Delaware
1870-1880
Agricultural Census
Volume 2

Linda L. Green

WILLOW BEND BOOKS
2006

WILLOW BEND BOOKS

AN IMPRINT OF HERITAGE BOOKS, INC.

Books, CDs, and more—Worldwide

For our listing of thousands of titles see our website
at
www.HeritageBooks.com

Published 2006 by
HERITAGE BOOKS, INC.
Publishing Division
65 East Main Street
Westminster, Maryland 21157-5026

International Standard Book Number: **0-7884-3829-8**

Introduction

This census names only the head of the household. Often times when an individual was missed on the regular U. S. Census, they would appear on this agricultural census. So you might try checking this census for your missing relatives. Unfortunately, many of the Agricultural Census records have not survived. But, they do yield unique information about how people lived. There are 52 columns of information on the 1870 census. I chose to transcribe only seven of the columns. There are 100 columns of information on the 1880 census. I chose to transcribe only sixteen of the columns. Below is a list of other types of information available on the 1870 and 1880 agricultural census.

Linda L. Green
13950 Ruler Court
Woodbridge, VA 22193

Other Data Columns 1870

Column/Title

7. Total Amount of Wages Paid During the Year including the Value of Board
8. Horses
9. Mules & Asses
10. Milch Cows
11. Working Oxen
12. Other Cattle
13. Sheep
14. Swine
16. Spring Wheat
17. Winter Wheat
18. Rye
19. Indian Corn
20. Oats
21. Barley
22. Buckwheat
23. Rice
24. Tobacco
25. Cotton
26. Wool
27. Peas and Beans
28. Irish Potatoes
29. Sweet Potatoes
30. Orchard Products
31. Wine
32. Produce of Market Gardens
33. Butter
34. Cheese
35. Milk Sold
36. Hay
37. Clover
38. Grass
39. Hogs
40. Hemp
41. Flax
42. Flax-seed
43. Silk Cocoons
44. Maple Sugar
45. Cane Sugar
46. Molasses
47. Bees Wax
48. Honey
49. Forest Products
50. Value of Home Manufacturers
51. Value of Animals Slaughtered or Sold for Slaughter
52. Estimated Value of All Farm Production, Including Betterment and Additions to Stock

Other Data Columns 1880

The 1880 Agricultural Census brought an increase in the number of columns due to further defining already existing categories. On this census, there are 100 columns of information. Columns 1-16 were transcribed. The other 84 columns are listed below.

17. Grasslands Acreage for 1879—Mown
18. Grasslands Acreage for 1879—Not Mown
19. Grasslands—Product Harvested in 1879—Hay
20. Grasslands—Product Harvested in 1879—Clover Seed
21. Grasslands—Product Harvested in 1879—Grass Seed
22. Horses of All Ages on hand as of June 1, 1880
23. Mules and Asses of All Ages on hand as of June 1, 1880
24. Neat Cattle and Their Products—on hand June 1, 1880—Working Oxen
25. Neat Cattle and Their Products—on hand June 1, 1880—Milch Cows
26. Neat Cattle and Their Products—on hand June 1, 1880—Other
27. Neat Cattle and Their Products in 1879—Calves Dropped
28. Neat Cattle and Their Products in 1879—Cattle of all ages—Purchased
29. Neat Cattle and Their Products in 1879—Cattle of all ages—Sold living
30. Neat Cattle and Their Products in 1879—Cattle of all ages—Slaughtered
31. Neat Cattle and Their Products in 1879—Cattle of all ages, Died, Strayed and Stolen not Recovered
32. Neat Cattle and Their Products in 1879—Milk Sold or Sent to Butter and Cheese Factories in 1879
33. Neat Cattle and Their Products in 1879—Butter Made on the Farm in 1879
34. Neat Cattle and Their Products in 1879—Cheese Made on the Farm in 1879
35. Sheep on hand June 1, 1880
36. Sheep Movement in 1879—Lambs Dropped
37. Sheep—Sheep & Lambs Purchased
38. Sheep—Sheep & Lambs—Sold Living
39. Sheep—Sheep & Lambs—Slaughtered
40. Sheep—Sheep & Lambs—Killed by Dogs
41. Sheep—Sheep & Lambs—Died of Disease
42. Sheep—Sheep & Lambs—Died of Stress or Weather
43. Sheep—Clip, Spring, 1880, Shorn and to be Shorn—Fleece
44. Sheep—Clip, Spring, 1880, Shorn to be Shorn—Weight
45. Swine—on hand June 1, 1880
46. Poultry on hand June 1, 1880 exclusive of Spring Hatchings—Barnyard
47. Poultry on hand June 1, 1880 exclusive of Spring Hatchings—Other
48. Eggs Produced in 1879
49. Cereals—Barley, 1879—Area Acres
50. Cereals—Barley, 11879—Crop in Bush.

51. Cereals—Buckwheat, 1879—Area Acres
52. Cereals—Buckwheat, 1879—Crop in Bush.
53. Cereals—Indian Corn, 1879—Area Acres
54. Cereals—Indian Corn, 1879—Crop in Bush.
55. Cereals—Oats, 1879—Area Acres
56. Cereals—Oats, 1879—Crop in Bush.
57. Cereals—Rye, 1879—Area Acres
58. Cereals—Rye, 1879—Crop in Bush.
59. Cereals—Wheat, 1879—Area Acres
60. Cereals—Wheat, 1879—Crop in Bush.
61. Pulse—Canada Peas (Dry), 1879—Bush.
62. Pulse—Beans (Dry), 1879—Bush.
63. Fiber—Flax, 1879—Acres in Crops
64. Fiber—Flax, 1879—Seed in Bush.
65. Fiber—Flax, 1879—Straw in Tons
66. Fiber—Flax, 1879—Fiber in pounds
67. Fiber—Hemp, 1879—Acres
68. Fiber—Hemp, 1879—Tons
69. Sugar, Sorghum, 1879—Area in Crops Acres
70. Sugar, Sorghum, 1879—Sugar in Lbs.
71. Sugar, Sorghum, 1879—Molasses in Gals.
72. Sugar, Maple, 1879—Sugar in Lbs.
73. Sugar, Maple, 1879—Molasses in Gals.
74. Broom Corn, 1879—Acres
75. Broom Corn, 1879—Lbs.
76. Hops, 1879—Area Acres
77. Hops, 1879—Crop in Lbs.
78. Potatoes, Irish, 1879—Area Acres
79. Potatoes, Irish, 1879—Crop in Bush.
80. Potatoes, Sweet, 1879—Area Acres
81. Potatoes, Sweet, 1879—Crop in Bush.
82. Tobacco, 1879—Area Acres
83. Tobacco, 1879—Crop in Lbs.
84. Orchards, 1879—Apple—Acres
85. Orchards, 1879, Apple Bearing Trees
86. Orchards, 1879, Apples, Bush.
87. Orchards, 1879—Peach Acres
88. Orchards, 1879 Peach Bearing Trees
89. Orchards, 1879 Peach, Bush.
90. Total Value of Orchard Products of all kinds Sold or Consumed
91. Nurseries Acres
92. Nurseries Value of Produce Sold in 1879
93. Vineyards, No. of Acres
94. Vineyards—Grapes Sold in 1879 Lbs.
95. Vineyards, Wine Made in 1879 Gals.
96. Market Gardens—Value of Produce Sold in 1879 in Dollars

97. Bees, 1879—Honey in Lbs.

98. Bees, 1879—Wax in Lbs.

99. Forest Products—Amount of Wood Cut in 1879 in Cords

100. Forest Products—Value of all Forest Products Sold or Consumed in 1879 in Dollars

Table of Contents

1870   **Pages**

Kent County   1

New Castle County   48

Sussex   91

Index   160

New Castle County Addition   189

New Castle County Addition Separate Index   195

1880

Kent County   197

New Castle County   252

Sussex County   303

Index   397

This agricultural census was filmed from original records in the Delaware State Archives in Wilmington Delaware by the Delaware State Archives Microfilm office.

There are some fifty-two columns of information on each individual. Only the head of household is addressed. I have chosen to use only seven columns of the information because I feel that this information best illustrates the wealth of the individuals. These are shown below:

1. Name of Owner
2. Acres of Improved Land
3. Woodland Unimproved
4. Other Unimproved Acreage
5. Cash Value of Farm
6. Value of Farm Implements and Machinery
15. Value of All Livestock

Thus, the numbers following the names represent columns 2, 3, 4, 5, 6, 15.

The following symbol is used to maintain spacing where information in a column is left blank (-). This symbol is used where letters, names or numbers are not legible (_).

NOTE: Beginning with the 1870 Agricultural Census the column names have changed. And, the names for this county appear as Last Name First.

Rash, Moses, 80, 180, -, 15000, 250, 550

Thomas, Joseph H., 4, -, -, 1200, 50, 85

Pearson, John W., 65, 25, -, 500, 95, 435

Pearson, Abraham, 60, 30, 10, 5000, 140, 555

Cohee, Sarah E., 12, -, 17, 2500, -, 145

Hured, Philip, 10, -, 5, 200, -, 15

Cohee, Vincen, 3, -, -, 600, -, 6

Thompson, Obediah, 60, 30, 10, 3000, 40, 285

Sammeser, Nehemiah, 4, 3, 2, 300, -, 40

Newmine, Wm. T., 19, -, 1, 800, -, 40

Boggs, David, 95, 95, -, 7000, 85, 400

Thompson, Obediah, 12, 8, -, 300, -, 80

Gibbs, Stephen, 7, 13, -, 600, 120, 255

Barber, Charles, 40, 10, 10, 160, 115, 335

Taylor, John, 25, 10,8, 1500, 65, 165

Thompson, Silas, 15, -, -, 60, 75, 165

Sylvester, John, 10, -, -, 200, -, 175

Taylor, John, 20, -, 5, 1000, 35, 204

Allen, James S., 16, -, -, 500, -, 50

Buckingham, Isaac, 162, 150, 30, 800, 75, 330

Callem, Benjamin, 3, -, -, 300, -, -

Martin, Benj. W., 180, 105, -, 8000, 175, 876

Johnson, John A., 100, 37, -, 6000, 35, 280

Bearman, Henry, 100, 45, -, 7000, 100, 558

Johnson, Thomas, 100, 50, -, 6000, 105, 450

Melvin, Truitt, 30, 17, 10, 1500, 35, 155

Rash, John, 153, 85, -, 13000, 275, 726

Rash, James, 75, 32, -, 7000, 190, 420

Pearson, Isaac H., 100, 35, -, 6000, 270, 685

Jones, Levin P., 15, 15, -, 2000, -, 46

Wright, Charles G., 100, 40, -, 6000, 110, -

Ware, John L., 50, 10, -, 3000, 50, 160

Ware, John T., 55, 20, -, 3000, 75, 330

Lord, John B., 12, 1, 10, 1000, 100, 180

Hurd, Joseph, 150, 30, 60, 9000, 280, 900

Shorts, John, 260, 40, 10, 10000, 140, 752

Mable, William, 45, 20, -, 3000, 25, 212

Carlile, Cassie, 8, -, -, 280, 12, 55

Collins, Manta, 26, -, -, 400, -, 10

Dyer, John H., 35, 55, -, 3000, 25, 124

Irland, Joseph, 250, 150, -, 16000, 435, 940

Wallace, Samuel C., 75, 30, -, 3000, 193, 370

John, Elias, 70, -, 20, 4000, 38, 280

Griffin, Charles, 7, 7, 6, 800, -, -

Pearson, Francis., 50, 5, 5, 2000, 70, 340

Mitchell, Joshua R., 45, 42, -, 2000, 70, 407

Hudson, John I., 40, -, -, 350, -, 40

Hallowell, William B., 60, 20, -, 2500, 65, 310

Clark, Jacob, 18, -, -, 800, -, -

Adams, Thomas H., 4, -, -, 200, -, -

Hammon William, 83, 10, -, 6500, 150, 618

Hirone, Samuel R., 80, 20, -, 3000, 115, 379

Hirone, William, 140, 40, -, 5000, 80, 545

Rash, Daniel, 75, -, 200, 13000, 50, 357

Pove, Benjamin, 10, -, -, 800, -, 40

Grenwell, Henry, 150, 40, 400, 900, 250, 790

Luich, Reiley, 180, 40, -, 10000, 320, 1168

Wilcuts, Robert H., 150, 40, 40, 9000, 188, 558

Wilkenson, Ruben, 80, -, 800, 2000, 35, 100

Davie, Truslin L., 33, -, -, 100, -, -, -

Harris, John W., 37, 10, -, 1500, 70, 120

Campbell, William, 10, -, -, 1000, 10, 385

David, John L., 17, -, -, 800, -, -

Fountain, Aaron, 3, -, -, 150, -, 53

Ford, Henry, 7, 1, -, 300, -, 5

Phillips, George, 113, 113, -, 6000, 175, 40

Darg, Samuel, 18, -, -, 600, -, -

Hoaney, Thomas, 3, -, -, 300, -, -

Gibbs, Richard A., 10, -, -, 150, -, 35

Danny, Samuel O., 35, -, -, 1000, 5, 45

Thompson, Obediah, 15, 3, -, 600, 25, 123

Jackson, Alexander, 10, -, -, 700, -, 65

Hawkins, Isaac, 50, 10, -, 1500, 50, 300

Farrow, Joseph D., 22, 20, -, 1000, 55, 190

Dear, George W., 20, 5, -, 600, -, 40

Barrett, George W., 20, 21, -, 800, 25, 93

Clayton, William, 32, 3, -, 600, 10, 135

Barrett, Isaac, 6, -, -, 200, 5, 145

Clark, John, 1, 29, -, 1000, -, -

Barrett, James, 30, -, -, 1000, -, -

Ford, Christopher, 20, 6, -, 1000, 60, 345

Harris, Lemuel, 136, 135, -, 4000, 135, 395

Richerson, James, 10, 2, -, 200, -, -

Poor, Thomas D., 60, 50, -, 2000, 125, 250

Lane, William H., 55, 5, 2500, 80, -, 220

Cregg, Samuel, 70, 70, -, 4000, 70, 485

Gibbs, Nathan, 75, -, -, 2000, 45, 239

Wright, Elisha, 31, 10, -, 2500, 75, 585

Hawkins, Geo. F., 8, -, -, 400, -, 90

Horsey, Thomas, 37, 20, -, 800, -, -

Burris, Edward, 42, -, -, 2500, -, 165

Hollingsworth, Ray__, 3, -, -, 200, -, -

Powell, Charles H., 40, 20, -, 2000, 20, 226

Morris, Beecham, 90, 50, -, 4000, -, 40

Jones, James P. 6, 12, -, 400, -, 6

Franklin, Zeddach, 10, -, -, 300, -, 20

Gibbs, Stephen L., 15, 65, -, 2000, -, 100

Williams, Waitman, 20, 10, -, 400, -, 50

Walls, James, 120, 60, -, 6000, 50, 306

Larrimoore, Ruth, 2, -, -, 100, -, 35

Webster, David H., 4, -, -, 600, -, 190

Johns, James, 135, 65, -, 8000, 315, 878

Sparks, Solomon, 4, -, -, 400, -, 72

Davis, John M., 125, 5, -, 3000, 85, 707

Nickerson, Joshua, 96, -, -, 2000, 11, 185

Luff, Thomas, 65, 75, -, 5000, 45, 240

Oakhall, Mary E., 25, 175, -, 1000, -, -

Stevens, John, 26, 25, -, 1500, 15, 155

Slaughter, Asbury, 14, -, -, 1500, -, -

Powell, Samuel R., 100, -, -, 3000, 90, 310

Hawkins, Wm. F., 125, 125, -, 8000, 170, 1184

Martion, John, 70, 30, -, 2500, 110, 225

Powell, Samuel C., 100, -, -, 4000, 145, 200

Rash, Isaac, 50, -, -, 150, 35, 85

Grenge, Samuel, 65, 15, -, 1500, 40, 283

Guessford, James H., 75, 25, -, 2500, 95, 140

Vible, Samuel, 15, 9, -, 800, 35, 333

Graham, William T., 40, 100, -, 2000, 55, 327

Powell, Michael, 20, 80, -, 1500, 25, 385

Morris, George, 25, 275, -, 5000, 30, 170

Stevens, Robert, 85, -, -, 25, 45, 188

Barnett, Joseph, 45, 15, 45, 2800, 110, 783

Johnson, William H., 17, -, -, 500, 17, 110

Scotten, Sarah A., 79, 20, -, 2500, 65, 384

Jones, William, -, 85, -, 2500, 41, 187

Scotten, Phil__, 90, 30, -, 4500, 165, 460

Hartwell, Lawrence, 4, -, 17, -, 300, 15, 40

Williams, William I., 90, 10, -, 1500, 50, 375

Downs, Daniel W., 35, 5, -, 2000, 50, 170

Scotten, Spencer, 90, 30, -, 3000, 90, 509

Scotton, Charles E., 70, 20, -, 2000, 20, 164

Greenwell, Francis, 75, 45, -, 1400, 60, 355

Marvel, Frances J., 110, 90, -, 10000, 495, 1232

Linch, Andrew G., 80, 70, -, 5000, 50, 180

Slaughter, William, 190, 41, -, 18000, 550, 1385

Boyles, William, 45, 20, -, 2000, 75, 583

Harrington, Thos. M., 75, 24, -, 1000, 44, 197

Kenton, Ely, 60, 5, -, 1200, 21, 163

Fell, John B., 150, 52, -, 6000, 275, 835

Perry, John C., 50, 35, -, 1300, 30, 331

Groff, Robert H., 25, 100, -, 4000, 365, 1153

Slay, George, 185, 185, -, 10000, 165, 827

Hawkins, Clayton F., 150, 25, 25, 6000, 85, 514

Minos, John W., 10, -, -, 30, -, 40

Williams, John, 150, 100, -, 6000, 140, 635

Day, Matthew, 95, 35, -, 3500, 195, 586

Powell, James E., 170, -, 40, 5000, 180, 410

Hartwell, Timothy, 17, -, -, 5000, -, 40

Hartwell, Edward, 12, -, -, 50, 25, 185

Patten, James, 50, -, 50, 2000, 20, 130

Montague, John, 83, -, -, 3000, 285, 700

Kersey, William, 60, -, 24, 3000, 51, 426

Boyer, James E., 88, -, -, 4500, 175, 648

Hall, William T., 60, -, -, 2500, 68, 291

Colescott, Wm. H., 20, -, -, 1200, 20, 212

McCollum, Samuel, 50, -, 5, 1200, -, 150

Meredith, William, 12, 3, -, 3000, 130, 305

Kersey, James, 87, -, -, 200, 135, 137

White, Benjamin, 22, -, -, 250, -, 20

Furgerson, Henry, 18, -, -, 200, 10, 100

Shahan, Jacob D., 60, -, -, 500, 15, 60

Faulkner, Thomas, 75, -, 10, 1000, 40, 300

Hollan, Isaac G., 35, 10, 11, 1200, 50, 172

Gear, William _., 20, 5, 5, 300, 52, 82

Crammer, Joseph, 70, 30, 19, 1500, 20, 483

Hayes, Bazzell, 20, 10, 20, 700, 20, 78

Boyles, James, 50, 20, -, 1000, -, 40

Quillen, Isaac, 10, 5, -, 400, -, 80

Jones, Garrett, 8, -, -, 300, -, -

Jones, Jesse M., 44, 10, -, 3000, 125, 416

Slaughter, Jonathan, 40, 13, -, 2500, 120, 205

Black, Frederick, 90, 10, -, 2500, 40, 269

Moor, Thomas H., 72, 10, -, 2000, 20, 228

Williams, James, 180, 25, 25, 6000, 303, 1141

Clark, James, 120, 20, -, 4500, 310, 478

Tribbett, Daniel, 30, 10, -, 1000, 15, 40

Simmons, James, 100, 25, 25, 1500, 20, 198

Bedwell, Caleb, 10, -, -, 150, -, 65

Thomas, Isaac, 60, -, 15, 1200, 50, 374

Briscoe, Isaac L., 18, 32, -, 600, 10, 145

Griffith, James W., 40, 75, 15, 3000, 100, 50

Fisher, John R., 90, -, 40, 2000, 75, 154

Thompson, William, 30, 10, -, 300, -, 56

Agnue, John, 150, 130, 8, 6000, 320, 899

Pearson, William W., 100, -, -, 1000, -, 241

Casson, John G., 80, 20, -, 4500, 284, 495

Jones, William H., 35, 25, 15, 2000, 15, 95

Jones, George F., 175, -, -, 5000, 215, 698

Jones, George, 25, -, -, 1500, 138, 352

Anderson, James, 150, 25, 25, 5000, 105, 685

Bowers, William, 150, 50, -, 5000, 180, 352

Pratt, Isaac S., 100, 50, 6, 5000, 125, 625

Thompson,William, 63, 30, -, 1500, 65, 290

Poor, Martin, 5, 5, -, 200, -, 45

Hurd, Rachael, 88, 30, -, 1000, -, -

Short, Embry, 12, 8, -, 500, -, 110

Voshell, Obediah, 43, 30, 2, 2500, 175, 170

Foreaker, Williams, 40, 46, -, 1000, -, -

Taylor, John R., 20, 5, -, 800, -, 75

Fasig, Samuel, 122, 50, -, 8000, 389, 920

Adams, William F., 70, 75, 5, 4000, 200, 492

Marvel, Henry D., 63, 70, -, 3500, 175, 150, 486

Crammer, James, 20, 30, 28, 1000, 20, 135

Tue (Tuc), Betsey, 5, -, -, 150, -, 30

Griffith, John F., 75, -, -, 1500, 25, 262

George, Thomas N, 125, 2, -, 3000, 200, 840

George, Joseph, 87, 10, -, 1500, 120, 310

Voshell, John J., 125, -, -, 5000, 160, 1147

Crampitt (Clampitt), James, 80, 20, 2000, 10, 360

Hall, James, 75, -, 25, 2100, 15, 488

Todd, William M., 13, -, -, 1500, -, 236

Myers, Abraham, 84, 6, -, 2800, 150, 636

Arron, Powell, 12, 2, -, 8000, 275, 855

Pratt, Henry Sr., 165, 75, -, 9000, 210, 745

Pratt, Henry Jr., 200, -, -, 8000, 695, 1475

Darling, William, 100, -, -, 4500, 170, 466

Nickerson, Capril (Cassril), 10, 8, -, 4500, 294, 445

Arthers, James, 40, 5, -, 1500, -, 60

Seaney, Thomas, 68, 20, -, 4000, 450, 493

William, Amos C., 60, 25, -, 3000, 290, 505

Green, James H., 80, 31, -, 4000, 120, 553

Foreaker, Masculine, 100, -, -, 1000, -, -

Willis, William M., 80, 5, -, 3000, 137, 686

Vurden, William, 200, 200, -, 12000, 527, 1849

Vurden, James, 200, -, 70, 9500, 205, 1175

Reed, Frederick, 38, -, 12, 1200, 55, 130

Lowdine, Joshua, 21, -, 4, 800, 30, 155

Ford, Moses, 180, 30, 30, 8500, 845, 1820

Green, Jonathan, R., 53, 52, -, 3000, 79, 252

Perry, Thomas, 1458, 30, 6500, 140, 156

Williams. Thomas, 44, 16, -, 1800, 130, 270

Slaughter, Solomon, 211, -, -, 5000, 85, 435

Hayden, Daniel, 100, -, -, 2500, 45, 297

Rash, Darling, 83, 58, -, 2000, 45, 191

Hubbard, Nuton, 80, 27, -, 3000, -, 380

Dodd, William A., 25, -, -, 800, 70, 200

Hutchins, John, 33, 67, -, 1500, 20, 360

Hutchins, Thomas, 60, 17, -, 1500, 25, 250

Clement, Thomas, 229, 45, -, 9000, 295, 740

Price, Thomas, 50, 20, -, 1500, 105, 318

Purnell, John S., 150, 50, -, 4000, 250, 515

Price, Solomon, 15, 6, 5, 600, -, 40

Shettler, Hezekiah, 55, 20, -, 1500, 145, 400

Marvel, Philip B., 110, 30, -, 3500, 145, 610

Smith, Samuel C., 100, 13, -, 2200, 15, 300

Montacue, Jesse, 75, 50, 10, 5600, 275, 215

Hubbard, Edward, 152, 20, 3, 5500, 295, 700

Kersey, Moses, 80, 20, -, 3000, 65, 359

Voshell, Samuel, 40, -, -, 1200, 25, 215

Wilkinson, James, 21, -, -, 800, 30, 180

Voshell, Peter, 33, -, -, 600, 50, 198

Connely, Tilghman, 113, -, -, 1600, 46, 110

Cregg, Jacob, 46, -, -, 400, 45, 125

Vincent, Nicholas, 17, -, -, 1000, 40, 148

Price, James J., 200, -, 40, 5000, 100, 350

Steel, Josiah, 150, 25, 350, 5500, 275, 929

Clark, James, 175, 25, 50, 6000, 230, 1214

Cooper, Samuel, 37, -, -, 500, -, -

Wooten (Worton), Manlove, 97, -, -, 800, -, 80

Potter, Richard L., 32, -, -, 500, -, 55

Marvel, Avary, 100, -, 6, 2000, 200, 445

Beddle, William P., 87, -, -, 1500, 60, 477

Parvis, William, 22, 15, -, 800, 25, 34

Hansberry, William, 58, -, -, 1500, 70, 164

Johnson, John, 32, -, -, 600, 60, 348

Johnson, Henry, 100, -, 17, 1700, 110, 292

Gooden, John, 200, 86, -, 10000, 685, 1580

Clark, James W., 18, -, -, 800, 40, 195

Rash, James, 34, -, -, 400, -, 140

Rash, Elizabeth, 7, -, -, 300, -, -

Voshell, Hannah, 3, -, -, 100, -, -

Voshell, John D., 100, 10, 10, 2500, 65, 416

Herring, Jeremiah, 4, -, -, 300, -, -

Jones, George W., 160, -, -, 2500, 65, 228

Frances, James, 21, -, -, 500, 75, 308

Smith, John, 27, -, 300, -, 73

Edge, Robert, 100, 30, 30, 2500, 75, 210

Thomas, Isaac, 235, 50, -, 550, 75, 314

Cook, James, 50, -, -, 1200, 15, 230

Carter, Ann, 30, -, -, 50, -, 45

Jarmon, George, 97, -, -, 2000, 130, 275

Callahan, Alexander, 15, -, -, 900, 100, 187

Rawley, William, 300, 10, -, 800, 40, 86

Clark, William H., 60, 15, -, 1100, 50, 515

Mathew, John, 37, -, -, 500, -, 80

Jones, George H, 100, 60, -, 2500, 60, 360

Richards, James, 100, 25, -, 1500, -, 150

Bedwell, Preston, 100, 50, 25, 4200, 100, 595

Moor, William S., 60, 40, -, 2000, 70, 363

Kersey, John, 37, -, -, 1000, 70, 150

West, Thomas C., 100, -, -, 3500, -, -

Whitby, Nathan, 22, -, -, 220, -, 60

Rash, Henry, 17, -, -, 200, -, 60

Vincen, James, 80, -, -, 1000, -, 260

Ford, Martin R., 120, 100, -, 6000, 280, 720

Callahan, Edward, 40, -, 5, 500, -, 122

Lofland, Isaac D., 81, 94, -, 8500, 135, 280

Johnson, William, 100, -, -, 2000, 80, 350

Moor, Thomas S., 80, 70, -, 4000, 110, 525

Green, James, 60, -, -, 1700, -, 251

Hawkins, Susan, 14, -, -, 1500, 140, 297

Massey, Sarah, 20, -, -, 400, -, 30

Slay, Mary, 30, -, -, 1800, 35, 295

Smith, Henry, 42, -, -, 550, -, 60

Kersey, Jonathan, 10, -, -, 200, -, -

Tribbett, John, 75, 20, -, 1500, 100, 337

Hudson, Nathaniel, 90, -, -, 2500, 50, 225

Foraker, James, 100, 30, 30, 1500, 35, 261

Anderson, Benjamin, 40, -, -, 600, -, 100

Cook, William, 100, -, 21, 3000, 225, 632

Marvel, Phillip, 140, 10, -, 3000, 315, 953

Cooper, Thomas B., 150, 50, -, 4000, 420, 1239

Clark, Nemieh, 125, 55, -, 3000, 20, 197

Powell, James, 100, -, -, 3000, 150, 302

Aron, David, 50, -, -, 500, 20, 50

Slaughter, William P., 60, -, -, 1800, 35, 375

Powell, Thomas, 50, -, 20, 2500, -, 40

Scuse, Thomas, 125, -, 250, 14000, 85, 610

Moore, Thomas W., 90, -, -, 3500, 105, 191

Hewes, Robert, 50, 50, -, 3500, 330, 780

Cregg, John, 224, 40, -, 1000, 372, 791

Proud, Levi, 100, - 40, 6500, 245, 410

Powell, James R., 100, 24, -, 3500, 690, 1015

Hollowell, Thomas, 130, 20, -, 3500, 120, 472

Corse, Jackson, 100, -, -, 200, 50, 150

Hutchison, Eliza, 50, -, -, 1000, 10, 75

Steffey, William, 100, 26, -, 6500, 150, 405

Wright, Robert, 85, 45, -, 7000, 210, 585

Wright, Robert B., 80, 83, -, 4500, 175, 640

Mosely, John, 30, -, 70, 4000, 50, 195

Collwell, Edward, 25, 5, -, 300, -, 52

Hargadine, Elizer, 75, 33, -, 3500, 125, 305

Stover, Peter, 50, 43, -, 5000, 155, 200

Winger, Moses, 50, 41, -, 4500, 100, 360

Cade, John W., 69, 69, -, 5000, 85, 290

Rash, Eugene, 100, 14, -, 5000, 110, 500

Cohee, Nehemiah, 50, -, -, 2000, 50, 80

Rash, Myers, 80, -, 500, 7000, 280, 470

Dennis, Mahalia, 10, 10, 5, 300, -, 85

Wharton, John B., 141, 29, 200, 14000, 275, 776

Cullen, John W., 119, -, 16, 22000, 968, 1435

Dennis, John, 100, -, 20, 15000, 205, 1098

Ridgway, William, 50, 45, -, 7000, 260, 810

Taylor, Augusta, 90, 5, -, 6000, 210, 390

Deats, George, 90, 40, -, 7000, 95, 233

Ennes, William H., 58, 9, -, 3500, -, 636

Vanslyk, Selden, 57, -, -, 4000, 155, 355

Tate, James, 100, -, -, 2500, 50, 270

Carsen, William, 75, -, -, 3500, 160, 305

Pickett, Lewis, 10, -, -, 300, -, 60

Cahoon, Wm. R., 50, -, 30, 4800, 235, 651

Hancock, Sallie E., 150, -, 50, 16000, 365, 976

Mosely, Wingate, 70, -, 90, 8000, 75, 266

Bobbett, Chas. C., 225, 27, -, 25000, 905, 1340

Kimmey, Alcorn, 157, -, 15, 27000, 319, 865

Ridgely, Eugene, 215, 85, -, 36000, 295, 1513

Hope, James M., 100, 6, -, 11000, 320, 685

Baum, Chester, 107, 8, -, 12000, 350, 1000

Mason, Samuel R., 100, 10, -, 12000, 500, 1130

Goodman, Jacob, 70, 8, -, 7800, 360, 900

Hamon, John, 229, 15, -, 25000, 225, 1405

Lewis, William A., 60, -, -, 6000, 90, 150

Hope, Samuel W., 100, 27, -, 9500, -, 385

Jones, Richard M., 137, -, -, 13200, 165, 1030

Saxton, Richard, 110, 30, -, 6000, 95, 340

Tucker, James, 96, -, -, 3500, 260, 470

Laws, Bazzel, 70, -, -, 2000, 50, 240

Todd, James H., 340, -, 50, 20000, 650, 1096

Hope, James M., 90, -, 10, 4500, -, 349

Sturdevant, Saml. T., 40, -, 28, 4000, 110, 186

Waples, James _., 170, 40, 5, 28000, 465, 1413

Morris, William R., 150, -, 20, 20000, 370, 867

Shockley, John B., 70, -, 41, 12000, 215, 570

Webb, William, 176, -, 20, 20000, 170, 1202

Garten (Gasten), William, 200, -, 100, 19000, 445, 1090

Wheatly, William, 185, -, 10, 15000, 308, 1070

Vandergriff, John D., 175, -, 10, 20000, 290, 1080

Day, Joseph, 108, -, -, 14000, 290, 1080

Mills, David, 360, -, 50, 3000, 535, 2000

Smith, Peter H., 150, -, 10, 14500, 335, 1306

Wildsmith, Harry, 100, -, 50, 6000, 115, 360

Slaughter, John, 115, -, 40, 6000, 355, 853

Coverdale, David, 150, -, 10, 18000, 430, 910

Longfellow, Wm. C., 335, 50, -, 58000, 500, 1770

Wilson, Robert, 70, 20, 46, 10000, 440, 680

Flick, Jacob R., 110, 10, -, 1000, 250, 635

Green, Levi, 15, 60, -, 2500, -, 20

Grey, William, 100, 50, 63, 9000, 250, 1292

Bolk, Peter, 16, 61, -, 2500, 50, 135

Farrow, Daniel, 80, 30, -, 3500, 35, 175

Barcus, Elisabeth, 11, -, -, 800, 28, 165

Todd, Henry, 80, 25, 6000, 405, 745

Green, John M., 100, 25, -, 5000, 80, 345

Moore, Abraham C., 48, -, 39, 6500, 200, 788

Moore, Joseph, 35, -, 35, 4000, 80, 298

Seim, Adams, 110, 56, -, 7000, 150, 600

Harrald, Williams D., 10, 25, 25, 4000, -, 238

Hammon, Joseph, 40, 10, -, 1200, 50, 115

King, Thomas W., 15, -, -, 600, -, 215

Cratser, William, C., 75, -, 236, 7000, 40, 172

Everett, John, 40, 20, -, 1500, 125, 332

Abbott, Nehemiah, 100, 50, 1000, 500, 70, 366

Moore, Henry, 80, 42, -, 5000, 80, 638

Hafnal, Saml. M., 15, -, -, 500, -, 41

Shahan, William E., 3, -, -, 350, -, 85

Hickey, Samuel, 10, 5, -, 300, -, 55

Green, Robert H., 225, 300, -, 22000, 315, 114

Edenfield, Thomas, 20, -, 2, 400, 20, 63

Mason, Edward, 90, -, 40, 4100, 40, 290

Register, Abriham, 18, -, -, 300, -, -

Hull (Hall), John, 15, -, -, 300, -, -

Dennis, George W., 50, -, -, 800, -, -

Boggs, Joseph, 80, -, 20, 4500, 150, 755

Paxson, George W., 50, -, 30, 4000, 175, 785

Clark, John M., 80, -, 29, 4000, 240, 950

Clendaniel, James, 100, -, 100, 7000, 120, 536

Johns, Zaceriah, 90, -, -, 8000, 5, 700

Reed, Charles, 100, -, -, 6000, 50, 305

Moore, James, G., 100, -, -, 5500, 100, 230

Moore, Barkley, 134, -, 80, 10500, 48, 350

McKee, William F., 175, -, -, 12000, 400, 685

Shorse, Mary, 16, -, -, 300, -, -

Reed, Allen, 20, -, -, 1000, -, 80

Denny, John M., 164, 251, 10000, 80, 1109

Denny, Charles, 112, 75, -, 14000, 650, 1670

Argo, Jorden, 80, 50, 50, 7500, 186, 870

Hardison, John, 50, 16, -, 3500, 125, 160

Emory, Samuel, 46, -, -, 2300, 40, 215

Reed, Thomas, 60, -, -, 4500, 45, 280

Be__, John H., 165, 25, -, 15000, 550, 800

Leonard, Dexter, 160, 45, -, 10500, 465, 664

Haman, Edward, 280, 32, -, 20000, 340, 1570

Moore, Samuel, 20, -, -, 2000, 25, 340

Eager, William, 74, -, -, 2000, 150, 370

Wells, Jacob, 40, 33, 2000, 170, 190

Lawson, David, 114, -, -, 3500, 100, 655

Cannon (Connor), Henry W., 140, 17, 100, 26000, 600, 1775

Vacant Farm, 200, -, -, 2000, -, -

Pearson, James W., 200, -, -, 4000,70, 270

Burnahm, Thomas, 200, 65, 50, 24000, 378, 1615

Murphy, Thomas G., 200, 50, -, 25000, 257, 255

Miluts, John, 205, -, -, 6000, 225, 1525

Eccles, Edwin O., 161, 25, -, 22500, 445, 1090

Daniphan, Asbury, 94, -, -, 7500, 170, 551

America, Moses, 24, -, -, 1600, 100, 400

Slaughter, George, 85, -, -, 4500, 1250, 620

Daws, William, 200, -, -, 25000, 465, 2780

Williams, Samuel, 197, 3, -, 25000, 415, 570

Denney, Thos. H., 164, 50, -, 25000, 405, 1513

Baker, John F., 150, -, -, 15000, 345, 435

Voshell, Draper, -, -, -, -, -, -

Bolden, Robert, 75, -, 10, 1600, -, 125

Clayton, David, 20, -, -, 5000, 40, 180

Bell, Isaac, 7, -, -, 300, -, 40

Bell, Joseph, 136, -, 155, 10000, 75, 410

Dackstader, Wm. H., 58, -, 4, 7500, 193, 744

Baker, Isaac, 305, -, 25, 20000, 475, 1604

Harrington, James, 216, 4, -, 15000, 1008, 2120

Cowgill, Daniel, 114, -, 6, 18000, 1145, 1030

Collins, Alexander, 50, 50, -, 3000, 400, 960

Hayt, Daniel, 40, -, -, 3000, 165, 110

Beker, Milliam H., 51, 2, 6, 3600, 160, 700

Bell, Alfred, 20, -, 10, 2000, -, 20

Lafferty, Rachell, 55, 5, -, 3000, 115, 305

Paruice, William, 15, -, 9, 900, 65, 238

Reed, Mary, 15, -, -, 400, -, 15

Mitchell, Elizabeth, 23, -, -, 700, 40, 145

Canno (Cannon), Bartholomew, 10, -, -, -, 600, -, -

Anderson, Charles, 55, -, -, 4000, -, 200

McGonnigal, Joshua, 10, -, -, 4000, -, 320

Hobson, Joseph, 140, -, 100, 12000, 215, 820

Slaughter, Thomas J., 85, -, 85, 8000, 117, 880

Postle, William G., 160, -, 10000, 625, 1100

Godwin, James, 125, 15, 10000, 160, 799

Postle, Thomas, 180, -, 20, 12000, 580, 1235

Manlove, William W., 125, 25, -, 11350, 150, 150

Hollie, Charles W., 50, 10, 5, 1000, 50, 295

Paradee, Edward, 183, 20, -, 12000, 625, 1960

Pepper, Ellie W., 175, -, 25, 16000, 155, 646

Biddle, William H., 300, 40, 10, 24200, 800, 1840

Rockwell, Daniel, 45, -, -, 5000, 800, 540

Wilson, Mary A., 200, -, -, 20000, 300, 850

Wharton, Joshua B., 300, -, 50, 30000, 355, 1155

Wharton, Ernest, 123, -, 118, 13000, 445, 1126

Parry, Thomas, 117, 18, -, 8000, 100, 190

Slaughter, Timothy, 146, 10, 6, 10000, 395, 1030

Raymond, James, 120, 20, 20, 12000, 140, 602

Sapp, Elias, 25, 10, 10, 500, -, 76

Davis, James A., 90, 20, -, 11000, 190, 810

Harrington, Harriett, 140, 10, 10, 13000, -, -

Reivewey, James D., 250, 20, -, 22000, 920, 2213

Wharton, Bolitha L., 151, -, -, 7500, 195, 874

Hook, James, 12, -, -, 2000, -, 225

Anderson, Curtis, 230, -, 290, 45000, -, -, -

Loomis, Alviras, 164, -, -, 26000, 440, 470

Milbourn, William T., 20, 5, 5, 1000, -, 135

Butler, William, 300, 20, 10, 30000, 300, 1205

McIlvain, Henry, 150, -, -, 8000, 75 275

Covard, Augustus, 18, -, -, 900, 20, 625

Moor, Henry, 100, -, -, 20000, 195, 590

Atkins, Deborah, 100, -, -, 1000, 40, 260

Rodney, George, 50, -, -, 1000, -, -

Bell, Alexander, 100, -, 60, 1500, -, 220

Anderson, William, 240, 216, 197, 8000, 50, 340

Anderson, William, J., 30, 9, 35, 900, 10, 200

Godwin, John, 40, 14, 6, 1000, -, 110

Huffington, Jacob, 27, 8, 4, 900, -, 190

Taylor, Nathaniel S., 150, -, 150, 24000, 200, 1528

Taylor, James, 19, -, -, 500, -, 193

Tinley, Draper D., 260, 40, -, 25000, 40, 410

Knight, Martin, 250, 25, 3000, 15000, 168, 1410

Slaughter, William, 9, -, -, 600, 40, 400

Slaughter, Andrew, 240, -, -, 15100, 240, 922

Cowgill, Palmer L., 200, 20, -, 16000, 270, 900

Hutchison, William, 50, 10, -, 4000, 200, 344

Lober, William, 31, 7, -, 700, 50, 775

Gidersleeve, Daniel S., 300, 4, -, 25000, 440, 1220

Voshell, Draper, 128, 26, -, 12500, 150, 540

Patten, William, 23,11, -, 800, 30, 540

Wharton, Charles M., 180, 20, -, 20000, 690, 1311

Sipple, John, 138, 12, -, 12000, 145, 940

Norris, John R., 29, 3, -, 500, -, 220

Knight, Robert, 10, 7, -, 500, 25, 160

Gibson, Thomas, 9, 5, -, 400, 80, 160

Knight George, 100, 10, 20, 3100, 300, 830

Anderson Charles H. 11, 5, -, 300, 25, 185

Fisher, George W., 20, 8, 3, 800, 65, 380

Reed, Calob, 140, 60, -, 12000, 275, 500

Jenkins, Jabez, 113, 150, 15, 15000, 280, 1400

Webb, Joseph, 20, 5, 9, 1000, -, 195

Stant, John M., 250, 60, 40, 27000, 550, 4135

Cullen, Hezekiah, 200, 10, 20, 12000, -, 500

Taylor, John, 287, 30, 579, 17000, -, 595

Ward, James, 200, 30, 30, 16000, 125, 1095

Tomlinson, Askin, 10, 10, 100, 500, 25, 310

Williams Joseph, 160, 30, 100, 15000, 260, 1492

Levick, Mary, 130, 20, 500, 10000, 115, 438

Lewis, Evan R., 320, 20, 30, 50000, 525, 2190

Hancock, David, 170, 10, 5, 20000, 175, 665

Heverin, Williams, 400, 90, 7000, 26000, 550, 3149

Lofland, William, 100, -, -, 2000, -, 340

Wicks, Joseph, 13, -, 2, 600, -, 12

Salisbury, Gove, 15, -, 3, 3000, -, 450

Stuart, Mary A., 9, -, 5, 500, -, -

Todd, Henry, 100, -, 10, 5000, -, 215

Wilds, William, 8, -, -, 400, -, 625

Nicholson, John A., 10, 5, -, 1000, -, 540

Cornagys (Comegys), Joseph P., 15, -, -, 5000, -, 550

Heverin, James L., 12, -, -, 2400, -, 500

Wilds, James P., 10, -, -, 500, -, 50

Bonnill, William G., 8, -, -, 1600, -, 458

Jester, Milliam M., 20, -, 5, 3000, -, 110

Wallace, James C., 9, -, -, 500, -, 50

Bradford, Thomas B., 20, -, -, 4000, 451, 1350

Marshal, Samuel D., 10, -, -, 500, -, 300

Culbreth, Thomas O., 12, -, -, 700, -, 100

Cowgill, James, 15, -, -, 2000, -, -

Anderson, Edward S., 10, -, -, 1000, -, 259

Tucker, Joseph I., 8, -, -, 400, -, 80

Kerbin, James M., 22, -, -, 4000, 220, 490

Jones, James, 6, -, -, 300, -, 100

Cowgill, Daniel, 8, -, -, 2000, -, 125

Knight, Haughett, 7, -, -, 400, 430, 2485

Allee, James F., 15, -, -, 600, -, 150

Jump, Robert P., 7, -, -, 400, -, 150

Cahoon, William R., 6, -, -, 600, -, 175

Schaich, George C., 8, -, -, 700, 415, 305

Reeve, John W., 10, -, -, 500, -, 80

Hoffaker, Daniel C., 5, -, -, 500, -, 180

Carpenter, Perry, 7, -, -, 600, -, 80

Clements, Charles, 5, -, -, 500, -, 150

Chambers, Joseph M., 5, -, -, 400, -, 200

Clifton, James A., 7, -, -, 500, -, 30

Day, Charles H. B., 5, -, -, 400, -, 310

Schwartz, William S., 7, -, -, 500, -, 3530

Lewis, Jacob, 5, -, 3, 400, -, 835

Cox, Holly, 10, -, -, 500, -, 15

Boone, William G., 6, -, -, 400, -, 138

Schindler, George, 9, -, -, 800, -, 130

McGonigal, Samuel, 11, -, -, 900, -, 160

Taylor, James, 5, -, -, 400, -, 10

McCollom, Hugh, 5, -, -, 500, -, 15

Smith, Charles, 7, -, -, 600, -, 20

King, Robert, 5, -, -, 500, -, 20

Miller, Rachael, 9, -, -, 400, -, 25

Day, George, 5, -, -, 400, -, 20

Draper, Thomas, 7, -, -, 500, -, 8

Pierce, Thomas, 8, -, -, 60, -, 60

Butler, Zadock, 6, -, -, 500, -, 40

Cattle, Charles, 5, -, -, 400, -, 115

Waller, John, 8, -, -, 700, -, 10

Griffin, Sylas, 5, -, -, 500, -, 10

Keirn, John, 7, -, -, 700, -, 60

Gieser, Lewis, 6, -, -, 500, -, 20

Morgan, Donbar, 5, -, -, 400, -, 50

Hews, Frazier, 8, -, -, 700, -, 15
Blizzard, Gideon, 10, -, -, 900, -, 210
Lewis, Luff, 5, -, -, 500, -, 20
Teat, William, 5, -, -, 500, -, 150
Casson, Myers, 6, -, -, 500, -, 250
Morgan, George W., 8, -, -, 600, -, 45
Perry, James, 5, -, -, 400, -, 42
Ennis, William B., 8, -, -, 700, -, 60
Thomas, William R., 5, -, -, 400, -, 90
Wilson, Robert, 9, -, -, 700, -, 325
Smith, James L., 13, -, -, 800, -, 300
Holston, Thomas, 5, -, -, 40, -, 93
Jones, John L., 7, -, -, 600, -, 160
Foreaker, Joseph C., 5, -, -, 400, -, 200
Lewis, John, 7, -, -, 500, -, 65
Crouch, Elijah, 11, -, -, 1000, 100, 100
Sharp, William, 9, -, -, 1100, -, 250
Wallace, William H., 3, -, -, 1000, -, 180
Salsbury, Eli, 10, -, -, 1500, -, 200
Noyes, Manlove, 45, -, -, 18000, 165, 560
Shakespier, William N., 100, -, -, 30000, 25, 2420
Ridgely, Henry, 30, -, -, 3000, -, 1728
Davis, Jacob, 4, -, -, 600, 10, 30
Downey, Robert, 42, -, 8, 6300, 500, 1500
Vanwinkle, Charles, 30, -, -, 3000, 12, 275
Reed, Ezekiel G., 10, 12, -, 450, 50, 260
Scout, Augustus, 13, -, -, 5000, 200, 250
Morris, Aaron D., 6, -, -, 900, 20, 150
Trainer, Willam, 85, 5, -, 8500, 125, 550
Reynolds, Joseph, 62, 18, -, 7500, 250, 475

Johnson, Lemuel T., 147, 17, -, 12000, 200, 1000
Johns. John K., 150, 10, -, 2000, 180, 860
Griffin, Antoynette D., 40, -, -, 5200, 75, 450
Garrison, Gamaliel, 100, 25, -, 8710, 60, 760
Severson, John, 160, 36, -, 10000, 500, 690
Boyer, William E., 107, 15, -, 11000, 150, 695
Wallace, Benjamin, 120, 15, -, 11000, 150, 695
Boyer, James S., 120, -, -, 9000, 50, 750
Dickson, Jacob P., 120, 20, -, 6000, 400, 50
Wheatman, Samuel, 120, -, -, 9000, 100, 750
Reece, John P., 100, -, -, 10000, 400, 520
Cavender, Wilson, 170, 10, -, 21000, 100, 1500
Reed, Clement, 166, 50, -, 20000, 1000, 800
Cowgill, Mary C., 280, 100, -, 20000, 200, 1200
Faison, Nathaniel, 100, -, 35, -, 4000, 25, 530
Baccus, Issac, 40, 30, -, 2000, 25, 75
Ford, Daniel, 100, 20, 30, 12000, 200, 1125
Truax, Henry, 100, 15, 16, 20000, 250, 1200
William, John H., 40, 15, -, 2000, 475, 980
Pleasanton, Alexander, 30, 30, -, 1500, 30, 420
Faison, Sarah, 10, 9, -, 3000, 125, 150
Register, Eliza, 300, 20, -, 25000, 500, 1600
Voshell, Obadiah, 160, 30, -, 14000, 100, 1058

Johnson, Daniel R., 85, 15, -, 8500, 150, 750

Slaughter, Henry, 220, -, -, 22000, 300, 2750

Young, James, 150, 10, -, 12000, 500, 1500

Collins, George W., 180, 20, -, 15000, 200, 1475

Farrell, William R., 200, 45, 5, 17000, 300, 1500

Hoffecker, William D., 165, 10, 300, 20000, 400, 1250

Rickards, Molten, 100, -, 300, 7000, 200, 620

Truax, Benjamin, 305, -, -, 21000, 225, 1300

Slaughter, John C., 150, -, -, 12000, 50, 900

Collins, John M., 200, 15, -, 15000, 400, 1400

Farron, John C., 200, -, -, 20000, 400, 500

Spear, Andrew, 263, -, -, 24000, 500, 2025

Collins, Robert P., 134, -, 25-, 10000, 150, 575

Lattamore, Benjamin F., 120, -, -, 8400, 50, 575

Fennell John, 200, 60, -, 2000, 500, 2000

Slaughter, Ezekiel, 200, 50, 500, 18000, 200, 1300

Reynolds, John, 200, 10, -, 20000, 225, 1300

Turner, Benjamin, 140, 100, 107, 11000, 50, 408

Rawley, Robert, 80, 20, 117, 8000, 250, 1000

Cummins, Philip, 180, 20, -, 4500, 10, 350

Rawley, John, 75, 8,-, 3000, 100, 470

Rench, Washington, 100, 45, -, 7000, 150, 800

Reynolds, Francis, 200, -, -, 18000, 300, 1300

Hubbard, John W., 160, -, -, 5000, 375, 1675

Moor, Sewall, 72, 8, -, 3200, 10, 175

Denney, Frank J., 200, -, 75, 15000, 700, 1600

Snow, Jackson, 150, -, -, 12000, 250, 1451

Jefferson, Charles H., 100, -, -, 6000, 40, 345

Snow, James P., 148, 16, -, 10000, 140, 980

Hill, Henry C., 175, -, -, 10000, 500, 1900

Staten, William H., 158, -, -, 8000, 500, 2000

Tombleson, James, 45, -, -, 1500, 15, 361

Garner, Levi, 100, -, -, 4000, 100, 560

Jones, George T., 140, 20, -, 12000, 200, 1290

Cleaver, Noah, 90, 10, -, 3000, 50, 570

Rawley, William, 125, -, -, 4800, 100, 550

Starling, James, 150, 50, -, 8000, 400, 1120

Mitchell, James W., 180, 15, -, 15000, 600, 2250

Frazier, Carey, 160, 75, -, 15000, 400, 1451

Griffin, Samuel S., 175, -, -, 13000, 300, 750

Quillen, John, 100, 50, -, 8000, 125, 530

Wheatman, Charles, 112, 48, -, 10000, 200, 750

Peterson, John H., 125, 25, -, 25000, 800, 1200

Savin, John R., 200, 17, -, 13000, 250, 1290

Poor, John T., 121, -, -, 12000, 200, 690

Allee, Johnathan, 150, -, -, 10000, 200, 603

James, John J., 250, 58, -, 11000, 450, 1400

Wilday, Jacob _., 94, 12, -, 3000, 30, 350

Thompson, John D., 85, 15, -, 4000, 300, 330

Truax, James, 300, -, -, 20000, 500, 1265

Truax, Isaac 200, 10, -, 16000, 500, 1200

Truax, James T., 150, 10, -, 10000, 300, 1070

Hoffecker, James, 60, 15, -, 5000, 100, 308

Jones, William, H., 40, 5, -, 2000, 30, 390

Hoffecker, William H., 251, 5, 80, 20000, 350, 900

Denney, Samuel W., 100, 50, -, 9000, 500, 400

Perry, Alexander, 60, -, 40, 3000, 50, 400

Walker, Ezekiel, 200, -, 1000, 120000, 125, 635

Jones, Richard, 200, 10, -, 14000, 500, 1000

Beddle, John, 300, -, -, 15000, 200, 1807

Vane, Jesse S., 56, -, -, 3000, 50, 688

Ford, Presley, 175, 15, -, 8000, 300, 2045

Griffin, Jacob, 100, -, -, 4000, 400, 720

Robison, James C., 200, -, 20, 15000, 300, 2189

Smith, Alfred H., 200, -, 1000, 15000, 200, 1695

Register, Charles H., 165, 100, -, 12000, 400, 1350

Register, Isaac J., 100, 34, 100, 7000, 50, 900

Cannon, John W., 120, 10, 50, 10000, 100, 650

Sutton. Thomas L., 720, -, -, 20000, 375, 850

Ford, Anderson, 135, 15, -, 5000, 50, 1105

Palmatary, Daniel, 90, 10, -, 7500, 100, 617

Ford, Mathew, 110, -, -, 3000, 150, 655

Bennett, Thomas H, 100, -, -, 2500, 50, 410

Hollis, John, 100, -, -, 4000, 50, 313

Williams, John, 72, -, -, 3500, 50, 390

Benn, John J., 38, -, -, 800, 10, 100

Goldsborough, Joseph, 125, 10, 165, 8000, 150, 620

Bacon, James L., 100, -, 400, 3000, 100, 450

Robison, George, 10, -, -, 300, 10, 100

Severson, John, 15, 15, -, 600, 20, 100

Anderson, Joseph, 80, 15, -, 2700, 100, 300

McNutt, James, 45, 20, -, 2000, 10, 100

Walker, Charles W., 100, 25, -, 3000, 50, 250

Shorts, John H., 161, 15, -, 3500, 125, 660

Moor, Nemiah, 250, 50, -, 8000, 400, 1550

Maloney, Thomas, 125, -, -, 3500, 40, 460

Reed, Austin, 100, -, -, 3000, 50, 500

Hobbs, Titus J., 200, 40, -, 18000, 150, 1733

Collins, William T., 160, 53, -, 10000, 200, 1138

Jones, Nemiah, 60, 250, 5000, 125, 425

Logan, John R., 300, -, -, 20000, 400, 2100

Brown, Abram, 300, -, -, 10000, 300, 1200

Loatman, Samuel, 300, -, -, 5000, 100, 450

Collins, James H., 150, -, 4000, 10000, 200, 100

Spruance, James H., 235, -, -, 21150, 500, 200

Goldsborough, John T., 200, -, 600, 15000, 300, 2305

Fortner, James, 100, -, 1000, 10000, 1000, 750

Forbutter, John, 250, -, -, 10000, 100, 400

Robison, Peter W., 50, -, -, 1500, 100, 115

Goldsborough, John, 100, -, 125, 4000, 100, 300

Shorts, George, 170, -, 40, 4500, 50, 200

Shorts, William T., 133, 20, 93, 4000, 50, 400

Cummins, James, 100, -, 5000, 3000, 50, 275

Collins, William B., 195, -, 36, 20000, 300, 1524

Allee, Jacob R., 160, 5, 15, 12000, 375, 800

Jefferson, Ephraim, 137, 25, -, 8000, 250, 540

Staten, Charles E., 275, 50, -, 47000, 1500, 3000

Wright, William P., 140, 20, -, 16000, 300, 520

Gott, Mathew H., 160, 10, -, 25000, 200, 1030

Webster, Henry D., 125, 40, 10000, 400, 750

Ransom, William M., 150, -, -, 30000, 800, 1625

Woodhull, Francis, 21, -, -, 4000, 100, 225

Gilmore, James, 120, -, -, 12000, 125, 250

Anthony, John, 240, -, -, 240000, 1500, 1800

Hudson, Alfred L., 245, 20, -, 33000, 1000, 2270

Spruance, Horance, 310, -, -, 40000, 1000, 3000

Palmantary, Timothy C., 50, -, -, 5000, 500, 100

Cummins, George W., 175, 51, -, 30000, 600, 2500

Collins, William, 10, -, -, 1000, 50, 500

Brown, George F., 290, 35, 20, 30000, 460, 1950

Clements, James R., 50, -, -, 8000, 100, 530

Stockley, Ayres, 30, -, -, 5000, 200, 440

Dulin, Charles H., 125, 6, 6, 1500, 500, 1000

Pernell, John, 75, 50, 20, 3000, 75, 500

Wiest, George, 170, 46, 30, 12000, 500, 1066

Crawford, Jas., 340, 45, 15, 24000, 250, 1068

Jobson, John, 140, -, 3, 8000, 150, 800

Rembers, Charles, 80, 11, 5, 4000, 50, 1100

Stevens, Westly, 100, 30, 5, 7000, 150, 615

Green, Isaac, 150, 15, 10, 9000, 300, 1370

Finn, John, 160, 2, 20, 10000, 150, 850

Marshel, Moses, 80, 50, -, 4000, 60, 150

David, Nehemiah, 175, -, 30, 4000, 50, 700

Hurlock, Benjamin, 196, 67, -, 18000, 400, 1200

Parrey, Joseph, 130, 18, 4, 25000, 500, 1400

Lightsop, Joseph W., 130, 15, -, 7500, 300, 1000

Pratt, Nathaniel S., 150, 50, -, 12000, 150, 380

Foxwell, Tilghman, 170, 20, -, 20000, 300, 1100

Griffin, James, 100, -, -, 10000, -, 965

Jones, Robert, 80, 30, 18, 12000, 60, 250

Buckmaster, Wm., 720, 10, 4, 13000, 300, 1630

Stevans, William, 155, 30, 15, 16000, 300, 950

Jones, David, 100, 40, -, 5600, 200, 700

Wilson, Peter, 155, -, 4, 15000, 156, 600

Reese, Elizabeth, 175, -, 25, 18000, 200, 600

Staats, John P., 145, 51, -, 12000, 300, 1300

Harrington, Benj., 150, -, 20, 18000, 200, 1063

Frazier, Job M., 120, 46, -, 12000, 200, 880

Reams, John G., 50, 10, 5, 2000, 25, 480

Burrows, Henry, 30, 80, -, 2500, 100, 620

Ross, Charles G., 1100, 100, 100, 18000, 700, 1500

Walker, George D., 200, 75, 25, 15000, 300, 1900

Shermer, Henry, 140, 25, 25, 8000, 300, 1075

Mabury, Thomas, 200, 30, 170, 12000, 50, 800

Pratt, John, 200, 25, 30, 6000, 200, 1000

Pratt, Henry, 146, -, 1250, 100, 300

Chefferis, Benj., 125, 15, 10, 3000, 150, 400

Robison, William, 30, 50, 20, 1100, 100, 200

Collins, John W., 125, -, -, 11000, 200, 934

Spruance, David S., 170, 30, -, 15000, 300, 1250

Cook, George, 100, -, 3, 7000, 200, 880

Streets, Eduard, 119, -, -, 8000, 100, 825

Frazier, William, M, 130, 30, -, 7000, 100, 350

Burrows, Wm. _., 125, 50, -, 12000, 300, 925

Wilds, David, S., 20, -, 15, 20000, 695, 1150

Gassaway, Wm, 200, 15, -, 10750, 200, 1700

West, Benjamin, 160, 20, 20, 10000, 200, 700

Burrows, Benj. D., 105, 20, -, 6000, 250, 688

Truax, Sam P., 120, 32, -, 11000, 200, 825

Knotts, George W., 130, 30, -, 8000, 250, 1040

Boyles, James S., 60, -, 10, 3000, 20, 330

Hollett, Joseph H., 100, 10, -, 300, 20, 300

Hudson, Henry, 10, 4, 8, 1050, 50, 80

Loper, Joseph, 25, 15, 7, 2000, 100, 225

Jones, James, 90, -, 30, 250, 300, 950

Helyard, William, 58, -, 2, 2000, 100, 500

Taylor, John, 800, -, 60, 50000, 800, 3000

Downs, Nehemiah C., 151, 14, -, 10000, 250, 1500

Arnold, George G., 150, 20, 10, 17000, 500, 1175

Williams, James, 160, 12, -, 20000, 1000, 1900

Welds, James F., 85, -, 16, 10000, 275, 1000

Baily, Mason, 130, -, 30, 10000, -, 673

Knotts, Frederick, 200, -, 52, 10000, 150, 1364

Coverdale, Luke W., 150, -, 20, 12000, 100, 480

Cleaver, John, 125, 20, 20, 6000, 400, 970

Deakyne, John, 200, 83, -, 6000, 50, 290

Lamb, Thomas, 200, -, -, 12000, 100, 300

Myres, John, 72, -, -, 3000, 150, 440

Pratt, Henry S., 175, 50, -, 7000, 300, 1500

Saulsbury, Thomas, 450, 200, 50, 21000, 1500, 950

Wright, Ebenezer, 65, -, 30, 8000, 175, 425

Conner, Joseph B., 125, 8, -, 5453, 295, 1600

Bell, Leuro M., 130, 20, 25, 11000, 500, 800

Pernel, Henry H., 25, -, 125, 3000, 50, 800

Burrows, Ebenezer, 50, 10, 4, 1500, 50, 535

Rash, Joseph _., 40, 75, -, 2000, 50, 750

Copes, Thomas, 25, 50, 50, 2000, 50, 200

Burrows, James, 30, 60, 50, 1500, 50, 120

Grady, Rodgers, 80, 40, -, 3000, 50, 320

David, John, 150, 30, 20, 6000, 50, 550

Burrows, Edward, 60, 50, 20, 2000, 100, 500

Attix, Timothy, 50, 20, 15, 4000, 150, 400

Attix, Thomas, 120, 50, 25, 6000, 200, 820

Earl, Dariel (Daniel), 75, 25, 30, 3000, 50, 230

Brinkly, Carson, 75, 5, 20, 2500, 5, 350

Reed, Daniel, 90, 38, 10, 3000, 75, 410

Clark, James D., 180, 175, 20, 10000, 500, 1625

Clark John N., 65, 75, 40, 6000, 200, 900

Marris (Morris), William W., 60, 40, -, 3000, 200, 600

Downs, Joseph, 40, 75, 25, 3500, 100, 770

Downs, Enoch, 150, 25, 20, 9000, 200, 640

Guesford, James, 150, 25, 20, 9000, 200, 640

Shahan, Mary, 120, 3, -, 700, -, 606

Taylor, William, 360, 45, 40, 12000, 500, 1800

Consealer, Edward, 100, 15, 15, 2000, 50, 840

Kemp, Thomas, 35, -, 5, 1000, 25, 225

Rash, John S., 40, 10, 22, 1500, 100, 205

Jacobs, William H., 100, 30, 44, 400, 100, 300

Moor, Robert D., 100, 30, 17, 3500, 200, 400

Adkinson, John, 40, 14, 100, 2500, 200, 300

Duhadaway, James, 50, -, -, 150, 50, 250

Short, Joshua, 160, 20, 40, 180, 100, 442

Hazel, Mathew, 130, 20, 25, 6000, 400, 900

Truet, John K., 120, 40, 10, 6000, 100, 600

Short, Isaac J., 100, 10, 100, 8000, 50, 400

Carey, William H., 40, 21, -, 1500, 100, 500

Martin, William, 50, 6, 60, 3700, 100, 450

Taylor, Thomas, 115, 14, 12, 3000, 200, 865

Husbands, Howell, 120, 30, -, 5000, 200, 900

Hutchenson, Samuel, 100, 16, -, 5000, 500, 960

Short, Greenbury H., 125, 25, 50, 6000, 400, 600

Boyer, William, 40, 40, 40, 1500, 50, 270

Saulsbury, A__y, 100, 20, -, 3500, 25, 325

Pleasanton, John, 125, 25, 25, 4000, -, 710

Short, James M., 86, -, -, 3000, 100, 550

Short, Isaac W., 20, 56, -, 2000, 50, 280

Short, William S., 100, 40, 10, 3000, 100, 625

Thompson, Joseph, 155, 37, -, 8000, 200, 1025

Lafferty, John W., 140, 10, 65, 8000, 250, 600

Durbrour, Armeull, 100, -, 14, 5000, 165, 650

Graham, Robert, 35, -, -, 1500, 50, 380

Hazel, Mary, 75, 15, 10, 300, 10, 400

Hazel, James P., 80, 20, -, 4000, 200, 625

David, Enoch, 100, 40, 15, 2500, 100, 800

Numbers, John, 90, 15, 3500, 120, 375

Hobson, James, F., 125, 50, -, 4000, 150, 450

Bailey, John, 110, -, 15, 2100, 50, 230

Holland, Williams, 53, -, -, 3000, 100, 215

English, Thomas, 75, 10, 17, 4000, 80, 250

Jones, Edward F., 60, 20, -, 240, 125, 800

Donovan, Thomas B., 100, 10, 54, 3000, 150, 325

Moor, Thomas, 60, -, 33, 3640, 150, 600

Pruett, William, 200, 100, 70, 10000, 200, 1500

Woodall, James, 100, 40, 20, 8000, 100, 400

Dodd, James M. 80, 30, 25, 6000, 50, 200

Carey, Cherrey, 62, 158, 4, 6500, 25, 550

Ford, Martin B., 50, 25, 25, 3000, 25, 300

Moor, Andrew F., 29, -, -, 4000, 25, 100

Jones, William M., 60, 30, 10, 2000, 20, 300

Bailey, Thomas E., 250, 130, -, 24000, 700, 2150

Douna, David O., 150, 30, 10, 11000, 350, 1250

Taylor, Jenifer S., 65, 60, -, 5000, 50, 300

Rees, Davis, 160, -, -, 15000, 350, 1000

Graham, John W., 70, 75, 35, 4000, 750, 775

Moor, John, 140, 65, -, 16000, 400, 1050

Merideth, Peter, 240, 60, -, 15000, 500, 1500

Hutchinson, Samuel, 108, -, 25, 9000, 100, 800

Cavender, James, 250, -, 20, 16000, 200, 1000

Boyd, Joshua, 140, 45, 15, 1000, 200, 900

Carney, William, 50, 87, 3000, 50, 400

Hall, John, 160, 25, 30, 10000, 50, 850

Register, Saml. M., 150, 24, 24, 14000, 100, 900

Daris (Davis), Thomas, 110, 25, 10, 13000, 1400, 950

Laphan, Isaac, 172, -, 28, 20000, 400, 650

Bilderback, John, 205, 20, -, 20000, 500, 1400

Garrison, Ebenezer, 140, 10, 30, 13000, 250, 730

Dear, Robert, 65, 10, 8, 4000, 100, 450

Bryant, Robert, 40, -, 5, 2000, 10, 250

Morgan, John, 175, 26, -, 16000, 150, 700

Morgan, William, 150, -, 2, 8000, 75, 800

Ringold, Lemuel, 300, 150, 50, 3000, -, 1000

Daniels, Samuel, 110, 40, 20, 8000, 300, 350

Green, John, 185, -, 12, 14000, 100, 1500

Smithers, William, 135, 8, 9, 12000, 300, 1300

Mason, Trustand, 150, 30, 20, 10000, 300, 806

Wright, John 161, 30, -, 12000, 140, 1220

Nelson, William, 190, 12, 25, 17000, 400, 1325

Jones, William S., 150, 30, -, 1250, 415, 1100

Lewis, Rees, 75, 3, -, 5000, 100, 300

Parson, Joseph, 70, -, 4, 350, 75, 600

Massey, Henry P., 125, -, 10, 12000, 500, 900

Underwood, Nathan F., 200, 38, -, 10000, 200, 800

Blackiston, Samuel, 80, 10, 5, 3000, 100, 250

Young, William F., 140, 15, -, 5000, 150, 700

Moor, James C., 160, 10, 30, 10000, 100, 650

Davis, Jesep W., 120, 28, 10, 10000, 100, 350

Blackiston, Bryan F., 300, 60, 26, 30000, 150, 1500

Parson, John, 35, -, 25, 1000, 50, 250

Smear, William, 150, 50, 100, 13000, 75, 1660

Jones, Samuel M., 140, 20, 40, 12000, 600, 800

Ford, James, 40, 10, 20, 4000, 100, 450

Wisher, John, 175, 15, 15, 20000, 300, 1400

Garrison, Ephraim, 165, 15, 43, 18000, 600, 1860

Jefferson, Elihu, 100, 60, -, 10000, 250, 600

Cexper, Richard, 80, 10, -, 6000, 100, 800

Lofton, Flary C., 5, -, -, 4000, 100, 400

Moor, Harvy H., 47, -, -, 6000, 100, 850

Bishop, John H., 80, 25, -, 8000, 500, 500

Finlow, Heran, 60, -, -, 6000, -, 420

Emerson, James W., 100, -, -, 6000, 60, 300

Boggs, David, 110, 25, -, 600, 100, 650

Sinn (Finn), John J., 175, 50, -, 14000, 100, 600

Reis, Frances, 75, 95, -, 7000, 100, 600

Wilson, Henry L., 40, 125, -, 6500, 150, 950

Spring, Merrit, 152, 40, -, 10000, 500, 968

Pleasant, Daniel, 100, 70, -, 4000, 125, 500

Dill, William, 200, 30, -, 10000, 200, 700

Farrow, Nathan, 760, -, -, 7000, 50, 1400

Potter, Faris, 55, 5, -, 300, 5, 800

Wheatby (Wheatly), Richard, 250, -, -, 18000, 150, 340

Ross Daniel, 90, 10, -, 6000, 200, 1200

Wilson, John C., 150, 12, -, 15000, 250, 750

Hasbands (Husbands), John, 60, 20, -, 5000, 60, 1569

Foster, Thomas _., 962, 31, -, 12500, 250, 420

Lawson, David, 100, -, -, 9000, 125, 4480-

Fox, Abraham, 60, 6, -, 6000, 25, 780

Donovan, Grover, 40, 20, -, 7000, 200, 110

Willis, William, 100, -, -, 8000, 50, 1270

Clements, Charles, 130, -, -, 10000, 175, 200

Hazel, Henry K., 260, 25, -, 16000, 150, 685

Carson, Robert E., 170, -, -, 10000, 500, 1838

Ham, Benjamin F., 140, 20, -, 15000, 220, 500

Pleasanton, Henry, 20, 11, -, 7000, 100, 618

Cowgill, William, 400, -, -, 14000, 550, 928

Gils, Levi, 160, 40, -, 15000, 200, 800

Moor, Isaac, 300, 20, -, 20000, 400, 500

Barcus, John, 60, 4, -, 5500, 150, 1550

Pleasanton, Stephen, 100, 15, -, 5000, 150, 300

McCauly, Truston, 40, 10, -, 6000, 125, 600

Moor, Abraham, 17, -, 1000, 50, 300

Hazel, William, 100, 3, -, 6500, 150, 613

Benson, Elza, 75, -, 20, 2500, 50, 450

Laws, Alexandria, 200, 14, -, 23000, 1500, 2500

Barton, Robert, 250, 25, 132, 12000, 50, 700

Marvel, Emory, 150, -, 150, 8000, 200, 710

Deges (Deger), Frederick, 75, -, 10, 3000, 50, 150

Rawley, John, 120, 15, 500, 6000, 200, 1200

Robison, James, 100, 10, 20, 12000, 200, 650

Lenard, Webster D., 125, 10, 20, 2500, 200, 600

Thomas, Robert, 125, -, -, 10000, 100, 562

Cowgill, Daniel C., 125, 25, -, 12000, 325, 1650

Brown, Timothy, 90, -, 20, 5000, 150, 800

Huffnel, Adam, 245, 50, -, 15000, 245, 2780

Slaughter, Isaac, 470, 15, 60, 17000, 300, 1500

Husband, Wm. 120, -, 30, 5000, 50, 500

Nowel, James S., 130, -, 15, 16000, 320, 1877

Codwright, Mary, 100, -, 22, 9000, 150, 700

Parris, George, 145, -, 5, 15000, 600, 1248

Hutchinson, Wm., 150, -, -, 12000, 150, 500

Berry, Isaac, 150, 15, -, 12000, 600, 1000

Cowgel, Joseph C., 180, 30, 20, 15000, 350, 1075

Blackiston, Jas., 150, -, 25, 12000, 150, 1150

Johnson, James D., 40, -, 30, 2000, 50, 200

Richison, James B., 28, -, -, 3000, 25, 670

Lober, Peter E., 100, -, 40, 8000, 250, 1000

Spencer, William, 60, 20, -, 3000, 50, 500

Kemp, John, 30, 10, 10, 2500, 50, 160

York, Samuel, 50, 13, 50, 4000, 50, 673

Cadmus, Michel S., 125, -, 123, 1000, 300, 485

Griffith, Charles, 260, 25, -, 40000, 400, 1600

Donovan, Burton, 160, 180, 20, 15000, 200, 1100

Keath, Thomas, 78, 27, 18, 7020, 200, 790

Durban, Benjamin, 100, 100, -, 2500, 30, 345

Smith, William J., 200, 10, 4, 13500, 50, 345

George, Thomas, 50, -, -, 300, 40, 430

Harper, Charles, 156, 20, 15, 8000, 100, 1000

Barber, Joseph M., 104, 50, 18, 8000, 100, 500

Keath, James, 100, -, 30, 8000, 150, 940

Slaughter, John, 100, -, 10, 15000, 100, 700

Satterfield, Chs., 180, 20, 40, 20000, 150, 1500

Cardwright, Theo., 201, 120, -, 5000, 100, 925

Seward, William, 300, -, 10, 30000, 600, 2230

Ennis, James, 80, -, 50, 4000, 100, 175

Vaughn, Charles, 100, -, 10, 8000, 50, 350

Griffith, Jas. T., 100, 31, -, 3000, 60, 1520

Gols, Richard, 250, 75, 400, 18000, 300, 1750

Denner, Joseph, 100, 10, -, 8000, 200, 600

Seward Joseph, 185, 15, -, 16000, 200, 1200

Hamm, Pleasanton, 8, -, -, 2500, 10, 225

Keys, Eliza, 80, -, -, 4000, 50, 585

Graverly, William H., 40, -, 9, 6000, 50, 670

Hutchinson, Mathew, 175, -, 70, 14000, 125, 1025

Hargadine, Samuel, 25, -, -, 10000, 300, 800

Smyth, William, 150, -, 20, 16000, 200, 1620

Clow, John, 96, -, -, 9000, 200, 400

Ranals, John C., 75, 15, 10, 8500, 100, 530

Riggs, Alex. S., 125, 12, 13, 7500, 10, 878

Boyer, Thos. R., 150, -, 15, 8000, 150, 692

Boyer, Thomas D., 150, -, 10, 600, 140, 1000

Stockwell, Lewis, 90, 10, 14, 6000, 100, 500

Dish, Joseph E., 130, 12, 40, 9000, 400, 1200

Denney, Isaac M., 100, 10, 50, 8000, 200, 900

Armstrong, George, 38, 3, 61, 3000, 50, 200

Hammond, John, 110, 40, -, 6000, 600, 850

Harrington, R., 120, 20, -, 5000, 200, 900

Robbins, D. H., 90, 60, -, 5000, 100, 600

Harrington, Jas., 45, 6, 30, 3000, 56, 450

Watkins, E. R., 50, 10, -, 2000, 100, 250

Voshel, Alexander, 80, 100, -, 6000, 27, 560

Holland, Wm. D., 175, 118, 100, 800, 135, 798

Martin, Wm. D., 50, 10, -, 2500, 50, 300

Allcott, Jared, 30, 25, -, 2000, 50, 175

Masten, Hezekiah, 140, -, -, 8000, 200, 700

Spencer, Wm., 140, 22, -, 5000, 200, 630

Tomlinson, J. D., 80, 8, -, 4000, 50, 150

Sapp, James, 50, -, -, 3000, 200, 625

Herrington, Martin, 124, 6, -, 10000, 375, 735

Davis, Joshia, 100, 30, -, 4000, 100, 500

Warrington, A., 100, 20, -, 5000, 150, 200

Jacobs, Molten, 160, 42, -, 6000, 150, 380

Wharton, Charles, 160, -, -, 6000, 209, 445

Huburt, J. L., 50, -, -, 1000, 15, 178

Morris, G. F., 35, 8, -, 1200, 40, 390

Marvel, Josiah, 100, 13, 22, 6500, 240, 475

Hill, John, 70, 5, -, 5000, 50, 349

Case, John, 240, -, -, 6000, 340, 533

Taylor, John, 70, 16, 15, 4000, 135, 445

Harrington, Jas., 150, 27, 20, 12000, 700, 1450

Draper, Anry, 130, -, -, 12000, 150, 685

Frazier, Jas., 150, 18, -, 12000, 680, 1445

Clarkson, Hester, 30, 7, -, 1200, 16, 40

Morgan, Wm., 80, -, -, 4000, 50, 346

Sharp, Clement, 160, 40, 70, 2000, 500, 670

Heathers, John J., 100, 40, -, 2500, 50, 100

Traider, W. R., 100, -, -, 2000, 40, 110

Taylor, Peleg, 75, 25, 25, 6000, 50, 170

Betts, Stephen, 70, 37, 37, 3000, 250, 300

Wilcox, Parker, 100, 36, -, 4000, 150, 200

Tomlinson, Wmn, 65, 35, 2500, 60, 200

McColly, Truston, 80, 30, -, 300, 10, 300

Dausen (Dawson), Wm. H., 80, 23, -, 2008, 25, 50

Emory, Charley, 100, 150, -, 4000, 75, 100

McColley, George, 80, 27, -, 2500, 200, 325

Vineyard, Henry, 80, 26, -, 3000, 75, 370

Marvel, H. H., 100, 100, 48, 6000, 150, 298

Simpson, W. J., 140, 70, -, 5000, 275, 70

Frader, Benj., 140, -, -, 2000, -, -

Ensly, James M., 10, 25, -, 1000, 50, 100

Dickerson, Charles, 70, 30, -, 2600, 150, 250

Letts, Randolph, 130, 30, -, 6000, 100, 250

Weeks, Lorenze B., 11, -, -, 2000, -, 50

Scott, Romey P., 5, -, -, 800, 50, 150

Vineyard, Wm., 65, -, -, 1000, 150, 230

Vineyard, Carter, 100, 45, -, 4000, 100, 175

Anderson, J. C., 38, 16, -, 1400, 75, -, 240

Wilson, Wm., 25, -, 9, 1000, 135, 165

Lippencott, Isaac, 160, 40, -, 5000, 95, 230

Lahe, Andrig, 65, 20, 15, 6000, 140, 200

Sapp, Wm., 100, 300, -, 1200, 100, 40

Herrington, M. A. 130, 40, -, 6000, 415, 528

Quillen, John, 4, -, -, 1000, 100, 545

Sharp, James, 150, 100, -, 10000, 775, 795

Macklin, A., 112, -, -, 10000, 100, 448

Hitch, Spencer, 75, 38, -, 3000, 125, 293

Register, David, 150, 52, -, 5000, -, -

Degraff, Elbert, 45, 9, -, 2000, 50, 75

Estes, D. C., 17, 3, -, 1000, 50, 100

Watson, Daniel, 188, 10, -, 5000, 50, 465

Austin, Wm. H., 140, 20, -, 8000, 150, 800

Riggs, James, 55, 5, -, 3000, 65, 270

Townsend, N. J., 40, 17, -, 1500, 354, 195

Townsend, C. B., 45, 12, -, 1000, -, 200

Jump, Lemuel, 50, 100, -, 2500, 10, 200

Armor, Samuel, 13, 44, -, 1200, 55, 125

Taylor, H. S., 40, 10, -, 2500, 25, 90

Scott, Ira, 10, -, -, 1500, 25, 125

Scott, David, 300, 200, -, 20000, 250, 600

Davis, Wm., 10, -, -, 1500, 50, 75

Smith, Nelson, 30, 5, -, 2000, 50, 175

Wilson, John D., 15, 145, -, 2500, 100, 130

Higgins, H., 50, 25, -, 1500, 30, 75

Sharp, Peter, 100, 80, -, 8000, 75, 95

Redden, Jeremiah, 16, -, -, 600, -, 60

Owens, Joseph, 135, 37, -, 6000, 65 435

Willis, Joseph, 75, 10, -, 3000, 60, 228

Philips, W. R., 51, -, -, 5000, 205, 295

Whitenack, Thomas, 125, 40, 9000, 150, 230

Meserle, George B., 53, 5, -, 6000, 175, 396

Lofland, John M., 35, -, -, 2500, 225, 341

Richards, James, 47, 5, -, 10000, 150, 1175

Redifor, Levon W., 145, 30, -, 6000, 170, 495

Griffin, Sarah A., 100, 10, -, 5000, 360, 740

Collins, John W., 100, 25, -, 3000, 20, 200

Harrington, J. K., 90, 40, 40, 6000, 175, 748

Lynch, Jas. H., 200, 40, -, 20000, 570, 1140

Willis, James H., 51, -, -, 2500, 65, 165

McColly, R. M., 150, 20, -, 8000, 25, 372

Vineyard, Carter, 120, 9, -, 8000, 275, -

Webb, John M. C., 100, 16, 85, 8000, 350, 1145

Webb, John M., 75, 25, 64, 4000, 200, 480

Sipple, James D., 75, 56, 265, 7000, 250, 1192

Tompson, Daniel, 70, 40, 44, 5000, 750, 607

Webb, Charles A., 30, 15, 143, 1500, 125, 445

Quillen, Jacob, 75, 30, 1500, 5000, 200, 1103

Kirby, Benjamin, 100, 5, 75, 6000, 117, 400

Mosely, Morris, 130, 65, 40, 7000, 89, 568

Master, Wm. T., 116, 30, 54, 8000, 175, 785

Kerby, Wesly, 100, 80, 213, 5000, 242, 564

Tompson, Levi, 8, -, -, 400, -, 65

Riley, Robert, 150, 10, -, 8000, 155, 495

Moore, John M., 88, 10, -, 5000, 160, 575

Emory, John, 192, 30, 60, 14000, 570, 1448

Burr, Charles M., 50, -, -, 6000, 180, 2110

Coleburn, John, 120, -, 100, 7000, 185, 928

James, Elias P., 130, 10, 50, 9000, 320, 385

Stephenson, R. H., 140, 40, -, 12000, 320, 764

Hitch, John, 100, 20, -, 7000, 77, 557

Bethards, W. H., 100, 50, 30, 8000, 333, 840

Ross, James, B., 100, 140, -, 3000, 160, 427

Willis, Hiram, 59, -, -, 2000, -, 15

Morris, John C., 160, 5, -, 8000, 746, 652

Dickerson, Benj, 150, 35, -, 8000, 200, 472

Marvel, James, 135, 15, -, 10000, 105, 521

Waler, Blake, 150, 323, -, 10000, 500, 675

Wyatt, George E., 40, 30, -, 3000, 50, 135

Herring, Wm. G., 183, 52, -, 8000, 372, 573

Rather, Robert C., 160, 46, -, 10000, 1055, 895

Ready, James, 100, 58, -, 6000, 220, 788

Homes, Nathaniel, 100, 47, -, 9000, 640, 10

Reynolds, Zacheriah, 195, -, -, 12000, 290, 755

Hammond, E. A., 180, 16, -, 9000, 75, 423

Hydon, Benj., 50, 26, -, 3000, 181, 365

Hinsley, Ephraigm, 20, 10, -, 1000, 35, 95

Morris, Francis, 40, 20, -, 2000, 10, 195

Milton, Matthew, 140, 30, -, 9000, 500, 986

Milton, Wm. B., 90, -, 100, 6000, 170, 635

Webb, George F., 60, 80, -, 5000, 95, 423

French, Nathaniel, 60, 8, -, 2000, 35, 335

Mills, David, 60, 8, 29, 2000, 30, 260

Milton, Jas. C., 70, 60, -, 3000, 150, 403

Hill, Absalom, 8, -, -, 2000, 175, 511

French, Wm. H., 54, 54, -, 1200, 75, 253

Bennett, Joshua, 142, 32, -, 8000, 387, 1060

Lindle, Jesse, 85, 25, -, 400, 60, 243

French, George, 100, 35, -, 6000, 95, 484

Watson, David, 50, 8, -, 3000, 275, 816

Holland, James, 100, 25, -, 5000, 115, 565

Webb, Thos. _., 60, 15, 40, 4000, 181, 781

Milton, Alexander, 60, 15, 20, 4000, 80, 415

Porter, William, 150, 20, 300, 12000, 145, 1340

Coverdale, Samuel, 30, -, -, 1000, 125, 285

Wadkins, J. R., 300, 300, 500, 10000, 800, 373

Hinsley, Cornelius, 60, 40, -, 2000, 75, 322

Cole, Manlove, 30, 12, -, 2500, 50, 350

Davenport, Nat, 30, -, -, 500, -, 100

Mason, D. H., 105, 55, 120, 7000, 337, 484

Lynch, Robert, 200, 11, -, 11000, 340, 667

Postles, Jas. H., 100, 35, -, 10000, 470, 706

Hynson, G. L., 118, 47, -, 12000, 368, 885

Short, John, 87, 20, -, 6000, 201, 410

Jackson, Levi H., 85, 12, -, 3000, 200, 550

Tompson, Selby, 68, 20, -, 3000, 170, 462

Hammond, J. H., 80, 50, 2, 5000, 220, 525

Tolbert, Joshua, 60, 4, 25, 4500, 215, 553

Neadles, Benj., 17, 30, -, 2500, 100, 225

Hall, Silas, 40, 7, 113, 3000, 300, 90

Townsend, W., 20, -, -, 500, 10, 70

Richards, W. H., 35, 45, -, 1500, 115, 266

FitzGarrell, Elijah, 9, 20, -, 800, 30, 105

Wyatt, Major, 15, 20, -, 700, 10, 30

Jester, Isaac, 25, 15, -, 1000, 97, 310

Foster, Burtin, 26, 4, -, 500, -, 40

Fowler, John, 50, 20, -, 1800, 95, 385

Hall, Draper, 10, -, 350, 18, 242

Evans, Joseph H., 92, 15, -, 3000, 140, 395

Parsons, Joseph, 8, 22, -, 1100, 20, 480

Hall, Charles P., 30, 6, -, 1000, 25, 95

Thomas, Andrew T., 35, 5, -, 2500, 162, 177

Hall, W. B., 30, 30, -, 1500, 45, 350

Hall, Wm. E., 100, 120, -, 3500, 25, 230

Hall, J. W., 80, 40, -, 2000, -, 12

Bennet, J. M., 40, 73, -, 2000, 170, 563

Scott, Robert, 20, 20, -, 800, 10, 125

Andrew, Lutheer 80, 50, 200, 5000, 150, 847

McColly, Wm. J. H., 40, 20, 127, 3000, 351, 721

Wilkins, Eli W., 26, -, 120, 2000, 40, 226

Mills, Richard, 100, 35, 75, 4000, 400, 1236

Atkins, Robert, 60, 80, -, 2000, 155, 200

Donovan, Gibson, 70, 90, 10, 2000, 40, 334

Scott, Nehemiah, 4, 3, -, 200, -, 100

Scott, Garrison, 4, 10, -, 200, -, 73

French, Joseph, 150, 80, -, 6000, 235, 677

FitzGarrel, George, 40, 33, -, 1500, -, 100

Coverdale, Elizabeth, 30, 39, -, 900, -, 180

Thomas, George, 25, 30, -, 1500, 254, 405

Herring, Saml., 22, -, -, 2000, 100, 250

Clifton. Daniel, 23, 20, -, 1500, 64, 575

Bechel, John A., 20, 60, 20, 5000, 400, 445

Thomas, N. B., 40, 20, -, 1500, 150, 500

Beswick, C. B., 80, 50, 130, 4000, 1000, 900

Beswick(Berwick), W. P., 75, 40, 40, 2000, 500, 875

Hall, J. W., 30, 60, -, 1500, 200, 550

Maloney, John, 50, 35, 70, 2500, 20, 400

Coverdale, George, 45, 60, 30, 1500, 10, 40

Cain, Mary B., 60, 100, 220, 5000, 300, 565

Thomas, J. M., 100, 100, -, 5000, 150, 125

Maloney, John, 12, 30, 100, 1200, 100, 450

Macklin, Elias, 30, 120, 30, 2000, 60, 15

Maloney, A. J., 200, 130, 150, 8000, 600, 900

Hudson, James, 150, 90, 200, 5000, 200, 700

Hughs, Benj., 50, 450, -, 2500, -, -

Torbert, Peter, 100, 65, 328, 8000, 160, 800

Johnson, J. M., 20, 10, -, 500, 20, 150

Davis, Jehu, 90, 100, 300, 5000, 350, 500

Wheder, Mary, 25, 10, 15, 1000, 60, 240

Hall, David, 30, 6, 40, 3000, 300, 700

Touett, George, 50, 10, -, 4000, 300, 408

Jester, Ann, 15, 5, -, 500, 100, 125

Burhance (Buchance), G. H., 108, 10, -, 9000, 300, 400

Hall, Hetty, 300, 100, 150, 15000, 250, 125

Bradley & Bennet, 140, 56, -, 1200, 125, 375

Jester, Isaac R., 20, 10, -, 1000, 300, 200

Smith, E., 30, 82, -, 2000, 300, 175

Mitten, Daniel, 44, 3, -, 800, 70, 255

Herring, Benj., 60, 80, -, 4000, 200, 250

Burr, Daniel, 100, 40, -, 10000, 250, 250

Houston, Joseph, 100, 15, -, 3000, 250, 350

Whitacre, Stephen, 40, 5, -, 1000, 40, 130

Hammond, N. D., 80, 70, -, 4000, 140, 225

Fisher, George, 80, 40, 15, 2000, 30, 350

Mills & Son, 100, 35, -, 5000, 400, 500

Jenkins, S. T., 130, 25, -, 9000, 350, 630

Meredith, R. G., 155, 25, -, 6000, 300, 375

Heveloe, Solomon, 45, 10, -, 1400, 100, 100

Andre, Elijah, 80, 30, -, 1800, 75, 100

Banks, N. J., 40, 24, -, 3000, 6, 310

Smith, A., 40, 16, -, 3000, 150, 490

Fountain, N., 100, 35, 30, 4000, 200, 500

Davis, George, 100, 60, 100, 8000, 100, 975

Meredith, D., 120, 180, 10, 6000, 200, 300

Collins, Gibson, 125, 50, 50, 2250, 125, 180

Cubbage, Saml., 90, 133, 50, 5000, 300, 670

Redden (Redder), John, 132, 50, 190, 7000, 250, 380

Cole, Nehemiah, 160, 125, 14, 6000, 700, 200

Davis, Wm. E., 50, 16, -, 3500, -, 250

Phelps, Richard, 18, -, -, 4000, 150, 200

Griffin, James, 48, 2, -, 5000, 175, 370

Henderson, Benj., 80, 50, -, 10000, 200, 475

Webb, John M., 100, 28, -, 4000, 130, 400

Murphy, G. O., 33, 3, 2000, 15, 50

Wroten, Robert, 50, 90, -, 1500, -, 250

Hill, Joshua, 66, -, -, 3000, 800, 900

Carpenter, Robt., 95, -, -, 1300, 400, 800

Davis, Joshua, 130, -, -, 4000, -, 500

Carlesie, M. R., 10, -, -, 1000, -, 450

Mitchel, J. R., 50, -, -, 2000, -, 240

Buell, Rowland, 120, 48, -, 7500, 175, 180

Gedwin (Godwin), Daniel, 20, -, -, 1000, -, 500

Pratt, Nath., 69, 40, 60, 5000, 150, 550

Davis, Nehemiah, 16, 250, -, 5000, 45, 600

Louery, Justus, 10, -, -, 1500, 100, 1500

Davis, James, 6, -, -, 700, 200, 360

Ross, Samuel, 5, -, -, 500, -, 60

Torbert, J. W., 6, -, -, 490, -, 125

Marshal, Wm., 7, -, -, 600, -, 300

Harris, Len, 18, -, -, 800, -, 800

Lofland, Joseph, 12, -, -, 100, -, 100

Postles, James, 11, -, -, 900, -, 200

Gough, John, 29, 113, 24, 40000, 2000, 10520

Grier, George S., 5, -, -, 500, -, 200

Norton (North), A. T., 102, 25, -, 9000, 700, 800

Lyndon, Caleb, 68, 8, -, 5500, 350, 525

Snell, Hiram, 55, 8, -, 10000, 430, 530

Rosa, John J., 50, 10, -, 15000, 500, 400

Wallace, Thos., 22, -, -, 8000, 200, 300

Dorsey, W. N. W., 175, 29, 35, 16000, 680, 1650

Griffith, Robert, 180, 10, -, 4000, 473, 1260

McIntosh, J. O., 80, 20, -, 5000, 200, 400

McColley, John, 80, 80, 110, 8000, 300, 700

Masten, D. R., 3, -, -, 500, 20, 40

Hadlock, P. J. 60, 30, -, 4000, 200, 350

Richard, Wm., 30, -, -, 1000, -, 30

Quillen, James, 170, 118, -, 13000, 412, 690

Needles, John F., 65, 10, -, 2000, 75, 300

Prettyman, William, 411, 15, -, 1000, 10, 200

Powell, Nathaniel C., 160, 138, -, 1450, 200, 700

Whitaker, Jacob, 72, 72, 6000, 65, 300

Tomlinson, William T., 25, 55, -, 2000, -, 200

Shockly, James W., 250, 50, -, 3000, -, 300

Ochletree, John V., 100, 71, -, 7400, 40, 300

Davis, Charles, 35, -, -, 100, -, 35

Sharp, Jesse, 100, 80, -, 7200, 40, 500

Coverdale, Jeremiah, 6, -, -, 60, -, 500

McCauley, Clement C., 100, 100, -, 4000, 20, 300

Ward, Lewis, 100, 36, -, 2700, 50, 200

Estes, George E., 38, 12, -, 1000, 20, 150

Manlove, Sarah A., 4, -, -, 80, 10, 400

Prettyman, John, 75, -, -, 900, 25, 200

Smith, Johnathan, 58, 6, -, 1900, 40, 200

Dickerson, Asa, 60, -, -, 1200, 10, 100

Wroten, Smith, 125, 25, -, 1500, 130, 200

Tatman, Saml. P., 210, 211, -, 7000, 50, 400

Griffith, Jane H., 45, 15, -, 900, 30, 100

Griffith, Albert, 116, 25, -, 2900, 90, 500

Holly, James, 150, 50, -, 5000, 200, 800

Messick, Covington, 100, 20,-, 1200, -, -

Hatfield, James T., 150, 50, -, 2400, 75, 400

Prettyman, Burton, 65, 55, -, 2400, 125, 400

Decker, Isaac, 4, -, -, 80, -, 20

Outin, Perry, 60, 10,-, 700, -, -

Willis, Nathan, 4, -, -, 60, -, 60

Cooper, Samuel C., 3, -, -, 60, -, 20

Ralston, Robert, 130, 110, -, 6000, 50, 700

Scott, Wesly, 85, 15, -, 2000, 60, 300

Hamilton, William, 75, 30, -, 2100, 20, 400

Lord, Robert, 125, -, -, 3200, 20, 400

Johnson, Alexander, 220, 80, -, 15000, 300, 1500

Cordry, Nath., 6, -, -, 90, -, 25

Prettyman, Alfred P., 60, 50, -, 1500, 40, 250

Prettyman, Jane B., 75, 25, -, 2000, 160, 600

Corday, Thomas H., 80, 60, -, 3000, 50, 1400

Rush, James, 66, 34, -, 2500, 80, 500

Morgan, James H., 75, 25, -, 1540, 300, 600

Booth, Elias T., 180, 96, -, 7500, 200, 1100

Hamilton, Johnathan, 75, 15, -, 1500, 15, 60

Raughly, Easter, 80, 75, -, 20000, 90, 115

Jones, Samuel O., 150, 57, -, 48000, 150, 485

Redden, John O., 60, 10, -, 12000, 35, 53

Hopkins, William N., 200, 50, -, 3000, 130, 300

Tyler, Stephen, 65, 30, -, 1000, 30, 63

Thomas, John R., 3, 2, -, 149, 5, 28

Raughley, John W., 70, -, 60, 1530, 55, 293

Butler, Willis W., 26, 14, -, 560, 50, 105

Corday, Charles H., 60, 40, -, 1000, 50, 135

Downs, Henry, 75, 30, -, 1200, 10, 350

Downs, William H, 4, -, -, 80, 25, 1125

Scott, Noah, 113, 113, -, 3390, 60, 511

Dickerson, Joshiah, 110, 37, -, 2200, 100, 360

Long, Isaac J., 3, 37, -, 400, -, 10

Davis, William T., 2, 1, -, 120, -, -

Shockly, George, 100, 37, -, 2000, -, 115

Fleming, George, 4, -, -, 60, -, 120

Willard, Norman, 25, 20, -, 1600, 30, 125

Hill, Daniel, 30, 15, -, 675, 100, 200

Johns, John H., 100, 25, -, 2500, 300, 500

Johnson, Nathaniel B., 70, 80, -, 2250, 50, 200

Messick, James B., 100, 25, -, 15000, 20, 270

Wood, James J., 300, 117, -, 8000, 20, 2500

Harrington, Charles, 100, 64, -, 3250, 120, 650

Willis, Joseph W., 70, 70, -, 2080, 20, 75

Walker, James J., 55, 56, -, 1332, 10, 92

Dickerson, William H., 140, 60, -, 4000, 200, 400

Bradly, Thomas, 90, 110, -, 1570, 25, 150

Corday, Jacob, 223, 223, -, 2956, 30, 600

Adkins, Thomas R., 60, -, -, 360, 100, 300

Fisher, Shwinger, 100, 100, -, 6000, 58, 300

Harrington, Johnathan, 104, 103, -, 305, 25, 300

Fleming, Beniah, 8, 100, -, 1400, 20, 200

Hammond, Mathias, 150, -, -, 3000, 200, 475

Griswold, Abner, 125, 155, -, 6000, 280, 417

Nelson, Daniel P., 188, 188, -, 7500, -, 300

Satterfield, John, 40, -, -, 800, 25, 130

Grier, John, 200, 100, -, 4500, 75, 400

Harrington, Samuel S., 263, 110, -, 1652, 100, 500

Bosworth, William C., 300, 150, -, 8208, 500, 1000

Welch, Jacob, 75, 161, -, 1092, 40, 200

Masten, Joseph A., 200, 150, -, 7000, 100, 486

Steel, Robert D., 10, -, -, 100, 5, 109

Graham, Joseph, 300, 70, -, 3720, 9000, 185

Wyatt, Isaac, 60, 15, -, 750, 40, 263

Thomas, Allen, 4, -, 1, 40, -, -

Miner, John, 40, 8, -, 480, 50, 184

Camper, William P., 60, 10, -, 700, 75, 260

Harris, Samuel, 110, 40, -, 2250, 40, 300

Moore, James A., 90, 70, -, 3000, 100, 200

Butler, William, 200, 100, -, 6000, 400, 1070

Harrington, William M., 143, 142, -, 5400, 75, 750

Wyatt, John, 54, 12, -, 1000, 75, 253

Mininer, Samuel, 35, 250, -, 770, 50, 300

Voshell, John H., 75, 8, -, 996, 10, 165

Sapp, Elizabeth, 125, 50, -, 2625, 400, 333

Welch, Thomas, 40, 10, -, 500, 100, 485

Jester, James G., 100, 20, -, 1270, 80, 105

Jester, Thomas A., 130, 70, -, 3000, 95, 500

Jester, William F., 60, 20, -, 800, 40, 300

Warren, Andrew, 10, 5, -, 180, 60, 365

Reed. S__, 75, 20, -, 1440, 75, 30

Hewes, Alexander W., 45, 30, -, 1125, 40, 400

Kenton, Nathaniel, 60, 30, 1000, 50, 95

Brown, Johnathan, 80, 10, -, 1200, 75, 230

Duniphen, James G., 60, 20, -, 960, 300, 145

Staten, John W., 25, 43, -, 680, 100, 230

Laramore, John F., 60, 10, -, 700, 30, 150

Levi, Jacob, 40, 20, -, 600, 150, 220

Chelooth, William, 80, 10, -, 900, 25, 162

Wyatt, Moses, 25, 25, -, 600, 10, 130

Hopkins, Samuel, 100, 82, -, 1400, 50, 130

Fleming, Mathew, 40, 15, -, 700, 15, 155

Reed, Edward, 75, 38, -, 3000, 60, 120

Hopkins, Waitman, 78, 100, -, 1142, 100, 530

Reed, John C., 125, 100, -, 2250, 20, 250

Fisher, Bird, 70, 70, -, 1400, 30, 200

Larramore, Joseph W., 20, 13, -, 264, 50, 130

Lofland, Elias F., 200, 100, -, 5000, 200, 200

Booth, Joseph, 135, 34, -, 1828, 150, 588

Booth, John F., 30, 10, 900, 30, 135

Callary, Cirtis, 700, 40, -, 1000, 40, 125

Frem, Charles G., 30, 20, -, 1000, 600, 50

Rathell, Thomas, 10, 5, -, 200, 10, 36

Shuridan, Philip K., 40, 10, -, 1000, 65, 100

Morris, Henrietta, 50, 30, -, 800, 200, 120

Friend, Isaac, 200, 100, -, 1500, 25, 129

Jerrold, William, 62, 20, -, 820, 100, 115

Parris, William, 40, 10, -, 600, 25, 110

Macklin, Curtis, 60, 10, -, 500, 10, 50

Thistlewood, Bengamin, 100, 76, -, 6000, 40, 197

Hall, John D., 100, 40, -, 1400, 30, 253

Anderson, James, 392, 186, -, 1176, 300, 573

Powell, Samuel, 6, 8, -, 140, -, -

Watters, Mahala, 6, 2, -, 80, -, 10

Plumer, James, 60, 10, -, 600, 50, 150

Satterfield, Levi, 40, 10, -, 800, 10, 35

Denham, George W., 230, 35, -, 7000, 250, 450

Moore, John A., 40, 10, -, 1000, 8, 97

Tygert, John E., 150, 50, -, 2000, 150, 200

Curtis, John R., 40, 64, -, 2600, 75, 101

Scott, Peter W., 140, 30, -, 2000, 30, 310

Michall, John H., 100, 30, -, 1300, 25, 452

Harrington, David, 182, 52, -, 3675, 80, 333

Dorman, Solomon, 100, -, -, 300, 40, 75

Hill, Ruth, 90, 68, -, 4000, 10, 200

Nowell, Peter, 70, 30, -, 2000, 60, 340

Harrington, Daniel, 60, 36, -, 1920, 75, 230

Tinsley, John, 175, 53, 5225, 300, 650

Jacobs, William A., 75, 25, -, 1000, 10, 108

Murrel, Theodore, 100, 30, -, 2000, 20, 181

Abbott, Warner, 20, 10, -, 300, 25, 80

Riller, William C., 50, 30, -, 2000, 115, 150

Harrington, Matilda, 100, 40, -, 2800, 140, 300

Pierce, Elias, 142, 125, -, 2670, 125, 190

Miner, Johnathan, 125, 65, -, 2600, 75, 220

Minner, Nathaniel, 12, -, -, 300, 60, 200

Abler (Ables), Thomas E., 60, -, -, 1000, 50, 170

Brown, Thomas, 172, 172, -, 5430, 120, 430

Brown, James, 80, 20, -, 1200, 60, 398

Morgan, Nath., 35, 35, -, 740, 20, 150

Tribbet, Jacob, 75, -, -, 600, 10, 275

Smith, David W., 100, -, -, 100, 65, 450

Anthony, Thomas G., 90, -, -, 800, 125, 265

Anthony, Daniel, 160, 50, -, 2940, 125, 224

McKinah, George H., 90, 50, -, 2160, 400, 1460

Hix, William H. 98, 85, -, 3800, 450, 940

Hopkins, Waitman, 60, 20, -, 1200, 30, 295

Hopkins, Henry, 45, 35, -, 960, 30, 467

Porter, James, 100, 8, -, 1944, 120, 480

Frazier, Alexander, 100, 20, -, 2000, 70, 512

Marine, Davis, 246, 247, -, 10000, 125, 655

McKnase, William, 40, 6, -, 300, 75, 317

Smith, James A., 164, 40, -, 5344, 240, 865

Smith, Edwin B., 95, 29, -, 3100, 220, 300

Harrington, Richard J., 235, 110, -, 4128, -, 446

Hix, Eliza, 75, 25, -, 2000, -, 70

Cain, Noah, 54, 25, -, 2376, 40, 235

Hickman, John L., 50, 50, -, 100, -, 63

Jester, William, 50, 25, -, 1125, 30, 200

Wyatt, Major M, 200, 50, -, 5000, 150, 1120

Raughly, Thomas H., 50, 40, -, 1800, 20, 2440

Anderson, Outen, 231, 100, -, 14067, 1600, 1044

Sipple, Noah, 60, 20, -, 1500, 200, 432

Porter, Edward, 40, -, -, 400, -, -

Anthony, John, 23, 23, -, 560, -, -

Parris, William, 25, 15, -, 600, 30, 132

Brown, John, 100, 7, -, 2040, 100, 853

Brown, Tilghman, 95, 20, -, 1800, -, 406

Raughly, Robert, 200, 150, 7080, 300, 1029

Johnson, George, 60, -, -, 1400, -, -

Demby, Henry, 120, 100, -, 3500, 60, 258

Fleming, Biniah, 75, 29, -, 2080, 85, 315

Smith, Charles A, 75, 50, -, 1500, 50, 450

Morris, Curtis, 50, 63, -, 1200, 10, 230

Brown, Mary, 80, 39, -, 1865, 25, 120

Cain, William, 60, 5, -, 1250, 30, 512

Ross, James, 8, -, -, 160, -, 15

Thomas, Philemon, 60, 10, -, 800, 40, 100

Reed, Nancy, 40, 36, -, 900, -, -

Reed, John W., 75, 30, -, 2100, 45, 140

Cain, Elizabeth, 105, -, 600, 30, 125

Cooper, Louisa, 40, 40, -, 800, 100, 210

Beetts, Wingate, 40, 20, -, 600, -, 30

Scott, William, 30, 40, -, 840, 10, 60

Graham, Henery, 40, 10, -, 1000, 70, 2156

Barlow, Aaron, 65, 35, -, 2000, 250, 130

Friend, James, 248, -, -, 3000, 30, 130

Herrington James Jr., 75, -, -, 800, 10, 145

Hanly, James, 90, 30, -, 1800, 30, 500

Harrington Benjamin, 75, 75, -, 5000, 100, 160

Barker, Joseph, 10, 40, -, 600, 10, 75

Brown, Isaac, 10, -, -, 100, -, 65

Wolcatt, Joseph, 68, 22, -, 3680, -, -

Cain, Calep, 80, 40, -, 1600, 125, 300

Scott, Samuel, 30, 10, -, 600, 80, 100

Stubs, John H., 40, 10, -, 600, 50, 120

Hewes, James, 50, 30, -, 800, 30, 200

Wheeler, William, 60, 27, -, 875, 40, 153

Welch, Jacob, 10, 5, -, 500, 40, 200

Travis, John, 75, 65, -, 2800, 30, 400

Barker, Thomas J., 200, 60, -, 3000, 900, 525

Masten, William L., 200, 150, -, 7000, 500, 1072

Kelly, James, 60, 30, -, 1800, 10, 146

Masten, William H., 120, 227, -, 5872, 200, 656

Hill, Joshia M., 100, 10, -, 1000, 50, 360

Cox, John, 200, 60, -, 2000, 175, 693

Porter, James, 90, 90, -, 2500, 15, 400

Harrington, Samuel, 58, 58, -, 1160, 10, 400

Porter, John, 40, 20, -, 600, 10, 196

Callans, John V., 105, 57, -, 4500, 150, 439

Morgan, Nancy, 59, 30, -, 840, 15, 210

Burnham, Andrew P., 40, 32, 1080, 10, 120

Grant, Andrew J., 30, 10, -, 600, 10, 125

Bowen, Levi, 60, -, -, 2000, 50, 439

Brown, Joseph, 130, 110, -, 2000, 75, 535

McKnatt, Nathan, 80, 10, -, 1000, 40, 200

Brown, David, 40, 20, -, 400, 25, 190

Fleming, Ezekiel, 38, 104, -, 1600, 50, 1200

Walker, E. W. H., 5, 2, -, 300, 10, 500

Sharp, M__ _., 150, 200, 3500, 40, 130

Tharp, John A., 270, 153, -, 14000, 20, 130

Short, Robert A., 40, 9, -, 3000, 150, 330

Cain, James M., 72, 25, -, 2200, 30, 300

Sh___, William, 300, 192, -, 1800, 610, 900

Severin, Gilliam K., 200, 150, -, 10500, 350, 300

Minner, Nimrod, 100, 125, -, 2500, 25, 100

Tharp, Samuels, 75, 30, -, 2400, 15, 200

Tharp, Beniah, 40, 10, -, 150, 10, 200

Cain, Sarah K., 72, 22, -, 1800, 400, 200

Carroll, William C., 72, 25, -, 2000, 100, 200

Tharp, William, 70, -, -, 1400, 10, 600

Anderson, Ezekiel, 60, 15, -, 1500, 10, -

Calaway, Hanes, 40, 10, -, 1000, 15, 100

Walker, John, 40, 10, -, 1000, 15, 100

Callaway, Curtis, 60, 32, -, 1400, 10, 100

Hutchins, Thomas M., 250, 50, -, 12000, 200, 800

Parris, Israel, B., 66, 20, -, 1800, 100, 400

Melvin, Riley, 140, 20, -, 4000, 300, 600

Stafford, Elizabeth, 74, 10, -, 1700, 10, 200

Bynard, Thomas H., 70, 34, -, 2000, 60, 200

Bynard, Ferdinand V., 280, 140, -, 7000, 80, 400

Raughly, Henry, 65, 35, -, 1200, 40, 200

Smith, Robert H., 100, 50, -, 5000, 60, 500

Garratt, Richard, 200, 70, -, 6000, 100, 400

Williams, William S., 40, -, -, 800, 10, 100

Barker, Alexander, 150, 50, -, 4000, 20, 400

Cahall, Ann, 10, 5, -, 300, 10, -

Morris, James R., 40, -, -, 800, 10, 200

Voss, James, 60, 35, -, 1900, 40, 600

Raughly, James, 96, 4, -, 2000, 40, 400

Melvin, James, 98, 8, -, 3200, 60, 300

Callary, Manlove, 40, 27, -, 3400, 80, 200

Lavanie, Thomas H., 15, 10, -, 400, 15, 300

Smith, Nathan Sr., 90, 40, -, 2000, 80, 300

Smith, Nathan Jr., 80, 45, -, 2500, 250, 800

Price, James H., 45, 30, -, 1400, 30, 100

Smith, Alexander, 60, 30, -, 1800, 50, 100

Herd, Benjamin L., 142, 100, -, 5000, 150, 800

Sipple, Zadoc, 100, 50, -, 3000, 300, 600

Callary, Johnathan, 100, 10, -, 3400, 200, 400

Baruros (Barurich), John J., 90, 25, -, 3000, 60, 300

Beader, William H., 110, 37, -, 3000, 400, 400

Porter, David, 66, 34, -, 4000, 30, 300

Spence, Anias E., 60, -, -, 1800, 200, 400

Hynerston (Hynupton), Clement A., 70, 50, 9100, 400, 700

Goshey, Mathew, 160, 20, -, 390, -, 50

Clark, Henry N., 160, -, -, 300, 60, 300

Lynch, Noah, 100, 50, -, 4500, 30, 1000

Gorden, Mathew, 125, 25, -, 3100, 10, 60

Lank, Robert J., 24, -, -, 7700, 10, 200

Morris, Thomas E., 100, 66, -, 3200, 24, 400

Anderson, Samuel, 60, 60, -, 2400, 10, 300

Anderson, David, 75, 30, -, 2100, 15, 300

Jester, Isaac, 100, 30, -, 2600, 30, 500

Jones, Joshua B., 250, 60, -, 1610, 10, 100

Hall, Samuel L., 60, 90, -, 2600, 10, 200

Booth, John, 80, 160, -, 500, 5, 100

Harrington, Peter D., 100, 40, -, 2400, 26, 300

Lewis, Nimrod, 50, 30, -, 1600, 15, 200

Atkinson, Elizabeth, 30, 10, -, 800, 5, 100

Seedens, William W., 200, 200, -, 6000, 20, 300

Williams, John, 100, 60, -, 3300, 40, 700

Callis, Frisby, 109, -, -, 2000, 5, -, 75

Jones, Ezekiel, 100, 30, -, 2600, 40, 600

Wrother, Eli, 100, 30, -, 2600, 40, 600

Porter, James, 125, 4, -, 3200, 60, 700

Porter, John F., 100, 20, -, 2400, 20, 200

Todd, John W., 75, 25, -, 200, 10, 200

Richards, Charles, 23, 23, -, 900, 30, 200

Adams, Charles M., 200, 100, -, 8000, 40, 900

Peters, Samuel, 60, 40, -, 200, 10, 200

Peters, William R., 140, 60, 4000, 15, 1100

Parris, Mathew, 100, 78, -, 3600, 10, 300

Alaband, Martin F., 100, 30, -, 2600, 30, 360

Jester, Isaac, 21, -, -, 800, -, 100

Graham, Mary, 31, -, -, 2000, 5, 2000

Callaway, Peter, 100, 60, -, 3200, 200, 500

Victor, John W., 100, 50, -, 3000, 10, 200

Callaway, Henry, 60, 20, -, 3300, 15, 300

Ross, Hooper B., 50, -, -, 1600, 20, 400

Hopkins, Elias P., 90, 14, -, 2000, 400, 400

Collins, Joseph, 80, 30, -, 1300, 30, 300

Melvin, Sidenham, 120, 100, -, 4000, 30, 600

Stickly, Jacob, 40, 40, -, 500, 10, 200

Melvin, Bascomb, 50, 14, -, 100, 10, 200

Graham, Sarah, 60, 14, -, 800, 8, 100

Taylor, David, 85, 40, -, 3400, 40, 300

Purnell, William, 60, 30, -, 800, 10, 200

Blades, Thomas, 40, 30, -, 600, 10, 200

McKnatt, John W., 40, 30, -, 600, 10, 300

Anthony, Robert, 65, -, 800, 15, 300

Laramer, Anna M., 130, 30, -, 1500, 15, 600

Simpson, William, 200, 50, -, 5000, 50, 600

Cain, John, 65, 28, -, 900, 30, 800

Brown, William, 75, 5, -, 800, 10, 300

Adams, Garrison, 60, 30, -, 800, 10, 300

Clark, Waitman, 100, 50, -, 1800, 40, 500

Harrington, Robert, 70, -, -, 80, 5, 580

Tutle, Samuel, 20, 10, -, 600, 15, 200

34

Simpson, Alexander, 100, 25, -, 4000, 140, 200

Cain, George W., 30, 20, -, 100, 30, 400

Salmins, Minus, 60, 30, -, 3000, 15, 300

Sapp, Curtis, 55, 12, -, 1500, 15, 400

Sapp, Elias, 60, 10, -, 1400, 10, 200

Lewis, Stephen, 150, 50, -, 3000, 30, 500

Hopkins, Robert G., 240, 230, -, 4000, 100, 800

Tharp, Benj. H., 120, 30, -, 300, 15, 300

Hill, Robert _., 40, 30, -, 200, 40, 300

Tharp, Lewellen, 100, 200, -, 600, 10, 500

Cahall, William, 75, 30, -, 2500, 10, 100

Hopkins, Zebulum, 700, 170, -, 8000, 10, 300

Redder, Stephen, 100, 44, -, 5000, 60, 500

Cordry, Jeremiah P., 80, 90, -, 3000, 40, 800

Taylor, David, 300, 60, 5200, 30, 800

Collins, George W., 80, 10, -, 1800, 60, 200

Merikan, Richard, 103, 102, -, 801, -, 500

Hobkins, Hooper, 250, 150, -, 8000, -, 800

Fountain, Daniel, 100, -, -, 3000, -, 400

Wroten, Eli, 80, 50, -, 2400, -, 600

Wroten, Charles, 130, 170, -, 6000, -, 300

Tatman, Henry, 75, 75, -, 3000, -, 100

Hopkins, John, 100, 86, -, 5000, -, 600

Jacobs, Isaac, 140, 30, -, 3400, -, 400

Fisher, James H., 130, 60, -, 4000, -, 600

Andrews, James, 100, 50, -, 2000, -, 300

Williamson, Charles, 140, 100, -, 500, -, 600

Smith, Elizabeth, 10, -, -, 400, -, 40

Thomas, Zebulum, 100, 30, -, 2600, -, 400

Fisher, Samuel, 128, 30, -, 310, -, 500

Burdick, Ryals, 150, 153, -, 1300, -, -

Lewis, Jacob F., 80, 20, -, 300, -, 400

Outin, Charles, 60, 10, -, 3300, -, 400

Graham, Jacob, 60, 22, -, 1700, -, 400

Fisher, Wheelty, 75, 70, -, 1800, -, 42

Roe, William, 70, 30, -, 1400, -, -

Cavender, Ruben, 100, 30, -, 3000, -, 200

Lewis, William, 80, 31, -, 1000, -, 300

Smith, Clement, 100, 37, -, 1800, -, 200

Jones, William, 100, -, -, 100, -, 100

Salisbury, Garrison, 100, 30, -, 3800, -, 400

Smith, Martin, 150, 44, -, 3200, -, 500

Hardesty, George, 40, 40, -, 1000, -, 300

Smith, Elizabeth, 90, 30, -, 1200, -, 100

Salisbury, Henry, 90, 30, -, 1200, -, 50

Layton, William, 150, 150, -, 3000, -, 300

Layton, Jane, 100, 100, -, 4000, -, -

Williams, Thomas, 60, 30, -, 9000, -, -

Outin, John H., 40, 30, -, 800, -, 200

Ferrins, Jesse, 30, 20, -, 600, -, 200

Dill, James W., 40, 30, -, 1000, -, 300

Webster, Benjamin, 60, 30, -, 1000, -, -

Redding, Eamils, 60, 40, -, 1000, -, 200

Ferrins, John, 100, 50, -, 1200, -, 200

Ferrins, Mary, 60, -, -, 600, -, -

Cahall, John A., 100, 83, -, 5000, -, 500

Thomas, Robert H., 80, 46, 200, 10, 413

Jones, George R., 60, 20, -, 3000, 46, 500

Harrington, Nimrod, 68, 67, -, 6800, 30, 400

Meleham, Edward D., 20, 23, -, 2200, -, 200

Simpson, Clement C., 150, 260, -, 10200, -, 1320

Tharp, Beniah, 100, 60, -, 3000, -, 300

Callary, James, 50, -, -, 1000, -, 500

Merrikin, Rebecca A., 100, 44, -, 2000, -, 200

Cahall, Luff, 140, 50, -, 3400, -, 600

Smith, John W., 140, 60, -, 2500, -, 400

Ward, Jesse, 60, 40, -, 2000, -, 400

Ward, Sarah, 40, 10, -, 1000, -, 100

Cain, La___, 55, -, -, 1100, -, 500

Wheatly, Baynard, 70, 7, -, 3200, 50, 600

James, Albert, 167, 25, -, 5000, 25, 265

Vincent, Burton, 30, -, -, 3000, 20, 110

Bancroft, Edward H., 40, -, -, 8000, 75, 500

Dager, John, 60, -, -, 12000, 60, 500

Mifflin, Samuel H. S., 215, 30, -, 20000, 100, 615

Jenkins, John, 200, -, -, 10000, 600, 1040

Donovan, Benjamin, 212, -, -, 5600, 20, 270

Lillie, William, 90, -, -, 3000, 10, 765

Lawrence, Alonzo, 63, -, -, 13000, 600, 550

Allen, John, 110, 40, -, 11250, 100, 350

Kilmer, Robert, 100, 60, -, 1000, 100, 625

Frear, Cornelius, 247, 315, -, 16100, 100, 1580

Frear, William, 175, 5, 20, 20000, 500, 550

Thompson, Levi, 35, 5, -, 3000, 50, 300

Barnard, Daniel P., 100, 35, -, 15000, 600, 875

Shockly, Thomas, 120, 67, 20, 7000, 50, 460

Draper, Thomas, 300, 60, 1000, 23000, 300, 635

Lodge, John, 95, 5, -, 10000, 100, 618

Pickering, Thomas J., 450, 150, -, 3000, 200, 1370

Owens, Daniel, 90, 10, -, 6000, 100, 260

Lewis, Caleb, 140, 60, -, 5000, 150, 600

Maier, Lemuel B., 40, -, -, 6000, 150, 250

Jackson, Edward, 97, 20, 5, 12000, 300, 670

Carey, Absolem, 100, -, -, 16000, 200, 500

Lofland, John G., 50, -, 8, 5000, 200, 380

Wharton, Samuel, 160, -, -, 9000, 200, 700

Wilson, John, 16, -, -, 1200, 25, 200

Frear, DeWitt, 205, -, -, 17000, 500, 1025

Greer, James, 175, 20, -, 10000, 200, 500

Brown, William, 105, 12, -, 2500, 500, 550

Deweese, Lamuel (Samuel), 250, -, -, 10000, 200, 600

Wright, John W., 190, -, 10, 20000, 300, 750

Lowber (Sowber), Charles, 200, -, -, 18000, 600, 1400

Moody, Dunbar, 65, -, -, 50000, 200, 260

Wilson, Henry, 95, 7, -, 17000, 800, 850

Lum, Wilson, 177, 35, -, 17000, 800, 850

Hairson, Thomas, 136, -, -, 20000, 300, 800

Sterne, Levi, 104, -, -, 10000, 100, 470

Jester, John H., 115, 10, -, 6000, 25, 380

Shorts, Charles, 40, 6, -, 3000, 100, 160

Briley, Joshua, 50, 4, -, 3000, 25, 160

Anderson, James, 85, 15, -, 5000, 50, 333

Davis, John, 50, 8, -, 3000, 50, 300

Ridgway, William, 210, 130, -, 10000, 500, 950

Townsend, Samuel, 270, 30, -, 15000, 200, 1636

Voshall, James, 125, 25, -, 8000, 100, 830

Gildersleeve, George, 347, 40, 15, 15000, 150, 964

Aldrich & Rhodes, 45, -, -, 10000, 500, 515

Postles, Henry, 35, -, -, 2000, 100, 250

Greer, James, 88, 2, 100, 8800, 150, 430

Dunn (Durna), John, 112, -, -, 8000, 150, 755

Brown, Jacob G., 200, -, -, 25000, 450, 1450

Lahr, William, 104, -, -, 8000, 100, 445

Conwell, Charles, 135, 15, 100, 8000, 150, 520

Maloney, William, 96, -, -, 5000, 60, 355

Dyer, James S., 60, 65, 15, 4500, 50, 425

Lindale, Turner P., 86, 10, -, 3500, 150, 340

Voshell, Daniel, 94, 15, -, 3500, 100, 250

Pratis, Henry, 23, 15, 3, 1500, 40, 150

Bailey, Nathan, 80, 60, -, 3500, 150, 285

Voshell, William, 85, 15, -, 4000, 100, 270

Holston, William, 30, -, 4, 1000, 25, 250

Gibbs, Henry, 37, -, -, 2000, 50, 265

Rickards, James, 60, -, -, 8000, 25, 570

Carpenter, Henry, 30, -, -, 3000, 25, 275

Evans, William K., 131, -, -, 12000, 500, 250

Kens, Philo A., 21, -, -, 4000, 250, 278

Blood, James, 130, -, -, 9000, 150, 558

Parvis, Joseph, 160, 40, -, 5000, 100, 690

Jonsin, Shade, 175, 75, -, 6000, 150, 668

Sipple, John, 100, 80, -, 4000, 100, 270

Frasher, Thomas, 75, 25, -, 2500, 100, 900

Griwell, Isaac, 175, 75, -, 5000, 150, 605

Threasher, James, 170, 80, -, 10000, 150, 825

Frasher, Robert, 130, 15, -, 7000, 00, 600

Steel, John W., 100, 25, 12, 3000, 150, 400

Smith, William, 130, 170, 20, 4000, 200, 975

Tinley, Jacob, 200, -, -, 10000, 200, 1200

Cohee, Benjamin, 207, 188, -, 20000, 400, 667

Cowgill, Ezekiel, 125, 25, -, 5000, 400, 1000

Carter, Henry, 34, 11, -, 1200, 150, 200

Ratlage, George, 100, 60, -, 300, 100, 300

Gordwin, William, 90, 55, -, 4500, 100, 600

Gordwin, Thomas, 80, 60, -, 4500, 275, 1156

Clark, Peter, 120, 110, -, 600, 150, 1015

Cathn, Edward, 25, -, -, 1000, 25, 80

Griswell, Peter, 136, 37, -, 2500, 150, 500

Frasher, Ezekiel, 100, 60, -, 5000, 200, 500

Moor, Nathan, 100, 25, -, 450, 100, 475

Killin, James, 125, 75, -, 6000, 400, 160

Robrson, Samuel, 65, 10, -, 2500, 500, 200

Carter, James H., 75, 125, -, 3000, 50, 385

Kemp, Thomas, 18, -, -, 700, 25, 300

Cooper, Peter L., 135, 50, -, 5000, 200, 885

Gordwin, Isaac, 115, 75, -, 6000, 100, 820

Cohee, John L., 120, 20, -, 5000, 150, 500

Reynolds, Sarah, 70, -, -, 3500, 250, 240

Edwards, Milliam, 125, 25, -, 6000, 50, 334

Clark, Ezekiel, 100, 50, -, 5000, 100, 260

Dill, Lutar M., 89, 89, -, 5000, 100, 580

Cubbage, William, 70, 20, -, 2000, 100, 528

Smith, Henry, 190, 10, -, 5000, 500, 600

Slaughter, Andrew, 100, 20, -, 2000, 50, 250

Smith, Thomas, 100, 25, -, 2000, 100, 400

Jackson, Thomas, 120, 66, 30, 9000, 216, 700

Snydar, George, 14, -, -, 2000, 25, 100

Allaland, William, 150, 50, -, 1000, 500, 825

Kinney, Edgar J., 52, 8, -, 5000, 250, 350

Dickson, William, 282, 15, -, 25000, 300, 2050

Brookway, Laurence, 131, 25, -, 10000, 100, 565

Richards, Charles M., 40, 12, -, 3000, 300, 310

Able, Davis, 35, 15, -, 3000, 200, 115

Cooper, Henry, 160, 5, -, 18000, 50, 950

Credis, William, 120, -, -, 7000, 150, 250

Achor, Charles, 25, -, 20, 1500, 110, 130

Postals, Joseph, 12, -, -, 1000, 75, 130

Presiden (Pursiden), Daniel B., 112, 2, -, 5000, 25, 108

Cox (Case), William B., 120, -, -, 7000, 150, 254

Carter, Edward, 220, 25, -, 30000, 500, 1450

Jones, Glaney J., 25, -, -, 3000, 700, 250

Woodle, John, 11, -, -, 1200, 50, 115

Jurman, Mathias, 12, -, -, 1800, 30, 300

Clark, James, 50, 50, -, 1500, 30, 150

Cook, Thomas, 260, 60, -, 12800, 300, 1100

Calk, William S., 148, 100, -, 10150, 200, 870

Murry, Henry, 240, 50, -, 12000, 300, 725

Jones, Charles, 34, 6, -, 1000, 40, 220

Downham, John M., 60, -, -, 1500, 10, 155

Harress, Isaac, 39, -, -, 1000, 50, 100

Killin, John, 70, 20, -, 1000, 10, 200

Hammond, Issac, 30, 270, -, 200, 30, 100

Langral, John, 40, 46, -, 1000, 20, 360

Andrew, Wilson, 43, 27, -, 1200, 50, 273

Greswell, Joseph, 147, 50, -, 4000, 150, 725

Painter, John, 50, 150, 50, 2000, 250, 125

Copes, John N., 120, 40, 2000, 200, 360

Sylvestor, Samuel, 175, 85, -, 15000, 250, 133

Shockley, John, 290, 50, -, 7000, 50, 245

Cooper, Samuel S., 149, 30, 5, 6000, 400, 580

Dill, Alexander, 152, 30, -, 6000, 226, 760

Greenly, William, 25, 25, -, 1000, 50, 156

Raughley, Peter, 125, 75, -, 6000, 200, 400

Chemmonds, Joshua R., 325, 55, -, 15000, 250, 1600

Cooper, Jonathan, 60, 40,-, 1500, 20, 140

Parvis, John N., 70, 20, -, 1500, 100, 880

Blizzard, David, 72, 8, -, 800, 25, 176

Marker, William, 70, 30, -, 1200, 25, 300

Marker, John, 8, 26, -, 1200, 20, 150

Billings, James, 90, 50, -, 3000, 75, 450

Frashier, Alexander, 80, 40, -, 5000, 125, 450

Frashier, William, 90, 12, -, 4000, 300, 655

Cohee, Samuel, 136, 30, -, 5000, 300, 575

Carr, Samuel, 92, 25, -, 16000, 250, 900

Clemmonds, Ezekiel B., 110, 40, -, 10000, 300, 700

Morris, Ambrose, 50, 59, -, 3000, 25, 200

Cubbage, Benjamin, 115, 25, -, 5000, 100, 750

Broodway, Ambrose, 115, 38, -, 3500, -, 900

Hill, John, 75, 25, -, 5000, 125, 150

Shaw, Edward, 100, 167, -, 5000, 25, 675

Ross, William, 44, -, -, 7500, 10, 200

French, Elias, 26, -, -, 1000, 50, 135

Burchnel, John, 40, 6, -, 5000, 150, 600

Jarrell, James, 30, 10, -, 700, 25, 150

Jarrell, Thomas, 40, -, -, 1000, 125, 250

Hubert, Peter, 80, 37, -, 4000, 150, 280

Jarrell, John, 75, 15, -, 3000, 110, 255

Hargadine, Henry, 135, 25, -, 10000, 250, 750

Catlin, Robert, 475, 75, -, 40000, 500, 245

Hargass, William B., 100, -, -, 10000, 350, 600

Smith, John L., 110, -, -, 7000, 100, 530

Calk, William, 112, 25, -, 7000, 150, 620

Roberts, Clayton, 65, 15, -, 2000, 100, 200

Odaniel, Samuel, 100, 40, -, 3000, 100, 300

Barnet, Isaac, 100, 21, -, 2500, 200, 640

Barcus, John, 70, 43, -, 1500, 25, 400

Lipkins, Laurens, 50, 90, -, 2000, 100, 105

Gordwin, John C., 110, 15, -, 4000, 25, 515

Kursey, Ann, 70, 80, -, 3000, 30, 100

Slaymaker, James B., 60, -, -, 5000, 150, 925

Whetsit, Aron E., 215, 15, -, 25000, 400, 1660

Dudley, Albert, 162, 2, 6, 20000, 200, 1000

Smith, Jacob, 140, -, -, 10000, 600, 662

Whitmire, John, 140, -, -, 10000, 300, 300

Gibbs, Samuel, 175, -, 250, 5000, 100, 125

Lewis, Robert, 22, -, 100, 2000, 100, 150

Varnarmon, Phalando, 139, 25, -, 10000, 300, 450

Fredell, Charles, 40,-, -, 2500, 25,300

Frashier, Mandly, 75, -, 5, 1600, 25, 200

Buckmaster, Charles, 40, 30, -, 2000, 250, 150

James, Clemment, 125, 120, -, 15000, 120, 500

Jonson, Edward, 70, 40, -, 3000, 100, 156

Willis, Joseph, 100, 50, -, 5000, 25, 400

Graham, Parrot P., 1112, 20, -, 6000, 175, 1324

Conner, Barret, 75, 20, -, 300, 75, 500

Burchnell, Joseph, 56, -, -, 3000, 50, 200

Linch, Joseph, 70, 20, -, 3500, 500, 320

Huffington, Henry, 10, 10, -, 8000, 100, 400

Cowgell, Henry, 61, -, -, 8000, 250, 400

Peters, John, 103, -, -, 6000, 500, 525

Ronsh, Charles, 61, -, -, 6000, 300, 450

Lysher, William, 63, 12, -, 12000, 500, 850

Saxton, Thomas, 125, 25, -, 5000, 200, 350

Jones, Jones H., 200, 100, -, 5000, 25, 300

Barcus, James W., 50, -, -, -, 2500, 185

Calley, Andrew J., 195, 45, -, 12000, 800, 700

Meed, Genes, 135, -, -, 5000, 150, 170

Cook, Nathan, 190, 70, -, 10000, 200, 1200

Wabruth, Sanford, 135, 15, -, 10000, 100, 570

Clever, John, 130, 30, -, 15000, 400, 1200

Cooper, Samuel B., 175, 8, -, 9000, 500, 555

Stubbs, William, 150, 25, -, 8000, 200, 600

Grey, William A., 112, 8, -, 10000, 200, 600

Lewis, Thomas B., 100, 25, -, 6000, 150, 450

Griffin, Asa, 55, 50, -, 4000, 300, 300

Lewis, Garret, 93, 10, -, 5000, 150, 30

Evans, William, 93, 10, -, 5000, 150, 30

Maginns, John H., 120, 20, -, 5000, 150, 300

Hensley, Annis, 135, 14, -, 6000, 115, 85

Bradley, John, 120, 25, -, 12000, 300, 900

Moore, John A., 20, 10, -, 7500, -, -

Needles, David, 32, 30, -, 3000, 30, 75

Vincent, Levy, 35, 8, 2, 3000, 50, 200

Needles, Catharine, 50, 20, 27, 3000, -, 75

Cullen, William W., 15, -, -, 2500, -, 75

King, David T., 85, 5, -, 10000, 130, 775

Simpson, William, 100, -, -, 7500, 150, 414

Jarrell, James, 125, 25, -, 15000, 250, 690

Carter, John W., 19, -, -, 600, 125, 165

Boynton, John, 40, 50, 100, 10000, 100, 145

Snell, William, 17, 2, -, 4000, 25, 110

Benson, Matthew, 26, -, 4, 2000, 10, 300

McMahan, James, 20, 9, -, 4000, 125, 490

Salisbury, Earl P., 41, -, -, 5000, 200, 175

Jollie, Philip A., 150, 50, 50, 10000, 300, 951

Satterfield, William M., 200, 100, -, 10000, 200, 1020

Hodgson, Robert, 30, -, 14, 2500, 60, 190

Kelley, Thomas H., 36, -, 4, 2500, 40, 173

Vickers, Elizabeth, 20, -, 14, 2500, 25, 160

Kidder, Hiram B., 58, 24, -, 3500, 100, 370

Bostick, James P., 51, 12, -, 2500, 50, 331

Harrington, Samuel, 20, -, -, 2000, 50, 145

Sipple, Thomas, 40, 20, 110, 5000, 5, 247

Jones, Robert, 90, 20, -, 10000, 250, 726

Downham, John, 150, 44, 5, 12000, 150, 662

Anderson, Letitia, 15, -, -, 600, -, 50

Bailey, Edmond, 130, 30, -, 16000, 600, 2475

McIlvain, William S., 60, 22, -, 3500, 75, 582

Meredith, William H, 120, 20, -, 5000, 25, 142

Anderson, John D., 60, 20, -, 5000, 200, 570

Judd, Hiram D., 35, 15, -, 3000, 150, 330

Sipple, James, 60, 20, 60, 4000, -, -

Hawkins, George W., 40, 10, -, 3500, 150, 320

Bendin, Harrison, 15, -, -, 1500, 100, 322

Young, John, 20, -, 10, 1500, 20, 205

Young, Thomas, 9, 1, -, 600, -, 60

Bell, Caleb S., 100, 180, 10, 6000, 50, 219

Cahall, Samuel R., 40, 20, 40, 3500, 25, 455

Berrian, A. J., 40, -, 1, 2000, -, -

Perry, William, 15, -, -, 500, 40, 120

Carter, Henry, 35, -, -, 4000, 300, 545

Carter, Henry, 53, -, -, 3000, -, -

Angell, Daniel M., 11, -, -, 3000, 100, 90

Whitaker, Henry, 46, 4, 3, 5300 200, 430

Whitaker, Henry, 45, 9, -, 4000, -, -

Emerson, John, 100, 70, -, 2000, 1200, 1381

Wilcutts, William J., 189, 30, -, 15000, 75, 1074

Herrington, Henry, 160, 42, -, 11100, 400, 920

Creadick, John C., 50, 25, 75, 4000, 100, 435

Creadick, William T., 140, 45, 10, 7000, 100, 825

Burchenal, Joseph, 12, -, -, 3500, 100, 150

West, John, 90, 20, 87, 10000, 50, 230

Smithers, Joseph, 108, -, -, 20000, 300, 225

Brown, Joseph, 200, 10, -, 8000, 100, 590

Clark, Henry, 145, -, 15, 12000, 500, 1045

Harrington, Isaac, 100, 10, -, 5000, 300, 330

Fountain, Stephen B., 20, -, 7, 1500, 75, 145

Harrington & Bro., 30, 80, -, 6000, 100, 405

Young, Thomas, 12, -, 3, 1000, 15, 106

Harrington, John, 92, 7, -, 8000, 300, 1485

Harrington, Caleb B., 14, -, 1000, 10, 418

Griffin, William, 30, -, 56, 3000, 20, 90

Clark, Benjamin V., 100, 46, -, 7000, 150, 500

Coursey, Thomas B., 73, 12, 5, 2000, 250, 1165

Bennett, Joshua A., 60, 40, -, 5000, 100, 310

James, Levin, 160, 40, -, 7000, 125, 240

Boone, Joseph H., 100, 50, 45, 4000, 50, 225

Green, John, 35, 10, -, 2500, 40, 240

Rice, Charles, 60, 15, -, 7000, 150, 580

Sliter, Charles, 60, -, 15, 4800, 100, 625

Martin, George W., 12, -, 12, 700, 50, 150

Cox, William, 133, 65, 20, 8000, 100, 720

Gibson, Samuel, 200, 59, -, 25000, 500, 1423

Cox, Charles C., 260, 40, -, 25000, 300, 1823

Spencer & Mason, 295, 6, 34, 26000, 200, 1215

Anderson, Joseph B., 40, -, -, 3000, 120, 195

Barratt, George, A., 200, 50, -, 15000, 150, 697

Evans, James H. 100, 50, -, 10000, 500, 783

Bradley, Joseph, 150, 30, -, 10000, 250, 736

Darby, Samuel W., 200, 200, -, 30000, 100, 2500

Lowber, Elizabeth, 80, -, -, 8000, -, -

Dill, Samuel C., 775, 50, -, 18000, 575, 1379

Emerson, John P., 60, 12, 18, 6000, 50, 415

Stephens, James, 16, 53, -, 2000, 15, 75

Everett, Macena, 130, 35, 25, 2000, 25, 124

Brown, William C., 100, 30, 45, 8000, 10, 927

Snell, Melville, 30, 25, -, 4000, 100, 55

Sherman, Orin, 38, 82, -, 4800, 200, 545

Reynolds, Thomas P., 200, 75, -, 12000, 100, 335

Carpenter, George W., 100, 37, -, 6000, 130, 655

Murry, Stansbury, 39, 4, -, 4000, 200, 675

Billings, William, 150, 50, 8, 10000, 100, 630

Bateman, John W., 60, 20, 20, 3000, 80, 690

Kersey, Thomas, 60, 20, 20, 3000, 75, 380

Lowber, William, 20, -, -, 1000, 65, 195

Perry, Heman, 24, -, -, 3500, 20, -

Needles, Andrew, 7, -, 2, 800, 10, 108

Fitts, Elmer, 80, 20, -, 4000, 25, 237

Clampitt, John, 30, 15, -, 4000, 50, 100

Savage, Joseph, 45, 15, -, 10000, 150, 765

Magill, Ebenezer E., 200, 40, -, 12000, 200, 870

Satterfield, John, 70, 30, -, 5000, 50, 440

Pratt, John L., 190, 18, -, 18000, 750, 1160

Truitt, Joseph C., 130, 20, -, 10000, 50, 175

Meredith, Isaac F., 70, 30, 50, 6000, 100, 280

Lollis, Thomas W., 100, 25, -, 4000, 50, 445

Pasco, Robert V., 80, 20, -, 8000, 100, 445

Moore, Samuel T., 150, 125, 25, 10000, 200, 971

McIntosh, John, 160, 14, -, 7000, 100, 746

Williams, Risden, 80, 45, -, 8000, 150, 550

Anderson, George W., 130, 20, -, 1000, 125, 633

Baker, Charles, 80, 30, -, 6000, 400, 565

Ells, Daniel S., 22, 4, -, 2000, 100, 150

Mosley, James, 60, 7, -, 2500, 10, 270

Johnson, John T., 70, 20, 10, 5000, 150, 655

Pritchet, Wormack, 200, -, -, 16000, 200, 690

Postles, Sarah A., 140, 15, -, 12000, 100, 451

Roach, Robert H., 100, -, -, 8000, 150, 560

Rogers, Hesekiah, 80, 40, -, 10000, 600, 585

Chrylar, James P., 50, -, 24, 5000, -, 135

Burton, Erasmus, 170, 15, -, 10000, 200, 1605

Satterfield, Aaron, 50, -, -, 2500, 15, 125

Satterfield, William E., 35, -, -, 2500, 200, 506

Williams, Benjamin, 170, 30, -, 16000, 250, 1010

Reed, Jehu M., 250, 10, -, 35000, 3000, 2336

Emory, Thomas, 140, 30, -, 15000, 250, 101

Barnes, Isaac W., 200, 4, -, 24000, 300, 690

Perkins, Joseph, 85, -, 15, 3000, 30, 180

Hansly, Nehemiah, 140, -, 10, 1500, 10, 75

Johnson, Aner, 75, -, -, 3000, 50, 285

Coverdale, William H., 100, 20, -, 3000, 20, 236

Short, Robert J., 70, -, -, 2000, 10, 2000

Salevan, James H., 152, 15, -, 12000, 300, 1340

Emory, Abner W., 195, 60, -, 15000, 160, 820

Russell, Elias, 150, 4, -, 18000, 400, 1343

Grier, James W., 143, -, -, 6300, 200, 750

Grier, John E., 150, 40, -, 12000, 100, 875

Grier, James, 225, 30, -, 21000, 1500, 1544

Wyatt, Thomas H., 80, 25, -, 5000, 100, 730

Wyatt, Robert, 40, -, -, 1200, 10, 90

Rash, John W., 80, 20, -, 3000, 20, 150

Sapp, William H., 80, 20, -, 5000, 150, 525

Lindale, James M., 60, 12, -, 3000, 100, 515

Robbins, William H., 150, 15, -, 10000, 150, 730

Gray, William H., 17, -, -, 2000, 100, 250

Grier, Robert K., 115, 25, -, 4500, 600, 865

Williams, Henry, 120, 20, -, 12000, 500, 922

William, Thomas C., 70, 30, -, 8000, 200, 710

Saxton, John, 160, 50, -, 10000, 500, 1055

Scanlon, Michel, 11, -, -, 1000, 10, 60

Scanlon, Jeremiah, 20, -, -, 1400, 50, 40

Lindale, William P., 100, -, -, 7000, 150, 500

Morris, William S., 90, 20, -, 4000, 150, 640

Reed, James H., 250, 100, -, 15000, 500, 1300

Cohee, Anthony, 60, 30, 10, 5000, 50, 220

Lean, Enos, 10, -, -, 700, -, 130

Scanlon, John, 10, -, -, 800, -, 185

Bradley, James, 40, 8, -, 4000, 75, 385

Warren, George R., 200, 30, -, 20000, 500, 1395

Coverdale, John, 200, 60, -, 25000, 200, 1891

Stephenson, James, 70, 20, -, 6000, 50, 260

Stephenson, Tinly B., 120, 10, -, 8000, 150, 865

Fulton, Andrew M., 110, 70, -, 12000, 100, 690

Wright, James H., 48, -, 3, 5000, 100, 520

James, Thomas, 250, 95, 50, 20000, 150, 1660

Binns, Benjamin F., 50, -, -, 450, 150, 265

Holt, Henry P., 75, 6, -, 5000, 60, 540

McIlvain, McIlvoy, 250, -, -, 24000, 600, 1400

Layton, Louder T., 136, -, -, 10000, 50, 380

Chambers, Benjamin D., 75, 6, -, 5000, 100, 570

Hanser, Cornelius, 67, 8, -, 4000, 100, 439

Jackson, Wesly, 60, 40, 20, 4000, 70, 440

Roe, Jonathan D., 150, -, -, 6000, 50, 360

Moore, William W., 125, 25, -, 9000, 150, 345

Callen, John, 75, 35, -, 3500, 50, 380

Conner, John J., 130, 20, -, 15000, 500, 1876

Lynch, John W., 160, -, 3, 10000, 50, 705

Lynch, Thomas, 250, 5, 10, 20000, 80, 1875

Kelley, Joseph H., 100, 8, -, 8000, 300, 760

Chambers, Mark G., 160, 40, -, 15000, 200, 660

Barnett, Henry, 60, -, -, 5000, 200, 655

Godwin, Isaac, 15, -, -, 2500, 150, 220

Jackson, Lewis, 50, -, 12, 3000, -, 145

Catts, James E., 80, -, 10, 4000, 150, 425

Jackson, James, 50, 10, 13, 2500, 25, 130

Ackerman, Ralph, 40, 5, -, 3000, 100, 180

Massey, Peter H., 60, 30, -, 6000, 125, 425

Faulkner, James, 48, 25, -, 4000, 30, 295

Sharrard, Samuel, 100, 12, 95, 4500, 35, 200

Wooters, Elijah, 34, 35, -, 2500, 20, 240

Stout, Edmonds, 80, 12, -, 15000, 300, 525

Johnson, James C., 80, -, -, 4000, 200, 540

Vanburkaloe, Wm., 106, 9, -, 10000, 200,720

Peck, Samuel L., 200, 200, -, 15000, 500, 995

Leach, Royal D., 50, 14, -, 5000, 100, 345

Kulp, Jacob, 100, 30, -, 7500, 200, 882

Klumpp, Charles, 80, 125, -, 5000, 200, 703

Bastirn, George M., 50, 82, -, 4000, 100, 492

Herring, Samuel S., 80, 20, -, 6000, 100, 345

Grennell, Virgil, 40, 20, -, 2500, 200, 260

Grennell, Myron G., 60, -, -, 3000, -, -

Downham, Isaac R., 125, 100, 35, 10000, -, 100, 615

Holden, Edmond, 100, 25, -, 5000, 200, 495

Townsend, Elias, 100, 120, -, 9000, 150, 455

Hopkins & Warner, 115, 186, 14, 18500, 250, 935

Dewey, Talman, 50, 60, -, 6000, 200, 553

Melbourn, Samuel W., 125, 25, -, 6000, 150, 475

Morrison, William 11, 55, -, 4000, -, -

Needles, James R., 20, 80, -, 3000, 50,2 58

Freedell, Jacob, 100, 63, -, 6500, 1200, 1025

Sheetts, Samuel, 40, 20, 35, 2500, 100, 640

Semans, Lewis, 30, 60, 10, 3000, 75, 342

Caldwell, Elijah S., 40, 20, -, 5000, 200, 250

Spence, Pierson C., 60, 100, 25, 3000, 100, 495

Johnson, George A., 60, -, 55, 2000, 150, 435

Black, James, 20, -, 20, 500, 40, 150

Jarrell, Joseph, 40, 5, -, 1500, 40, 225

Grayham, George, 75, 50, 25, 4000, 75, 315

Morris, Alexander, 60, 70, -, 2000, 50, 200

Hamish, Christian, 225, 49, -, 13000, 500, 935

Cook, Risden, 175, 20, -, 8000, 200, 1278

Moore, George, 80, 20, -, 3000, 50, 170

Knotts, William, 100, 30, 100, 3000, 150, 833

Tinley, Johnathan, 70, 100, 30, 8000, 150, 564

Marker, Henry, 25, 15, 10, 800, 15, 170

Hopkins, John, 100, 25, 75, 4000, 50, 440

Beakling, John S., 125, 50, 75, 5000, 100, 557

Kemp, Mathew, 100, 200, 50, 7000, 90,500

Perry, David, 60, 10, 45, 3000, 100, 280

Perry, Spencer, 90, 30, -, 4000, 50, 357

Kemp, George, 40, -, 20, 1500, 10, 220

Goodin, Benjamin, 150, 100, 9, 10000, 300, 1540

Gruwell, John, 190, 150, -, 12000, 200, 710

Cohee, Elizabeth, 120, 8, -, 4000, -, 60

Cook, John, 75, 50, -, 6000, 200, 915

Reynolds, Robert, 200, 175, -, 10000, 300, 1413

Reed, Benjamin L., 15, 37, -, 15000, 40, 178

Meredith, Ezekiel C., 60, 59, -, 2500, 100, 673

Cooper, John W., 30, 30, 20, 2000, 10, 150

Sipple, Uriah, 75, 73, -, 5000, 200, 575

Price, Edward T., 60, 59, -, 2500, 30, 217

Dill, Catharine, 40, 25, 20, 2500, 25, 356

Pierson, William, 35, 20, -, 1000, 50, 230

Killen, Thomas, 80, 50, 33, 3000, 20, 235

Moore, Griffin, 125, 75, -, 10000, 100, 680

Hubbard, Peter T., 90, 100, -, 8000, 150, 370

Meredith, William W., 40, -, -, 2000, 20, 95

Meredith, Peter K., 110, 15, -, 4000, 400, 1260

Clark, Nath., 100, 40, -, 4000, 150, 505

Christopher, John W., 100, 50, -, 2000, 60, 315

Conner, Samuel D., 65, 100, 25, 4000, 70, 520

Smith, David S., 13, 10, -, 800, 40, 345

Longfellow, Jonathan, 100, 68, -, 4000, 125, 475

Clark, Lemuel, 20, 30, -, 5000, 100, 460

Minner, Thomas, 20, 15, -, 1000, 90, 360

Minner, Samuel, 50, 50, -, 2000, 50, 220

Minner, William, 30, 70, -, 1500, 50, 425

Dill, Abner, 150, 100, 16, 4000, 150, 758

Smith, Samuel W., 40, 16, 4, 1500, 50, 140

Dodd, Silas, 52, 9, 9, 1500, -, -

Vickery, William 100, 30, -, 2500, 30, 305

Dodd, Silas, 77, 20, -, 2000, 50, 327

Scabinger, John, 100, 45, -, 5000, 150, 625

Warren, John, 200, 300, 100, 15000, 300, 1283

Herd, Ann, 30, 30, 10, 1500, 30, 206

Sipple, Robert H., 80, 26, -, 2700, 90, 564

Frazier, William, 90, 96, -, 4000, 130, 746

Hoyd, John, 100, 100, -, 6000, 300, 935

Wyatt, John J., 75, 25, -, 2000, 30, 420

Sapp, William M., 60, 15, -, 1500, 35, 425

Moore, Benjamin, 80, 10, -, 2000, 80, 265

Dill, William H., 38, 19, -, 800, 30, 350

Hues, Eben, 50, 20, 25, 2000, 200, 460

Melvin, Hinson, 47, 16, -, 1500, 60, 360

Melvin, James A., 100, 50, 100, 3000, 50, 380

Greenlee, Robert, 75, 100, 25, 3000, -, 135

Jump, William C., 75, 30, -, 2000, 125, 630

Morris, Elisha, 20, 3, 30, 700, 30, 123

Hues, Samuel, 10, 8, -, 1000, 50, 255

Hurd, James, 100, 30, -, 4000, 200, 845

Cooper, Samuel B., 80, 20, -, 3000, 30, 330

Dill, Peter, 80, 45, -, 2500, 50, 215

Draper, John L., 12, 11, -, 300, 10, 100

Edwards, William, 125, 160, 15, 5000, 110, 400

Clark, Caleb, 60, 30, 10, 1500, 30, 175

Edwards, John W., 20, 20, -, 500, 30, 112

Reynolds, George, 100, 49, -, 2250, 25, 160

Hurd, James K., 185, 100, 15, 6000, 40, 360

Irvin, John, 80, -, 20, 3000, 100, 560

Montague, Daniel, 160, 5, -, 3500, 150, 412

Brown, William H., 110, 90, -, 5000, 75, 660

Green, Thomas, C., 280, 25, 20, 15000, 800, 2300

Carter, Philemon C., 225, 75, 15, 12000, 300, 1800

Bell, Thomas H., 16, -, -, 800, 50, 300

Dill, Samuel L, 100, 10, -, 1000, 100, 320

Dill, Andrew, 75, -, -, 1000, 40, 200

Lewis, Jacob A., 100, -, -, 2500, 80, 385

Baker, John H., 40, 20, -, 1000, 10, 80

Hurd, William, 45, 15, 4, 1000, -, 15

Cooper, John S., 120, 30, -, 3000, 40, 490

Cooper, John W., 200, 130, -, 20000, 1000, 1520

Hues, William, 100, 200, 280, 3000, 300, 320

Pinder, Edward, 75, 85, -, 1600, 40, 315

Hurd, John, 75, 75, -, 2000, 80, 567

Voshel, William H., 10, 7, -, 700, 60, 355

Draper, Benjamin H., 25, 22, -, 1200, 60, 395

Tribbet, Stephen, 60, 60, -, 2000, 50, 324

Faulkner, William, 16, 12, -, 600, 30, 175

Faulkner, Samuel, 40, 10, -, 1000, 10, 159

Clark, John, 45, 15, -, 1000, 25, 150

Cohee, James, 60, 30, 10, 2500, 150, 375

Clark, Mary, 40, 40, -, 2000, -, 50

Barcus, William J., 100, 30, -, 2600, 25, 320

Edwards, Philemon, 10, 7, -, 500, 30, 225

Hurd, Thomas G., 50, 15, -, 1200, 20, 100

Meredith, John R., 100, 100, 30, 2000, 100, 510

Luff, George W., 50, -, 12, 1200, 80, 560

Bernite, James K., 60, -, -, 12000, 250, 1550

Bernite, James K., 70, 109, 30, 8000, -, -

Bernite James K., 10, 39, -, 1000, -, -

Bernite, James K., 50, -, 10, 2500, -, -, -

Kelley, Silas B., 75, 25, 15, 10000, 300, 670

Killen, George W., 50, 6, -, 3500, 150, 461

Holden, Susan, 130, 97, -, 4000, 75, 665

Mack, Flavel C., 40, 22, -, 5000, 125, 255

New Castle County Delaware
1870 Agricultural Census

This agricultural census was filmed from original records in the Delaware State Archives in Wilmington Delaware by the Delaware State Archives Microfilm office.

There are some fifty-two columns of information on each individual. Only the head of household is addressed. I have chosen to use only seven columns of the information because I feel that this information best illustrates the wealth of the individuals. These are shown below:

1. Name of Owner
2. Acres of Improved Land
3. Woodland Unimproved
4. Other Unimproved Acreage
5. Cash Value of Farm
6. Value of Farm Implements and Machinery
15. Value of All Livestock

Thus, the numbers following the names represent columns 2, 3, 4, 5, 6, 15.

The following symbol is used to maintain spacing where information in a column is left blank (-). This symbol is used where letters, names or numbers are not legible (_).

NOTE: Beginning with the 1870 Agricultural Census the column names have changed. And, the names for this county appear as Last Name First.

Townsend, Richard, 100, -, -, 16000, 500, 1800

Laffoneur, Levi W., 80, 50, -, 7800, 400, 1400

Ennis, George W., 4, -, -, 1500, 25, 30

Budd, James, 160, 4, -, 13000, 300, 1000

Shears, William T., 50, -, -, 1500, 100, 400

Taylor, James T., 12, -, -, 1200, 100, 400

Ginn, Benjamin F., 150, 50, -, 1000, 300, 800

Thompson, John, 14, -, -, 1000, 50, 100

Dought__, Wilson, 100, 80, -, 10000, 200, 800

Vannagner, Joel, 4, 4, -, 300, 25, 75

Manering, James, 26, 8, -, 1500, 75, 400

Welch, Eli, 60, 120, -, 5000, 200, 600

Collins, John P., 40, -, -, 16000, 100, 700

Thornton, John, 100, 60, -, 3000, 50, 300

Boone, Skinner, 30, 40, -, 3000, 50, 300

Watson, Cornelous, 25, 8, -, 13000, 300, 1800

Staats, John F., 150, 86, -, 10000, 200, 1500

Lewis, John R., 220, 12, -, 12000, 150, 1600

Tatman, Cyrus, 139, -, -, 10000, 230, 1000

Finley, Archabald, 40, 17, -, 5000, 200, 560

Money, Benjamin, 100, 30, -, 10000, 300, 700

Roberts, Samuel, 130, 20, -, 10000, 300, 1000

Manering, Suel, 150, -, -, 15000, 250, 975

Deakyne, John W., 100, 80, -, 10000, 100, 900

Silcox, Charles P., 15, -, -, 1500, 50, 200

Rust, George F., 110, 40, -, 15000, 100, 900

Jones, Pernel, 170, 20, -, 18000, 500, 2065

Silcox, Edward, 15, -, -, 1500, 50, 200

Fleming, Alexander, 275, 25, -, 20000, 300, 1500

Henry, Joseph, 4, -, -, 250, 25, 150

Brister, Scharloot, 5, 4, -, 400, 25, 100

Hutchison, John, 20, -, -, 1400, 100, 600

Roberts, Joseph, 170, 55, -, 18000, 300, 1800

Newton, Charles, 200, 20, -, 20000, 200, 1000

Lynum, William W., 23, 50, -, 2500, 25, 300

Maloner, Thomas, 15, -, -, 1000, 50, 150

Townsend, Job, 80, 50, -, 5000, 100, 700

Shaw, William, 100, -, -, 4000, 100, 800

Lind, William H., 80, 30, -, 6000, 40, 800

Weldon, William, 260, 40, -, 15000, 300, 1500

Watson, Anthony, 9, -, -, 500, 80, 300

Fennimore, Lewis, 150, 60, -, 7000, 100, 800

Brockson, Richard, 100, 50, -, 8000, 140, 600

Ginn, Martin, 100, 200, -, 8000, 100, 700

Danields, William J., 250, 50, -, 15000, 200, 1500

Moore, James, 180, -, -, 10000, 200, 800

Lord, Frisby, 8, 7, -, 500, 50, 100

Frances, William, 150, 50, -, 1000, 100, 1000

Frances, James H., 100, 54, -, 8000, 100, 300

Segars, David W., 52, -, -, 4000, 50, 150

Davis, Nehemiah Jr., 70, 30, -, 15000, 200, 1500

Appleton, Henry H., 200, -, -, 30000, 650, 2000

Crouding, Jacob, 100, 20, -, 1000, 200, 900

Goldsborough, Andrew J., 100, 50, -, 6000, 100, 250

Fortner, Absolam, 30, 7, -, 3000, 50, 150

Barlow, Gideon E., 60, 140, -, 6000, 50, 400

Appleton, John M., 230, 60, -, 35000, 400, 2500

Tilman, Elias, 210, -, -, 15000, 200, 1000

Moore, Elias, 135, 35, -, 25000, 200, 700

Kennard, Thomas, 165, 10, -, 20000, 200, 800

Ratledge, William, 110, 10, -, 10000, 100, 800

Faulkner, William, 150, 50, -, 12000, 300, 1500

Prettyman, Perry, 300, -, 100, 35000, 400, 2000

Lister, Edward, 190, 17, -, 15000, 400, 1500

Ryley, William E., 300, 50,-, 30000, 500, 2000

Edward, Joseph, 200, 20, -, 18000, 500, 1500

Warren, Samuel R., 100, 25, -, 7000, 300, 1500

Rose, David C., 150, 90, -, 10000, 400, 1000

Lord, Theadore, 135, 5, -, 6000, 100, 600

Armstrong, Richardson, 50, 120, 55, 10000, 500, 1500

Rothwell, Gideon E., 300, 35, 25, 30000, 400, 2500

Thompson, Isaac, 200, 500, -, 20000, 200, 1000

Spearman, Pen P., 60, 110, -, 6000, 100, 500

Danields, George, 150, 50, -, 8000, 400, 1200

Richard, Zacarias, 300, -, 100, 25000, 300, 25000

Athly & Brothers, 200, -, 140, 25000, 300, 2000

Perry, Frances T., 200, -, -, 3000, 300, 2500

Darby, Alexander, 170, 10, -, 25000, 300, 4100

Townsend, George L., 135, 15, 15000, 800, 1000

Vandyke, William S., 250, -, 130, 25000, 500, 500

Walker, Wilson W., 155, -, 25000, 600, 1500

Boylishitts, Michal, 120, -, -, 25000, 500, 800

Thomas, Lydia, 400, 10, 200, 5000, 1000, 3500

Fennimore, Edward, 200, -, 500, 30000, 800, 2500

Fennimore, Samuel, 250, -, 140, 30000, 500, 1500

Buris, Samuel, 190, -, -, 20000, 500, 2000

Reynolds, Aaron, 200, -, -, 25000, 800, 1500

Whitlock, Henry, 270, 30, 3000, 40000, 1000, 2500

Tinly, Stringer L., 150, -, -, 20000, 500, 2000

McCoomie, John, 180, -, -, 2000, 500, 1000

Bratton, Abraham, 125, 75, -, 5000, 100, 400

Ingram, Abraham, 5, -, -, 500, 50, 200

Denny, Benjamin, 175, 75, -, 15000, 300, 1200

Hardiway, Chatfield, 100, 60, 20, 8000, 200, 800

Collins, James R., 300, 50, 50, 30000, 500, 2500

Collins, Andrew J., 300, -, 30, 30000, 500, 2700

Lattemus, Kooms, 100, 50, -, 5000, 300, 800

Buckson, James, 150, -, 150, 10000, 175, 1200

David, Frances, 60, -, 60, 3000, 50, 200

David, Benjamin, 120, -, 80, 6000, 200, 1000

Staats, Isaac, 120, -, 200, 5000, 300, 1100

Johnson, Robert, 100, 529, -, 10000, 50, 300

Otterson, Elias, 300, -, 100, 30000, 1000, 3000

Derrickson, Robert, 160, -, 100, 6800, 300, 2000

Haden, Abraham, 300, 15, 100, 20000, 400, 2000

Hden, Abraham Jr., 102, -, -, 5000, 100, 500

Ennis, Annias, 80, 40, 40, 12000, 500, 1500

Mathews, Wingate, 105, 45, -, 12000, 500, 1000

Lewis, James H., 50, 54, -, 6000, 400, 500

Collins, Catharine, 80, -, 160, 6000, 200, 800

Statts, Jacob, 120, 40, -, 8000, 200, 600

Staats, Jacob Jr., 200, 100, -, 12000, 300, 800

Weller, Frederick H., 100, -, -, 4000, 200, 500

Staats, Isaac R., 100, 25, -, 6000, 300, 1000

Reynolds, William, 85, -, -, 3000, 100, 400

Deakyne, Alexander, 95, 46, -, 6000, 300, 800

Derrickson, John, 100, 15, -, 6000, 300, 800

David, James L., 125, -, -, 5000, 200, 800

Bungy, Auther, 125, 10, -, 5000, 200, 500

Staats, George, 186, 10, -, 8000, 400, 1000

Staats, William H., 85, -, -, 3300, 100, 400

Staats, James R., 97, -, -, 3500, 100, 300

Lyrman (Lyman), David, 140, -, -, 4000, 200, 400

Staats, David, 40, -, -, 1500, 100, 300

Gibbins, John, 53, -, -, 2000, 100, 200

Lambson, Thomas, 200, 50, -, 8000, 200, 800

Young, John, 80, -, -, 4000, 200, 400

Staats, Samuel J., 46, -, -, 2500, 200, 300

Bratten, Thomas, 22, -, -, 1000, 100, 250

Hoffecker, Philip, 160, 20, -, 7000, 40, 500

Johnson, George W., 150, 30, -, 10000, 200, 400

Ford, Christofer, 200, 80, -, 16000, 300, 1000

Roberts, James, 150, -, -, 10000, 300, 1000

Herd, George, 75, 125, -, 6000, 200, 500

Thompson, James, 60, 40, -, 4500, 200, 400

Dolson (Polson), Larrance, 80, 120, -, 5000, 50, 280

Pierson, Daniel, 90, 36, -, 5000, 100, 256

Macy, Isaac, 150, 50, -, 8000, 100, 600

Crowley, Owen, 36, 10, -, 1500, 50, 500

Dunaway, John, 30, 40, -, 3000, 100, 250

Stevens, Abraham, 200, 25, -, 10000, 200, 1000

Gardner, William, 300, 50, 100, 15000, 300, 800

Termman (Lermmons), Jesse, 10, 8, -, 800, 100, 500

Lockerman, John, 100, 20, 30, 4500, 100, 250

Deakyne, Thomas, 200, -, -, 10000, 500, 1200

Harteuss, Thomas, 150, -, 20, 10000, 500, 800

Lattamous, James, 80, -, -, 5000, 200, 850

Garner, James, 43, -, -, 1500, 100, 350

Woodrufus, Henry, 43, -, -, 1500, 600, 400

Staats, Edwin, 200, 25, 30, 15000, 300, 1000

Huggins, Roberts, 80, 20, 25, 4000, 100, 350

Craig, William, 65, -, -, 2000, 100, 250

Rickards, Joseph, 150, -, 4000, 18000, 300, 1000

Deakyne, Napolian, 80, -, 100, 6000, 200, 500

Jarrell (Javrel), Charles, 130, -, 30, 11000, 150, 350

Deakyne, George, 130, 47, 30, 7000, 200, 600

Deakyne, William _., 150, -, 100, 8000, 200, 800

Cooper, John, 60, -, -, 3000, 100, 300

Javrel, John, 9, -, -, 1000, 50, 200

Warran, George, 160, -, 100, 10000, 300, 1000

Cooper, Henry, 200, 40, 20, 15000, 300, 1000

Deakyne, Jacob, 140, -, 100, 10000, 400, 1200

Barrat, Elizabeth, 100, 80, -, 7000, 200, 400

Alston, William, 120, 60, -, 8000, 200, 510

Hickman, Martin, 70, 30, -, 4000, 100, 400

Pierce, Charles, 90, 10, -, 3000, 100, 700

King, Edward, 100, 25, -, 3000, 100, 500

Hill, Jacob, 185, 15, -, 10000, 200, 1200

Powell, Robert, 40, 120, -, 4000, 100, 450

Powell, Thomas, 100, 100, -, 4000, 100, 200

Blackston, John, 40, 113, -, 5000, 50, 200

Parker, Joseph, 70, 38, -, 4000, 100, 300

Blidt, Michal, 50, 60, -, 4000, 100, 400

Bennit, Samuel, 80, -, -, 2500, 75, 350

Keen, Henry, 14, -, -, 1400, 25, 200

Macy, Peter, 115, 20, -, 4000, 100, 400

Davis, Cornelious, 250, -, 100, 10000, 2500, 1800

Huggins, Samuel, 200, 50, -, 10000, 250, 100

Fowler, Stephen, 100, 10, -, 5000, 150, 1200

Brown, Jonathan, 180, -, -, 9000, 200, 1500

Danields, Jacob, 250, 25, -, 5000, 100, 800

Barber, John, 200, 15,-, 700, 150, 1000

Walker, Isaac P., 40, -, 20, 4000, 50, 276

Nailer, William, 80, 43, -, 5000, 100, 600

Gordin, George, 175, 25, -, 10000, 300, 1000

Dill, Philomon, 25, 28, -, 2000, 25, 100

Allen, John, 5, -, -, 400, 25, 150

Wright, John, 240, 15, -, 16000, 200, 1200

Warner, Peter, 130, 30, 10, 6500, 100, 800

Deavon(Dearon), Bayard, 10, 60, -, 2000, 25, 150

Dearon, John, 75, -, -, 3000, 100, 300

Jerman, John, 200, 50, -, 18000, 300, 1200

Middleton, Thomas, 160, 20, -, 16000, 300, 800

Middleton, Benjamin, 140, -, 80, 7000, 200, 600

Johnson, James, 250, -, 50, 15000, 300, 1000

Kenneda, Joseph, 80, -, -, 3000, 100, 600

Crow, Owen, 160, 20, -, 15000, 300, 1500

Cavender, Theodore, 150, 10, -, 15000, 300, 800

Winford, John, 200, 75, -, 15000, 250, 800

White, Clement, 70, -, -, 3500, 100, 451

Trustee, Jacob, 10, 10, -, 800, 25, 150

Spear, John, 180, 50, -, 12000, 200, 850

Nolan, James, 70, 30, -, 3000, 100, 400

Buckanan, James, 34, -, -, 2000, 50, 300

Armstrong, Samuel, 250, 5, -, 18000, 300, 1000

Armstrong, John, 180, 80, -, 10000, 300, 1000

Coursey, James, 32, -, -, 1000, 120, 125

Wells, Joseph, 200, -, -, 10000, 300, 1000

Prier, Jeremiah, 40, 3, -, 4000, 100, 400

Stephenson, Joseph, 240, -, -, 18000, 300, 800

Broadway, William, 60, -, -, 3000, 50, 250

Donoho, William, 130, 4, -, 13000, 100, 103

Donoho, Sallie, 10, 4, -, 800, 25, 150

Townsend, John, 175, -, -, 30000, 1300, 1600

Danields, William, 127, -, -, 13000, 300, 1080

Vinyard, James, 200, 100, -, 15000, 500, 1200

Atwell, Edward, 175, 125, -, 15000, 200, 1000

Danields, Stockeley, 23, 4, -, 2000, 25, 300

Caulk, Wesley, 9, -, -, 500, 25, 150

Hanson, Joseph H., 125, 75, -, 6000, 100, 400

Rothwell, Washington, 300, 40, -, 22000, 700, 1700

Brooks, James, 100, 25, -, 12000, 300, 1000

Rothwell, James, 130, 20, -, 15000, 500, 1500

Rothwell, Moody, 212, 20, -, 20000, 600, 2000

Deakyne, Ward, 30, 20, -, 1200, 50, 350

Prettyman, George, 100, 125, -, 8000, 100, 500

Wilson, Manlove, 180, 20, -, 18000, 400,1 500

Warran, Kendal, 100, -, -, 4000, 50, 200

Brister, Jeremiah, 175, 45, -, 10000, 100, 1000

Alfry, William, 150, 110, -, 32000, 600, 2500

Hand, Jonathan S., 150, 27, -, 8000, 200, 1500

Gill, Benjamin, 160, 15, -, 12000, 250, 900

Harris, John B., 30, -, 18000, 200, 1500

Hutchison, Joseph, 100, -, -, 10000, 150, 1000

Ginn, Lorenza, 75, 25, -, 10000, 150, 400

Roberts, John B., 250, 20, -, 18000, 300, 1000

Couchran, Robert, 325, 40, -, 40000, 1500, 3500

Davis, Manlove, 260, -, -, 30000, 600, 2000

Naudain, Arnold, 200, -, -, 18000, 500, 1200

Davis, Isaac, 150, -, -, 15000, 250, 1000

West, John, 120, 20, -, 10000, 300, 1000

Davis, Mark, 150, -, -, 20000, 300, 1400

Kanely, Benjamin, 200, 25, -, 20000, 300, 1500

Kanely, James C., 200, -, -, 20000, 400, 1600

Hanson, Richard, 275, 25, -, 20000, 300, 1800

Redmiles, Thomas, 90, 10, -, 4000, 100, 600

Knotts, George, 200, 50, -, 15000, 200, 800

Hevern, John, 7, 9, -, 400, 75, 200

Allen, Isaac, 30,70, -, 3500, 50, 250

Landers, Philip, 100, 200, -, 10000, 200, 1200

Jones, William, 100, 50, -, 5000, 154, 800

Sweatman, Thomas, 40, 30, -, 3000, 100, 500

Vandyke, Jacob C., 4, 17, -, 4000, 100, 500

Browning, George, 75, 25, -, 4000, 100, 500

Warrran, Samuel, 70, 14, -, 6000, 100, 400

Hollis, Eli, 10, 45, -, 1500, 25, 300

Jester, Isaac, 200, -, -, 15000, 200, 800

West, Joseph, 165, -, 232000, 700, 1000

McCrone, Mary, 350, 50, -, 30000, 500, 1000

Hanson, Joseph, 200, 50,-, 20000, 500, 1200

Wilson, William, 200, -, -, 20000, 500, 2000

Toreum, John, 125, 75, -, 7000, 200, 400

Crawford, James, 332, 10, -, 30000, 500, 1600

Beddle, George, 170, 65, -, 7000, 200, 500

Wilson, William, 160, -, -, 20000, 500, 1500

Aibh, Benjamin, 200, -, -, 20000, 500, 1500

Knotts, William 250, -, -, 22000, 500, 1500

Lockerman, Elias, 110, 100, -, 10000, 100, 1000

Townsend, Joshua, 7, -, -, 300, 25, 100

Harris, Jacob, 40, 5, -, 1500, 50, 300

Bartley, Samuel, 38, 5, -, 3000, 25, 150

Davis, Nehemiah, 400, 100, -, 20000, 300, 2000

Clayton, Isaac, 14, 6, -, 1000, 25, 200

Colyer, George, 130, 82, -, 10000, 520, 1000

Reynolds, Walter, 19, 14, -, 3000, 50, 200

Lattamus, Alexander, 100, -, -, 4000, 100, 400

Clayton, John, 64, -, -, 2500, 75, 300

McCarter, James, 50,80, -, 3500, 100, 400

McKay, Benjamin, 80, 40, -, 3000, 100, 500

Reed, James, 20, -, -, 5000, 50, 250

Philips, Samuel, 10, 5, -, 700, 25, 100

Tibbit, Daniel, 15, 40, -, 400, 25, 150

Clayton, Jacob, 30, 36, -, 1200, 50, 300

Holt, James, 100, 103, -, 4000, 150, 600

Carrow, James, 100, 200, -, 5000, 200, 450

Beck, Samuel, 50, 71, -, 2500, 100, 500

Prier, James, 50, 30, -, 2000, 75, 300

Dixon, Cooper, 50, 30, -, 2000, 57, 300

Clark, John, 30, 10, -, 400, 50, 250

Gankle, Israel, 30, 120, -, 2500, 75, 200

Gesford, John, 60, 120, -, 3000, 100, 300

Bennet, Thomas, 18, 4, -, 800, 50, 300

Brockson, James, 240, 40, -, 18000, 300, 1500

Sena, James, 50, 80, -, 4000, 50, 150

Smallwood, Washington, 30, -, -, 1000, 25, 200

Cox, Price, 150, 100, -, 10000, 100, 500

Gott, John, 100, -, -, 4000, 100, 450

Nailer, William 200, 8, -, 15000, 400, 1300

Porter, Abel J., 65, -, -, 6000, 250, 700

Lockerman, William, 120, 30, -, 8000, 200, 800

Crocket, James, 4, 1, -, 400, 25, 260

Cullen (Culler), David, 50, 25, -, 5000, 50, 200

Collins, George, 100, 35, -, 9000, 150, 600

Thomas, Alexander, 15, 8, -, 500, 25, 150

Pierce, Noah, 12, -, -, 300, 25,75

View, Joseph, 5, -, -, 200, 25, 150

Clayton, John M., 15, 7, -, 600, 25, 100

Davis, Peter, 11, -, -, 350, 15, 125

Greer, Charles, 50, -, -, 200, 10, 75

Ford, Jefferson, 200, 100, -, 10000, 200, 1000

Young, John, 150, 50, -, 8000, 200, 1000

Murphy, Charles, 150, 100, -, 6000, 200, 1000

Webster, Andrew, 140, 83, -, 7000, 200, 800

Webster, James, A., 40, 37, -, 2000, 25, 350

Graham, Samuel, 75, -, 500, 20, 150

Moore, James R., 14, -, -, 700, 25, 150

Melvin, James, 65, -, -, 2500, 75, 250

Simmons, John, 50, 65, -, 2000, 75, 230

Vail, Thomas, 30, 40, -, 1200, 25, 200

Reynolds, James, 36, 100, -, 2000, 75, 300

Reynolds, William, 45, 30, -, 1000, 25, 300

Banks, John, 25, 15, -, 800, 50, 250

Ford, William, 60, 36, -, 1500, 50, 250

Sathell, Henry, 50, 46, -, 2000, 50, 300

Birch, Peter, 80, 47, -, 2500, 150, 400

Jones, Miles T., 80, 150, -, 5000, 150, 350

Webb, Isaac, 85, 20, -, 2000, 100, 400

Ford, John W., 75, 75, -, 3000, 150, 650

Hoofman, Joseph, 90, 50, -, 3500, 200, 700

Clayton, Daniel, 4, 3, -, 400, 25, 240

Gewman, John, 90, 15, -, 3500, 300, 1000

Schockly, Daniel, 85, 225, -, 6000, 200, 800

Marten, James, 50, 15, -, 1600, 100, 400

Budd, William, 18, 25, -, 400, 75, 340

Hanifee, Patrick, 140, -, -, 10000, 150, 750

Donovan, James, 200, 100, -, 12000, 200, 600

Riggs, William, E., 180, -, -, 18000, 200, 1000

Clouds, David, 140, 25, -, 10000, 200, 700

Caul, William, 15, -, -, 700, 25, 150

Lee, David, 10, -, -, 450, 25, 150

Moffett, William, 100, 30, -, 6000, 100, 500

Thomas, John, 35, 35, -, 2000, 25, 300

Wright, Roberts, 80, 100, -, 6000, 75, 1000

Sparks, Samuel, 130, 20, -, 5000, 100, 750

Powell, James, 65, 18, -, 4000, 250, 700

Marine, Thomas, 75, 100, -, 4500, 25, 350

Deakyne, John, 100, 100, -, 5000, 100, 800

Keefer, Nicholas, 40, 55, -, 2500, 50, 350

Raymond, Henry, 12, 25, -, 1000, 50, 300

Reed, John, 70, 500, -, 8000, 25, 250

Tush, William, 95, 5, -, 4000, 75, 500

Anderson, William 4, 4, -, 300, 20, 150

Lee, William, 70, -, -, 2000, 100, 500

Gesford, Nathaniel, 60, 230, -, 5000, 100, 800

Fleetwood, Alexander, 40, 98, -, 2500, 75, 150

Morgan, Nathan, 150, 250, -, 10000, 300, 1500

Rash, John, 20, 5, -, 500, 25, 250

Nucum, John, 70, 40, -, 1500, 50, 300

Hall, Israel, 60, 100, -, 5000, 75, 400

Townsend, Samuel, 120, 10, -, 40000, 400, 2000

Carney, Abraham, 6, 10, -, 8000, 25, 200

Townsend, Joshua, 6, -, -, 400, 25, 100

Scheggs, John, 60, 88, -, 5000, 50, 3500

Scheggs, William, 70, 90, -, 6000, 50, 600

Roberts, Zacaras Mc., 300, 100, -, 15000, 200, 1200

Furgerson, Cole, 10, -, -, 1500, 25, 250

Geston, Robert, 96, -, -, 4000, 100, 700

Gale, Jeremiah, 22, -, -, 1200, 25, 300

Caulk, Isaac, 200, 100, -, 10000, 200, 1000

Beason (Benson), David H., 88, 2, -, 16000, 300, 780

McCoy, George W., 172, 3, -, 35000, 700, 4210

Beason, John T., 115, 45, -, 16000, 300, 1848

Zebley, Jacob, 99, 45, -, 16000, 300, 1310

Sellars, George H., 60, -, -, 18000, 150, 1400

Husbands, Andrew, 15, -, -, 3300, 150, 700

Forward, Miller, 80, 15, -, 12000, 200, 1200

Thomas, Class, 60, 5, -, 8000, 200, 700

Perry, Oliver H., 48, 4, -, 10000, 250, 840

Miller, Martin, 59, 6, -, 9000, 300, 850

Talley, John, 35, 5, -, 6000, 150, 600

Weldon, Joseph, 28, -, -, 3000, 75, 700

Thompson, George, 15, -, -, 8000, 100, 300

Smith, David, 230, -, -, 50000, 800, 5055

Mendenhall, Eli, 92, -, -, 14000, 300, 1690

Satwell, Edward, 60, 25, -, 77000, 300, 700

Hance, George W., 91, -, -, 27000, 250, 2000

Wilson, Edmond, 86, -, -, 20000, 300, 1583

Swift, James, 6, -, -, 1500, 250, 186

Ragan, Patrick, 38, -, -, 9000, 250, 800

Todd, James, 127, -, -, 20000, 300, 2575

McDonald, Alexander, 50, -, -, 75000, 250, 920

Naylor, Isaac, 81, -, -, 13000, 200, 1190

Sturn, Isac, 10, -, -, 5000, 200, 475

Forward, Anna, 52, -, -, 5000, 150, 455

Nelder, William, 86, -, -, 12000, 200, 850

Robinson, Mary, 140, -, -, 28000, 300, 2280

Weldon, William R., 34, -, -, 5000, 120, 404

Weldon, Levi M., 60, -, -, 6000, 150, 690

Louderman, Isaac T., 74, -, -, 12000, 150, 1210

Welsh, Patrick, 160, -, -, 24000, 300, 1650

Moore, John W., 80, -, -, 13000, 200, 1052

Beason, John, 130, 24, -, 15000, 75, 768

Hipple, Jacob, 20, -, -, 6000, 100, 275

Robinson, Hanson, 51, -, -, 60000, 150, 1350

McCullough, William, 65, 3, -, 9000, 150, 900

Gest, Alfred, 33, -, -, 9000, 100, 775

Mahoney, Meshal, 90, -, -, 25000, 500, 2420

Oan (Ocen, Oau), Robert, 84, -, -, 10000, 300, 1300

Beason (Benson), Robinson, 59, 7, -, 6000, 200, 580

Congleton, John B., 33, 10, -, 4300, 200, 230

Elewright, Zachariah, 32, 6, -, 5000, 50, 135

Hopkins, Joseph, 45, 6, -, 4000, 50, 410

Righton, Evins, 19, 3, -, 3500, 200, 460

Peina (Prince), Henry, 11, -, -, 1600, 75, 145

Meholson, Rebecca, 36, 5, 3000, 200, 550

Hanby, William 80, 20, -, 1000, 300, 775

Hanby, William E., 50, 15, -, 7000, 150, 585

Benson, George T., 100, 30, -, 12000, 200, 1070

Hanby, Samuel, 119, 25, 34, 13500, 500, 1548

Langly, Joseph, 90, 10, -, 12500, 500, 1370

Sharp, Jesse, 720, 10, -, 32000, 500, 2560

Way, Marcer (Mercer) G., 145, 20, -, 25000, 300, 2455

Veale, John, 80, 20, -, 10000, 100, 580

Veale, George, 82, 15, -, 13000, 100, 925

Bird, William, 86, 14, -, 12000, 300, 1220

Phillips, Thomas, 100, 60, -, 2000, 200, 1500

Haney, Jeremiah, 75, 25, -, 15000, 300, 1450

Lodge, Samuel G., 51, 9, -, 11000, 300, 750

McKay, James, 85, 15, -, 18000, 300, 1163

Vernon, William, 46, 4, -, 10000, 100, 750

Rice, Joseph, 80, 4, -, 9500, 100, 700

Goodley, Thomas, 40, 4, -, 6000, 150, 830

Ennis, Francis, 19, 5, -, 2000, 75, 195

Carpenter, Joseph, 81, 10, -, 10000, 250, 880

Prince, John M., 76, 30, -, 10000, 400, 1390

Prince, Walter, 40, 2, -, 5000, 200, 1180

Bunting, Joshua, 65, 6, -, 9000, 400, 650

Hanby, James, G., 1000, 8, -, 10000, 300, 1300

Petetedemange, John, 108, 12, -, 10000, 300, 150

Cloud, George L., 50, -, 10, 6000, 100, 710

Cloud, Lott, 135, 25, -, 13000, 300, 1500

Grubb, Thomas B., 25, 5, -, 3000, 50, 195

Fally (Talley), William A., 40, 2, -, 4200, 100, 500

Peirce, Alfred D., 30, 3, -, 3700, 75, 435

Philips, Lewis, 30, -, -, 3000, 100, 300

Herbey, Charles, 50, 10, -, 9000, 200, 670

Talley, Curtis M., 51, 10, -, 6000, 250, 570

Ferris, Thomas M., 68, 25, -, 50000, 1200, 1060

Price, James, 52, -, -, 9000, 200, 900

Churchman, C. & M. H., 60, -, 6, 12000, 100, 1445

Ford, Franklin, 30, -, -, 1500, 100, 150

Ford, Franklin Jr., 120, -, -, 25000, 640, 1500

Cozyens, John, 100, -, 100, 30000, 250, 1430

Danley, Alfred T., 25, -, -, 24000, 200, 550

Sage, James M., 61, 5, -, 20000, 250, 935

Lodge, Isaac W., 28, 4, -, 3200, 200, 505

Gest, James H., 35, 20, -, 6700, 200, 810

Mayne, Theophilus, 50, 30, -, 14000, 100, 720

Talley, Jesse, 41, 4, -, 7000, 250, 565

Osborne, James, 100, 25, -, 15000, 350, 1880

Grubb, James, 55, -, -, 14000, 200, 490

Jones, Joseph H., 83, 17, -, 20000, 1500, 2950

Price, Marcellus, 200, 70, -, 35000, 600, 3640

Cassey, William, 200, -, -, 200000, 500, 350

Kimber, Thomas, 29, -, -, 20000, 400, 900

Elliott, John C., 90, 6, -, 19000, 200, 230

Anderson, John, 111, 13, -, 9000, 300, 1130

Goodwin, John, 30, 20, -, 10000, 200, 1350

Magilegan, James, 92, 8, -, 13000, 200, 975

Husbands, Thomas, 28, 5, -, 13000, 500, 1020

Ewing, Henry, 9, 1,-, 3800, 100, 300

Murphy, Alfred, 65, 10, -, 11000, 300, 755

Husbands, William, 40, 20, 9000, 300, 675

Husbands, Abraham, 29, -, -, 5500, 300, 785

Petetedemange, Frank P., 87, 20, -, 20000, 250, 652

Evins, George B., 11, 3, -, 2500, 75, 200

Derrick, Samuel, 8, -, -, 2500, -, 221

Hall, Thomas, 61, 11, -, 13000, 600, 875

Sterling, Hugh, 25, -, -, 10000, 300, 700

Elliott, Benjamin, 9, 28, -, 25000, 250, 1630

Lentz, Joseph, 5, -, -, 1000, 700, 75

Bettey, Robert, 70, 20, -, 10000, 150, 1070

Stedham, Isaac, 70, 10, -, 10000, 100, 670

Miller, George L., 62, 5, -, 9000, 300, 750

Welden, George W., 59, 10, -, 10000, 200, 935

Kink, Marish, 5, -, -, 1200, 5, 100

Dixon, Samuel H., 75, 20, -, 20000, 600, 1115

Bringhurst, Edward, 38, 8, 12000, 800, 1230

Welden, Jacob R., 222, 24, -, 22400, 400, 2180

Hewitt, Smith, 155, 50, -, 20000, 300, 1189

Sharpley, William, 35, 10, -, 6600, 100, 680

Newell, Michal, 87, 3, -, 10000, 250, 1550

Chapman, William H., 125, 50, -, 20000, 300, 830

Plank, Mariah, 25, 3, -, 3000, 100, 295

Goodyear, Robert, 40, -, -, 6000, -, 75

Dayly, Thomas, 50, 10, -, 6000, 100, 945

Rotherise, William, 72, 8, -, 11000, 500, 1380

Talley, Elihu, 153, 8, -, 15000, 300, 1900

Talley, William, 111, 12, -, 15000, 300, 1694

Banks, Robert, 150, 50, -, 20000, 800, 845

Bullock, John M., 20, 1, -, 2000, 50, 150

Trimble, John, 34, 6, -, 5000, 500, 355

Peirce, Irwin, 22, -, -, 2400, 50, 110

Judd, Henry B., 25, -, -, 12000, 100, 380

Pattison, William, 9, -, -, 2000, 250, 120

Canby, Samuel, 218, 2, -, 10000, 500, 1078

Eldrige, Levi, 100, 11, -, 18000, 150, 720

Wilson, William, 44, 1, -, 4500, 200, 435

Deakin, John, 100, 27, -, 8000, 300, 1080

Husbands, John, 138, 72, -, 14500, 300, 1270

Proud, John, 100, 18, -, 20000, 500, 1588

Hoonby, James E., 45, 20, -, 6500, 200, 1020

Righter, John, 42, 4, 25, 3000, 200, 700

Polhouse (Rolhouse), John, 70, 8, -, 11000, 1500, 880

Cloud, William, 65, -, -, 10000, 200, 835

Lentz, David, 4, -, -, 1000, 500, 200

Barlow, Malachi, 45, 2, 20, 9000, 100, 682

Talley, Lewis, 45, 20, 8000, 200, 840

Peirce, Joseph N., 54, 10, -, 6400, 200, 1075

Vance, Thomas, 24, 4, -, 3000, 100, 400

Peirce, Uriah, 32, 18, -, 5000, 75, 710

Peirce, Clark, 65, 20, -, 7000, 75, 450

Casey, Robert, 68, 4, -, 7200, 200, 1300

Foulk, James K., 55, 5, -, 8000, 150, 760

Zebley, Sarah A., 30, -, 5, 3500, 75, 385

Ween (Weer), William, 5818, -, 10750, 75, 1085

Zebley, Owen Sr., 30, -, -, 2500, 50, 475

Duffy, Neal, 40, -, 7, 4000, 50, 275

Day, John M., 90, 30, -, 19000, 500, 1200

Dunn, John, 81, 15, 4, 10000, 200, 1345

Mousley, Joseph, 18, -, -, 3500, 100, 348

Robinson, Rebecca, 22, -, -, 2000, 25, 150

Tally, Curtis, 90, 10, -, 7000, 100, 1180

Dougherty, George, 40, 2, -, 4000, 50, 440

Houghey, Peter, 50, 2, -, 6000, 150, 830

Berry, John T., 43, 1, -, 7000, 250, 670

Forward, Ann T., 38, 8, -, 4600, 200, 485

Talley, John, 8, -, -, 2500, 50, 340

McCalister, James, 40, 20, -, 7500, 150, 680

Talley, John, 32, -, -, 1500, 100, 200

Stafford, James, 140, 20, -, 12000, 150, 520

Grubb, Isaac, 33, 2, -, 4500, 100, 400

Journey, Mosses, 30, 40, -, 1000, 200, 450

Talley, Samuel M., 57, 5, -, 7750, 300, 1000

Prince (Peirce), Alexis, 31, 8, -, 4000, 150, 630

Mullen, Michal, 20, 50, -, 1500, 90, 210

Wilson, David J., 12, 10, -, 2000, 50, 320

Twaddle, James, 88, 25, -, 13000, 300, 1300

Talley, William T., 141, 49, -, 2000, 200, 1560

Chandler, John, 11, -, -, 3000, 200, 305

Moore, Annie, 16, -, -, 3000, 100, 115

Malin, Harry, 7, -, -, 2000, 50, 180

Leach, Charles, 70, 10, -, 10000, 250, 1015

Ramsey, Henry, 120, 10, -, 8000, 150, 1100

Hawkins, Thomas, 130, 11, -, 10000, 200, 1297

Greenfield, Benjamin, 15, -, -, 1200, 100, 200

Graves, John, 70, 6, -, 10000, 500, 1165

Talley, Amer, 7, -, -, 2800, 100, 585

Kellum, John, 90, 7, -, 9000, 250, 1200

Henkson, Minsil, 70, 20, -, 9000, 400, 890

Logan, Samuel, 11, -, -, 2500, 150, 280

Talley, Abner, 62, 10, -, 10000, 200, 1100

Webster, Henry, 70, 24, -, 10000, 200, 965

Talley, George M., 298, 2, -, 30000, 300, 6150

Talley, John, 66, 10, -, 10000, 125, 870

Miller, Joseph, 60, 6, 10000, 400, 878

Paschal, Charles, 74, 6, -, 12000, 300, 1250

Talley, James, W., 56, 2, -, 7250, 100, 570

Rowland, Richard, 47, 35, -, 8000, 50, 365

Lemons, Charles, 21, -, -, 3500, 250, 440

Pierce, Joseph, J., 63, 35, -, 8000, 300, 815

Lober, George, 32, -, 10, 3000, 150, 515

McCarety (McConty), Denis, 43, 5, -, 5000, 50, 590

Hand, Rachal, 22, -, -, 2500, 100, 410

Lenderman, Isaac G., 40, 2, -, 3000, 20, 242

Stoops, Albert T., 40, 5, -, 3000, 200, 820

Henk, Lewis L., 38, 2, -, 4000, 50, 335

Perkins, James A., 74, 5, -, 6500, 75, 770

Talley, Thomas L., 137, 3, -, 12600, 300, 1440

Nickolson, Emory, 16, -, -, 2000, 60, 345

Day, Thomas R., 95, 9, -, 10000, 300, 1060

Shole, John, 42, -, -, 14000, 150, 150, 290

Clark, William H., 50, -, -, 4000, 150, 585

Smith, George W., 46, 5, -, 5000, 250, 600

Beason (Benson), Edward, 55, 50, -, 7500, 100, 395

Almand, John, 33, 4, -, 3500, 150, 400

McCafferty, John, 280, 70, -, 35000, 600, 3430

Wilson, Hannah, 50, 9, -, 6500, 200, 665

Orr, William T., 60, -, -, 8000, 150, 1060

Forward, John, 160, 40, -, 2000, 300, 194

Peirce, Joseph, 14, -, -, 2400, 100, 340

Casiday, Peter, 40, 20, -, 6200, 100, 520

Talley, Penrose, 90, 40, -, 16000, 300, 1000

McCollum, Frances, 139, 17, -, 16000, 400, 1725

Hanby, Joseph, 84, 6, -, 10000, 300, 1110

Welden, Stephen G., 75, 40, -, 10000, 300, 1310

Sharpley, Jacob, 44, 2, -, 7700, 200, 850

Day, John, 20, -, -, 2500, 100, 455

Mousley, George R., 20, 8, -, 4500, 200, 605

Webster, Clark, 77, 20, -, 10000, 300, 761

Webster, Isaac, 42, 5, -, 6500, 250, 716

Talley, Thomas, 90, 10, -, 15000, 300, 670

Forwood, Valentine, 53, 15, -, 6800, 100, 555

Bird, Lewis, 40, -, -, 6000, 200, 448

Dougherty, George, 37, 10, -, 14000, 300, 1200

Smith, James A. B., 56, 4, -, 7000, 300, 835

Perkins, Esaw, 11, 16, -, 1800, 500, 1730

Blackwell, Steaphen, 25, -, -, 5000, 100, 650

Lodge, Clark, 30, 6, -, 4500, 100, 3600

Valentine, Engle, 51, -, -, 8000, 75, 760

Lodge, William P., 127, 20, -, 22000, 400, 1815

Huchson, Thomas, 90, 9, -, 20000, 500, 1315

McSorley, Frank, 20, -, -, 4000, 200, 665

Perkins, Christian, 40, -, -, 6000, 100, 792

Philips, William 50, -, -, 7000, 200, 920

Lodge, Martin, 50, -, -, 7000, 75, 996

Forward, Amon, 153, 9, -, 26000, 300, 1450

Begger, Elwood, 110, 50, -, 20000, 200, 1400

Edwards, Richard, 36, -, -, 4000, 700, 735

Talley, Henry, 81, -, -, 13000, 300, 1050

Barlow (Banlow), Henry, 100, 7, -, 8000, 200, 1245

Talley, Patten, 47, 5, -, 8000, 300, 750

O'Byrne, John, 46, -, -, 12000, 200, 900

Elliott, Isaac T., 235, 25, -, 41000, 500, 2780

Bennett, John J., 80, 10, -, 8000, 200, 1150

Shepley, Sarah, 84, 117, -, 40000, 1000, 1150

Lecarpentier, Edward, 7, 5, -, 5000, 100, 600

Lecarpentier, Charles, 100, 25, 105, 23000, 200, 1000

Wilson, Norris, 80, -, 20, 4500, 203, 1000

Dalton, John, 80, 24, -, 10000, 300, 500

Tompson, Isaac, 65, 9, 3, 7000, 300, 400

Husbands, Adulpheus, 100, 25, 35, 15000, 1000, 800

Green, William, 25, 5, -, 2500, 200, 3500

Ringold, Robert, 3, -, -, 400, -, 40

Ely, George, 75, 5, 600, 14000, 700, 2000

Wilson, Thomas, 80, 5, -, 9000, 600, 1500

Milson, Robert, 30, -, -, 4000, -, 600

Chandler, Marshael, 100, 30, 27, 15000, 800, 1500

Palmer, Morris, 180, 25, -, 20000, 500, 200

Clark, Jones, 90, 10, 15, 12000, 250, 550

McDonnell, Bridgett, 4, -, 3, 700, -, 150

Bullock, Chockley, 30, 10, -, 140, -, 300

Hendrickson, Adaline, 95, 10, -, 10000, 450, 1000

Clark, William, 75, 5, 5, 10000, 450, 1000

Bakey, Petre, 8, -, 2, 1000, -, 125

Kane, Thomas, 10, -, -, 2000, 75, 100

Holley, Daniel, 14, -, -, 2000, -, 180

Negendank, Charles, 2, -, 5, 2000, 200, 250

Negendank, Lewis, 100, 30, -, 12000, 500, 1000

Jobson, Francis, 75, 15, 15, 10000, 500, 1000

Swayne, Henry, 60, 10, 10, 10000, 250, 800

Thompson, William, 100, 20, 5, 12000, 500, 1500

Chandler, Poulson, 150, 59, -, 20000, 700, 1800

Oaks, Alban, 7, 3, -, 2000, 100, 275

Dalton, Joel, 36, 4, -, 4000, 200, 500

Bogan, Hugh, 60, 10, 8, 4800, 20, 390

McCullough, Joshua, 105, 10, -, 12000, 400, 1100

Bogan, Mary, 25, 2, -, 2000, 25, 100

Barry, Thomas, 39, 7, -, 5000, 50, 300

Ritchie, James, 20, 4, -, 3300, 150, 200

Taylor, David, 42, 63, -, 5000, -, 200

Kent, Henry, 8, 5, -, 2000, 40, 275

Thompson, John, 110, 20, -, 18000, 400, 1900

Wilmot, William, 137, 75, 2, 30000, 600, 1800

Cloud, Joseph, 50, 15, -, 6000, 300, 800

Seal, William, 26, 2, -, 5000, 250, 250

Moore, Cromwell, 24, 8, 44, 6000, 50, 450

Wright, Jane, 50, 15, 20, 7000, 100, 800

Nickelis, Elis, 50, 5, -, 8000, 150, 350

Chandler, Philema, 82, 4, -, 12000, 500, 1000

Chandler, Joseph P., 60, 7, -, 10500, 300, 600

Nickelis, Catharine, 14, 2, -, 6000, 100, 200

White, James, 35, 5, -, 10000, 300, 300

Kilroy, Daniel, 4, -, -, 1000, -, 100

Scot, Jane, 25, 2, -, 4000, 100, 275

Parkin, William, 30, 6, -, 6000, 250, 200

Peters, William, 125, 8, -, 15000, 100, 1500

Armstrong, Milliam, 6, -, -, 4000, 200, 250

Walker, Ann, 40, 10, -, 8000, 100, 175

Dixon, Alexandria, 3, -, -, 5000, -, 320

Delaplane, James, 8, 4, -, 10000, -, 250

Lancaster, James, 33, 7, -, 15000, 150, 350

Bartran, Benjamin, 28, -, -, 12000, 300, 800

Kelly, John, 90, 22, -, 14500, 200, 800

Faivre, Zepharine C., 73, 7, -, 20000, 300, 500

Carpenter, Alfred, 14, 2, -, 4000, 100, 300

Garrett, David, 73, 6, -, 70000, 300, 600

Carpenter, James, 125, 25, -, 20000, 700, 1500

Dilworth, Levi, 135, 14, 17, 18000, 800, 2500

Clemet, Canby, 110, 25, -, 9000, 300, 1200

Hendrickson, William H. 85, 15, -, 12000, 300, 1400

Kane, Michael, 14, -, -, 3000, 50, 250

Bugless, Jethro, 25, -, 2, 3000, 150, 225

Pasmore, William, 133, 40, -, 25000, 300, 2000

Fred, Elizabeth J., 35, -, -, 6000, 500, 500

Cloud, Harlin, 130, 35, -, 20000, 300, 2500

Baldwin, George, 120, 24, -, 6500, 400, 1800

Klair, Frederick, 105, 7, -, 15500, 500, 1500

Nickelis, Ellis, 60, 8, -, 8000, 200, 500

Toben, Matthias, 7, -, 2, 2500, -, 100

Dilworth, James, 155, 40, -, 20000, 1000, 2500

Dilworth, French, 50, 20, -, 8000, 200, 700

Gambol, Samuel, 54, 6, -, 10000, 200, 900

Springer, Jeremiah, 60, 12, 15, 8000, 100, 280

Sharpless, Benjamin, 55, 5, -, 6000, 600, 1300

Sharpless, John W., 5, -, -, 1000, -, 150

Kane, Harry, 28, 7, 15, 5000, 150, 400

Davis, Harman, 35, 10, -, 6000, 200, 700

Walker, Abiha, 30, 6, -, 4000, 100, 400

Jacobs, Samuel, 107, 9, -, 14000, 600, 1500

Graves, Hays, 12, -, -, 5000, 100, 300

Armstrong, Samuel, 20, 20, -, 4000, 200, 225

Way, Jacob, 60, -, -, 7000, 200, 650

Snodgrass, David, 65, 10, 15, 9000, 200, 1000

Way, Edward, 85, 10, -, 8000, 200, 450

Murry, Sylvestor, 20, -, -, 3000, 100, 150

Holmes, Jackson, 64, 7, -, 9000, 600, 900

Pyle, John, 160, 30, -, 30000, 400, 2500

Durham, Iseral, 12, -, -, 3000, -, 300

Green, Elwood, 60, 24, -, 10500, 375, 1700

Press, William, 8, -, -, 1000, -, 300

Caleher, Denis, 6, -, -, 1500, 40, 200

Lowther, David & Thomas, 88, 12, -, 10000, 500, 1400

Thatcher, Edward, 100, 37, -, 17000, 500, 1350

Pyle, Garrett, 50, 6, 3, 6000, 200, 300

Hanegan, James, 25, 10, 5, 4000, 50, 400

Green, Parker, 85, 20, -, 10500, 450, 1100

Morrow, Joseph, 95, 10, 5, 11000, 400, 800

Mackey, Daniel, 40, 11, -, 4000, 150, 500

Jackson, Marshell, 85, 10, -, 8000, -, 700

Brannan, John, 60, 8, -, 5500, 100, 750

Dougherty, Patrick 40, 10, -, 5000, 150, 500

Brown, James, 15, -, -, 2000, 200, 600

Vernon, Ottey (Otley), 100, 20, -, 16000, 1200, 1750

Grady, Michael, 11, -, -, 1500, -, 150

Sharpless, Anna & Jehu, 165, 15, 5, 21000, 1000, 3000

Speakman, John, 20, 10, 10, 4000, 175, 400

Leach, Jones, 80, 15, -, 9000, 1000, 1500

Kilcanon, Martin, 55, 6, 4, 5000, 75, 500

Leeds, Joseph, 8, 5, -, 3000, 20, 200

Pyle, Reese, 12, -, -, 2000, -, -

Patten, Hugh, 4, -, -, 1500, -, 150

Alcorn, James, 15, 4, -, 1500, 25, 200

Murphy, Patrick, 27, 7, -, 3000, 50, 1000

Curry, Thomas, 65, 15, -, 8000, 200, 600

Cayton, John, 6, -, -, 1600, -, 150

Graves, Lewis, 110, 30, -, 10000, 400, 800

Lowther, William, 57, 13, 15, 9000, 250, 2100

Lowther, John, 11, -, -, 3000, 100, 800

Frame, Joseph, 90, 16, -, 5000, 150, 700

Woodward, Abner, 75, 18, -, 10000, 500, 1100

Rolster, Mathew, 23, 2, -, 3000, 25, 200

Springer, Levi_, 40, 40, -, 8000, -, 700

Biderman, Harman, 6, -, -, 1500, 100, 200

Trainer, Patrick, 5, 7, -, 1200, 50, 200

Starit, Jones, 50, 23, -, 5000, 50, 200

Hand, James A., 30, 10, -, 2600, -, 300

Magrillis, James, 70, 10, -, 8000, 250, 500

Dougherty, Joseph, 35, 21, -, 6000, 50, 500

Magrillis, James, 10, -, -, 2600, 50, 120

Dilworth, Thomas, 80,-, -, 12000, 200, 1000

Green, Charles, 57, -, -, 1500, 400, 800

Bartram, Elwood, 90, 10, -, 20000, 800, 1450

Welsh, Michael, 40, 5, -, 8000, 175, 500

Corard (Conard), Anthony, 20, 5, 15, 8000, 250, 700

Beaty, Benjamin, 50, 44, -, 10000, 50, 480

Ball (Bull), John, 75, 5, 10, 13500, 500, 800

Taylor, Caleph, 10, -, -, 10000, 400, 475

Banning, John, 110, 10, -, 18000, 400, 1000

Collins, Petre, 5, -, -, 5000, -, 100

Lynam, George, 88, 15, 10, 14000, 400, 900

Robinson, William P., 100, 40, -, 18000, 175, 900

Hendrickson, Charlott, 80, 30, 10, 20000, 400, 1500

Ford, Abraham, 130, 30, 5, 20000, 600, 1300

Haley, Benjamin, 200, 11, -, 30000, 500, 3200

Travis, Daniel, 20, 10, -, 3000, 10, 100

Brinton, George W., 22, -, -, 4000, 100, 350

Ely, Oliver, 29, 1, -, 15000, 300, 450

Wells, George W., 100, 12, -, 3000, 800, 1000

West, Christopher, 12, 5, 23, 5000, 100, 800

Chandler, Thomas, 90, 19, -, 11000, 400, 1000

Jackson, Charles, 40, 10, 14, 7000, 400, 700

Snyder, Henry, 95, 10, 4, 20000, 250, 1200

Hofficker, James H., 94, 5, -, 2400, 300, 1350

White, Henry, 103, -, 6, 15000, 500, 1500

Stidham, Gilpen P., 14, -, -, 6000, 150, 150

Jordon, John, 95, 15, -, 1000, 400, 600

Lynum, Milliam, 100, 13, -, 12000, 500, 800

Armstrong, James P., 90, 20, 10, 15000, 1000, 1600

Armor, James, 62, -, 4, 7000, 500, 600

Conley, James, 76, 10, 48, 14000, 150, 500

Richardson, John, 55, 35, -, 20000, 500, 1200

Richardson, Joseph P., 75, 37, -, 25000, 500, 1500

Lynum, James H., 150, 10, -, 17000, 100, 200

Lynum, Abert A., 95, 5,-, 10000, 200, 975

Cranston, Samuel, 160, 3, 37, 20000, 500, 2000

Flinn, William, 95, 5, -, 15000, 500, 2500

Lynum, Joseph S., 95, -, 5, 10000, 400, 800

Flinn, Franklin, 130, 20, -, 20000, 500, 6000

Rothwell, Benjamin, 36, -, 15, 5000, 150, 600

Irons, Alexandria, 17, -, -, 3500, -, 250

Kilgore, Joseph, 6, -, -, 15000, 150, 100

Lynum, Robert F., 40, -, -, 8000, 250, 400

Cranston, Francis W., 75, -, 5, 15000, 800, 1000

Everson, Alex. W., 147, -, -, 25000, 800, 1500

Flinn, Howard E., 20, -, -, 3000, 200, 300

Flinn, Robert B., 130, 10, 20, 20000, 500, 2500

Flinn, Isaac N., 40, -, -, 8000, 600, 600

Lynum, John R., 100, -, 30, 10000, 500, 600

Lynum, Thomas P., 100, -, -, 10000, 500, 600

Lakens, Isaiah, 20, -, -, 10000, 100, 400

Updite, Isaac, 46, -, 4, 100, -, 100

Grubb, George W., 106, -, -, 15000, 1500, 1600

Grigg, Rufus, 70, 40, -, 15000, 300, 1200

Philips, Bernard, 4, -, -, 800, 50, 100

Petetdemange, Joseph, 100, 25, 32, 20000, 1000, 2000

Stidham, James, 80, -, -, 12000, 300, 1300

Calahan, Joseph, 70, 25, -, 15000, 300, 1000

Lynum, David, 100, -, -, 20000, 500, 1000

Richardson, Ashton, 58, 2, -, 15000, 500, 800

Robinson, Robert L., 60, 20, -, 16000, 1000, 1600

Tucker, Patrick, 100, -, 30, 250000, 800, 1600

Brown, John A., 50, -, -, 20000, 150, 700

Langley, Joel, 20, -, -, 6000, 100, 250

Limon, George, 16, -, -, 11500, -, 250

Scot, William, 12, -, -, 6000, -, 125

Tatum, John R., 100, 30, -, 31000, 300, 1285

Ware, James R., 90, 5, 35, 20000, 200, 500

Kellam, Thomas, 30, 10, 3, 10000, -, 200

Harris, Thomas C., 5, -, -, 2500, -, 100

Mahoney, Denis, 70, 15, -, 16000, 200, 900

Pyle, Joseph, 12, -, -, 20000, 200, 250

Pugh, Lewis H., 9, -, -, 4000, 125, 200

Ferris, William, 33, -, -, 30000, 300, 450

Wood, John, 11, -, -, 5000, -, 125

More, Eliza M., -, 105, -, 4000, -, -

Martin, Emily O., -, 150, 61, 4000, -, -

Deupont, Margaret E., 40, -, -, 20000, 1500, 1000

Mchugh, Bridgett, 15, -, -, 25000, -, 200

Hicks, John G., 20, -, -, 8000, 100, 300

Bradford, Milliam, 6, -, -, 8000, 100, 200

Weldon, George, 17, -, -, 7000, 100, 250

Banning, Henry G., 30, -, -, 40000, 500, 100

Lynum, Thomas, 75, 25, -, 20000, 500, 1000

McDaniel, Delaplain, 8, -, -, 20000, 500, 1000

Foreman, Isaac, 75, 15, -, 15000, 250, 300

Richardson, Samuel, 100, 20, 12, 15000, 200, 500

Richardson, William P., 300, 47, -, 31000, 1000, 3000

Crabb, John, 60, 25, 5, 8500, 200,1000

Walter, Enos, 116, 10, 9, 25000, 600, 1500

Brown, James, 70, -, -, 14000, 1000, 1800

Derickson, Milliam Z., 60, -, 15, 8000, 400, 900

Foreman, Joseph S., 100, 20, -, 12000, 500, 1000

Woodward, Aaron, 110, 4, -, 16000, 400, 1500

Hollingsworth, Abner, 138, 8, 4, 25000, 500, 1500

Coleman, William, 153, 32, 8, 40000, 500, 1400

Armstrong, R. Lewis, 101, 24, -, 25000, 650, 2500

Cranson, James, 110, 15, -, 20000, 475, 1000

Flinn, John J., 195, -, 8, 20000, 500, 1500

Price, David M., 8, -, -, 2000, 100, 150

Armstrong, Milliam, 85, 6, 14, 8000, 150, 1100

Woodward, Joseph, 130, 10, -, 20000, 300, 900

Brackin, William H., 100, 10, -, 16000, 300, 1000

Armstrong James, 100, 30, -, 18000, 400, 700

Bracken, Thomas, 90, 26, -, 12000, 300, 800

Stewart, John, 50, 6, -, 8000, 500, 600

Riddle, James, 60, 30, 40, 25000, 350, 240

Bancroft, Joseph 25, 16, 17, 10000, 400, 700

Mercer, William _., 4,-, -, 4500, -, 150

Miller, James, 4, -, -, 5000, 50, 200

Brinkley, Julian, 35, -, 15, 40000, 200, 800

Thompson, William 40, 20, -, 10000, 200, 400

Law, William 20, -, 8, 3000, 100, 300

Toy, James, 16, -, -, 10000, 100, 550

Dougherty, Charles, 10, -, -, 5000, -, 250

Fleming, William P., 65, -, -, 32000, 175, 1000

Gregg, Samuel, 75, 25, 15, 50000, 400, 1200

Cornog, Augustus, 60, 10, -, 5000, 200, 700

McKee, John, 100, 10, -, 13000, 500, 1000

Woodward, George K., 100, 18, -, 20000, 400, 1200

Deupont, Charles _., 58, 7, -, 18000, 600, 800

Palmer, Abraham, 87, 7, 2, 10500, 700, 4000

E. I. Deupont-deNemours, 600, 400, -, 100000, 2000, 17550

Stinson, James, 22, 1, -, 300, 150, 385

James, Jesse, 38, 2, -, 4000, 200, 530

Eastburn, Oliver, 90, 17, -, 10700, 350, 13500

Thompson, Lewis, 95, 12, -, 13750, 1000, 1200

Walker, William H., 80, 20, -, 10000, 600, 1320

Walker, Thomas M., 75, 25, -, 8000, 200, 750

Ocheltree, John, 54, 7, -, 6100, 400, 718

Eastburn, David, 85, 15, -, 9000, 400, 965

Buckingham, Richard, 75, 5, -, 8000, 400, 1000

Bideman, John, 87, 10, -, 9500, 500, 1000

Fell, Jonathan, 88, 10, -, 8000, 1000, 1500

Fell, Ezra, 60, 23, -, 8000, 500, 1000

Walker, Rebecca, 90, 20, -, 11000, 250, 785

Chambers, John _., 97, 3, -, 8000, 400, 787

McDowell, James, 53, 7, -, 3000, 300, 350

Howet, Jackson, 150, 50, -, 10000, 50, 400

McCormick, Nathan, 56, 10, -, 4000, 50, 450

Sentman, Joseph, 11, 1, -, 1500, -, 100

Seal, Thomas, 85, 15, -, 11000, 1000, 1000

Moore, Isaac, 40, -, -, 4000, 10, 350

Thompson, George, 105, 19, -, 12040, 400, 1050

Walker, Robert, 70, 20, -, 8100, 250, 950

Bell, Elizabeth, 178, 2, -, 14560, 400, 1040

Aikin, James, 79, 4, -, 4150, 250, 410

Craig, Walker, 52, 3, -, 2750, 200, 430

Davis, Wm. & Eli, 100, 11, -, 8325, 600, 1200

Collins, Edward, 75, 5, -, 5000, 100, 715

Campbell, Ann, 8, -, 3, 690, -, 75

Moore, Levi A., 110, 10, -, 10000, 300, 1255

David, Robert J., 85, 15, 9000, 400, 1325

Hopkins, Abel J., 153, 10, -, 16030, 500, 2000

Harkness, William, 173, 30, -, 13000, 500, 1390

Hanna, Robert, 25, -, -, 1000, 40, 100

Dennison, James, 75, -, -, 6500, 300, 458

Hanna, Joseph L., 55, 7, -, 5000, 300, 529

Klair, Aaron, 118, 17, -, 13500, 700, 1110

Whiteman, Israel, 118, 12, -, 10000, 600, 1075

Whiteman, Henry, 95, 6, -, 8000, 500, 950

Whiteman, Andrew J., 21, -, -, 3000, 200, 275

Bailey, John, 19, 29, 90, 6500, 15, 70

Nivan, David, B., 125, 25, -, 10000, 200, 820

Russel, Samuel, 35, 5,-, 4000, 100, 345

Rankin, Sarah, 100, 48, -, 10000, 500, 1310

Rankin, Joseph, 110, 40, -, 10000, 500, 1096

Jacobs, Sarah, 75, 20, -, 4500, 25, 440

Barlow, John, 21, -, -, 2500, 10, 310

Russel, George, 4, -, -, 700, -, 70

Mote, Coats E., 56, 10, -, 7000, 200, 552

Higgins, Thomas, 114, 6, -, 8000, 500, 1024

McCormick, Levi, 52, -, -, 4500, 110, 320

Worrall, Samuel, 108, -, -, 9000, 200, 1040

Whiteman, Charles, 95, 19, -, 6500, 400, 880

Little, James, 105, 12, -, 8000, 400, 875

Mitchel, James, 62, 6, -, 4300, 200, 480

Mote, Eli, 114, 12, -, 10000, 150, 890

Harkness, Samuel, 45, 4, -, 4000, 300, 410

Guthrie, John, 10, -, -, 1000, -, 42

Chambers, Sarah, 10, -, -, 1000, 50, 100

Macklem, Andrew, 10, -, -, 14000, 75, 250

Jazuett, James, 16, -, -, 2000, -, 200

Guthrie, Joseph, 35, 9, -, 3000, 50, 266

Taylor, Robert, 107, 55, -, 9000, 300, 760

Barris, Elizaabeth, 5, -, -, 1000, 75, 145

Brooks, Thomas, 171, 30, -, 10000, 400, 1007

Eastburn, Sarah B., 3, -, -, 1000, 50, 175

Woodward, John H., 85, 15, -, 8000, 400, 1250

Eastburn, Samuel, 74, 27, -, 12000, 200, 1047

Eastburn, Joseph Jr., 50, -, -, 5000, 150, 700

Buckingham, Alan, 68, 7, -, 7500, 400, 810

Morrison, Benjamin, 9, -, -, 1500, 10, 100

Rankins, Robert, 155, 25, -, 9000, 200, 930

Fell, Samuel, 5, -, -, 1000, 10, 145

Eastburn, William 180, 20, -, 16000, 300, 2575

Oliver, Joseph G., 17, -, -, 1000, -, 100

Robinson, John, 12, -, -, 1500, -, 150

Eastburn, Joseph, 150, -, -, 18750, 805, 2618

Appleby, Richard, 241, 355, -, 22040, 1000, 1890

Drake, Uriah, 100, 13, -, 8000, 300, 1000

Drake, Uriah Jr., 100, 12, -, 7000, 200, 1100

Maree, William, 93, 7, -, 7000, 400, 500

Bonsall, Levi, 29, 2, -, 2500, 150, 290

Yarnall, James, 8, 2, -, 1500, 6, -

Brown, Ann, 8, 1, -, 100, -, 60

Macklem, Eliza, 11, -, -, 1200, 20, 10

Gary, Robert, 39, 6, -, 3000, 75, 385

Fitzsimins, Simmons, 40, 39, -, 5000, 100, 390

Agram, Joseph, 7, -, -, 1500, 50, 165

Whitman, Henry M., 66, 4, -, 3500, 100, 595

Johnson, Hiram J., 105, 20, -, 8000, 100, 1050

Chrillas, David, 150, 57, -, 25000, 150, 1390

Curtis, S. M., 22, -, -, 2500, 100, 610

Reynolds, John A., 95, 5, -, 10000, 1000, 1400

Crossan, James, 72, -, -, 4000, -, 160

Morrison, William, 132, 75, -, 17000, 500, 1290

Jackson, Peter, 86, 20, -, 7000, 150, 1000

Mote, William H., 100, 25, -, 6000, 100, 1000

Lewis, Evan, 90, 9, -, 7000, 100, 600

McKewan, Robert, 120, 20, -, 8000, 100, 800

Russel, John, 66, 6, -, 4000, 10, 280

Currender, William, 91, 10, -, 9000, 100, 875

Greenwalt, John L., 47, -, -, 6000, 200, 614

Medill, George, 88, 3, -, 9000, 300, 790

Greenwalt, Joseph, 49, -, -, 4000, 100, 160

Ridgway, John, 136, -, -, 800, 100, 1600

Little, William F., 28, -, -, 1800, 200, 295

Richards, Nathaniel, 53, 10, -, 2500, 150, 470

Ball, Joseph, 30, -, -, 1500, 15, 100

Kelly, Jefferson, 48, 12, -, 2500, 100, 276

Taylor, Thomas, 10, -, -, 3000, -, 100

Foot, Benjamin, 17, -, -, 2000, 100, 2110

Garret, George A., 4, -, -, 3000, -, 140

Harman, Edmond, 80, 5, -, 6000, 500, 810

Anthony, William, 4, -, -, 2500, -, 160

Woods, James, 23, -, -, 1500, 15, 120

Taylor, Samuel, 68, -, -, 5000, 400, 1035

Klair, Egbert, 115, 10, -, 8750, 300, 1185

Whiteman, King J., 125, 25, -, 8750, 300, 1357

Scott, Owen, 240, 35, -, 20000, 1000, 2220

Woolaston, Albert, 93, 12, -, 7600, 300, 1169

Satterthwait, Reuben, 108, -, -, 11000, 500, 2019

Bratlett, Henry, 56, -, -, 15000, 75, 435

Banks, Jabez, 287, -, -, 20000, 600, 2300

Holland, John, 101, -, -, 10000, 600, 820

Brown, James Jr., 175, -, -, 13000, 450, 288

McCoy, Robert, 132, 15, -, 8000, 100, 1070

McCalister, David, 100, 10, -, 7000, 100, 1000

Taylor, Robert, 25, 24, -, 2500, 200, 630

Welsh, Wilson, 62, -, -, 8000, 100, 760

Trenden, Joseph, 48, 10, -, 4500, 100, 775

Spencer, Nathaniel, 110, 20, -, 8000, 100, 900

Paxton, Alfred, 57, -, -, 7000, 500, 526

Hammell, Joseph, 98, 3, -, 8000, 100, 730

Permock, Lewis, 95, 20, -, 7000, 1000, 1190

McFarlin, Robert, 50, 22, -, 6000, 200, 500

McKee, Andrew, 85, 3, -, 6000, 200, 1070

Cannon, Charles, 50, -, -, 6000, 150, 485

Walker, Quinby, 58, -, -, 5000, 200, 590

Foot, James, 66, 5, -, 8000, 1000, 1010

Reubencame, Jacob, 110, 10, -, 12000, 500, 1200

Nurlin, Samuel, 119, 20, -, 12000, 900, 1068

Hazlett, William, 46, -, -, 8520, 400, 460

Cranston, Edward, 52, -, -, 4000, 200, 690

Permock, William H., 12, -, -, 3000, 50, 480

Derrickson, Joseph, 84, 20, 5, 8000, 200, 850

Ball, Reuben, 130, 30, 40, 10000, 500, 1735

Jones, Lemuel, 14, -, -, 2700, 100, 295

McClear, George P., 3 ½, -, -, 1500, 50, 255

Allcorn, James, 12, -, -, 3000, 25, 790

Chandler, Abraham, 9, -, -, 3000, -, 278

Ball, John, 58, -, -, 5800, 600, 930

Brackin, Jane, 4, -, -, 1000, 40, 140

Lynam, Robert T., 94, 2, -, 10000, 300, 1540

Cranston, Joseph, 132, 18, -, 9400, -, 1100

Smith, Pasly P., 30, 1, -, 3000, 30, 310

Springer, Allen, 140, 10, -, 10000, 200, 1307

Naudain, Arnold, 140, 10, -, 12000, 500, 2010

Jester, Janus N., 145, 25, 20, 17000, 200, 890

Eastburn, May J., 126, 20, -, 8500, 300, 1030

Parker, John, 24, -, -, 2400, 259, 330

Miney, Patrick M., 15, -, -, 1500, 20, 210

Thompson, John, 144, 40, -, 14720, 1000, 915

Mitchell, Joseph, 175, 25, -, 20000, 400, 1438

Mitchell, Abner, 83, 12, -, 9000, 350, 1000

Derrickson, Aquilla, 916, 14, -, 11000, 500, 1405

Derrickson, Zachariah, 100, 30, -, 10400, 500, 920

Foot, James, 50, 12, -, 5500, 300, 720

McElwee, Lewis, 72, 10, -, 6000, 200, 1030

Yearsley, Samuel, 20, -, -, 3000, 100, 313

Yearsley, McCoy, 47, 8, -, 5500, 200, 740

Pierce, Henry, 90, 7, -, 12000, 500, 1410

Claranon, James, 55, 15, -, 5000, 200, 560

Gregg, Benjamin, 130, 20, -, 12000, 500, 620

Gregg, Beeson, 29, 7, -, 3300, 200, 315

Crosson, Hannah, 21, 7, -, 3300, 200, 315

Crossen, John R., 58, 9, -, 4500, 200, 335

McElwee, Jane, 48, 18, -, 3300, 150, 400

Steward, Hugh, 24, -, -, 3000, 100, 320

Woodward, Abner, 106, 20, -, 13000, 500, 1200

Mitchel, Stephen, 90, 12, -, 10200, 200, 870

Woodward, Frederick, 45, 14, -, 8000, 300, 750

Baldwin, Thomas L. J., 63, 4, -, 4000, 100, 470

Davis, George, 95, 75, -, 11000, 300, 462

Springer, John, 78, 8, -, 6000, 300, 775

Davis, Alran, 27, 4, -, 3100, 25, 250

Taylor, John, 40, 6, -, 3000, 200, 420

Rearden, Patrick, 32, -, -, 3000, 150, 470

Phillip, John, 88, 10, -, 6500, 20, 1290

Ahern, David, 7, -, -, 700, 10, 135

Derrickson, Lewis, 104, 12, -, 9000, 400, 1480

Chambers, Elizabeth, 26, -, -, 2500, 50, 340

Dennison, Samuel, 80, 16, -, 7680, 300, 830

Dennison, John, 118, 20, -, 200, -, 900

Fisher, John, 118, 20, -, 200, -, 900

Klair, George, 108, 12, -, 12000, 100, 200

Durnall, Harvey, 52, 20, -, 5000, 200, 230

Peirson, Hiram, 100, 10, -, 11000, 500, 1200

Dewey, Christian, 3, -, -, 500, 10, 75

Hanna, John, 90, 6, -, 9800, 400, 1470

Gebhart, Benjamin F., 57, 30, -, 8700, 500, 1395

Jackson. James, 78, -, -, 15000, 500, 930

Brackin, Bertha, 44, 3, -, 5000, 250, 660

Corwell (Connell), Peter, 45, 6, -, 5000, 400, 670

Chandler, Thomas J., 30, 3, -, 5000, 200, 475

Laferty, John, 12, -, -, 2500, -, 60

Rumford, Francis, 63, -, -, 63000, 250, 565

Foot, George W., 65, 10, -, 6500, 200, 540

Currender, Washington, 100, 6, -, 10000, 500, 1315

Brackin, William, 36, -, -, 6000, 150, 360

Lindsey, Joseph, 140, 60, -, 25000, -, -, -

Wirt, Charles, 75, 10, 19, 6000, 250, 530

Montgomery, William A., 9, -, -, 3500, 40, 50

Lynam, Evan, 96, 6, -, 7500, 500, 1210

Cheltin, John, 63, -, -, 6300, 300, 753

Chandler, Souther M., 4, -, -, 3500, -, 350

McCrossen, Denis, 4 ¾, -, -, 1000, 50, 175

Ferguson, Margaret, 10, -, -, 2400, 75, 160

Fell, Franklin, 110, 40, -, 40000, 400, 2150

Graves, David, 73, -, -, 7300, 400, 745

Clark, Henry, 50, -, -, 5000, 200,760

Bailey, Samuel A., 66, 4, -, 7100, 200, 980

Bailey, John, 57, 3, -, 5500, 150, 800

Bramble, George, 6, -, -, 1850, 250, 270

Robinson, John B., 75, 5, -, 15000, 100, 975

Philip, Isaac D., 25, -, -, 3900, 75, 325

Philip, William G., 10, -, -, 6000, -, 350

Highfield, John J., 25, -, -, 2500, -, 100

Marshall, Calvin P., 52, -, -, 7800, 313, 1000

Duross, Bernard, 40, 10, -, 3000, 100, 670

Duncan, Benjamin T., 136, 35, -, 12000, 500, 1200

Bower, Stacy, 6, -, -, 2000, 40, 300

Cranston, James, 140, -, -, 16800, 500, 1893

Derrickson, Cornelius, 87, 10, -, 9700, 200, 550

Fisher, Richard, G., 84, 84, 8, -, 9000, 500, 1500

McElwee, John E., 120, 12, -, 10560, 400, 1415

Pierson, George, 98, 15, -, 11300, 500, 1560

Brown, Joseph, 38, 4, -, 4200, 200, 300

Shakespeare, James, 100, 60, -, 12500, 400, 960

Moore, William, 95, -, 5, -, 6000, 300, 1140

Adker, Frank D, 128, 20, -, 12800, 500, 1780

Porter, James F., 27, -, -, 4000, 200, 225

Lynam, David, R., 295, -, 5, 30000, 1000, 2975

Cranston, Benjamin, 97, -, -, 150000, 300, 1295

Shakespeare, Benjamin, 114, 12, -, 15600, 400, 2130

Hayes, Joseph, 6, -, -, 1000, -, 175

Kerns, Edward, 25, -, -, 4000, 100, 390

Lobb, George, 175, 15, -, 15000, 200, 1642

Swinderman, Lawrence, 7, -, -, 1500, 50, 200

Romeled, John, 40, 7, -, 5000, 500, 910

Gregg, Edwin, 70, 8, -, 8000, 400, 1300

McElwee, Barton, 85, 17, -, 7500, 500, 795

Crossen, Calvin, 93, 7, -, 1000, 150, 950

Gibson, Robert, 4, -, -, 1000, -, 100

Jackson, Pusey, 12, -, -, 5000, -, 660

Peoples, William, 65, -, -, 8000, 400, 1155

Graves, Lemuel, 45, 5, -, 6500, 200, 500

Crossen, Isaac, 89, 12, -, 9000, 200, 670

Dewe, Henry, 66, 6, -, 6000, 150, 600

Nickendank, William 42, -, -, 8000, 300, 700

Righfield, Calvin, 9, -, -, 2500, 25, 160

Hollingsworth, John, 107, 9, -, 1600, 400, 1000

Whiteman, Frank, 45, 3, -, 5000, 150, 850

Foust, John, 17, 1, -, 4000, 100, 370

Margargal, Ephraim, 54, 6, -, 7500, 400, 680

Souden, Edward, 70, 36, 20, 9440, 200, 850

Bowan, Joseph, 50, 26, -, 7600, 300, 2120

Dixon, Samuel, 100, 44, -, 14440, 250, 2580

Veal, Phebe, 3, -, -, 1200, -, -

Sharpless, John & Amos, 115, 15, -, 13000, 1000, 1760

Mason, Jonathan, 95, 25, -, 8400, 200, 1200

Hanna, Lewis, 80, 20, -, 8000, 200, 750

Armstrong, John, 75, 23, -, 7840, 500, 954

Marshall, Hannah, 18, 2, -, 1500, 100, 250

Baker, Joshua B., 55, 10, -, 6500, 300, 820

Bowman, Jacob G., 1, -, 14, 700, -, 25

Chandler, Jacob, 70, 30, 10, 6500, 300, 930

Vandever, Thomas, 75, 25, 7620, 500, 900

Graves, Robert, 26, -, -, 4000, 100, 500

Leach, Joseph, 103, 25, -, 10000, 350, 1000

Hulett, John H., 68, 6, -, 7400,1000, 140

Green, Violetta, 15, -, -, 1400, -, -

Reece, William 34, 1, -, 5400, 250, 400

Wilson, Stephen, 140, 40, -, 1800, 500, 1410

Hoops, Thomas, 52, 5, -, 7000, 250, 450

Springer, Stephen, 42, 4, -, 5000, -, 660

Graves, Samuel, 98, 10, -, 13400, 500, 1215

Miller, Lewis, 92, 10, -, 9000, 250, 1000

Ralston, William, 25, 2, -, 3000, 100, 275

Wilson, James, 60, 15, -, 9000, 500, 500

Hobson, William, 90, 7, -, 6000, 300, 1000

Grace, Walker, 38, -, -, 4000, 100, 450

Lamborn, Chandler, 120, 30, -, 13500, 400, 1225

Pool, Thomas H., 81, 8, -, 8101, 400, 765

Wilson, Stephen, 100, 12, -, 11120, 300, 875

Pool, John, 130, 20, -, 15000, 300, 1600

Muldoon, Patrick, 4, -, -, 1700, 10, 120

Sharpless, Samuel, 140, 10, -, 13500, 500, 2660

Hanna, Jacob, 102, 12, -, 15000, 800, 2450

Bartholomew, Edward, 85, 12, -, 16400, 300, 1635

Dillworth, John D., 103, 17, -, 12000, 400, 1500

Dixon, Isaac, 135, 15, -, 15000, 350, 1595

Chandler, Spencer, 125, 20, -, 17800, 600, 1880

Collins, George, 80, -, -, 8000, 150, 770

Gess, Palmer, 38, -, -, 5760, 50, 605

Cleadin, Daniel, 29, -, -, 4000, 130, 555

Mitchell, John, 444, 12, -, 16500, 400, 1690

McVaughn, Frank, 29, -, -, 4250, 300, 450

Dewees, Chandler, 3, -, -, 2000, -, 175

Highfield, Eli, 75, -, -, 8000, 200,800

Flinn, Westley J., 106, 10, -, 14000, 300, 660

Heald, Caleb, 10, -, -, 6000, 200, 300

Porter, Solomon, 21, 4, 2500, 25, 200

McCarty, Cornelius, 35, 5, -, 3000, 100, 700

Yetter, John H., 106, 10, -, 8500, 300, 870

Calvin, Hall J., 100, 16, -, 14500, 1000, 1200

Lamborn, Lewis, 20, 4, -, 4000, 100, 315

Harrington, Jeremiah, 10, -, 830, 3000, 55, 200

Loudon (London), James, 100, -, -, 1500, 20, 80

Pierson, William, 30, -, -, 4500, 100, 150

Hoops, Enoc C., 86, 6, -, 8000, 200, 830

McGorem, Edward, 25, 1, -, 2800, 50, 1160

McGorem, John, 3, -, 6, 800, -, 95

Hyde, David, 2, -, 6, 700, -, 75

Gay, Howard L., 47, 6, -, 7500, 300, 425

McVaughn, Lakins, 72, -, 5, 7500, 250, 910

Hulett, Josiah G., 55, 10, -, 5250, 413, 593

Springer, George, 65, 8, -, 7300, 500, 1570

Thompson, Cyrus, 123, 12, -, 13500, 600, 860

Wilson, Ephraim, 120, 8, -, 12800, 300, 1560

Dixon, Joshua H., 76, 20, -, 9000, 200, 725

Little, William, 95, 8, -, 15000, 300, 900

Lacy, George, 29, -, -, 3600, -, 160

Jackson, John G., 95, -, -, 28600, 200, 1200

Robinson, William, 16, 2, -, 1800, -, 390

Mendenhall, J. N., 70, 20, -, 6000, 200, 480

McCormick, John, 90, 10, -, 10000, 300, 760

London, Jarvis Jr., 6, -, -, 800, -, 175

Moore, Thomas J., 70, 6, -, 6740, 200, 830

Appleby, Oliver H., 236, 1, -, 28000, 1200, 3500

Bolton, James L., 90, -, -, 8000, 300, 1010

Grimes, Thomas, 300, -, -, 25000, 500, 1800

Warren, David, 270, -, 40, 32000, 700, 2800

Lodge, Eurastus, 250, -, 250, 30000, 700, 200

Barnaby, Joseph, 110, -, -, 10000, 100, 700

Tybout, George Z., 360, -, -, 43000, 1500, 7640

Tybout, George M., 200, -, -, 30000, 800, 1720

Armstrong, Spencer, 100, -, -, 1200, 500, 1600

Hawthorn, George, 164, 4, 30, 41000, 300, 1390

Lank, Milliam J., 130, -, 20, 13000, 150, 700

Sterling, Ephraim, 65, -, -, 5000, 30, 600

Sutherland, Roderick, 25, -, -, 3000, 400, 1100

Massey, Alfred, 220, -, 30, 12000, 1000, 1200

Rickards, John, 60, 40, -, 8000, 200, 550

Smith, John, 3, -, -, 400, 200, 30

Gray, Henry, 3, -, -, 300, 50, 50

Poole, Thomas, 20, 10, -, 2000, 20, 60

Keegan, James, 175, 25, -, 24000, 500, 2100

Jacobs, John, 187, -, -, 10000, 450, 850

Johns, John, 8, -, -, 8000, -, 650

Douglass, James C., 300, -, -, 45000, 500, 1465

King, Ethan B., 125, -, -, 16000, 300, 1100

McMullen, Henry, 100, -, -, 20000, 1200, 1600

McCrone, Rebecca, 6, -, -, 1000, 100, 200

Gray, Henry, 3, -, -, 400, 125, 120

Friel, John, 3, -, -, 400, 125, 200

McCoy, William B., 100, 30, -, 7000, 500, 1100

McMullen, John, 190, 10, -, 10000, 300, 800

Paynter, George, 155, -, -, 14000, 300, 4100

Wiswell, John, 140, -, -, 12000, 300, 700

Pennington, Ashbury, 70, -, -, 4000, -, 650

Thompson, William, 140, 10, -, 12000, 450, 1100

Watterhouse, Joseph, 212, -, -, 20000, 800, 2000

Appleby, Rachel A., 130, 50, -, 15000, 400, 2000

Crossin, Eli, 185, 16, -, 14000, 400, 500

Pennington, Francis, 120, 40, -, 14000, 300, 600

Downward, Thomas, 90, 22, -, 11000, 200, 1200

Pyle, Newton, 172, 8, -, 22000, 600, 2000

McKee, Elwood B., 150, 30, -, 20000, 600, 1500

Stroup, Eugene, 220, 46, -, 25000, 1000, 1800

Diehl, John, 254, -, 33000, 2000, 3500

Bowl, George, 100, 16, -, 11000, 400, 1800

Brown, Joseph T., 230, -, -, 28000, 500, 5500

Lofland, Reuben P., 170, 30, -, 18000, 400, 1100

Lee, George A., 225, -, -, 25000, 800, 2400

Biddle, Deborah, 400, -, -, 30000, 800, 4000

Hays, John, 100, -, -, 11000, 300, 700

Lewden, Josiah, 150, 100, -, 15000, 250, 1400

Allen, George, 50, -, 100, 4500, 50, 300

Groves, Andrew, 120, -, -, 11000, 200, 1000

Smith, John, 108, -, -, 12000, 400, 1200

Gemmill, David W., 220, 44, -, 2500, 200, 600

Morrison, John C., 126, -, -, 13000, 300, 800

Clock, Philip R., 285, -, 30000, 500, 2000

George, Jonathan, 165, -, -, 20000, 300, 1500

McCrone, George, 142, 20, -, 25000, 300, 9000

Booth, Benjamin, 290, 50, -, 42500, 800, 3500

Robinson, Westleigh, 130, 20, -, 14000, 300, 1300

Tush, Jacob, 70, 26, -, 8000, 70, 300

Ward, William, 150, 30, -, 18000, 200, 700

Vandergrift, George, 104, -, -, 1000, 100, 650

Vandergrift, George, 195, 15, -, 20000, 500, 1800

Pennington, Albert, 120, -, -, 12000, 200, 600

McFarland, Edward, 120, -, -, 12000, 600, 1400

Morrison, George W., 101, -, -, 10000, 1000, 1100

Welden, Lewis, 350, -, -, 45000, 800, 9000

Bradford, John, 90, -, -, 10000, 300, 600

Everson, John, 280, -, -, 25000, 300, 2000

Slock, Thomas, 190, 10, -, 20000, 200, 800

Marley, Benjamin, 125, -, -, 12000, 200, 800

Biggs, Alexander, 102, -, -, 12000, 100, 1200

Halcomb, Bankson, 170, -, -, 37000, 600, 2800

Halcomb, Bankson, 35, 35, -, 45000, 600, 2800

Jester, James, 85, 25, -, 11000, 300, 900

McFarland, John H., 141, 20, -, 20000, 300, 1700

Jackson, George, 200, -, -, 40000, 1000, 3710

Casey, Mary, 70, -, -, 8000, 150, 700

White, William, 136, -, -, 20000, 900, 1200

Peach, William, 80, 40, -, 11000, 200, 500

McCafferty, Edward, 125, -, -, 25000, 500, 1460

Welch, Thomas, 270, 45, -, 50000, 600, 3500

King, George 95, -, -, 22000, 200, 1000

Davis, Jason, 240, -, 40, 35000, 600, 3000

Stroup, William, 130, -, -, 20000, 500, 2000

McCoy, Douglas, 87, -, -, 10000, 300, 1100

Peter, Randolph, 200, -, -, 35000, 500, 1200

White, George, 300, 11, -, 40000, 600, 3000

Hirst, Samuel, 14, -, -, 2800, 20, 200

Pugh, John, 119, -, -, 10000, 200, 1300

Waters, James, 20, -, -, 5000, 200, 200

Flynn, James, 15, -, -, 4000, 150, 250

Pugh, James, 10, -, -, 3000, 150, 150

Fox, John, 17, -, -, 5000, 150, 300

Reynolds, Lewis, 82, 14, -, 12000, 200, 1000

Downey, Edward, 205, -, -, 30000, 500, 1500

Miller, Adam, 9, -, -, 1000, 50, 400

Jackson, Richard, 200, -, -, 35000, 500, 2200

Callahan, Mary, 140, -, -, 20000, 500, 1500

Goodley, William, 318, -, -, 40000, 1000, 1600

Boyle, John, 100, -, -, 10000, 250, 600

Townsend, Solomon, 215, -, -, 65000, 800, 300

Alrich, Lucas, 120, -, -, 18000, 500, 1500

Jackson, Samuel, 130, -, -, 20000, 500, 1500

Cann, Francis, 128, -, -, 20000, 500, 1600

Lefevre, John, 120, -, -, 15000, 300, 900

Getty, John, 50, -, -, 7500, 200, 300

Hunter, Joseph, 165, -, -, 20000, 300, 1100

Lambson, William H, 198, -, -, 30000, 500, 1800

Lefevre, Joseph, 53, -, -, 7000, 250, 400

Burr, Horace, 23, -, -, 10000, 200, 400

Hanson, George, 93, -, -, 14000, 500, 1160

Smith, William, 230, -, -, 50000, 1000, 2800

May, George S., 14, -, -, 13000, 1100, 250

Lofland, Elias, 210, -, -, 30000, 800, 1800

Russell, Sarah, 97, -, -, 1500, 600, 1100

Eckles, John S., 160, -, -, 20000, 600, 1200

Russell, James, 170, -, -, 30000, 700, 1600

Hirst, James, 80, -, -, 15000, 400, 1500

Morgan, David, 200, 50, -, 25000, 400, 1400

Edwards, George, 400, 50, -, 54000, 1000, 4650

Nivin, David, G., 150, 40, -, 20000, 500, 2000

Higgins, Joseph, 75, -, -, 10000, 250, 700

Janvier, Julian, 172, -, -, 34400, 1000, 3000

Lobdell, G. G., 200, -, -, 50000, 1000, 10000

Moore, Hannah, 170, -, -, 50000, 1000, 2000

Townsend, Sylvester, 300, -, 16, 40000, 1000, 2000

White, Henry, 200, -, 10, 20000, 500, 1400

Landers, John, 10, -, -, 4000, 300, 250

White, Thoams, 200, 8, -, 25000, 800, 1600

Fols, Ezekiel, 88, -, -, 10000, 400, 1200

Moore, Robert, 160, -, 40, 16000, 800, 1800

Banks, William, 120, -, 30, 15000, 800, 1700

McCoy, John, 175, -, -, 17000, 800, 1800

Smith, Azariah, 130, 40, -, 15000, 600, 1200

Bell, Benjamin, 28, -, -, 11000, 400, 500

Bolton, William, 140, -, -, 14000, 400, 900

Gooding, Jesse, 110, -, -, 10000, 300, 800

Clark, Elmer, 230, -, -, 35000, 600, 2000

Taggard, Joseph, 150, -, -, 20000, 500, 1500

Newlove, John, 162, -, -, 20000, 600, 1500

Cleaver, Mark M., 102, -, -, 10000, 500, 700

Marvin, Ebenezer, 30, -, -, 6000, 300, 300

Broad, Jacob S., 50, -, -, 12000, 600, 1100

Alexander, Jessie, 27, -, -, 6000, 400, 250

Miller, Peter, 14, -, -, 3000, 500, 200

Getts, Benjamin, 10, -, -, 2500, 200, 250

Goodman, Alexander, 150, -, -, 35000, 800, 1700

Lentz, Henry, 7, -, -, 1400, 150, 200

Luinn (Quinn), Jones, 11, -, -, 2000, 150, 100

Janvier, Ferdinand, 190, 20, 10, 15000, 500, 1835

Tetter, John D., 85, 2, 15, 7000, 400, 636

Vansant, William, 90, 30, -, 75000, 50, 486

George, James H., 144, -, 6, 8000, 100, 700

Stafford, John, 100, 30, 22, 65000, 200, 1066

Moss, Robert D., 18, -, - 1500, 50, 225

Whitakre, Conrad, 84, -, -, 25000, -, 105

Calhoun, Wm. W., 55, -, 45, 6000, 200, 395

Ellison, Thos. B., 95, 3, 5, 9000, -, 590

Racine, F. P., 119, 25, -, 1000, 100, 515

Johnston, John, 125, 160, 15, 13000, 150, 700

Portham, James, 173, 50, -, 20000, 900, 985

Ford, Wm. B., 10, -, -, 5000, 500, 470

Cann, James, 100, 15, 26, 6000, 50, 685

Mills, Joseph, 62, 5, -, 4000, 300, 355

Kettlewood, Mathew, 122, 8, 10, 9800, 200, 775

Porter, Thos. G., 60, 10, -, 5000, 200, 520

Moore, A. S., 80, 13, 7, 7000, 200, 580

Stradley, Wm. H., 175, 30, 25, 15000, 200, 1025

Porter, Samuel, 30, 6, 30, 1000, 100, 256

Racine, John G., 140, 5, 5, 6000, 200, 800

Tendel, D. M., 40, 2, 5, 3500, 1150, 310

Mitchel, Robert, 150, 10, 3, 16300, 400, 1040

Stewart, S. W., 120, 259, 12000, 300, 788

Walter, Thomas, 125, 25, 10, 10000, 75, 1340

Reed, B. A., 100, -, 40, 6000, 50,720

King, Robert, 10, -, 3, 1000, 50, 350

Stewart, James, 100, 95, 12, 16000, 275, 810

Cunningham, S., 100, -, 50, 4000, 345, 955

McMullen, James, 100, 7, 20, 5000, 40, 535

Bays(Hays), Henry, 130, -, 170, 10500, 300, 800

Cann, Alister, 175, -, 250, 12000, 350, 904

Adair, Wm. P., 35, 10,-, 2500, -, 190

Reed, Robert, 68, -, 30, 3000, 104, 530

Farris, D. Brainard, 80, 15, 5, 7000, 350, 1060

Frazer, John, 112, 11, -, 725, 100, 443, 310

Clark, Anna, 53, 400, 40, 80000, 400, 335

Barton, Miller, 30, 40, -, 14000, 150, 360

Cullen, D. H., 38, 5, 8, 3000, 200, 440

Bacon. Edward, 29, 1, -, 3000, 100, 355

Graves, George W., 60, -, 15, 6000, 200, 706

Davis, John T., 41, -, -, 4000, 25, 490

Harman, Andrew, 90, -, -, 7000, 100, 950

Mattie, Marcelius, 11, -, -, 1000, 100, 190

Davis, J. W., 86, 2, 3, 9000, 100, 200

Kylee, Fannie, 18, -, -, 2000, -, -

Scott, John, 16, -, -, 1750, 100, 230

Boulden, Jessee, 13, -, 2, 4000, 200, 110

Dean, Jacob, 60, 30, 10, 15000, 350, 1380

Nelson, James, 135, 5, -, 6000, 150, 800

Boulden, George, 250, 125, -, 22500, 500, 1985

Boulden, George Jr., 130, 95, -, 12395, 10, 953

Tindle, Thomas, 100, 50, -, 8000, 300, 648

Hog, John, 60, -, 20, 2500, 20, 200

McIntire, Samuel 100, 15, 55, 8500, 300, 850

McIntire, James, 80, 50,10, 10000, 500, 730

Raimond, James, 160, 20, -, 15000, 280, 1030

Sapp, Johnathan, 50, 15, 15, 3600, 100, 230

Ellison, Curtis B., 108, -, -, 16000, 450, 1130

Ellison, C. B., 92, 20, -, 40000, 825, 575

Vail, David C., 400, 100, -, 50000, 400, 214

Ingram, N. R., 143, -, -, 15000, 200, 600

Biggs, Wm. P., 192, 4, 16, 13200, 340, 964

Biggs, Wm. P., 114, 8, 30, 10000, -, -

Keely, Michael, 66, 33, 32, 7500, 300, 590

Cooch, J. Wilkins, 250, 100, 150, 35000, 2000, 2025

Cooch, J. Wilkins, 60, 20, -, 10000, 200, 380

Stanton, Isaac B., 70, -, 10, 6000, 200, 415

Woodward, Joel, 40, 45, 21, 6900, 300, 1240

Cavender, William, 250, 50, -, 18000, 400, 1710

Veasy, James T., 70, -, 6, 8000, -, 650

Veasy, James L., 20, 20, -, 1600, -, 75

Paxon, Owen, 150, 40, -, 15000, 100, 680

Hinson, James A., 85, 15, -, 5000, 150, 300

Henether, __, 18, 1, 67, 2000, 50, 330

Paxon, Morris H., 375, 25, -, 40000, 875, 1490

Ellison, Johnathan, 250, 8, 42, 30000, 65000, 1490

McMahan, Andrew, 90, 70, 20, 7000, 100, 1030

Wright, Stephens, 98, 7, -, 6300, -, 660

Beggs, Sewell R., 150, 26, 10, 17240, 600, 1210

Lum, Charles A., 222, 31, -, 22770, 340, 2890

Eliason, J. D., 378, -, 25, 40300, 100, 2790

Price, John T., 200, -, 3, 20000, 990, 2265

McIntire, Thos., 200, 20, 6, 22000, 1000, 1300

Harbert, Wm. K. H., 151, -, 20, 213750, 150, 1250

Harbert, Timothy, 151, -, 20, 213750, 150, 650

Reynolds, J. W., 184, -, -, 28400, 300, 1950

Cotts, Elvin, 241, -, -, 25000, 1000, 2400

Veasy, James L., 95, -, 5, 5000, 400, 665

Sutton, Samuel, 75, 15, -, 3500, 100, 275

Walker, Mary C., 32, -, -, 2000, 50, 330

McMahan, Thos., 100, 25, 15, 40600, 100, 365

Dickerson, Levi, 250, -, 50, 14000, 550, 1625

Holton, Spencer S., 130, 15, 12, 12500, 140, 1352

Pierson, Ross D., 125, 8, 23, 3000, 100, 735

Wiley, Stephen, 50, -, 1, 13000, 100, 400

Wright, Samuel, 100, 20, -, 14000, 500, 1275

Dayett, Wm. T., 50, 10, 9, 10000, 200, 810

Black, Robert M., 160, 40, 55, 17850, 176, 1060

Frazer, Samuel, 180, 15, 5, 1300, 200, 1038

Ward, John, 60, -, 15, 4500, 125, 345

Cavender, L. A., 147, 15, 15, 20000, 350, 1950

Boys, Jacob, 5 ½, -, -, 1800, 183, 560

Nicholson, James, 113, -, -, 11300, 650, 1553

Pleasanton, Edward, 175, -, 5, 14400, 300, 1175

Gray, Francis, 320, 50, -, 37000, 300, 1487

McCoy, Thomas, 200, -, 200, 18100, 250, 1263

Davidson, John, 200, 12, -, 20000, 3000, 2240

Lecompt, Charles, 910, -, -, 73250, 3500, 1500

Gibbs, Henry, 6, -, 6, 800, -, -

Doucan, George A., 30, -, 27, 30000, 100, 175

Porter, Henry, 8, -, 4, 900, 40, 145

Math__, Lewis, 10, -, -, 700, 55, 125

McCoy, James, 120, -, -, 7000, 100, 580

Stephenson, Daniel, 15, -, 2, 1000, 25, 182

Roy(Ray), Reuben, 28, -, 6, 2000, 145, 230

Boulden (Boulder), James, 70, 65, 30, 8000, 200, 610

Hance, Edward, 200, 41, 9, 2000, 900, 1840

Nickols, Isaac W., 30, -, 10, 1200, 60, 270

Nickols, Charles, 45, -, 15, 1800, 50, 1800

Est of Causdon, J., 18, -, -, 1800, -, -

Christian, Adam, 90, 16, -, 26, 6000, 150, 435

Fetter, Alfred B., 30, 33, -, 16330, 400, 2480

Murry, Samuel, 8, -, 9, 350, -, 90

Pierce, Edward T., 100, -, 24, 6000, 200, 820

Williams, Jas. H., 16, -, 4, 800, 100, 212

Williams, Jas. Jr., 25, -, 1, 1300, -, -

Greenage, William, 4, -, -, 300, 55, 400

Hockinstien, Goodfrey, 20, -, 25, 1100, 35, 225

Gallops, William, 125, -, 60, 8000, 74, 480

Murphy, Hugh, 24, -, 4, 950, 50, 230

Simpson, Alexander, 26, 4, 47, 3000, 75, 172

Dale, Samuel T., 65, 65, -, 10000, 200, 780

Walton, Samuel C., 76, -, 300, 10000, 200, 600

Stewart, Samuel, 48, -, 41, 2300, 110, 220

Green, Levi, 6, -, 19, 375, 150, 121

Lum, Elizabeth, 30, 10, -, 2000, 300, 450

Lewis, George, 20, -, -, 1000, -, -

Wilson, Washington, 30, 25, 5, 3600, 50, 475

Collins, William, 12, 18, -, 900, 40, 220

McDaniel, William, 108, -, 15, 1800, 100, 360

Pennington, Joseph, 9, -, -, 600, -, 110

Lum, Elias, 25, 10, 14, 2000, 40, 370

Armstrong, Robert, 15, -, -, 400, 60, 196

Dutton. George, 102, 30, 20, 7530, 200, 660

Champion, Benjamin, 14, -, -, 500, 50, 199

Miller, William, 72, 2, 2, 780, 25, 60

Howard, William, 5, -, -, 300, -, 140

Jazzard, Thomas, 13, -, -, 1200, 20, 120

Reece, Thomas, 125, 20, 10, 15000, 2000, 1275

Powel, W. G., 75, -, 25, 6000, -, 150

Hall, John, 100, -, 40, 10000, 2000, 610

Conely, Samuel, 135, 33, 35, 18000, 100, 515

Chandles, Thomas C., 70, 15, 15, 6000, 500, 440

Johnson, Godfrey, 43, -, -, 43000, 165, 415

Chandles, D. W., 115, 5, -, 10000, 100, 567

McCorns, John, 185, 35, -, 22000, 1000, 1325

Griffith, Mary E., 216, 10, 4, 23000, 800, 2500

Carnog, Wm. D., 151, 14, -, 20000, 972, 2805

Bradly, H. H., 13, -, 4, 13000, 100, 700

Frazer, William, 75, 15, 15, 6000, 50, 555

Morrison, J. R., 61, -, 16, 5375, 400, 1000

Mote, William, 176, 12, 10, 20000, 350, 1253

Burk, Patric, 20, 5, 15, 2500, 25, 185

McConaughy, Thos., 150, -, 100, 18750, 600, 550

Hall, William, 40, 20, -, 3000, 20, 550

Sillitoe, William, 107, 15, -, 7000, 250, 545

Bowers, William, 20, 50, 30, 30000, 1000, 2420

Macey James, 78, 16, 6, 7600, 256, 598

McConaughy, Wm., 150, 26, -, 12000, 800, 1605

Keely, Michael, 50, -, 5, 2700, -, 165

Cole, Elias, 15, -, -, 1500, 50, 275

Melburn, Robert, 60, 10, -, 2000, 275, 260

McCluskey, Arthur, 80, 10, 40, 7000, 100, 506

Megget, Peter, 76, -, 4, 4000, 380, 650

Slack, William, 94, 30, 26, 15000, 150, 1065

Gellis, Paul, 20, -, 23, 2500, 75, 179

Walton, Elijah, 40, -, 10, 2500, 100, 210

Frazer, W. W., 45, -, 20, 5000, 500, 630

Rambo, Enos, 75, 75, 10, 6000, 200, 515

Stroud, Edward, 142, 18, -, 16000, 650, 914

Rupp, John, 36, -, -, 1800, 700, 280

Walton, Charles, 70, -, 33, 5000, 350, 1395

Ott, Andrew, 17, -, 8, 1500, 200, 140

Moody, John, 120, 12, 8, 11000, 300, 650

Ford, David, 150, -, 50, 9000, 500, 1118

Loans, James, 9, -, 1, 1000, 20, 100

Sapp, Joseph, 200, -, 90, 17000, 300, 1320

Benson, Jas. A., 130, 20, 5, 12000, 600, 1505

Clendenin, S. H., 20, -, 20, 2500, 25, 332

Alexander, George, 50, 15, 22, 3000, 84, 360

Conelly, Peter, 4, -, 8, 500, -, 60

O'Roark, Jones (James), 20, -, 85, -, 1500, 20, 170

O'Roark, Timothy, 17, -, 7, 1200, 20, 175

Lopers, Samuel, 20, 20, 3, 2500, 50, 186

James (Jones), Joseph, 13, -, -, 1300, -, 80

James, David, 5, -, -, 300, -, 90

James, Joseph H., 10, -, -, 300, -, 80

Lewis, Dolphin, 5, -, -, 300, -, 20

Green, Benjamin, 1, -, -, 500, -, 35

O'Roark, Bartholomew, 35, -, 15, 1400, 25, 279

Whitaker, G. P., 30, -, 140, 7900, 25, 1540

Roach, Patric, 12, -, 8, 1200, 30, 100

O'Roark, Timothy, 14, -, 9, 1800, 15, 280

Sulivan, Patric, 20, -, 2, 1000, 100, 340

Coemis, James, 3, -, -, 500, -, 80

Walker, William, 10, -, -, 300, -, 41

Jones, Jane, 3, -, -, 150, -, 9

Seeny, William, 20, -, -, 400, 60, 150

Kenether, Martin, 44, -, 39, 2760, 200, 517

Kendal, Henry, 60, -, 35, 2500, 60, 505

Lyman, Joseph, 40, 40, 19, 2500, 200, 400

Brown, John, 40, -, 50, 3000, 150, 418

Brown, Thomas, 23, -, 2, 1500, -, 190

Dayett, Adam, 60, -, 40, 400, 250, 1120

Crawford, Joseph, 11, -, -, 1000, 50, 200

Sheldon, George, 108, 8, 22, 7500, 150, 445

Beck, Walter, 140, -, 157, 10000, 300, 1350

Norris, William, 7, -, 40, 1000, 14, -

Johnston, Henry, 7, -, 41, 14000, 40, 135

Batten, Mahlon, 90, 14, 26, 9500, 400, 1090

Dickenson, Saml., 350, 40, 50, 50000, 1000, 3450

Colburn, Arthur, 100, -, 180, 7000, 2500, 16510

Beck, William, 220, -, 100, 20000, 1000, 2377

Lester, Henry S., 28, -, -, 5000, 100, 500

Hewes, Clement, 175, -, 100, 30000, 500, 1500

Beck, John, 90, -, 70, 20000, 150, 750

Davidson, Alexander, 230, -, 60, 20000, 150, 2000

Clark, Thomas J., 80, -, 10, 75000, 500, 500

Deputy, Solomon, 250, -, -, 50000, 500, 1835

Clark, James H., 209, -, 60, 47400, 450, 4500

Clark, William H. H., 300, -, -, 60000, 600, 6587

Clark, Elizabeth, 50, -, -, 10000, -, 800

Davidson, John W., 200, -, 40, 20000, 500, 2400

Reybold, Anthony, 350, -, -, 50000, 2000, 6540

Reybold, William, 300, -, 100, 60000, 3000, 8650

Longland, Benjamin, 200, -, -, 17000, 300, 700

Reybold, John F., 240, -, 40, 35000, 675, 2955

Carpenter, John K., 150, 20, 35, 30000, 100, 300

Ochletree, Robert, 200, -, 35, 20000, 500, 984

Jones, William, 66, 6, -, 7000, 400, 550

Campbell, Joseph, 80, -, 20, 8000, 150, 600

Pennington, Benjamin, 720, -, 30, 15000, 600, 800

McCall, Samuel, 125, -, -, 15000, 1000, 1500

Montgomery, James, 290, -, -, 41000, 631, 2800

Smith, George H., 20, -, 5, 6500, -, 260

Aspriel, John A., 300, -, 20, 36200, 1500, 1580

Padley, James, 175, -, 125, 24000, 300, 1700

Taylor, Richard, 175, -, 100, 24000, 1000, 1638

Lester, Henry, 10, -, -, 1300, -, 150

Smith, Job, 740, -, 20, 20000, 200, 984

Sutton, Samuel B., 180, -, 20, 20000, 1000, 2000

Hurlock, Thomas, 200, -, 10, 20000, -, 1200

Rickards, Joshua, 200, -, 28, 22800, 800, 2329

Cleaver, Peter, 180, -, -, 16000, 800, 1750

Carr, Richard T., 500, -, 80, 30000, 400, 1400

Olston, Abner, 710, -, -, 11000, 200, 700

Taylor, Henry, 200, -, -, 25000, 1200, 3190

Gray, Montgomery J., 175, -, -, 17500, 200, 800

Stuckart, William _., 260, -, 15, 25000, 1000, 3670

Belville, John P., 190, -, 23750, 500, 2500

Evans, Mitchel, 170, -, 20, 30000, 900, 1852

Hurlock, Timothy, 200, -, 20, 20000, 350, 600

Clark, John C., 270, -, 3, 28000, 1500, 4098

Lee, James A., 250, -, 50, 15000, 200, 870

Gray, Richard, 192, -, 8, 20000, 400, 2100

Belville, Thomas W., 121, -, -, 15000, 350, 1670

Reybold, Barry, 66, -, -, 16500, 300, 1420

Harrison, Roberts, 190, -, -, 19282, 500, 1985

Biddle, Alexander M., 112, -, 30, 12000, 300, 1073

Casperson, William, 250, -, 30, 25000, 200, 1406

Hill, John, 260, -, -, 20000, 300, 2040

Denny, Michael, 6, -, -, 1000, 100, 240

Massey, John, 70, -, 30, 7500, 150, 1205

Dempsey, John, 140, -, 20, 12000, 200, 913

Calhoun, John, 100, -, -, 12000, 250, 935

Corbit, Charles, 200, -, 2, 50000, 800, 2700

Brady, George F., 160, -, 65, 30000, 1000, 1750

Clark, Theodore, 280, -, -, 42500, 500, 1871

Davidson, Amos E., 108, -, 100, 16000, 500, 1646

McCall, John, 80, -, 6, 6000, -, 680

Clark, William D., 255, -, -, 30000, 1300, 7500

Walter, H. C. 165, -, 125, 20000, 1500, 800

Alrich, J. B., 80, -, 62, 12000, 500, 600

Vandergrift, L. G., 313, 5, 32, 30000, 500, 3720

Diehl, Wm. B., 148, 10, 7, 15000, 400, 1550

Moore, Richard agt., 100, 2, 40, 9000, 100, 450

Hardcastle, Peter, 150, -, 50, 15000, 150, 955

Jefferson, Margaret, 152, 20, -, 15000, 400, 1225

Graham, Philip, 60, 40, 40, 4500, 500, 630

Higgin, Samuel, 150, 10, -, 13000, 250, 900

Foard, Thomas J., 380, 100, 75, 45000, 100, 3000

Vandergrift, J. C., 128, 6, 15, 15000, 300, 1190

Spear, Wm. B., 318, 10, -, 60000, 800, 2820

Shallcross, S. F., 525, -, 10, 85000, 2000, 4000

Shallcross, S. F., 525, -, 10, 85000, 2000, 4000

Shallcross, J. F., 50, -, 20, 8000, 500, 1000

Shallcross, S. F., 200, -, -, 30000, -, -

Shallcross, S. F., 171, -, -, 27000, 500, 1500

Cochran, R. W., 150, -, -, 30000, 500, 1600

Janvier, James J., 225, 15, -, 33000, 500, 2545

Boyd, John M., 50, -, -, 6500, 250, 300

Lord, Smieson, Jr., 80, -, 30, 12000, 400, 575

Lewis, Samuel, 140, 10, 140, 17000, 600, 1000

Cleaver, S. B., 7, -, -, 1400, 10, 300

Cleaver, Dauab, 125, -, 3, 20000, 500, 1835

Zacheris, John, 12, -, -, 7200, 60, 100

Stewart, David, 200, -, 40, 30000, 1000, 150

Jefferson, Samuel, 90, -, -, 12000, 500, 820

Cleaver, Isaac Jr., 128, -, -, 20000, 500, 1250

McClane, Hugh, 4, -, -, 2000, 50, 225

Cleaver, Joseph, 275, -, 100, 40000, 1000, 5170

Cleaver, William, 200, 51, -, 50000, 1000, 3050

McMullen, Wm., 20, 15, 50, 23500, 800, 2100

Dilworth, Thos. F., 260, 20, 63, 30000, 500, 1600

Price, Henry, 286, -, -, 40000, 1000, 1820

Carpenter, Thomas S., 228, -, -, 30000, 500, 2850

Green, Wilson, 12, -, -, 4600, 100, 360

Cleaver, Geo. G., 161, -, 2, 17000, 200, 500

Cleaver, Geo. G., 200, -, 110, 18000, 2000, 3300

Cleaver, William, 160, -, 160, 16000, 500, 1265

Nelson, John B., 85, 2, 40, 10000, 300, 1100

Cleaver, John A., 80, -, -, 8000, 200, 1000

Vandergrift, A. J., 200, 40, -, 40000, 800, 2100

Vandergrift, Christopher, 140, -, -, 12000, 400, 1300

Ellison, Joseph, 180, 25, 60, 21000, 400, 1500

Lofland, Trusten, 90, -, 150, 7000, 300, 1300

Burges, Geo. O., 100, 14, 100, 10000, 400, 1500

Segars, Samuel, 100, -, -, 10000, 500, 1500

Vandergrift, Wilson E., 275, -, -, 31000, 1000, 2450

Fowler, Robert, 100, 25, 60, 12000, 400, 712

Townsend, Sisters, 135, 30, 20, 12000, 400, 800

Averill, Alex, 110, 40, 30, 11000, 400, 1200

Foard, Geo. H., 200, 12, 208, 25000, 500, 1800

Stevens, Edward S., 240, 30, 50, 30000, 400, 2400

Foard, Chas. T., 160, 5, 41, 25000, 700, 2600

Foard, Chas. T., 88, -, -, 13000, -, -

Watkins, Gasaway, 31, -, -, 10000, 300, 550

Shallcross, Jacob F., 210, -, -, 40000, 1000, 1800

Perkins, Henry A., 150, 19, -, 25000, 800, 2075

Garner, James, 50, -, -, 10000, 400, 1040

McWhorter, Francis S., 3210, 37, -, 40000, 800, 2850

Bryan, Michael H., 30, -, -, 5000, 100, 350

McWhorter, Thos. S., 167, -, -, 25000, 1000, 1000

Otterson, Samuel T., 230, -, -, 30000, 700, 1600

Osborne, Hafert G., 272, 22, -, 50000, 800, 2300

Simkins, Wm., 200, -, 40000, 800, 2800

Cochran, Julian, 200, -, -, 30000, 500, 2000

Holston, Isaac, 210, -, -, 30000, 1000, 3300

Roberts, Samuel U., 150, 3, -, 13000, 300, 1100

Burris, Nehemiah, 250, 50, -, 25000, 600, 2000

Brady, William, 200, 15, -, 42000, 800, 1200

Vannekle, Fredis, 300, 80, -, 3000, 400, 2000

Cochran, Edward R., 250, 25, -, 55000, 1000, 3400

Readwig, Henry P., 200, 28, -, 22800, 300, 900

Harman, Alston, 140, 12, -, 25000, 400, 1100

Certs, Lewis, 120, 30, -, 10000, 300, 800

Noring, Elwood, 120, -, -, 10000, 300, 950

Eton, Richard, 190, -, 7, 20000, 600, 2255

Davis, Manlove, 400, 50, -, 40000, 800, 3500

Bearster, Chas., 150, -, -, 30000, 400, 1200

Meade (Medde), Robert H., 55, 17, -, 2500, 50, 400

Eliason, Wm. J., 100, -, -, 8000, 50, 600

Lecompte, James, 260, -, -, 32000, 600, 2900

Boggs, Samuel, 170, 12, -, 18000, 500, 2100

Lore, Williams, 250, 20, -, 32000, 600, 4800

Egee, Joseph, 313, -, -, 40000, 600, 2300

Hudson, John P., 300, 10, -, 37000, 600, 2820

Robson, John, 75, -, -, 12000, 300, 1100

McWhorter, John F., 140, -, -, 24000, 500, 1605

McWhorter, William, 112, -, 12000, 300, 1000

Pont, James T., 100, -, -, 12500, 500, 1045

Newner & Burris, 100, 40, -, 14000, 200, 800

Eliason, Andrew, 225, 10, -, 30000, 1000, 2500

Eliason, Andrew Jr., 300, 50, -, 25000, 1000, 3000

Rothwell, Samuel, 250, -, -, 25000, 800, 2100

Eliason, Ebenezer, 155, -, -, 20000, 1200, 1500

Boulden, Edward F., 100, -, -, 10000, 500, 1115

Loflund, Alford, 295, -, -, 40000, 700, 2300

Lynch, Pernell J., 300, 28, -, 32000, 700, 2200

Lynch, Pernell J., 130, 10, 15, 18000, 100, 550

Houston, Wm. H., 370, 30, -, 40000, 200, 2775

Clayton, Richard, 200, -, 5, 25000, 1000, 1900

Clayton, Joshua J., 133, -, 5, 20000, 500, 2000

Clayton, Charles E., 133, -, -, 20000, 500, 2000

Jones, John A., 350, -, -, 70000, 1000, 5000

Clayton, Henry, 150, -, -, 20000, 600, 1300

Clayton, Henry, 30, -, -, 5000, 700, 750

Gray, James, 400, 100, -, 50000, 1200, 4000

Paradee, Charles, 180, -, -, 18000, 150, 600

Clayton, Thomas, 190, -, 10, 25000, 800, 2000

Jones, Samuel, 300, 50, 10, 21000, 300, 600

Stoops, William, 220, 30, -, 25000, 1000, 1200

Green, Lewis, 150, 12, -, 16000, 300, 750

Armstrong, Benj., 160, 15, 22, 20000, 500, 1200

Ratledge, Robt. D., 180, 60, -, 20000, 600, 1675

Wood, William, 350, 50, -, 40000, 1200, 3000

Cavender, Thomas, 390, 60, -, 15000, 1200, 3890

Dodson, James, 240, -, -, 24000, 800, 2300

Hanson, Benj. Jr., 280, -, -, 30000, 600, 2000

Holton, Randolph, 150, 55, -, 16500, 500, 1500

Cleaver, Isaac, 210, 10, -, 25000, 500, 1515

Bennett, John R., 200, 5, 35, 30000, 600, 1640

Bennett, Wm. H., 100, 10, -, 10000, 500, 10000

Craven, Thomas J., 225, -, -, 28000, 1500, 4435

Craven, Thomas J., 75, -, -, 10000, -, -

Craven, Thomas J., 100, -, -, 12000, -, -

Townsend, Geo. W., 87, -, -, 15000, 500, 700

McMullen, James, 195, -, -, 23000, 500, 1270

Vandergrift, James M., 187, -, -, 29000, 400, 2200

Vandergrift, James M. 150, -, -, 15000, 150, 940

Karsner, Geo. W., 30, -, -, 9000, 75, 1185

Rickards, Ezekiel, 70, -, -, 12000, 300, 300

McVay, Samuel, 150, 25, -, 20000, 600, 1500

Clair, James F., 182, -, -, 18300, 200, 2415

Vail, Samuel C., 152, -, -, 19000, 500, 1600

Vandergrift, Jacob F., 120, 8, -, 15000, 600, 10000

Vandergrift, Abram, 225, 10, 100, 25000, 300, 1300

Longland, John, 170, -, -, 12000, 500, 2480

Longland, Zenos, 250, -, -, 10000, 300, 1400

Fogg, Richard, 380, -, 7, 55000, 2000, 8000

Vail, John C., 200, 15, -, 25000, 500, 1800

Garman, James, 143, -, -, 17875, 500, 1140

Vail, William, 275, 25, -, 30000, 500, 2300

Swan, John, 250, 50, 10, 20000, 1000, 1800

Vandergrift, Isaac _., 260, 40, -, 35000, 1000, 3171

Houston, John, 54, -, -, 10000, 250, 300

Houston, Thomas J., 120, -, -, 10000, 300, 800

McVay, Henry H., 55, -, -, 8500, 400, 600

Hopkins, Thos R., 177, -, -, 30000, 200, 850

Templeman, Geo. S., 88, -, -, 800, 500, 1000

Jones, Samuel, 20, -, -, 1500, 100, 150

Jones, Abram, 25, -, -, 2500, 100, 400

Burnham, Elizabeth, 165, -, -, 25000, 300, 2000

Cochran, Thomas, 550, 25, -, 100000, 1200, 4900

Hopkins, Levin, 200, -, -, 15000, 300, 1100

Riley, Thomas P., 200, 10, -, 20000, 400, 2000

Money, James, 812, 10, -, 20000, 400, 2000

Rothwell, Saml. T., 135, -, -, 13500, 500, 700

Jester, John A., 100, 10, -, 5000, 200, 400

Lord, Joseph A., 6, -, -, 1000, 100, 200

Mailby, Richard L., 245, -, -, 50000, 500, 2500

Naudain, Geo. W., 11, -, -, 2500, 100, -

Corbin, David Jr., 69, -, -, 8970, 100, 1140

Corbit, John, 65, -, -, 17000, 500, 500

Polk, William, 185, -, -, 20000, 150, 2500

Polk, William, 150, 15, -, 16000, 500, 1750

Polk & Hyatt, 40, -, -, 10000, 500, 300

Brown, Jane J., 160, -, 36, 25000, 400, 1925

Gould, Thomas H., 180, 20, -, 20000, 500, 1800

Lloyd, Horatio G., 150, -, 60, 25000, 800, 1200

Parker, Wm. C., 150, -, 50, 20000, 500, 1300

Walker, Martin E., 225, -, 25, 17000, 800, 2000

Dale, William, 20, -, -, 1500, 50, 250

Racine, Geo. N., 20, -, -, 6000, 150, 1100

Jones, Abram, 75, -, -, 9000, 200, 500

Nowland, A. J., 340, -, 50, 50000, 1000, 4500

Pennington, F. J., 270, -, 80, 40000, 1000, 200

Cochran, W. R., 200, -, 70, 41000, 1000, 3000

Vail, Alexander H., 172, -, -, 30000, 500, 1500

Merritt, Thomas S., 166, -, 20, 20000, 500, 1200

Merritt, Thomas S., 58, -, -, 15000, 300, 500

Barton, James, 40, -, -, 4000, 100, 500

Williams, Jonathan, 300, -, 26, 40000, 1500, 4358

Pennington, Samuel, 90, -, -, 18000, 250, 1100

Green, William, 285, -, 15, 60000, 1000, 2780

Derrickson, Charles, 260, 35, -, 40000, 1000, 2700

Willetz, Merrit, 320, -, -, 60000, 2000, 3000

Cochran, Richard R., 200, 20, -, 35000, 1500, 2700

Hoffecker, James R., 144 15, -, 16000, 500, 1810

Crockett, Alfred, 125, -, -, 15625, 800, 1265

Hushabeck, Andrew H., 8, -, -, 2400, 100, 200

Derrickson, Geo., 200, 15, -, 40000, 700, 2200

Cochran, Wm. R., 300, -, -, 50000, 2000, 2000

Cochran, John, 200, -, -, 40000, 1500, 2500

Lynch, Amos W., 300, -, -, 60000, 1500, 300

Hanson, Benjamin F., 160, -, -, 32000, 1000, 1800

Price, John, 150, 50, -, 20000, 500, 1500

Reynolds, Geo., 125, 15, -, 20000, 400, 1100

Jones, Geo., 130, 20, -, 15000, 500, 1500

Cochran, Chas. P., 375, 40, -, 50000, 1000, 4000

Murphy, Thomas, 158, -, -, 20000, -, 2100

Ray, James H., 456, 65, -, 40850, -, 200

Pelling, John, 9 ½, -, -, 9000, 50, 350

Williams, George, 6, -, -, 4000, 30, 200

Besser, William P., 6, -, -, 1500, 150, 350

Wright, Samuel B., 7, -, -, 8000, 10, 295

Husler, George M., 15, -, -, 2200, 10, 100

Glenn, William, 5, -, -, 3500, 20, 140

Williams, John F., 11, -, -, 4200, 50, 500

Pennington, William 4 ½, -, -, 2600, 100, 200

Choate, John W., 5, -, -, 1000, -, 350

Miller, John, 5, -, -, 1000, 5, 70

Curtis, Frederick A., 5, -, -, 4000, -, 175

Drennon, Jonathan, 4, -, -, 2000, -, 60

Evans, John W., 9, -, -, 4000, 20, 100

Wilson, Samuel Y., 12, -, -, 4000, -, -

Evans, George G., 10, -, -, 4000, -, 100

Laws, Alexander, 140, 10, -, 15000, 1000, 1600

Porter, Edward D., 65, 8, -, 12000, 1500, 1500

Haines, Eric W., 76, -, -, 9000, 50, 800

McColough, George, 160, 28, 15, 21527, 500, 1272

Kerr, George, 131, 18, -, 12000, 1000, 1930

Casho, Jacob, 35, -, -, 10000, 50, 669

Casho, George A., 80, -, 2, 5000, 357, 840

Morrison, Thomas H., 150, 60, -, 12000, 1000, 1528

McKeowan, Thomas, 38, -, -, 3000, 50, 370

Worrel, Nimrod, 35, -, 4000, 500, 560

Holland, Margaret, 50, -, -, 2500, -, 525

Lewis, Albert G., 100, -, -, 12000, 500, 1445

Waller, Frances A., 89, -, -, 10000, -, 150

Fisher, Levi, 7, -, -, 3000, 500, 480

Murphey, David J., 135, -, -, 20000, 1200, 2250

O'Donnell, John, 8, -, -, 1200, -, 150

Blandy, Charles W., 39, 3, -, 10000, 400, 485

McLaughlin, Constantine, 13, -, -, 12000, 100, 410

Edmundson, William, 90, 5, -, 9000, 300, 895

Crow, James, 58, 2, -, 5500, 500, 540

Crow, George, 33, 4, -, 2000, 100, 350

Steele, John, 120, 10, -, 10000, 500, 1398

Steele, Thomas, 20, -, -, 6000, 200, 325

Robinson, Joshua, 82, -, -, 9000, 300, 1100

Courtney, Henry B., 110, 10, -, 11000, 500, 760

Donnell, Andrew, 138, 10, -, 2000, 840, 1977

Hossinger, Joseph, 130, 30, -, 11000, 800, 1426

Heisler, William E., 140, 11, -, 20000, 1000, 860

Deputy, Samuel, 70, 10, -, 8500, 300, 730

McKowean, John, 45, 6, -, 6000, 100, 830

Baer, Aaron, 85, 15, -, 10000, 500, 1190

Simmons, Richard, 50, 46, -, 7000, 150, 1120

Sergent, Robert, 7, -, -, 1000, 50, 60

Lindsey, Samuel, 228, 70, -, 24840, 1000, 3550

Blackwell, Ephraim, 5, -, -, 2500, -, 250

Smith, William, 100, 20, -, 9000, 250, 1325

Oliver, William, 11, -, -, 1200, 50, 145

Robinson, William, 20, -, -, 3000, 50, 295

Radclift, Thomas, 5, -, -, 1150, 10, 14

Waid, Elizabeth, 70, -, -, 5390, 250, 500

Palmer, Elizabeth, 30, 12, -, 3000, 15, 180

McLaughlin, John, 40, 10, -, 3000, 15, 50

Pimberton, John, 20, 13, -, 1000, 25, 720

Clark, William, 89, 8, -, 7000, 200, 475

Mackey, James H., 24, 1, -, 4000, 40, 270

Wright, John E., 5, -, -, 1000, 50, 200

Conlsey, Jonathan, 6, -, -, 1500, 100, 265

Teaf, John L., 45, 5, -, 5000, 300, 480

Gregg, William, 39, 5, -, 5000, 300, 480

Mote, James H., 89, 10, -, 8000, 300, 1360

Travers, John M., 21, 4, -, 2500, 20, 446

Johnston, John T., 67, 5, -, 7500, 1000, 1250

Evans, Owen, 109, 31, -, 18000, 600, 3400

Chambers, Mary J., 130, 20, -, 16000, 400, 1625

Crosson, James L., 160, 15, -, 13000, 300, 1150

Twed, Manswel, 85, 40, -, 15000, 800, 1350

Pyle, Lamborn, 25, 14, 6, 7000, 300, 825

Thompson, Joel B., 5, 25, -, 25000, 1000, 4245

McClelland, William, 84, 16, -, 15000, 300, 950

Lamb, Thomas, 118, 12, 1, 15000, 600, 1760

Lawthorn, Thomas, 153, 60, 12, 18000, 125, 478

Scott, Samuel J., 28, -, 5, 2500, 50, 238

Ruth, Theodore, 66, -, 34, -, 7500, 200, 800

Dean, William, 126, 40, -, 42500, 1000, 3240

Rothwell, Abraham, 54, 6, -, 9000, 300, 1280

Pritchard, Geo & Joseph, 103, -, -, 12500, 500, 1255

Cavender, John, 20, -, 8, 1400, 50, 230

Moore, Hiram, 7, 116, 3, 9000, 300, 510

Sayers, Sarah, 50, 10, -, 6000, 300, 815

Sheppard, Carter W., 66, -, 6, 6000, 500, 950

Cling, Robert A., 131, 8, -, 11000, 500, 1450

Smith, Edmond, 20, -, -, 2000, 50, 380

Campbell, William, 35, -, -, 4000, 75, 470

Clay, William, 20, -, -, 4000, 70, 160

Mansfield, James H., 12, -, -, 2500, 25, 170

Webbe, Thomas, 148, -, -, 12000, 700, 1910

Egbert, Daniel, 5, -, -, 2000, 25, 150

Deugan, Joseph, 23, 10, -, 1500, 50, 250

Duff, Thomas, 27, -, -, 3000, 150, 400

Appleby, David, 142, 14, -, 20000, 562, 1705

Price, James R., 56, -, 40, 3600, 250, 730

Starr, William, 22, -, -, 1000, 10, -

Fenton, John, 19, -, -, 1200, 25, 350

Eastburn, Franklin, 73, 10, -, 8000, 250, 990

Singer, John A., 58, -, 20, 4585, 100, 350

Cavender, Frederick, 20, 5, -, 600, 10, 100

Oldham, Alexander, 100, 10, 10, 7000, 250, 650

Gibons, William, 12, -, 5, 1500, 250, 360

James, Eber, 100, -, -, 9000, 500, 1115

Newlin, Charles, 130, 50, 50, 16000, 800, 1625

Morris, Louis, 30, 20, -, 3000, 350, 440

Morrison, William, 299, 40, -, 33900, 1000, 1940

McBride, William, 142, 12, 28, 18000, 1000, 1495

Alrich, Sarah A., 82, 20, -, 15000, 500, 1150

Maree, Margaret, 55, 20, -, 6000, 500, 640

Ruth, Levi, 88, 10, -, 8000, 500, 740

Wright, James, 180, 150, -, 28500, 1000, 2760

James, William, 150, -, -, 15000, 800, 2055

Hawthorn, Thomas, 100, 25, -, 12500, 350, 980

Cartey, George, 14, -, -, 800, 100, 250

Noolin, Dennis, 10, -, -, 400, 50, 165

Stroud, William J., 100, -, 20, 12000, 500, 1240

Roomer, Henry, 5, -, -, 1000, 250, 260

Ayers, Jefferson, 435, 15, 8, 30000, 800, 2350

Hawthorn, William M., 67, 35, 10, 10000, 100, 450

Churchman, Henry L., 590, -, -, 59000, 1568, 17431

Johnson, Samuel M., 208, -, -, 2000, 500, 2030

Reynolds, Francis, 110, -, -, 10000, 300, 1678

Draper, Daniel, 110, -, -, 10070, 300, 850

Rylott, Mathew, 235, 15, -, 25000, 1000, 4556

Peters, Benjamin, 200, 50, 103, 17500, 350, 930

Hill, James, 14, 7, -, 2000, 57, 275

Carslile, Samuel, 175, -, 70, 20000, 400, 1850

Peters, Henry W., 57, -, -, 2500, 100, 160

Bowing, Robert, 125, 7, -, 6500, 300, 650

Smalley, William F., 70, 10, -, 16000, 200, 1550

Wright, Abraham E., 132, -, -, 10000, 400, 1140

Groves, Jonathan, 112, -, -, 7500, 500, 970

Palmer, George W., 80, 20, 47, 5000, 200, 690

Whitten, Thomas, 200, 15, 45, 25000, 1000, 1986

Brooks, George T., 105, 25, -, 9000, 200, 780

Lynam, William, 136, -, 4, 14000, 900, 2200

Armstrong, Robert, 120, 15, 21, 15000, 1000, 1575

Smalley, James H., 60, 41, 10, 7000, 200, 610

Switcher, Isaac, 52, 15, -, 4000, 200, 340

McNee, Thomas, 20, 10, -, 2000, 100, 800

Martin (Morton), George, 25, 5, -, 2100, 50, 300

Pogue, David & Joseph, 135, 22, 25, 19000, 1500, 1390

Brooks, William W., 125, 15, 15, 13000, 1000, 1210

Brooks, Henry, 100, 6, 29, 12000, 300, 940

Comly, Robert, 64, 3, 8, 5000, 300, 468

Weldon, John, 75, -, 75, 8000, 150, 375

Bostic, Mary E., 18, 2, -, 600, -, 10

Leman (Lemon), William T., 135, -, 5, 15000, 600, 1000

Cavender, William W., 135, 10, -, 14500, 500, 1010

Roberts, Samuel, -, 3, 15, 400, 100, 120

Motherall, William 252, 20, 10, 38000, 800, 2972

Platt, Franklin, 150, -, 18, 15000, 700, 1566

Johnston, George, 40, 10, 20, 3000, 100, 520

Morrison, Samuel W., 125, -, -, 12500, 500, 1320

Crouch, Isaac L., 30, -, -, 15000, 1000, 1200

Sussex County Delaware
1870 Agricultural Census

This agricultural census was filmed from original records in the Delaware State Archives in Wilmington Delaware by the Delaware State Archives Microfilm office.

There are some fifty-two columns of information on each individual. Only the head of household is addressed. I have chosen to use only seven columns of the information because I feel that this information best illustrates the wealth of the individuals. These are shown below:

1. Name of Owner
2. Acres of Improved Land
3. Woodland Unimproved
4. Other Unimproved Acreage
5. Cash Value of Farm
6. Value of Farm Implements and Machinery
15. Value of All Livestock

Thus, the numbers following the names represent columns 2, 3, 4, 5, 6, 15.

The following symbol is used to maintain spacing where information in a column is left blank (-). This symbol is used where letters, names or numbers are not legible ( ).

There are some missing pages near the end of this county. Missing pages appear from page 8 of Lewes & Rehoboth Hundred to Page 21 Little Creek Hundred Subdivision 22. It is impossible to determine how many pages after 8 in the same Hundred and possibly another eleven pages in the next section. There are five pages throughout the county where the census taker recorded first name first. All of the rest of this county last name first is used.

Warrington, David M., 6, -, -, 200, -, 250
McColum, John R., 80, 11, -, 1000, 30, 810
Brasure Jacob H., 100, 100, 25, 2500, 150, 775
Showell, Robert, 50, 100, -, 1500, -, 25
Truitt, Andrew C., 10, -, 5, 300, -, 50
Truitt, Jane, J., 12, -, 3, 300, 20, 173
Truitt, Irwin, 50, 40, -, 600, -, 75
Rogers, Americus, 50, 6, 4, 700, -, 200

Collins, Josiah, 100, 100, -, 2400, 14, 263
Cropper, William, 90, 20, -, 1300, 50, 275
Derickson, Ananias D., 118, 38, -, 5000, 313, 927
Rickards, Eliza, 60, 40, -, 1100, -, -
Layton, John, 75, 67, -, 2000, -, 270
Cooper, Benjamin P., 34, 48, -, 2050, -, 162
Thomas, Littleton, 40, 98, -, 1600, -, 60

Hudson, John H., 80, 80, -, 3000, 205, 583

Tingle, Ananias R., 40, 20, -, 900, -, -

Bunting, John, 3, 1, -, 200, -, 60

Holloway, Aaron, 9, -, -, 500, 20, 163

Evans, John W., 40, 60, -, 1500, -, 107

Latchen, Levin C., 75, 175, -, 3500, 50, 437

Collins, James B., 20, 30, -, 700, -, 151

Taylor, Thomas, 14, 6, -, 600, 30, 310

Bunting, James, 87, 36, 6, 1300, -, -

Brasure, James A., 42, 48, -, 2000, 50, 273

Atkins, Thomas, 50, 38, -, 1800, 50, 520

Williams, Ezekiel C., 20, 4, -, 1200, 20, 438

Hudson, Josiah, 30, 40, -, 1000, -, 180

Magee, John W., 20, 20, -, 400, -, 130

Bourlin, James S., 20, 17, -, 800, 50, 112

Savage, Joseph, 10, 1, -, 150, -, 50

Oliver, William, 36, 14, -, 1000, 50, 430

Harrison, Rouse, 100, 50, -, 3000, 50, 472

Daisey, Joseph N., 50, 50, -, 1000, 50, 270

Brasure, Joshua, 40, 10, -, 1000, 25, 260

Hudson, William J., 35, 15, -, 900, 25, 220

Brasure, James L., 30, -, -, 900, 25, 213

Brasure, James, 45, 25, -, 1400,2 5, 300

Hudson, James, 4, 4, -, 100, -, 45

Holloway, David, 48, 15, -, 600, -, 30

Megee, Auther L., 16, 21, -, 600, 50, 150

Dickerson, James B., 52, 40, -, 900, -, 90

McCabe, Edward N., 45, 55, -, 2000, 150, 312

Tracey, John H., 15, 30, -, 500, -, 95

Truitt, William R., 50, 50, -, 2500, -, 160

Murrey, Kendal, H., 60, 100, -, 4000, 30, 32

Evans, Henry J., 30, 120, -, 3000, -, 150

Timmons, Cyrus, 60, 90, -, 3000, 50, 180

Rogers, Joshua, 30, 26, -, 1100, -, 215

Lynch, Alfred, 60, 40, -, 3500, 200, 900

Eshun, Joseph, 47, 100, -, 3000, 50, 220

Bunting, Ezekiel W., 50, 50, -, 3000, 50, 225

Murrey, Laben H., 60, 64, -, 3000, 200, 280

Lynch, Lambert, 25, 55, -, 2000, -, 200

Lynch, Ezekiel, 65, 25, -, 2000, 50, 250

Bunting, Charles, 40, 28, -, 2000, -, 150

Hudson, Charles, 40, 80, -, 2000, -, 50

Timmons, Henry, 32, 18, -, 1200, -, 160

Derickson, Joseph, 30, 30, -, 1500, -, 80

Lynch, Aaron, 50, 15, -, 2500, 40, 200

Hickman, Caleb E., 15, -, -, 1000, -, 180

Hickman, Nathaniel, 30, 30, -, 2000, 30, 225

Furman, Edwin J., 50, 30, -, 1500, 150, 735

McCabe, Ebe, 25, 28, -, 2000, 100, 200

Rickards, Robert, 70, 25, -, 2500, 200, 285

McCabe, Garritson, 30, 40, -, 2000, 125, 260

Davis, Samuel C. S., 40, 50, -, 2000, 10, 20

Bunting, Edward D., 30, 30, -, 2000, 60, 200

Derickson, Peter M., 72, 28, -, 4000, 100, 200

Warrington, George, W., 32, 68, -, 2000, -, 100

Morris, Mitchel, 40, 66, -, 3000, 30, 210

Carey, John L. B., R., 56, 94, -, 4000, -, 510

Morris, Jackson, 13, 18, 32, -, 1200, -, 145

Hudson, David, 20, 60, -, 1600, -, 50

Hudson, Levin, 60, 40, -, 3000, 100, 416

Timmons, Josiah, 40, 46, -, 3800, 150, 590

Warrington, Richard, 28, 80, -, 2000, -, 100

Morris, Armwell, 25, 28, -, 1500, -, -

Morris, Armwell, 20, 12, -, 500, -, 250

Lynch, John B., 24, -, -, 700, -, 150

Bunting, Peter, 40, 15, -, 1500, -, 275

Bunting, Milby, 20, 4, -, 1000, -, 100

Hickman, William, 40, 80, -, 2500, -, 100

Bunting, Elijah, 50, -, -, 2000, -, 175

Herrin, Charles W., 20, 50, -, 1500, -, -

Evans, Ganage(George), 25, 50, 4, 2500, -, 125

Stephens, Henry, 50, 50, -, 1700, -, 100

Murrey, Washington, 25, -, -, 800, -, 250

McCabe, Elijah, 30, 50, -, 800, -, 180

McCabe, Elisha, 6, -, -, 1000, -, 170

McCabe, Isaac, 13, -, -, 3000, 125, 350

McCabe, Lebo, 60, 10, -, 3500, -, 248

McCabe, William S., 16, 20, 22000, 175, 370

McCabe, Amos, 42, 42, -, 2000, 100, 425

Murray, Milborn, 60, 40, -, 2500, 60, 303

Campbell, Eli, 3, 8, -, 400, -, -

Murray, Albert J., 26, 29, -, 1000, 100, 240

McCabe, William B., 50, 50, -, 3000, 20, 324

Murray, Stephen W., 48, 3, -, 2000, -, 200

Morris, Isaac H., 21, -, -, 1000, -, 210

Stephens, Joshua, 100, 200, -, 6000, -, 500

Brasure, William T., 10, 6, -, 2000, 25, 175

Tubbs, Samuel, 35, 27, -, 2500, -, 10

Hickman, William B, 4, 5, -, 1500, -, 55

West, James D., 3, -, -, 2000, -, 175

Lynch, Joshua T., 20, 12, -, 1000, 25, 250

Lynch, Jacob, E., 7, 3, -, 1000, 20, 200

Lynch, Alward W., 25, 5, -, 1000, -, 150

Tubbs, William R., 50, -, -, 1400, -, 130

Watson, Henry H., 11, -, -, 1500, -, 200

Lacey, Samuel W., 5, -, -, 1500, -, 100

McCabe, Levin T., 8, 12, -, 1000, 25, 259

Murray, Joseph G., 40, 30, -, 3000, -, 300

Hudson, Joseph G., 40, 14, -, 1200, 50, 150

Sharp, William, 30, -, -, 1200, 60, 220

McCabe, Josiah, 23, 20, -, 1000, -, -
Harrison, Joseph G., 53, 15, -, 2700, 50, 360
Hudson, William L., 60, 30, -, 3000, 50, 464
Lynch, Levi, 70, 10, -, 3000, -, 150
Lynch, John D., 90, 30, -, 4500, 100, 400
Lynch, Caleb M., 5, -, -, 500, -, 140
Evans, Burton R., 38, 10, -, 1000, -, 225
Lynch, Joshua, 10, 2, -, 800, -, 270
Lynch, Caleb W., 40, 4, -, 1600, -, 150
Lynch, David, 46, 4, -, 1500, -, 100
Hill, John B., 20, -, -, 800, -, 100
Lynch, Isaiah, 50, 39, -, 2500, 30, 350
Lynch, William A., 68, 5, -, 2000, 50, 400
Davidson, Joseph, 40, 10, -, 2000, -, 300
Bishop, Joshua W., 55, 24, -, 2500, 100, 378
Wilgus, Joshua, 24, 16, -, 1500, -, -
Murray, Jackson B., 70, 60, -, 3000, -, 150
Evans, George C. W., 35, 25, -, 2000, -, 250
Murray, Caleb, 30, 20, -, 2000, 500
Williams, Henry Jr., 45, 10, -, 1500, 25, 250
Wilgus, James, 26, 10, -, 2000, -, -
Wilgus, Jacob A., 14, 9, -, 4000, -, 600
Wilgus, John, 3, -, -, 1700, -, 175
Stephens, Thomas, 128, 40, -, 8000, 133, 617
Brasure, Littleton, 48, 32, -, 2500, -, 200
Murrey, Joseph, 4, 8, 60, -, 2500, -, 200
Eshun, Kendal, 128, 40, -, 6000, 50, 595
Murrey, Joshua B., 58, 12, -, 2500, 160, 483

Beckman, John, 24, 36, -, 600, -, -
Holloway, Thomas J., 45, 10, -, 1500, 50, 395
Magee, Jesse P., 44, 6, -, 1500, 50, 317
Layton, William, 28, 22, -, 1200, -, 200
Handy, George, 32, 30, -, 2000, -, 125
Murray, Sacker, 45, 15, -, 1500, -, 286
Moore, Eli, 50, 74, -, 1500, -, 286
Carey, Joseph S., 80, 200, -, 5600, 200, 910
Murray, Lot, 140, 70, -, 4000, -, 617
Murray, John B., 52, -, -, 1500, -, 175
Baker, Jonathan, 95, 47, 50, 3000, 80, 595
Murray, William C., 30, 30, -, 1500, -, 200
Hastings, William M., 120, 40, -, 4000, -, 408
Long, Joshua P., 40, 10, -, 1000, -, 500
Clog, John, 45, 5, -, 1000, -, 100
Harris, Joseph G., 100 67, -, 3000, 150, -
Long, Stephen L., 64, 64, -, 3000, -, -
Holloway, Jacob, 100, 155, 53, 4000, -, -
Long, Lemuel, 64, 36, -, 2000, -, -
Hickman, Handy, 10, -, 100, 800, -, -
Long, Charles H., 36, 10, 20, 1000, -, -
Rogers, Soloman, 80, -, 200, 1500, -, -
Stephens, Charles, 128, 30, 20, 4000, -, -
Hudson Joshua, 64, 60, -, 2000, -, -
Hudson, John H., 20, -, -, 400, -, -
Cambell, Lambert, 75, 45, -, 4000, 50, -
Hudson, Elisha, 73, -, 92, 2000, -, -,
Campbell, George, 30, 20, -, 2500, -, -

Long, Seth, 80, 20, -, 3000, -, -
Hudson, Peter W., 56, 50, -, 2000, -, -
Hudson, Isaac, 10, -, -, 400, -, -
Hudson, Noah, 80, 50, -, 5000, -, -
Lockwood, John, 36, 24, -, 2500, -, -
Long, Isaiah, 45, 35, -, 3000, 75, -
Rogers, William, 72, 20, 4000, -, -
Long, Zeno P., 60, 40, -, 4000, -, -
McNeal, Wolsey B., 60, 240, -, 6000, 200, -
Hudson, Jerry, 76, 70, -, 2500, -, -
Rogers, John, 80, 70, -, 4000, -, -
Jones, Ebe, 60, 30, -, 2000, -, -
Campbell, Eli, 60, 30, -, 2000, -, -
Hastings, Joshua, 48, 38, -, 2000, 40, -
Hudson Isaiah, 80, 30, -, 4500, -, -
Holloway, Peter, W., 60, 50, -, 3000, 100, -
Hudson, David, C., 56, 66, -, 3000, 100, -
Bunting, Joseph B., 50, 47, -, 3000, -, -
Hudson, Lemuel, 60, 15, -, 2500, -, -
Hudson, Joseph W., 52, 57, -, 3000, 150, -
Jacobs, James, 20, 40, -, 1500, -, -
Evans, William 20, 30, -, 800, -, -
Bunting, Milby, 75, 360, -, 12000, 200, -
Bunting, George, 4, 8, -, 400, -, -
Carey, James A., 75, 75, -, 2000, -, -
Holloway, Thomas, 25, 5, 90, 800, -, -
Clark, William, 6, -, -, 300, -, -
Holloway, Elisha, 50, 38, -, 3000, 15, 260
Andrews, Benjamin, 52, 28, -, 2000, -, 200
Hickman, John of N., 80, 134, -, 6000, 150, 705
Rickards, Luke, 30, 70, -, 2000, -, 50
Campbell, Wilson, 7, 2, -, 650, -, 50
Johnson, Maggie, 32, 35, -, 3000, -, 125

Godwin, Charles, 56, 40, -, 2500, -, -
Tingle, John, 60, 38, -, 2000, -, 700
Lockwood, Henry, 40, 50, -, 1500, -, -
Dingle, Isaac, 30, -, -, 800, -, -
Long, Henry W., 135, 165, 4, 3000, 160, 745
Bunting, Mitchel, 75, 75, -, 4500, -, -
Waples, John S., 48, 10, -, 4000, -, 175
Hickman, Henry W., 44, 100, -, 4000, 150, 500
Evans, Nathaniel, 25, 22, -, 2000, -, 200
Hudson, John H., 32, 60, -, 2500, -, 200
Murrey, Ananias, 50, 30, -, 2500, -, 200
Hickman, Selby, 28, 10, -, 1000, -, 225
Rickards, Isaac, 32, 8, 3, 1200, -, 250
Derickson, George T., 40, 40, -, 3000, -, 225
Gray, Mitchel, 50, 25, -, 2000, 75, 245
Tingle, Charles C., 94, 39, -, 2000, -, 315
Dukes, Thomas, 100, 30, -, 4500, 150, 645
Evans, Jacob, 28, 30, -, 2500, -, 250
Taylor, John, 20, 26, -, 2000, -, 300
Welburn, George, 40, 10, -, 800, -, 150
Williams, Lemuel S., 60, 30, -, 1500, -, 500
Megee (Mcgee), William, 100, 180, -, 2500, -, 300
Rickards, Isaac, 25, 32, -, 1800, 40, 200
Chamberlin, B___ J., 150, 50, -, 2500, 50, 450
Tiro, Mitchel, 30, 170, -, 2500, -, 100
Daisey, Henry, 62, 24, -, 2000, -, 190
Daisey, Thomas of P., 60, 40, -, 2000, 200, 600

Townsend, Zadoc, 18, 30, -, 2000, -, 250

Welden, Abner, 80, 100, -, 3000, -, 350

Taylor, John, 9, -, -, 300, -, -

Tunell, Nathaniel, 110, 50, 20, 4000, -, 760

Williams, Milby, 160, 100, -, 2000, -, 200

Vickers, Elisha, 24, 26, -, 1000, -, 200

Lekits, Samuel B., 90, 50, -, 4000, -, 130

Timmons, James, 20, 40, -, 1800, -, 50

Tingle, James, 13, 2, -, 1200, -, 350

Jefferson, Mary L., 9, -, -, 200, -, -

Dealon, William M., 13, -, -, 800, -, 250

Quillin, Nathan, 29, 10, -, 500, -, 200

Derickson, Robert, 50, 8, -, 500, -, 40

Lewis, Jonathan, 19, -, -, 200, -, 200

Evans, Lemuel, 30, 30, -, 1500, -, 60

Lewis, George, 28, -, -, 600, -, 300

Evans, William, 35, 4, -, 800, -, 250

Evans, David, 9, 9, -, 600, -, 120

Knox, George W., 16, -, -, 600, -, 90

Evans, Stephen W., 44, 6, -, 1200, -, 260

Quillin, Peter, 50, 130, -, 3000, -, 150

Hall, William, 70, 8, -, 1000, -, 700

Evans, Elisha, 70, 30, -, 3500, -, 390

Herrin, Alfred W., 50, 20, -, 2500, -, 170

Collier, Peter, 85, 40, -, 6500, 150, 750

Hocker, Jacob, 15, 15, -, 1000, -, 350

Hudson, Henry, 40, 12, -, 3000, 150, 520

Hudson, Parker M., 25, 15, -, 1200, 50, 200

Williams, Lemuel, 4, -, -, 700, -, 50

Hudson, William, 25, 15, -, 1000, -, 150

Daisey, Thomas, 55, 56, -, 4000, 150, 740

Daisey, Jonathan, 55, 55, -, 2000, -, 120

Turner, William 20, 10, -, 1000, -, 175

Jones, John W., 50, 50, -, 4000, 100, 785

Jones, John W., 60, 36, -, 5000, -, -

Evans, Stephen, 40, 10, -, 800, -, -

Turner, Isaac, 5, 8, -, 400, -, 35

Quillin, Ebe D., 18, 15, -, 1000, -, 260

Wharton, Isaac, 25, 47, -, 2500, 50, 780

Wharton, John B., 16, 30, -, 1500, -, 150

Coard, Charles, 8, 4, -, 360, -, -

Knox, James M., 10, 5, -, 500, -, 75

Dickerson, Littleton, 75, 79, -, 3500, -, 320

Furman, Mary, 8, 20, -, 500, -, 100

Hudson, William, 28, 230, -, 2000, -, 296

Williams, John W., 30, 13, -, 1200, -, 200

Williams, Edward, 25, 25, -, 1500, -, 150

Evans, John A., 12, 21, -, 1200, -, 445

Hickman, James A., 56, 14, -, 2500, -, 180

Grice, John G., 60, 90, -, 3500, 200, 552

Evans, William T., 40, 10, -, 1000, -, 150

Evans, James, 30, 30, -, 1500, -, 225

Daisey, James W., 45, 5, -, 1500, -, 390

Dawson, Jasper, 35, 15, -, 1500, -, 120

Daisey, Eli R., 225, 15, -, 800, 50, 200

Daisey, Eber A., 40, 20, -, 1500, 50, 250

Lynch, Reuben, 80, 50, -, 3000, 100, 510

Hickman, Richard, 50, 6, -, 2000, -, 350

Hudson, Jerry, 36, 48, -, 3000, 200, 400

Derickson, Sarah, 36, 48, -, 3000, 200, 400

Rickards, Joseph, 40, 30, -, 2000, -, 225

Hickman, Richard, 36, 174, -, 6000, -, 300

Hudson, David, 80, 35, -, 5000, -, 400

Rickards, Stephen, 34, 14, -, 2500, 50, 300

Evans, Joshua, 26, -, -, 300, -, 25

Anderson, James, 40, 60, -, 3600, 125, 620

Godwin, David, 60, 39, -, 3000, -, 250

Lockwood, Benjamin, 40, 30, -, 2500, -, 240

Williams, George L., 64, 33, -, 3000, 100, 500

West, Isaac C., 50, 40, -, 2500, 100, 252

Hall, George, 50, -, -, 1500, 50, 225

Taylor, John N., 40, 16, -, 2000, -, 325

Hall, Charles, 40, 60, -, 2500, -, 350

Sermon, Samuel, 24, 76, -, 2500, -, 200

Melson, John, 20, 15, -, 1500, -, 250

Rickards, William A., 80, 80, -, 7000, -, 440

Barnet, Luke, 14, 14, -, 1500, -, 235

Melson, James, 8, -, -, 500, -, 50

Magee, John, 28, 17, -, 2200, -, 300

Miller, Jenkins, 8, 12, -, 1000, -, 50

Hall, Charles, 12, 118, -, 4000, -, -

Hudson, William, 20, -, -, 2000, -, 450

Short, John, 20, 26, -, 2000, 150, 800

James, Isaac W., 100, 160, -, 7000, 200, 1000

Evans, Selby, 45, 55, -, 1800, -, 350

Hall, Sarah, 21, 21, -, 600, -, 50

Tunnell, James A., 7, -, -, 1000, -, 120

Hall, Joseph, 200, 100, 50, 6000, 100, 440

Hall, Hetty, 60, 20, 15, 2000, -, 80

Evans, Sylvester, 40, 60, -, 1500, -, 20

Wharton, William, 30, -, -, 1000, -, 160

Hudson, David, 30, -, -, 700, -, 250

Burbage, Ananias, 16, -, -, 800, -, 150

Derickson, Lemuel, 35, -, -, 1000, -, 300

Burton, Elizabeth, 260, 53, -, 4000, -, 625

Derickson, William, 8, -, -, 300, -, 50

Hudson, James, 30, -, -, 700, -, 175

Gray, Milby, 70, -, -, 2500, -, -

Quillin, Hetty, 90, 10, -, 2000, -, 350

Rickards, John, 70, 30, -, 1500, -, 250

Davis, James, 75, 15, -, 1500, -, 100

Rickards, Lemuel, 35, -, -, 1200, -, 250

Wharton, John, 24, -, -, 700, -, -

West, George H., 250, -, -, 5000, -, 480

Williams, William L., 12, 12, -, 600, -, 150

Daisey, David, 20, -, -, 500, -, -

Lynch, James H., 75, 5, -, 1800, -, 300

Derickson, Elizabeth, 20, 10, -, 1000, -, 200

Evans, Stephen, 80, 32, -, 4000, 400, 530

Derickson, Stephen, 20, 18, -, 1000, -, 250

Derickson, Elizabeth, 15, 10, -, 1000, -, 50

Miller, John, 20, 20, -, 1000, -, 50

Townsend, James, 80, 80, -, 2500, -, 125

Godwin, David, 60, 40, -, 3000, 150, 500

Lynch, Levin, 32, 10, -, 1500, -, 250

Rickards, William, 29, -, -, 1500, -, 200

Murrey, Henry B., 85, 110, -, 6000, -, 600

Steel, Thomas N. Jr., 15, 11, -, 1000, -, 150

Steel, Thomas N. Sr., 20, 18, -, 1000, -, -

Hudson, George, 28, 22, -, 2000, -, 350

Furman, Lemuel, 20, 30, -, 3000, -, 340

Furman, James, 70, 30, -, 3000, -, 450

Messick, John, 50, 40, -, 1200, -, 200

Townsend, Peter, 30, 69, -, 1800, -, 500

Holt, James F., 45, 20, -, 3000, 100, 500

Holt, James F., 45, 20, -, 2000, -, -

Williams, Henry, 9, 5, -, 300, -, -

Morris, Wesly W., 30, 40, -, 2500, -, 200

Betts, James, 80, 50, -, 3000, -, 700

Derickson, George J., 20, 20, -, 1000, -, 700

Evans, Isaac H., 25, 15, -, 1500, -, 150

Betts, John J., 20, 15, -, 1000, -, 160

Quillin, Robert, 40, 15, -, 1200, -, 270

Steel, Thomas R., 75, 136, -, 2500, 150, 560

Banks, Henry, 25, 25, -, 1000, -, 140

Aydelott, David, 30, 100, -, 1200, -, 300

Murrey, Caleb, 50, 127, -, 2000, -, 300

Gray, Joseph, 200, 300, -, 6000, -, 575

Wells, Cannon, 6, 14, -, 300, -, -

Dolby, Peter, 120, 180, -, 4000, -, 80

Banks, Joshua, 50, 100, -, 1500, -, 70

Evans, George, 30, 40, -, 900, -, 100

Townsend, Luke, 25, 60, -, 1000, -, 130

Walter, George, 200, 120, 25, 4000, -, 200

Townsend, Joshua C., 30, 90, -, 5000, 200, 450

Cannon, Eleanor, 15, 35, -, 1000, -, 100

Steel, Isaac, 3, 7, -, 600, -, -

Townsend, Isaac, 20, -, -, 600, -, 200

Dale, Peter, 25, 25, -, 1000, 30, 225

Collins, Stephens, 52, 52, -, 2500, -, 300

Parsons, John B., 60, 80, -, 2800, -, 100

Melson, Stephen, 60, 40, -, 1500, -, 100

Derickson, John D., 65, 14, -, 1500, -, 350

Daisey, James T., 150, 100, -, 2200, -, 263

Brown, Selby, 70, 10, -, 1000, -, 100

Walter, Ebe, 250, 50, -, 6000, 520, 970

Aydelott, Stephen C., 100, 120, -, 5000, 100, 525

West, Nancy W., 20, 35, -, 2000, -, 225

Ellis, Isaiah, 100, 100, -, 3000, -, 400

Bennett, Joshua R., 40, 80, -, 2500,-, 500

Wharton, William, 50, 90, -, 1500, -, 40

Rickards, William, 80, 120, -, 4000, 40, 165

Townsend, Isaac, 70, 30, -, 1200, -, 350

Steel, John, 4, 13, -, 800, -, 2020

Williams, Daniel, 40, 50, -, 3000, -, 200

Steel, John R., 50, 60, -, 3000, 60, 380

Steel, John Sr., 23, 30, -, 600, -, 20

Townsend, John, 30, 70, -, 2000, -, -

Rickards, Mitchel, 30, 50, -, 1000, -, 250

Townsend, Major, 10, 10, -, 200, -, -

Steel, Mias B., 60, 10, -, 2500, 100, 300

Carey, Cornelius, 9, 11, -, 500, -, 250

Walker, Daniel, 25, 25, -, 1000, -, 300

Derickson, Benjamin, 30, 49, -, 1000, -, 70

Hall, Henry, 32, 40, -, 1000, -, 125

Lynch, Jacob, 36, 30, -, 2500, -, 200

Halloway, Ebenezer, 40, 50, -, 4000, 60, 400

Rickards, Hiram B., 8, 12, -, 100, -, 125

Tunnel, Henry, 30, 10, -, 1600, -, 300

Bennett, John, 70, 60, -, 3500, 100, 480

Wilgus, Thomas, 40, -, -, 1500, -, 175

Hudson, Ann, 14, -, -, 500, -, -

Hudson, Jacob H., 50, -, -, 3000, -, -

Rickards, James, 48, 15, -, 2100, -, 225

Hickman, Selby, 30, 8, -, 1500, -, 200

Derickson, William, 40, 25, -, 1500, -, 275

Bennett, John S., 45, 40, -, 4000, -, 500

Lynch, Elijah, 20, 35, -, 2500, -, 550

Lynch, Lemuel, 20, 25, -, 2000, -, 150

Derickson, Levin H., 70, 41, -, 2000, -, 200

Bennett, John, 65, 56, -, 4000, -, 525

Bennett, John, 55, 95, -, 2500, -, 250

Bennett, Levin H., 70, 70 -, 4000, 300, 532

Rickards, Kendal, 310, 130, 115, 5000, 100, 850

Davis, Josiah, 190, 10, 70, 200, -, 200

Law, James H., 50, 3, -, 3000, 125, 250

Hudson, George, 60, 20, -, 2000, 50, 200

Tunnell, Daniel, 100, 100, -, 6000, -, 400

Evans, Jedadiah, 50, 30, -, 3000, -, 400

Evans, Abigail, 20, 60, -, 800, -, -

Derickson, Robert, 12, -, -, 200, -, -

Tunnell, James, 60, 20, -, 800, -, 200

Johnson, George, 20, 28, -, 1000, 25, 485

Tunnel, John, 15, 15, -, 1000, -, 30

Turner, James, 30, 70, -, 1200, -, 100

Bull, John, 15, 15, -, 400, -, 100

West, Philip, 25, 45, -, 1000, -, 40

Calhoon, Ephraim, 12, -, -, 200, -, 160

Dalae, James M., 70, 30, -, 1000, -, 300

Layton, William, 68,46, -, 1700, 75, 500

Johnson, Barton, 30, 30, -, 800, -, -

Derickson, Isaiah J., 50, 30, -, 3600, 50, 600

Derickson, Henry, 30, 32, -, 2500, -, 175

Poleti, William 70, -, -, 3000, 20, 400

Lynch, Joseph I., 45, 31, -, 3500, 100, 360

Furman, George, 40, 10, -, 1500, 30, 250

Townsend, Ebe, 65, 75, -, 4000, 80, 800

Evans, Levin, 24, 45, -, 1000, -, 150

Calhoun, John, 80, 70, -, 2000, 50, 350

Nalls, Absalum, 40, 27, -, 1200, -, 350

Gray, William T., 10, 20, -, 500, -, 100

Chamberlin, Sophia, 5, -, -, 500, -, 50

Johnson, John A., 35, 5, -, 1000, -, 200

Pool, Robert, 37, 30, -, 1200, -, 300

Pool, Keturah, 125, 125, -, 6000,
150, 300
Lynch, Caleb, 40, 23, -, 1500, 50,
200
Hudson, John, 48, 15, -, 2000, -, 300
Bull, Josiah, 20, -, -, 500, -, 150
Evans, Lemuel, 70, 20, -, 2500, -,
200
Evans, Charles, 50, -, -, 1000, -, 200
Lynch, Edward, J., 41, 70, -, 1500, -,
200
Hudson, Benjamin, 40, 10, -, 1500,
50, 325
Gray, William H., 55, 30, -, 3500,
50, 350
Howard, George, 45, 25, -, 1000, -,
200
Cloy, David, 40, 30, -, 2500, -, 200
Derickson, William A., 35, 5, -,
1400, -, 220
Howard, Charles, 60, 40, -, 2500, -,
350
Banks, John, 60, -, -, 1000, -, -
Godwin, John, 80, 50, -, 2500, -, 400
Roberts, James, 100, 100, -, 4000,
100, 400
Moore, Peter, 30, 9, -, 1000, -, 100
Hickman, John, 35, 20, -, 1200, 30,
200
Hill, Henry, 40, 30, -, 2500, -, 250
Chericks, James, 30, 30, -, 1000, -,
250
Hill, Clement, 30, 70, -, 1500, -, -
Johnson, Henry R., 80, 65, -, 3500,
100, 730
Banks, John R., 50, -, -, 800, -, -
Hudson, Hannah, 6, -, -, 200, -, -
Moore, Mias B., 60, -, -, 2500, -, 400
Hudson, Nathaniel, 30, 10, -, 2500,
80, 200
Hubbard, James, 10, 2, -, 1000, 100,
120
Collier, Z. P., 3, 1, -, 500, 50, 175
Moore, B. J., 175, 40, -, 5000, 100,
200

Wootten, Reitter S., 210, 80, -, 6000,
25, 500
Hearn, Jonathan, 165, 150, -, 10000,
50, 425
Parker, Ebenezer, 50, 30, -, 800, -, 60
Callany, Luren S., 20, -, -, 5000, 100,
110
Daleny, James H., 33, -, -, 930, 50, -
Collier, Joshua J., 50, 10, -, 3000, -,
250
Sermon, Isaac, 20, 14, -, 2000, -, 80
Bacon, Joseph, 266, 201, -, 6460,
100, 150
Riggin, Thomas, 90, -, -, 6000, -, 150
Cannon, Thomas, 300, 400, -, 10000,
100, 300
Cannon, Burton, 60, 53, -, 1500, -, -
Remmey (Kenney), Samuel, 160,
158, -, 5000, -, 1000
Wootten, Nathaniel, 233, 125, -,
6400, -, 160
Windsor, William 100, 250, -, 5000,
-, 25
Hearn, John, 15, 11, -, 1500, -, 60
Collins, Jonathan, 50, 10, -, 8000, -,
200
Collins, Mary, 600, 800, -, 4200, -,
368
Wright, John T., 700, 450, -, 17000,
100, 260
Smith, O. P. 52, 52, -, 3500, -, 450
Mauck, Aaron, 117, 117, -, 8000, -,
150
Hearn, Daniel, 313, 243, -, 1600,
100, 50
Jones, Joseph, 20, 45, -, 5000, -, 100
Taylor, Edward, 75, 115, -, 1400, -,
235
James, William, 80, 54, -, 1800, -,
175
Hearn, Henry, 175, 65, -, 12000, 40,
200
Moore, Louther, 11, 3, -, 5000, -, -
Smith, George, 50, 50, -, 1000, -, -
Horsey, Thomas, 130, 30, -, 2800, -,
300

Wheatly, William 75, 50, -, 3000, -, 300

Bacon, John S., 28, -, -, 2000, -, 25

Wootten, John W., 100, 53, -, 3500, -, 75

Wolfe, William, 50, 10, -, 3000, -, 125

Phillip, George B., 11, 2, -, 1200, -, 100

Cullaway, William, 42, -, -, 4000, -, 1000

Ward, Jacob P., 110, 45, -, 4650, -, 250

Hearn, Martin, 84, 60, -, 4000, -, 100

Moore, Josiah, 75, 60, -, 2100, -, -

Maull, Loren J., 10, 13, -, 150, -, 25

Anderson, William, 20, 30, -, 1000, 25, 125

Lewis, Henry C., 250, 150, -, 1600, -, 1700

Hearn, Elijah, 11, -, -, 1000, -, 100

Parker, Asbery W., 54, 10, -, 1300, -, 50

Hearn, Rissey, 40, 20, -, 500, -, 100

Giles, Thomas, 50, -, -, 1500, -, 200

Moore, William, 90, 27, -, 300, -, 150

Wheatly, Stansbury S., 120, 80, -, 2000, -, 200

Delaney, James, 10, -, -, 400, 25, -

Andrew, J. Harey, 100, 30, -, 3000, -, 400

Maull, Joseph W., 37, 13, -, 1000, -, 100

Hitch, Benjamin, 140, 100, -, 11800, -, 130

Ebzery, James, 75, 25, -, 6000, -, 550

Hitch, William S., 15, 15, -, 400, -, 300

Hearn, Thomas, 15, 75, -, 700, 100, 150

Oliphant, William, 60, 140, -, 4500, 100, 200

Callaway, Handy, 150, 217, -, 9000, -, 235

Fooks, Benjamin, 320, 26, -, 13900, -, 1025

Adams, Isaac, 150, 70, -, 2460, 25, 200

Elliott, William, 120, 40, -, 1920, -, 150

Maull, James B., 220, 80, -, 6000, -, 300

Hitchens, John, 80, 20, -, 1500, -, 400

Legat, Peerry, 160, 40, -, 2500, -, 700

Alling, William, 325, 75, -, 8000, 100, 500

Welley, George, 200, 40, -, 4800, -, 250

Phillip, Henry W., 60, -, -, 1200, 50, 200

Marel, Joshua H., 15, -, -, 6000, -, 100

Runouls, Wilson, 30, -, -, 3000, -, 100

Frost, Sidney B., 7, 7, -, 175, -, 50

Moore, Daniel, W., 85, 15, -, 5000, 100, 300

Spicer, William 90, 30, -, 3000, -, 100

Moore, David, 40, 44, -, 1600, -, 300

Mitchel, Scot, 40, 40, -, 800, -, 200

Jerman, Isaac S., 60, 60, -, 2400, -, 50

Baker, Bryard, 35, 25, -, 900, -, 200

Wilson, John R., 81, 90, -, 10000, 500, 1000

Parmer, Charles, 175, 50, -, 4000, -, 300

West, Mary A., 100, 29, -, 2000, 100, 150

Gordy, John, 50, 50, -, 1500, 50, 300

Collins, Lamberson, 70, 27, -, 2000, 100, 250

Hall, John, 46, 4, -, 1200, 25, 140

English, Levinsco, 75, 25, -, 1000, 50, 25

English, James, 50, 53, -, 1000, 100, 300

Collins, William, 100, 42, -, 2500, 50, 100

West, Able W., 40, 20, -, 700, -, 10

Smith, Sallie, 50, 50, -, 1000, -, 150

Rogers, Thomas T., 60, 46, -, 4000, -, 350

Truitt, Burton, 50, 48, -, 1000, -, 200

Truitt, Burton, 30, 30, -, 600, -, -

Phillip, Parker N., 75, 49, -, 1500, -, 250

Phillip, Sarah R., 60, 36, -, 1000, 50, -

Lowe, James W., 50, 10, -, 600, -, 100

Collins, Joseph S., 30, 16, -, 460, -, 200

Littleton, John D., 25, 15, -, 500, -, 100

Mathews, Catherine, 100, 100, -, 2000, -, 200

Downs, Wright, 100, 98, -, 2000, 50, 500

Selbey, George, 50, 40, -, 900, -, -

West, Isaac C., 30, 22, -, 500, -, 75

Cannon, Jacob, 25, 10, -, 300, -, 50

Lengo, Minus B., 40, 46, -, 1300, -, 200

Wells, Joseph, 20, 10, -, 250, -, 10

Collins, Josiah, 100, 100, -, 3000, -, 400

Collins, Ephraim, 50, 50, -, 1500, -, 200

Jones, John, 100, 30, -, 2000, -, 200

Collins, Elijah, 100, 20, -, 1000, 50, 250

Timmons, Noble, 50, 50, -, 800, -, 150

Timmons, Nancy, 50, 50, -, 800, -, 200

Megee, Sallie, 100, 20, -, 1500, -, 150

Mitchel, Samuel, 100, 20, -, 1500, -, 150

Short, Elizabeth, 100, 50, -, 1500, -, 200

Short, Elijah, 25, 25, -, 500, 30, 200

Truitt, Joseph, 55, 47, -, 1200, 20, 50

Jones, George W., 25, 75, -, 1000, 310, 200

Mitchel, Rufus, 80, 67, -, 1800, 50, 400

Jones, Jacob S., 80, 114, -, 1500, 100, 300

Jones, Isaac S., 50, 25, -, 2500, 200, 700

Louis, George W., 35, 31, -, 1000, 30, 150

Hickman, John, 100, 10, -, 1000, 70, -

Gray, Joshua, 45, 21, -, 1600, 20, 100

Truitt, Mathew, 100, 60, -, 1000, 10, 400

Collins, Elius, 40, 10, -, 400, 50, 150

Collins, Sallie, 30, 10, -, 400, 25, 200

West, H. B., 100, 30, -, 4000, 35, 500

West, Sallie, 50, 25, -, 2000, 25, 200

Short, Mirus, 100, 45, -, 2500, 100 500

Wootten, Edward, 50, 50, -, 500, -, 200

Short, Isaac B., 100, 73, -, 3000, 200, 735

Moore, Robert, 55, -, -, 500, 20, 20

West, Burton, 200, 80, -, 3000, -, 100

Downs, Noah, 100, 25, -, 2000, 10, 300

Mitchel, Thomas, 10, 10, -, 200, 30, 10

Evans, Lemuel, 10, 10, -, 200, -, 50

Cary, Phillip, 42, 5, -, 500, -, 25

West, Wingate, 40, -, -, 800, -, 150

Timmons, Ezekiel, 100, 100, -, 3000, 50, 425

Truitt, Benjamin, 20, 20, -, 500, 30, 150

Burris, Serena, 40, 50, -, 500, -, 200

Mitchel, William, 75, 29, -, 2000, 150, 300

Hitchens, John, 100, 12, -, 1200, 50, 600

Hitchens, Eduard, 60, 20, -, 600, 50, 100

Oneal, David, 128, 128, -, 5000, 100, 300

Boyce, James, 100, 17, -, 2300, 50, 225

Hitchens, Noah, 50, 50, -, 1000, 50, 200

Melson, Isaac W., 50, 27, -, 1000, 50, 250

Givens, Isaac, 50, 53, -, 1000, 30, 200

Mathus, Isaac, 50, 30, -, 800, 25, 100

Mather, Nutter, 100, 100, -, 2000, 25, 300

Truitt, Mary, 50, 20, -, 700, -, 50

Cannon, Jacob, 100, 60, -, 2000, 100, 300

Burton, Joseph, 80, 63, -, 2000, 50, 200

Hitchens, Elijah, 50, 50, -, 1000, 50, 100

Mathius, Hezkiah, 100, 100, -, 2000, -, 500

Mathews, Hezekiah, 55, 175, -, 1200, 100, 500

West, Mary A., 100, 100, -, 4000, 200, 500

Wright, John W., 50 80, -, 2000, 250, 75

Parsons, Mary A., 50, 38, -, 1200, 150, 200

Phillip, John, 30, 20, -, 1000, 100, 200

Wootten, William, 75, 58, -, 2000, -, 100

Elliott, Benjamin, 50, 10, -, 1000, 100, 350

Hastings, Benjamin, 100, 10, -, 1500, 150, 200

Mitchel, Nathaniel, 60, 50, -, 1000, 100, 300

Workman, Joshua, 100, 100, -, 2400, 24, 350

Cergy, Nelley, 100, 50, -, 2000, -, 200

Mitchel, Dinnard, 30, 16, -, 500, 50, 50

Ward, James H., 50, 38, -, 1500, 50, 250

Baker, Joseph, 40, 30, -, 1800, 50, 500

Mitchell, Nancy, 20, 16, -, 360, 60, 50

Baker, Samuel, 200, 110, -, 3500, 50, 200

Baker, Sallie, 13, 10, -, 200, 20, 50

Baker, Sarah, 36, 10, -, 300, -, 100

Baker, Noble, 150, 20, -, 800, -, 100

Jerman, Isaac L., 30, 15, -, 600, -, 150

Dunaway, Henry, 100, 25, -, 1000, -, 200

Baker, William, 50, 54, -, 1400, 40, 50

Moore, Henry, 100, 40, -, 2000, -, 200

Tingle, Albert, 75, 25, -, 2000, -, 100

Moore, Luren, 150, 75, -, 2000, -, 250

Ake, John S., 52, 10, -, 1000, -, 200

Ake, Joseph, 100, 20, -, 800, -, 150

Ake, Thomas D., 50, 25, -, 1500, 50, 150

Jerman, James, 100, -, -, 1500, 30, 100

Hearn, James S., 150, 40, -, 2000, -, 1110

Gumby, David, 40, 16, -, 1000, -, 75

Gumby, Jacob, 25, 45, -, 1000, -, 200

Gumby, John B., 60, 48, -, 1500, 50, 550

Gumby, Stephen, 14, 14, -, 500, -, 150

Wells, Thomas 8, 3, -, 300, -, 250

Short, Uriah, 50, 50, -, 3000, -, 400

Short, Shederick, 200, 95, -, 4000, 400, 1000

Baker, Seth W., 25, 15, -, 600, 60, 150

Truett, Thomas J., 100, 96, -, 1500, 50, 200

Pennel, Selious, 75, 65, -, 2800, -, 250

Hickman, John W., 80, 45, -, 2500, -, 785

Mitchell, Samuel, 35, 92, -, 2500, -, 200

Windsor, Joseph, 25, 23, -, 500, 50, 100

Taylor, Lius, 44, -, -, 500, 50, -

Taylor Lius (Lins), 100, 72, -, 2580, 30, 150

Bacon, John L., 100, -, -, 3000, 50, 200

Moore, William 100, 40, -, 2000, 30, 150

Moore, Joseph, 50, 24, -, 1500, 50, 200

Wiley, John, 200, 50, -, 300, 50, 350

Phillip, Henry, 100, 20, -, 2000, 30, 200

Oneal, William, 150, 100, -, 5000, 50, 400

Mills, John, 200, 50, -, 3000, 50, 250

Jobarts, John J., 70, 76, -, 3500, 50, 200

Smith, Becker (Baker), 15, 25, -, 300, -, -

Phillip, William, 75, 25, -, 2000, -, 400

Marell, James, 50, -, -, 500, 30, -

Waller, John W., 15, -, -, 450, 50, 350

Whaley, Isaac T., 40, -, -, 800, 50, 150

Workman, William, 25, 40, -, 1500, 30, 400

Mathew, Cathrine, 75, 25, -, 1000, 30, 100

Workman, John, 25, -, -, 500, 50, 50

Pepper, Henry, R., 100, 25, -, 1500, 50, 200

Boyce, William, 100, 30, -, 2600, -, 460

Harmin, George, 100, 80, -, 1800, -, -

Boyce, James, H., 20, 25, -, 450, -, -

Roburts, Aray, 100, 60, -, 3000, 30, 530

Roberts, George, 50, 29, -, 590, -, -

Whealey, James, 20, 25, -, 450, 50, -

Whealey, William, 100, 25, -, 1500, 50, 400

West, George W., 100, 140, -, 3000, -, 200

Elliott, Elisha, 11, -, -, 1000, -, 200

King, Bengamin, 100, 175, -, 3000, 100, 200

Kinikin, Isak, 100, 40, -, 1500, -, 10

Pennell, Lemuel, 10, 39, -, 350, -, -

King, Noble A., 100, 33, -, 1330, 100, 200

Tingle, David, B., 100, 10, -, 2200, -, 300

Truitt, William, 12, 50, -, 100, -, 50

Short, John J., 15, 25, -, 1200, 50, 350

King, Nathaniel, 60, 14, -, 2000, 50, 350

Hamilton, George, 110, 25, -, 1800, -, 300

Truitt, George, 100, 40, -, 1800, -, -

Beetts, Charles K(R.), 65, 25, -, 1200, 50, 200

Parsons, James, 96, 100, -, 2500, 50, 30

Baker, Solathel, 40, 75, -, 2500, -, 300

Baker, Seth, 40, 50, -, 1000, 30, -

West, William, 68, -, -, 1500, -, 250

West, Cornelious, 100, 56, -, 5000, -, 700

Legates, Edward, 32, -, -, 1000, 50, -

Otwell, William, 100, 100, -, 2500, -, 200

West, Payter (Payton), 100, 20, -, 500, -, 150

Truett, James, 100, 130, -, 2500, -, 600

English, James W., 50, 61, -, 1000, -, 300

Truett, Henry, 100, 50, -, 1200, 25, 4100

Shirden, John W., 60, 18, -, 1200, -, 150

Wingat, Rachel, 6, -, -, 250, -, 20

Elensworth, Bovern, 100, 40, -, 2000, -, 300

Cannon, Wingate, 100, 100, -, 2000, -, 100

Vinson, Henry, 40, 40, -, 500, -, 10

Cannon, John, 50, 46, -, 1000, -, 100

Fer, George, 50, 40, -, 900, -, 50

Rodney, Robert, 110, 70, -, 2000, 50, 400

Rodney, William, 28, -, -, 300, -, 100

Gordy, Sarah, 200, 200, -, 4000, 50, 400

Rodney, John, 100, 50, -, 2000, -, 150

Hastings, Z. P., 60, 33, -, 1200, -, 300

James, Rubon, 30, 30, -, 100, -, 30

Chipman, John, 80, 120, -, 3000, 100, 650

Chipman, Charles, 100, 100, -, 2000, 100, 95

Workman, Thomas, 100, 90, -, 4000, 25, 350

Workman, Edward, 40, 30, -, 1000, -, 150

Morris, John, 100, 100, -, 2000, 50, 30

Warrington, William, 75, 22, -, 2000, 30, 200

Hamilton, Talbert, 50, 50, -, 1000, -, 150

Carman, Burton, 160, 160, -, 3000, -, 600

Shields, John, 75, 25, -, 1000, 40, 150

Carmean, James, 40, 53, -, 500, -, 50

Worten, William 25, 50, -, 500, -, 200

Records, William D., 40, 25, -, 2000, 50, 450

Elensworth, Roberson, 100, 40, -, 2000, -, 100

Spicer, Edward, 20, 10, -, 500, -, 75

Melson, Benjamin, 100, 40, -, 1500, -, 10

Pussey, Henry, 75, 35, -, 1200, 50, 10

Hopkins, John W., 75, 34, -, 1500, -, 150

Phillips, Elijah W., 150, 50, -, 3000, 100, 500

Burris, Serern, 50, 25, -, 1000, -, 100

Pussey, John, 25, 75, -, 1000, -, 20

Melson, Nathaniel, 100, 30, -, 2000, 50, 250

Hitchens, H. C., 30, 25, -, 1000, 10, 200

Watman, O. Brine, 300, 200, -, 8000, 100, 500

Givins, Colwell, 150, 156, -, 4000, 30, 200

Burris, Edward, 200, 75, -, 5000, 40, 400

Gordy, Peter, 300, 200, -, 2500, -, 300

Workman, Charles, 100, 15, -, 1000, -, 100

James, Brancen, 160, 61, -, 4000, 150, 800

Spicer, William 40, 20, -, 600, 30, 300

Workman, Jesse, 100, 40, -, 1000, 40, 150

Jones, Noah, 150, 50, -, 3000, 100, 400

Hitchens, Minus, 20, 100, -, 1500, 50, 300

Hitchens, Guren S., 75, 40, -, 1000, 40, 300

Hitchens Guren, 20, 10, -, 240, -, 75

Watten, Henry, 70, 45, -, 600, -, 75

Fleetwood, John, 50, 50, -, 800, 40, 100

Hitchens, Edmon, 80, 160, -, 1600, -, 300

James, Joshua, 100, 20, -, 1400, -, 100

Lamden, Robert, 150,100, -, 2000, -, 800

Ducks, John, 100, 40,-, 1000, -, 200
Watson, Talbert, 100, 47, -, 1500, -, 400
Hearn, Mickle, 100, 50, -, 6000, 100, 500
Truett, Miles, 25, 100, -, 3000, -, 200
Hitchens, George, 95, 50, -, 200, -, 30
Messick, Claton 75, 55, -, 1500, -, 150
Hitchens, George, 20, 100, -, 1500, -, 50
Mathew, Phillip, 100, 121, -, 3000, -, 20
Muthen, Phillip, 43, 100, -, 1500, 100, 500
Huston, Joseph, 100, 60, -, 2000, 50, 200
Messick, Nemiah, 40, 40, -, 600, -, 50
Horsey, Samuel, 50, 50, -, 1000, -, 5
Legates, Elijah, 25, 50, -, 900, -, 100
Hearn, Clemsth, 150, 50, -, 3000, 50, 530
Baker, Bryant, 25, 25, -, 500, -, 25
West, Payter (Payton), 30, 30, -, 600, 30, 40
Cilierr, Chalres, 100, 25, -, 1000, 100, 250
Pollet, Lewis, A., 75, 25, -, 1500, 50, 100
Records, Isaac, 100, 50, -, 2000, -, 100
Husten, Joseph, 100, 50, -, 2500, -, 300
Lynch, William, 74, -, -, 1000, 50, 175
Hosea, John H., 200, 94, -, 6000, -, -
Purnell, John, 100, 85, -, 1500, -, -
Conner, James D., 50, 50, -, 1000, -, 100
Gordey, Petter, 60, 390, -, 4500, -, -
Ward, Joseph, 100, 17, -, 1400, -, -
Legates, James R., 100, 90, -, 2000, 50, 400

Ward, William, 70, 30, -, 2000, -, 150
Morris, John, 100, 50, -, 2500, 100, 450
Elliott, James R., 100, 100, -, 3000, 30, 150
Legates, Alexandria, 100, 95, -, 1500, 50, 150
Morris, John, 50, 40, -, 800, -, -
Morris, John, 50, 50, -, 2000, -, -
Hearn, Joseph, 100, 81, -, 2500, -, 10
Ellis, Thomas H., 55, 58, -, 700, -, -
Faskey, James B, 15, 15, -, 300, -, 5
Mcfaden, James P., 30, -, -, 800, -, 200
Jones, Jacob R., 75, 58, -, 1500, 150, 400
Powell, Heyram, 15, 15, -, 400, 25, 100
Hearn, Joseph, 30, 15, -, 450, -, 100
Hearn, Joseph, 10, -, -, 200, -, -
Brimer, Joshua, 10, -, -, 100, -, 25
Bearn(Hearn), Bengamin, 80, 10, -, 4000, 30, 500
Hearn, Isaac T., 30, 15, -, 400, -, 100
Jerman, Isaac, 50, 10, -, 400, 25, 200
Brittingham, Hyram, 60, 55, -, 1000, -, 200
Hearn, George H., 200, 50, -, 3000, 50, 600
Carmean, William, 100, 18, -, 2500, -, 500
Carmean, William, 60, 30, -, 1800, -, 200
Cannon, Joseph, 20, 43, -, 630, 40, 150
Brittingham, Benjamin, 20, 70, -, 900, -, 50
Cannon, William, 100, 50, -, 1100, 30, -
Callaway, William, 40, 36, -, 760, -, 200
Hudson, Joseph, 25, -, -, 250, 100, 250
Hudson, David, 20, 20, -, 1000, 50, 800

Wheatley, William, 200, 100, -, 6000, 150, 300

Freeney, Eligah (col), 200, 50, -, 2700, 100, 250

Morris, Isaac, 100, 33, -, 3325, 200, 175

Wheatley, James B., 250, 250, -, 10000, 200, 50

Callaway, Jonathan, 175, 95, -, 5000, -, 100

Culben, Elijah, 100, 50, -, 3250, -, 25

Hastings, Elihue, 300, 75, -, 15000, 100, 200

Hearn, William, 50, -, -, 1000, 50, 40

Mills, John, 100, 25, -, 3750, -, 300

Mills, Isham, 80, 50, -, 1200, -, 200

Hastings, Cyrus, 120, 30, -, 3500, -, 400

Hearn, Jonathan, 150, 87, -, 4000, 100, 300

West, Isaac, 175, 155, -, 4000, -, 400

Dun, Burton, 100, 34, -, 2500, 150, 250

Horsey, George W., 200, 125, -, 14000, 200, 1000

Henry, John, 80, 20, -, 2500, 150, 100

Hitche, Lenn, 30, 10, -, 1000, 200, 1000

Hitch, George, 30, 30, -, 1200, 100, 175

Hearn, William 29, 11, -, 1000, 100, 150

Mills, John, 100, 25, -, 3750, -, 200 (may be repeat)

Mills, John, 30, 50, -, 1200, 100, 300

Mills, John, 10, -, -, 100, 50, 200

Evans, Nathaniel, 60, 51, -, 1000, 100, 200

Hearn, William, 175, 125, -, 2400, 50, 350

Cannon, James M., 72, 25, -, 800, -, 75

Oneal, Joseph T., 12, 2, -, 300, -, 600

Hearn, George H., 50, 155, -, 2000, 50, 900

Oneal, George, 65, 65, -, 2000, -, 500

Huston, Elisha, 100, 50, -, 2000, -, 300

Riggin, William, 175, 45, -, 5000, -, 30

Warrington, Elijah, 105, 55, -, 2000, -, 35

Marel, James A., 37, 12, -, 7500, -, 50

Marel, James A., 25, -, -, 500, -, 30

Cannon, Cyrus W., 125, 77, -, 4000, 200, 400

Hitch, William, 150, 150, -, 6000, 150, 400

Messick, George M., 90, -, -, 1800, 100, 150

Hastings, James, 100, 25, -, 2500, 50, -

Jester, Burton, 100, -, -, 2000, 40, -

Elliott, Hetty, 200, 107, -, 3000, -, 500

Hastings, Colwell, 30, 14, -, 500, 40, 100

Ennis, Jesse J., 115, 115, -, 5100, 50, 100

Prettyman, William, 60, 40, -, 1000, 10, 200

Blizzard, Henry C., 35, 175, -, 400, 10, 125

Ennis, David J., 115, 35, -, 3000, 90, 5475

Veasey, Josiah, 70, 70, -, 1400, 40, 430

Veasey, William 85, 80, -, 1750, 110, 120

Warrington, Peter, 60, 120, -, 1800, 20, 125

Johnson, William D., 40, 50, -, 900, 10, 125

Martin, John D., 89, 18, -, 970, 40, 400

Tompson, John C., 100, 100, -, 5100, 150, 1100

Virden, Joseph B., 100, 20, -, 2400, 50, 40

Dickerson, Richard, 60, 20, -, 3000, 80, 270

Dickerson, Samuel, 80, 50, -, 1300, 95, 190

Hopkins, Margaret, 40, 40, -, 1800, 20, 120

Pettyjohn, Zachariah, 50, 75, -, 20000, 100, 1100

King, Cornelious, 50, 43, -, 2000, 25, 585

King, James A., 130, 25, -, 4000, 100, 400

Dutton, George H., 60, -, -, 1200, 50, 300

White, Benjamin, 95, 75, -, 3000, 100, 350

White, William N., 125, 85, -, 840, 50, 490

White, Henry H., 70, 20, -, 3000, 200, 550

Lank, Peter C., 80, 17, -, 2900, 65, 500

Reynolds, Burton, 40, 47, -, 1200, 25, 100

Burton, Benjamin, 70, 30, -, 2500, 150, 300

Holland, Elisha, 150, 100, -, 5000, 200, 500

Brittingham, Smith, 75, 50, -, 1200, 50, 170

Carpenter, Benjamin, 125, 40, -, 1600, 20, 205

Carpenter, Benton, 110, 50, -, 1600, 60, 415

Chace, Isaac, 40, -, 40, 2200, 70, 570

Bennum, George W., 70, 45, -, 2500, 12, 85

Wilson, Thomas B., 40, 25, -, 1000, 10, 220

Blizzard, John T., 70, 15, -, 1700, 15, 350

Dutton, Robert, 110, 5, -, 700, 20, 500

Wilson, Ruben, 49, -, -, 400, 15, 200

Wilson, John W., 50, 20, -, 1000, 25, 150

Walls, John W., 70, 70, -, 6000, 20, 140

Akins, Sarah A., 40, 40, -, 1000, 18, 150

Coulter, James, 40, 50, -, 1500, 50, 325

Johnson, Minus, 25, 25, -, 6000, 40, 300

Vent, William 40, 56, -, 1920, 30, 250

Walker, Thomas, 75, 94, -, 33380, 100, 650

Simples (Simpler), James, 75, 10, 90, 2100, 30, 200

Holland, Andrew, 125, 100, -, 4500, 40, 40

Veasey, William W., 65, 15, -, 2500, 75, 400

Prettyman, Nathan, 210, 60, -, 2000, 10, 115

Holland, John S., 50, 57, -, 3000, 100, 490

Lemon, Sallie, 60, 70, -, 1300, 20, 180

Carpenter, Gideon, 50, 50, -, 1500, 80, 250

Conwell, Asa, 80, -, -, 300, 75, 340

Simpler, John R., 80, 40, -, 1200, 20, 160

Burton, Benjamin H., 80, 70, -, 1500, 15, 230

Carey, James H., 125, 125, -, 4000, 75, 325

Messick, George, 75, 20, 20, 2100, 20, 30

Maull, Purnall, 35, 8, -, 860, 110, 2116

Mcgee (Megee), Moses, 78, 15, -, 1000, 40, 290

Hood, Henry, 50, 50, -, 1000, 40, 130

Coffin, Matilda, 60, 40, -, 800, 30, 310

Clifton, William, 70, 70, -, 3000, 30, 300

Clifton, George A., 40, 50, -, 3000, 20, 80

Sharp, Henry, 100, 40, -, 2000, 20, 300

Walker, William, 60, 100, -, 200, 25,150

Clark, Norris, 30, 30, -, 1000, 25, 75

Hudson, Henry, 40, 90, -, 3000, 40, 186

Warrington, Nathaniel, 10, -, 40, 200, 20, 90

Dutton, John T., 60, 3000, -, 4800, 40, 150

White, Wallace W., 75, 80, 25, 1650, 100, 560

Black, Thomas, 40, 30, 26, 3000, 25, 250

Burton, Daniel, 16, -, -, 7000, 5, 995

Dutton, Jesse, 50, -, -, 1000, 10, 400

Palmer, Jackson, 70, 130, -, 6000, 25, 205

Robins, James, 100, 30, -, 3900, 25, 350

Hudson, Walker, 100, 40, 100, 4800, 75, 495

Downing, James B., 50, -, 40, 1800, 15, 110

King, Isaac, 100, 60, -, 2500, 50, 240

Harmon, Eli, 12, -, -, 460, 6, 100

King, Charles F., 40, 57, -, 1800, 65, 500

King, Sarah, 90, 35, -, 1800, 40, -

Martin, William, 60, 50, -, 1800, 40, 175

Burton, George, 20, 7, -, 800, 30, 150

Warrington, Alfred, 40, 65, -, 2000, 40, 250

Russel, David, 30, 40, -, 120, 15, 290

Richards, Mary, 20, 10, -, 900, 15, 200

Warrington, Coard, 75, 20, 30, 2000, 110, 1110

Holland, Cyrus, 20, -, -, 800, 10, 150

Holland, Elsey, 25, -, -, 500, 10, 130

Fisher, John, 40, 43, 100, 2500, 50, 500

Fisher, James, 40, 60, -, 2500, 50, 520

Holland, Jeremiah, 10, -, -, 300, 20, 200

Martin, Ebe, 40, 10, -, 1500, 15, 170

Fisher, Myres, 50, 6, 130, 2000, 25, 460

Howard, Elijah, 250, 20, 50, 8000, 15, 410

Dean, John, 8, 4, -, 400, 30, 280

Warrington, Rowland P., 40, 10, 40, 1880, 20, 126

Chase, James, 50, 50, 70, 3000, 40, 100

Holland, Joseph, 50, 3, -, 1500, 20, 330

White, Alfred, 70, 35, -, 2100, 20, 1810

Palmer, Silvester, 50, 50, -, 2000, 40, 180

Rust, Absolom, 60, 37, -, 2000, 100, 300

Rust, William, 10, 47, -, 1100, 12, 100

Milby, William, 30, 10, -, 1000, 20, 200

Simpler, William H., 175, 100, -, 4000, 30, 360

Wilson, Elisabeth, 20, 360, -, 4000, 30, 280

Bennum, William H., 100, 25, -, 3000, 40, 400

Perry, Thomas, 102, 50, 30, 6000, 140, 635

Hazel, William, 100, 100, -, 6000, 30, 661

Joseph, John W., 125, 175, -, 5000, 100, 570

Hopkins, Peter J, 100, 125, -, 2280, 25, 400

Reed, Peter, 100, 80, -, 4000, 50, 270

David, John B., 50, 40, -, 1000, 25, 380

Lank, Levin J., 15, 10, -, 500, 25, 100

Stevenson, Sarah H., 75, 5, -, 3000, 5, 200

Carey, Cornelius, 80, -, 20, 2000, 30, 305

Carey, Lemuel, 60, -, 3, 1000, 5, 70

Reynolds, James, 80, 10, 3, 2000, 10, 200

Merck, Matthew J., 95, 5, -, 1000, 50, 260

Martin, Samuel J., 100, 10, -, 3000, 100, 500

Wilson, Thomas R., 16, 2, -, 900, 30, 160

Brittingham, Moses, 40, 12, 25, 2000, 20, 225

Betts, Robert, 44, 6, 30, 2500, 75, 590

Reed, Alfred, 80, 10, -, 1500, 30, 590

Workman, John, 75, 40, 5, 3000, 10, 75

Robbins, David, 125, 100, -, 5000, 80, 44

Conwell, William A., 60, 57, 12, 2000, 50,3 80

Lofland, James, 60, 40,-, 2000, 30, 590

Johnson, Greenbury P., 40, 60, -, 2000, 30, 310

Heavelow, Morris, 110, 90, -, 10000, 10, 630

Reed, Benjamin, 100, 10, -, 2000, 75, 410

Russel, David, 80, 20, 12, 4000, 400, 500

Conwell, Hester B., 90, 10, -, 4000, 10, 220

Reed, William D., 75, 20, 25, 10000, 100, 600

Russel, Alfred, 40, 20, -, 3100, 140, 470

Carey, Jane T., 80, -, -, 3000, 200, 375

Robinson, John S., 120, 40, -, 4000, 50, 360

Heavelow, William J., 100, 20, -, 6000, 1000, 970

Wiltbank, David A., 120, 30, 90, 6000, 200, 810

Robinson, Thomas, 90, -, 96, 3000, 20, 820

Roach, 100, -, 40, 3000, 15, 355

Roach, James, 100, -, 40, 3000, 15, 355

Reed, Philip R., 40, 10, 30, 2000, 20, 480

Craige, William, 100, -, 20, 4000, 10, 250

Vaughan, Charles, 52, 10, -, 1350, 1200, 946

Robbins, James C., 80, -, 30, 2750, 40, 600

Reed, William, 60, 30, 15, 5000, 125, 460

Morris, John, 100, 30, 100, 10000, 130, 620

Nailer, David H., 80, 12, -, 5000, 30, 640

Wilson James, 75, 18, -, 3000, 50, 278

Jones, Kersey, 72, 20, -, 3000, 50, 275

William, John, 100, 25, -, 3000, 35, 248

Lofland, Samuel, 175, 50, 70, 2900, 35, 445

Heavelow, Joshua _., 52, -, -, 1200, 20, 400

Heavelow, Joseph P, 36, 5, 15, 2000, 10, 50

Reed, Abraham P., 70, 30, -, 2000, 20, 410

Wilson, Mark P., 80, 40, -, 3000, 30, 200

Walls, John H., 160, 40, -, 3000, 40, 134

Heather, Horatio, 40, -, 7, 1000, 10, -

Johnson, David M. P., 40, -, -, 1200, 5, 240

Reed, James, 62, 4, -, 1200, 55, -, 160

Conwell, James T., 100, 45, -, 6000, 100, 790

Rust, Sylvester H., 60, 84, -, 4000, 50, 661

Johnson, Henry M., 75, 55, -, 2000, 15, 104

Donovan, Mary J., 70, 70, 30, 2000, 20, 110

Spicer, William, 40, 60, -, 1000, 15, 100

Pettyjohn, Benjamin, 60, 140, -, 1200, 15, 120

Dickerson, James, 40, 240, 100, 3000, 12, 194

Workman, Philip T., 90, 100, -, 3200, 30, 390

Jester, Isaac, 50, 40, -, 1000, 20, 200

Messick, Henry T., 25, 32, -, 550, 5, 100

Workman, Philip, 80, 50, -, 1800, 25, 405

Hopkins, William, 80, 30, -, 1200, 30, 270

Abbott, James, 75, 140, -, 1800, 25, 254

Abbott, Alfred, 75, 25, -, 3000, 40, 400

Donovan, William H., 100, 170, -, 5000, 200, 490

Warren, John R., 40, 28, -, 500, 15, 160

Messick Levy, 50, 200, -, 2000, 100, 285

Carey, Maria, 70, 130, -, 1000, 10, 60

Heavelow, Mary, 150, 100, -, 2500, 20, 200

Donovan, Alfred, 100, 200, -, 3000, 20, 400

Peter, Mitchel, 125, 90, -, 2000, 25, 200

Walls, Robert J., 40, 100, -, 1400, 75, 230

Warren, Ebenezer, 50, 75 -, 2000, 150, 300

Fowler, William, 60, 90, 100, 2500, 100, 175

Clendaniel, Jacob, 130, 70, -, 2000, 20, 160

Workman, Robert B., 210, 60, -, 1000, 50, 200

Dutton, James, 60, 10, -, 1000, 15, 110

Johnson Purnal, 50, 100, -, 2000, 20, 400

Hayes, George, 80, 110, -, 2000, 40, 230

Mosley, Elsey, 20, 20, -, 600, 25, 220

Morris, Thomas, 40, 40, -, 300, 25, 350

Clifton, John O., 40, 50, -, 900, 25, 175

White, Wallace _., 80, 20, -, 2500, 100, 425

Ellingsworth, James, 40, 60, -, 1000, 10, 150

Johnson, Isaiah, 40, 20, -, 600, 10, 130

Prettyman, William, 20, -, -, 900, 30, 320

McClin, Virden, 20, 30, -, 600, 10, 440

Prettyman, Elisha, 20, 5, -, 800, 40, 200

Welsh, Nehemiah D., 9, -, -, 1000, 60, 175

Black, Joseph, 10, -, -, 1000, 40, 320

Hazzard, John C., 44, -, -, 2200, -, 200

Hazzard, John A., 175, 175, -, 2000, 25, 150

Betts, Joseph A., 55, -, -, 2500, 10, 170

Williams, Thomas, 125, 25, 60, 4000, 110, 195

Roach, James, 40, 21, -, 1200, 15, 95

Collier, William, 40, 10, -, 800, 20, 380

Morris, Robert R., 130, 80, -, 6300, 150, 900

Johnson, Barton H., 60, 15, -, 4000, 100, 460

Collier, Eli, 50, 50, -, 1000, 15, 435

Clendaniel, Luke, 30, 100, -, 1000, 20, 75

Jones, Elizabeth, 45, 10, -, 1000, 20, 180

Reynolds, Matilda, 60, 40, -, 1000, 20, 80

Morris, William, 50, 50, -, 1500, 50, 200

Tharp, Elisha, 300, 100, -, 5000, 20, 287

Betts, James, 15, -, -, 300, 100, 135

Salimons, Robert, 40, 100, -, 1500, 25, 420

Morris, Robert, 60, 50, -, 1650, 25, 150

Lindal, John H., 30, 120, -, 2000, 20, 100

West, Thomas P., 60, 60, -, 1200, 25, 147

Ellingsworth, Noble, 60, 15, -, 5000, 100, 700

Morris, William, 60, 40, -, 1000, 50, 175

Morris, Edward, 60, 40, -, 1500, 40, 150

Ellingsworth, James, 60, 40, -, 1500, 100, 200

Craige, James, 60, 40, -, 1500, 30, 275

Pepper, Thomas, 100, 75, -, 3000, 16, 150

Abbott, George F., 75, 125, -, 4000, 60, 390

Bryan, Robert B., 75, 70, -, 4000, 40, 485

Betts, Isaac S., 60, 10, -, 300, 100, 340

Mason, William S., 100, -, 6, 10000, 60, 590

Chase, Prissilla, 70, 70, -, 4000, 20, 270

Bruington, Cade, 30, 5, -, 800, 15, 50

Ingram, Nathaniel, 20, -, -, 500, 20, 178

Calhoon, Gideon, 45, 95, -, 1400, 20, 250

Dodd, John, 45, 22, -, 2000, 75, 317

Lindal, Joshua, 40, 60, -, 1200, 15, 80

Megee, George W., 50, 8, -, 1000, 20, 250

Reed, James T., 10, 30, -, 600, 100, 280

Reed, Somerset, 120, 120, -, 2400, 30, 467

Reed, Peter, 100, 200, -, 4500, 60, 4000

Mosley, Levy, 40, 37, -, 1000, 87, 418

Walker, Jesse, 30, 50, -, 800, 70, 140

Donovan, Nancy, 50, 150, -, 2000, 10, 155

Wilson, William W., 80, 520, -, 3200, 100, 380

Hazzard, William, 100, 75, -, 8753, 300, 700

Wiltbank, John, 20, -, -, 1000, 40, 225

Hall, Hairston, 20, -, -, 25, 30, 20

Johnson, Benton, 70, 30, -, 200, 30, 300

Jester, James M., 12, 15, -, 400, 20, 160

Short, Alfred, 140, 80, -, 1000, 400, 600

Hemmons, Wm. J., 6, 34, -, 500, 20, 50

Short, William, 60, 40, -, 4000, 200, 325

Reed, Elias B., 91, 14, -, 3000, 200, 70

Shockley, Purnal, 50, 20, -, 1500, 50, 50

Lawson, Jehue, 200, 100, -, 3000, 50, 250

Lawson, Daniel L., 50, -, -, 500, 20, 100

Betts, Edward D., 50, 30, -, 2500, 100, 450

Clendaniel, John H., 115, 35, -, 4000, 100, 500

Welch, George H., 61, 50, -, 1500, 20, 200

Coverdale, Tobias, 80, 100, -, 1500, 50, 400

Shockley, Elias, 140, 60, -, 3000, 140, 550

Clendaniel, Ahab, 55, 55, -, 600, -, -

Clendaniel, John, 50, -, -, 400, 50, 100

Pettyjohn, Dosdine, 40, 20, -, 1000, 50, 500

Smith, W.W., 200, 250, 50, 3000, 300, 500

Clendaniel, Samuel, 175, 100, -, 3500, 100, 500

Smith, Joseph A., 100, 100, -, 2000, 100, 100

Warren, Robt. H., 50, 30, -, 1000, 50, 30

Wilson, Levi H., 120, 50, -, 2500, 75, 300

Betts, Jonth. T., 205, 130, 30, 3000, -, -

Betts, Johnson J., 70, 6, -, 1000, 30, 150

Shepherd, James, 30, 10, -, 800, 25, 60

Warren, Stephen, 98, 25, -, 2000, 25, 200

Paswaters, Boaz, 98, 25, -, 2000, 25, 260

Warren, David S., 30, 30, -, 600, -, 26

Clendaniel Arey, 100, 50, 25, 1700, 100, 500

Clendaniel, James, 50, -, -, 500, 25, 175

Truitt, Nehemiah, 100, 55, 20, 1200, 20, 250

Hellens, Jacob, 200, 115, 25, 1800, 25, 200

Hellens, Alex., 70, -, -, 500, 40, 200

Banning, Thos. C., 55, 20, -, 800, 100, 450

Williams, Jonathan H., 45, 35, -, 400, 25, 150

Vauss, Salley, 35, 50, -, 400, 20, 70

Hemmons, Jas. T., 30, 20, -, 300, 20, 50

Fleetwood, Purnal, 50, 50, -, 1000, 25, 75

Deputy, Jas. B., 160, 180, 60, 4000, 100, 500

Williams, Whittington, 250, 150, -, 10000, 300, 600

Welch, Nathl, 50, -, -, 600, 75, 250

Tucker, David S., 60, -, -, 1000, -, 80

Hudson, Jas. H., 300, 100, 28, 11000, 150, 600

Austin, Henry, 300, 250, 50, 9000, 10, 500

Stuart, Henry W., 135, 121, -, 2500, 125, 350

Banta, John, 125, 15, -, 6000, 300, 1100

Hartwell, Thompson, 90, 15, -, 7500, 250, 350

Macklin, John S., 150, 50, -, 4000, 250, 250

Fisher, Joseph, 150, 150, -, 3000, 40, 30

Jester, Benj. E., 14, 30, -, 1500, -, 200

Macklin, Henry D., 30, 40, -, 1000, -, 100

Milman, Jonathan, 30, -, -, 400, 50, 200

Shew, Putnam, 40, 40, -, 1000, 50, 100

Milby, Peter, 60, 64, -, 1500, 50, 200

Clendaniel, George, 200, 175, -, 5000, 300, 700

Haskins, Joseph W., 100, 30, -, 3000, 50, 150

Murphy, Albert, 60, 80, -, 1200, 20, 5

Arnold, Erastus, 100, 100, -, 2000, 50, 310

Abbott, Jas. P., 70, 60, -, 2800, 50, 75

Betts, Wm. H., 60, 70, -, 2000, 125, 300

Warren, Joseph, 100, 75, -, 1500, 30, 450

Donovan, Peter, 60, 90, -, 1500, 40, 100

Clendaniel, Saml. H., 300, 200, -, 10000, 100, 300

Warren, Isaac R., 50, 60, -, 2000, 50, 200

Tatman, Petty, 40, 50, -, 400, -, 30

Milman, Joseph, 25, 10, -, 400, 150, 200

Burris, Jesse, 30, -, -, 500, 30, 200

Fay, Cyrus M., 50, 25, -, 1000, 50, 130

Causey, John W., 37, 56, -, 5000, 500, 1450

Causey, Peter F. Sr., 30, -, -, 3000, -, 350

Lewis, Hetell, 186, 60, -, 10000, 400, 600

Davis, Mark H., 166, -, 150, 8000, 50, 300

Draper, Lemuel, 150, 40, 80, 7000, 100, 1200

Walls, Purnal, 20, -, -, 1000, 50, 200

Truitt, Alex, 20 -, -, 1000, 25, 150

Davis, Nutter L., 50, -, -, 1000, -, 450

Russel, George, 168, 50, -, 8700, 200, 500

Holland, David H., 175, 24, 200, 10000, -, 350

Ratcliff, W. H., 100, 60, -, 6000, 200, 450

Stephens, George L., 36, -, -, 2000, -, -

Betts, Isaac, 1130, 570, -, 34000, 50, 485

Truitt, Catharine, 400, 170, -, 8500, -, 50

Davidson, Lemuel J., 100, -, -, 4000, 200, 450

Reed, Jos. B., 60, 20, -, 2000, 50, 160

Truitt, Joshua W., 30, 50, -, 1000, 50, 80

Williams, Richard, 50, 25, -, 2000, 50, 100

Truitt, Mary, 100, 30, -, 3000, 100, 320

Abbott, Nehemiah, 40, 40, -, 1000, 100, 300

Derrickson, Nathl., 70, 100, -, 2500, 100, 350

Macklin, John N., 45, -, -, 1000, 25, 80

Macklin, Barkley, 40, -, -, 1000, 75, 150

Sharp (Tharp), Nehemiah, 30, -, -, 800, 30, 150

Gaylord (Gaylow), H. J., 10, -, -, 4000, 50, 270

Dutton, Silas, 100, -, -, 2000, 50, 200

Prettyman, D. H., 50, 70, -, 7500, -, 50

Rister, M. R., 150, -, -, 3000, 100, 100

Hall, Alfred K., 54, 25, -, 4000, -, 75

Houston, James, 200, 100, -, 9000, 400, 300

Warren, Isaac F., 160, 160, -, 6400, 200, 650

Truitt, B. W., 18, -, -, 2000, 250, 50

Townsend, W. A., 20, -, -, 4000, 40, 240

Pullin, Alex., 23,-, -, 3000, 150, 350

McColly, Rustin P., 616, 800, 300, 75820, 265, 1700

Pierce, William, 50, 6, -, 1000, 50, 100

Davis, Thos. J., 567, 250, 633, 18800, 100, 480

Hopkins, Wm. E., 24, 6, -, 6000, 400, 250

Bennum, N. W., 125, 125, -, 7000, -, 300

Fowler, George, 30, -, -, 500, 20, 170

Fowler, Noah B., 59, -, -, 1000, 50, 60

Warren, Wm. Jr., 60, 60, -, 1500, 50, 250

Lynch, David, 20, 17, -, 1000, 50, 225

Chambers, Joshua, 80, -, -, 2000, 200, 250

Shockley, Elias of L., 40, 10, -, 1800, 50, 560

Morris, Bevens (Berens), 45, 28, -, 1000, 25, 150

Jester, Samuel L., 100, -, -, 2000, 50, 200

Clendaniel, John H., 75, 70, -, 2000, 50, 250

Russel, Charles H., 50, -, -, 1000, 50, 300

Webb, John, 100, 30, -, 2000, 125, 675

Johnson, Benj., 100, 30, -, 2000, 60, 250

Johnson, Alexr., 40, -, -, 600, 50, 300

Macklin, George, 50, 45, -, 1200, 50, 100

Dawson, John O., 100, 70, -, 1700 50, 250

Dawson, Isaac J., 50, 120, -, 1700, 50, 50

Warren, Stephen, 75, 25, -, 1700, 100, 400

Lofland, Littleton M., 60, 90, -, 1500, 40, 200

Morris, Isaac P., 70, 70, -, 1000, 40, 400

Jones, Erasmus, 80, 100, -, 1500, 100, 400

Purnell, Thos. C., 30, 20, -, 1500, 50, 150

Abbott, John, 60, 70, -, 1500, 30, 300

Roach, Robert, 100, 40, -, 1800, 50, 250

Carey, Eli B., 35, 30, -, 1200, 40, 150

Roach, Elizabeth, 50, 50, -, 1000, 40, 100

Dodd, Jesse, 50, 100, -, 1500, 50, 325

Donovan, W. H., 125, 125, -, 5000, 100, 400

Jefferson, Jas. P. 57, -, -, 2000, 50, 300

Roach, David K., 90, 100, -, 3000, 150, 300

Roach, Wm. W., 70, 100, -, 1500, 60, 150

Truitt, John, 80, 70, -, 3000, 80, 300

Shockley, Charles, 130, 60, 200, 4000, 100, 500

Messick, Nehemiah J., 70, 20, -, 2000, 80, 550

Cirwithen (Cirvirthen), Allen A., 50, -, 60, 1500, 50, 250

Wilson, Riley C., 50, 25, 25, 2000, 50, 200

Stephenson, Peter, 100, -, -, 1000, 50, 200

Waples, Benj. F., 100, 100, -, 3000, 200, 600

Wilson George, 300, 70, 40, 10000, 300, 800

Conwell, David, 40, 10, -, 1000, 50, 150

Reynolds, Myers, 125, 75, -, 10000, 250, 1000

Abbott, George C., 50, -, 30, 2000, 60, 400

Roach, Theodore, 30, -, -, 600, 40, 200

Warrington, Stephen H., 180, -, -, 2500, 100, 600

Roach, Edward, 100, 115, 50, 2000, 80, 300

Roach, Thomas, 100, 30, 100, 2000, 50, 500

Roach, Thomas, 50, 100, 100, 1500, 50, 300

Wilkins, George, 50, -, -, 1000, 50, 1100

Shockley, Wilson, 100, -, -, 1500, 50, 500

Shockley, Jos. M., 25, -, -, 600, 25, 150

Coffin, David, 40, -, -, 800, 30, 300

Joseph, Jesse C., 50, 20, 100, 2000, 25, 400

Dickerson, James, 70, 30, -, 1500, 50, 450

Draper, Joseph H., 300, 20, 200, 10000, 500, 2000

Jones, John H., 100, -, 200, 2000, 50, 350

Davis, Purnal T., 112, -, -, 3500, 300, 700

Davis, John W., 100, -, 100, 2000, 100, 200

Boyce, John W., 60, -, 50, 2500, 75, 200

Dickerson, Andrew J., 65, -, -, 2500, 60, 200

Young, David, 70, -, 100, 1500, 40, 150

Bennett, Riley W., 750, 150, 2000, 18000, 300, 1800

Bennett, David, 57, 60, -, 3000, 60, 350

Draper, Henry C., 135, 10, 15, 2500, 50, 220

Deputy, Sarah, 70, 45, -, 2000, 50, 250

Draper, George, 125, 75, -, 3000, 250, 700

Draper, Laurence R., 140, 50, 110, 6000, 100, 700

Abbott, Wm. L., 150, 50, -, 4000, 100, 600

Davis, Purnal, 40, 10, -, 1000, 50, 1000

Davis, Joseph M., 70, -, 75, 2500, 200, 500

Scott, James, 70, 20, -, 1500, 50, 150

Roach, Jas. H., 100, 60, -, 2000, 100, 200

Dickerson, Peter, 60, 30, -, 1500, 70, 250

Hudson, Robert, 40, 10, -, 800, 50, 60

Shockley, George W., 37, -, -, 500, 50, 125

Shockley, Jas. M., 30, -, -, 500, 50, 100

Young. Charles, 12, 20, -, 1000, 50, 150

Deputy, Zachariah, 65, 25, 10, 3000, 75, 200

Clendaniel, Joshua, 150, 100, -, 3000, 100, 500

Clendaniel, Isaac, 200, 50, -, 3000, 100, 500

Houston, Clement, 100, 100, -, 7000, 100, 400

Cuykendall, Elias, 80, -, -, 10000, 200, 550

Owens, George A., 100, 50, -, 4000, 100, 500

Johnson, John H., 120, 70, -, 2500, 50, 350

Perdee, John, 75, 100, -, 4000, 50, 400

Lecount, George H., 50, 20, -, 1000, 50, 200

Hammons, Samuel, 90, 30, -, 5000, 100, 250

Walter (Walton), Stuart, 100, 75, -, 10000, 200, 250

Deputy, Jas. H. (rZ), 100, 50, 100, 3000, 100, 250

Webb, Joshua, 200, 100, -, 10000, 200, 650

Houston, Curtis S., 125, 125, -, 6000, 150, 300

Milman, Michael, 75, 75, -, 1000, 50, 200

Truitt, Andrew V., 40, 40, -, 800, 20, 100

Watson, David, 70, 80, -, 1500, 50, 300

Hatfield, W. P., 100, 80, -, 1000, 30, 100

Lofland, Jesse, 85, 25, -, 1000, 400, 100

Warren, Spicer, 40, 38, -, 1500, 40, 150

Stott, Wm. B., 100, 34, 1500, 100, 100

Ingram, John, 100, -, -, 2000, 75, 250

Ingram, John B., 50, 50, -, 1200, 50, 150

Burton, Jacob, 15, -, -, 500, 30, 100

Coulter, Thos. A., 250, 50, -, 5000, 200, 500

Cirvirthen (Cirwithen), Philip, 40, 10, -, 1000, 50, 60

Shockley, Robinson, 40, 10, -, 1000, 50, 200

Davis, Henry, 16, -, -, 500, 30, 200

Vincent, Wm., 10, -, -, 400, 30, 200

Watson, Jacob, 40, -, -, 1000, 40, 200

Young, Nathan, 36, -, -, 1000, 40, 350

Shockley, Salomon, 44, -, -, 500, 40, 150

Wilkins, James, 120, -, 18, 3500, 300, 800

Jefferson Wm. P., 120, 80, -, 2500, 200, 500

Hazzard, Robt., 50, -, -, 2000, 50, 175

Holstein, Jas. J., 50, -, -, 1000, 40, 75

Warren, Elihu, B., 60, -, 140, 3000, 200, 300

Williams, Saml. J., 40, -, -, 1000, 100, 250

Warren, Bennett, 100, 10, -, 3000, 150, 300

Shepherd, Joseph B., 100, 100, 36, 3000, 150, 300

Davis, John S., 100, 10, 100, 3000, 100, 400

Clendaniel, W. H., 60, -, -, 1200, 60, 300

Benston, James, 60, 30, -, 1500, 50, 100

Fountain, W. H., 75, 50, -, 2000, 75, 300

Milman, David H., 50, 20, 50, 2000, 100, 350

Argo, John, 100, 50, 50, 2500, 150, 450

Argo, Joseph, 80, -, 50, 2500, 200, 600

Bennett, James, 80, -, 60, 3000, 100, 350

Bennett, John W., 75, 25, -, 2500, 80, 300

Davis, Lot W., 70, 60, -, 3000, 150, 900

Bennett, Henry, 65, 50, -, 2000, 50, 200

Cirwithen, Isaac, 120, 50, 40, 2500, 250, 650

Warrington, Robt., 40, -, -, 1000, 40, 150

Darvu (Darvee), Wm. Jones (James), 24, -, -, 800, 40, 400

Beidienar, Wm., 79, 40, -, 3000, 100, 400

Shockley, Wm. V., 55, 55, -, 2000, 50, 350

Wilson, Thos. B., 100, 17, -, 1800, 30, 225

Conwell, Robt. R., 50, 50, -, 1000, 25, 175

Calhoon, Joseph A., 80, 20, -, 2500, 75, 300

Calhoon, Wm. F., 60, -, -, 1000, 25, 130

Calhoon, Peter, 140, 130, -, 6000, 100, 500

Ingram, Anthony, 150, 150, -, 8000, 100, 425

Young, Joseph, 10, -, -, 300, 20, 125

Warren, Francis A., 60, 40, -, 2000, 25, 150

Hazel, Robert, 60, -, -, 1000, 30, 180

Higman, James, 100, -, -, 1500, 120, 425

Watson, Henry S., 70, 15, 45, 4000, 75, 200

Lindall, David, 50, 10, -, 1000, 30, 70

Jefferson, Saml. B., 100, 10, -, 1600, 50, 300

Donovan, Ann R., 80, 50, -, 2000, 60, 250

Swain, George, 100, 100, -, 2500, 80, 2500

Bebee, Ezra, 60, -, -, 3000, 50, 325

Stayton, Andrew J., 40, 60, -, 1000, 30, 180

Clifton, Pemberton, 60, -, -, 2000, 80, 800

Tucker, Benjamin, 100, 40, -, 1000, 25, 200

Clifton, Jane W., 40, 50, -, 1000, 180, 300

Wroton, Jane, 25, -, -, 500, 30, 200

Johnson, Alexr., 60, -, -, 800, 30, 225

Clifton, Asa, 35, -, -, 1000, 100, 375

Clifton, John W., 1000, 400, -, 40000, 50, 350

Hayes, John, 100, 40, -, 2000, 60, 280

Hare, Alexr., 90, 60, -, 5000, 75, 450

Titus, Platt, 110, 65, -, 8500, 150, 400

Campbell, John F., 125, 40, -, 10000, 200, 350

Gilman, Robt. H., 60, -, -, 6000, 200, 250

Turner, John, 80, 80, -, 1000, 40, 100

Wilkinson, John, 50, 20, -, 2000, 75, 70

Underhill Robt. B., 130, 55, -, 5000, 300, 250

Fenn, George J., 80, -, -, 6500, 50, 325

Stayton, David B., 11, -, -, 2000, 100, 350

Coshell(Voshell), Thos., 200, 200, -, 4000, 100, 250

Webb, Emery C., 100, 200, -, 3000, 50, 180

Owens, George, 30, -, -, 1000, 25, 140

Humphrey, Thos., 60, -, -, 3000, 40, 250

Wiswell, George G., 60, 15, -, 2500, 100, 500

Simpler, Samuel M., 100, 50, -, 3000, 75, 250

Reed, Jas. of J., 85, 25, 17, 5000, 75, 500

Delovan, Orland, 60, -, -, 4000, 125, 450

Clark, Allen, 100, 20, -, 6000, 100, 560

Donovan, Thos. W., 60, 30, -, 2500, 100, 400

Watson, Wm. P., 60, 40, -, 2500, 75, 150

Warren, David C., 40, 10, -, 2000, 50, 150

Wapler, Richard, 80, -, -, 2000, 40, 200

Potter, John W., 100, 50, -, 3000, 75, 275

Fountain, Solomon, 50, -, -, 1000, 40, 200

David, Robt. H., 575, 75, 200, 20000, 225, 1000

Pierce, Wm., 120, 15, 30, 5000, 200, 600

Townsend, Wm .H., 100, 50, 50, 5000, 100, 950

Fitzgerald, Ezekiel, 50, -, -, 1000, 50, 600

Hyman, Elijah, 80, -, -, 2000, 100, 200

Spencer, Henry B., 100, 30, -, 5000, 100, 500

Draper, Samuel, 40, -, -, 1000, 40, 175

Clendaniel, Wm. H., 12, -, -, 1000, 40, 200

Daniel, Moulton, 85, 15, -, 2000, 50, 300

Deputy, Jas. G., 75, 100, -, 3000, 50, 170

Potter, Benj. E., 100, 32, 179, 4500, 150, 600

Hill, John, 8, -, -, 100, 20, 100

Mills, David, 100, 100, -, 3000, 100, 550

Greenly, Elisha, 80, -, 100, 2000, 80, 425

Burton, Daniel, 40, -, -, 1000, 30, 100

Roberts, Robt. D., 15, -, -, 2000, 30, 100

Hill, Joseph, 50, -, -, 1000, 25, 100

Ennis, Stephen M., 80, 20, -, 2500, 75, 250

Boyce, Jas. H., 60, -, -, 1500, 50, 275

Spencer, Joseph, 100, -, 100, 4000, 125, 400

Evans, James, 50, -, -, 1000, 40, 200

Pierce, Henry J., 100, 23, 10, 5000, 200, 600

Cannon, Jacob, 140, 27, -, 4000, 40, 2810

McCally, Stephen, 120, 100, -, 3000, 100, 600

Boyce, Samuel C., 40, -, -, 1000, 60, 150

Watson, Wm., 150, 70, -, 6000, 150, 600

Vankirk, Wm., 100, 50, -, 2500, 50, 250

Watson, Thos., 50, -, -, 1000, 50, 160

Prettyman, Elizabeth, 100, 20, 50, 2000, 75, 400

Parks, Wm. H., 57, -, -, 1500, 50, 350

Wells, Joshua B. K., 93, -, -, 6000, 100, 400

Grosurvisch, Edgar, 38, -, -, 2500, 50, 100

Sharp, Roland P., 60, -, -, 2000, 60, 275

Sharp, Wm. H., 60, -, -, 2000, 60, 300

Tease, John, 60, -, 40, 2000, 30, 160

Ingram, Thos. H., 50, -, -, 1000, 30, 175

Sackett, George, 60, 20, -, 4000, 50, 250

Shockley, Lemuel, 60, 30, -, 1500, 25, 175

Shockley, Elias, 15, -, -, 500, 25, 100

Morgan (Moyer), Elizabeth, 100, -, -, 1500, 40, 150

Vreeland, Henry, 160, 20, -, 7000, 300, 450

Shockley, Wm., 110, 20, -, 6000, 150, 300

Titus, Joseph, 150, 50, -, 4000, 150, 700

Titus, Joseph V., 100, 30, -, 4000, 150, 900

Daniel, Elias, M., 112, 8, -, 4000, 50, 280

Carpenter, Burton, 120, -, -, 5000, 80, 350

Wood, Stephen J., 60, -, -, 12000, 200, 400

Allendorf, Lewis, 20, -, -, 2000, 30, 300

Wilkins, William, 140, -, -, 3000, 80, 400

Hill, William 100, -, -, 1500, 30, 250

Daniel, W. B., 75, 20, -, 2000, 50, 200

Benston, Wm., 70, -, -, 1500, 30, 300

Mason, Joseph, 100, 20, -, 3000, 100, 400

Benston, John, 1120, 100, -, 3000, 50, 200

Fitzgerald, George, 80, -, -, 2000, 75, 250

Simpson, Isaac, 200, 100, -, 8000, 200, 400

Solomon, Sarah, 100, 10, -, 1600, 100, 202

Mills, Miles T., 150, 50, -, 12000, 150, 850

Collins, Samuel, 135, 10, -, 1000, 200, 450

Miller, Wm., 100, -, -, 1500, 50, 175

Walls, W. B., 100, 30, -, 3000, 100, 500

Smith, Wm. H., 90, 34, -, 1600, 80, 400

Lofland, Parker, 150, 40, -, 10000, 100, 450

Deputy, Solomon, 200, 400, -, 10000, 150, 700

Deputy, Henry, 300, 700, -, 10000, 150, 1000

Webb, A. M., 60, -, -, 1000, 100, 150

Shockley, Charles, 40, 20, -, 500, 20, 100

Timmons, Wesly, 100, -, -, 3000, 50, 600

Hudson, Levin of C., 40, 80, -, 3000, -, 200

Parsons, Lemuel, 80, 70 -, 3000, -, 100

Houston, John M., 50, 260, -, 7500, -, -

Truitt, George, 34, 130, -, 2000, 25, 200

Tingle, Solomon, 50, 50, -, 3000, 25, 500

Wagner, Benjamin, 60, 50, -, 3000, 100, 340

Hudson, Isaiah, 34, 100, -, 3000, -, 200

Hudson, Alfred, 100, 100, -, 3000, 25, 300

Moore, Shepard, 50, 100, -, 2000, -, 150

Dukes, Hiram, 20, 50, -, 700, -, 90

Hudson, Seth, 50, 14, -, 1000, -, 150

Hudson, Clayton M., 160, 100, -, 4000, 100, 700

Wolfer, George F., 40, 160, -, 3000, -, 200

Hazzard, John C., 40, 110, -, 1500, -, 250

Steen, Hazard, 15, 60, -, 700, -, -

McCabe, Isaac, 40, 25, -, 2000, -, 450

Davidson, George, 15, 15, -, 600, -, 100

Davidson, Joseph, 30, 30, -, 500, -, -

Brasure, Perry, 40, 60, -, 700, -, 160

Joins, Robert, 20, 80, -, 530, -, 100

Tolbert (Talbert), Joseph, 20, 65, -, 425, -, 30

Ander, James, 25, 25, -, 500, -, 200

Hudson, Burton, 50, -, -, 2000, -, 40

McCray, Joseph, 40, 40, -, 1600, -, 100

McCray, Lemuel, 60, 40, -, 2500, -, 50

Cannon, E. W. & G. W., 180, 375, -, 12000, 250, 1430

Rickards, James, 40, 30, -, 1000, -, 80

Tooney, William, 40, 260, -, 3000, -, 200

Tompson, Isaac J., 17, -, -, 200, -, 175

Barton, George H., 50, 30, -, 640, -, 75

Dukes, James, 60, 70, -, 600, -, 170

Marvil, Alfred B., 100, 100, -, 1200, -, 250

Rogers, Jacob, 60, 40, -, 1000, -, 80

Burton, John, 50, 50, -, 1500, -, 250

Tingle, John C., 20, -, -, 400, -, 120

Ellis, William R., 80, 70, -, 4500, -, 200

Burton, Eli, 100, 100, -, 5000, 100, 600

Marner, William, 50, 100, -, 2000, -, 50

Moore, Isaac C., 76, 76, -, 3000, 75, 200

Derickson, Joshua, 60, 80, -, 4000, 125, 500

Williams, Isaac, 30, 20, -, 600, -, -

Derickson, Joshua, 50, 80, -, 1200, -, -

Derickson, Joshua, 40, 150, -, 2000, -, -

Short, Wingate, 20, -, -, 500, -, 30

Derickson, James, 100, 140, -, 4000, 50, 200

Morris, William L., 14, 26, -, 400, -, -

Helm, Peter M. B., 125, 175, -, 1500, 50, 310

Christopher, Armwell, 35, 45, -, 640, -, 76

Parkhurst, William J., 15, -, -, 1000, -, 100

Steen, William H., 20, -, -, 1000, -, 100

Hobkins, David R., 40, 75, -, 1000, -, 20

Truitt, Henry H., 20, 90, -, 1000, -, -

Timmons, John, 40, 17, -, 450, -, 150

Parkhurst, Daniel, 17, -, -, 800, -, 200

Short, Elias, 180, 150, -, 3000, 40, 350

Short, Elias, 25, 42, -, 1000, -, -

Godwin, Jacob R., 6, -, -, 1150, -, 175

Wise, Mitchel D., 12, -, -, 500, 25, 200

Hickman, Jepthah, 60, 60, -, 1100, -, 50

Dukes, John H., 50, 180, -, 3000, -, 200

Savage, Wesly W., 54, 75, -, 2000, -, 150

Timmons, William, 25, 25, -, 400, -, 100

Simpler, Merian L., 100, 20, -, 1000, -, 250

Wingate, Connor, 75, 75, -, 1500, 50, 300

Wingate, Hezakiah, 80, 80, -, 3000, -, -

Steen, James, 25, 35, -, 480, -, 100

Timmons, Henry, 25, 62, -, 500, -, 75

Simpler, Andrew, 40, 20, -, 600, -, -

Ingram, Peter, 40, 200, -, 1650, -, -

Johnson, Tilghman, 20, -, -, 1000, -, 300

Elliott, John, 55, 125, -, 1500, -, -

Burton, John H., 45, 150, -, 7000, -, 75

Morris, Joseph, 120, 30, -, 5000, 100, 380

McCray, Spencer, 50, 64, -, 1000, -, -

Moore, James, 95, -, -, 3000, -, -

Adams, Daniel, 25, 75, -, 800, -, 100

Dukes, Littleton, 18, 25, -, 400, -, 150

Moore, John, 8, 9, -, 500, -, 140

Fosque, William, 25, 9, -, 500, -, 10

Gray, Alfred L., 40, 50, -, 1500, 60, 250

Vickers, Nathaniel, 40, 60, -, 1000, -, 100

Brasure, George, 33, -, -, 200, -, 25

Bunting, John, 140, 176, -, 5000, 50, 250

Bunting, Marshal, 60, 40, -, 2700, -, 20

Hudson, Levin, 75, 75, -, 3000, 25, 125

Carey, Jonathan, 72, 120, -, 5000, 150, 740

Reed, John T., 6, -, -, 2000, -, 250

Gum, Menain, 50, 7, -, 20000, 200, 450

Lockwood, Burton W., 50, 50, -, 2000, -, 110

Long, Alfred, 50, 50, -, 1500, -, -

Hudson, Robert G., 50, 50, -, 3000, -, 300

Hudson, William B., 85, -, -, 3000, 50, 200

Long, Eber, 30, 83, -, 2000, 40, 200

Long, Elisha, 50, 40, 200, 2000, 100, 400

William, Joshua, 25, 5, -, 500, -, 130

Floyd, William, 60, 25, -, 4000, 100, 450

Timmons, William, 100, 100, 35, 6000, -, 200

Quillin, William, 50, 46, -, 1500, -, -

Timmons, Aaron, 100, 40, -, 2000, -, -

Hosier, Riley, 10, -, 100, 900, -, -

Brasure, John, 120, 100, -, 4000, -, 400

Justis, Isaac, 50, 50, -, 1500, -, 75

Long, George M., 64, 60, -, 2000, -, 200

Atkins, Elijah, 50, 50, -, 1500, -, 100

Daisey, William, 30, 25, -, 1000, -, 450

Davidson, William, 70, 100, -, 4000, -, 300

Long, William, 60, 500, -, 4000, -, 250

Hudson, Eber L., 50, 50, -, 2000, -, 150

Davidson, Elijah, 20, 15, -, 1000, -, 125

Muniford, John, 50, 200, 100, 5000, -, 600

Mumford, Charles, 40, 60, -, 300, -, 200

Hudson, Robert, 37, 13, -, 2000, -, 300

Moore, William, 20, 80, -, 1800, -, 100

Moore, John J., 50, 50, -, 2500, -, 250

Wilkinson, James, 56, 60, -, 3000, -, 200

Carey, Peter, 60, -, -, 2000, 100, 450

Littleton, Henry, 15, 220, -, 3000, -, -

Houston, John M., 120, 130, -, 6500, 300, 1200

Atkins, Joshua L., 20, 10, -, 500, -, 60

Parsons, George, 90, 138, -, 2500, 100, 200

Atkins, Willard, 80, 127, -, 1500, -, 100

James, Burton, 52, 53, -, 1000, -, 30

West, John C., 40, 70, -, 1100, -, 120

Short, John, 40, 50, -, 1200, -, 200

Cormine (Carmine), John, 60, 200, -, 3000, 75, 300

Cormine, Henry T., 30, 20, -, 500, -, 150

Hood, James, 30, 40, -, 800, -, 160

Tompson, Milla, 20, 50, -, 700, -, -

Walls, Edward, 10, 180, -, 1000, -, -

White, Lyddie, 25, 70, -, 2000, -, -,

Hanes, Francis, 20, 50, -, 1500, -, 30

Derickson, John F., 22, 40, -, 2000, 150, 600

Long, Benjamin, 10, -, -, 600, -, 100

Toomey, Thomas R., 60, 170, 150, 3000, -, 100

Mitchel, Lemuel, 60, 67, -, 3000, 100, 200

Philops, John H., 50, 50, -, 2000, -, 500

Mitchel, Henry B., 25, 90, -, 2000, -, 400

Mitchel, Robert, 35, 100, -, 2000, -, 250

Hastings, William B., 50, 50, -, 2000, -, 150

Lewis, Joseph, 75, 25, -, 2500, 200, 742

Lewis, Charles R., 35, 45, -, 1600, -, 150

Bowden, James, 10, 300, 90, 2000, -, 250

Short, Reuben J., 36, 240, -, 2500, -, 200

Mitchel, James, 24, 200, -, 2500, -, 150

Baker, Levin, 50, 50, -, 1500, -, 150

Mitchel, Samuel, 36, 100, -, 1500, -, 70

Bowden, James, 36, -, 100, -, 1500, -, -

Smith, Sampson, 20, 80, -, 1500, -, 100

McCalister, William, 75, 100, -, 2000, -, 300

Moore, Henry H., 80, 20, -, 2000, -, 400

Dunaway, William, 30, 20, -, 1000, -, 150

Ennis, John L., 100, -, -, 1500, -, 300

Wills, William, 60, 40, -, 3000, -, 450

Baker, Minos, 48, -, -, 1000, -, 250

Philops, William, 100, 150, -, 3500, -, 200

Baker, Abisha, 75, 20, -, 2500, 15, 150

George, Lewis W., 40, -, 10, 500, -, 130

Brittingham, Isaac, 60, 12, -, 2000, -, 300

Dunaway, Elhovser B., 6, 150, -, 600, -, -

Dunaway, Thomas, 70, 30, -, 1500, 50, 200

Brimer, Caleb, 80, 20, -, 2000, 100, 300

Baker, Joseph E., 40, 70, -, 2000, -, 350

Hudson, Joshua, 100, 125, -, 4000, 100, 400

Chanler, William, 75, 200, -, 2700, -, 300

Johnson, Miles, 10, 115, -, 1250, -, -

Carey, Elijah W., 130, 150, -, 2800, 100,700

Crampfield, Isaac, 32, 168, -, 4000, -, 150

Short, William H., 70, 56, -, 3000, -, 120

Carey, Robert, 50, 30, -, 3600, -, 300

Shockley, Peter D., 50, 100, -, 1700 -, 400

Shockley, Thomas R., 15, 10, -, 300, -, 120

West, Eli, 60, 20, -, 800, -, 120

Steen, Ephraim, 75, 75, -, 1500, -, 150

Otwell, Eliza, 30, -, -, 1000, -, 150

Otwell, John C., 75, 25, -, 2000, -, 400

Parmer, Isaac J., 100, 200, -, 4500, -, 600

Truitt, Hamilton, 150, 100, -, 2000, 75, 500

Philips, Burton, 150, 87, -, 2000, 100, 650

Philops, John B., 100, 30, -, 1300, -, 150

Heron, William, 46, -, -, 800, -, 50

Matthews, Josiah, 100, 100, -, 1500, -, 160

Messick, Nathan, 35, 75, -, 1000, 25, 200

Allen, Levin, 50, 72, -, 800, -, -

Bailey, John, 60, 17, -, 700, - 160

Philips, Elihu, 80, 61, -, 1400, 50, 700

Philips, Joseph, 150, 50, -, 2500, 70, 660

Conaway, Miles, 40, 20, -, 700, -, 60

Philops, Harriet, 140, 10, -, 1500, -, 25

Short, Philip, 100, 60, -, 1000, 50, 450

Mears, Robinson, 50, 100, -, 1500, -, 60

Laws, Saxey, 50, 50, -, 1000, -, 50

Prettyman, Robert, 100, 70, -, 2000, 100, 500

Johnson, Purnel, 50, 100, -, 700, -, -

Short, William, 23, 10, -, 300, -, -

Short, Mevain, 75, 70, -, 3000, 50, 400

Hitchens, Dalbey, 50, 30, -, 800, -, -

Short, Stephen, 100, 200, -, 2000, 100, 480

Mathews, Wingate T., 70, 28, -, 1200, -, 125

Short, Eli S., 50, 30,-, 800, 25, 200

Brion, Joshua B., 100, 116, -, 2000, 50, 500

Smith, Francis, 80, 40, -, 1500, -, 160

Short, Thomas W., 103, 50, -, 2000, 100, 400

Littleton, Minos, 35, 15, -, 500, -, -

Short, John W., 74, 75, -, 3000, 100, 500

Gunby, John, 100, 100, -, 2000, -, 200

Derickson, Thomas, 50, 50, -, 2000, 40, 250

Steen, James, 40, 20, -, 1000, -, 160

Short, Peter, 100, 50, -, 2000, 100, 650

Short, Leonard, 60, 50, -, 800, -, 200

Rogers, Minos, 75, 30, -, 1500, 100, 300

Scott, Curtis W., 60, -, -, 1200, -, 150

Prettyman, John C., 40, -, -, 800, -, 25

Jones, William, 18, 12, -, 500, -, 20

Fooks, Cyrus Q., 75, 75, -, 1800, -, 200

Mathews, David, 75, 75, -, 1800, -, 450

Scott, George, 75, 25, -, 1500, 100, 600

Prettyman, Cornelius, 50, 20, -, 1000, 50, 100

Bulcher, Noah, 13, 10, -, 200, -, 70

Rogers, John W. Sr., 60, 100, -, 3000, -, 500

Rogers, John W. Jr., 80, 75, -, 3000, -, 200

Rogers, Curtis, 100, 40, -, 2000, -, 300

Short, Mattford, 70, 70, -, 1200, -, 100

Rogers, James, 40, 40, -, 700, -, -

Ennis, John P., 25, 17, -, 800, -, 150

Jones, Zachariah, 50, 50, -, 1000, -, 50

Johnson, Asa, 100, 150, -, 1500, -, 250

Short, William O., 200, 204, -, 8000, 200, 1500

Elliott, Wingat, 80, 46, -, 1700, -, 312

Rogers, Daniel W., 50, 10, -, 1000, -, 70

Holloway, William, 200, 200, -, 4000, 50, 250

Carpenter, Noah J., 20, 30, -, 250, -, 150

Derickson, Elizabeth, 40, 200, -, 1000, -, -

Steen, Thomas, 60, 30, -, 400, -, 80

Thoroughgood, William N., 60, 80, -, 1000, 40, 200

Short, Henry W., 40, 55, -, 700, -, 90

Wainwright, Isaac, 30, 100, -, 700, -, -

Hitchens, Menain, 50, 50, -, 800, -, 50

Marvel, William, 100, 130, -, 2000, -, 130

Philips, John W., 75, 50, -, 2500, 100, 950

Betts, Joseph, 20, 4, -, 600, -, 60

Jones, William P., 30, 35, -, 1800, 50, 400

Moore, John, 150, 83, -, 2300, -, 300

Waples, Anna, 10, 13, -, 500, -, 25

Johnson, Joseph, 30, 20, -, 500, -, 100

Heron, George W. C., 200, 400, -, 5000, 150, 1000

Morris, Maranda, 7, -, -, 600, -, 60

Burton, Isaac, 40, -, -, 500, -, 250

Prettyman, William, 25, 24, -, 400, -, -

Morris, Benjamin, 75, 375, -, 4000, -, 30

Joseph, Elihu, 75, 75, -, 1000, 30, 145

Morris, Joshua S., 100, 75, -, 3000, 100, 450

Simpler, Isaac, 4, -, -, 500, -, 160

Wilson, Elsey, 150, 20, -, 700, -, 40

Morris, Joseph, 24, 75, -, 1500, 30, 175

Warren, Jacob S., 50, 80, -, 1000, -, 100

Pepper, Joshua W., 80, 80, -, 4000, 120, 600

Morris, Mary, 5, 75, -, 400, -, -

Lawson, George, 50, 50, -, 1000, 10, 100

Ennis, Adam, 40, 12, -, 800, -, 200

Jefferson, William, 100, 95, -, 3000, 40, 400

Jefferson, Thomas H., 60, 40, -, 1800, -, 100

Rogers, Philip, 75, 25, -, 2000, 50, 600

Rogers, Harrison, 16, -, -, 400, 50, 400

Rogers, John T., 80, -, -, 500, -, 100

Rogers, George M., 20, 6, -, 500, 20, 120

Rogers, Joel N., 50, 24, -, 800, -, 800

Ennis, Aaron B., 30, 30, -, 1000, -, 250

Simpler, Henry, 8, 7, -, 200, -, -

Rogers, George, 30, -, -, 500, 20, 200

Prettyman, Robert, 70, 200, -, 2500, 50, 350

Marvel, Joseph, 150, 50, -, 2000, 60, 520

Johnston, Ranford, 90, 160, -, 6000, 100, 238

Marvel, John P., 50, 105, -, 1000, -, 100

Johnson, Minos, 40, 90, -, 800, -, 70

Prettyman, John W., 70, 170, -, 2000, -, 150

Derickson, Alfred, 50, 100, -, 1500, -, 125

Spiser, William E., 50, 40, -, 900, -, 400

Davidson, James, 60, 140, -, 2500, -, 180

Mcgee, George, 100, 100, -, 2000, -, 40

Marvel, George W., 40, -, -, 500, -, -

Marvel, Nutter, 100, 116, -, 3000, 60, 380

Roach, Benjamin, 50, 22, -, 2000, 50, 200

Waples, Cornelious, 100, 100, -, 3000, -, 250

Rogers, John M., 30, 50, -, 1500, -, 45

Melson, William, 50, 50, -, 1500, 10, 150

Johnson, William S., 40, 60, -, 1000, 40, 370

Johnson, Thomas W., 60, 60, -, 1200, 50, 250

Holstein, Elisha, 50, 300, -, 3000, -, 20

Atkins, Elijah, 4, -, -, 400, -, 20

Wilson, William, 50, 20, -, 1800, -, 150

Wilson, Jonathan, 50, 25, -, 500, -, 50

Tingle, Henry, 4, 100, -, 500, -, -

Philops, John of J., 100, 129, -, 1200, -, 100

Webb, James R., 50, 400, -, 2000, -, 50

Evans, James R., 57, 75, -, 800, -, 60

Greenly, John P., 100, 130, -, 2000, -, 100

Philops, Joshua, 125, 75, -, 5000, 100, 350

Philops, Nathaniel, 100, 55, -, 3000, 100, 550

Houston, Robert B., 175, 225, -, 8000, 200, 1200

William, James, 100, 130, -, 5000, 100, 700

Lewis, Hillery, 50, 40, -, 900, -, -

Atkins, Noah, 60, 30, -, 1000, -, 70

Brasure, William, 25, 75, -, 1500, -, -

Hudson, George, 25, 25, -, 1000, -, 125

Floid, John W., 20, 40, -, 800, -, -

Timmons, George, 60, 20, -, 800, -, 125

Thompson, Wollis, 150, 50, -, 5000, 100, 840

Halleck, Robert, 40, 100, -, 3500, -, 175

Halleck, Shepard, 60, 100, -, 4000, -, 100

Halleck, Joseph, 150, 150, -, 7500, 200, 1500

Sudler, Joseph B., 40, 80, -, 1200, -, 300

Day, Levin B., 110, 40, -, 4000, 150, 350

Donohoe, Peter, 110, 50, -, 1500, 15, 350

Rogers, Daniel, 70, 60, -, 1000, 20, 250

Conaway, Dixon, 16, 2, -, 600, 20, 150

Pepper, Luther, 85, 85, -, 2000, 25, 150

Martin, James, 100, 95, -, 2000, 40, 350

Pepper, Asbury C., 150, 72, -, 4000, 300, 1550

Prettyman, Jacob, 150, 40, -, 2700, 50, 200

Legatts, Cyrus, 100, 80, -, 2700, 50, 560

Morris, William, 20, 25, -, 1000, 20, 50

Smith, Prettyman, 80, 26, -, 2000, 50, 225

Rogers, Wingat, 80, 26, -, 2000, 50, 200

Pepper, David M., 120, 80, -, 2500, 40, 450

Pepper, Charles T., 120, 90, -, 3000, 50,830

Wilson, Edward R., 40, 30, -, 1600, 30, 100

Hudson, Elizabeth, 40, 20, -, 1500, 25, 200

Messick, Daniel, 30, -, -, 370, 10, 100

Smith, Mitchel, 50, 20, -, 4000, 50, 353

Lynch, Greensbury, 50, 14, -, 5000, 40, 435

Pepper, Greensbury, 70, 45, -, 1500, 20, 350

Warren, John, 20, 20, -, 500, 20, 75

Dickerson, Russel, 80, 25, -, 6000, 40, 256

Dickerson, Charles, 30, 50, -, 6000, 15, 225

Wilson, Major W., 50, 40, -, 3600, 40, 150

Marvel, Dagewthey, 90, 60, -, 3000, 30, 150

Marvel, Maria, 40, -, -, 800, 20, 75

Marvel, Thomas R., 60, 40, -, 2000, 40, 200

Lindal, John, 40, 40, -, 1000, 50, 250

Dickerson, Jonathan, 20, 40, -, 400, 20, 75

Walls, George, 100, 60, -, 1920, 75, 200

Dickerson, William, 10, 20, -, 700, 10, 75

Hudson, Ann, 40, 30, -, 800, 40, 285

Lam, William, 111, -, -, 1650, 300, 830

Day, John R., 45, 10, -, 1200, 30, 200

West, Joseph, 50, 30, -, 1200, 20, 150

Conaway, Thomas, 50, 30, -, 1500, 40, 225

West, Stockley, 60, 75, -, 1460, 40, 200

Swain, Theophilus, 50, 7, -, 2000, 10, 200

Messick, John D., 90, 30, -, 2000, 26, 300

Hurt, Peter, 45, 32, -, 2945, 20, 125

Dickerson, George, 100, 50, -, 2000, 40, 200

Reed, Alexander, 70, 25, -, 1500, 25, 230

King, John, 45, 320, -, 3650, 60, 260

Lindal, Thomas, 45, 20, -, 1000, 30, 150

Cooper, Theodore, 24, 20, -, 580, 20, 216

Lecate, Job, 50, 100, -, 1500, 10, 250

Donovan, Wingat, 25, 25, 25, 700, -, -

Salmon, Wingat, 20, 30, -, 500, 5, 150

Torbert, Warren, 30, 6, -, 600, 40, 148

Joseph, Peter P., 60, 20, -, 4000, 45, 260

Wilson, Peter J., 40, 10, -, 1000, 10, 100

Pettyjohn, Pinkey, 50, 70, -, 1200, 25, 50

Wilson, Hiram, 60, 14, -, 900, 50, 202

Sharp, Josiah, G., 45, 115, -, 2000, 40, 245

Wilson, Edward R., 25, 105, -, 1300, 30, 150

Wilson, George, 60, 22, -, 1000, 30, 310

Vickers, Joseph K., 50, 40, -, 1800, 25, 212

Pepper, Joshua, 35,10, -, 1200, -, -

Wilson, George F., 80, 70, -, 1800, 50, 290

Wilson, Thomas, 30, 40, -, 840, 20, 125

Wilson, Nathaniel, 70, 80, -, 1620, 90, 235

Messick, John P., 60, 30, -, 1000, 10, 175

Pepper, Henry M., 180, 30, -, 5000, 150, 528

Morris, Wingate, 100, 30, -, 3000, 100, 300

Calhoon, George W., 9, 31, -, 350, 30,75

Agens, Ella, 66, 66, -, 1000, -, -

Morris, George F., 14, -, -, 500, 10, 200

Handcock, William, 60, 44, -, 2080, 50, 400

Martin, William, 30, 9, -, 300, 30, 45

Stuart, David A., 25, 29, -, 1080, 45, 180

Atkins, Peter E., 100, 105, -, 2600, 5, 75

Bennum, Henly, 40, 65, -, 3000, 25, 220

Dodd, Peter P., 75, 130, -, 2150, 200, 400

Baker, Joseph, 100, 60, -, 2410, 125, 400

Rust, Peter, 70,75, -, 2250, 40, 350

Atkins, Kindal, 50, 100, -, 1250, 75, 400

Burres, Joseph T., 70, 55, -, 2000, 25, 300

Rust, Absalom, 10, 20, -, 2500, 20, 300

Mcolley, Edward, 100, 75, -, 3000, 50, 200

Collins, Josiah A., 50, 50, -, 1000, 30, 100

Collins, M. B., 40, 20, -, 600, 20, 200

King, Robert, 50, 50, -, 2000, 25, 150

Salmons, David, 50, 25, -, 1500, 15, 125

Scott, Thomas, 125, -, -, 3500, 50, 410

McColley, James, 40, -, -, 800, 25, 100

Russel, Isaac, 20, 5, -, 500, 50, 175

Smith, Lemuel, 22, -, -, 300, 40, 150

Steel, James, 100, 50, -, 3000, 10, 250

Wilson, Theodore, 15, 55, -, 1000, 30, 120

Macklin, Byard, 25, 35, -, 770, 60, 300

Hart, Peter J., 8, 7, -, 300, 20, 35

Reynolds, David, 40, 50, -, 3000, 50, 200

McConaughey, Jonathan, 25, 371, 50, 8000, 120, 1100

Short, Isaac, 55, 50, -, 1600, 25, 150

Sharp, John, 100, 131, -, 2350, 40, 190

West, Huett W., 60, 20, -, 700, 20, 50

Evans, James A., 55, 45, -, 5000, 115, 200

Davis, James A., 80, 40, -, 1200, 20, 200

Sharp, John W., 100, 50, -, 2000, 30, 230

Carey, Albert, 70, 30, -, 1000, 50, 241

Wilkins, James B., 30, 15, -, 2500, 80, 340

Chase, Mary R., 40, 10, -, 1200, 18, 44

Dickerson, James of L., 150, 15, -, 1950, 30, 350

Megee (Mcgee), Levin J., 40, 30, -, 800, 30, 200

Calhoon, James K., 50, 57, -, 950, 10, 150

Pettyjohn, William, 100, 155, -, 2500, 80, 528

Britenham, Smith, 40, 78, -, 90, 20, 110

Warren, Robert, 45, 55, -, 1000, 20, 140

Workman, John, 60, 15, -, 1200, 50, 300

Spicer, William, 50, 38, -, 1000, 30, 200

Calhoon, Thomas L., 50, 213, -, 4200, 50, 150

Pettyjohn, George W., 85, 10, 5, 900, 30, 235

Greenley, Robert, 90, 95, -, 1851, 30, 150

Sharp, Asa, 60, 60, -, 500, 10, 100

Wilkins, Joseph, 45, 100, -, 1450, 20, 160

Sharp, Kincey, 75, 180, -, 2500, 125, 395

King, John, 25, 100, 25, 1500, 20, 150

Reed, Job, 50, 40, -, 800, 10, 150

Dutten, Lewis P., 50, 75, -, 1200, 70, 450

Donovan, Mary, 50, 90, -, 1200, 20, 190

Palmer, Gremsley (Greensley), 50, 40, -, 1600, 15, 150

Donovan, Zachariah, 60, 100, -, 1080, 10, 160

Donovan, Riley, 100, 100, -, 1000, 25, 200

Donovan, William, 100, 150, -, 5000, 20, 140

Donovan, George, 45, 25, -, 700, 20, 130

Donovan, Ruben, 75, 55, -, 2000, 20, 238

Davis, Solomon, 60, 80, -, 1440, 50, 450

Middlton, Henry, 40, 55, -, 1410, 70, 55

Robins, Joseph, 80, 50, -, 1500, 20, 195

Lynch, Joshua A., 80, 50, -, 1500, 50, 300

Burton, Benjamin D., 50, 57, -, 1500, 100, 281

Pettyjohn, Robert, 40, 62, -, 420, 10, 160

Brown, Robert, 40, 100, -, 1400, 20, 300

Brown, Robert, 40, 62, -, 420, 10, 160

Donovan, John, 75, 77, -, 1200, 35, 380

Abbott, Riley, 100, 100, -, 2000, 30, 200

Pettyjohn, William, 20, 150, -, 1700, 20, 170

Lynch, Kindal, 50, 40, -, 800, 20, 180

Donovan, Gibson, 40, 100, -, 1400, 20, 110

Macklin, John, 160, 100, -, 5200, 200, 670

Saterfield, William W., 35, 50, -, 1500, 50, 240

Lynch, Sarah, 50, 50, -, 1000, 30, 150

Pettyjohn, Lenny, 20, 40,-, 600, 20, 250

West, Thomas, 30, 50, -, 800, 15, 70

Torbert, George, 60, 34, -, 1880, 40, 450

Sastings, Solomon T., 60, 21, -, 2000, 20, 150

Pepper, David, 300, 200, -, 10000, 300, 1521

Harris, George, 12, -, -, 600, 10, 60

Ellegood, William, 13, -, -, 1300, 200, 400

Kimmey, Jacob, 20, 5, -, 2000, 125, 580

Conaway, Isaac, 35, 50, -, 1600, 100, 300

Hatfield, Thomas, 40, 2, -, 1600, 20, 200

Downing, Hiram T., 9, -, -, 3000, 30, 300

Layton, Samuel, 40, 85, -, 3000, 40, 470

Willen, George W., 100, 75, -, 5000, 120, 548

Anderson, James, 200, 40, -, 6000, 100, 1000

James, William, 10, -, -, 600, 35, 200

Tunnel, Charles, 140, 60, -, 450, 50, 250

Wilkins, John, 125, 75, -, 8000, 60, 520

Moore, Jacob, 60, 65, -, 5000, 100, 250

Pepper, Thomas, 75, 100, -, 8000, 50, 380

Lynch, James W., 11, -, -, 1100, 20, 390

Pepper, Edward G., 10, -, -, 1000, 70, 300

Ewings, Adolphus P., 40, 117, -, 3140, 80, 500

Jorden (Sorden), John, 14, -, -, 700, 20, 300

Wolf, Westley, 50, 30, -, 1600, 20, 240

Conaway, Curtis, 40, 70, -, 3500, 70, 700

Short, Edward, 345, 10, -, 10000, 60, 1435

Roach, Eli, 30, 8, -, 600, 40, 100

Skiner, George, 45, 50, -, 1940, 10, 175

Messick, Joseph, 50, 15, -, 1700, 20, 200

Pettyjohn, Able J., 50, 50, -, 1000, 30, 320

Wright, Gardener W., 60, 60, -, 10000, 150, 400

Workman, Jacob, 40, 100, -, 1500, 25, 150

Culley, Charles _., 67, 15, -, 1000, 200, 850

Serman, John, 40, 60, -, 1000, 40, 300

Hurdle, Jacob F., 50, 10, 1400, 50, 235

Davidson, Henry L., 20, -, -, 240, 50, 100

Coffin, David H., 5, 20, -, 250, 25, 150

Johnson, Robert, 50, 50, 20, 1200, 37, 235

Collins, Thomas, 58, 50, 39, 1476, 50, 450

Harmon, Whitteton, 75, 50, -, 1200, 50, 250

Johnson, John D., 62, 75, -, 1100, 30, 200

Wells, Johnathan, 60, 20, -, 1000, 30, 157

Atkins, Edward, 100, 100, 54, -, 2540, 36, 160

Thoroughgood, John B., 100, 100, 40, 23000, 75, 425

Waples, Jane, 140, 100, 20, 2600, 50, 405

Waples, Cornelious, 60, 70, -, 1300, 50, 365

Johnson, James, 20, 40, -, 1000, 50, 160

Stephenson, James, 18, 9, -, 600, 20, 125

McIlvain, Lewis W., 70, 50, -, 5000, 75, 280

Warrington, Wm. T., 200, 268, -, 13700, 250, 1160

Johnson, Whittleton, 60, 40, -, 2500, 110, 400

Mcgee (Megee), John W., 275, 355, -, 6300, 40, 525

Lawson, Robert, 50, 50, 20, 1500, 60, 260

Harmon, John, 50, 60, -, 1100, 21, 15

Joseph, John H., 85, 50, -, 1350, 60, 250

Joseph, David H., 100, 20, -, 1440, 50, 220

Green Lydia, 35, -, -, 600, 50, 150

Burton, Theodore H., 50, 25, 25, 2000, 75, 375

Norwood, Samuel B., 75, 80, 27, 2000, 25, 210

Warrington, Silus M., 125, 200, 40, 5000, 50, 376

Hopkins, John M., 40, 60, -, 1500, 25, 295

Joseph, Nehemiah, 75, 175, -, 2000, 50, 330

Johnson, John E., 55, 100, -, 1550, 30, 140

Prettyman, Charles H., 40, -, -, 400, 25, 150

Carey, Wooley B., 50, 68, 40, 950, 40, 184

Johnson, Nathaniel, 40, 20, 15, 750, 4, 275

Simpler, Josiah, 75, 100, 32, 2000, 25, 280

Megee (Mcgee), Wm. T., 70, 45, -, 1205, 30, 110

Pettyjohn, Truitt, 150, 300, 50, 4000, 50, 335

Wilson, Ann, 75, 80, -, 7800, 50, 210

Walls, Ann, 175, 90, -, 2000, 25, 150

Carey, Shepard W., 65, 48, -, 110, 25, 150

Walls, Gideon, 60, 40, -, 1200, 32, 245

Johnson, David H., 40, 15, -, 800, 20, 115

McDowel, Eli, 45, 100, 50, 1600, 20, 60

Prettyman, Bagwell, 60, 50, -, 1100, 75, 135

Davidson, James H., 45, 92, -, 1376, 65, 400

Joseph, Paynter, 40, -, -, 480, 30, 130

Joseph, Willard, 35, 30, -, 975, 28, 180

Davidson, Samuel, 75, 124, -, 2000, 50, 277

Prettyman, James, 50, 60, -, 2000, 25, 425

Vaughn, Samuel M., 50, 100, 16, 1300, 50, 306

Lynch, Peter R., 75, 75, 25, 2000, 50, 355

Norwood, Stephen, 35, 50, -, 800, -, 100

Coffin, James, 45, 50, -, 1000, 25, 125

Frame, Elizabeth, 200, 100, -, 6000, 100, 825

Rust, Thomas, 224, 441, -, 5238, 30, 300

Rust, James, 275, 300, 37, 6000, 75, 525

Rust, Wm. B., 100, 100, 10, 1680, 25, 225

Rust, Wm. H., 50, -, -, 500, 22, 100

Rust, Thomas B., 30, -, -, 300, 15, 15

Megee (Mcgee), Wm. C., 60, -, -, 700, 20, 176

Simpler, Jas. B., 80, 70, 35, 1500, 40, 260

Johnson, Wm. M., 60, 40, -, 9000, 20, 68

Johnson, Hazel, 30, 30, 20, 640, 22, 175

Hunter, Joseph, 110, 200, 70, 4000, 50, 718

Perry, John M., 50, 112, -, 1500, 20, 218

Martin, Shepard P., 175, 125, -, 13000, 75, 600

Joseph, Jesse E., 75, -, -, 1000, 40, 200

Johnson, John S., 50, 100, 17, 1670, 30, 288

Drain, Jacob, 55, 155, -, 1050, 25, 260

Gosler, Wm. W., 66, 32, -, 1000, 50, 410

Davidson, Samuel J., 75, 25, -, 9000, 30, 250

Lingo, Wm., 40, 62, -, 800, 20, 160

Burton, John B., 85, 80, 26, 1910, 100, 320

Burton, William H., 30, -, -, 300, 20, 125

Coffin, Nehemiah, 60, -, -, 1500, 25, 160

Miller, Major, 60, -, -, 600, 20, 175

Clark, Robert, 75, -, -, 1000, 45, 440
Bennum, Henry O., 80, 50, 25, 1800, 40, 540
Johnson, Albert J., 50, 64, -, 1320, 50, 275
Walls, John C., 78, 12, -, 900, 30, 240
Walls, John 40, 11, -, 1000, 25, 200
Blizzard, Stephen E., 30, 50, -, 1000, 50, 110
Blizzard, Levin C., 50, 16, -, 792, 20, 150
Rust, Peter W., 100, 60, -, 1920, 50, 380
Carey, John T., 50, 25, -, 1000, 20, 225
Carey, Robert F., 48, 40, -, 700, 20, 125
Verdier, James H., 70, 40, -, 1100, 25, 200
Jacobs, William K., 50, 50, -, 1800, 50, 250
Johnson, Annanias, 50, 71, -, 2500, -, 200
Jones, Benjamin, 100, 50, -, 1000, -, 300
Lingo, Wm. _., 75, 100, 75, 2500, 50, 260
Dodd, Samuel _., 40, 160, -, 2000, -, 20
Robinson, Sarah, 25, 175, -, 3000, -, 190
Dory, George, 50, 25, -, 900, 18, 90
Lawson, Selby, 75, 75, -, 1500, 20, 150
Joseph, Wm. C., 100, 150, 75, 3000, 50, 350
Barker, Thomas R., 66, 34, -, 1200, 75, 400
Frame, Henry C., 150, 272, -, 4642, 100, 525
Pusey, Jehu (John) S., 40, 100, -, 1425, 25, 127
Rust, Peter A., 60, -, -, 600, 30, 250
Vickers, Obed., 56, 70, -, 1100, 20, 35

Thoroughgood, James F., 75, 80, -, 1450, 35, 200
Lawson, Robert T., 60, 15, 25, 1200, 20, 70
Joseph, John M., 75, 25, 50, 1500, 20, 200
Collins, Stephen S., 80, 132, -, 1800, 25, 145
Coffin, James B., 60, 75, -, 1200, 20,3 00
Thoroughgood, William N., 100, 100, 90, 2610, 40, 250
Thoroughgood, John E., 75, 100, 75, 200 (2000), 20, 275
Davis, Henry, 35, 180, 55, 3000, 25, 150
Rust, Charles H., 48, 100, -, 1500, 18, 125
Lingo, Henry L., 10, 30, -, 400, 18, 150
Lingo, Alfred B., 61, 100, 40, 1800, 45, 325
Johnson, George, 44, 20, -, 600, 30, 300
Johnson, Sylvester, 45, 100, 55, 2000, 22, 130
Harmon, Isaac, 100, 104, -, 3400, 50, 600
Atkins, James A., 75, 100, 25, 2500, 25, 325
Harmon, Garrison, 68, 50, 81, 1400, 25, 175
Thoroughgood, Simeon W., 50, 50, 51, 1000, 35, 300
Thompson, John, 53, 45, -, 450, 18, 200
Street, Theophilus, 75, 75, -, 2000, 28, 350
Lingo, McClain, 50, 25, -, 6000, 20, 125
Harmon, Wesley, 45, 50, -, 760, 15, 125
Simpler, Elizabeth, 19, -, -, 285, -, 48
Davidson, Lemuel, 200, 300, 50, 6000, 75, 200

Burton, Wm. C., 300, 214, -, 7200, 150, 1615

Burton, Thomas W., 150, 150, -, 6000, 100, 800

Burton, Nathaniel W., 190, 175, 29, 4628, 75, 694

Dorman, Peter W., 70, 100, -, 2000, 15, 350

Lingo, Geo. F., 50, 59, -, 1100, 18, 190

Dorman, Abraham, 53, 83, -, 1350, 20, 328

Collins, Samuel C., 100, 108, -, 2400, 400, 800

Stockley, John, 25, 72, -, 1000, 15, 85

Warrington, Robert, 60, 94, -, 1540, 15, 220

Burton, Emily, 100, 93, -, 2895, 40, 375

Burton, John C., 143, 143, -, 2220, 28, 270

Lingo, Daniel C., 200, 300, 50, 7700, 50, 1060

Steel, John W., 20, 43, -, 450, 25, 270

Bales, Benjamin M., 80, 100, -, 2200, 30, 370

Lingo, John A., 75, 100, 24, 2000, 120, 400

Wilson, Jacob E., 63, 17, -, 1500, 40, 180

Wilson, Geo. F., 141, 34, -, 2000, 25, 150

Green, John O., 85, 200, 15, 4000, 30, 450

Baker, Wm. T., 75, 75, 25, 2100, 40, 400

Warrington, Wm. F., 25, 5, -, 300, 25, 135

McCray, Edward, 60, 26, -, 860, 30, 210

Lynch, Elisha, 51, 22, -, 720, -, 200

Clark, James H., 40, 17, -, 570, -, 110

Hanson, Nancy, 25, -, -, 300, -, -

Lingo, Paynter, 186, 150, -, 3360, 40, 325

Hanson, Eli, 35, 100, 25, 1500, -, 125

Cornmean, Nathaniel, 40, 20, -, 480, 40, 200

Burton, Joshua R., 100, 300, 500, 4500, 50, 500

Burton, John R., 100, 350, 136, 9600, 125, 1800

Street, Isaac M., 50, 25, -, 1000, 32, 300

Harrison, Theodore, 50, 100, 75, 1800, 15, -

Goslee, Peter E., 313, 39, 20, 6242, 50, 750

Barton (Burton), Lewis, 175, 45, 25, 2450, 15, 40

Lewes, Barton (Burton) C., 165, 50, 30, 2000, 15, 20

Warrington, Kindal J., 100, 50, 500, 2500, 30, 462

Street, Daniel W., 200, 25, 40, 3000, 40, 530

Massey, H. P. W., 76, -, -, 1200, 50, 250

Burton, John E., 75, 125, 100, 3500, 50, 600

Warrington, Wm. F., 50, -, -, 550, 25, 115

Joseph, Zachariah, 55, 75, 20, 1600, 20, 210

Lawson, James L., 48, 50, -, 1200, 25, 175

Scott, Mitchell, 20, 5, -, 325, 25, 225

Rogers, Nathaniel, 45, 30, -, 800, 20, 119

Burton, John S., 45, 30, -, 800, 20, 250

Lynch, William 200, 250, -, 4000, 34, 230

Harmon, Cord, 57, 23, -, 1000, 37, 175

Hood, Henry E., 100, 155, -, 2000, 38, 530

Strafford, Joseph, 85, 100, -, 4680, 30, 388

Steel, James, 75, 785, 25, 1750, 25, 125

Massey, John E., 100, 143, -, 2000, 20, 225

Brereton, James, 100, 175, -, 3000, 40, 293

Maull, James H., 40, -, -, 800, 20, 350

Joseph, Thomas A., 60, 53, -, 1200, 45, 345

Brittingham, Nathaniel, 50, 84, -, 600, 20, 120

Marvel, Mamaen B., 75, 25, -, 1000, 27, 300

Robinson, Peter, 100, 340, -, 3400, 30, 290

Hart, Samuel R., 26, -, -, 300, 10, 125

Johnson, Perry, 40, 20, -, 600, 20, 200

Long, John D., 125, 162, -, 3444, 40, 500

Miller, Isaac, 40, 60, -, 1500, 20, 100

Hurdle, Joseph C., 100, 25, 125, 1200, 18, 150

Parsons, John, 40, -, -, 800, 40, 350

Robinson, Parker, 30, 15, -, 900, 35, 375

Hazzard, Richard, 100, 100, 25, 2190, 25, 275

Robinson, Thomas, 100, 200, 50, 3850, 50, 611

Hazzard, William R., 40, 100, 10, 1500, 18, 100

Simpler, Amos, 50, 16, -, 700, 30, 304

Robinson, William, 100, 100, 45, 2456, 60, 350

Miller, Roland, 40, 100, 20, 1200, 25, 85

Hazzard, Robert, 100, 100, 19, 2190, 23, -

Robinson, George, 100, 100, -, 2400, 60, 467

Allen, Burton, 50, 50, 18, 1460, 20, 90

Prettyman, Wm. F., 100, 60, -, 2140, 75, 550

Marsh, James P. W., 80, 50, 62, 2605, 75, 600

Waples, Elizabeth, 57, 50, 50, 1884, 30, 227

Wilson, Daniel B., 100, 100, -, 3600, 125, 642

Miller, Nehemiah, 50, 75, 120, 3000, 30, 192

Hazzard, Alphonso, 95, 110, -, 3000, 100, 365

Hazzard, John E., 54, 80, -, 2010, 80, 437

Hazzard, Robert, 150, 150, 36, 1632, 60, 275

Stockley, John M., 37, -, -, 300, 20, 165

Lingo, Joseph B., 50, 40, 60, 3530, 30, 460

Harris, William, 55, 50, 20, 1200, 20, 275

Snyder, Henry, 140, 10, -, 2500, 25, 220

Burton, Lemuel P., 110, 200, 90, 6000, 75, 675

Wilson, Thomas P., 70, 50, 50, 1700, 50, 210

Vessels, Miras V., 100, 100, 85, 3000, 75, 500

Drain, Abraham, 72, -, -, 864, 20, 235

Webb, John, 50, 50, 20, 1200, 18, 85

Waples, James, 75, 79, -, 1840, 35, 241

Drain, Daniel, 50, 9, -, 725, 25, 300

Wright, Frederick, 54, -, -, 650, 20, 223

Walls, Elie, 115, 45, -, 4000, 50, 500

McIlvain, John S., 5, 58, 65, -, 1600, 30, 300

Burton, Peter W., 50, 50, 40, 1680, 25, 175

Joseph, Zachariah S., 45, 2, -, 1000, 40, 180

Walls, Asa, 40, 75, 25, 2000, 20, 40

Mustard, John, 92, 345, 50, 5000, 50, 650

Carey, Joseph, 100, 75, 25, 2000, 25, 200

Johnson, Henry R., 100, 200, -, 3000, 25, 550

Harmons, Agenish, 37, 20, -, 600, -, 205

Johnson, Noah, 40, 25, 35, 900, 20, 200

Warrington, Silas M., 70, 30, 22, 1600, 30, 230

Prettyman, David M., 50, 50, 50, 1000, 45, 265

Stevenson, Robert D., 115, 200, 25, 6000, 75, 666

Stevenson, Jesse _., 85, 15, -, 1500, 75, 275

Lingo, Mary W., 100, 100, 50, 2500, 40, 225

Street, David, 100, 100, 44, 2440, 75, 442

Britingham, James, 40, 50, -, 900, 20, 100

Davidson, Samuel, 75, 100, 25, 1600, 30, 200

Hutson, Cyrus, 65, 75, 60, 2000, 20, 100

Walls, Gilly S., 50, 75, 30, 1500, 20, 195

Burton, Peter R., 700, 400, 200, 16000, 150, 1500

Hurdle, Wm. W., 75, 75, -, 2000, 75, 575

Joseph, Hezekiah, 60, 18, 40, 1200, 30, 225

Warren, James H., 75, 85, -, 1500, 40, 250

Morris, George, 75, 45, -, 1400, 40, 185

Pride, James, 60, 60, 30, 1500, 45, 225

Prettyman, B. C., 75, 125, -, 8000, 75, 450

Robinson, Thomas, 55, 100, -, 1600, 40, 200

Warrington, D., 50, 50, -, 2000, 40, 185

Wright, Philop, 65, 75, -, 1500, 31, 225

Clark, George, 60, 70, -, 1800, 20, 175

Hopkins, William, 75, 50, 15, 1600, 75, 650

Steel, Thomas, 75, 80, -, 1500, 20, 200

Croadal, George, 64, 84, 10, 2000, 20, 100

Martin, James F., 100, 100, 200, 2500, 30, 336

Martin, Robert, 40, 14, -, 1000, 80, 175

Martin, James, 20, -, -, 500, 15, 150

Hitch, William J., 75, 125, -, 6000, 30, 340

Hutson, Milford, 50, -, -, 500, 20, 90

Johnson, John, 60, 52, -, 1250, 25, 300

Marsh, John A., 45, 75, -, 2000, 20, 150

Mantoon, Peter, 50, 50, -, 1000, 25, 300

Coursey, Nancy, 50, 75, 50, 2000, 20, 100

Lank, Samuel J., 85, 15, -, 1800, 30, 210

Waples, Burton, 50, 20, -, 1000, 28, 200

Lank, John C., 50, 20, -, 1400, 40, 215

Reynolds, Eldayman, 75, 20, 25, 2400, 35, 220

Johnson, Isaac, 25, 5, -, 500, 20, 90

Clark, Nathaniel, 50, 50, 50, 1500, 22, 225

Lingo, Samuel W., 79, 79, -, 1600, 75, 320

Lank, James, 110, 40, -, 3000, 50, 400

Palmore, Samuel, 95, 75, 10, 1500, 30, 175

Wells, Eli, 100, 100, 70, 3000, 30, 200

Carson, Alfred, 60, 40, 20, 2000, 40, 315

Lankford, John W., 300, -, 100, 4000, -, -

Russel, Robert, 160, 160, 180, 10000, 125, 1100

Waples, David, 100, 100, 70, 4000, 75, 500

Prettyman, Joel, 125, 26, 25, 10000, 300, 1000

Lodge, John, 22, -, -, 1320, 38, 150

Lodge, Samuel, 125, 48, 60, 7000, 85, 725

Lingo, Benjamin, 75, 25, 30, 2000, 80, 400

Webb, Charles, 75, -, -, 1000, 60, 400

Kelley, Charles, 150, 275, 175, 280000, 200, 375

Webb, George, 75, -, -, 800, 20, 250

Maull, Henry, 50, 20, -, 700, 20, 140

Maull, Orange, 20, -, -, 400, 20, 175

Wright, David, 75, 25, -, 2500, 50, 325

Norwood, William 100, 130, -, 3600, 20,300

Metcalf, John, 40, -, -, 3000, 75, 1160

McIlvain, David, 75, -, -, 5000, 80, 275

Orr, William, 25, -, -, 1250, -, 450

Hall, David, 50, -, -, 1500, -, 585

Maull, Thomas, 25, -, -, 1500, 50, 425

Russel, Edward, 100, -, 600, 5000, 75, 610

Phillips, Benjamin, 50, -, -, 1250, 75, 450

Dunning, Charles, 40, -, -, 1000, 30, 175

Tindle, Samuel, 16, -, -, 640, 40, -

Salmons, John, 50, -, 25, 1500, 50, 225

Lyons, Laban L., 60, -, -, 3000, 50, 575

Hitchens, Edwad, 100, -, 21, 6000, 75, 325

Burton, Joshua S., 100, -, 30, 6000, 110, 870

Hickman, Harbison, 610, 410, 78, 40000, 400, 4000

Paynter, Wm. D., 52, -, -, 1200, 75, 350

Hevalow, Moses, 50, 30, -, 700, 25, 225

Marsh, Joseph, 75, 100, 25, 3000, 75, 680

Lank, William, 75, -, -, 1000, 20, 190

Turner, Thomas, 100, 40, 40, 3600, 75, 450

Jones, George W., 54, 70, 20, 1000, 38, 225

Maull, James E., 58, 25, 25, 1080, 30, 281

Wood, John & Wm., 150, 89, -, 10000, 100, 480

Wolf, Wm. S., 50, 10, 40, 1200, 40, 325

Simpler, Lacey, 95, 25, 50, 2500, 50, 265

Wright, Maria, 28, -, -, 1000, 25, 125

Prettyman, James, 96, 35, -, 4000, 50, 470

Prettyman, Gideon, 55, -, -, 3000, 50, 375

Coursey, Eliza, 80, -, -, 60, 25, 140

Wilson & Bro., John, 100, 75, 35, 4000, 75, 410

Warrington, Silas, 75, 100, 25, 2800, 50, 285

Burton, John, 88, -, -, 1760, 75, 400

Goselee, Samuel, 75, 25, 50, 2000, 25, 175

King, David, 50, 20, 20, 1000, 25, 175

Joseph, Daniel, 60, 10, -, 1400, 30, 225

Lyons, Rodney, 90, 60, -, 4500, 100, 680

Roach, John, 65, -, -, 1300, 45, 300

Welch, Joseph, 150, 100, -, 6000, 100, 810

Fine, Phillip, 67, -, -, 2000, 50, 400

Fisher, Hiram C., 64, 100, -, 3000, 50, 275

Pride, Wingat, 50, 75, 28, 3500, 40, 260

Burton, Catharine, 50, 50, -, 2000, 50, 150

Cannon, Rease, 5, -, -, 125, 20, 230

Goldsborough, Charles, 100, 100, 45, 2500, 35, 300

Burton, Lemuel, 104, -, -, 1800, 25, 150

Fletcher, Joseph F., 114, 16, -, 2500, 50, 325

Warrington Charles R., 140, -, -, 5000, 75, 450

Fletcher, John M., 50, 50, -, 2000, 40, 200

Fisher, Robert, 75, 25, -, 2000, 50, 290

Burton, David B., 40, -, -, 800, 20, 250

Monix, Richard, 60, -, -, 1200, 40, 345

Palmore, John, 50, 100, -, 1500, 40, 170

Marsh, Peter, 75, 75, 100, 3500, 55, 515

Dunaphan, B., 125, 150, 175, 5500, 75, 645

Burton, Lemuel P., 100, 100, -, 4000, 75, 735

Morris, James, 23, -, -, 660, 20, 187

Hudson, Thomas, 50, -, -, 1000, 25, 35

Coursey, James, 60, -, -, 1800, 40, 140

Hevalow, John, 15, -, -, 300, 35, 190

Lynch, Theodore, 100, 40, -, 3000, 35, 365

Hudson, Peter, 22, -, -, 440, 30, 125

Dodd, William A., 285, 110, 135, 10600, 250, 1729

Miller, Edward, 50, -, 55, 2500, 25, 155

Palmore, Lemuel, 60, 75, -, 1600, 20, 140

Burton, Wm., 100, -, 2000, 20, 150

White, Shepard S., 50, 75, -, 120, 30, 300

Marsh, Lemuel, 75, 100, 25, 3150, 50, 310

Stockley, Woodman, 100, 62, -, 3000, 50, 310

Stockley, Henry, 40, 50, 10, 1100, 20, 120

Robinson, Robert, 100, 100, 72, 5440, 75, 592

Thompson, Radz, 100, 30, -, 3000, 50, 410

Burton, Eliza, 40, 13, -, 260, 25, 200

Dodd, Jas. A., 250, 10, -, 4700, 200, 1038

Thompson, Wm. 50, 47, -, 2800, 100, 900

Martin, Lounson D., 75, -, -, 1500, 40, 220

Hephaon, John, 60, -, -, 800, 30, 140

Marsh, Andrew, 100, 100, -, 4000, 60, 350

Harmon, Purnal, 60, -, -, 1200, 40, 250

Hutson, James, 100, 60, -, 4000, 75, 655

Holland, Comfort, 175, 100, 159, 10000, 150, 1300

King, William H., 100, 100, 1200, 4500, 75, 6000

Dodd, Joseph, 350, 20, 130, 15000, 250, 1975

Walls, John, 100, 100, 25, 4000, 55, 240

Wolfs, Daniel, 100, 100, 175, 7500, 75, 575

Waples, David, 75, 100, 25, 3500, 60, 275

Wolfe, William, 100, 100, 100, 6000, 75, 500

Paynter, Richard, 50, 25, -, 1500, 40, 183

Hart, Thomas R., 80, -, -, 2400, 50, 620

Collins, Jonathan, 35, -, 10, 1000, 50, 400

Marsh, John, 55, 100, 50, 4000, 30, 235

Palmore, Henry, 80, -, -, 3000, 75, 525

Walker, Thomas, 140, 60, -, 8000, 125, 950

Hood, John H., 100, 35, 35, 7000, 125, 850

Harrison, William, 150, 100, 30, 8000, 50, 365

Little, Henry, 75, -, -, 3000, 41, 240

Burton, Edward, 150, 50, 50, 6000, 75, 590

Marshall, John P., 110, 15, -, 9000, 50, 450

Stockley, Henry, 100, 33, -, 3500, 50, 195

Hevalow, Stafford, 75, -, -, 2250, 40, 370

Houston, Shepherd P., 200, 200, -, 15000, 125, 421

Hearn, Jonathan, 80, 20, -, 2000, 80, 175

Windsor, James H., 40, 20, -, 1200, 40, 150

Game, George, 100, 70, -, 1600, 50, 400

Gordery, William 150, 50, -, 4000, 100, 500

Gordery, William 100, 9, -, 2500, -, 200

Gordery, Benjamin, 100, 96, -, 2000, 50, 300

Gordery, Aaron, 100, 96, -, 2000, 100, 30

Gordery, Joseph, 100, 96, -, 2000, 50, -

Cordry, John, 100, 95, -, 3000, 100, 300

Baker, Jacob, 100, 40, -, 3000, -, 200

Anderson, James W., 135, 100, -, 2360, 150, 600

Marine, Jacob, 100, 90, -, 1000, -, 800

Marin, Jacob, 100, 80, -, 2800, -, 50

Collins, Josiah, 100, 20, -, 900, -, 75

Hastings, Joshua, 40, 50, -, 900, -, 100

Hastings, Joshua, 10, 8, -, 500, -, 100

Selbey, William, 67, 30, -, 1800, -, 60

Wirtten, (Wrotten), George, 100, 50, -, 1000, 75, 50

Elliott, Mary A., 100, 100, -, 1800, -, 100

Collins, John C., 150, 30, -, 2600, 50, 12

Elliott, Nathaniel, 100, 50, -, 1500, -, 300

Elliott, William 75, -, -, 1500, -, 50

Horney, William, 100, -, -, 1500, 50, 250

Hastings, Ezekiel, 100, 50, -, 2500, -, 300

Lerve (Lowe), Selbey M., 175, 25, -, 2500, -, 250

Jerman, John, 50, -, -, 1000, -, -

King, William, C., 160, 165, -, 6525, -, 450

King, William C., 75, 42, -, 2540, -, 150

Cordery, James H., 20, 22, -, 420, -, 150

Elliott, Edward, 100, 47, -, 2470, -, 150

Collins, Sarah, 100, 10, -, 900, -, 30

Melson, John W., 40, 35, -, 700, 100, 50

Melson, John W., 60, 20, -, 600, -, 30

Serman, William, 100, 100, -, 200, -, 400

Hastings, Daniel, 100, 25, -, 1200, 75, 150

Daris (Davis), John D., 20, 10, -, 600, -, 50

Morris, James, 60, -, -, 800, -, 100

Morris, Isaac, 100, 25, -, 1200, 100, 200

James, Minus, 100, -, -, 1600, -, 200

James, George, 103, -, -, 1600, -, 200

Williams, William A., 100, -, -, 1200, 40, 200

Pallott, Lewis, 75, 25, -, 1500, 30, 350

Hearn, Isaac, 100, 13, -, 2500, -, 200

Loue, Selbey, 100, 50, -, 3000, -, 300

Elliott, Joshua, 200, 12, -, 2500, -, 400

Cannon, Isaac, 80, -, -, 800, -, -

Cannon, Isaac, 50, 33, -, 1600, -, 100

Game, Daniel, 70, 30, -, 2000, -, 300

Collins, John, 20, -, -, 2000, -, 100

Lynch, William 74, -, -, 1000, -, 150

Vinson, Jinsey, 75, 25, -, 1200, 30, 200

Vinson, Minus, 50, 50, -, 1200, 40, 200

Green, Charles, 75, 50, -, 1000, -, 150

Hosa, John H., 100, 70, -, 3500, -, 200

Davis, (Daris), William, 50, 50, -, 2000, -, 200

Calhoon, Ephraim, 100, 100, -, 2000, -, 300

Harmon, George, 50, 50, -, 800, -, -

Hitchens, John, 50, 50, -, 1000, -, 20

Cannon, Phillip M., 100, 70, -, 2500, 100, 370

Gordey, Benton, 185, 40, -, 2250, -, 500

Elliott, Jacob, 95, 25, -, 1000, -, 200

Triggs, Cyrus, 60, -, -, 800, -, 250

Umphry, Ward, 40, -, -, 500, -, 100

Colman, Thomas, 50, 50, -, 1200, -, 200

Calloway, Wingat, 50, 30, -, 1000, -, 200

Swain, Bengiman, 30, 50, -, 500, -, 30

Callaway, Levi, 120, 110, -, 2300, 40, 600

Ward, Thomas, 75, 25, -, 1000, 50, 100

Carmean, William, 30, 20, -, 1000, -, 200

Carmean, Elijah, 100, 75, -, 1500, -, 50

Wingat, John, 100, 50, -, 500, -, 150

Baker, John, B., 75, 26, -, 1200, -, 400

Baker, John B., 33, -, -, 400, -, -

Hastings Joshua, 100, 46, -, 1500, 200, 200

Ward, Lenard, 75, 30, -, 1600, -, 200

Davis, Josiah, 60, 8, -, 1300, -, 150

Ward, Sallie, 100, 40, -, 2000, 40, 100

Ward, John, 150, 150, -, 4500, 100, 500

Ward, John, 50, 20, -, 1100, -, 300

Ward, Cyrus, -, 41, -, 500, -, -

Hearn, William G., 60, 40, -, 2000, -, 150

Hearn, William G., 80, 94, -, 2500, -, 200

Ralph, William _., 70, 30, -, 1000, -, 360

Records, Joseph C., 40, 20, -, 800, -, 150

Cannon, Joshua C., 205, 50, -, 4000, 50, 75

Elliott, Burton, 30, 5, -, 1500, -, -

Cordey, Elizabeth, 40, 10, -, 1000, -, 4

Penuel, Elijah, 150, 50, -, 400, 50, 300

Penul, Hyram F., 60, 55, -, 2300, -, 300

Elliott, William T., 100, 100, -, 4000, -, 275

Hastings, William 70, 10, -, 800, -, 260

Flowers, Owens, 5, 14, 7, -, 700, -, 30

Hastings, Ezekiel, 30, 50, -, 1200, -, 250

Jackson, James H., 70, 12, -, 3000, 40, 60

Elzey, Lewis C., 75, 25, -, 1200, -, 150

Phillip, Thomas, 100, 75, -, 3500, -, 300

Hitchens, Jeremiah, 30, 43, -, 900, 100, 200

Moore, Welfield, 75, 25, -, 800, -, 200

Elliott, Risden, 10, 5, -, 2000, -, 40

Moore, Whit, 4, 2, -, 300, -, 20

Duran, Noah, 4, 1, -, 200, -, 30

Wofford, William, 6, 3, -, 200, -, 10

Spencer, Samuel, 40, 20, -, 300, -, 150

Spencer, Bengiman, 4, 20, -, 300, -, -

Phillip, Isaac G., 200, 100, -, 6000, 200, 500

English, Charity, 30, -, -, 300, -, -

English, Thomas, 75, 25, -, 1000, -, 50

Rinals, Isaac H., 10, 90, -, 1200, -, 25

Waller, William 40, 10, -, 500, -, 150

Waller, James, 40, 17, -, 500, -, 5

Waller, John, 50, -, -, 1000, -, 300

Cooper, William, 140, 140, -, 2800, 25, 500

Henderson, Isaac, 40, 40, -, 1600, -, 100

Cooper, William B., 25, 25, -, 500, 50, 700

Ellis, Stephen, W., 50, 50, -, 2000, -, 300

Ellis, Stephen, 50, 23, -, 800, -, 400

Furbush, Ananias, 40, 45, -, 1000, -, 25

Fletcher, John, 50, 50, -, 2500, -, 200

Culer, Wilmar, 100, 100, -, 1000, 50, 200

Bradly, Samuel, 40, 31, -, 1500, 40, 60

Adams, Henry, 250, 250, -, 6000, 80, 600

Collins, Loren L., 100, 45, -, 1200, -, 200

Adams, Jeremiah, 200, 75, -, 3500, -, 350

Walston, Julia, 60, 40, -, 1200, 75, 300

Walston, Lambert, 75, 50, -, 1200, -, -

Walston, Thomas, 75, 40, -, 1200, -, -

Collins, James, 84, 45, -, 1500, 50, 75

Collins, Henry, 40, 63, -, 1100, -, 150

Colling, Loren A., 40, 10, -, 500, -, 100

Ellis, Joseph, 140, 105, -, 3000, 150, 900

Ellis, Josephus, 75, 55, -, 1560, 50, 200

Owens, Isaac W., 100, 50, -, 1800, 50, 200

Owens, Hamilton, 75, 20, -, 1300, -, 300

Bradly, Freeman R., 40, 50, -, 1000, -, 50

Owens, Josiah, 90, 50, -, 1500, -, 50

Owens, Josiah, 30, -, -, 600, -, -

Nickelson, Benjamin. 50, 10, -, 600, -, 300

Bradly, Allen R., 40, 10, -, 60, -, 50

Tuilley, Robert, 140, 80, -, 3000, 200, 700

Major, Bradly, 35, 22, -, 800, 100, 200

Carmine, R___, 75, 25, -, 1200, -, 300

Wright, Loren, S., 40, 75, -, 3000, 80, 601

Bradly, Flabius, 40, 30, -, 700, 40, 200

Bradly, Jean A. D., 125, 55, -, 1800, 30, 450

Bird, James (Jane), 100, 90, -, 1500, -, 75

Phillips, Roger, 50, 60, -, 1500, 50, 250

Cooper, Noah, 75, 78, -, 3500, 100, 300

Phillip, Mary, 60, 40, -, 800, 50, 100

Winder, Samuel, 40, 10, -, 1000, -, 100

Phillip, Elish, 100, 100, -, 2000, -, 150

Phillip, Samuel, 100, 50, -, 2000, 75, 200

Elzey, Charles, 50, 50, -, 1000, 40, 250

Phillip, Samuel, 150, 110, -, 6500, 30, 900

Howard, James, 50, 20, -, 700, -, 75

Corler, Isaac, 100, 380, -, 4800, -, 300

Phillips, Samuel, 40, 20, -, 600, -, -

Rinney, George W., 58, 58, -, 1000, 50, 400

Cooper, John, 100, 100, -, 3000, 75, 500

Taylor, Jacob, 61, -, -, 600, -, -

Hastings, Elihu, 100, 75, -, 4000, 175, 400

Hastings, Eli, 200, 25, -, 2000, 100, 300

Culser, Salathel, 100, 50, -, 5050, -, 300

Workman, William, 140, 60, -, 3000, -, 200

Rinekin, Gilley, 175, 25, -, 3000, -, 300

Hitch, Eligah, 100, 50, -, 2500, -, 500

Dickerson, Cathersy, 100, 25, -, 1500, -, 530

Dickerson, Cathersy, 50, 25, -, 1000, -, -

Giles, Isaac, 250, 150, -, 15000, 300, 1200

Ellis, Mathews, 60, 24, -, 2200, -, 300

Giles, Isaac, 100, 56, -, 3000, 200, -

Rinekin, John T., 100, 50, -, 1800, -, 200

Culser, Henry P., 100, 100, -, 2000, -, 300

Bailey, Jonathan, 200, 40, -, 6000, 100, 1000

Bailey, Jonathan, 160, -, -, 1500, -, -

Phillips, Mary E., 300, 300, -, 12000, 300, 1000

Phillips, Eliza, 100, 50, -, 3000, -, 150

Pippin, Robert, 75, 25, -, 1000, 200, 150

Moore, Charles N., 180, 80, -, 4000, 30, 500

Moore, Loren, 100, -, -, 1000, -, 150

Moore, Loren, 100, 63, -, 1200, -, 200

Ralph, Sarah, 50, 27, -, 1000, -, 300

Hill, John, 100, 20, -, 1000, -, -

Rinny, Elijah, 90, 7, -, 1445, -, 400

Rinny, William, 90, -, -, 2000, 300, 150

Ellis, Noah, 50, 40, -, 900, -, 300

Collins, Antiney, 60, 20, 800, -, 300

Ellis, Joseph, 80, 17, -, 900, -, 300

Hastings, Elezy, 60, 20, -, 1000, -, 25

Culser, Louis, 100, 20, -, 1000, 40, 40

Ellis, Martin M., 150, 50, -, 4000, 100, 1000

Loue, Ebenezer M., 125, 75, -, 2400, 40, 500

Callaway, Joseph, 100, 120, -, 2000, 50, 200

Ellis, James, 75, 85, -, 2000, 500

Loue, James C., 80, 45, -, 2000, -, 350

Henry, Isaac, 100, 70, -, 1700, -, 300

Loue, James, 12, -, -, 500, -, -

Hearn, Burton, 60, 65, -, 1250, -, 200

Phillip, Thedora, 60, 40, -, 1000, 100, 300

Beach, Barnabus, 100, 80, -, 1800, 50, 300

Ounes, James C., 60, 73, -, 1320, 40, 220

Collins, Joah W., 30, 38, -, 700, 100, 200

Tuelley, James E., 60, 50, -, 1100, -, 300

Waller, John F., 20, 50, -, 600, -, 100

Phillip, Dervons (Dervoies), 60, 74, -, 1600, -, 400

Tuilley, Loren P., 100, 200, -, 3600, 100, 500

Tuilley, Robert O., 100, 60, -, 1200, -, 250

Waller, Ebenezer, 75, 20, -, 2200, 100, 300

Waller, Ebenezer, 40, 20, -, 1200, 100, 400

Records, William, 100, 25, -, 4250, 50, 1000

Gottler, Griffith, 75, 15, -, 450, -, 100

Collins, John B., 50, 10, -, 1000, -, 50

Walston, Charles, 75, 95, -, 3000, 100, 500

Walston, Charles, 30, 30, -, 600, -, 50

Moore, Colwell, 100, 70, -, 3400, 100, 250

Rinekin, Stephen, 75, 25, -, 1600, -, 250

Rinekin, John, 25, -, -, 500, -, 175

Rinekin, William B., 50, 5, -, 1000, -, 150

Morris, Thomas P., 60, 40, -, 2000, -, 200

Hill, John, 100, 25, -, 1200, -, 50

Collins, Eben, 120, 100, -, 2200, -, 200

Hastings, Margart, 100, 20, -, 1200, -, 200

Rhods, Solomon, 150, 50, -, 2000, -, 250

Collins, Daughty, 150, 50, -, 1200, 25, 350

Warrington, Elijah, 100, 50, -, 3500, -, 750

Records, Edward, 80, 80, -, 1800, -, 150

Phillip, William, 201, 10, -, 4530, -, 200

Horsey, Nathaniel, 125, 125, -, 5000, 150, 800

Hastings, Elzey, 30, 40, -, 1500, 40, 100

Hitch, George, 30, 70, -, 1000, -, 300

Megee, John W. D., 175, 35, -, 2500, -, 500

Ellis, James E., 150, 51, -, 4000, -, 500

Dunn, Thomas, 200, 70, -, 4000, -, 400

Wingat, Mathew, 100, 100, -, 2000, 50, 500

Culser, Hardy, 200, 100, -, 6000, 40, 500

Boyce, Daniel, 200, 100, -, 5000, 100, 500

Hearn, Luthe, 200, 50, -, 4000, 30, 50

Cordery, Burton, 100, 50, -, 3000, 40, 50

Rinekin, Stephen, 60, 30, -, 800, 50, 200

Rinney, Eligah C., 175, 77, -, 5000, 40, 450

Records, Jonathan, 40, 10, -, 1200, 100, 150

Records, Thomas S., 45, 17, -, 1500, -, 150

Callaway, Sallie, 50, 20, -, 1500, -, 150

Game, Joseph, 150, 85, -, 4700, 50, 100

Beach, Isariah, 60, 30, -, 1600, 100, 400

Hill, Loren J., 100, 50, -, 2500, 40, 500

Game, John G., 100, 38, -, 2700, -, 250

Bacon, Mary, 200, 100, -, 6000, -, 100

Bacon, Samuel, 5, 25, -, 500, -, -,

Bacon, Thomas, 130, 23, -, 3060, -, 760

Bacon, John L., 75, 39, -, 3000, -, 200

Hearn, Rendle B., 70, 20, -, 1000, -, 400

Hearn, Rendle B., 30, 90, -, 1000, -, 200

Hearn, William, 100, 70, -, 1000, -, 500

Callaway, Jobe, 80, 70 -, 1500, -, 200

Culser, Cherlotte, 700, 10, -, 1000, -, 300

Legates, Peggey, 23, 50, -, 500, -, 100

Serman, Jobe, 50, 15, -, 650, -, 200

Serman, Jobe, 22, 10, -, 300, -, 400

Hearn, William, 70, 90, -, 1500, -, 400

Elliott, John M., 130, 70, -, 4000, -, 700

Hastings, Eliza, 100, 30, -, 2600, -, 200

Hastings, Joseph, 70, 30, -, 2000, -, 300

Gordey, Benjamin, 100, 100, -, 2000, -, 500

Callaway, Burton, 90, 68, -, 2000, 100, 1300

Oliphant, Elizah, 112, 100, -, 2500, 100, 600

Lynch, Samus (James), 100, 10, -, 1200, 50, 250

Hester, Lynch, 100, 10, -, 1000, -, 200

Lynch, John, 100, 10, -, 1200, -, 150

Lynch, James, 100, 10, -, 1200, -, 200

Wilkins, James, 75, 100, -, 2000, -, 200

Adams, Alfred, 110, 70, -, 4300, -, 700

Adams, Alfred, 40, 50, -, 700, -, 300

Wirtten, John, 75, 26, -, 1600, 560, 150

Ring (King), George, 100, 100, -, 2000, 30, 50

Minas, Jam__, 100, 94, -, 2000, -, 200

Game, Daniel, 40, 50, -, 800, -, 200

Philips, Eliza, 40, 40, -, 1000, 50, 250

Short, James, 15, 65, -, 1000, 40, 200

Ellis, George W., 60, 140, -, 2500, -, 600

Hastings, Elihu, 22, -, -, 500, 31, 50

Hastings, Josses, 100, 25, -, 1600, 40, 200

Culser, Daniel, 40, 20, -, 850, -, 200

Waller, Jonathan, 120, 47, -, 3000, 100, 800

Waller, Jonathan, 175, 55, -, 4600, -, 500

Waller, Jonathan, 75, 65, -, 1700, -, -

Whit, Peter, 100, 50, -, 1200, -, -

Dunn, Thomas, 110, 35, -, 1600, 40, 250

Ellis, Ridgway, 200, 100, -, 200, -, 2000, 40, 350

Hastings, Elie, 11, 50, -, 1150, -, -

Hastings, Winder W., 150, 50, -, 4000, -, 400

Whit, Winder, 100, 90, -, 4000, -, 620

Cole, Washington, 40, 25, -, 1000, -, 150

Hastings, William, N., 50, 40, -, 2000, -, 150

Freeney, Elijah, 60, 55, -, 3200, -, 200

Andrew, H___, 29, -, -, 490, -, -

Hearn, Isaac, 200, 20, -, 4000, 75, 500

Callaway, John, 100, 60, -, 2600, -, 200

Ellis, James of E., 30, 26, -, 1200, -, 300

Dunn, Luther B., 100, 48, -, 3500, -, 600

Morris, Nehmiah, 100, 50, -, 3500, -, 200

Callaway, Washington, 160, 100, -, 4600, 50, 820

Hastings, John, 70, 20, -, 1800, -, -

Hearn, William, 60, 100, -, 3200, -, 700

Beach, Jonathan, 50, 25, -, 1500, -, 600

Callaway, Sallie, 100, 30, -, 2600, -, 150

Waller, Hamilton, 50, 30, -, 1600, 100, 400

Hastings, John H., 100, 100, -, 4000, -, 300

McCearly, Auther, 50, 30, -, 1000, -, 150

Hastings, James, 120, 120, -, 1600, -, 250

Elles, Angelin, 50, 24, -, 1500, -, 300

Louie, John P., 100, 100, -, 4000, -, 300

Rinney, Samuel, 150, 170, -, 3600, 40, 260

Rinekin, Gilley, 170, 70, -, 1600, 100, 300,

Davis, Daniel, 80, 40, -, 2640, 50, 150

Graham, Phillop, 100, 70, -, 1800, -, -

Phillips, Robert, 100, 100, -, 2000, 320, 500

Graham, William, 100, 111, -, 2000, 150, 200

Lingo, William, 100, 50, -, 4000, 200, 300

Messick, Samuel, 100, 50, -, 4000, 175, 400

Hitchens, Phillop, 100, 50, -, 4500, 300, 175

Hitchens, John, 40, 10, -, 1000, -, 100

Outten, Thomas, 100, 40, -, 1200, -, 50

Morgan, Elender, 4, 20, -, 400, -, 50

Culser, Rilley, 25, 27, -, 800, -, 5

Culser, Asa, 30, 23, -, 500, -, 300

Culser, Asa, 15, -, -, 300, -, 25

Culser, Asa, 40, 50, -, 1000, -, 50

Scott, Evans, 10, 100, -, 500, -, 400

Morgan, James, 100, 100, -, 1000, 50, 600

Renuls, Jacob, 200, 100, -, 7000, -, -

Waller, George, 30, 20, -, 1500, -, 300

Chipman, Thomas, 100, 56, -, 5000, 100, 300

Floyd, George S., 80, 100, -, 2500, 50, 300

Husten, Mait__, 100, 130, -, 2000, -, 250

Husten, Martin, 40, 4, -, 400, -, -

Penten(Penter), James, 20, 10, -, 300, -, -

Chase, Siner, 5, 3, -, 600, -, 60

Wallas, Stephen, 50, 50, -, 2000, 150, 200

Conley, Eliza, 60, 110, -, 1500, -, 200

Carmean, Greensly, 100, 50, -, 5000, 200, 500

Layton, Jesse, 70, 90, -, 5000, -, 100

Morgan, Jacob, 4, 4, -, 400, -, 100

Morgan, Jacob, 50, 30, -, 500, -, -

Collins, Grus, 50, 20, -, 400, -, -

Faylan, Solomon, 10, 22, -, 300, -, -

Scott, William, 30, 40, -, 800, -, 75

Scott, William, 100, 20, -, 400, -, -

Moore, Luther, 100, 10, -, 500, -, 200

Moore, Loren, 100, 40, -, 500, -, 200

Riggin, Purnell, 55, 5, -, 2400, -, 200

Serman, Loren S., 40, 60, -, 1800, -, 458

Serman, Loren S., 40, 10, -, 450, 250, 200

Lamden, Joshua, 150, 150, -, 2000, 30, 600

Truitt, Wm., 400, 300, -, 7000, 100, 500

Giles, William, 100, 100, -, 4000, 50, 600

Hitchens, George, 100, 50, -, 3000, -, -

Moore, Nancy, 65, 65, -, 1500, -, 50

Collins, Bengiman, 100, 50, -, 1800, 50, 100

Cannon, Johnson, 100, 100, -, 2000, 40, 100

Records, Isaac, 100, 100, -, 2000, 50, 50

Hitchens, Nathaniel, 40, 20, -, 1200, -, -

Messick, Miles, 400, 20, -, 1200, -, -

Dolbey, Sarah, 100, 125, -, 2500, -, 300

James John, 40, 40, -, 800, -, 100

Messick, Joel H., 50, 50, -, 1000, -, -

Messick, Nehemiah, 50, 46, -, 600, -, 100

Messick, John, 20, 40, -, 600, -, 200

Messick, Clayton, 40, 20, -, 700, -, -

Tindal, Gorden, 100, 100, -, 2000, -, 600

Morgan, William W., 200, 92, -, 6000, 50, 1200

Plumer, Hudson, 60, 50, -, 1800, 40, 300

Parmer, John, 200, 20, -, 1800, -, 39

Hitchens, Gillis, 40, 90, -, 2000, -, 150

Boyce, David, 100, 50, -, 3000, -, 400

Pussy, Morris, 100, 60, -, 3100, 50, 800

Collins, John A. D., 100, 60, -, 3500, 100, 400

Boyce, David, 60, 29, -, 1800, 100, 300

Owens, James, 70, 30, -, 1800, -, 400

Baker, Purnell, 100, 25, -, 1800, -, 350

Messick, Charles, 100, 100, -, 2500, -, 500

Lingo, Elizabeth, 50, 70, -, 1000, -, 450

Waller, William, 100, 100, -, 2000, -, 200

Allen, William, 100, 150, -, 4000, 200, 400

Fooks, Nathaniel, 50, 60, -, 1600, -, 60

Vinson, Joseph W., 50, 30, -, 1600, -, 550

Green, Thomas, 40, -, -, 2000, -, -,

Jones, Henry, 40, 55, -, 3000, -, 200

Taylor, William, 40, 28, -, 1200, -, 100

Graham, Phillip, 100, 74, -, 1500, 200, 300

Graham, William, 100, 70, -, 1500, -, 20

Callaway, William 100, 65, -, 1500, 50, 50

Wootten, Eligah, 50, 25, -, 750, 50, 150

Elliott, Joshua, 60, 25, -, 1500, -, 375

Elliott, John L., 50, 20, -, 1000, -, 250

Hearn, Isaac F., 40, 20, -, 600, -, 300

Elliott, Burton, 20, 20, -, 600, -, 100

Ellis, Joseph, 75, 20, -, 3000, -, 300

Hearn, William, 200, 97, -, 3000, -, 500

Ellis, Thomas, H., 75, 57, -, 1000, -, 475

Hearn, William T., 100, 50, -, 2000, 300, 200

Madox, James, 100, 60, -, 2500, -, 600

Melson, William, 100, 15, -, 1200, 50, 300

Elliott, Stokley, 150, 25, -, 1200, 100, 150

West, Melbey, 50, 25, -, 600, 50, 600

Figgs, Cyrus, 50, 20, -, 600, 40, 150

Elliott, Elias, 785, 20, -, 900, 10, 250

Elliott, Joseph S., 100, 7, -, 1000, -, 100

Smith, Marshal, 200, 62, -, 3000, 100, 600

Adams, Isaac, 100, 100, -, 4000, -, 200

Elliott, Jacob, 50, 50, -, 1000, -, 300

Cormean, Joseph C., 50, 60, -, 1000, -, 200

White, Joseph G., 8, -, -, 1000, -, 100

White, James G., 40, 40, -, 800, -, 100

White, Spicer M., 30, -, -, 800, -, 200

White, Spicer, 60, -, -, 600, -, 50

Truitt, Sothey, 3, -, -, 600, -, 200

Foskey, James, 100, -, -, 1200, -, 200

Brittingham, Minus, 3, -, -, 500, -, 400

White, William S., 30, 20, -, 1000, 100, 200

Hastings, Sarah, 20, 15, -, 1000, -, 150

White, William B, 60, 35, -, 2000, 50, 300

White, William, 31, -, -, 800, -, 50

Parson, John W., 100, 47, -, 1000, 50, -

Melson, John, 100, 47, -, 1000, 40, -

Elliott, William, 100, 10, -, 1000, 20, 30

O___ Quillen, 10, 4, -, 300, 40, 300

Megee, George, 100, 50, -, 2000, 80, 300

Riggin, William, 200, 100, -, 3000, 70, 200

Bell, William, 100, 100, -, 2000, 100, 500

Bell, Edward, 100, 100, -, 200, -, 250

Bell, Bose (Base), 100, 100, -, 2000, -, 200

NEXT PAGE IS FIRST NAME FIRST

James Maston, 12, 15, -, 700, 40, 100

Mary Hopkins, 15, 10, -, 1000, 100, 70

William Hearn, 53, 15, -, 1000, 50, 30

George Phillips, 60, 40, -, 1500, 30, 100

Jonathan Moore, 40, 30, -, 2000, 40, 200

Leven, Hitch, 290, 40, -, 20000, -, 100

Jones, Wainwright, 27, 15, -, 500, -, 300

Eccestor Moore, 150, 150, -, 7500, -, 550

George York, 75, 45, -, 2400, -, 50

Joseph Moore, 50, 20, -, 1400, -, -

Llyman Spicer, 55, 76, -, 2500, 30, 350

Meils S. Holt, 300, 100, -, 4000, 100, 300

Harry S. Moore, 300, 50, -, 3000, 200, 200

Elie Clifton, 60, 10, -, 1600, -, 250

Leren Lank, 40, 160, -, 1200, -, 200

John Boyce, 10, 100, -, 500, -, 100

John Boyce, 50, -, -, 250, -, -

John Boyce, 75, 5, -, 275, -, 50

Samuel Ceol, 6, 5, -, 200, -, 200

Thomas Renouls, 50, 50, -, 1000, -, 125

James Renouls, 50, 50, -, 500, -, 5

George Moore, 75, 85, -, 2500, -, 200

Dennis Phillips, 24, 100, -, 6000, 200, 500

Elijah Morgan, 200, 500, -, 4000, -, 400

Morgan Morgan, 100, 40, -, 1500, -, 100

Jeremiah Eskridge, 100, 68, -, 3000, 100, 200

Isaac Boman (Bornan), 100, 100, -, 2000, -, 50

Hester Belle (Beul), 100, 100, -, 2000, 50, 200

NEXT PAGES AGAIN IN LAST NAME FIRST

Fisher, Alexandria, 60, 20, -, 1500, 300, 700

Fisher, Samuel, 60, 10, -, 1600, 300, 600

Hastings, Samuel, 60, 40, -, 1500, 100, 100

Harmon, Ennis, 200, 200, -, 4000, 60, 400

Sraswood (Traswood), Samuel, 160, 25, -, 2000, 100, 360

Sharp, Joshua, 60, 40, -, 1000, 60, -

Metter, Zebedee, 100, 300, -, 4000, -, -

Raordin, Benjamin, 460, -, -, 3000, 100, 1200

Owens, John, 260, 140, -, 6000, 200, 700

Brown, George F., 60, 70, 20, 2000, 100, 335

Brown, George F., 12, -, -, 100, -, -

Samuels, Edward, 25, -, -, 700, 100, 35

Samuels, John W., 100, 75, 25, 1500, 40, 160

Dyer, Thomas D., 100, 40, -, 1600, 20, 60

Jewell, William 90, 90, -, 1400, 20, 100

Brown, Tilghman, 36, 14, -, 500, 10, 90

Reynolds, George, 60, 100, -, 1500, 25, 100

Cannon, Thomas, 175, 100, -, 6000, 200, 900

Russel, William, 25, 50, -, 1000, -, 75

Elliott, James, 6, -, 11, -, 300, 100

Reynolds, William 100, 75, -, 2000, 60, 300

Stewart, Meiles, 4, -, -, 300, 20, 30

Stewart, George H., 275, 100, -, 2700, 160, 600

Laws, Benjamin, 5, -, -, 400, -, 30

Hill, George, 150, 92, -, 4000, 150, 300

Spicer, Thomas, 200, 50, 20, 5000, 160, 360

Boys, James, 130, 70, -, 3000, 100, 400

Conaway, William, 60, 140, -, 2200, 60, 420

Crampfield, Jacob, 40, 33, -, 1800, 25, 200

Spicer, Charles, 50, -, 16, 600, 15, 100

Willey, Tilghman, 200, 125, -, 6500, 175, 900

Owens, Edward, 80, 22, -, 2000, 50, 600

Isaacs, John W., 200, 40, -, 3000, 100, 400

Short, George, 64, 44, -, 1200, 40, 160

Lofland, John, 100, 158, -, 1200, 20, 200

Prettyman, Josiah, 125, 75, -, 3000, 75, 375

Russel, James, 70, 20, -, 1200, 20, 200

Rust, John, 80, 71, -, 3500, -, 280

Johnson, Bayard, 100, 30, -, 1820, 40, 461

Morgan, Martin, 200, 100, -, 5000, 200, 800

Conaway, Curtis, 150, 50, -, 4000, 100, 430

Short, John M., 63, 32, -, 1500, 25, 320

Short, Samuel, 63, 32, -, 1500, 25, 250

Short, John C., 80, 45, -, 5000, 150, 700

Short, Alfred, 45, 69, -, 2000, -, -

Short, Margaret, 25, 15, -, 1000, 30, 500

Coverdale, James, 75, 25, -, 1500, 20, 200

Swain, Spencer, 40, 25, -, 800, -, 90

Swain, John A., 37, -, -, 700, -, 100

Swain, John, 18, 12, -, 500, -, 200

Fooks, Johnathan, 75, 195, -, 3500, 100, 460

Swain, Walter, 76, 30, -, 2000, 35, 421

Day, John, 100, 50, -, 3000, -, 380

Short, Samuel _., 100, 71, -, 1300, -, 240

Short, Gilly C., 75, 20, -, 2000, -, 190

Swain, John B., 96, -, -, 600, 20, -

Smith, David _., 60, 71, -, 6000, 100, 425

Smith, James W., 60, 70, -, 1200, 25, 210

Nutter, Zebedee, 7, -, -, 300, 20, 300

Ed__son, Nehemiah, 60, 61, -, 2500, 50, 560

Eskridge, James, 100, 40, -, 1400, 100, 400

Tucker, Robert, 48, 47, -, 1800, 20, 140

Jones, Sarah, 82, 33, -, 1000, 10,40

Spicer, Charles, 100, 60, -, 1600, 20, 200

Turner (Garner), Charles, 17, -, -, 400, -, -

Oliphant, Isaac, 80, 110, -, 2500, 100, 300

Jones, Thomas, 95, 15, -, 1500, 150, 400

Prichard, Mariah, 50, 50, -, 1000, 30, 180

Conaway, Nancy, 100, 100, 2000, 75, 580

Fleetwood, Curtis, 40, -, -, 2000, 100, 100

Dawson, Sally, 100, 196, -, 2000, 40, 220

Johnson, Thomas, 100, 100, -, 2800, 25, 190

Messick, James, 100, 52, -, 1500, -, 900

Dolbey, Andrew _., 200, 100, -, 4000, 200, 600

Giben, Julio, 30, 10, -, 400, -, -

Spicer, Edward, 40, -, -, 600, 25, 110

Jones, Charles, 30, 4, -, 300, -, 270

Conaway, Louis, 50, 38, -, 700, -, 200

Joseph, Gideon, 100, 25, -, 1200, -, -

Conaway, Purnell, 78, -, -, 100, -, -

Tindal, Benton, 150, 50, -, 2000, 100, 600

Conaway, Wingat, 80,10, -, 800, 20, 300

Tindal, James, 100, 30, -, 1000, 100, -

Tindal, Abrams, 150, 100, -, 3000, 75, 350

Messick, Phillip, L., 60, -, -, 800, 20,180

Elliott, John H., 110, 40, -, 1000, 20, 210

Teague, Mary, 50, 15,-, 500, - , 60

Tindal, Johnathan, 40, 40, -, 600, -, -

Messick, Mary, 76, -, -, 1000, 20,140

Jefferson, Cyrus, 60, -, -, 500, 10, 40

Dreddon, Stephen, 6, -, -, 300, 20, 200

Lambden, Sovrin, 180, 180, -, 3600, 50, 340

Harris, Salomon, 60, 98, -, 1000, 40, 400

Fleetwood, Phillip, 125, 125, -, 2000, 50, 380

Conaway, Nathaniel, 90, 40, -, 400, -, -

Conaway, Nathaniel, 10, -, -, 400, 100, 220

Tindal, Johnathan, 100, 50, -, 1500, 110, 330

Fleetwood, William, 120, 40, -, 1600, 160, 600

Lollace, James, 32, -, -, 300, -, -

Conaway, Levin, 140, 60, -, 2000, 100, 260

Hastings, James, 100, 65, -, 2000, 30, 300

Conaway, Isaac, 100, 100, -, 2000, 60, -

Warrington, Samuel, 200, 200, -, 8000, 1000, 461

Conaway, Minos T., 250, 250, -, 5000, 200, 180

Bryant, Burton, 180, 60, -, 2500, 100, 690

Tindal, George, 150, 50, -, 2000, 40, 270

Williams, William, 100, 50, -, 1200, 20, 200

Hutchens, Alfred, 9, -, -, 400, -, -

Philips, Warren, 100, 50, -, 1000, 30, 180

Dalby, Johnathan, 136, 70, -, 3000, 140, 600

Smith, Frank E., 75, 25, -, 1000, 30, 120

Barr, David, 125, 75, -, 2000, 150, 500

Barr, Robert, 200, 50, -, 2500, 150, 480

Fooks, Esther, 190, 10, -, 4500, 100, 700

Marvel, Josiah P., 160, 40, -, 2400, 300, 600

Joseph, Thomas, 150, 100, -, 5000, 20, 125

Hill, David, 50, -, -, 600, 40, 200

Hill, Johnathan, 50, -, -, 500, 40, 160

Swain, Abe, 100, 30, -, 1000, 20, 100

Donophon, Robert, 140, 60, -, 2500, 100, 500

Conaway, Noble, 260, 100, -, 3000, 200, 700

Conaway, Purnell, 80, 80, -, 1000, 20, 120

Tindal, Peter, 120, 50, -, 1800, 75, 300

Fleetwood, Cyrus, 150, 150, -, 3000, 100, 360

Hill, Manna, 100, 50, -, 1800, 100, 310

Hitchens, Gideon, 60, 40, -, 1500, 30, 300

Hearn, Benjamin, 100, 10, -, 1500, 40, 340

Prettyman, Louisa, 30, 25, -, 600, -, -

Conaway, Elizabeth, 7, 4, -, 100, -, -

Calloway, William, 120, 40, -, 1200, 40, 200

Carey, Theodore, 100, 25, -, 1000, 40, 290

Tindal, Isaac, 70, 70, -, 1000, 75, 400

Hurley, William, 80, 20, -, 1000, 20, 180

Lynch, Peter, 61, -, -, 1000, 20, 140

Davidson, John, 85, 80, -, 1500, 40, 500

Coffin, Thomas, 40, 100, -, 130, 10, 60

Cary, John, 40, 50, -, 900, 10, 70

Hurley, Theophilus, 100, 10, -, 1100, 20, 180

Wilkins, James, 40, 40, -, 900, 10, 130

Wilkins, Thomas, 100, 116, -, 2400, 100, 430

Smith, Gideon, 80, 30, -, 1000, 10, 300

Wilson, David, 80, 30, -, 100, 10, 300

Taylor, William, 150, 250, -, 3000, 70, 660

Short, Wingate, 90, 22, 1100, 50, 200

Short, John, 40, 77, -, 900, 50, 130

West, James, 50, 100, -, 1500, 10, 40

Chorman, Alfred, 40, -, -, 200, 10, 190

Dickson, Benton, 20, -, -, 100, 10, 60

Dill, Edward, 100, 80, -, 1200, 10, 50

Smith, William C., 80, 35, -, 1000, 40, 220

Smith, Sally A., 100, 20, -, 1500, 40, 200

McCauley, Eliza, 100, 40, -, 1400, 30, 160

McCauley, Daniel, 75, 105, 40, 2500, 100, 1400

Swain, John H., 100, 50, -, 1600, 20, 300

Isaacs, Noah, 100, 250, -, 4500, 100, 440

Isaacs, James O., 100, -, -, 1000, 20, 110

Cornwell, Baptist, 100, 20, -, 4500, 50, 410

Maxwell, Elias, 100, 100, -, 3000, 50, 300

Perkins, Purnell, 100, 25, -, 1300, 10, 51

Hurley, James, 150, 100, -, 2500, 40, 300

Conaway, Gilly(Gelby, Gellis) J., 25, 25, -, 500, 10, 240

McDowell, Zach, 15, 35, -, 500, 25, 100

Reynolds, David, 110, 20, -, 800, 50, 200

Lofland, Cornelius, 175, 100, -, 2700, 20, 400

Short, James E., 70, 40, -, 1000, 25, 450

Wilson, Joseph, 75, 25, -, 2000, 50, 310

Isaacs, Joseph, 140, 160, -, 3500, 40, 300

Isaacs, Minos T., 120, 250, -, 1875, 50, 500

Isaacs, Hiram, 90, 60, -, 1500, 40, 250

Samuels, William, 80, 120, -, 2000, 10, 80

Turner, Lewis, 40, 40, -, 700, 10, 160

Smart, Alexandria, 60, 60, -, 1000, -, 30

Beebe, John, 55, 145, -, 2000, 30, 100

Sharp, Benton, 150, 124, -, 2740, 75, 400

Sharp, Theophilus, 100, 60, 60, 2300, 25, 150

Conaway, Elizabeth, 140, 60, -, 2000, 20, 400

Sharp, Joseph, 60, 57, -, 1500, 75, 258

Sharp, William W., 150, 150, -, 5000, 200, 800

Dickerson, Benton, 60, 40, -, 1000, 10, 270

Smith, William, 80, 37, -, 1000, -, 280

Turner, Isaac, 75, 30, -, 1000, -, 280

Griffin, Charles, 100, 50, -, 1500, 20, 200

Griffin, James, 100, 90, -, 1900, 20, 250

Passwaters, Sebastian, 65, 140, -, 2000, 10, 90

Passwaters, James, 40, 20, -, 400, 10, 100

Passwaters, Samuel, 35, 15, -, 300, 10, 110

Banner, Thomas, 100, 80, -, 1500, 20, 90

Smith, Chalton, 150, 75, -, 3000, 60, 500

Miller, James, 30, 10, -, 400, 20, 140

Brace, George, 48, 78, -, 4000, 50, 300

Webb, James, 150, 50, -, 2000, 25, 370

Griffith, Jeremiah, 80, 40, -, 2000, 40, 390

Johnson, Charles, 27, 26, -, 636, 50, 280

Davis, John, 100, 25, -, 2000, 50, 200

Satterfield, John H., 100, 33, -, 2300, 50, 270

Bainbridge, Robert, 80, 57, -, 2000, 50, 190

Bayley, John S., 150, -, -, 4000, 30, 340

Coats, Thomas, 80, 80, -, 4000, 120, 350

Dawson, William, 120, 120, -, 2500, 40, 400

Bullock, John H., 90, 40, -, 2500, 10, 180

Lynch, John W., 125, 16, -, 1700, 100, 210

Dempsey, George S., 125, 50, -, 2000, -, 230

Vincent, James, 100, 40, -, 2500, 80, 410

Elliott, Kingsbury, 56, -, -, 3000, 90, 290

Carpenter, James, 5, -, -, 700, 10,40

Jones, Edward, 120, 180, -, 8000, 100, 500

Carlisle, Joseph, 200, 150, -, 14000, 200, 800

Willey, George, 100, 100, -, 3000, 40, 200

Willey, John, 100, 50, -, 2000, 30, 380

Staton, Amos, 100, 30, -, 2000, 25, 300

Lynch, David, 115, 40, -, 2000, 30, 430

Marine, Charles, 100, 200, -, 1800, 10, 130

Rion, John L., 80, 20, -, 1000, 20, 158

Tatman, John, 175, 75, -, 2500, 40, 400

Owens, Robert, 160, 60, -, 7000, 100, 640

Tatman, Sally, 80, 71, 2500, 80, 459

Tatman, Charles P., 80, 73, -, 1900, 20, 420

Smith, Michael, 158, 65, -, 2000, 100, 500

Murphy, Lizzie, 4, -, -, 100, 40, 161

Petrigen (Petrigru), George, 70, 90, -, 1600, 10, 120

Willey, Samuel, 50, 83, -, 1380, 20, 130

Hatfield, Jones, 75, 75, -, 1500, 20, 160

Hemmons, Joshua, 70, 70, -, 1400, 10, 70

Carlisle, Charles, 80, 80, -, 1500, 10, 100

Rion, Thomas, 160, 240, -, 3000, 10, 400

Tatman, Carlton, 90, 10, -, 1000, 30, 200

Haye, John, 20, 46, -, 500, 10, 160

Spanish, John, 90, 50, -, 2000, 30, 310

Williams Benjamin, 70, 100, 95, 1800, 20, 200

Petrigen (Petrigru), 60, 40, -, 1200, 10, 200

Willey, Weightman, 60, 40, -, 1200, 10, 110

Hersey, John, 75, 25, -, 1600, 20, 295

Willey, Ann, 50, 35, -, 500, 30, 60

Willey, Margaret, 70, 15, 10, 4000, 10, 40

Lynch, William, 100, 80, 34, 1200, 40, 280

Lynch, Alexander, 80, 40, -, 1000, 15, 200

Welch, James, 94, -, -, 2500, 45, 480

Webb, Isaac, 125, 200, -, 4000, 75, 500

Webb, James P., 200, 236, -, 2000, 10, 150

Paswaters, Clement, 100, 200, -, 1500, 10, 100

Sharp, James, 50, 51, -, 1000, 15, 500

Webb, William, 75, 225, -, 2000, 10, -

Hagyard (Haggard), William, 200, 200, -, 4000, 100, 400

Macklin, Johnathan, 100, 75, -, 3000, 50, 800

Spicer, James, 125, 100, -, 3000, 100, 270

Fowler, Margaret, 80, 50, -, 2000, 20, 220

Owens, Isaac D., 70, 50, -, 2000, 100, 300

Owens, Johnathan, 200, 68, -, 2000, 40, 240

Sharp, Joshua, 300, 200, -, 17000, 200, 1100

Willey, James, 300, 20, -, 10000, 70, 540

Lyons, John H., 200, 40, -, 18000, 150, 800

Carlisle, Samuel, 106, 105, -, 4100, 60, 600

Rion, William, 140, 50, -, 3800, 30, 590

Willey, John, 200, 100, -, 8000, 75, 750

Hollace, Silas, 275, 185, -, 10000, 200, 1000

Todd, Luther, 200, 100, -, 3000, 40, 300

Todd, Jacob, 600, 200, -, 8000, -, -

Scott, Alexandria, 15, -, -, 500, 60, 340

Layton, Thomas, 170, 130, -, 6000, 121, 500

Johnson, John Q., 40, 60, -, 2500, 27, 100

Buley, George, 150, 50, -, 4000, 80, 180

Jordan, William, 15, 15, -, 1500, 60, 170

Rawlins, Thomas, 200, 50, -, 5000, 160, 520

Graham, Uriah, 85, 10, -, 2000, 40, 450

Collins, Willliam, 165, 70, -, 5500, -, -

Smith, George, 40, 165, -, 2000, 10, 161

Reddon, George, 150, 100, 50, 3000, 40, 400

Adams, Nicholas, 75, 175, -, 1500, 40, 225

Phillips, George H., 75, 90, -, 3000, 40, 200

Ellegood, Robert, 250, 200, -, 13500, 200, 500

Calloway, Samuel, 125, 25, -, 3000, 30, 390

Penniwell, John, 175, 75, -, 5000, 40, 290

Allen, Edward, 130, 45, - , 3500, 20, 180

Calhound, Thomas, 100, -, -, 2000, 70, 140

Calhound, John, 100, 100, -, 3000, 25, 310

Jones, Ezekiel, 120, 110, -, 2800, 50, 150

Gosllin, James, 140, 35, -, 2000, 40, 320

Conaway, Henry, 60, 30, 900, 40, 180

Johnson, Lewis, 20, 50, -, 1000, 40, 190

Johnson, David, 90, 20, -, 1600, 40, 200

Outen, James, 20, -, -, 2000, 40, 160

Mathews, John H., 40, -, -, 800, 20, 170

Jones, Thomas, 90, 10, -, 2000, 40, 200

Bennett, Nicholas, 130, 50, -, 2500, 50, 260

McDowell, James, 4, -, -, 200, 10, 40

Spicer, Jacob, 52, -, -, 600, 15, 180

Walker, John, 75, 97, -, 3000, 200, 300

Owens, Isaac D., 70, 50, -, 2000, 60, 340

Owens, Alexandria, 60, 17, -, 1500, 40, 400

Radcliff, James H., 60, 17, -, 1000, -, -

Messick, Miles, 343, 100, -, 10000, 180, 900

Hindson, John, 12, -, -, 200, 25, 250

Russel, Theodore, 60, 60, -, 1000, 20, 160

Johnson, William D., 90, 40, -, 1200, -, 200

Russel, Martin, 75, 175, -, 2000, -, 190

Carey, Albert, 300, 75, -, 10000, 500, 1200

Nelson, Daniel, 150, 50, -, 1800, 40, 150

Isaacs, Louis, 100, 30, -, 1500, -, 200

Sudler, John R., 318, 40, 30, 35000, 300, 1200

Lord, Luther, 57, -, -, 2500, 40, 400

Sullivan, James, 38, -, -, 1800, 40, 200

Cannon, Philip L., 1156, -, -, 12000, 500, 800

Willen, Thomas, 30, -, -, 2000, -, 160

Myers, David S., 109, -, -, 8000, 600, 400

Hatfield, David, 19, -, -, 300, -, 40

Vandeburg, John H., 50, -, -, 4000, 100, 190

Carey, William, 80, -, -, 4000, 40, 150

Dale, John, 200, 60, 20, 15000, 500, 1100

Cannon, John, 225, 75, -, 15000, 100, 800

Walls, Lafayatte, 40, -, -, 4000, 20, 210

McIlvain, Robert, 150, -, -, 6000, 160, 600

Sharp, Jesse, 200, 50, -, 3000, 30, 200

Records, John R., 150, 50, 40, 10000, 50, 500

Jones, John F., 150, 40, -, 10000, 250, 600

Ridgway, Tilghman, 100, 35, -, 2000, 25, 240

Allen, Zaddock, 125, 133, -, 3000, 15, 240

Cordray, Robert, 75, 17, -, 2000, 100, 350

Jones, Albert, 250, 30, -, 2240, 40, 300

Collison, John M., 129, 64, -, 6000, 190, 700

Lord, Andrew, 51,-, -,2500, 100, 440

Hayes, John, 180, 20, -, 10000, 400, 1000

Barwick, Nathaniel, 70, 42, -, 2000, 150, 400

Mims, Beechum, 300, 210, -, 5400, 150, 1000

Collison, Robert, 130, 234, -, 3750, 25, 200

Collins, George, 100, 60, -, 3000, 200, 500

Pratt, Henry, 60, 27, -, 1200, 50, 400

Hastings, Joseph, 100, 100, -, 2000, 10, 160

Hamilton, John, 90, 70, -, 2000, 75, 500

Edgell, William, 130, 90, -, 2200, 100, 740

Maurice, Johnathan, 40, -, -, 500, 10, 180

Maurice, Hezekiah, 84, -, 41, 2000, 50, 400

Maurice, William, 100, 60, -, 1900, 50, 400

Lane, John, 100, 100, -, 4000, 100, 350

Oday, Charles, 40, 50, -, 2000, 15, 200

Staton, Nehemiah, 130, 80, -, 6000, 175, 600

Gray, Peter, 125, 35, -, 5000, 100, 500

Ellingsworth, John, 125, 75, -, 3000, 40, 380

Bradley, William, 100, 40, -, 2000, -, 310

Adams, Garrison, 80, 100, -, 1000, 20, 300

Pennwell, Simeon, 500, 200, 100, 20000, 500, 1100

Stephens, John W., 94, 26, -, 2000, 75, 601

Flemming, John, 100, 30, -, 2000, 80, 500

Smuller, Whittington, 110, 80, -, 1500, 30, 300

Williams, E. R., 152, 89, -, 3500, 1000, 1000

Willey, Mary A., 100, 30, -, 5000, 100, 300

Rawley, Samuel, 100, 40, -, 3000, 350, 400

Oday, Solomon, 100, 50, -, 3600, 50, 390

Oday, James, 100, 35, -, 2900, 40, 371

Jones, Alexander, 150, 54, -, 3500, 100, 700

Conway, Job, 200, 150, -, 5000, 25, 370

Cartrell, S. B., 145, 81, -, 4000, 50, 231

Knox. Samuel, 125, 20, -, 1450, 100, 300

Gray, Thomas, 100, 130, 2500, 40, 260

Scott, John W., 80, 20, -, 2000, 20, 375

Carlisle, William, 166, 20, -, 4000, 35, 400

Wright, William, 120, 37, -, 2000, 20, 140

Ross, William, 40, 40, -, 1000, 10, 185

Petrigren (Petigru), William, 130, 20, -, 3000, 36, 300

Collison, Alexandria, 150, 89, -, 2500, 60, 250

Jones, John, 150, 89, -, 2590, 60, 250

Tuttle, Reuben, 52, -, -, 2000, -, -

Fisher, David W., 80, 17, -, 1800, 50, 600

Todd, John H., 130, 70, -, 3600, 100, 700

Blanchard, Henry, 160, 52, -, 2540, 200, 900

Jones, Charles, 100, 40, -, 1500, 50, 100

Kinnerman, Ambrose, 50, 100, -, 3000, 50, 300

Currey (Carrey), Thomas, 40, 5, -, 500, 10, 200

Hallowell, George, 100, 40, 17, 3580, 50, 341

Adkinson, Gove, 80, 50, 30, 250, 40, 300

Chafing (Chassing), Matthews, 100, 100, -, 3000, 50, 400

Adams, William, 150, 50, -, 2500, 200, 500

Adams, Rogers, 175, 125, -, 8000, 50, 500

Adams, Edward, 100, 100, -, 3600, 170, 530

Reynolds, Rod, 50, 250, -, 3000, 40, 200

Pattin, Wellington, 100, 150, -, 5000, 200, 250

Chafing, John, 60, 60, 37, 1000, 40, 360

Hignut, John, 200, 50, -, 2000, 35, 400

Hastings, Rendle, 165, 165, -, 2500, 25, 200

Lord, Thomas, 130, 80, -, 10000, 160, 900

Oday, William, 300, 200, -, 8000, 100, 458

Willey, Isaac, 55, 6, -, 1800, 60, 400

Maurice, Jeremiah, 63, -, -, 2500, 100, 300

William, Ark, 15, 15, -, 500, 20, 340

Nichols, Umphers, 300, 200, -, 8000, 100, 100

McCauley, Edmond, 100, 100, -, 2000, 40, 400

Richards, John E., 120, 100, -, 5500, 400, 700

Blades, Isaiah, 250, 100, -, 7000, 75, 500

Hill, Isaac, 150, 150, -, 2400, 16, 390

Lord, John, 200, 200, -, 5000, 50, 300

Gamben, Matthew, 15, 15, -, 50, 10, 200

Collison, William, 150, 58, -, 4000, 40, 1000

Lankford, Benton, 75, 100, -, 1700, 10, 190

Hignut, George, 110, 30, -, 1400, 10, 210

Viland, Jasond, 75, 35, -, 1000, 15, 300

Adams, Daniel C., 180, 80, -, 2100, 151, 600

Stokes, James, 70, 70, -, 1000, 10, 161

Reed, Ekiel, 60, 29, -, 700, 15, 150

Willoughby, Job, 135, 138, -, 3000, 50, 380

Ross, William, 80, 70, -, 4500, 125, 500

Ross, Thomas G., 100, 90, -, 2500, 25, 400

Bradley, Minos, 125, 70, -, 2400, 10, 100

Woodcock, Arthur, 4, 56, -, 800, 50, 300

Hill, William, 60, -, -, 1500, 100, 240

Tredfield, Bradly, 40, 59, -, 2000, 300, 360

Robins, Samuel, 150, 136, -, 4000, 165, 310

Gullet, A. W., 150, 100, -, 3000, 50, 200

Collison, William, 75, 50, -, 1500, -, -

Hignut, Ellender, 90, 70, -, 1400, -, -

Swann, James, 10, 40, -, 500, -, -

Layton, Beechum, 40, 35, -, 700, -, 280

Hignut, Margaret, 35, 15, -, 500, -, -

Reed, William, 60, 50, -, 1000, -, 190

Adkison, Burton, 40, 46, -, 1000, -, -

Kerr, Lorenzo, 100, 60, -, 4000, -, 280

Lawthe, Levi, 85, 15, -, 3000, 75, 300

Harris, Frank, 50, 12, -, 1000, -, -

Trout, George T., 175, 25, -, 8000, 500, 600

McMillen, M. W., 89, 39, -, 5000, 500, 400

Brown, Daniel, 5, 20, -, 2000, 100, 281

Harris, William, 150, 50, -, 7000, 150, 500

Wade, Henry, 6, -, -, 400, -, -

Reed, Stansberry, 175, 50, -, 4000, 100, 700

Culver, George A., 100, 50, -, 2000, 40, 300

Coats, Mark A., 125, 125, -, 4000, 400, 500

Coats, William J., 100, 90, -, 4000, 160, 600

Leadenham, Noah, 140, 23, -, 3000, 200, 700

Reegum, Mitchell, 25, 137, -, 2500, 20, 257

Warren, George, 105, 56, -, 3200, 40, 80

Kinder, Jacob, 106, 80, -, 5280, 150, 300

Kinder, Luis, 100, 60, -, 2000, 100, 460

Hickman, John, 200, 200, -, 4500, 80,600

Parker, William, 200, 355, -, 5500, 40, 260

Ward, John W., 60, 40, -, 1000, -, 40

Willoughby, Richard, 100, 166, -, 2500, 100, 300

Fratt, Nicholas, 159, 237, -, 11000, 250, 700

Wright, Louis N., 120, 173, -, 8000, 200, 600

Wright, Isaac K., 100, 74, -, 3480, 125, 465

Kinder, Isaac A., 100, 74, -, 3480, 125, 465

Kinder, John, 130, 156, -, 6800, 300, 900

Swain, Cornelius, 100, 91, -, 1500, 80, 400

Laramore, Coel (Joel), 160, 200, -, 4000, 10, 160

Short, Daniel B., 120, 120, -, 4800, 200, 500

Coats, Raymond, 150, 75, -, 5000, 150, 500

Whitney, Isaac A., 100, 107, -, 1500, 100, 300

Hastings, Dewit N., 10, -, -, 200, -, 20

Brown, Charles, 100, 100, -, 1000, -, -

Johnson, C. S., 8,20,-, 280, 200, 210

Melson, Benjamin, 200, 160, -, 3600, 100, 800

Marshal, James, 75, 118, -, 1550, 40, 361

Elligood, Charles, 100, 300, -, 3000, 10, 280

Campbell, John, 90, 30, -, 2500, 100, 470

Ray, John, 75, 133, -, 2500, 200, 300

Burton, James, 80, 60, -, 2200, 10, 100

Swain, Robert, 125, 64, -, 7580, 125, 500

Swain, James, 175, 100, -, 11000, 500, 700

Hitch, Henry, 149, 75, -, 2400, 100, 600

Hitch, Henry, 100, 100, -, 2000, -, -

Ward, George, 30, -, -, 300, 10, 200

Cannon, Louvenia, 9, -, -, 250, 10, 10

Rickets, John T., 110, 140, -, 3000, 800, 350

Brown, William, 96, -, -, 1500, 25, 250

Maurice, Tyrus, 100, 125, 125, 2500, -, 260

Clarkson, F. M., 80, 20, -, 1000, 500, 1000

Brown, Joshua, 90, 60, -, 1000, 40, 300

Cannon, Abraham, 100, 50, -, 2000, 60, 280

Kinder, Tilghman, 120, 45, -, 3500, -, 500

Jacobs, Thomas, 160, 80, 20, 5000, 19, 1950

Horsey, John, 109 26, -, 7000, 100, 400

Kinder, Daniel D., 130, 231, -, 5000, 150, 700

Wright, Twiford, 65, 75, -, 1500, 75, 500

Ressum, Rylie, 100, 31, -, 2000, 16, 280

Parker, Joseph, 80, 20, -, 2000, -, 100

Rust, Catesby F., 150, 80, -, 4000, 200, 600

Shepherd, John, 38, 5, -, 1000, 100, 400

Townsend, Robert, 20, 15, -, 200, -, 50

Williams, William, 20, 15, -, 230, -, 80

Jacobs, Henrietta, 300, 100, -, 6000, 200, 400

Kinder, John H., 180, 60, -, 3500, 150, 420

Corbin, William, 100, 25, -, 2000, 40, 390

Noble, Johnathan, 120, 67, -, 2805, 120, 700

Pussum, Plymouth, 50, 35, -, 800, 40,2 00

Cannon, Jery, 60, 70, -, 3000, 40, 380

Bradly, Peter, 30, 65, -, 1200, 60, 240

Corbin, Stephen, 100, 60, -, 3500, 200, 540

Rickets, Elijah, 135, 135, -, 3000, 10, 20

Corbin, Stephen, 65, 41, -, 2000, -, 41

Bradly, Anthony, 11, -, -, 200, 10, 10

Williams, Jesse, 94, 14, -, 3000, 40, 300

Williams, William, 100, 80, -, 2000, 20, 200

Hill, Osmos D., 112, 116, -, 3000, 100, 360

Sullivan, Meredith, 95, 37, -, 3000, 100, 300

Williams, Robert, 40, -, -, 500, 30, 200

Conley, Daniel, 100, 41, -, 2000, 10, 290

Prawl, William, 100, -, -, 3000, 20, 140

Meilmon, Johnathan, 200, 200, -, 12000, -, 560

Layton, Garret S., 53, 15, -, 1000, 250, 400

Satterfield, Archa, 87, 43, -, 4000, 60, 250

Richards, Philip H., 200, 180, -, 10000, 200, 600

Sedgwick, John, 30, -, -, 300, -, 200

Jacobs, Loxley R., 180, 120, -, 7500, 100, 90

Dilworth, John D., 180, 210, -, 10000, 420, 800

Carroll, Jerry, 11, -, -, 200, -, 50

Gordon, Elisha, 100, 87, -, 4000, 200, 700

Ross, Sanders, 110, 40, -, 2000, 10, 120

Brown, Isaac, 150, 150, -, 3000, 15, 100

Jacobs, Alexander, 150, 130, -, 6000, 180, 700

Moore, Thomas, 260, 145, -, 6000, 225, 500

Bryant, William, 130, 135, -, 5300, 100, 600

Lord, Henry, 192, 95, -, 3000, 75, 400

Jacobs, Nathaniel, 175, 75, -, 8000, 250, 900

Records, Joseph, 166, 84, -, 6250, 200, 1100

Cary, James, 85, 80, -, 1000, 60, 1100

Bullock, Thomas, 100, 80, -, 3000, -, 450

Melson, Thomas, 85, 45, -, 1000, -, 280

Williams, David, 90, 30, -, 2000, -, 120

Jones, William T., 70, 40, -, 1800, -, 240

Richard Dodd, 80, 30, -, 5000, 100, 500

FIRST NAME FIRST TO END

Joshua Obier (Obien), 400, 200, -, 8000, 300, 1250

Mathew Davis, 150, 70, -, 2000, 40, 150

Major Allen, 300, 200, -, 8000, 250, 900

William Medford, 150, 50, -, 2000, 50, 300

James Fleetwood, 75, 25, -, 4000, 250, 220

Joseph Neal, 200, 315, -, 12000, 300, 1025

George Rickards, 50, -, -, 1000, 50, 100

William Cannon, 100, 50, -, 2000, 100, 250

Jacob Kinder, 169, 40, -, 2430, 350, 500

Alfred Fisk, 110, 36, -, 4000, 500, 500

John Twiford, 300, 100, -, 2000, 150, 300

Silas Smith, 50, -, -, 500, 50, 100

Ann Brookfield, 5, -, -, 50, -, 2

Thomas Dawson, 267, -, -, 8000, 200,700

Joseph Jones, 88, 20, -, 3000, 100, 100

David Hearn, 47, -, -, 500, 50, 250

Augustus Obien, 184, -, -, 1500, 175, 250

James Moore, 180, -, -, 1200, 300, 300

Samuel Eskridge, 100, 56, -, 4000, 100, 300

Leven Lecate, 100, 30, -, 2000, 100, 600

Perry Darby, 150, 50, -, 5000, 100, 600

David Williams, 70, 10, -, 1200, 100, 200

Edward, Willen, 100, 62, -, 3000, 400, 400

Lilby (Lilly) Williams, 70, 30, -, 2500, -, 300

Perry Wright, 5, -, -, 150, -, 25

James Wainwright, 30, 5, -, 800, 10, 100

Jesse Allen, 76, 20, -, 3000, 200, 275

Milmer Williams, 110, 30, -, 3000, 20, 250

Tyrus Phillips, 10, -, -, 600, 5, 100

Henry Tull, 40, -, -, 500, 25, 200

James Lines, 150, 50, -, 5000, 100, 600

Martin Lankford, 100, 50, -, 1000, 200, 500

James Oday, 40, 25, -, 2500, 200, 500

Leven Allen, 90, 10, -, 1500, 100, 400

Jacob Borons, 387, 262, -, 8000, 1000, 800

George Brown, 20, -, -, 200, 25, 150

Robert Hastings, 100, 50, -, 2000, 25, 25

George Collins, 200, 75, -, 3000, 50, 150

Peter Tull, 50, -, 15, 700, 20, 200

Robert Tull, 27, 6, -, 800, -, 200

William Penton, 75, 100, -, 3000, 100, 200

James Moore, 6, -, -, 200, -, 25

James T. Ward, 100, 40, -, 2000, 50, 200

Henry Bell, 100, 20, -, 130000, 100, 400

Leven Brown, 150, 50, -, 4000, 300, 240

Mary Obien (Obier), 150, 75, -, 5000, 100, 175

Charles Benston, 45, 5, -, 1500, -, 200

John Vandever, 50, 25, -, 2500, -, 200

Warren, Feantom, 150, 50, -, 4000, 100, 400

George Foster, 20, 10, -, 600, -, 5

Edward Gillis, 80, 20, -, 5000, 300, 600

Leven Nicholis, 80, 20, -, 2250, 100, -

Luke Messick, 37, 12, -, 1000, 25, 250

Silas Wainwright, 160, 40, -, 2000, 40, 350

Robert Knowles, 70, 70, -, 1400, 150, 150

Littleton Lankford, 300, 100, -, 8000, 120, 600

Samuel Messick, 100, 150, -, 3000, 200, 350

Elijah Colbern 150, 150, -, 3000, -, 400

William Allen, 100, 130, -, 3000, 600, 500

Thomas Allen, 100, 50, 25, 1500, -, 125

Edward Calloway, 36, -, -, 300, -, -

Andrew Vinsin, 150, 75, -, 3000, 300, 200

Thomas Brown, 85, 89, -, 3500, 1000, 600

William Cannon, 200, 100, -, 5000, 1000, 600

Hugh Brown, 70, 30, -, 3000, 300, 200

Samuel Jackson, 4, -, -, 400, -, 200

Ira King, 70, 37, -, 2500, 500, 429

John Cannon, 60, 90, -, 4000, 40, 125

Peter Waller, 100, 40, -, 6000, 150, 250

Levi Cannon, 16, 2, -, 1000, -, 10

William Allen, 100, 50, -, 3000, 100, 200

William Collins, 100, 10, -, 2000, 100, 200

Timothy Skidmore, 55, 20, -, 1500, 30, 175

William Messick, 140, 163, -, 6000, 200, 600

Halsey Mastien, 150, 75, -, 5000, 200, 400

Ezekiel Reed, 70, 24, -, 6000, 50, 228

Samuel Collins, 5, -, -, 200, 100, 160

Gilly Ellis, 16, 8, -, 500, -, 140

Abraham Lewis, 125, 75, -, 5300, 100, 150

George Jones, 75, 60, -, 1300, -, 150

Edward Wheatly, 65, 5, -, 200, -, 200

William Husten, 250, 250, -, 7000, 200, 600

Josiah Vaughn, 300, 300, -, 8000, 150, 400

Samuel Neal, 20, 180, -, 2000, 100, 50

John Helm, 100, 50, -, 3000, 100, 300

Elijah Ellis, 80, 20, -, 5000, 50, 280

William Elliott, 100, 20, -, 4000, 50, 400

George Ellis, 47, 100, -, 11000, -, 230

William Ellis, 140, 150, -, 15000, 200, 1000

Theodore White, 100, 50, -, 2000, 10, 200

Jacob Nicholson, 310, 325, -, 21950, 750, 633

Jacob Allen, 200, 200, -, 5000, 150, 500

Curtis Cannon, 40, 51, -, 1500, -, 225

William Admore, 6, -, -, 300, -, 20

Charles Flowers, 180, 120, -, 12000, 300, 425

Alfred Wright, 6, -, -, 500, -, 100

James Trewett, 150, 50, -, 5000, 200, 6000

Robert Frame, 200, 60, -, 6000, 200, 1000

Warren Kinder, 600, 300, -, 30000, 1000, 630

John Collins, 10, -, - 1000, 100, 500

James Layton, 500, 100, 10900, 100, 600

William Bettis, 6, -, -, 300, -, 100

Jesse Wright, 100, 150, -, 4000, -, 20

William Allen, 50, 27, -, 3000, 100, 200

Emale Meligain, 100, 25, -, 3000, 150, 200

Joseph Allen, 25, 70, -, 4000, -, 200

Robert Brown, 150, 57, -, 5000, 400, 500

Burton Allen, 40, 20, -, 2000, 200, 700

James A. Willey, 100, 1, -, 1800, 1000, 125

John Williams, 250, 140, -, 14000, 200, 600

Thomas Anderson, 50, -, -, 1000, 25, 80

Elizabeth Cannon, 100, 25, -, 3000, 100, 500

Charles Hydrick, 170, 194, -, 2000, 340, 1030

Abraham Bivens, 125, 100, -, 14000, -, 300

Jeremiah Adams, 4, -, -, 4000, -, 300

Lewis Shue, 100, 91, -, 4000, 200, 600

William Handy, 200, 150, -, 5000, 150, 300

John Allenson, 90, 20, -, 2000, 100, 40

Eliza Wright, 400, 100, -, 34000, 500, 1300

Turpin Adams, 150, 50, -, 5000, -, 200

William Cannon, 700, 150, -, 5000, 500, 1000

James Brown, 125, 200, -, 7000, -, 700

Batte Cannon, 175, 225, -, 3040, -, 250

Elisah Dickison, 75, 25, -, 3000, -, 400

Leven Waller, 75, 25, -, 1800, -, 150

Nathaniel Lewis, 125, 75, -, 3000, -, 110

James Reynolds, 75, 25, -, 2000, -, 255

Thomas Loyd, 200, 50, -, 4000, -, 252

John Simpson, 100, 50, -, 3000, -, 250

Hugh Taylor, 150, 125, -, 5000, -, 300

William Neal, 160, 100, -, 3000, -, 75

Jerome Layton, 157, 100, -, 500, -, 200

Alece Kinder, 250, 100, -, 6000, -, 500

Abraham Taylor, 120, 20, -, 4000, -, 200

Julie Hall, 100, -, -, 1500, -, 450

William Wright, 150, 50, -, 6000, -, 460

Joshua Butler, 10, 21, -, 500, 30, 125

John Loyd, 100, 40, -, 4000, -, 180

Joseph Cannon, 50, 42, -, 2500, -, 300

Isaac Cannon, 75, 37, -, 4000, -, 300

Leven Williams, 100, 55, -, 4000, -, 450

Batte Adams, 200, 100, -, 8000, -, 650

Burton Hurly, 150, 50, -, 4000, -, 600

Jacob Williams, 180, 45, -, 5000, -, 450

Alead Dawson, 6, -, -, 3400, -, 175

Edward Martin, 225, 100, -, 33000, 150, 1000

Anna Chambers, 130, 10, -, 5600, -, 300

William Dulaney, 250, 225, -, 12000, -, 600

William Brown, 150, 500, -, 2000, -, 150

William Ross, 650, 200, -, 100000, 500, 1800

William Kaevlis, 200, 100, -, 3000, -, 450

Frank Brown, 50, 100, -, 2000, -, 200

William Rogers, 130, 60, -, 8000, -, 600

Jane Brown, 75, 25, -, 2500, -, 400

Newton Williams, 160, 163, -, 5000, -, 300

Wesley Perkins, 100, 50, -, 2000, -, 400

William Spicer, 200, 100, -, 1500, -, 100

Abbott, 9, 31, 111-112, 114-116, 128
Able, 39
Abler, 31
Ables, 31
Achor, 39
Ackerman, 44
Adair, 77
Adams, 2, 5, 34, 101, 121, 139, 142,
145, 151-153, 158-159
Adker, 71
Adkins, 29
Adkinson, 18, 153
Adkison, 154
Admore, 158
Agens, 127
Agnue, 5
Agram, 68
Ahern, 70
Aibh, 54
Aikin, 67
Ake, 103
Akins, 108
Alaband, 34
Alcorn, 64
Aldrich, 37
Alexander, 76, 80
Alfry, 53
Allaland, 38
Allcorn, 69
Allcott, 22
Allee, 12, 14, 16
Allen, 1, 36, 52-53, 74, 123, 133,
144, 151-152, 156-158
Allendorf, 119
Allenson, 158
Alling, 101
Almand, 60
Alrich, 75, 82, 89
Alston, 52
America, 10
Ander, 120
Anderson, 5, 7, 10-12, 15, 23, 30-31,
33-34, 37, 41-43, 55, 58, 97, 101,
129, 137, 158

Andre, 27
Andrew, 26, 39, 101, 142
Andrews, 35, 95
Angell, 41
Anthony, 16, 31, 34, 69
Appleby, 68, 73-74, 88
Appleton, 49
Argo, 9, 117
Armor, 24, 64
Armstrong, 22, 50, 52-53, 62-64, 66,
72, 74, 79, 85, 89
Arnold, 17, 113
Aron, 7
Arron, 5
Arthers, 5
Aspriel, 81
Athly, 50
Atkins, 11, 26, 92, 121-122, 125,
127, 129, 131
Atkinson, 34
Attix, 18
Atwell, 53
Austin, 23, 113
Averill, 83
Aydelott, 98
Ayers, 89
Baccus, 13
Bacon, 15, 77, 100-101, 104, 142
Baer, 87
Bailey, 19, 37, 41, 67, 71, 123, 140
Baily, 17
Bainbridge, 149
Baker, 10, 43, 47, 72, 94, 101, 103-
104, 106, 122-123, 127, 132, 137-
138, 144
Bakey, 62
Baldwin, 63, 70
Bales, 132
Ball, 64, 69
Bancroft, 36, 66
Banks, 27, 55, 59, 69, 76, 98, 100
Banlow, 61
Banner, 149
Banning, 64, 66, 133

Banta, 113
Barber, 1, 22, 52
Barcus, 9, 21, 39-40, 47
Barker, 32-33, 131
Barlow, 32, 49, 59, 61, 67
Barnaby, 73
Barnard, 36
Barnes, 43
Barnet, 39, 97
Barnett, 3, 44
Barr, 148
Barrat, 52
Barratt, 42
Barrett, 2-3
Barris, 68
Barry, 62
Bartholomew, 73
Bartley, 54
Barton, 21, 77, 86, 120, 132
Bartram, 64
Bartran, 62
Barurich, 33
Baruros, 33
Barwick, 152
Bateman, 42
Batten, 81
Baum, 8
Bayley, 149
Bays, 77
Be__, 9
Beach, 141, 143
Beader, 33
Beakling, 45
Bearman, 2
Bearn, 106
Bearster, 84
Beason, 56-57, 60
Beaty, 64
Bebee, 118
Bechel, 26
Beck, 54, 81
Beckman, 94
Beddle, 6, 15, 54
Bedwell, 4, 7
Beebe, 149
Beetts, 32, 104

Begger, 61
Beggs, 78
Beidienar, 117
Beker, 10
Bell, 10-11, 18, 41, 47, 67, 76, 145, 157
Belle, 145
Belville, 82
Bendin, 41
Benjamin, 4, 24
Benn, 15
Bennet, 26-27, 54
Bennett, 15, 25, 42, 61, 85, 98-99, 116-117, 151
Bennit, 52
Bennum, 108-109, 114, 127
Benson, 21, 56-57, 60, 80
Benston, 117, 119, 157
Bernard, 71
Bernite, 47
Berrian, 41
Berry, 21, 59
Berwick, 26
Besser, 87
Beswick, 26
Bethards, 24
Bettey, 58
Bettis, 158
Betts, 23, 98, 110-114, 124
Beul, 145
Biddle, 10
Bilderback, 19
Biddle, 74, 82
Bideman, 67
Biderman, 64
Biggs, 75, 78
Billings, 39, 42
Binns, 44
Birch, 55
Bird, 57, 61, 140
Bishop, 20, 94
Bivens, 158
Black, 4, 45, 79, 109, 111
Blackiston, 20-21
Blackston, 52
Blackwell, 61, 87

Blades, 34, 153
Blanchard, 153
Blandy, 87
Blidt, 52
Blizzard, 13, 39, 107-108, 131
Blood, 37
Bobbett, 8
Bogan, 62
Boggs, 1, 9, 20, 84
Bolden, 10
Bolk, 9
Bolton, 73, 76
Boman, 145
Bonnill, 12
Bonsall, 68
Boone, 12, 42, 48
Booth, 28, 30, 34, 75
Bornan, 145
Borons, 157
Bostic, 90
Bostick, 41
Bosworth, 29
Boulden, 77-79, 84
Boulder, 79
Bourlin, 92
Bowan, 72
Bowden, 122
Bowen, 32
Bower, 71
Bowers, 5, 80
Bowing, 89
Bowl, 74
Bowman, 72
Boyce, 103-104, 116, 119, 141, 144-145
Boyd, 19, 83
Boyer, 4, 13, 19, 22
Boyle, 75
Boyles, 4, 17
Boylishitts, 50
Boynton, 41
Boys, 79, 146
Brace, 149
Bracken, 66
Brackin, 66, 69, 71
Bradford, 12, 66, 75

Bradley, 27, 29, 40, 42, 44, 152, 154
Bradly, 80, 139-140, 155
Brady, 82, 84
Bramble, 71
Brannan, 63
Brasure, 91-94, 120-121, 125
Bratlett, 69
Bratten, 51
Bratton, 50
Brereton, 133
Briley, 37
Brimer, 106, 123
Bringhurst, 58
Brinkley, 66
Brinkly, 18
Brinton 64
Brion, 123
Briscoe, 4
Brister, 49, 53
Britenham, 128
Britingham, 134
Brittingham, 106, 108, 110, 123, 133, 145
Broad, 76
Broadway, 53
Brockson, 49, 54
Broodway, 39
Brookfield, 156
Brooks, 53, 68, 89
Brookway, 38
Brown, 15-16, 21, 30-32, 34, 36-37, 42, 47, 52, 63, 65-66, 68-69, 71, 74, 81, 86, 98, 128, 146, 154-159
Browning, 54
Bruington, 112
Bryan, 83, 112
Bryant, 20, 148, 156
Buchance, 26
Buckanan, 52
Buckingham, 1, 67-68
Buckmaster, 17, 40
Buckson, 50
Budd, 48, 55
Buell, 27
Bugless, 63
Bulcher, 124

Buley, 151
Bull, 64, 99-100
Bullock, 59, 62, 149, 156
Bungy, 51
Bunting, 57, 92-93, 95, 121
Burbage, 97
Burchenal, 41
Burchnel, 39
Burchnell, 40
Burdick, 35
Burges, 83
Burhance, 26
Buris, 50
Burk, 80
Burnahm, 10
Burnham, 32, 85
Burr, 24, 27, 76
Burres, 127
Burris, 3, 84, 102, 105, 114
Burrows, 17-18
Burton, 43, 97, 103, 108-109, 117, 119-121, 124, 128-130, 132-137, 155
Butler, 11-12, 29-30, 159
Bynard, 33
Cade, 8
Cadmus, 21
Cahall, 33, 35-36, 41
Cahoon, 8, 12
Cain, 26, 31-36
Calahan, 65
Calaway, 33
Caldwell, 45
Caleher, 63
Calem, 1
Calhoon, 99, 112, 117, 127-128, 138
Calhoun, 77, 82, 99
Calhound, 151
Calk, 38-39
Callahan, 7, 75
Callans, 32
Callany, 100
Callary, 30, 33, 36
Callaway, 33-34, 45, 101, 106-107, 138, 140-144
Callen, 44
Calley, 40

Callis, 34
Calloway, 138, 148, 151, 157
Calvin, 73
Cambell, 94
Campbell, 2, 67, 81, 88, 93-95, 118, 155
Camper, 29
Canby, 59
Cann, 75, 77
Cannon, 10, 15, 69
Cannon, 98, 100, 102-103, 105-107, 119-120, 136, 138, 144, 146, 152, 155-158
Cardwright, 22
Carey, 18-19, 36, 93-95, 99, 108, 110-111, 115, 121-123, 127, 130-131, 134, 148, 151-152
Carlesie, 27
Carlile, 2
Carlisle, 150-151, 153
Carman, 105
Carmean, 105-106, 138, 143
Carmine, 122, 139
Carney, 19, 56
Carnog, 80
Carpenter, 12, 27, 37, 42, 57, 62, 81, 83, 108, 119, 124, 150
Carr, 39, 82
Carrey, 153
Carroll, 33, 156
Carrow, 54
Carsen, 8
Carslile, 89
Carson, 21, 135
Carter, 6, 21, 38, 41, 47
Cartey, 89
Cartrell, 153
Cary, 102, 148, 156
Case, 23, 38
Casey, 59, 75
Casho, 87
Casiday, 60
Casperson, 82
Cassey, 58
Casson, 5, 13
Cathn, 38

Catlin, 39
Cattle, 12
Catts, 44
Caul, 55
Caulk, 53, 56
Causdon, 79
Causey, 114
Cavender, 13, 19, 35, 52, 78-79, 85, 88-90
Cayton, 64
Ceol, 145
Cergy, 103
Certs, 84
Cexper, 20
Chace, 108
Chafing, 153
Chamberlin, 95, 99
Chambers, 12, 44, 67-68, 70, 88, 115, 159
Champion, 79
Chandler, 60-62, 64, 69, 71-73
Chandles, 80
Chanler, 123
Chapman, 58
Chase, 109, 112, 127, 143
Chassing, 153
Chefferis, 17
Chelooth, 30
Cheltin, 71
Chemmonds, 39
Chericks, 100
Chiman, 105
Chipman, 143
Choate, 87
Chorman, 148
Chrillas, 68
Christian, 79
Christopher, 46, 120
Chrylar, 43
Churchman, 58, 89
Cilierr, 106
Cirvirthen, 115, 117
Cirwithen, 115, 117
Clair, 85
Clampitt, 5, 42
Claranon, 70

Clark, 2-4, 6-7, 9, 18, 33-34, 38, 42, 46-47, 54, 60-62, 71, 76, 81-82, 88, 95, 109, 118, 131-132, 134
Clarkson, 23, 155
Clay, 88
Clayton, 3, 10, 54-55, 84
Cleadin, 73
Cleaver, 14, 17, 76, 82-83, 85
Clement, 6
Clements, 12, 16, 21
Clemet, 62
Clemmonds, 39
Clendaniel, 9, 111-118
Clendenin, 80
Clever, 40
Clifton, 12, 26, 108-109, 111, 118, 145
Cling, 88
Clock, 74
Clog, 94
Cloud, 57, 69, 62-63
Clouds, 55
Clow, 22
Cloy, 100
Coard, 96
Coats, 149, 154
Cochran, 83-86
Codwright, 21
Coemis, 81
Coffin, 108, 116, 129-131, 148
Cogwell, 40
Cohee, 1, 8, 38-39, 44-45, 47
Colbert, 57
Colburn, 81
Cole, 25, 27, 80, 142
Coleburn, 24
Coleman, 66
Colescott, 4
Collier, 96, 100, 111-112
Colling, 139
Collins, 2, 10, 14-17, 24, 27, 34-35, 48, 50, 54, 64, 67, 73, 79, 91-92, 98, 100-102, 119, 127, 129, 131-132, 137-141, 143-144, 151-152, 157-158
Collison, 152-154
Collwell, 7

Colman, 138
Colyer, 54
Comegys, 12
Comly, 89
Conard, 64
Conaway, 123, 125-126, 128-129, 146-149, 151
Conelly, 80
Conely, 80
Congleton, 57
Conley, 65, 143, 155
Conlsey, 88
Connell, 71
Connely, 6
Conner, 18, 40, 44, 46, 106
Connor, 10
Consealer, 18
Conway, 153
Conwell, 37, 108, 110-111, 115, 117
Cooch, 78
Cook, 6-7, 17, 38, 40, 45
Cooper, 6-7, 28, 32, 38-40, 45-47, 52, 91, 126, 139-140
Copes, 18, 39
Corard, 64
Corbin, 86, 155
Corbit, 82, 86
Corday, 28-29
Cordery, 137, 141
Cordray, 152
Cordry, 28, 35, 137
Corler, 140
Cormean, 145
Cormine, 122
Cornagys, 12
Cornmean, 132
Cornog, 66
Cornwell, 149
Corse, 7
Corwell, 71
Coshell, 118
Cotts, 78
Couchran, 53
Coulter, 108, 117
Coursey, 42, 53, 134-136
Courtney, 87

Covard, 11
Coverdale, 9, 17, 25-26, 28, 43-44, 113, 146
Cowgel, 21
Cowgill, 10-13, 21, 38
Cox, 12, 32, 38, 42, 54
Cozyens, 58
Crabb, 66
Craig, 51, 67
Craige, 110, 112
Crammer, 4-5
Crampfield, 123, 146
Crampitt, 5
Cranson, 66
Cranston, 65, 69-71
Cratser, 9
Craven, 85
Crawford, 16, 54, 81
Creadick, 41
Credis, 39
Cregg, 3, 6-7
Croadal, 134
Crocket, 54, 86
Cropper, 91
Crossan, 68
Crossen, 70, 72
Crossin, 74
Crosson, 70, 88
Crouch, 13, 90
Crouding, 49
Crow, 52, 87
Crowley, 51
Cubbage, 27, 38-39
Culben, 107
Culbreth, 12
Culer, 139
Cullaway, 101
Cullen, 8, 11, 41, 54, 77
Culler, 54
Culley, 129
Culser, 140-143
Culver, 153
Cummins, 14, 16
Cunningham, 77
Currender, 68, 71
Currey, 153

Curry, 64
Curtis, 31, 68, 87
Cuykendall, 116
Dackstader, 10
Dager, 36
Daisey, 92, 95-98, 121
Dalae, 99
Dalby, 148
Dale, 79, 86, 98, 152
Daleny, 100
Dalton, 61-62
Daniel, 26, 118-119
Danields, 49-50, 52-53
Daniels, 20
Daniphan, 10
Danley, 58
Danny, 2
Darby, 42, 50, 156
Darg, 2
Daris, 19, 138
Darling, 5
Darvee, 117
Darvu, 117
Dausen, 23
Davenport, 25
David, 2, 16, 18-19, 50-51, 67, 109, 118
Davidson, 79, 81-82, 94, 114, 120, 122, 125, 129-131, 134, 148
Davie, 2
Davis, 3, 11, 13, 19-20, 22, 24, 26-29, 37, 49, 52-55, 63, 67, 70, 75, 77, 84, 93, 97, 99, 14, 116-117, 127-128, 131, 138, 143, 149, 156
Daws, 10
Dawson, 23, 96, 115, 147, 156, 159
Day, 4, 8, 12, 60-61, 125-126, 147
Dayett, 79, 81
Dayly, 58
Deakin, 59
Deakyne, 18, 49, 51-53, 55
Dealon, 96
Dean, 77, 88, 109
Dear, 2, 19
Dearon, 52
Deats, 8

Deavon, 52
Decker, 28
Deger, 21
Deges, 21
Degraff, 23
Delaney, 101
Delaplane, 62
Delovan, 118
Demby, 32
Dempsey, 82, 149
DeNemours, 66
Denham, 30
Denner, 22
Denney, 10, 14-15, 22
Dennis, 8-9
Dennison, 67, 70
Denny, 9, 50, 82
Deputy, 81, 87, 113, 116, 118, 120
Derickson, 66, 91, 93, 95-100, 120, 122-125
Derrick, 58
Derrickson, 50-51, 69-71, 86, 114
Deugan, 88
Deupont, 65-66
Dewe, 72
Dewees, 73
Deweese, 36
Dewey, 45, 70
Dickenson, 81, 108, 111, 116
Dickerson, 23, 25, 28-29, 78, 92, 96, 126-127, 149
Dickison, 158
Dickson, 13, 38, 148
Diehl, 74, 82
Dill, 20, 35, 38-39, 42, 46-47, 52, 148
Dillworth, 73
Dilworth, 62-64, 83, 156
Dingle, 95
Dish, 22
Dixon, 54, 58, 62, 72-73
Dodd, 6, 19, 46, 112, 115, 127, 131, 136, 156
Dodson, 85
Dolbey, 144, 147
Dolby, 98

Dolson, 51
Donnell, 87
Donoho, 53
Donohoe, 125
Donophon, 148
Donovan, 19, 21, 26, 36, 55, 111-112, 114-115, 117-118, 126, 128
Dorman, 31, 132
Dorsey, 28
Dory, 131
Doucan, 79
Dougherty, 59, 61, 63-64, 66
Dought__ 8
Douglass, 74
Douna, 19
Downey, 13, 75
Downham, 39, 41, 45
Downing, 109, 128
Downs, 3, 17-18, 29, 102
Downward, 4
Drain, 130, 133
Drake, 68
Draper, 12, 23, 36, 46-47, 89, 114, 116, 118
Dreddon, 147
Drennon, 87
Ducks, 106
Duff, 88
Duffy, 59
Duhadaway, 18
Dukes, 95, 120-121
Dulaney, 159
Dulin, 16
Dun, 107
Dunaphan, 136
Dunaway, 51, 103, 122-123
Duncan, 71
Duniphen, 30
Dunn, 37, 59, 141-143
Dunning, 135
Duran, 139
Durban, 22
Durbrour, 19
Durham, 63
Durna, 37
Durnall, 70

Duross, 71
Dutten, 128
Dutton, 79, 108-109, 111, 114
Dyer, 2, 37, 146
Eager, 10
Earl, 18
Eastburn, 67-68, 70, 88
Ebzery, 101
Eccles, 10
Eckles, 76
Ed__son, 147
Edenfield, 9
Edge, 6
Edgell, 152
Edmundson, 87
Edward, 50
Edwards, 38, 46-47, 61, 76
Egbert, 88
Egee, 84
Eldrige, 59
Elensworth, 105
Elewright, 57
Eliason, 78, 84
Ellegood, 128, 151
Elles, 143
Elligood, 155
Ellingsworth, 111-112, 152
Elliott, 58, 61, 101, 103-104, 106-107, 121, 124, 137-139, 144-147, 150, 158
Ellis, 14, 98,, 106, 120, 139-143, 158
Ellison, 77-78, 83
Ells, 43
Ely, 61, 64
Elzey, 139-140
Emerson, 20, 41-42
Emory, 9, 24, 43
English, 19, 101, 104, 139
Ennes, 8
Ennis, 13, 22, 48, 50, 57, 107, 119, 122, 124-125
Ensly, 23
Ephraim, 20
Eshun, 92, 94
Eskridge, 145, 147, 156
Estes, 23, 28

Eton, 84
Evan, 102
Evans, 26, 37, 40, 42, 82, 87-88, 92-100, 107, 119, 125, 127
Everett, 9, 42
Everson, 65, 75
Evins, 58
Ewing, 58
Ewings, 129
Faison, 13
Faivre, 62
Fally, 57
Farrell, 14
Farris, 77
Farron, 14
Farrow, 2, 9, 20
Fasher, 38
Fasig, 5
Faskey, 106
Faulkner, 4, 44, 47, 49
Fay, 114
Faylan, 143
Feantom, 157
Fell, 4,67-68, 71
Fenn, 118
Fennell, 14
Fennimore, 49-50
Fenton, 88
Fer, 105
Ferguson, 71
Ferrins, 35-36
Ferris, 57, 65
Fetter, 79
Figgs, 144
Fine, 136
Finley, 49
Finlow, 20
Finn, 16, 20
Fisher, 5, 11, 27, 29-30, 35, 70-71, 87, 109, 113, 136, 145-146, 153
Fisk, 156
Fitts, 42
FitzGarrel, 26
FitzGarrell, 26
Fitzgerald, 118-119
Fitzsimins, 68

Fleetwood, 56, 105, 113, 147-148, 156
Fleming, 29-30, 32, 49, 66
Flemming, 152
Fletcher, 136, 139
Flick, 9
Flinn, 65-66, 73
Floid, 125
Flowers, 139, 158
Floyd, 121, 143
Flynn, 75
Foard, 82-83
Fogg, 85
Fols, 76
Fooks, 101, 124, 144, 146, 148
Foot, 69-71
Foraker, 7
Forbutter, 16
Ford, 2-3, 5, 7, 13, 15, 19-20, 51, 55, 58, 64, 77, 80
Foreaker, 5, 13
Foreman, 66
Fortner, 16, 49
Forward, 56, 59-61
Foskey, 145
Fosque, 121
Foster, 20, 26, 157
Foulk, 59
Fountain, 2, 27, 35, 42, 117-118
Foust, 72
Fowler, 26, 52, 83, 111, 114-115, 150
Fox, 21, 75
Foxwell, 16
Frader, 23
Frame, 64, 130-131, 158
Frances, 6, 49
Franklin, 3
Frasher, 37
Frashier, 39-40
Fratt, 154
Frazer, 77, 79-80
Frazier, 14, 17, 23, 31, 46
Frear, 36
Fred, 63
Fredell, 40

Freedell, 45
Freeney, 107, 142
Frem, 30
French, 25-26, 39
Friel, 74
Friend, 30, 32
Frost, 101
Fulton, 44
Furbush, 139
Furgerson, 4, 56
Furman, 92, 96, 98-99
Gale, 56
Gallops, 79
Gamben, 153
Gambol, 63
Game, 137-138, 141-142
Gankle, 54
Gardner, 51
Garman, 85
Garner, 14, 51, 83, 147
Garratt, 33
Garret, 69
Garrett, 62
Garrison, 13, 19-20
Garten, 8
Gary, 68
Gassaway, 17
Gasten, 8
Gay, 73-74
Gaylord, 114
Gaylow, 114
Gear, 4
Gebhart, 70
Gedwin, 27
Gellis, 80
Gemmill, 74
George, 5, 22, 45, 74, 77, 122
Gesford, 54-55
Gess, 73
Gest, 57-58
Geston, 56
Getts, 76
Getty, 76
Gewman, 55
Gibbins, 51
Gibbs, 1-3, 37, 40, 79

Giben, 147
Gibons, 89
Gibson, 11, 42, 72
Gidersleeve, 11
Gieser, 12
Gildersleeve, 37
Giles, 101, 140, 144
Gill, 53
Gillis, 157
Gilman, 118
Gilmore, 16
Gils, 21
Ginn, 48-49, 53
Givens, 103
Givins, 105
Glenn, 87
Godwin, 10-11, 27, 44, 95, 97-98, 100, 121
Goldsborough, 15-16, 49, 136
Gols, 22
Gooden, 6
Goodin, 45
Gooding, 76
Goodley, 57, 75
Goodman, 8, 76
Goodwin, 58
Goodyear, 58
Gorden, 33
Gordery, 137
Gordey, 106, 138, 142
Gordin, 52
Gordon, 156
Gordwin, 38, 40
Gordy, 101, 105
Goselee, 135
Goshey, 33
Goslee, 132
Gosler, 130
Goslin, 151
Gott, 16, 54
Gottler, 141
Gough, 27
Gould, 86
Grace, 72
Grady, 18, 63

Graham, 3, 19, 29, 32, 34-35, 40, 55, 82, 143-144, 151
Grant, 32
Graverly, 22
Graves, 60, 63-64, 71-72, 77
Gray, 43, 74, 79, 82, 84, 95, 97-100, 102, 121, 152-153
Grayham, 45
Green, 5, 7, 9, 16, 20, 42, 47, 61, 63-64, 72, 79-80, 83, 85-86, 129, 138, 144
Greenage, 79
Greenfield, 60
Greenlee, 46
Greenley, 128
Greenly, 39, 118, 125
Greenwalt, 68
Greenwell, 4
Greer, 36-37, 55
Gregg, 66, 70, 72, 88
Grenge, 3
Grennell, 45
Grenwell, 2
Greswell, 39
Grey, 9, 40
Grice, 96
Grier, 27, 29, 43
Griffin, 2, 12-16, 24, 27, 40, 42, 149
Griffith, 5, 21-22, 28, 80, 149
Grigg, 65
Grimes, 73
Griswell, 38
Griswold, 29
Griwell, 37
Groff, 4
Grosurvisch, 119
Groves, 74, 89
Grubb, 57-59, 65
Gruwell, 45
Guesford, 18
Guessford, 3
Guillen, 4
Gullet, 154
Gum, 121
Gumby, 103
Gunby, 123

Guthrie, 68
Hadlock, 28
Hafnal, 9
Haggard, 150
Hagyard, 150
Haines, 87
Hairson, 37
Halcomb, 75
Haley, 64
Hall, 4-5, 9, 19, 25-27, 30, 34, 56, 58, 80, 96-97, 99, 101, 112, 114, 135, 159
Halleck, 125
Halloway, 99
Hallowell, 2, 153
Ham, 21
Haman, 9
Hamilton, 28-29, 104-105, 152
Hamish, 45
Hamly, 32
Hamm, 22
Hammell, 69
Hammon, 2, 9
Hammond, 22, 25, 27, 29, 39
Hammons, 116
Hamon, 8
Hanby, 57, 61
Hance, 56, 79
Hancock, 8, 12
Hand, 53, 60
Handcock, 127
Handy, 94, 158
Hanegan, 63
Hanes, 122
Haney, 57
Hanifee, 55
Hanna, 67, 70, 72-73
Hansberry, 6
Hanser, 44
Hansly, 43
Hanson, 53-54, 76, 85-86, 132
Harbert, 78
Hardcastle, 82
Hardesty, 35
Hardison, 9
Hardiway, 50

Hare, 118
Hargadine, 7, 22, 39
Hargass, 39
Harkness, 67-68
Harman, 69, 77, 84
Harmin, 104
Harmon, 109, 129, 131-132, 136, 138, 146
Harmons, 134
Harper, 22
Harrald, 9
Harress, 39
Harrington, 4, 10-11, 17, 22-24, 29-32, 34, 36, 41-42, 73
Harris, 2-3, 27, 29, 53-54, 65, 94, 128, 133, 147, 154
Harrison, 82, 92, 94, 132, 137
Hart, 127, 133, 137
Harteuss, 51
Hartwell, 3-4, 113
Hasbands, 20
Haskins, 113
Hastings, 94-95, 103, 105, 107, 122, 137-143, 145-147, 152-154, 157
Hastirn, 45
Hatfield, 28, 116, 128, 150, 152
Hawkins, 2-4,7, 41, 60
Hawthorn, 74, 89
Hayden, 6
Haye, 150
Hayes, 4, 71, 111, 118, 152
Hays, 74, 77
Hayt, 10
Hazel, 18-19, 21, 109, 117
Hazlett, 69
Hazzard, 111-112, 117, 120, 133
Hden, 50
Heald, 73
Hear, 100
Hearn, 100-101,103, 106-107, 137-138, 140-145, 148, 156
Heather, 110
Heathers, 23
Heavelow, 110-111
Heisler, 87
Hellens, 113

Helm, 120, 158
Helyard, 17
Hemmons, 112-113, 150
Henderson, 27, 139
Hendrickson, 62, 64
Henk, 60
Henkson, 60
Hennum, 131
Henry, 49, 107, 140
Hensley, 40
Hephaon, 136
Herbey, 57
Herd, 33, 46, 51
Heron, 123-124
Herrin, 93, 96
Herring, 6, 25-27, 45
Herrington, 22-23, 32, 41
Hersey, 150
Hester, 142
Hevalow, 135-137
Heveloe, 27
Heverin, 12
Hevern, 53
Hewes, 7, 30, 32, 81
Hewitt, 58
Hews, 13
Hickey, 9
Hickman, 31, 52, 92-100, 102, 104, 121, 135, 154
Hicks, 66
Higgin, 82
Higgins, 24, 67
Highfield, 71, 73
Higman, 117
Hignut, 153-154
Hill, 14, 23, 25, 27, 29, 31-32, 35, 39, 52, 82, 89, 94, 100, 118-119, 140-141, 146, 148, 153-155
Hindson, 151
Hinsley, 25
Hinson, 78
Hipple, 57
Hirone, 2
Hirst, 75-76
Hitch, 23-24, 101, 107, 134, 140-141, 145, 155

Hitche, 107
Hitchens, 101-103, 105-106, 123-124, 135, 138-139, 143-144, 148
Hix, 31
Hoaney, 2
Hobbs, 15
Hobkins, 35, 120
Hobson, 10, 19, 72
Hocker, 96
Hockinstien, 79
Hodgson, 41
Hoffaker, 12
Hoffecker, 14-15, 51, 86
Hofficker, 64
Hog, 78
Holden, 45, 47
Hollace, 151
Hollan, 4
Holland, 19, 22, 25, 69, 87, 108-109, 114, 136
Hollett, 17
Holley, 62
Hollie, 10
Hollingsworth, 3, 66, 72
Hollis, 15, 54
Holloway, 92, 94-95, 124
Hollowell, 7
Holly, 28
Holmes, 63
Holstein, 117, 125
Holston, 13, 37, 84
Holt, 44, 54, 98, 145
Holton, 78, 85
Homes, 25
Hood, 108, 122, 132, 137
Hoofman, 55
Hook, 11
Hoonby, 59
Hoops, 72-73
Hope, 8
Hopkins, 29-31, 34-35, 45, 57, 67, 85, 105, 108-109, 111, 114, 130, 134, 145
Horney, 137
Horsey, 3, 100, 106-107, 141, 155
Hosa, 138

Hosea, 106
Hosier, 121
Hossinger, 87
Houghey, 59
Houston, 27, 84-85, 114, 116, 120, 122, 125, 137
Howard, 79, 100, 109, 140
Howet, 67
Hoyd, 46
Hubbard, 6, 14, 46, 100
Hubert, 39
Huburt, 23
Huchspn. 61
Hudson, 2, 7, 16-17, 26, 84, 92-100, 106, 109, 113, 116, 120-123, 125-126, 136
Hues, 46-47
Huffington, 11, 40
Huffnel, 21
Huggins, 51-52
Hughs, 26
Hulett, 72-73
Hull, 9
Humphrey, 118
Hunter, 76, 130
Hunting, 95
Hurd, 2, 5, 46-47
Hurdle, 129, 133-134
Hured, 1
Hurley, 148-149
Hurly, 159
Hurlock, 16, 82
Hurt, 126
Husband, 21
Husbands, 18, 20, 56, 58-59, 61
Hushabeck, 86
Husler, 87
Husten, 106, 143, 158
Huston, 106-107
Hutchenson, 18
Hutchins, 6, 33, 148
Hutchinson, 19, 21-22
Hutchison, 7, 11, 49, 53
Hutson, 134, 136
Hyatt, 86
Hyde, 73

Hydon, 25
Hydrick, 158
Hyman, 118
Hynerston, 33
Hynson, 25
Hynupton, 33
Ingram, 50, 78, 112, 117, 119, 121
Irland, 2
Irons, 65
Irvin, 47
Isaac, 151
Isaacs, 146, 148-149
Jackson, 2, 25, 36, 38, 44, 63-64, 68, 71-73, 75, 139, 157
Jacobs, 18, 23, 31, 35, 63, 67, 74, 95, 131, 155-157
James, 4, 15, 24, 36, 40, 42, 44, 66, 80, 89, 97, 100, 105, 110, 122, 129, 138, 144
Janvier, 76, 83
Jarmon, 6
Jarrell, 39, 41, 45, 51
Javrel, 51-52
Jazuett, 68
Jazzard, 79
Jefferson, 14, 16, 20, 82-83, 96, 115, 117, 124, 147
Jenkins, 11, 27, 36
Jerman, 52, 101, 103, 106, 137
Jerrold, 30
Jesse, 57
Jester, 12, 26-27, 30-31, 34, 37, 54, 70, 75, 86, 107, 111-114
Jewell, 146
Jobarts, 104
Jobson, 16, 62
John, 2
Johns, 3, 9, 13, 29, 74
Johnson, 2-3, 6-7, 13-14, 21, 26, 28-29, 32, 43-45, 50-52, 68, 80, 89, 95, 99-100, 107-108, 110-112, 115-116, 118, 121, 123-125, 129-131, 133-134, 146-147, 149, 151, 155
Johnston, 77, 81, 88, 90, 125
Joins, 120
Jollie, 41

Jones, 2-8, 12-17, 19-20, 29, 34-36, 38-51, 53, 55, 58, 69, 81, 84-86, 95-96, 100, 102, 105-106, 112, 115-116, 124, 131, 135, 144-145, 147, 150-153, 156, 158
Jonsin, 37
Jonson, 40
Jordan, 151
Jorden, 129
Jordon, 64
Joseph, 109, 116, 124, 126, 129-134, 136, 147-148
Journey, 59
Judd, 41, 59
Jump, 12, 24, 46
Jurman, 38
Justis, 121
Jutson, 134
Kaevlis, 159
Kane, 62-63
Kanely, 53
Karsner, 85
Keath, 22
Keefer, 55
Keegan, 74
Keely, 78, 80
Keen, 52
Keirn, 12
Kellam, 65
Kelley, 41, 44, 47, 135
Kellum, 60
Kelly, 32, 62, 69
Kemp, 18, 21, 38, 45
Kendal, 81
Kenether, 81
Kennard, 49
Kenneda, 52
Kenney, 100
Kens, 37
Kent, 62
Kenton, 4, 30
Kerbin, 12
Kerby, 24
Kerns, 71
Kerr, 87, 154
Kersey, 4, 6-7, 42

Kettlewood, 77
Keys, 22
Kidder, 41
Kilcanon, 63
Kilgore, 65
Killen, 46-47
Killin, 38-39
Kilmer, 36
Kilroy, 62
Kimber, 58
Kimmey, 8, 128
Kinder, 154-156, 158-159
King, 9, 12, 41, 42, 58, 74-75, 77, 104, 108-109, 126-128, 135-137, 142, 157
Kinikin, 104
Kinnerman, 153
Kinney, 38
Kirby, 24
Klair, 63, 67, 69-70
Klumpp, 45
Knight, 11-12
Knotts, 17, 45, 53-54
Knowles, 157
Knox, 96, 153
Kulp, 45
Kursey, 40
Kylee, 77
Lacey, 93
Lacy, 73
Laferty, 71
Lafferty, 10, 19
Laffoneur, 48
Lahe, 23
Lahr, 37
Lakens, 65
Lam, 126
Lamb, 18, 88
Lambden, 147
Lamborn, 72-73
Lambson, 51, 76
Lamden, 105, 143
Lancaster, 62
Landers, 53, 76
Lane, 3, 152
Langley, 65

Langly, 57
Langral, 39
Lank, 34, 74, 108, 110, 134-135, 145
Lankford, 135, 153, 157
Laphan, 19
Laramer, 34
Laramore, 30, 154
Larramore, 30
Larrimoore, 3
Latchen, 92
Lattamore, 14
Lattamous, 51
Lattamus, 54
Lattemus, 50
Lavanie, 33
Law, 66, 99
Lawrence, 36
Laws, 8, 21, 87, 123, 146
Lawson, 10, 20, 112, 123, 129, 131-132
Lawthe, 154
Lawthorn, 88
Layton, 35, 44, 91, 94, 99, 129, 143, 151, 154, 156, 158-159
Leach, 45, 60, 63, 72
Leadenham, 154
Lean, 44
Lecarpentier, 61
Lecate, 126, 156
Lecompt, 79
Lecompte, 84
Lecount, 116
Lee, 55, 74, 82
Leeds, 63
Lefevre, 76
Legat, 101
Legates, 104, 106, 142
Legatts, 126
Lekits, 96
Leman, 90
Lemon, 90, 108
Lemons, 60
Lenard, 21
Lenderman, 60
Lengo, 102
Lentz, 58-59, 76

Leonard, 9
Lermmons, 51
Lerve, 137
Lester, 81
Letts, 23
Levi, 30
Levick, 12
Lewden, 74
Lewes, 132
Lewis, 8, 12-13, 20, 34-36, 40, 47-48, 50, 68, 79-80, 83, 87, 96, 101, 114, 122, 125, 158
Lightsop, 16
Lillie, 36
Limon, 65
Linch, 4, 40
Lind, 49
Lindal, 112, 126
Lindale, 37, 43-44
Lindall, 117
Lindle, 25
Lindsey, 71, 87
Lines, 157
Lingo, 130-135, 143-144
Lipkins, 40
Lippencott, 23
Lister, 49
Little, 68-69, 73, 137
Littleton, 102, 122-123
Lloyd, 86
Loans, 80
Loatman, 15
Lobb, 71
Lobdell, 76
Lober, 11, 21, 60
Lockerman, 51, 54
Lockwood, 95, 97, 121
Lodge, 36, 57-58, 61, 73, 135
Lofland, 7, 12, 24, 27, 30, 36, 74, 76, 83-84, 110, 115-116, 119, 146, 149
Lofton, 20
Logan, 15, 60
Lollace, 147
Lollis, 43
London, 73
Long, 29, 94-95, 121-122, 133

Longfellow, 9, 46
Longland, 81, 85
Loomis, 11
Loper, 17
Lopers, 80
Lord, 2, 28, 49-50, 83, 86, 151-153, 156
Lore, 84
Louderman, 56
Loudon, 73
Loue, 138, 140
Louery, 27
Louie, 143
Louis, 102
Lowber, 37, 42
Lowdine, 5
Lowe, 102, 137
Lowther, 63-64
Loyd, 159
Luff, 3, 47
Luich, 2
Luinn, 76
Lum, 37, 78-79
Lyman, 51, 81
Lynam, 64, 70-71, 89
Lynch, 24-25, 33, 44, 84, 86, 92-94, 97-100, 106, 115, 126, 128-130, 132, 136, 138, 142, 148-150
Lyndon, 27
Lynum, 49, 64-66,
Lyons, 135-136, 150
Lyrman, 51
Lysher, 40
Mable, 2
Mabury, 17
Macey, 80
Mack, 47
Mackey, 63, 88
Macklem, 68
Macklin, 23, 26, 30, 113-115, 127-128, 150
Macy, 51-52
Madox, 144
Magee, 92, 94, 97
Magilegan, 58
Magill, 43

Maginns, 40
Magrillis, 64
Mahoney, 57, 65
Maier, 36
Mailby, 86
Major, 139
Malin, 60
Maloner, 49
Maloney, 15, 26, 37
Manering, 48-49
Manlove, 10, 28
Mansfield, 88
Maree, 68, 89
Marel, 101, 107
Marell, 104
Margargal, 72
Marin, 137
Marine, 31, 55, 137, 150
Marker, 39, 45
Marley, 75
Marner, 120
Marris, 18
Marsh, 133-137
Marshal, 12, 27, 155
Marshall, 71-72, 137
Marshel, 16
Marten, 55
Martin, 1, 18, 22, 42, 65, 89, 107, 109-110, 125, 127, 130, 134, 136, 159
Martion, 3
Marvel, 4-7, 21, 23, 25, 124-126, 133, 148
Marvil, 120
Marvin, 76
Mason, 8-9, 20, 25, 42, 72, 112
Massey, 7, 20, 44, 74, 82, 132-133
Masten, 22, 28-29, 32
Master, 24
Mastien, 157
Maston, 145
Math__, 79
Mather, 103
Mathew, 7, 104, 106
Mathews, 50, 102-103, 123-124, 151
Mathius, 103

Mathus, 103
Matthews, 123
Mattie, 77
Mauck, 100
Maull, 101, 108, 133, 135, 145
Maurice, 152-153, 155
Maxwell, 149
May, 76
Mayne, 58
McBride, 89
McCabe, 92-94, 120
McCafferty, 60, 75
McCalister, 59, 69, 122
McCall, 81-82
McCally, 119
McCarety, 60
McCarter, 54
McCarty, 73
McCauley, 28, 148, 153
McCauly, 21
McCearly, 143
McClane, 83
McClear, 69
McClelland, 88
McClin, 111
McCluskey, 80
McColley, 23-24, 28, 127
McCollom, 12
McCollum, 4, 61
McColly, 23, 26, 114
McColough, 87
McColum, 91
McConaughey, 127
McConaughy, 80
McCoomie, 50
McCormick, 67, 73
McCorns, 80
McCounty, 60
McCoy, 56, 69, 74-76, 79
McCray, 120-121, 132
McCrone, 54, 74-75
McCrossen, 71
McCullough, 57, 62
McDaniel, 66, 79
McDonald, 56
McDonnell, 61

McDowel, 130
McDowell, 67, 149, 151
McElwee, 70-72
Mcfaden, 106
McFarland, 75
McFarlin, 69
Mcgee, 95, 108, 125, 127, 129-130
McGonigal, 12
McGonnigal, 10
McGorem, 73
Mchugh, 66
McIlvain, 11, 41, 44, 129, 133, 135, 152
McIntire, 78
McIntosh, 28, 43
McKay, 54, 57
McKee, 9, 66, 69, 74
McKeowan, 87
McKewan, 68
McKinah, 31
McKnase, 31
McKnatt, 32, 34
McKowean, 87
McLaughlin, 87-88
McMahan, 41, 78
McMillen, 154
McMullen, 74, 77, 83, 85
McNeal, 95
McNee, 89
McNutt, 15
Mcolley, 127
McSorley, 61
McVaughn, 73
McVay, 85
McWhorter, 83-84
Meade, 84
Mears, 123
Medde, 84
Medford, 156
Medill, 68
Meed, 40
Megee, 92, 95, 102, 108, 112, 127, 129-130, 141, 145
Megget, 80
Meholson, 57
Meilmon, 156

Melbourn, 45
Melburn, 80
Meleham, 36
Meligain, 158
Melson, 97-98, 103, 105, 125, 137, 144-145, 155-156
Melvin, 2, 33-34, 46, 55
Mendenhall, 56, 73
Mercer, 66
Merck, 110
Meredith, 4, 27, 41, 43, 45-47
Merideth, 19
Merikan, 35
Merrikin, 35
Merritt, 86
Meserle, 24
Messick, 28-29, 98, 106-108, 111, 115, 123, 126-127, 129, 143-144, 147, 151, 157
Metcalf, 135
Metter, 146
Michall, 31
Middleton, 52
Middlton, 128
Mifflin, 36
Milbourn, 11
Milby, 109, 113
Miller, 12, 56, 58, 60, 66, 72, 75-76, 79, 87, 97, 119, 130, 133, 136, 149
Mills, 8, 25-27, 77, 107, 107, 118-119
Milman, 113-114, 116-117
Milson, 61
Milton, 25
Miluts, 10
Minas, 142
Miner, 29, 31
Miney, 70
Mininer, 30
Minner, 31, 33, 46
Minos, 4
Mitchel, 27, 68, 70, 77, 101-103, 122
Mitchell, 2, 10, 14, 70, 73, 103-104
Mitten, 27
Moffett, 55
Money, 49, 86

Monix, 136
Montacue, 6
Montague, 4, 47
Montgomery, 71, 81
Moody, 37, 80
Moor, 4, 7, 11, 14-15, 18-21, 38
Moore, 7, 9, 10, 24, 30, 40, 43-46, 49, 55-56, 60, 62, 67, 71, 73, 76-77, 82, 88, 94, 100-104, 120-122, 124, 129, 139-141, 143-145, 156-157
More, 65
Morgan, 12-13, 20, 23, 28, 31-32, 56, 76, 119, 143-146
Morris, 3, 8, 12, 18, 23, 25, 30, 32-34, 39, 44-46, 89, 93, 98, 105-107, 110-113, 115, 120-121, 124, 126-127, 134, 136, 138, 143
Morrison, 45, 68, 74-75, 80, 87, 89-90
Morrow, 63
Morton, 89
Mosely, 7-8, 24
Mosley, 43, 111-112
Moss, 77
Mote, 67-68, 80, 88
Motherall, 90
Moulton, 118
Mousley, 59, 61
Moyer, 119
Muldoon, 72
Mullen, 59
Mumford, 122
Muniford, 122
Murphey, 87
Murphy, 10, 27, 55, 58, 64, 79, 86, 113, 150
Murray, 93-94
Murrel, 31
Murrey, 92-95, 98
Murry, 38, 42, 63, 79
Mustard, 134
Muthen, 106
Myers, 5, 152
Myres, 18
Nailer, 52, 54, 110
Nalls, 99

Naudain, 53, 70, 86
Naylor, 56
Neadles, 25
Neal, 156, 158-159
Needles, 28, 40-42, 45
Negendank, 62
Nelder, 56
Nelson, 20, 29, 77, 83, 151
Nenether, 78
Newell, 58
Newlin, 89
Newlove, 76
Newmine, 1
Newner, 84
Newton, 49
Nicholis, 157
Nichols, 153
Nicholson, 12, 79, 158
Nickelis, 62-63
Nickelson, 139
Nickendank, 72
Nickerson, 3, 5
Nickols, 79
Nickolson, 60
Nivan, 67
Nivin, 76
Noble, 155
Nolan, 52
Noolin, 89
Noring, 84
Norris, 11, 81
North, 27
Norton, 27
Norwood, 129-130, 135
Nowel, 21,
Nowell, 31
Nowland, 86
Noyes, 13
Nucum, 56
Numbers, 19
Nurlin, 69
Nutter, 147
O'Byrne, 61
O'Donnell, 87
O'Roark, 80-81
Oakhall, 3

Oaks, 62
Oan, 57
Oau, 57
Obien, 156-157
Obier, 156-157
Ocen, 57
Ochletree, 28, 67, 81
Odaniel, 39
Oday, 152-153, 157
Oldham, 89
Oliphant, 101, 142, 147
Oliver, 68, 88, 92
Olston, 82
Oneal, 103-104, 107
Orr, 60, 135
Osborne, 58, 84
Ott, 80
Otterson, 50, 83
Otwell, 104, 123
Ounes, 141
Outen, 151
Outin, 28, 35
Outten, 143
Owens, 24, 36, 116, 118, 139, 144,
146, 150-151
Padley, 81
Painter, 39
Pallott, 138
Palmantary, 16
Palmatary, 15
Palmer, 61, 66, 88-89, 109, 128
Palmore, 135-137
Paradee, 10, 84
Parker, 52, 70, 86, 100-101, 154-155
Parkhurst, 120-121
Parkin, 62
Parks, 119
Parmer, 101, 123, 144
Parrey, 16
Parris, 21, 30-31, 33-34
Parry, 11
Parson, 20, 26, 145
Parsons, 98, 103-104, 120, 122, 133
Paruice, 10
Parvis, 6, 37, 39
Paschal, 60

Pasco, 43
Pasmore, 63
Passwaters, 149
Paswaters, 113, 150
Patten, 4, 11, 63
Pattin, 153
Pattison, 59
Paxson, 9, 78
Paxton, 69
Paynter, 74, 135, 137
Peach, 75
Pearson, 1-2, 5, 10
Peck, 45
Peina, 57
Peirce, 57, 59-60
Peirson, 70
Pelling, 87
Pennel, 104
Pennell, 104
Pennington, 74-75, 79, 81, 86-87
Penniwell, 151
Pennwell, 1252
Penten, 143
Penter, 143
Penton, 157
Penuel, 138
Penul, 138
Peoples, 72
Pepper, 10, 104, 112, 124-129
Perdee, 116
Perkins, 43, 60-61, 83, 149, 159
Permock, 60, 69
Pernel, 18
Pernell, 16
Perry, 4, 6, 13, 15, 41-42, 45, 50, 56,
109, 130
Peter, 75, 111, 127
Peters, 34, 40, 62, 89
Peterson, 14
Petertedemange, 57
Petetdemange, 65
Petetedemange, 58
Petigru, 153
Petrigen, 150
Petrigren, 153
Petrigru, 150

Pettyjohn, 108, 111, 113, 126, 128-130
Phelps, 27
Philip, 71
Philips, 24, 54, 57, 61, 65, 123-124, 142, 148
Phillip, 70, 101-104, 139-141
Phillips, 2, 57, 105, 135, 140, 143, 145, 151, 157
Philops, 122-123, 125
Pickering, 36
Pickett, 8
Pierce, 12, 31, 52, 55, 60, 70, 79, 144, 118-119
Pierson, 46, 51, 71, 73, 78
Pimberton, 88
Pinder, 47
Pippin, 140
Plank, 58
Platt, 90
Pleasant, 20
Pleasanton, 13, 19, 21, 79
Plumer, 30, 144
Pogue, 89
Poleti, 99
Polhouse, 59
Polk, 86
Pollet, 106
Polson, 51
Pont, 84
Pool, 72, 99-100
Poole, 74
Poor, 3, 5, 14
Porter, 25, 31-34, 54, 71, 73, 77, 79, 87
Portham, 77
Postals, 38
Postle, 10
Postles, 25, 27, 37, 43
Potter, 6, 20, 118
Pove, 2
Powel, 80
Powell, 3-4, 7, 28, 30, 52, 55, 106
Pratis, 37
Pratt, 5, 16-19, 27, 43, 152
Prawl, 156

Presiden, 38
Press, 63
Prettyman, 28, 49, 53, 107-108, 111, 114, 119, 123-126, 130, 133-135, 146, 148
Price, 6, 33, 46, 57-58, 66, 78, 83, 86, 88
Prichard, 147
Pride, 134, 136
Prier, 53-54
Prince, 57, 59
Pritchard, 88
Pritchet, 43
Proud, 7, 59
Pruett, 19
Pugh, 65, 75
Pullin, 114
Purnell, 6, 34, 106, 115
Pursiden, 38
Pusey, 131
Pussey, 105
Pussum, 155
Pussy, 144
Pyle, 63, 65, 74, 88
Quillen, 14, 23-24, 28, 145
Quillin, 96-98, 121
Quinn, 76
Racine, 77, 86
Radcliff, 151
Radclift, 88
Ragan, 56
Raimond, 78
Ralph, 138, 140
Ralston, 28, 72
Rambo, 80
Ramsey, 60
Ranals, 22
Rankin, 67
Rankins, 68
Ranson, 16
Raordin, 146
Rash, 1-3, 6-8, 18, 43, 56
Ratcliff, 114
Rathell, 30
Rather, 25
Ratlage, 38

Ratledge, 49, 85
Raughley, 29, 39
Raughly, 31-33
Rawley, 7, 14, 21, 152
Rawlins, 151
Ray, 79, 86, 155
Raymond, 11, 55
Readwig, 84
Ready, 25
Reams, 17
Rearden, 70
Records, 105-106, 138, 141, 144, 152, 156
Redden, 24, 27, 29
Redder, 27, 35
Redding, 36
Reddon, 151
Redifor, 24
Redmiles, 53
Reece, 13, 72, 79
Reed, 5, 9-11, 13, 15, 18, 30, 32, 43-45, 54-55, 77, 109-110, 112, 114, 118, 121, 126, 128, 154, 158
Reegum, 154
Rees, 19
Reese, 17
Reeve, 12
Register, 9,13, 15, 19,
Reis, 20
Reivewey, 11
Rembers, 16
Remmey, 100
Rench, 14
Renouls, 145
Renuls, 143
Ressum, 155
Reubencame, 69
Reybold, 81-82
Reynolds, 13-14, 25, 38, 42, 45-46, 50-51, 54-55, 68, 75, 78, 86, 89, 108, 110, 112, 115, 127, 137, 146, 149, 153, 158
Rhodes, 37
Rhods, 141
Rice, 42, 57
Richard, 28, 50

Richards, 7, 24-25, 34, 38, 69, 109, 153, 156
Richardson, 65-66
Richerson, 3
Richison, 21
Rickards, 14, 37, 51, 74, 82, 85, 91, 93, 95, 97-99, 120, 156
Rickets, 155
Riddle, 66
Ridgely, 8, 13
Ridgway, 8, 37, 69, 152
Riggin, 100, 107, 143, 145
Riggs, 22, 24, 55
Righter, 59
Rightfield, 72
Righton, 57
Riley, 24, 85
Riller, 31
Rinals, 139
Rinekin, 140-141, 143
Ring, 142
Ringold, 20, 61
Rinney, 140-141, 143
Rinny, 140
Rion, 150-151
Rister, 114
Ritchie, 62
Roach, 43, 80, 110-111, 115-116, 125, 129, 136
Robbins, 22, 43, 110
Roberts, 39, 49, 51, 53, 56, 84, 90, 100, 104, 119
Robins, 109, 128, 154
Robinson, 56-57, 59, 64-65, 68, 71, 73, 75, 87-88, 110, 131, 133-134, 136
Robison, 15-17, 21
Robrson, 38
Robson, 84
Roburts, 104
Rockwell, 10
Rodney, 11, 105
Roe, 35, 44
Rogers, 43, 91-92, 94-95, 102, 120, 123-126, 132, 159
Rolhouse, 59

181

Rolster, 64
Romeled, 71
Ronsh, 40
Roomer, 89
Rosa, 28
Ross, 17, 20, 24, 27, 32, 34, 39, 153-154, 156, 159
Rotherise, 58
Rothwell, 50, 53, 65, 84, 86, 88
Roughly, 29
Rowland, 60
Roy, 79
Rumford, 71
Runouls, 101
Rupp, 80
Rush, 28
Russel, 67-68, 109-110, 114-115, 127, 135, 146, 151
Russell, 43, 76
Rust, 49, 109, 111, 127, 130-131, 146, 155
Rustin, 114
Ruth, 88-89
Ryley, 50
Rylott, 89
Sackett, 119
Sage, 58
Salevan, 43
Salimons, 112
Salisbury, 12, 35, 41
Salmins, 35
Salmon, 126
Salmons, 127, 135
Salsbury, 13
Sammeser, 1
Samuels, 146, 149
Sapp, 11, 22-23, 30, 35, 43, 46, 78, 80
Sastings, 128
Saterfield, 128
Sathell, 55
Satterfield, 22, 29-30, 41, 43, 149, 156
Satterthwait, 69
Satwell, 56
Saulsbury, 18-19

Savage, 43, 92, 121
Savin, 14
Saxton, 8, 40, 44
Sayers, 88
Scabinger, 46
Scanlon, 44
Schaich, 12
Scheggs, 56
Schindler, 12
Schockly, 55
Schwartz, 12
Scim, 9
Scot, 62, 65
Scott, 23-24, 26, 28-29, 31-32, 69, 77, 88, 116, 123-124, 127, 132, 143, 151, 153
Scotten, 3-4
Scout, 13
Scuse, 7
Seal, 62, 67
Seaney, 5
Sedgwick, 156
Seedens, 34
Seeny, 81
Segars, 49, 83
Selbey, 102, 137
Sellars, 56
Semans, 45
Sena, 54
Sentman, 67
Sergent, 87
Serman, 129, 137, 142-143
Sermon, 97, 100
Severin, 33
Severson, 13, 15
Seward, 22
Sh__, 33
Shahan, 4, 9, 18
Shakespeare, 71
Shakespier, 13
Shallcross, 82-83
Sharp, 13, 23-24, 28, 32, 57, 93, 109, 114, 119, 126-128, 146, 1490150, 152
Sharpless, 63, 72
Sharpley, 58, 61

Sharrard, 44
Shaw, 39, 49
Shears, 48
Sheetts, 45
Sheldon, 81
Shepherd, 113, 117, 155
Shepley, 61
Sheppard, 88
Sherman, 42
Shermer, 17
Shettler, 6
Shew, 113
Shields, 105
Shirden, 105
Shockley, 8, 39, 112-113, 115-117, 119-120, 123
Shockly, 28-29, 36
Shole, 60
Shorse, 9
Short, 5, 18-19, 23, 25, 32, 43, 97, 102-104, 112, 120-124, 127, 127, 142, 146-149, 154
Shorts, 2, 15-16, 37
Showell, 91
Shue, 158
Shuridan, 30
Sidenham, 34
Silcox, 49
Sillitoe, 80
Simkins, 84
Simmons, 4, 55, 87
Simpler, 108-109, 118, 121, 124-125, 130-131, 133, 135
Simples, 108
Simpson, 23, 34-36, 41, 79, 119, 159
Singer, 89
Sinn, 20
Sipple, 11, 24, 31, 33, 37, 41, 46
Skidmore, 157
Skiner, 129
Slack, 80
Slaughter, 3-4, 6-7, 9-11, 14, 21-22, 38
Slay, 4, 7
Slaymaker, 40
Sliter, 42

Slock, 75
Smalley, 89
Smallwood, 54
Smart, 149
Smear, 20
Smith, 6-8, 12-13, 15, 22, 24, 27-28, 31-33, 35-40, 46, 56, 60-61, 70, 74, 76, 81, 87-88, 100, 102, 104, 113, 119, 122-123, 126-127, 145, 147, 148-151, 156
Smithers, 20, 42
Smuller, 152
Smyth, 22
Snell, 27, 41-42
Snodgrass, 63
Snow, 14
Snydar, 38
Snyder, 64, 133
Sober, 37
Solomon, 119
Sorden, 129
Souden, 72
Spanish, 150
Sparks, 3, 55
Speakman, 63
Spear, 14, 52, 82
Spearman, 50
Spence, 33, 45
Spencer, 21-22, 42, 69, 118-119, 131
Spicer, 101, 105, 111, 128, 145-147, 150-151, 159
Spiser, 125
Spring, 20
Springer, 63-64, 70, 72-73
Spruance, 16-17
Sraswood, 146
Staat, 50
Staats, 17, 48, 51
Stacy, 71
Stafford, 33, 59, 77
Stant, 11
Stanton, 78
Starit, 64
Starling, 14
Starr, 88
Staten, 14, 16, 30

Staton, 150, 152
Statts, 51
Stayton, 118
Stedham, 58
Steel, 6, 37, 29, 98-99, 127, 132-134
Steele, 87
Steen, 120-121, 123-124
Steffey, 7
Stephens, 42, 93-94, 114, 152
Stephenson, 24, 44, 53, 79, 115, 129
Sterling, 58, 73
Sterne, 37
Steveans, 17
Stevens, 3, 16, 51, 83
Stevenson, 110, 134
Steward, 70
Stewart, 66, 77, 79, 83, 146
Stickly, 34
Stidham, 64-65
Stinson, 66
Stockley, 16, 132-133, 136-137
Stockwell, 22
Stokes, 154
Stoops, 60, 85
Stott, 117
Stout, 44
Stover, 7
Stradley, 77
Strafford, 133
Street, 17, 131-132, 134
Stroud, 80, 89
Stroup, 74-75
Stuart, 12, 113, 127
Stubbs, 40
Stubs, 32
Sturdevant, 8
Stuckart, 82
Sturn, 56
Sudler, 125, 151
Sulivan, 81
Sullivan, 152, 155
Sutherland, 74
Sutton, 15, 78, 82
Swain, 118, 126, 138, 146-148, 154-155
Swan, 85

Swann, 154
Swayne, 62
Sweatman, 53
Swift, 56
Swinderman, 71
Switcher, 89
Sylvester, 1
Sylvestor, 39
Taggard, 76
Talbert, 120
Talley, 56-61
Tally, 59
Tate, 8
Tatman, 28, 35, 49, 114, 150
Tatum, 65
Tavis, 64
Taylor, 1, 5, 8, 11-12, 17-19, 23, 24, 34-35, 48, 62, 64, 68-70, 81-82, 92, 95-97, 100, 104, 140, 144, 148, 159
Teaf, 88
Teague, 147
Tease, 119
Teat, 13
Templeman, 85
Tendel, 77
Termman, 51
Tetter, 76
Tharp, 32-33, 35-36, 112, 114
Thatcher, 63
Thistlewood, 30
Thomas, 1, 4, 6, 13, 21, 26, 29, 32, 35-36, 50, 55-56, 91
Thompson, 1-2, 5, 15, 19, 36, 48, 50-51, 56, 62, 66-67, 70, 73-74, 88, 125, 131, 136
Thornton, 48
Thoroughgood, 124, 129, 131
Threasher, 37
Tibbit, 54
Tilghman, 155
Tilman, 49
Timmons, 92-93, 96, 102, 120-121, 125
Tindal, 144, 147-148
Tindle, 78, 135
Tingle, 92, 95-96, 103-104, 120, 125

Tinley, 11, 37, 45
Tinly, 50
Tinsley, 31
Tiro, 95
Titus, 118-119
Toben, 63
Todd, 5, 8-9, 12, 34, 56, 151, 153
Tolbert, 25, 120
Tombleson, 14
Tomlinson, 12, 22-23, 28
Tompson, 24-25, 61, 107, 120, 122
Toomey, 122
Tooney, 120
Torbert, 26-27, 126, 128
Toreum, 54
Touett, 26
Townsend, 24-25, 37, 45, 48-50, 53-54, 56, 75-76, 83, 85, 96-99, 114, 118, 155
Toy, 66
Tracey, 92
Traider, 23
Trainer, 13, 64
Traswood, 146
Travers, 88
Travis, 32
Tredfield, 154
Trenden, 69
Trewett, 158
Tribbet, 31, 47
Tribbett, 4, 7
Triggs, 138
Trimble, 59
Trout, 154
Truax, 13-15, 17-18
Truet, 18
Truett, 103-104, 106
Truitt, 43, 91-92, 102-104, 113-116, 120, 123, 144-145
Trustee, 52
Tubbs, 93
Tuc, 5
Tucker, 8, 12, 65, 113, 118, 147
Tue, 5
Tuelley, 141
Tuilley, 139, 141

Tull, 157
Tunell, 96
Tunnel, 99, 129
Tunnell, 97, 99
Turner, 14, 96, 99, 118, 135, 147, 149
Tush, 55, 75
Tutle, 34
Tuttle, 153
Twaddle, 60
Twed, 88
Twiford, 156
Tybout, 73
Tygert, 31
Tyler, 29
Umphry, 138
Underhill, 118
Underwood, 20
Updite, 65
Vacant Farm, 10
Vail, 55, 78, 85-86
Valentine, 61
Vanburkaloe, 44
Vance, 59
Vandenburg, 152
Vandergift, 82
Vandergriff, 8
Vandergrift, 75, 82-83, 85
Vandever, 72, 157
Vandyke, 50, 53
Vane, 15
Vankirk, 119
Vannagner, 48
Vannekle, 84
Vansant, 77
Vanslyk, 8
Vanwinkle, 13
Varnarmon, 40
Vaughan, 110
Vaughn, 22, 130, 158
Vauss, 113
Veal, 72
Veale, 57
Veasey, 107-108
Veasy, 78
Vent, 108

Verdier, 131
Vernon, 57, 63
Vessels, 133
Vible, 3
Vickers, 41, 96, 121, 127, 131
Vickery, 46
Victor, 34
View, 55
Viland, 153
Vincen, 7
Vincent, 6, 36, 40, 117, 150
Vineyard, 23-24
Vinsin, 157
Vinson, 105, 138, 144
Vinyard, 53
Virden, 107
Voshall, 37
Voshel, 22, 47
Voshell, 5-6, 10-11, 13, 30, 37, 118
Voss, 33
Vreeland, 119
Vurden, 5
Wabruth, 40
Wade, 154
Wadkins, 25
Wagner, 120
Waid, 88
Wainwright, 124, 145, 157
Waler, 25
Walker, 15, 17, 29, 32-33, 50, 52, 62-63, 67, 69, 78, 81, 86, 108-109, 112, 137, 151
Wallace, 2, 12-13, 28
Wallas, 143
Waller, 12, 87, 104, 139, 141-144, 157-158
Walls, 3, 108, 110-111, 114, 119, 122, 126, 130-131, 133-134, 136, 152
Walston, 139, 141
Walter, 66, 77, 82, 98, 116
Walton, 79-80, 116
Wapler, 118
Waples, 8, 95, 115, 124-125, 129, 133-135, 147

Ward, 75, 79, 101, 103, 106, 138, 154-155, 157
Ware, 2, 12, 28, 36, 65
Warner, 45, 52
Warran, 52-54
Warren, 30, 44, 46, 50, 73, 111, 113-118, 124, 126, 128, 134, 154
Warrington, 23, 91, 93, 105, 107, 109, 115, 117, 129-130, 132, 134-136, 141, 147
Waters, 75
Watkins, 22, 83
Watman, 105
Watson, 23, 25, 48-49, 93, 106, 116-119
Watten, 105
Watterhouse, 74
Watters, 30
Way, 57, 63
Webb, 8, 11, 24-25, 27, 55, 115-116, 118, 120, 125, 133, 135, 149, 150
Webbe, 88
Webster, 3, 16, 35, 55, 60-61
Weeks, 23
Ween, 59
Weer, 59
Welburn, 95
Welch, 29-30, 32, 48, 75, 113, 136, 150
Welden, 58, 61, 75, 96
Weldon, 49, 56, 66, 90
Welds, 17
Weller, 51
Welley, 101
Wells, 10, 53, 64, 98, 102-103, 119, 129, 135
Welsh, 56, 69, 111
West, 7, 17, 41, 53-54, 64, 93, 97-99, 101-104, 106-107, 122-123, 126-128, 144, 148
Whaley, 104
Wharton, 8, 11, 23, 36, 96-98
Whealey, 104
Wheatby, 20
Wheatley, 107
Wheatly, 8, 20, 36, 101, 158

Wheatman, 13-14
Wheder, 26
Wheeler, 32
Whetsit, 40
Whit, 142
Whitacre, 27
Whitaker, 28, 41, 80
Whitakre, 77
Whitby, 7
White, 4, 52, 62, 64, 75-76, 108-109, 111, 122, 136, 145, 158
Whiteman, 67-69, 72
Whitlock, 50
Whitman, 68
Whitmire, 40
Whitnack, 24
Whitney, 154
Whitten, 89
Wicks, 12
Wiest, 16
Wilcox, 23
Wilcuts, 2
Wilcutts, 41
Wilday, 15
Wilds, 12, 17
Wildsmith, 9
Wiley, 78, 104
Wilgus, 94, 99
Wiliamson, 35
Wilkenson, 2
Wilkins, 26, 115, 117, 119, 127-129, 142, 148
Wilkinson, 6, 118, 122
Willard, 29
Willen, 129, 152, 156
Willetz, 86
Willey, 146, 150-153, 158
William, 5, 13, 110, 121, 125, 153
Williams, 3-4, 6, 10, 12, 15, 17, 33-35, 43-44, 79, 86-87, 92, 94-98, 111, 113-114, 117, 120, 138, 148, 150, 152, 155-159
Willis, 5, 21, 24-25, 28-29, 40
Willoughby, 154
Wills, 122
Wilmot, 62

Wilson, 9, 11, 13, 17, 20, 23-24, 36-37, 53-54, 56, 59-61, 72-73, 79, 87, 101, 108-110, 112, 115, 124-127, 130, 132-133, 135, 148-149
Wiltbank, 110, 112
Winder, 140
Windsor, 100, 104, 137
Winford, 52
Wingat, 105, 138, 141
Wingate, 121
Winger, 8
Wirt, 71
Wirtten, 137, 142
Wise, 121
Wisher, 20
Wiswell, 74, 118
Wofford, 139
Wolcatt, 32
Wolf, 129, 135
Wolfe, 101, 137
Wolfer, 120
Wolfs, 136
Wood, 29, 65, 85, 119, 135
Woodall, 19
Woodcock, 154
Woodhull, 16
Woodle, 38
Woodrufus, 51
Woods, 69
Woodward, 64, 66, 70, 78
Woolaston, 69
Wooten, 6
Wooters, 44
Wootten, 100-103, 144
Workman, 103-105, 110-111, 128-129, 140
Worrall, 67
Worrel, 87
Worten, 105
Worton, 6
Wright, 2-3, 7, 16, 18, 20, 36, 44, 52, 55, 62, 78, 87, 89, 100, 103, 129, 133-135, 139, 153-155, 157159
Wroten, 27-28, 35
Wrother, 34
Wroton, 118

Wrotten, 137
Wyatt, 25-26, 29-31, 43, 46
Yarnall, 68
Yearsley, 70
Yetter, 73
York, 21, 145
Young, 14, 20, 41-42, 51, 55, 116-117
Zacheris, 83
Zebley, 56, 59

New Castle County Delaware
1870 Agricultural Census
Additional Section Located

This agricultural census was filmed from original records in the Delaware State
Archives in Wilmington Delaware by the Delaware State Archives Microfilm
office.

These additional pages were found in the middle of the Kent County Delaware
1880 Agricultural Census. They have been added after Sussex Co. 1870 and a
separate index created for these pages.

There are some fifty-two columns of information on each individual. Only the
head of household is addressed. I have chosen to use only seven columns of the
information because I feel that this information best illustrates the wealth of the
individuals. These are shown below:

1. Name of Owner
2. Acres of Improved Land
3. Woodland Unimproved
4. Other Unimproved Acreage
5. Cash Value of Farm
6. Value of Farm Implements and Machinery
15. Value of All Livestock

Thus, the numbers following the names represent columns 2, 3, 4, 5, 6, 15.

The following symbol is used to maintain spacing where information in a column
is left blank (-). This symbol is used where letters, names or numbers are not
legible (_).

**NOTE: Beginning with the 1870 Agricultural Census the column names have
changed. And, the names for this county appear as Last Name First.**

Janvier, Ferdinand, 190, 20, 10,
15000, 500, 1905
Titter (Titler), John D., 85, 2, 15,
7000, 400, 636
Vansant, Wm., -, -, -, -, 50, 486
George, James H., 144, -, 6, -, 8000,
100, 700
Stafford, John, 100, 30, 22, 6500,
200, 1066
Moss, R. D., 18, -, -, 1500, 50, 225

Whitaker, Conrad, 8, 2, -, -, 2500,
45, 105
Calhoun, W. W., 55, -, 45, 6000,
200, 395
Ellison, T. B., 95, 3, 5, 9000, -, 590
Racine, T. P., 119, 24, -,
1000(10000), 100, 515
Johnston, John, 125, 160, 15, 13000,
150, 700
Portman, James, 173, 50, -, 20000,
900, 965

Ford, Wm. B., 10, -, -, 5000, 5000, 470

Cann, James, 100, 15, 20, 6000, 50, 685

Mills, Joseph, 62, 5, -, 4000, 300, 355

Kettlewood, M., 122, 8, 10, 988, 200, 775

Portis, Thos. G., 60, 10, -, 5000, 200, 520

Moore, A. S., 80, 13, 7, 7000, 200, 580

Stradley, Wm. H., 175, 30, 25, 15000, 200, 1025

Porter, Samuel M., 30, 6, 30, 4000, 100, 256

Racine, John G., 40, 5, 5, 6000, 200, 800

Tindall, D. M., 40, 2, 5, 6500, 150, 310

Mitchell, Robert, 150, 10, 3, 6300, 400, 1040

Stewart, S. W., 120, 25, 9, 11000, 300, 988

Wullers, Thos., 125, 25, 10, 1500, 75, 1340

Reed, B. A., 100, -, 40, 6000, 50, 770

King, Robert, 10, -, 3, 1000, 50, 350

Stewart, James, 100, 75, 12, 1600, 275, 810

Cunningham, S., 100, -, 50, 4000, 345, 955

McMullen, James C., 100, 7, 20, 5000, 40, 535

Boy, Henry, 130, -, 170, 10500, 300, 800

Cann, Allebone, 175, -, 25, 12000, 350, 1904

Adair, Wm. J., 35, 10, -, 2500, -, 190

Reed, Robert, 68, -, 30, 3600, 104, 530

Harris, D. Brainerd, 80, 15, 5, 7000, 350, 1060

Frazer,John, 112, 11, -, 9725, 100, 310

Clark, Anna, 556, 400, 40, 80000, 400, 3350

Burton, Miller, 30, 40, -, 14000, 150, 361

Cullen, D. H., 38, 5, 8, 3000, 200, 440

Bacon, Edward, S., 29, 1, -, 3000, 100, 355

Groves, Groves W., 60, -, 15, 6000, 200, 7000

Davis, Jehu T., 41, -, -, 4000, 120, 490

Harmon, Andrew, 490, -, -, 7000, 100, 250

Matten, Marcelus, 11, -, -, 1000, 100, 190

Davis, J. W., 86, 2, 3, 9000, 100, 200

Kyle, Fanny, 18, -, -, 2000, -, -

Scott, John, 16, -, -, 1750, 100, 230

Boulden, Jesse, 13, -, 2, -, 4000, 200, 110

Dean, Jacob, 160, 30, 10, 1500, 350, 1380

Nelson, James, 135, 5, -, 6000, 150, 800

Boulden, George, 250, 125, -, 2250, 500, 1985

Boulden, George Jr., 130, 95, -, 12375, 100, 953

Tindall, Thomas, 100, 50, -, 8000, 300, 645

Hogg, John, 60, -, 20, 2500, 20, 200

McIntire, Samuel, 100, 15, 55, 8500, 300, 850

McIntire, James, 80, 60, 10, 10000, 500, 730

Raimond, James, 160, 20, -, 15000, 280, 741

Sapp, Jonathan, 50, 15, 15, 3600, 100, 230

Ellison, Curtis B., 570, 108, -, 16000, 450, 1030

Ellison. C. B., 92, 20, -, 40000, 825, 575

Vail, David C., 400, 100, -, 5000, 400, 2116

Ingram, N. R., 143, -, -, 15000, 200, 600

Biggs, Wm. P., 192, 4, 16, 13200, 340, 964

Biggs, Wm. P., 144, 8, 30, 10000, -, -

Keely, Michael, 66, 33, 32, 7500, 300, 590

Cooch, J. Wilkins, 250, 100, 150, 35000, 2000, 2025

Cooch, J. Wilkins, 60, 20, -, 10000, 200, 380

Stanton, Isaac B., 70, -, 10, 6000, 200, 415

Woodward, Joel, 40, 45, 21, 6900, 300, 1240

Cavender, William 250, 50, -, 18000, 400, 1710

Veasy, James T., 70, -, 6, 8000, -, 650

Veasy, James L., 95, -, 6, 10000, 500, 825

Veasy, James L. Farm No. 2, 70, 20, -, 1600, -, -

Paxon, Aron, 150, 40, -, 1500, 100, 650

Hinson, Jas. A., 85, 5, -, 5000, 150, 300

Harman, Alston, 150, 25, 23, 12000, 600, 990

Kenether, Valentine, 15, 1, 67, 2000, 50, 250

Paxon, Johnathan, 250, 8, 43, 30000, 600, 490,

McMahan, Andrew, 90, 20, 20,7000, 100, 1030

Dickinson, Samuel, 360, 40, 50, 50000, 1000, -

Wright, Stephen, 98, 7, -, 6300, -, 660

Biggs, Sewel C., 150, 26, 10, 16740, 600, 1210

Lum, Charles A., 222, 31, -, 22771, 340, 2390

Ellison, John D., 378, -, 25, 10300, 100, 2970

Price, John F., 200, -, 3, 20000, 990, 2265

McIntire, Thos., 200, 20, 6, 22000, 100, 1300

Harbert, Wm. H. H. 131, -, 20, 21375, 250, 1250

Harbert, Timothy, 157, -, 20, 21375, 150, 650

Reynolds, J. W., 185, -, -, 18400, 300, 1950

Catts, Lavin, 246, -, -, 25000, 1000, 2400

Veasy, Jas. L., 95, -, 5, 5000, 400, 665

Sutten, Samuel, 75, 15, -, 3500, 100, 375

Walker, Mary C., 32, -, -, 2000, 50, 330

McMahon, Thos., 100, 25, 15, 4600, 100, 365

Dickerson, Levi, 350, -, 50, 40000, 550, 1625

Holton, Spencer S., 130, 15, 12, 12560, 140, 1352

Parson, Ross L., 12, 58, 23, 15000, 100, 735

Wagley, Stephen, 50, -, 1, 3000, 100, 400

Wright, Samuel, 100, 20, -, 19000, 500, 1275

Dayett, W. Thomas, 800, 10, 9, 10000, 200, 810

Black, Robert M., 160, 40, 55, 17850, 175, 1060

Frazer, Samuel, 150, 15, 5, 13000, 200, 1038

Ward, John, 60, -, 15, 4500, 125, 345

Cavender, L. A., 147, 15, 15, 20000, 350, 1950

Boys, Jacob, 5 ½, -, -, 1800, 183, 560

Nickolson, James, 173, -, -, 11300, 650, 1553

Pleasanton, Edward, 175, -, 5, 14400, 300, 1175

Gray, Francis, 320, 50, -, 37000, 300, 1484

McCoy, Thomas, 20, -, 20, 18400, 250, 1263

Donelson, John, 200, -, 20, 18400, 250, 1263

Lecompt, Charles, 310, -, -, 23250, 350, 1500

Gibbs, Henry, 6, -, 6, 800, -, -

Deakyne, George A., 30, -, 27, 3000, 200, 175

Porter, Henry, 8, -, 4, 900, 40, 145

Money, Lewis, 10, -, -, 700, 55, 125

McCoy, James, 120, -, -, 7200, 100, 580

Stevenson, Daniel, 15, -, 2, 1000, 25, 182

Roy, Reuben, 28, -, 7, 2000, 45, 230

Boulden, Jas., 70, 65, 30, 8000, 200, 610

Hance, Edward, 200, 41, 9, 2000, 900, 1840

Nickols, Isaac W., 30, -, 10, 1200, 60, 270

Nichols, Charles, 45, -, 15, 1800, 50, 180

Est. Johnson, Causdon, 18, -, -, 1800, -, -

Christian, Adam, 90, 16, 20, 6000, 150, 425

Tilton (Titton), Aflred B., 300, 33, -, 16330, 400, 2480

Murry, Samuel, 8, -, 9, 350, -, 90

Pierce, Edward T., 100, -, 24, 6000, 200, 800

Williams, James W., 16, -, 4, 800, 100, 212

Williams, Jas., 25, -, 1, 1300, -, -

Greenage, William, 4, -, -, 300, 55, 400

Harkenstien, Godfrey, 20, -, 25, 1100, 35, 420

Hockenstien, G. 120, -, 60, 8000, 75, 488

Gallops, William, 24, 4, 750, 50, 230

Murphy, Hugh, 20, 4, 47, 3000, 75, 172

Simpson, Alexander, 65, 65, -, 10000, 200, 780

Dale, Samuel T., 760, -, 300, 10000, 200, 600

Walton, Samuel C., 48, -, 41, 2300, 110, 220

Stewart, Samuel, 6, -, 19, 375, 156, 121

Greer, Levi, 30, 10, -, 2000, 300, 450

Lum, Elizabeth, 20, -, -, 1000, -, -

Lewis, George, 30, 25, -, 2600, 50, 475

Wilson, Washington, 12, 18, -, 900, 40, 920

Collins, William J., 105, -, 15, 4500, 100, 360

McDaniel, William, 9, -, -, 600, -, 110

Pennington, Joseph, 25, 16, 140, 200, 40, 370

Lum, Elias, 15, -, -, 400, 60, 196

Armstrong, Robert, 102, 30, 20, 7550, 200, 660

Dutton, George, 14, -, -, 500, 50, 109

Champion, Benjamin, 22, 2, 2, 780, 25, 60

Miller, William, 5, -, -, 300, -, 140

Howard, William, 13, -, -, 1200, 20, 122

Rees, William R., 125, 200, 16, 15000, 2000, 1295

Jaggard, Thomas, 75, -, 25, 6000, -, 150

Powel, W. G., 100, -, 40, 10000, 200, 610

Hall, John, 135, 33, 35, 15000, 100, 515

Comly, Samuel 70, 15, 15, 6500, 500, 940

Chandler, Thos. C., 43, -, -, 4300, 160, 415

Chandler, D. W., 115, 5, -, 10000, 100, 567

McCorn, John, 185, 35, -, 22000, 1000, 1325

Griffith, Mary C., 216, 10, 4, 23000, 800, 2500

Corney, Wm. L., 151, 14, -, 20000, 972, 2805

Bradley, H. H., 130, -, 4, 13000, 100, 701

Frazer, William, 76, 15, 15, 6000, 50, 355

Morrison, R. J., 11, -, 16, 5775, 500, 1000

Mote, William, 178, 12, 10, 20000, 50, 1253

Berk, Pakie, 20, 5, 15, 2500, 25, 185

McConaughey, Thomas, 150, -, 100, 18750, 600, 550

Hall, William, 40, 20, -, 3000, 20, 550

Lilliloe, W. W., 107, 15, -, 7000, 250, 545

Bowers, William, 226, 50, 30, 30000, 1000, 2420

Murry, Jas., 78, 16, 6, 7600, 200, 595

McConaughey, William, 150, 26, -, 12000, 800, 1605

Keely, Michael, 50, -, 5, 2700, -, 165

Cole, Elias, 15, -, -, 1500, 50, 275

Melvin, Robert, 60, 10, -, 2000, 275, 200

McClusky, Arthur, 80, 10, 40, 7000, 100, 500

Megget, Peter, 66, -, 4, 5000, 350, 650

Slock, William, 94, 30, 26, 15000, 150, 1065

Gellis, Paul, 20, -, 23, 2500, 75, 179

Walton, Eligah, 40, -, 10, 2500, 100, 210

Frazer, W. W., 45, -, 20, 5000, 500, 630

Rambo, Enos, 75, 15, 10, 6000, 200, 515

Stroud, Edward, 142, 18, -, 16000, 650, 974

Rupp, John, 36, -, -, 1800, 700, 280

Walton, Charles, 90, -, 33, 5000, 350, 1395

Ott, Andrew, 17, -, 8, 1500, 200, 140

Moody, John, 120, 12, 8, 10000, 300, 650

Ford, David, 150, -, 40, 9000, 500, 1118

Loans, James, 9, -, 1, 1000, 20, 100

Sapp, Joseph, 200, -, 90, 17000, 300, 1328

Benson, Jas. A., 130, 20, 5, 12000, 600, 1505

Clendenin, S. H., 20, -, 20, 2500, 25, 331

Alexander, George, 30, 15, 22, 3000, 84, 360

Conely, Peter, 4, -, 8, 500, -, 60

O'Rouk, James, 20, -, 5, 1500, 20, 190

O'Rouk, Timothy, 97, -, 7, 1200, 20, 175

Sobers, Samuel, 20, 20, 32500, 50, 186

James, Joseph, 13, -, -, 1300, -, 80

James, David, 5, -, -, 300, -, 90

James, Joseph, 10, -, -, 300, -, 80

Lewis, Delphin, 50, -, -, 300, -, 20

Greer, Benjamin, 11, -, -, 500, -, 35

O'Rouk, Barthol., 35, -, 150, 1400, 25, 269

Whitaker, G. P., 30, 160, -, 7900, 25, 1540

Roach, Patric, 12, -, 8, 1200, 30, 160

Sulivan, Patric, 20, -, 2, 1000, 100, 340

Corns (Coms), James S., 3, -, -, 500, -, 80

Walker, William, 10, -, -, 300, -, 40

James, James, 3, -, -, 150, -, 9

Sceny, William, 20, -, -, 400, 60, 150

Kernether, Martin, 44, -, 39, 2760, 200, 519

Kendal, Henry, 60, -, 35, 2500, 60, 505

Lymer, Joseph, 40, 40, 19, 2500, 200, 410

Brown, John, 40, -, 50, 3000, 150, 415

Brown, Thomas, 23, -, 2, 1500, -, 190

Dayett, Adam, 60, -, 40, 4000, 250, 1126

Crawford, Joseph, 11, -, -, 1000, 50, 200

Shelden, George, 100, 8, 22, 7500, 150, 445

Beck, Walter, 140, -, 157, 10000, 300, 1350

Norris, William 7, -, 40, 1000, 14, -

Johnston, Henry, 7, -, 41, 1400, 400, 135

Batten, Maddon, 90, 14, 26, 9500, 400, 1060

Dickenson, Samuel, 360, 40, 50, 50000, 1000, 3450

Johnston, H. J., -, 100, 40, 9500, -, -

Adair, 190
Alexander, 193
Armstrong, 192
Bacon, 190
Batten, 194
Beck, 194
Benson, 193
Berk, 193
Biggs, 191
Boulden, 190, 192
Bowers, 193
Boy, 190
Boys, 191
Bradley, 193
Brown, 194
Burton, 190
Calhoun, 189
Cann, 190
Catts, 191
Cavender, 191
Champion, 192
Chandler, 192
Christian, 192
Clark, 190
Clendenin, 193
Cole, 193
Collins, 192
Comly, 192
Coms, 193
Conely, 193
Cooch, 191
Corney, 193
Corns, 193
Crawford, 194
Cullen, 190
Cunningham, 190
Dale, 192
Davis, 190
Dayett, 191, 194
Deakyne, 192
Dean, 190
Dickenson, 194

Dickerson, 191
Dickinson, 191
Donelson, 192
Dutton, 192
Ellison, 189-191
Ford, 190, 193
Frazer, 190-191, 193
Gallops, 192
Gellis, 193
George, 189
Gibbs, 192
Graves, 190
Gray, 191
Greenage, 192
Greer, 192-193
Griffith, 193
Hall, 192-193
Hance, 192
Harbert, 191
Harkenstien, 192
Harman, 191
Harmon, 190
Harris, 190
Hinson, 191
Hockenstien, 192
Hogg, 190
Holton, 191
Howard, 192
Ingram, 191
Jaggard, 192
James, 193
Janvier, 189
Johnson, 189, 192
Johnston, 194
Keely, 191, 193
Kendal, 193
Kenether, 191
Kernether, 193
Kettlewood, 190
King, 190
Kyle, 190
Lecompt, 192

Lewis, 192-193
Lilliloe, 193
Loans, 193
Lum, 191-192
Lymer, 193
Matten, 190
McClusky, 193
McConaughey, 193
McCorn, 192
McCoy, 192
McDaniel, 192
McIntire, 190-191
McMahan, 191
McMahon, 191
McMullen, 190
Megget, 193
Melvin, 193
Miller, 190, 192
Mills, 190
Mitchell, 190
Money, 192
Moody, 193
Moore, 190
Morrison, 193
Moss, 189
Mote, 193
Murphy, 192
Murry, 192-193
Nelson, 190
Nickols, 192
Nickolson, 191
Norris, 194
O'Rouk, 193
Ott, 193
Parson, 191
Paxon, 191
Pennington, 192
Pierce, 192
Pleasanton, 191
Porter, 190, 192
Portis, 190
Portman, 189

Powel, 192
Price, 191
Racine, 189-190
Raimond, 190
Reed, 190
Rees, 192
Reynolds, 191
Roach, 193
Roy, 192
Rupp, 193
Sapp, 190, 193
Sceny, 193
Scott, 190
Shelden, 194
Simpson, 192
Slock, 193
Sobers, 193
Stafford, 189
Stanton, 191
Stevenson, 192
Stewart, 190, 192
Stradley, 190
Stroud, 193
Sutten, 191
Tilton, 192
Tindall, 190
Titler, 189
Titter, 189
Titton, 192
Vail, 190
Vansant, 189
Veasy, 191
Wagley, 191
Walker, 191, 193
Walton, 192-193
Ward, 191
Whitaker, 189, 193
Williams, 192
Wilson, 192
Woodward, 191
Wright, 191
Wullers, 190

This agricultural census was filmed from original records in the Delaware State Archives in Wilmington Delaware by the Delaware State Archives Microfilm office.

There are some 100 columns of information on each individual. Only the head of household is addressed. I have chosen to use only the first 16 columns of the information because I feel that this information best illustrates the wealth of the individuals. In columns 2-4 the number 1 is used as an indicator only and represents no numeric value. These are shown below:

1. Name of Individual
2. Owns
3. Rents for Fixed Money Rental
4. Rents for Shares of Products
5. Improved Tilled Land
6. Improved Permanent Meadows, Pastures, Orchards, and Vineyards
7. Unimproved Woodland & Forest
8. Unimproved including Old Fields, and Not-Growing Wood
9. Farm Value of Farm including Land, Fences, and Buildings
10. Farm Value of Farming Implements and Machinery
11. Value of Livestock
12. Fences—Cost of Building and Repairing in 1879
13. Cost of Fertilizers Purchased in 1879
14. Labor—Amount Paid for Wages for Farm Labor during 1879, including Value of Board
15. Labor—Weeks paid labor in 1879 including Value of Board
16. Estimates Value of all Farm Productions (sold, consumed or on hand for   1879)

Thus, the numbers following the names represent columns 2, 3, 4, 5, 6, 7, 8, 9, 10,11, 12, 13, 14, 15, 16.

The following symbol is used to maintain spacing where information in a column is left blank (-). This symbol is used where letters, names or numbers are not legible (_). This census was split between two reels of microfilm. The first reel I transcribed columns 1-11 & 16, the second I transcribed 1-16.

Snow, Joseph, 1, -, -, 145, 6, 45, -, 6000, 350, 900, 1850
Cummins, William A., -, -, 1, 200, 12, 25, -, 18000, 750, 2150, 2400
Johnson, Lenard F., -, -, 1, 175, 1050, -, 10000, 500, 850, 1550
Raughley, Robert, -, -, 1, 170, 30, 20, -, 12000, 1540, 1100, 2620
Boyed, Jorlena, 1, -, -, 113, 12, 5,35, 6000, 620, 800, 1900
Desch, Joseph E., 1, -, -, 85, -, 6, -, -, 5000, 480,700, 1670
Mitchell, William C., -, -, 240, 60, -, 150, 15000, 650, 1485, 2850

Rice, John, 1, -, -, 955, -, -, 6500, 160, 600, 1200
Evans, Abel, -, -, 1, 177, 13, 10, -, 7000, 420, 700, 1968
Jefferson, Charles H., -, -, 1, 166, 10, -, -, 7500, 360, 960, 1710
Speakman, Gideon, 1, -, -, 115, 20, 30, -, 25000, 800, 750, 3560
Stacling, James, -, -, 1, 155, 26, -, 5, 14000, 490, 1485, 3040
Cummins, W. Polk, -, -, 1, 198, 104, -, 4, 25000, 1100, 1580, 8491
Hoffecker, Walter O., -, -, 1, 156, 46, 122, 18000, 700, 1635, 2735
Robinson, John R., -, -, 1, 144, 6, 26, -, 8000, 250, 655, 2105
Bleakley, Thomas W., 1, -, -, 93, -, 12, 20, -, 7000, 300, 580, 1060
Everett, Charles D., 1, -, -, 100, 10, 15, -, 6000, 450, 750, 1156
Goodwin, Alexander, -, -, 1, -, 60, 5, 15, -, 2000, 125, 160, 300
Cousin, Henry W., -, -, 1, 26, -, 10, -, 1200, 50, 175, 290
Robinson, Joseph C., -, -, 1, 175, 25, 5, 10, 7500, 360, 560, 2345
Farrell, John, -, -, 1, 176, 24, 20, 20, 12000, 450, 1620, 3035
Spear, Andrew, -, -, 1, 2100, -, -, 40, 2000, 1500, 2500, 4490
Pleasanton, Alexander, -, -, 1, 198, -, 2, -, 20, 12000, 1235, 1885
Denney, J. Frank, 1, -, -, 200, -, -, 35, 15000, 3175, 1975, 3700
Moore, Martian, -, -, 1, 133, -, 42, 10, 15, 12000, 225, 600, 30, 2480
Jones, John W., -, -, 1, 162, 38, -, 100, 20000, 450, 940, 2700
Condright, Isaac H., -, -, 1, 48, 54, -, 175, 4000, 380, 475, 1250
Hoffecker, Mary, 1, -, -, 160, 30, 10, 1000, 12000, 500, 1700, 2750
Anderson, John W., -, -, 1, 171, 29, 200, -, 16000, 265, 582, 1560
Williams, Thomas, -, -, 1, 178, 22, 10, 20, 10000, 480, 700, 2650
Garrison, John P., -, -, 1, 140, 1, -, 35, 8000, 300, 960, 1730
Turner, James J., 1, -, -, 25, 15, -, -, 3200, 100, 500, 260
Stephenson, John W., -, -, 1, 95, 34, -, -, 7500, 200, 515, 1820
Downs, Enoch S., -, -, 1, -, 40, 30, 60, -, 3500, 150, 225, 1380
Bishop, John H., 1, -, -, 72, 17, -, -, 4000, 100, 300, 900
Shorts, John H., -, -, 1, 170, 30, 25, -, 10000, 550, 910, 3060
Griffin, John S., -, -, 1, 455, 15, -, 2500, 150, 366, 300
Truax, Isaac, 1, -, -, 215, 35, 25, -, 14000, 650, 1760, 2950
Ford, Daniel, -, -, 1, 146, 4, 20, 100, 10000, 460, 1020, 1500
Kimmey, Elsy, -, -, 1, 35, -, 25, -, 2000, 60, 310, 500
Hutchinson, Mathew, 1, -, -, 106, 1, 25, -, 5000, 340, 400, 840
Frazier, Carey, -, -, 1, 160, 5, 40, 15, 9000, 400, 660, 1945
Wheatman, Samuel, -, -, 1, 75, 23, 25, -, 6000, 300, 420, 1256
Mclany, Jacob S., 1, -, -, 55, -, 15, -, 1800, 250, 330, 465
Voshell, John M., 1, -, -, 11, 5, -, -, 4600, 250, 410, 925
Ford, William H., 1, -, -, 105, 3, 16, -, 4000, 260, 475, 700
Jerman, John M., 1, -, -, 150, 10, 36, -, 6000, 280, 560, 1650
Riggs, William E. Jr., -, -, 1, 252, 12, 125, -, 10000, 325, 1000, 2050
Riggs, Abel S., -, -, 1, 150, 25, 5, -, 12000, 400, 780, 2360
Truax, Samuel P., -, -, 1, 204, 2, 14, -, 12000, 275, 1150, 3440
Silbey, Benjamin, -, -, 1, 94, -, 6, -, 3000, 120, 450, 380
Cary, Nehemiah, -, -, 1, 117, 5, 8, -, 3000, 155, 605, 550
Raughly, John, 1, -, -, 63, 15, 5, -, 4000, 240, 622, 635

Reynolds, John J., -, -, 1, 100, -, 15, 500, 5000, 120, 550, 670
Reynolds, John _., -, -, 1, 140, 20, 5, 200, 9000, 260, 530, 1660
Pearse, John F., -, -, 1, 120, -, 4, 50, 107, 2500, 75, 390, 780
Reynolds, Joseph R., -, -, 1, 113, 13, -, 150, 10000, 230, 900, 1510
Raughley, William, -, -, 1, 200, -, 40, 600, 9000, 320, 1150, 1975
Blake, James, -, -, 1, 130, -, -, -, 2500, 140, 520, 520
Robinson, James C., -, -, 1, 155, 25, 20, 15, 12000, 620, 1120, 3441
Raughley, John W., -, -, 7, 74, 15, 8, -, 4500, 100, 310, 545
Cary, George S., 1, -, -, 121, 4, 35, -, 8000, 250, 908, 1500
Lynch, George B., -, -, 1, 277, 10, 59, -, 15000, 490, 1516, 3089
Crow, Eugene, 1, -, -, 94, 46, 70, -, 5000, 160, 475, 935
Truax, James Sr., -, -, 1, 290, 30, 6, 50, 30000, 360, 1540, 3860
Ford, Presly, 1, -, -, 108, -, 4, -, 3000, 300, 565, 950
Kirby, Thomas, -, -, 1, 230, 35, -, 5, 100, 15060, 330, 1653, 2445
Roccass, Demot C., -, 1, -, 180, 8, 100, 2100, 6000, 60, 210, 6000
Cummins, Andrew, -, -, 1, 75, -, 3, 30, 45, 2000, 10, 75, 170
Cummins, Joseph, -, -, 1, 65, -, 1, 20, 30, 200, 35, 260, 700
Holtz, Mathew, 1, -, -, 45, -, -, 155, -, 1000, 60, 100, 260
Loatman, Samuel, l, -, -, 60, -, 340, 3000, 300, 690, 380
Johnson, Jacob, 1, -, -, 120, 150, -, 180, 4000, 300, 420, 540
Millington, George A., 1, -, -, 70, 10, 23, 320, 15000, 300, 1275, 700
Slaughter, George, 1, -, -, 146, -, 4, 50, 550, 4500, 150, 530, 985
Blades, John C., -, -, 1, 155, 49, -, 30, 6000, 325, 500, 1120
Keiser, William, -, -, 1, 58, -, 2, 30, -, 2000, 140, 335, 585
Robinson, George, 1, -, -, 10, -, -, -, 1000, 80, 145, 130
George, Henry, 1, -, -, 72, 35, -, 2000, 90, 310, 280
Sewerson, John, 1, -, -, 16, -, 13, -, 1000, 65, 92, 210
Rench, Washington D., -, -, 1, 120, 7, 10, 150, 8000, 220, 770, 1550
Goldsborough, Jeremiah R., 1, -, -, 60, -, -, 140, 2200, 175, 200, 590
Newcomb, Christopher C., 1, -, -, 41, 1, -, 10, 2400, 90, 165, 160
Connard, Isaac W., 1, -, -, 12, -, 50, -, 1000, 20, 95, 190
Butcher, Robert, 1, -, -, 10, -, 3, -, 850, 25, 75, 100
Schafee, Henry, 1, -, -, 180, -, -, 57, 8000, 140, 650, 1330
Denney, Robert, 1, -, -, 10, -, 3, -, 850, 25, 75, 100
Register, Charles H., -, -, 1, 180, -, -, 5, 7, 8000, 145, 1330
Turner, Enoch, -, 1, -, 12, -, -, 3, 1100, 75, 100, 260
Jones, George T., -, -, 1, 166, 14, -, 20, 10000, 310, 1005, 2165
Frazier, James, -, -, 1, 82, 3, 7, 8, 2500, 60, 305, 685
Anderson, Joseph H., -, -, 1, 14, 73, 8, -, 6500, 375, 1150, 1507
Spruance, William E., -, -, 1, 86, 14, -, 7000, 230, 540, 1628
Cole, Samuel J., -, -, 1, 40, 30, -, 5, 30, 1800, 50, 385, 190
Griffith, Charles H., -, -, 1, 250, -, 47, 10, 12000, 600, 1594, 2915
Vane, Jesse S., 1, -, -, 54, 2, -, -, 3500, 200, 445, 816
Short, William T., -, -, 1, 120, -, 35, 40, 3000, 75, 590, 431
Williams John H., -, -, 1, 105, 20, 6, 19, 10000, 260, 640, 1455
Masmon, Theodore, -, -, 1, 250, 20, 5, 75, 20000, 35, 1150, 2885

Miller, Edward, 1, -, -, 95, -, 10, -, 3500, 80, 270, 770
Raughley, John, 1, -, -, 105, 10, 40, 35, 4000, 100, 535, 1120
Cummins, George W., 1, -, -, 113, 62, 50, -, 30000, 1200, 1600, 6570
Scout, Augustus, 1, -, -, 16, 7, -, -, 8000, 500, 210, 1335
Carrow, Enoch Y., -, -, 1, 64, 2, 10, -, 3500, 125, 270, 608
Ford, Matthew, 1, -, -, 90, 10, 5, 5, 3500, 120, 833, 1075
Ford, Charles, -, -, 1, -, 100, -, -, 14, 2500, 60, 1010
Slaughter, Ezekiel, 1, -, -, 6, -, -, -, -, 300, 70, 150
Ford, Anderson, 1, -, -, 114, 36, 5, 10, 4000, 7, 5, 885
Newton, William D., -, -, 1, 138, 12, 5, 400, 3200, 70, 415, 965
Snyder, Cyrus K., 1, -, -, 100, 2, 40, 74, 3500, 120, 548, 1082
Williams, John, 1, -, -, 70, -, -, -, 2000, 60, 160, 554
Cooper, Hary, -, -, 1, 155, -, 15, 15, 15000, 355, 1530, 2180
Logan, John A., 1, -, -, 88, 2, 10, -, 4500, 345, 570, 1080
Moore, Sewell, -, -, 1, 122, 2, 5, 9, 4000, 110, 404, 705
Simpson, John H., -, -, 1, 113, 23, -, -, 4500, 530, 517, 760
Palmatory, Daniel, 1, -, -, 95, 1, 4, 6, 3800, 100, 230, 603
Truax, William G., -, -, 1, 163, 22, 15, -, 12000, 250, 955, 1975
Truax, James F., -, -, 1, 260, 42, 105, -, 20000, 340, 1085, 3650
Windal, Joseph S., 1, -, -, 44, -, -, 5, 1200, 80, 120, 305
Jones, John P., 1, -, -, 135, 10, 15, -, 9000, 380, 883, 1810
Snow, A. Jackson, 1, -, -, 150, 16, 5, 15, 7000, 190, 860, 1324
Hobbs, Titus J., -, -, 1, 175, 25, 5, 40, 14000, 800, 1435, 2668
Hobbs, Titus J., -, -, 1, 147, -, -, 3, 6500, 60, 901, 2195
Hobbs, Titus J., -, -, 1, 146, 2, -, 4, 6000, 150, 951, 2100
Cary, John P., -, -, 1, 127, 23, -, 10, 16000, 300, 840, 1964
Roo (Rov, Roe), Robert, -, -, 1, 110, -, 150, 525, 4500, 150, 445, 750
Carson, George, -, -, 1, 40, 4, -, 350, 2000, 120, 200, 400
Cummins, B. Frank, -, -, 1, 118, -, -, -, 4000, 120, 490, 910
Stockley, Ayres, 1, -, -, 32, -, -, -, 7000, 275, 400, 860
Morris, George S., 1, -, -, 12, -, -, -, 400, 30, 130, 145
Slaughter, Henry, -, -, 1, 250, 1, 30, 30, 20000, 620, 2450, 3925
Slaughter, Henry, -, -, 1, 80, 6, 20, -, 3000, 200, 550, 600
Morris, Aaron, 1, -, -, 6, -, -, -, 600, 55, 40, 236
Limberger, Jacob, 1, -, -, 24, 1, -, 5, 4500, 100, 455, 580
Jerman, J. T., -, 1, -, 16, -, -, -, 2400, 20, 215, 400
Tomlinson, ___, 1, -, -, 50, -, -, -, 1000, 25, 40, 90
Lewis, Hamilton, -, -, 1, 11, -, -, -, 1400, 25, 120, 373
Bailey, John H., -, 1, -, 10, -, -, -, 1000, 25, 25, 220
Cole, Ananias, -, -, 1, 15, -, -, -, 1400, 15, 50, 245
Moore, Robt. J., -, 1, -, 5, -, -, -, 500, 20, 100, 120
Bowden, Joseph, -, 1, -, 8, -, -, -, 600, 50, 220, 220
Lewis, John G., -, 1, -, 17, 10, -, -, 2200, 150, 250, 1000
Spruance, Horace, 1, -, -, 600, 145, 75, -, 30000, 1500, 1300, 10000
Anthony, John, 1, -, -, 155, 85, 44, -, 22000, 3000, 1260, 3510
Hudson, Alfred L., 1, -, -, 152, 158, 11, -, 30000, 1500, 1530, 7400

Woodhull, Francis, 1, -, -, 20, -, -, -, 4500, 150, 270, 350
Driggs, Lemuel B., 1, -, -, 37, 10, 10, -, 5000, 250, 250, 830
Boyles, James S., -, -, 1, 100, 20, -, -, 6500, 400, 600, 1800
Arrington, H. E., 1, -, -, 79, 32, -, -, 5000, 400, 400, 1250
Lowe, Ephraim, 1, -, -, 102, 20, 16, -, 8000, 600, 650, 180
Baker, Isaac, -, -, 1, 102, -, 15, -, -, 6000, 250, 500, 1150
Dickson, Jacob R., 1, -, -, 180, 10, 10, -, 15000, 300, 450, 2100
Goldsborough, Joseph, -, -, 1, 120, 60, 8, -, 11000, 400, 1250, 2660
Thompson, George T., -, -, 1, 150, 42, -, -, 20000, 400, 1100, 1930
Severson, John, 1, -, -, 170, 37, 15, -, 10000, 300, 765, 1900
Wheatman, Samuel, -, -, 1, 78, 42, 60, -, 6000, 400, 600, 2910
Griffin, Samuel, -, -, 1, 125, 25, 34, -, 12000, 500, 1043, 2315
Faison, Sarah V., 1, -, -, 10, -, 10, -, 1200, 100, 150, 215
Garrison, Gamalish G., -, -, 1, 183, 90, 10, -, 10000, 400, 1050, 4537
Truax, Henry F., 1, -, -, 18, 8, -, -, 2000, 250, 125, 440
Jones, John J., -, -, 1, 186, 47, 15, -, 7000, 700, 895, 2185
Trumpeller, David W., 1, -, -, 128, 30, 10, -, 8500, 600, 500, 1245
Jones, Ebenezer, -, -, 1, 198, 12, -, -, 5800, 100, 280, 585
Stockley, Ayres, 1, -, -, 15, -, -, -, 1800, 75, 160, 390
Collins, John Edwin, 1, -, -, 3, 5, -, -, 800, 50, -, 80
Cummins, William, 1, -, -, 16, -, -, -, 1600, 50, 100, 230
Wallace, Benjamin, -, -, 1, 144, 36, 2, -, 10000, 500, 1167, 1980
Ranson, William N., -, -, 1, 159, 31, -, -, 13000, 700, 835, 3832
Ranson, William N., -, -, 1, 26, -, -, -, 2500, 200, 125, 937
Jones, Caleb, -, -, 1, 14, -, -, -, 1400, 100, 225, 338
Davis, Mary J., 1, -, -, 23, -, -, -, 2500, 50, 100, 333
Raughley, Joshua, -, -, 1, 100, 20, -, -, 5000, 200, 1210
Mustard, John, 1, -, -, 40, -, -, -, 4000, 200, 270, 750
Cummins, Robt. L., 1, -, -, 25, -, -, -, 2400, 100, 50, 849
Wright, William, -, -, 1, 8, -, -, -, 800, 125, 180, 410
Reynolds, Samuel, 1, -, -, 20, -, -, -, 2000, 130, 175, 416
Levick, Parris M., -, -, 1, 50, 24, 15, -, 3000, 150, 200, 300
Postles, Zadoc, -, -, 1, 120, 25, 35, -, 10000, 500, 1000, 2500
Gartin, William R., -, -, 1, 100, 10, 10, -, 10000, 200, 800, 1600
Gartin, Henry, -, -, 1, 125, 14, 6, -, 10000, 200, 600, 2300
Webb, William D., -, -, 1, 140, 14, 12, -, 12000, 250, 100, 2000
Butler, William, -, -, 1, 135, 125, 60, -, 15000, 800, 1500, 3500
Moore, Thomas, -, -, 1, 40, 35, 25, -, 7000, 100, 400, 800
Conard, Augustus, -, -, 1, 70, 60, 12, -, 4000, 100, 200, 600
Morris, William R., 1, -, -, 75, 75, -, -, 20000, 300, 250, 2500
Slaymaker, James, 1, -, -, 9, 25, 3, -, 4500, 250, 500, 200
Downham, John W., -, -, 1, 150, 50, 20, -, 20000, 200, 1000, 3500
Cowgill, Daniel Jr., 1, -, -, 30, 30, 20, -, 17000, 200, 350, 300
Sheats, John, -, -, 1, 30, 30, -, -, 2000, 75, 150, 100
Collins, Alexander, -, -, 1, 100, -, 17, -, 8000, 800, 700, 1250
Givin, John S., 1, -, -, 30, 20, 5, -, 3400, 400, 100, 300

Denney, Thomas H., 1, -, -, 300, 100, 15, -, 4000, 1000, 3500, 12000
Eccles, Edwin O., 1, -, -, 100, 60, 25, -, 15000, 800, 600, 1300
Ladoux, William B., -, 1, -, 60, 20, -, -, 10000, 200, 1000, 1000
Smith, Sarah A., 1, -, -, 12, 3, -, -, 6000, 300, 400, 600
Sulivan, Noah, -, -, 1, -, 6, -, -, 300, 25, 75, 35
Kemp, John T., -, -, 1, 120, 15, 50, -, 10000, 130, 500, 1000
Farrow, Francis A., -, -, 1, 30, 4, 8, -, 4000, 200, 500, 600
Rash, Daniel, 1, -, -, 100, 70, 25, -, 5000, 300, 800, 1800
Wharton, Bolithat, 1, -, -, 115, 20, 105, -, 6000, 500, 400, 1300
Minones, William, -, 1, -, 12, -, 8, -, 500, 25, 150, 170
Wharton, Joshua B., 1, -, -, 300, 21, 30, -, 16000, 200, 1000, 2500
Jacobs, Oliver, 1, -, -, 87, 4, 10, -, 5000, 200, 300, 500
Jackson, Gilder, -, -, 1, 250, 2, 60, -, 16000, 600, 1200, 2500
Catts, Stephen, -, 1, -, 25, 50, 8, -, 2200, 150, 300, 400
Philips, Warren J., -, -, 1, 50, -, 5, -, 1500, 100, 100, 200
Gibson, Thomas, -, -, 1, 150, -, 12, -, 10000, 250, 800, 100
Gruman, John (overseer), 1, -, -, 150, -, -, -, 1000, 200, 12000, 1300
Slaughter, James W., -, -, 1, 150, -, 4, -, 10000, 100, 500, 1200
Argo, David, -, -, 1, 35, -, 10, -, 10000, 1000, 500, 1200
Slaughter, Thomas, -, -, 1, 150, 25, 25, -, 12000, 500, 500, 1000
Broom, William, -, -, 1, 40, -, -, -, 1200, 15, 100, 200
Wharton, John B., -, -, 1, 200, 40, 35, -, 15000, 500, 4000, 2500
Pardee, Charles H., -, -, 1, 250, 40, 30, 300, 16000, 350, 800, 1800
Cullen (Culler), William T., -, -, 1, 140, 40, 30, 18, 10000, 600, 700, 1500
Pardee, Edward Y., -, -, 1, 200, -, 20, 200, 10000, 1000, 800, 3000
Wilson, Robert, 1, -, -, 275, 40, 6, 200, 20000, 800, 2000, 3000
Caulk, John, -, -, 1, 125, 30, -, -, 8000, 300, 600, 600
Postles, William G., -, -, 1, 220, 8, 45, -, 16000, 1000, 2000, 3000
Collins, George, 1, -, -, 180, 30, 50, 100, 13000, 500, 1000, 2500
Patten, William, -, -, 1, 90, 1, 10, 25, 5000, 100, 300, 300
Conner, John J., 1, -, -, 120, -, 30, -, 6000, 50, 500, 1500
Dawes, William 1, -, -, 130, -, 15, -, 7000, 300, 1000, 2500
Priestly, John H., -, -, 1, 95, 12, 1, -, 5000, 200, 350, 510
Cowgill, Jacob, 1, -, -, 75, 10, 8, -, 6000, 150, 500, 600
Hutchinson, William, 1, -, -, 40, 10, 5, 5, 5000, 400, 2500, 500
Harrington, Isaac, -, 1, -, 230, -, 20, 250, 10000, 300, 1500, 2500
Johnston, George, -, -, 1, 180, 20, 15, 300, 16000, 400, 600, 1000
Slaughter, William, 1, -, -, 27, -, -, -, 500, 150, 200, 600
Taylor, Raymond J., -, -, 1, 200, 20, 60, -, 8000, 500, 1000, 1900
Anderson, Charles H. (col), -, -, 1, 32, -, -, -, 800, 100, 100, 200
Harrington, Harriet, -, -, 75, 30, 15, -, 5000, 250, 600, 1500
Lewis, Eben, -, -, 1, 210, 2, -, -, 10000, 1000, 1100, 3000
Wharton, Earnest, -, -, 1, 150, 20, 20, -, 5000, 300, 900, 1200
Raymond James, 1, -, -, 120, 10, 20, -, 5000, 50, 500, 500
Slaughter, Nathaniel, 1, -, -, 110, 5, 8, -, 8000, 100, 800, 1200
Catts, James E., -, -, 1,80, 1, 19, -, 3000, 100, 500, 500

Slaughter, Timothy, 1, -, -, 35, 2, 25, -, 10000, 500, 1000, 1200
Postles, Thomas, 1, -, -, 140, -, 13, -, 5000, 150, 300, 500
Taylor, James H., -, -, 1, 130, 14, 3, -, 10000, 100, 300, 1200
Davis, James A., 1, -, -, 138, 12, -, -, 5000, 600, 500, 1200
Bell, Alexander (col), -, -, 1, 32, 6, 6, -, 400, 60, 75, 100
Anderson, John (col), -, -, 1, 25, -, 10, -, 500, 75, 150, 150
Martin, Samuel, -, 1, -, 40, 10, -, 3, 400, 75, 150, 200
Knight, Morten, 1, -, -, 80, 3, -, 200, 1500, 100, 500, 600
Laffer (Loper, Lasser), William (col), -, -, 1, 25, 2, 1, -, 500, 50, 100, 300
Anderson, Emily, -, -, 1, 75, -, 10, 150, 5000, 50, 70, 200
Anderson, William 1, -, -, 60, -, 6, 60, 1000, 25, 125, 150
Dyer, Willliam, 1, -, -, 80, -, -, -, 4000, 100, 250, 1000
Slaughter, Andrew, 1, -, -, 95, 2, 5, -, 4000, 150, 500, 1000
Parvis, William, 1, -, -, 35, -, 5, -, 1600, 100, 200, 150
Lafferty, Rachel, 1, -, -, 59, 2, 1, -, 2500, 200, 250, 400
Knotts, William K., -, -, 1, 30, 4, 6, -, 6000, 150, 175, 700
Dyer, William, 1, -, -, 140, 30, 5, -, 12000, 500, 1200, 2800
Hayes, Manlove, 1, -, -, 140, 30, 8, -, 12000, 500, 1200, 2800
Paskins, Daniel, -, -, 1, 20, -, 10, -, 600, 56, 100, 200
Peak, Charles E., 1, -, -, 125, 10, 2, -, 8000, 300, 400, 1000
McClain, Abraham, -, -, 1, 125, 20, 5, -, 7500, 300, 100, 400
Welch, Robert, 1, -, -, 45, 7, 2, -, 3000, 50, 400, 400
Holt, Levi J., 1, -, -, 25, 5, -, -, 3000, 100, 150, 500
Fowler, William D., 1, -, -, 120, 25, 5, -, 7000, 300, 300, 1000
Knotts, William -, -, 1, 186, 4, -, 1, 12000, 2000, 1000, 1200
Bell, Isaac (overseer), 1, -, -, 457, 15, -, -, 5100, 100, 500
Harrington, James, -, -, 1, 140, 100, 16, -, 10000, 2500, 600, 1600
Denney, Charles, -, -, 1, 150, 100, -, -, 10000, 400, 100, 3500
Hevern, William S., -, -, 1, 500, 600, 50, -, 30100, 3000, 4000, 6000
Carter, William Q., -, -, 1, 110, -, 100, -, 20000, 300, 300, 3000
York, James T., -, -, 1, 75, 8, -, -, 5000, 200, 200, 600
Kimmey, Aleade, -, 1, -, 25, 13, 10, -, 5000, 150, 100, 600
Carrey, James, 1, -, -, 55, 20, -, -, 4000, 200, 300, 500
Blackston, Charles, -, -, 1, 29, 1, -, -, 3000, 100, 70, 200
James, Elias, -, -, 1, 208, 42, 12, -, 12000, 300, 900, 1500
Parvis, Joseph, -, -, 1, 144, 6, 30, -, 12000, 400, 1500
Rias, James, -, -, 1, 20, 10, 4, -, 500, 150, 100, 2
Voshell, Elisha, -, -, 1, 93, 1, -, -, 4000, 700, 300, 400
Danner, Hiram H., 1, -, -, 125, 15, -, 10, 12000, 300, 300, 800
Burnham, Thomas A., -, -, 1, 190, 15, 25, -, 12000, 700, 800, 2000
Wilson, Thomas W., 1, -, -, 50, 25, 5, -, 5000, 50, 100, 10
Walton, Staton, 1, -, -, 1, -, -, -, 500, 50, 100, 300
Sapp, Elias, 1, -, -, 30, -, -, -, 300, 100, 150, 300
Truitt, J. K., -, -, 1, 42, 5, -, -, 2000, 150, 400, 400
Cannon, Wilson L., 1, -, -, 120, 20, 70, 33, 15000, 250, 1200, 1000
Miller, Ollis P., -, 1, -, 131, 20, -, -, 10000, 500, 1200, 1000

Smith, James L., 1, -, -, 20, ½, -, -, 4000, 100, 150, 200
Saxton, Daniel, -, 1, -, 21, -, -, 1, 1000, 50, 100, 150
Moore, Samuel, 1, -, -, 20, -, -, 1, 1000, 75, 200, 200
Marvell, Theodore, -, -, 1, 80, 30, 20, -, 1500, 50, 100, 300
Downham, William, -, -, 1, 49, -, 31, -, 2000, 50, 100, 300
Berry, John, -, -, 1, 250, 30, 15, -, 17000, 400, 200, 3000
Carney, William, -, -, 1, 29, -, 10, -, 3000, 100, 350, 300
Farrow, George, -, -, 1, 45, 6, 10, -, 1500, 50, 100, 100
Moore, Joseph P., -, -, 1, 300, 36, 60, -, 20000, 300, 1000, 25000
Emory, Samuel, 1, -, -, 30, 3, 7, -, 1600, 25, 150, 300
Hancock, David, -, -, 1, 64, -, 2, -, 7000, 100, 250, 500
Jones, John L., 1, -, -, 12, -, -, -, 1200, 100, 100, 300
Lewis, Jacob G., -, -, 1, 250, 3, 20, 10, 8000, 200, 400, 500
Walker, William, 1, -, -, 20, 7, 160, -, -, 30000, 200, 500, 4000
Bowers, John C., 1, -, -, 110, 14, 12, -, 6500, 300, 350, 1400
Larned, H. D., -, -, 1, -, 160, 30, 5, 10000, 200, 1000, 1800
Larned, H. D., 1, -, -, 140, 20, 25, 20, 9000, 300, 600, 1250
Reed, Edward, -, -, 1, 65, 20, 20, -, 5000, 100, 300, 900
Emerson, Julian, 1, -, -, 100, 22, 10, -, 6000, 50, 200, 900
Hirons, Robert W., -, -, 1, 80, 15, 25, -, 4000, 300, 400, 600
Ridgway, Tilghman, -, -, 1, 20, 1, 7, -, 400, 25, 50, 160
Fulton, J. Alexander, 1, -, -, 44, 20, -, -, 5000, 100, 200, 650
Williams, Frederick A., 1, -, -, 14, 14, -, 2, 4000, 800, 50, 1200
John M. Downs, -, -, 1, 160, 26, 50, -, 2940, 300, 510, 980
Thomas, Jerald, -, -, 1, 150, -, 70, -, 1600, 75, 225, 250
Nancy May, 1, -, -, 93, -, 6, -, 2000, 50, 150, 250
Williams, Amos C., 1, -, -, 85, -, 25, -, 3000, 500, 300, 503
Powell, William, -, -, 1, 75, -, 25, -, 2500, 225, 275, 400
Marshall, Moses, -, -, 1, 60, -, 86, -, 800, 20, 100, 200
Legar, John, 1, -, -, 48, -, 12, -, 14000, 100, 225, 500
Cooper, Samuel, -, 1, -, 40, -, 10, -, 500, 130, 115, 75
Munce, Isiah, -, -, 1, 100, -, -, 1500, 130, 265, 200
Roe, William H., -, -, 1, 136, -, 50, -, 3000, 150, 250, 200
Darnell, Henry, -, 1, -, 16, -, -, -, 150, -, 40, 100
Summers, Jonathan, 1, -, -, 8, -, 8, -, 315, -, 10, 100
Voshall, John, -, -, 1, 180, -, -, 50, 5000, 50, 150, 450
Marvel, Philip, 1, -, -, 100, -, 50, -, 1800, 160, 400, 600
Cook, William, 1, -, -, 96, -, 25, -, 4000, 400, -, 1000
Paor, John H., 1, -, -, 5, -, -, -, 200, 40, 40, 400
Woodall, Thomas -, -, 1, 30, -, 50, -, 600, 100, 190, 160
Pronerel, Alphonso F., -, -, 1, 70, -, 20, -, 900, 60, 100, 150
Paor (Paon), Thomas, -, 1, -, 200, -, -, -, 200, 100, 185, 200
Philips, Samuels, -, -, 1, 700, 10, -, 10, 2000, 300, 400, 520
Everts, Honey, -, 1, -, 35, -, 75, -, 500, 175, 300, 400
Jackson, William, 1, -, -, 14, -, -, -, 300, 50, 200, 180
Shahan, John W., 1-, -, 14, -, 4, -, 320, 50, 150, 100

Caleal, William, 1, -, -, 55, -, 145, -, 1400, 100, 200, 300
Jackson, Andrew F., -, -, 1, 10, -, -, -, 100, 60, 100, 150
Morris, John, -, -, 1, 9, -, 16, -, 440, 50, 200, 190
Powel, Hiram C., -, 1, -, 24, -, -, -, 500, 50, 200, 300
Vible, Samuel, 1, -, -, 18, -, 16, -, 700, 50, 150, 225
Morris, Isaac, -, -, 1, 69, -, 15, -, 1000, 100, 200, 375
Roberts, James, -, -, 1, 18, -, 19, -, 4000, 100, 250, 119
Richardson, James, 1, -, -, 12, -, -, -, 800, 50, 200, 200
Richardson, William, 1, -, -, 18, -, 15, -, 125, 100, 150, 225
Morris, George, 1, -, -, 16, -, 25, -, 650, 150, 200, 150
Todd, Solomon, -, -, 1, 11, -, 14, -, 200, 50, 75, 100
Hammond, George, 1, -, -, 70, -, 10, 70, 300, 150, 300, 350
Greenage, Samuel, -, -, 1, 100, -, 30, -, 300, 75, 200, 350
Greenage, John, -, -, -, 40, -, 10, -, 50, -, 125, 260
Thompson, James, -, -, 1, 60, -, 175, 450, 75, 260, 175
Powell, W. R., 1, -, -, 30, -, 50, 10, 1000, 75, 275, 175
Powell, Michael, -, -, 1, 50, -, 16, 20, 1000, 100, 200, 200
Powell, Charles H., 1, -, -, 75, -, 20, -, 2000, 100, 350, 400
Davis, George, -, 1, -, 18, -, -, -, 225, 10, 50, 225
Morris, John, -, 1, -, 18, -, 4, -, 120, -, -, 150
Hay, John W., -, -, 1, 25, -, -, 55, 900, 50, 275, 150
Coznegys, William J., -, 1, -, 78, -, -, -, 200, 40,75, 200
Ortist(Artist), William, -, -, 1, 130, -, 25, 25, 2000, 100, 300, 500
Conly, William H., -, -, 1, 10, -, -, -, 300, 50, 100, 100
Johnson, William H., 1, -, -, 17, -, -, -, 700, 50, 75, 150
Cox, Nathan H., -, 1, -, 16, -, -, -, -, 150, -, 75
Nickerson, Joshua, 1, -, -, 50, -, -, -, 1500, 25, 250, 200
Barrett, Joseph, 1, -, -, 48, -, 10, -, 1000, 150, 400, 300
Caldwell, Jacob, -, 1, -, 78, -, 2, -, 1000, 50, 200, 400
Gibbs, Charles E., -, -, 1, 34, -, -, -, 2000, -, 400, 604
Hall, James, -, 1, -, 100, 15, 25, -, 3000, 150, 500, 760
Parker, Hall, 1, -, -, 15, -, -, -, 500, 40, 150, 140
Hawkins, William, 1, -, -, 125, 8, 67, -, 2000, 300, 500, 100
Jones, George, -, -, 1, 50, -, -, 50, 100, 75, 230, 150
Reynolds, George B., -, 1, -, 10, -, -, 50, 600, 75, 175, 137
Craig, Samuel, 1, -, -, 80, -, 57, 3, 3000, 400, 200, 700
Jacobs, Morton, -, -, 1, 70, -, -, 10, 1200, 25, 125, 10
Enich, James H., -, 1, -, 85, -, 65, 50, 1000, 50, -, 150
Fountain, Aaron, 1, -, -, 12, -, -, -, 500, 75, 100, 100
Ford, Henry, 1, -, -, 8, -, -, -, 275, 50, 100, 111
Oliver, Edward, -, -, 1, 10, -, -, -, 400, -, 100, 100
Gibbs, Richard, -, -, 1, 11, -, -, -, 500, -, 20,175
Delany, Henry, 1, -, -, 5, -, -, -, 125, 10, 50, 100
Shorts, Kinssey (Kinsey), -, -, 1, 34, -, 6, -, -, -, -, 150
Crammer, Samuel, -, 1, -, 10, -, -, -, 5, 40, 25, 100, 200
Clayton, William, -, -, 1, 12, -, 16, -, 450, 50, 150, 150

Wallace, William R., 1, -, -, 40, -, 8, -, 1500, 50, 50, 280
Campbell, William, 1, -, -, 125, -, -, 6, 1600, 150, 800, 700
Farrow, Joseph D., 1, -, -, 24, -, -, 20, 1000, 100, 300, 200
Mosely, Purnell, -, -, 1, 60, -, 10, -, 950, 40, 150, 200
Jacobs, Eli, -, -, 1, 100, -, 18, 18, 1000, 200, 300, 300
Faulkner, Andrew J., 1, -, -, 27, -, 40, -, 2000, -, 400, -
Pleasants, Jones, 1, -, -, 32, -, -, -, 320, 100, 500, 250
Derrah, Samuel, -, -, 1, 70, -, 5, -, 800, 50, 100, 175
Taylor, John, -, 1, -, 27, -, -, 3, 500, 40, 100, 150
Hammond, Harriet, 1, -, -, 15, -, 3, -, 150, -, 100, 140
Carter, George T., -, -, 1, 22, -, 3, -, 800, 75, 150, 150
Durham, Daniel, -, -, 1, 35, -, -, 10, 1100, 175, 300, 300
Williams, William, 1, -, -, 8, -, -, -, 80, 50, 75, 100
Willey, Francis, -, 1, -, 20, -, 30, -, 1500, 50, 225, 170
Collins, Benjamin T., -, -, 1, 30, -, -, 30, 1100, 50, 100, 150
Taylor, Martin, 1, -, -, 9, -, -, 1, 150, 25, 50, 50
Jewel, John, -, -, 1, 170, -, 120, 50, 3000, 100, 800
Carrow (Farrow), William, -, -, 1, 200, -, 75, 225, 10000, 300, -, 2000
Siere, Adam, 1, -, -, 16, -, -, -, 800, 56, 75, 150
James, Henry, -, -, 1, 8, -, -, 3, 80, 50, 75, 190
Furguson, Elizer, 1, -, -, 30, -, -, 4, 500, -, -, 150
Graham, Benjamin, -, -, 1, 85, 42, 43, -, 2800, 100, 600, 600
Newnum(Newsum), William T., 1, -, -, 28, -, -, -, 1000, 75, 75, 600
Smith, Blanchurd, -, -, 1, 9, -, -, -, 1200, 55, 125, 300
Rash, Eugene, 1, -, -, 30, -, 95, -, 40000, 150, 400, 880
Evert, John H., 1, -, -, 15, -, 14, -, 500, 60, 150, 200
Taylor, Robert S., 1, -, -, 20, -, 10, -, 600, 200, 50, 175
Moon, Henry, 1, -, -, 70, -, 60, -, 5000, 400, 50, 2000
Jackson, Alex, -, -, 1, 120, -, 80, -, 4000, 200, 300, 1400
Shahan, William E., 1, -, -, 12, -, 5, -, 1000, 375, 200, 400
Mclain, James, -, -, 1, 100, -, 80, 120, 1000, 40, 75, 100
King, Thomas M., 1, -, -, 15, -, 8, -, 800, 100, 100, 300
Bishop, Job, 1, -, -, 40, -, 10, -, 1700, 50, 150, 200
Pearson, James W., -, -, 1, 125, -, 75, -, 3000, 100, 225, 400
Ables, H. Nehemiah, -, -, 1, 125, -, 70, -, 3000, 500, 800, 1100
Wheatman, Charles, -, -, 1, 50, 4, 37, -, 2510, 500, 100
Simmons, Prince, -, 1, -, 40, -, 10, -, 1200, 50, 400, 200
Moore, Joseph, 1, -, -, 70, -, 35, -, 1750, 200, 300, 500
Hughs, John, -, -, 1, 150, -, 30, -, 8000, 300, 750, 600
Hefnal, Samuel M., -, -, 1, 110, -, -, -, 3300, 300, 700, 500
Gibbs, Nathan, -, -, 1, 75, -, -, 25, 1500,75, 200, 400
Thompson, Obediah, -, -, 1, 10, -, -, -, 400, 10, 150, 200
Bell, John, -, -, -, 1, 75, -, -, 36, 2000, 300, 200, 350
Voshall, Draper, 1, -, -, 42, -, -, -, 852, 30, 200, 150
Clow, William H., -, 1, -, 15, -, -, -, 300, 50, 150, 100
Melson, John H., -, 1, -, 15, -, -, -, 300, 50, 150, 100

Melvin, Trewett, 1, -, -, 55, -, -, -, 2000, 50, 200, 600
Lindale, Peter M., 1, -, -, 57, -, 10, -, 1000, 50, 150, 500
Gullett, Edward, 1, -, -, 140, -, 10, -, 5000, 250, 400, 600
Mosely, James, -, 1, -, 10, -, -, -, 300, 150, 200, 200
Wallace, Samuel C., 1, -, -, 90, -, 15, -, 2000, 100, 150, 700
Thompson, William, -, -, 1, 20, -, 6, -, 1000, 40, 200, 300
Bearman, Henry, 1, -, -, 100, -, 50, -, 4000, 200, 500, 600
Bearman, Henry D., -, -, 1, 38, -, 37, -, 1500, 75, 200, 300
Ireland, Joseph, -, -, 1, 200, -, 150, -, 4000, 200, 375, 600
Johnson, Isaac, 1, -, -, 100, -, 100, 35, 2500, 700, 500, 800
Mastin, Benjamin W., 1, -, -, 125, -, 95, -, 5000, 150, 100, 800
Moore, Edward, 1, -, -, 120, -, 6, -, 2500, 100, 200, 500
Driggars, Robert, -, -, 1, 24, -, -, -, 400, 20, 20, 100
Register, Abraham, 1, -, -, 18, -, -, -, 300, 40, 15, 100
Lord, John B., 1, -, -, 15, -, -, -, 1100, 300, 150, 350
Mosley, Nehemiah, 1, -, -, 28, -, 4, -, 200, -, 50, 45
Wiley, John, -, -, 79, -, 9, -, 6000, 500, 600, 1200
Shorts, Joseph M., -, -, 1, 100, -, 9, -, 2500, -, 400, 1000
Pearson, Truett H., -, -, 1, 36, -, 4, -, 500, -, 200, 300
Heye, David, 1, -, -, 70, -, 30, -, 200, -, 400, 1000
Hanger, Cornelius, -, -, 1, 50, -, 10, -, 200, -, 500, 300
Ridgway, Alfred, -, 1, -, 5, -, -, 45, 800, 40, 160, 180
Moore, Abraham C., 1, -, -, 9, -, -, -, 1900, 200, 250, 300
Slaughter, E. B., -, -, 1, 40, -, -, -, 2500, 65, 275, 250
Heard, Joseph, -, -, 1, 75, -, 25, -, 2500, 50, 250, 400
Rash, Mark, -, 1, -, -, -, -, -, -, 50, 300, 500
Nichols, Alexis D., -, 1, -, 175, -, 104, -, 10000, 150, 300, 1000
Driggis, George, -, -, 1, 7, -, -, -, -, 400, 100, 100
Driggers, Fordine, -, -, 1, 35, -, -, -, 1000, 100, 200, 300
Mosely, Isaac, -, -, 1, 15, -, -, -, 150, 40, 150, 75
McKee, William, 1, -, -, 130, -, 47, -, 12000, 800, 600, 2500
Ridgway, William, -, -, 1, 100, -, -, -, 3500, 150, 500, 600
Moseley, John M., -, -, -, 1, 10, -, 20, 80, 2200, 200, 400,800
Greer, Isaac, -, 1, -, 60, -, 40, -, 500, -, 1500, 300
Reville, Henry W., 1, -, -, 76, -, 10, -, 3100, 250, 200, 500
Whitenock, Cornelius, -, -, 1, 30, -, -, 44, 3500, 20, 200, 250
Legar, James, -, -, 1, 109, -, 50, 241, 10000, 300, 1300, 1100
Aron, William -, -, 1, 100, -, 14, -, 5000, 200, 760, 800
Wetherby, Joseph R., -, -, 1, 100, -, 10, 79, 5000, 600, 650, 800
Spence, R. A., 1, -, -, 40, -, 25, 28, 4000, 100, 175, 500
Brothers, William 1, -, -, 50, -, 42, -, 4300, 200, 400, 800
Hargadine, Eliza, 1, -, -, 70, -, 20, 18, 3000, 100, 400, 600
Johnson, Thomas, -, 1, -, 5, -, -, -, 250, 50, 80, 100
Smith, William J., -, -, 1, 800, -, 20, -, 4000, 150, 400, 820
Cohoon, William R., 1, -, -, 70, -, 10, -, 5000, 300, 700, 1600
Teet, James, 1, -, -, 60, -, -, 6, 2500, 100, 250, 400

Beers, James L., 1, -, -, 90, -, 15, -, 6000, 150, 300, 400
Hybedo, George, 1, -, -, 8, -, -, -, 500, 25, 100, 190
Jolly, D. A., -, -, 1, 90, -, -, 6, 5000, 200, 400, 750
Dennis, John, -, -, 1, 122, -, 10, -, 10000, 600, 1050, 1500
Rash, Moses, 1, -, -, 75, -, 42, -, 6000, 320, 400, 500
Pearson, Aron, 1, -, -, 32, -, -, -, 700, -, 800, 200
Thomas, Joseph, 1, -, -, 6, -, -, -, 700, 25, 190, 400
Pearson, Abraham, 1, -, -, 90, -, 1300, -, 3500, 500, 800, 1800
Pearson, William, 1, -, -, 106, -, 25, -, 1700, 100, 250, 300
Wilson, James R., -, -, 1, 100, -, 38, -, 5000, 600, 700, 1000
Rash, Joseph T., 1, -, -, 80, -, -, 20, 10, 5000, 200, 375, 600
Jones, Levin P., 1, -, -, 30, -, -, -, 1000, 50, 190, 100
Hurd, Robt. E., -, -, 1, 137, -, 200, -, 6000, 150, 200, 600
Newton, John T., -, 1, -, 30, -, -, -, 600, 200, 200, 375
Carpenter, Henry, 1, -, -, 10, -, -, -, 140, 50, 100, 200
Rash, Nehemiah, -, 1, -, 75, 25, 33, -, 3000, 300, 458, 1500
Shorts, John, -, -, 1, 150, 40, 60, 20, 5000, 200, 800, 1100
Sherard, Samuel, -, 1, -, 125, -, -, 20, 800, 125, 700, 700
Paulsbury (Saulsbury), Thomas J., -, -, 1, 190, -, 60, -, 11000, 500, 1500, 2500
Weslon, Lewis, 1, -, -, 18, -, -, -, 1000, 75, 100, 300
Kersey, Jonathan, 1, -, -, 38, -, -, -, 3000, 200, 350, 575
Voshall, John T., -, 1, -, 33, -, -, -, 900, 50, 300, 250
Corse, William, -, 1, -, 40, -, -, -, 1000, 50, 100, 200
Dodd, William, A., -, 1, -, 10, -, -, -, 600, 75, 150, 125
Slaughter, Thomas, -, 1, -, 200, -, -, -, 6000, 300, 725, 800
Faulkner, Thomas, 1, -, -, 106, -, 11, 2, 1500, 100, 400, 1000
Hutson, Henry, -, -, 1, 120, -, 75, -, 4000, 350, 500, 350
Johnis, James, 1, -, -, 170, -, 30, -, 6150, 150, 300, 1000
Webster (Welston), David H., 1, -, -, 13, -, -, -, 500, 60, 260, 250
Mearks, Solomon D., 1, -, -, 14, -, -, -, 250, 40, 150, 123
Scotter, Sarah A., 1, -, -, 80, -, 20, -, 2000, 100, 225, 300
Jones, William, 1, -, -, 79, -, 8, -, 1000, 100, 300, 400
Milbourne, Thomas H., 1, -, -, 60, -, -, -, 1400, 100, 250, 350
Milbourne, Elisabeth, -, -, 1, 60, -, -, -, 1600, 50, 175, 250
Downes, Daniel W., 1, -, -, 42, -, 5, -, 1500, 150, 200, 350
Boyer, Charles E., -, 1, -, 65, -, -, -, 1000, 150, 200, 250
Stevens, James, 1, -, -, 24, -, -, -, 400, 75, 160, 400
Hall, Charles E., 1, -, -, 80, -, 12, -, 1400, 150, 350, 410
Harnet, Timothy F., 1, -, -, 44, -, -, -, 1500, 50, -, 300
Harnet, Edmond, 1, -, -, 15, -, -, -, 550, 40, 150, 200
Roberts, Jacob, 1, -, -, 95, -, -, -, 1100, 100, 400, 350
Black, Fredric, -, -, 1, 100, -, -, -, 1000, 100, 300, 250
Hartnet, Lawrence, 1, -, -, 14, -, -, -, 255, -, 150, 100
Harnet, Lawrence F., 1, -, -, 40, -, -, -, 400, -, 100, 200
Scott__, Spencer, 1, -, -, 85, -, 45, -, 4000, 300, -, 400
Wickett, Robert, 1, -, -, 66, -, 10, -, 1500, 146, 150, 250

Greenwell, Francis, 1, -, -, 112, -, 8, -, 3500, 200, 1000, 500
Perry, George, -, -, 1, 35, -, 15, -, 1000, 25, 338, 200
Cornegys, Albert, -, -, 1, 100, -, 30, -, 2000, 100, 6000, 900
Bishop, Henry, -, 1, -, 40, -, -, 10, 700, 50, 75, 150
Moore, Enoch, -, 1, -, 100, -, -, -, 1500, 10, 125, 400
Boyer, Jerrel, 1, -, -, 90, -, 20, -, 5000, 300, 400, 1000
Hall, William D., 1, -, -, 62, -, -, 13, 2500, 100, 285, 250
Livingston, James, 1, -, -, 100, -, 40, -, 5000, 250, 450, 250
Kersey, George W., -, -, 1, 160, -, 10, -, 4000, 160, 475, 600
Moore, George D., -, -, 1, 77, -, 75, -, 2000, 150, 235, 610
Perry, Thomas C., -, -, 1, 90, -, -, -, 2000, 100, 225, 400
Roberts, Stephen, 1, -, -, 60, -, -, 10, 950, 20, 200, 225
Hackett, William, 1, -, -, 20, -, -, -, 200, 25, 50, 100
Course, Solomon, -, 1, -, 5, -, -, -, 150, 25, 100, 10
Virdin, John W., -, -, 1, 175, -, 75, -, 5000, 200, 750, 1675
Spenceneer, Stephen, 1, -, -, 17, -, -, 3, 240, 25, 70, 75
Parker, John B., 1, -, -, 85, -, -, -, 1200, 100, 300, 300
Perry, John C., -, -, 1, 85, -, -, -, 1500, 125, 300, 400
Cohol (Cahal), John H., -, -, 1, 150, -, 52, -, 1800, 75, 220, 400
Stephens, John, 1, -, -, 60, -, 2, -, 1000, 20, 400, 400
Stephens, Enoch, -, -, 1, 22, -, -, -, 500, 10, 100, 150
Moore, William, 1, -, -, 60, -, 46, -, 2000, 75, 900, 500
Kersey, John B., 1, -, -, 3, -, 25, -, 2500, 100, 250, 245
Rash, Henry, 1, -, -, 35, -, 15, -, 600, 85, 200, 195
Thompson, William Thomas, -, -, 1, 100, -, 55, -, 3800, 100, 200, 400
Whelby, Nathan, 1, -, -, 46, -, -, -, 300, 50, 40, 50
Muller, Henry, 1, -, -, 13, -, -, -, 500, -, 130, 75
Taylor, Enoch, 1, -, -, 8, -, -, -, 300, 40,75, 80
Craig, Jacob, 1, -, -, 28, -, -, -, 350, 25, 185, 50
Esra, William, 1, -, -, 20, -, -, -, 500, 75, 145, 200
Johnson, Bengamon, -, 1, -, 23, -, -, -, 450, 25, -, 360
Scott, Clement, 1, -, -, 17, -, -, -, 400, 30, 250, 100
Callahan, William, -, -, 1, 40, -, -, -, 1000, -, 160, 150
Jerman, George W., 1, -, -, 40, -, -, -, 500, 25, 100, 100
Ford, Martin F., 1, -, -, 200, -, 100, -, 13000, 600, 1200, 200
Pratt, Jonathan, 1, -, -, 27, -, 50, -, 1500, 20, 120, 300
Moore, Thomas S., 1, -, -, 100, -, -, -, 4000, 200, 400, 401
Moore, Thomas E., -, 1, -, 18, -, -, -, 1100, 150, -, 300
Smith, John of Lurnal, 1, -, -, 30, -, -, -, 800, 4, 930, 150
Griffith, John T., -, 1, -, 40, -, -, -, 1500, 50, 750, 500
Minos, John W., -, -, 1, 100, -, 100, -, 5000, 150, 40, 400
Crawly, Timothy, -, 1, -, 15, -, -, -, 400, 50, 120, 210
Hutchins, Daniel V., -, 1, -, 4, -, -, -, 600, -, -, -, 250
Moore, Thomas, 1, -, -, 82, -, -, -, 2000, 150, 350, 500
Slaughter, Jonathan, 1, -, -, 53, 5, -, -, 2500, 100, -, 400
Jones Samuel C., -, -, 1, 55, -, -, -, 3500, 100, 400, 750

Williams, William J., -, -, 1, 92, -, 18, -, 3000, 200, 500, 1000
Jones, Garrett, 1, -, -, 8, -, -, -, 400, -, 20, 410
Gier, William T., -, -, 1, 50, -, -, -, 1000, 100, 250, 200
Sparks, Samuel J., 1, -, -, 60, -, -, -, 1500, 100, 200, 370
Gomery, Jesse, 1, -, -, 20, -, -, -, 600, 50, 150, 140
Shaher (Shahan), Jacob, 1, -, -, 54, -, -, -, 1500, 100, 250, 500
Griffith, James W., 1, -, -, 75, -, 40, -, 2000, 100, 500, 300
Shaher, Jonathan, -, -, 1, 18, -, -, -, 450, 140, 150, 230
Cahal, James W., -, -, 1, 150, -, -, -, 2000, 75, 150, 400
Voshall, Obediah, -, -, 1, 60, -, -, -, 900, 150, 600, 375
Bedwell, Caleb, -, 1, -, 25, -, -, -, 500, -, 150, 700
Marnes, Thomas J., 1, -, -, 120, 6, 70, -, 6000, 500, 1500, 1100
Taglibet, Daniel, -, -, 1, 100, -, 24, -, 1800, 150, 300, 400
Stevens, James Sr., -, -, 1, 24, -, -, -, 900, 50, 200, 150
Jones, James, -, -, -, 1, 40, -, -, -, 700, 50, 250, 250
Ridgley, Henry, 1, -, -, 385, -, -, -, 50000, 500, 950, 2500
Jones, Francis E., -, 1, -, 167, -, -, -, 16700, 600, 1725, 2561
Hooker, James, -, 1, -, 35, -, -, -, 700, 150, 200, 600
Thomas, Even, 1, -, -, 12, -, -, -, 2000, 50, 150, 440
Culler, John W., 1, -, -, 135, -, -, -, 1400, 300, 200, 2200
Marvel, Emery, -, -, 300, -, -, -, 15000, 900, 1359, 2000
Carsons, William, -, -, 1, 75, -, 10, 15, 2500, 100, 375, 785
Staber, Magor, -, -, 1, 75, -, 15, -, 3000, 200, 300, 800
Mosely, John, 1, -, -, 200, -, -, -, 500, 25, 100, 100
Hews, Sue H., -, -, 1, 75, -, -, 30, 200, 200, 200, 200
Warl, John L., 1, -, -, 25, -, -, -, 100, 50, 185, 180
Miller, David, -, -, 1, 200, -, 40, -, 5500, 250, 625, 1010
Robinson, Samuel, -, -, 1, 107, -, 20, -, 3000, 250, 500, 800
Pearson, Thomas, 1, -, -, 60, -, 10, -, 1500, 150, 350, 450
Yolt (Colt, Holt), William N., 1, -, -, 55, -, 40, -, 1800, 100, 165, 300
David, Herran (Henson), 1, -, -, 75, -, 10, -, 2700, 300, 300, 550
Sapp, William T., -, -, 1, 160, -, -, 30, 11500, 100, 1000, 1975
Harmer, Bowen, 1, -, -, 72, -, -, -, 4500, -, 400, 400
Willis, John, -, -, 1, 67, -, -, -, 2500, 150, 300, 300
Willis, William A., 1, -, -, 60, -, -, -, 1800, 100, 645, 300
White, Clement, -, -, 1, 80, -, -, -, 3000, 100, 220, 150
James, Richard M., 1, -, -, 135, -, 2, -, 15000, 175, 640, 953
Kniverse, John J., 1, -, -, 100, -, 10, -, 9000, 450, 600, 1600
Hope, Samuel W., 1, -, -, 100, -, 27, -, 10000, 150, 250, 1100
Wheatly, William, -, -, 1, 160, -, -, -, 14400, 700, 1100, 2300
Violet, Dierty, -, -, 1, 100, -, -, -, 8000, 300, -, 700
Wharton, Stanly, 1, -, -, 410, -, -, -, 25000, 1000, 1390, 2100
Coffin, Joseph _., -, -, 1, 174, -, -, -, 12000, 200, 824, 2400
Coverdale, David, 1, -, -, -, 210, -, -, -, 12000, 400, 930, 2000
Hope, James M., 1, -, -, 100, -, -, 6, 1000, 150, -, 877
Baum, Collinston, 1, -, -, 107, -, -, 8, 17000, 500, -, 2200

Bellman, Daniel, 1, -, -, 78, -, -, -, 8000, 80, 457, 800
Duhadaway, James, -, -, 1, 60, -, -, -, 5500, 1175, 500, 400
Laws, Bozzle, 1, -, -, 85, -, 15, -, 2000, 75, 150, 190
Slaughter, John, 1, -, -, 115, -, 40, -, 4350, 250, 750, 2500
Bobbett, Charles C., 1, -, -, 222, -, 30, -, 18900, 600, 1000, 4400
Turner, Thomas, 1, -, -, 100, -, 33, -, 4500, 415, 278, 400
Kirk, William, -, -, 1, 50, -, 80, -, 3000, 250, 225, 500
Sturdivant, Samuel T., 1, -, -, 67, -, -, -, 3500, 50, 250, 450
Jones, George T., -, -, 1, 72, -, 18, -, 3000, 100, 250,800
Hatfield, Henry W., 1, -, -, 100, -, 55, -, 6000, 350, 400, 1200
Atkinson, Q. Charles, 1, -, -, 38, -, -, -, 2000, 65, 150, 200
Wright, Robert W., 1, -, -, 75, -, 55, -, 6000, 250, 600, 500
Baudlett, William H., -, -, 1, 110, -, 20, -, 4000, 175, 360, 795
Ste__, William, 1, -, -, 100, -, 26, -, 4300, 300, 325, 900
Abbott, Jonathan, 1, -, -, 50, -, 26, -, 5, 2000, 100, 300, 400
Farrow, Daniel, -, -, 1, 60, -, 25, 40, 4000, 150, 525, 450
Dreggis, Samuel, -, -, 1, -, -, -, -, -, 50, 150, 75
Covell, Francis A., 1, -, -, 21, -, -, -, 1000, 50, 115, 175
Shorts, Jacob M., -, -, 1, 20, -, -, -, 2800, 65, 200, 200
Skase, William, -, -, 1, 86, -, -, -, 1204, 100, 275, 350
Aron, Powell, 1, -, -, 122, -, -, -, 1000, 1100, 100, 1800
Greenly, James P., -, -, 1, 150, -, -, -, 2400, 175, -, 300
Wright, William, -, -, 1, 97, -, -, -, 3500, 30, 200, 670
Powell, James R., 1, -, -, 89, -, -, -, 5000, 600, 700, 850
Laubach, E. M., 1, -, -, 137, -, 7, -, 6000, 200, 350, 1860
Bonwill, Peter F., 1, -, -, 270, -, 20, -, 15000, 500, 1800, 1850
Massey, Elisha, -, -, 1, 47, -, 6, -, 2350, 50, 25, 456
Massey, Alfred, 1, -, -, 93, -, -, -, 4000, 300, 600, 409
Hutchinson, Henry P., 1, -, -, 100, -, 30, -, 2000, 125, 345, 575
Hirons, Samuel, -, -, 1,75, -, -, -, 1875, 200, 500, 265
Hirons, William, 1, -, -, 165, -, -, -, 4025, 350, 100, 690
Hurd, Lewis, -, -, 1, 40, -, -, -, 800, -, 100, 475
Boggs, John, - -, 1, 9, -, -, -, 450, 75, 100, 200
Kersey, James, 1, -, -, 30, -, -, -, 900, 100, 145, 284
Pearson, Charles H., 1, -, -, 18, -, -, -, 700, 150, 207, 268
Casson, Sarah, 1, -, -, 117, -, -, -, 3500, 220, 300, 600
Reed, Samuel L., 1, -, -, 46, -, -, -, 1000, 50, 125, 250
Clark, Auther, 1, -, -, 6, -, -, -, 500, 30, 140, 837
Sirse, Thomas, -, -, 1, 185, 50, -, -, 8000, 300, 465, 460
Coppage, Edward, -, -, 1, 70, -, -, -, 800, 1500, 400, 475
Pearse (Pearson), John W., 1, -, -, 63, -, -, -, 2000, 50, 350, 500
Downs, Francis W., -, -, 1, 109, -, -, -, 1500, 50, 330, 275
Pearson, Isaac, 1, -, -, 132, -, -, -, 5000, 200, 520, 786
Stevens, Emery, 1, -, -, 23, -, -, -, 600, 100, 300, 400
Bochwell (Bedwell), Preston, -, -, -, 1, -, -, -, -, -, 200, 615, 1105
Mayberry, John P., -, -, 1, 300, 40, 60, -, 10000, 1100, 1200, 1850

Gibbs, Stephen, -, -, 1, 219, -, -, -, 6800, 100, 426, 508
Laws, Noah, -, -, 1, 21, -, -, -, 800, 40, 100, 139
Clark, John, -, -, 1, 24, -, -, -, 750, 75, 200, 236
Wright, Elisha, 1, -, -, 90, -, 10, -, 2000, 200, 600, 1215
Morris, Elisha, -, -, 1, 15, -, -, -, 400, 50, 75, 200
Grant, E. R., 1, -, -, 53, -, -, -, 1500, 300, 221, 333
Morris, Bracham, & Cunningham, Samuel, 1, -, -, 80, -, 90, -, 1800, 200, 300, 340
Virdin, Alex, -, -, 1, 140, -, 70, 30, 4800, 325, 825, 1500
Braklin, Samuel, -, -, 1, 120, -, 15, 5, 2800, 30, 225, 840
Pratt, Henry Jr., 1, -, -, 142, -, 25, -, 5000, 480, 900, 3110
Gooden, Isaac K., -, -, 1, 175, -, 25,-, 4000, 350, 475, 1128
Seward, Luther, 1, -, -, 100, -, -, -, 3000, 350, 466, 980
Hutchins, James A., -, -, 1, 160, -, 20, 20, 4000, 300, 590, 580
Johnson, John H., 1, -, -, 150, -, 50, -, 1500, 200, 350, 300
Pratt, Isaac, 1, -, -, 110, -, 40, -, 4000, 200, 500, 3000
Paw__ly, Warren B., 1, -, -, 160, -, 15, -, 500, 325, 670, 1250
Morflight, Joseph, -, -, 1, 70, -, 60, -, 2500, 85, 220, 206
Hurd, Rachel, 1, -, -, 40, -, 40, -, 1200, 50, 105, 200
Zaylon (Taylor), John, -, -, 1, 15, -, -, 200, 40, -, 45, 250
Paon, Martin, 1, -, -, 6, -, -, 10, 250, 40, 100, 110
Cosden (Gooden), William, -, 1, -, 100, -, 80, -, 3300, 150, 395, 304
Thompson, William, -, -, 1, 100, -, 7, -, 1500, 75, 506, 494
Horsey, William, -, -, 1, 80, -, 20, -, 1800, 50, 185, 506
Hinsley, Mathew, -, -, 1, 60, -, 20, -, 1000, 50, 95, 270
Thompson, Obediah, -, -, 1,190, -, 100, -, 5800, 150, 500, 850
Fisher, John, 1, -, -, 110, -, 20, 7, 3000, 75, 560, 230
Holland, Isaac G., 1, -, -, 20, -, 5, 16, 500, 70, 50, 230
Thomas, William, -, -, 1, 75, -, 5, -, 2500, 300, 425, 735
Clark, James, 1, -, -, 160, -, 30, 40, 3000, 865, 40, 2000
Harrington, David, 1, -, -, 88, -, 5, -, 3500, 175, 450, 804
Marshall, Samuel C., -, -, 1, 75, -, 31, -, 1000, 25, 200, 200
Voshal, John J., 1, -, -, 158, -, 30, -, 3000, 300, 700, 770
Marvel, Philip D., 1, -, -, 100, -, 67, -, 3500, 250, 832, 1675
Marvel, Thomas Sr., -, -, 1, 50, -, -, -, 1100, 100, 115, 300
Miller, George W., 1, -, -, 38, -, -, -, 380, 30, 240, 178
Knapp (Knopp), Harlan, 1, -, -, 56, -, -, -, 1200, 100, 200, 159
Dill, Ezkel, -, -, 1, 100, -, 50, -, 1700, 50, 321, 340
Clements, Thomas, 1, -, -, 219, -, 50, -, 4035, 300, 694, 1635
Hutchens, John, 1, -, -, 60, -, 20, -, 12000, 75, -, 186
Reed, Charles, -, 1, -, 18, -, -, -, 500, 45, 235, 189
Voshall, Joseph, -, 1, -, 29, -, -, -, 2000, 45, 190, 500
Scotter(Scotten), Delilemer, -, -, 1, 146, -, -, 40, 5000, 40, 103, 384
Cox, Nathan, 1, -, -, 31, -, -, -, 620, 20, 770, 285
Burris, Harrison, -, 1, -, 180, -, -, -, 960, -, 440, 128
Hubbard, Newton, 1, -, -, 197, -, -, -, 1100, 300, 475, 700
Hubbard, Edward, 1, -, -, 175, -, -, -, 3500, 300, 600, 1530

Dill, John W., -, -, 1, 89, -, -, -, 1000, 50, 70, 670
Montague, Jesse, 1, -, -, 88, -, -, -, 2000, 150, 300, 340
Penise, John, 1, -, -, 100, -, -, -, 1000, 100, 221, 266
Marvel, David J., 1, -, -, 195, -, -, -, 4000, 645, 1125, 2855
Ford, Moses K., 1, -, -, 200, -, 50, -, 4000, 300, 570, 974
Ford, George E., -, -, 1, 90, -, -, -, -, 150, 262, 157
Williams, Bruce, 1, -, -, 77, -, -, -, 1200, 75, 200, 167
Clark, Williams, -, -, 1, 75, -, -, -, 1500, 100, 120, 400
Hutson, Nathaniel, 1, -, -, 30, -, 2, -, 500, 50, 185, 250
Honey, Thomas, -, 1, -, 31, -, -, -, 50, 173, 190
Moore, Thomas, -, -, 1, 60, -, -, -, 800, 50, 110, 129
Clark, John, -, 1, -, 90, -, -, -, 1100, 100, 135, 290
Wilkinson, John, -, -, 1, 14, 6, -, -, 700, 45, 90, 178
Slay, John, 1, -, -, 296, 50, 30, 20, 7735, 400, 620, 1296
Philips, Samuel, -, -, 1, 107, -, 21, -, 2540, 115, 350, 530
Green, James H., 1, -, -, 80, -, 31, -, -, 360, 50, 887
Walls, James, -, 1, -, 75, -, -, -, 900, 75, 200, 390
Williams, John B., 1, -, -, 120, -, -, -, 1000, 200, 315, 1504
Benwell, John, -, -, 1, 60, -, -, -, 800, 100, 125, 142
Burris, Edward, 1, -, -, 50, -, -, -, 500, 500, 220, 120
Voshall, John, 1, -, -, 928, -, 30, -, 3500, 200, 603, 547
Voshall, Charles, -, 1, -, 10, -, 5, -, 600, 300, 400, 200
Vincent, Nicholas, 1, -, -, 7, -, -, -, 300, -, 180, 120
Gooden, John, 1, -, -, 250, -, -, -, 10000, 100, 1461, 2154
Clark, James, -, -, 1, 250, -, 150, -, 6000, 500, 1216, 1029
Steel, Josiah, 1, -, -, 100, -, 100, -, 5000, 5000, 1275, 1100
Parvis, Wm., 1, -, -, 37, -, -, -, 600, 50, 125, 160
Johnson, John, 1, -, -, 34, -, -, -, 1500, 100, 400, 632
Johnson, Thomas, -, -, 1, 100, -, 100, -, 2000, 150, 250, 641
Becklin, John, -, -, 1, 119, -, 36, -, 3100, 300, 585, 670
Bedwell, Ann, 1, -, -, 88, -, -, -, 1000, 75, 250, 224
Marvel, Luther, -, -, 1, 105, -, -, -, 2100, 150, 610, 900
Virdin, William Jr., -, -, 1, 116, -, 60, -, 3500, 100, 600, 815
Denner, Henry P., -, 1, -, 10, -, -, -, 50, 500, 50, 150, 250
Cunningham, David, 1, -, -, 45, -, 15, -, 800, 75, 175, 150
Johnson, Henry, -, -, 1, 97, -, -, -, 2000, 150, 200, 280
Voshal, George T., 1, -, -, 35, -, -, -, 1000, 100, 110, 120
Rash, Thomas, 1, -, -, 35, -, -, -, 500, 50, 110, 200
Patten, Zaddock, -, 1, -, 22, -, -, -, 500, 75, 260, 250
Scott, Nathaniel, -, -, 1, 150, 25, 50, 25, 3300, 180, 500, 800
Sudler, Joshua, 1, -, -, 50, -, 20, -, 1000, 100, 500, 350
Swallow, William, 1, -, -, 119, -, -, -, 2000, 60, 150, 275
Thompson, John W., -, -, 1, 90, -, -, -, 1500, 150, 328, 510
Calahan, Jesse & Andrew, 1, -, -, 32, -, -, -, 1200, 100, 100, 290
Tolstone, James, -, -, 1, 100, -, -, -, 1600, 100, 400, 257
Thomas, Wesley John, 1, -, -, 140, -, -, -, 1400, 200, 400, 375

Thomas, Isaac, 1, -, -, 140, -, -, -, 1200, 108, 500, 350
Montage, Seth, -, 1, -, 100, -, 1000, 17, 1500, 100, 190, 200
Moore, John W., 1, -, -, 200, -, 85, -, 4750, 400, 640, 650
Patten, Lydia, 1, -, -, 39, -, -, -, 600, 40, 100, 150
Price, Thomas, 1, -, -, 21, -, -, -, 300, 40, 100, 175
Voshall, Joseph, 1, -, -, 87, -, -, -, 1000, 40, 450, 180
Caldwell, Daniel, -, -, 1, 40, -, -, -, 500, 75, 175, 125
Williams, James, 1, -, -, 173, -, 25, -, 4000, 500, 1100, 1896
Raughly, Joshua B., -, 1, -, 300, -, 200, -, 5000, 300, 1190, 2900
Barens, John, -, -, 1, 150, -, -, -, 1000, 50, 100, 120
Powel, William, 1, -, -, 10, -, -, -, 500, 25, 75, 75
Darling, William, -, -, 1, 150, -, 50, -, 3300, 150, 533, 355
Short, James, -, -, 1, 25, -, -, -, 510, 10, 125, 210
Calahan, Nathan, -, -, 1, 60, -, -, -, 150, 50, 698, 160
Zerratt, William, 1, -, -, 66, -, -, -, 2500, -, 100, 536
Nickerson, John B., 1, -, -, 110, -, -, -, 2500, 280, 530, 870
Whitaker, John, 1, -, -, 244, -, -, -, 15000, 900, 1380, 3922
Todd, James H., 1, -, -, 110, 90, 30, 30, 12000, 500, 2000, 1000
Demar, Joseph, -, -, 1, 120, -, -, 100, 5000, 100, 450, 1250
Saxton, Austin, -, -, 1, 80, -, 40, 80, 2000, 20, 50, 185
Saxton, Henry C., -, -, 1, 100, -, 40, 30, 300, 50, 220, 530
Handy, Parker, 1, -, -, 7, -, -, -, 600, 12, 100, 30
Denney, John P. M., 1, -, -, 76, 12, -, -, 6000, 50, 400, 100
Barcus, John, 1, -, -, 100, 1, 60, -, 2500, 25, 640, 625
Harrington, William, 1, -, -, 120, 17, -, 3500, 125, 625, 500
Hazel, Henry K., 1, -, -, 100, 3, 30, -, 3000, 75, 300, 650
Fenimore, John W., 1, -, -, 100, 4, -, -, 4000, 200, 1100, 1270
Taylor, Thomas K., 1, -, -, 145, 25, 45, -, 8600, 180, -, 1700
Grasly, William H., 1, -, -, 50, -, -, -, 3500, 50, 300, 850
Clendaniel, James, -, 1, -, 125, -, 15, -, 2000, 100, 500, 500
Quillon, John, -, -, 1, 105, 35, 5, -, 5500, 950, 600, 600
Ennis, Presly, 1, -, -, 65, -, 10, 25, 2500, 50, 550, 600
Seward, William P., -, 1, -, 300, 25, -, 150, 20000, 1200, 2760, 4200
Vaughn, Joseph E., -, -, 1, 137, -, 3, 25, -, 8000, 200, 100, 1200
Moore, William F., -, -, 1, 90, -, ½, 6, 10, 3000, 100, 350, 925
Moore Owen, -, -, 1, 45, 80, 15, 300, 3000, 75, 100, 190
Green, Washington, 1, -, -, 8, 10, -, -, 600, -, 10, 40
Rickards, Charles, 1, -, -, 23, -, 4, -, 600, -, 100, 30
Smith, George W., -, -, 1, 100, 5, 36, 10, 4000, 130, 600, 1100
Keith, Thomas Sr., 1, -, -, 80, 6, 38, -, 4000, 90, 625, 750
Senn, John Jacob, 1, -, -, 140, -, 100, -, 4000, 25, 100, 100
Cowgill, Daniel E., 1, -, -, 120, 5, 25, -, 6000,200, 650, 1540
Smith, Eli, -, -, 1, 100, 2, 25, -, 3000, 150, 600, 1600
Hurly, William, -, -, 1, 360, -, 40, -, 15000, 500, 1700, 2500
Davis, Manlove, 1, -, -, 10, -, 26, -, 600, 10, 75, 100
Holden, Edmond, -, -, 1, -, 5, -, -, -, 200, 600, 850

Smithers, William H., -, -, 1, 150, 150, 25, -, 8000, 200, 900, 2670
Pleasanton, John, -, 1, -, 15, 2, -, -, -, 25, 100, 150
Taylor, John, -, -, 1, 238, 3, -, 20000, 565, 1240, 4400
Stout, Emanuil J., 1, -, -, 260, 125, -, 800, 150, 1300, 1590
Seward, Joseph, -, -, 1, 190, 6, 15, -, 900, 303, 930, 2560
Richardson, James B., 1, -, -, 20, -, 8, -, -, 1000, 10, 125, 100
Green, Robert H., -, -, 1, 180, 15, -, 20, 15000, 250, 900, 3900
Wilson, Daniel M., 1, -, -, 200, 2, 25, -, 10000, 500, 1000, 4100
Wilson, Thomas W., Jr., -, 1, -, 155, -, 5, 15, -, 10000,700, 1100, 3250
Wilson, Thomas W. Sr., 1, -, -, 137, -, 2, -, -, 8000, 200, 425, 2560
Smith, William, -, -, 1, 130, 20, -, 20, 10000, 60, 800, 1500
Moore, Joseph S., 1, -, -, 80, -, -, -, 5000, 75, 525, 1000
Ennis, John, -, -, 1, 75, -, 5, 10, 1500, 50, 480, 450
Ennis, William H, 1, -, -, 22, -, 2, -, 1500, 500, 835, 1250
Wilson, Andrew J., 1, -, -, 125, 8, 70, 6, 3000, 50, 400, 600
Cole, Abner R., 1, -, -, 66, 3, 4, -, 1500, 50, 200, 450
Wilson, Henry L., 1, -, -, 130, 4, 20, -, 3000, 65, 850, 650
Coverdale, George W., -, -, 1, 75, -, -, -, 2000, 50, 200, 370
Wall, Thomas, -, -, 1, 90, -, 25, 20, 2500, 200, 720, 1500
Holston, Thomas, -, -, 1, 35, -, 40, -, 1000, 100, 320, 400
Ward, James, -, -, 1, -, -, -, -, -, 400, 750, 900
Lowber, Samuel M., -, -, 1, 90, 1, 15, 300, 3000, 50, 400, 1000
Williams, Henry A., -, -, 1, 160, 35, 7, -, 10000, 250, 1150, 2800
Carson, Robert E., -, -, 1, 230, -, 26, -, 10000, 250, 800, 2350
Pleasanton, Edward Jr., -, -, 1, 70, -, 4, -, 1500, 50, 250, 300
Hamm, Benjamin F., 1, -, -, 125, -, 35, -, 8, 8000, 100, 980, 750
Vaughn, James, B., -, -, 1, 130, -, -, 100, 3000, 125, 600, 900
Berry, Isaac M., -, -, 1, 200, 2, 50, 140, 7500, 180, 965, 1400
Moore, Abram, 1, -, -, 90, 2, 2, 3, 4000, 100, 600, 1200
Fox, Abraham, -, -, 1, 200, 20, 40, -, 8000, 285, 1150, 2000
Ennis, James M., -, -, 1, 70, ½, -, -, 3000, 125, 1000, 800
Pleasanton, David, 1, -, -, 135, 21, -, -, 8000, 800, 1475, 2000
Lowber, Peter E., -, -, 1, 135, 15, 2, 25, 10000, 50, 800, 2500
Keith, James, -, -, 1, 133, 17, 50, -, 700, 350, 900, 1650
Lafferty, James, -, -, 1, 150, 25, 25, -, 8000, 250, 1075, 2000
Smith, John, W., -, -, 1, 110, 9, 6, -, 4000, 125, 600, 975
Denney, Isaac M., 1, -, -, 100, 10, 25, 15, 6000, 200, 875, 1450
Raymond, John C., 1, -, -, 90, 3, 15, 12, 5000, 75, 400, 880
Miller, David, -, -, 1, 160, 5, 10, -, 5600, 160, 750, 955
George, Thomas R., 1, -, -, 125, 1, 8, -, 4000, 75, 730, 1050
Wilson, Thomas W., 1, -, -, 175, -, 3, 35, 15, 4000, 150, 1115, 800
Harper, Henry R., -, -, 1, 112, 12, 50, -, 4000, 100, 450, 900
Freeston, Matilda, 1, -, -, 74, 1, -, -, 5000, 60, 350, 500
Moore, Joseph S., 1, -, -, 573, -, 10, 2500, 40, 150, 250
Garrison, Ephraim S., 1, -, -, 132, 20, 10, 40, 14000, 500, 900, 2350
Jefferson, Elihu, 1, -, -, 26, 4, -, -, 3000, 50, 150, 200

Morgan, James, -, -, 1, 76, 1, -, -, 4000, 50, 400, 400
Cooper, Richard M., 1, -, -, 65, 15, 4, -, 4000, 200, 500, 1120
Smith, Frank M., 1, -, -, 51, -, 2, -, -, 2500, 100, 320, 200
Carney, Sherry, -, -, 1, 56, 15, -, -, 1500, 40, 125, 170
Hetsh__, John, -, -, 1, 126, 45, -, 10, 12000, 500, 1400
Ewing, Elizabeth, -, -, 1, 52, 2, -, -, 2000, 50, 250, 300
McGonigal, John R., 1, -, -, 100, -, 5, 100, 3000, 350, 325, 700
David, William T., -, 1, -, 65, -, -, -, 1000, 25, 170, 150
Harrington, John M., -, -, 50, -, 25, 50, 2000, 50, 400, 360
York, Samuel C., 1, -, -, 35, 12, 15, 6, 4000, 206, 575, 600
Hendricson, John H., 1, -, -, 54, 5, 12, 22, 5000, 50, 500, 500
Norris (Morris), John K., -, -, 1, 70, 1, 30, -, 3000, 100, 800, 1200
Vaughn, Charles R., -, -, 1, 100, 25, 25, 5, 4000, 200, 700, 800
Hamm, Pleasanton 1, -, -, 19, 8, -, -, 1500, 20, 150, 500
Laws, Alexander, 1, -, -, 323, 60, 40, 260, 23000, 200, 2500, 6970
Clements, William, -, -, 1, 100, 15, 25, 12, 3000, 50, 950, 1075
Clements, Charles S., -, -, 1, 200, 20, 25, 10, 10000, 260, 1200, 2250
Slaughter, Isaac C. -, -, 1, 78, 2, 6, 4, 5000, 250, 400, 450
McFann, John, -, -, 1, 130, 20, -, 150, 6000, 300, 800, 1265
Berry, Wilson P., -, -, 1, 130, 6, 25, 15, 2000, 30, 200, 375
Pleasanton, Edward, -, -, 1, 200, -, 30, 15, 6000, 300, 1000, 1100
Snech, Moses, 1, -, -, 148, 6, 30, 20, 4000, 300, 670, 470
Smith, John L., -, -, 1, 166, 21, 60, -, 10000, 200, 1330, 1235
Wilson, Elizabeth, -, -, 1, 135, 14, 40, -, 6000, 150, 695, 1240
Cummins, George, -, -, 1, 50, -, 30, -, 2000, 40, 225, 375
Perry, John H., -, -, 1, 93, 3, -, -, 2500, 50, 275, 700
Argae (Argo), Jorden, -, -, 1, 104, 1, 80, 8, 4000, 150, 570, 1100
Bishop, Resdon H., 1, -, -, 20, -, 6, 4, 2500, 50, 60, 310
Bishop, John H., 1, -, -, 85, 15, 12, 10, 4000, 200, 400, 485
McNutt, James, -, -, 1, 100, 50, 30, 3, 2500, 75, 300, 400
Williams, William, 1, -, -, 25, 2, 18, -, 1000, 10, 325, 200
Denney, William T., 1, -, -, 6, 4, -, -, 1000, 10, 80, 160
Sanders, Isaac, -, -, 1, 80, -, 110, -, 3500, 60, 300, 565
Boggs, David, -, -, 1, 85, 12, 40, 3, 5400, 100, 1075, 800
Clouds, David M., 1, -, -, 50, 10, -, -, 2500, 25, 200, 700
Emmerson, James, W., 1, -, -, 100, 23, 40, -, 6000, 100, 250, 1000
Stockwell, Louis, 1, -, -, 62, 18, 12, 22, 4000, 150, 550, 1014
Cudney, Albert S., -, -, 1, 20, 1, -, -, 1000, 20, 200, 700
Boyer, George, 1, -, -, 140, 1, -, 10, 6000, 250, 1280, 1000
Boyer, Charles, -, -, 1, 190, 10, 30, 70, 6000, 40, 1000, 1157
Harper, Charles F., -, -, 140, 10, 40, -, 6000, 200, 670, 1425
Lamb, James P., 1, -, -, 80, 8, 20, -, 3000, 200, 425, 575
Conaway, Minos, 1, -, -, 100, 30, 10, 10, 7500, 500, 1000, 1880
Morris, Ambrose, -, -, 1, 70, -, 6, -, 2500, 150, 530, 815
SOME NAMES ARE IN FIRST NAME FIRST
Garret L. Hynson, -, -, 1, 120, 10, 35,-, 10000, 200, 800, 2000

Davis H. Mason, -, -, 1, 100, 2, 25, -, 6000, 20, 600, 1050
Reuben Harrington, 1, -, -, 120, 3, 15, -, 6000, 200, 800, 1300
James Quillon, 1, -, -, 80, 2, 40, -, 4500, 200, 875, 1915
George Armstrong, 1, -, -, 40, 1, 60, -, 2000, 100, 125, 400
John W. Lynch, -, -, 1, 160, 8, 6, -, 10000, 200, 500, 2000
Joseph Frazier, 1, -, -, 150, 6, 28, -, 12000, 300, 1000, 2900
Mary C. Tomlinson, 1, -, -, 75, 10, 10, -, 3500, 200, 500, 1000
George B. Taylor, -, -, 1, 125, 10, 25, -, 4000, 50, 300, 900
William Spencer, -, -, 1, 10, 23, -, -, 7000, 400, 600, 11400
John S. Harrington, 1, -, -, 156, 1, 28, -, 10000, 700, 1350, 2980
Orlean DeHitt, 1, -, -, 86, 7, -, -, 5000, 365, 625, 1500
Henry H. Fisher, -, -, 1, 80, 5, 15, -, 3000, 25, 350, 400
James Sapp, 1, -, -, 56, 11, 2, -, 4000, 175, 400, 1400
Edward B. C. McColley, 1, -, -, 100, -, 15, -, 2500, 350, 300, 450
George Johnson, -, 1, -, 8, -, 6, -, 300, 5, 35, 30
James H. Postles, 1, -, -, 95, 20, 21, -, 6000, 400, 515, 1240
Wm. H. Mason, -, -, 1, 160, 15, 10, -, 6000, 200, 700, 1600
John T. Johnson, -, -, 1, 173, -, 225, -, 9000, 400, 410, 1650
George W. Holmes, 1, -, -, 30, 12, 5, -, 2500, 100, 250, 600
END FIRST NAME FIRST
Warren, Thos. W., -, -, 1, 100, 10, 20, -, 5000, 150, 350, 800
Clark, Jos. V., -, -, 1, 90, 5, -, -, 7000, 400, 600, 1340
Postles, James B., -, -, 1, 65, 4, 30, -, 1800, 165, 265, 475
Masten, David R. Sr., 1, -, -, 3, ½, -, -, 300, 25, 10, 50
Smith, James L., 1, -, -, 15, 21, -, -, 3000, 50, 100, 200
Hitch, Spencer, 1, -, -, 70, 2, 41, -, 2500, 75, 150, 450
Hammond, John W., -, -, 1, 135, 12, 55, -, 8000, 500, 800, 2350
Masten, Hezekiah, 1, -, -, 130, 3, 25, -, 6000, 450, 600, 1150
Lofland, Parran T., -, -, 1, 75, 1, 25, -, 4000, 100, 200, 450
Jacobs, Wm. A., 1, -, -, 30, -, 4, -, 1000, 50, 125, 300
Jacobs, Levi C., 1, -, -, 25, -, 5, -, 1000, -, 150, 740
Jacobs, John, 1, -, -, 20, -, -, 20, 1200, 100, 125, 500
Morris, Geo. T., -, -, 1, 27, -, 15, -, 1200, 20, 40, 200
Hill, John, 1, -, -, 60, 3, 15, -, 3300, 100, 250, 550
Thistlewood, Albert, -, -, 1, 37, -, -, -, 1200, 100, 150, 450
Melvin, Riley, 1, -, -, 40, -, 3, -, 800, -, -, 75
Harrington, Theodore C., -, -, 1, 120, 330, -, 6000, 300, 375, 700
Turner, Albert, -, -, 1, 75, 4, 50, -, 3500, 250, 250, 2600
Harrington, Charles, -, -, 1, 105, 25, -, -, 700, 400, 850, 3200
Quillen, John, 1, -, -, 51, -, 20, -, 3500, 250, 350, 900
Davis, William E., -, -, 1, 165, -, 100, -, 10000, 200, 600, 2400
Vinyard, Curtis, 1, -, -, 130, -, -, -, 6000, 225, 700, 1000
Lofland, John M., 1, -, -, 130, 35, 5, -, 5000, 100, 400, 650
Postles, Mark Q., 1, -, -, 100, 25, 20, -, 6000, 100, 250, 1000
Cannon(Connor), Denis, 1, -, -, 130, 5, 10, -, 7000, 300, -, 1700
Betts, Stephen K., 1, -, -, 75, 2, 32, -, 4000, 255, 450, 1100

Dixon, David L., 1, -, -, 80, 10, 20, -, 4500, 150, 200, 900
Sanders, Daniel, 1, -, -, 25, 9, 15, -, 1000, 75, 175, 225
Wilson, William, 1, -, -, 20, 1, 2, -, 1200, 125, 100, 200
Elliott, Caleb, -, -, 1, 70, -, 13, -, 1200, 30, 140, 600
Jacobs, James H., -, -, 1, 7, 1, -, -, 500, 50, 35, 105
Marvel, Josiah, 1, -, -, 100, 6, 25, -, 5000, 300, 400, 1000
Moore, Samuel, 1, -, -, 30, -, 30, -, 1000, 25, 150, 300
McColley, George, 1, -, -, 80, 10, 10, -, 2500, 100, 175, 500
Vinyard, Henry, 1, -, -, 90, 6, 40, -, 3000, 125, 500, 800
Kesler, Josiah, -, -, 1, 70, 18, 24, -, 3000, 50, 200, 500
Woodrow, Robert, 1, -, -, 110, 20, 20, -, 3500, 15, 50, 500
Tomlinson, Wm. C., 1, -, -, 70, -, 40, -, 2000, 50, 75, 250
Foreacre, Elmer, -, -, 1, 80, 10, 5, -, 3000, 100, 200, 650
Sapp, John M., -, -, 1, 150, 3, 150, -, 6000, 150, 320, 850
Harrington, Major A., 1, -, -, 130, 8, 40, -, 5000, 400, 600, 1900
Lewis, Haskell, 1, -, -, 160, 30, 60, -, 10000, 600, 1200, 2200
Hammond, Eli A., -, -, 1, 200, 40, 60, -, 12000, 150, 500, 2400
Alexander, Thomas, 1, -, -, 70, 20, 50, -, 5000, 300, 450,1600
Wilcuts, John, -, -, 1, 120, -, 30, -, 5000, 150, 1000, 800
Evans, Joseph, 1, -, -, 40, 6, 5, -, 2000, 100, 75, 400
Bye, J. Hart, 1, -, -, 80, 5, -, -, 4000, 300, 250, 1000
Davis, George F., -, -, 1, 150, 9, 15, -, 8000, 300, 800, 1600
Wilbur, Jeptha C., 1, -, -, 6, 13, -, -, 5000, 150, 60, 1200
Colborn, William J., -, -, 1, 35, 1, 40, -, 2000, 100, 200, 500
Bennett, John W., 1, -, -, 85, -, 20, -, 3000, 200, 350, 1000
Hill, Joshua D., -, -, 1, 65, 5, 20, -, 3000, 225, 400,750
Wales, Blake, 1, -, -, 90, 50, 40, -, 7000, 200, 400, 1600
Bennett, John H., 1, -, -, 60, 20, 15, -, 2000, 300, 600, 950
Bell, Caleb S., 1, -, -, 85, 2, 30, -, 3500, 125, 350, 600
Trader, Wm. R., 1, -, -, 60, 15, 4, -, 1000, 75, 100, 300
Owens, Joseph H., 1, -, -, 140, 4, 35, -, 6000, 150, 150, 900
Macklin, Curtis, -, 1, -, 30, 4, 6, -, 1000, 40, 250, 400
Burris, John, 1, -, -, 43, 2, 5, -, 500, 100, 400, 400
Hayes, Charles, -, -, 1, 40, 15, 10, -, 1200, 75, 160, 500
Anderson, John, -, 1, -, 35, -, 5, -, 1000, 75, 300, 400
Vinyard, William 1, -, -, 42, 2, 20, -, 1200, 75, 200, 700
Melvin, Joseph, -, -, 1, 17, 2, 1, -, 400, 150, 90, 200
Marvel, Henry N., 1, -, -, 150, 2, 100, -, 5000, 100, 125, 450
Alexander, Isaac D., -, 1, -, 57, 6, 3, -, 1600, 200, 200, 850
Scott, Ira, 1, -, -, 17, 3, -, -, 1000, 75, 100, 540
Watson, David B., -, 1, -, 21, 12, -, -, 1500, 125, 150, 325
Scott, David, 1, -, -, 41, 30, 200, -, 7000, 150, 425, 500
Wilson, Levi H., -, -, 1, 35, -, -, -, 900, 100, 250, 700
Richardson, Willard, 1, -, -, 100, 26, 24, 20, 5000, 100, 300, 800
Higgins, Charles, 1, -, -, 3, -, -, -, -, 150, 50, 40
Appleman, Alfred, 1, -, -, 47, 8, 28, 10, 2200, 75, 80, 600

Johnson, John H., 1, -, -, 20, -, 10, -, 1000, 100, 200, 400
Jump, James L., -, -, 1, 38, 6, 8, -, 1000, 40, 50, 250
Wilson, Wm. C. J., -, -, 1, 80, 1, 220, -, 3000, 50, 175, 100
Satterfield, Bruffet S., -, -, 1, 120, 3, 175, -, 3000, 50, 100, 416
Insley, James M., 1, -, -, 30, 2, 10, -, 800, 100, 150, 175
Hammond, George P., -, -, 1, 80, 2, 20, -, 2500, 100, 100, 350
Coverdale, Jeremiah, -, -, 1, 80, -, 25, -, 2000, 30, 35, 250
Emory, Charles, -, 1, -, 75, 8, 135, -, 1500, 125, 300, 275
Dorsey, William N., 1, -, -, 170, 35, 100, -, 4000, 150, 150, 450
Griffith, David R., 1, -, -, 180, 20, 100, -, 4000, 100, 295, 1025
Dickerson, William J., -, -, 1, 60, -, -, -, 900, 100, 300, 600
Connor, James, -, -, 1, 12, -, -, -, 200, 30, -, 200
Foreacre, Masculine, -, -, 1, 100, 12, 20, -, 1500, 60, 100, 300
Armor, Samuel H., 1, -, -, 30, 3, 10, -, 1200, 100, 100, 400
Wharton, Charles W., -, -, 1, 169, 20, 40, -, 4800, 200, 425, 1400
Townsend, William J., 1, -, -, 48, 1, 5, -, 1500, 100, 150, 400
Townsend, Charles B., 1, -, -, 54, 2, 15, -, 1500, 100, 175, 450
Hitchens, Samuel J., 1, -, -, 30, 25, 15, -, 2500, 125, 100, 750
McColley, Robert M., 1, -, -, 2, 16, 2, -, 400, 50, 150, 250
Wheatly, Henry, 1, -, -, 7, -, -, -, 250, 20, 150, 275
Mills, George W., -, -, 1, 175, 3, 40, -, 4000, 100, 150, 400
Chipman, William, 1, -, -, 50, 10, 20, -, 2000, 75, 85, 300
Isleib, George, 1, -, -, 30, 25,10, -, 2000, 100, 200, 750
Riggs, Levi B., -, -, 1, 40, -, 25, -, 1200, 25, 65, 225
Ackerman, A. J., 1, -, -, 87, 20, 15, -, 2500, 1300, 200, 300
Henry, Alexander, 1, -, -, 80, 15, 6, -, 3000, 200, 175, 900
Rudy, Frank, 1, -, -, 120, 14, 45, -, 4000, 200, 600, 1650
Kenney, H. R., -, 1, -, 20, 52, -, -, 3500, 200, 160, 600
Snell, Hiram, 1, -, -, 46, 17, 10, -, 7000, 300, 242, 600
Rash, John J., 1, -, -, 55, 35, 1, -, 9000, 1800, 450, 5500
Gardner, J. W. C., 1, -, -, 12, 6, 4, -, 8000, 200, 200, 1000
Pullen, Alexander, -, 1, -, 55, 10, -, -, 5000, 400, 375, 6000
Clendaniel, Wm. H., -, -, 1, 150, -, 10, -, 5000, 75, 450, -, 550
Williams, C. T. W., 1, -, -, 125, 32, 35, -, 2700, 250, 175, 550
Woodruff, Aaron B., 1, -, -, 100, 40, 20, -, 6000, 500, 500, 1500
Hollis, Charles W., -, -, 1, 40, 2, 25, -, 1800, 75, 260, 425
Phillips, Wm. R., 1, -, -, 60, 30, 20, -, 7000, 150, 300, 400
Van Vorst, Cornelius, 1, -, -, 100, 65, 10, 10000, 125, 1200, 1200
Serven, Benj. R., 1, -, -, 75, 10, 15, -, 5000, 75, 100, 650
Caverly, Samuel L., 1, -, -, 35, 25, 10, -, 8000, 50, 60, 600
Torbert(Tolbert), A. T. A., 1, -, -, 195, 35, 30,-, 20000, 800, 2200, 4000
Redifer, George, -, 1, -, 125, 10, 28, -, 5000, 30, 100, 200
Griffith, Sarah A., 1, -, -, 140, 12, 10, -, 3000, 495, 425, 500
Hill, Joshua H., -, -, 1, 120, -, 20, -, 4000, 100, 200, 500
Van Vorst, Wm., 1, -, -, 250, 170, 50, -, 50000, 2000, 600, 6600
Barker, Charles, -, 1, -, 110, 75, 15, -, 8000, 500, 875, 3350

Smoot, Henry H., -, -, 1, 160, 30, 30, -, 20000, 500, 1200, 400
Collins, Emma, 1, -, -, 63, 2, -, -, 6000, 250, 323, 500
Henderson, Mary A. S., 108, 2, 20, -, 4000, 125, 425, 450
Cubbage, Louisa, 1, -, -, 85, 15, 10, -, 1000, 300, 850, 1300
Cubbage, John H., -, -, 1, 50, 4, 6, -, 1200, 25, 85, 300
Webb, Emory, -, -, 1, 80, 2, 40, -, 3000, 100, 400, 425
Cole, Nehemiah, 1, -, -, 80, 6, 20, -, 2500, 50, 180, 500
Davis, James L., -, 1, -, 244, 16, 40, -, 7500, 75, 450, 850
Richards, Wm. H., 1, -, -, 40, -, 100, -, 4000, 50, 325, 400
Dorsey, Wm. N. W., 1, -, -, 150, 50, 28, -, 10000, 1000, 1150, 2100
Tolbert, William, -, -, 1, 65, 35, 40, -, 4000, 50, 200, 600
Redden, John W., 1, -, 1, -, 70, 34, 80, -, 5000, 150, 150, 725
Paisly, Edwin, 1, -, -, 11, 3, -, -, 1000, 5, -, 120
Hayes, Samuel, -, 1, -, 20, 1, 5, -, 1200, 50, 125, 325
Martin, James H., 1, -, -, 61, 1, 10, -, 2500, 300, 350, 700
Soffle, James D., 1, -, -, 80, 200, 120, -, 6000, 400, 1000, 1350
Roach, George, 1, -, -, 85, -, -, -, 3000, 125, 270, 1025
Short, John, 1, -, -, 10, 4, 48, -, 3000, 100, 15, 350
Bethards, William H., -, -, 1, 100, 30, 50, -, 5500, 300, 840, 1500
Booth, John T., -, -, 1, 110, 1, 30, -, 5000, 350, 500, 1100
Stevenson, Robert H., 1, -, -, 134, 16, 30, -, 9000, 600, 600, 1700
Robbins, Eli W., -, -, 1, 59, -, -, -, 150, 50, 250, 525
Colbourn, George W., -, -, 1, 40, -, -, -, 1500, 150, 200, 525
Davis, George H., -, -, 1, 110, -, 75, -, 4000, 100, 400, 800
Allcott, Jared B., 1, -, -, 62, 8, 10, -, 1600, 150, 450, 450
Mason, Alexander L., 1, -, -, 45, -, 15, -, 1000, 30, 125, 175
Scott, John, -, 1, -, 11, -, 4, -, 300, 10, -, 75
Jenkins, Silas T., 1, -, -, 100, 205, 40, -, 7000, 400, 500, 2350
Webb, James M., -, -, 1, 75, 34, 60, -, 5000, 200, 465, 825
Sanders, Frederick, 1, -, -, 43, 4, 3, -, 1500, 100, 270, 488
Watson, Elias S., -, -, 1, 35, -, -, -, 800, 40, 175, 275
Bickle, John D., 1, -, -, 42, 12, 10, -, 2400, 200, 225, 825
Andre, Elizabeth, 1, -, -, 20, -, 10, -, 1000, 30, 100, 110
Corsa, Harry C., 1, -, -, 37, 3, 20, -, 1000, 125, 200, 425
Hammond, Nicholas D., -, -, 1, 175, 12, 30, -, 8000, 400, 800, 1500
Griffith, James H., 1, -, -, 43, 3,4, -, 2500, 300, 450, 900
Davis, John H., 1, -, -, 40, 1, -, -, 2000, 125, 180, 325
Chapman, John, 1, -, -, 42, 6, -, -, 2400, 175, 225, 900
Lewis, Arch P., 1, -, -, 95, 23, 10, -, 6000, 300, 500, 965
Mills, Charles, 1, -, -, 10, -, 5, -, 600, 85, 135, 100
Mills, Davis S., 1, -, -, 60, 1, 25, 15, 2000, 100, 300, 500
Jester, Isaac R., 1, -, -, 20, 5, 10, -, 1000,80, 150, 300
Moore, John H., -, -, 1, 65, 72, -, 2000, 50, 200, 400
Burry, Serena, 1, -, -, 55, 15, 30, -, 2500, 125, -, 420
Needles, Benj & Theodore, -, -, 1, 50, 2, 10, -, 1200, 200, 300, 475
Milten, Daniel, 1, -, -, 14, -, 2, -, 500, 80, 250, 300

Jackson, Frank B., -, 1, -, 12, -, 22, -, 400, 60, 60, 175
Morris, Charles, -, -, 1, 20, 2, 13, -, 700, 40, 160, 300
Needles, Benj. P., -, -, 1, 16, 1, 40, -, 1000,50, 140, 100
Tolbert, Joshua, 1, -, -, 80, 4, 10, -, 4000, 300, 500, 650
Jackson, John & James, -, -, 1, 100, -, 10, -, 2500, 125, 225, 795
Taylor, Charles P., -, -, 1, 125, 125, 50, -, 6000, 150, 375, 1150
Hall, Charles P., -, -, 1, 60, -, 200, -, 5000, 80, 175, 350
Milten, Matthew F., -, -, 1, 150, 140, 50, -, 10000, 200, 600, 1500
Milten, William B., -, -, 1, 100, -, -, -, 4000, 75, 350,795
Wadkins, Eli R., -, -, 1, 68, 2, 20, -, 1800, 40, 100, 475
Virden, Thomas, -, 1, -, 35, -, 10, -, 1000, 100, 145, 300
Harrington, Jacob & James, -, -, 1, 100, 25, 30, -, 5000, 100, 400, 1050
Richards, William J., -, -, 1, 30, -, -, -, 100, 24, 200, 200
Masten, Mrs. Eliza, 1, -, -, 100, 55, 50, -, 6000, 20, 155, 665
Mosely, Morris, -, -, 1, 150, 50, 100, -, 6000, 100, 500, 1200
Kirby, Benjamin F., 1, -, -, 105, 30, 10, -, 8000, 250, 600, 1100
Hudson, John W., -, 1, -, 80, 100, 40, -, 5000, 400, 500, 1325
Kirby, John W., 1, -, -, 112, 204, 20, -, 4000, 450, 800, 1500
Kirby, James H., 1, -, -, 58, 68, -, -, 1500, 100, 500, 610
Short, Edward, 1, -, -, 40, 40, 15, -, 3000, 155, 400, 825
Webb, John W., 1, -, -, 125, 125, 18, -, 10000, 355, 700, 1520
Webb, Charles A., -, -, 1, 30, 50, -, -, 1800, 150, 300, 650
Webb, John M., 1, -, -, 60, -, 20, -, 2000, 200, 350, 625
Hall, William E., 1, -, -, 30, -, 5, -, 1000, 100, 500, 425
Thompson, Daniel, -, 1, -, 60, 40, 90, -, 3000, 900, 700, 850
Moseley, Joseph C., -, -, 1, 60, 6, 15, -, 1200, 50, 175, 600
Macklin, George, -, -, 1, 50, -, 50, -, 2500, 108, 170, 410
Thomas, George, 1, -, -, 35, 2,23, -, 1500, 100, 230, 450
Thomas, Aaron B., 1, -, -, 12, -, 12, -, 400, 2, 55, 100
Thomas, Andrew T., 1, -, -, 30, 9, 5, -, 1200, 50, 125, 520
French, Joseph, 1, -, -, 31, 1, 8, -, 800, 30, 75, 320
Hall, Winlock B., 1, -, -, 33, 6, 6, -, 1000, 70, 250, 300
Hall, David B., -, -, 1, 150, 5, 50, -, 8000, 250, 1200, 1250
Bennett, Joshua S., 1, -, -, 48, 2, 5, -, 2500, 250, 450, 900
Bennett, Joseph C., 1, -, -, 75, 100, 25, -, 2000, 200, 400, 700
Watson, Daniel, -, 1, -, 120, 150, 5, -, 7000, 200, 1150, 600
Holland, James, -, -, 1, 165, 500, 250, 150, 10000, 100, 700, 600
McColley, Wm. J. H., 1, -, -, 80, 233, 15, -, 2500, 250, 900, 550
Evans, Thomas & John, -, -, 1, 75, -, 50, -, 3000, 125, 550, 600
Hinsly, Cornelius, -, -, 1, 70, -, 35, -, 2500, 75, 350, 450
Wadkins, John R., 1, -, -, 60, -, 80, -, 1200, 75, 600, 400
Scott, Daniel, 1, -, -, 12, -, -, -, 250, 30, 130, 160
Jester, Benjamin F., -, -, 1, 40, -, 25, -, 1200, 100, 200, 300
French, William H., 1, -, -, 40, -, 40, -, 2000, 150, 300, 630
French, Nathaniel, -, -, 1, 50, -, 100, -, 2000, 150, 250, 400
Jester, George W., -, -, 1, 50, -, 18, -, 800, 600, 150, 100

Andrew, John W., 1, -, -, 45, -, -, 15, -, 1200, 80, 250, 40
Bennett, John M., 1 -, -, 60, -, 1, 20, -, 2000, 125, 225, 425
French, Samuel, -, -, 1, 18, -, 1, 2, -, 505, 25, 100, 125
Mills, Richard, 1, -, -, 110, 2, 40, -, 4000, 400, 750, 1500
Bennett, Joshua, 1, -, -, 120, 15, 32, -, 6000, 230, 1155, 1800
Hill, Absalom, 1, -, -, 120, 15, 32, -, 1200, 275, 600, 600
Watson, David K., -, -, 1, 80, -, 45, -, 3000, 15, 400, 1700
Hill, Samuel, -, -, 1, 75, -, 100, -, 1000, 19, 75, 100
Lindle, Thomas, 1, -, -, 89, -, 15, -, 1500, 75, 450, 850
Loper, John F., 1, -, -, 33, -, 3, -, 500, 20, 150, 300
Jester, Isaac, 1, -, -, 13, -, 12, -, 400, 25, 200
Ingram, Thos. H., -, -, 1, 16, -, 14, -, 450, 50, -, 180
Meredith, Edward, -, -, 1, 20, -, 20, -, 700, 125, 320, 350
Fitzgerald, Eligah, 1, -, -, 14, -, 16, -, 500, 20, -, 100
Richards, William H., 1, -, -, 40, -, 40, -, 2000, 75, 325, 400
Hopkins, Clarence, -, 1, -, 35, 3, 20, -, 1400, 60, 300, 300
Mills, James E., -, -, 1, 30, 1, 90, -, 2000, 75, 200, 400
Jackson, Levi H., 1, -, -, 69, 2,14, -, 2500, 125, 300,680
Hydom, Benjamin, 1, -, -, 46, 10, 20, -, 2500, 175, 325, 900
Beswick, William P., 1, -, -, 60, 42, 10, -, 2500, 300, 500, 800
Carpenter, Robert, -, -, 1, 95, 5, -, -, 8000, 300, 500, 1500
Clifton, Daniel, 1, -, -, 20, 1, 24, -, 1200, 50, 250, 210
Bickel, John A., 1, -, -, 130, 155, 75, -, 5000, 450, 600, 1400
Evan, James, -, -, 1, 30, 75, 40, -, 4000, 50, 275, 325
Webb, Thomas A., -, -, 1, 100, -, -, -, 4000, 175, 50, 800
Abbott, William L., 1, -, -, 54, 1, 20, -, 1500, 150, 300, 425
Cheasman, David, 1, -, -, 47, 2, 10, -, 1000, 65, 200, 320
Lynch, James, 1, -, -, 30, -, 70, -, 1000, 80, 350, 310
Coverdale, Joshua, 1, -, -, 20, -, 40, -, 900, 40, 100, 135
Thomas, Nathaniel B., -, -, 1, 150, 30, 150, -, 6000, 50, 600, 1100
Scott, Garrison, 1, -, -, 8, -, 4, -, 150, 20, 50, 90
Meredith, Dickerson M., -, -, 1, 100, 105, 70, -, 3600, 250, 700, 725
McColley, Nathaniel, -, -, 1, 23, -, 5, -, -, 600, 80, 220, -, 550
Walls, Jesse W., -, -, 1, 40, 10, 25, -, 800, 40, 100, 140
Johnson, Louis, -, -, 1, 14, -, -, -, 150, 15, 140, 65
Voshell, Alexander, -, -, 1, 100, 100, 40, -, 5500, 100, 350, 600
Hickman, Caleb, -, 1, -, 10, -, 50, -, 900, 25, 120, 150
Bradley, James, -, 1, -, 13, -, 2, -, 150, 20, 100, 130
Holliger, George, -, -, 1, 10, -, -, 2, 140, 10, 100, 85
Truitt, William G., 1, -, -, 100, 100, 40, -, 4000, 150, 540, 1600
Scott, Eliza, 1, -, -, 25, -, 14, -, 800, -, -, 100
Hall, John W., 1, -, -, 77, -, 10, -, 3000, 100, 350, 525
French, George, -, -, 1, 80, 75, 20, -, 5000, 125, 400, 725
Thomas, George, -, -, 1, 50, 25, 25, -, 3000, 100, 160, 625
Tribbitt, John, -, -, 1, 30, 50, 15, -, 1000, 75, 350, 360
Macklin, Elias, -, -, 1, 20, 4, 5, -, 450, 30, 120, 210

Webb, George F., -, -, 1, 50, 60, 40, -, 2000, 80, 200, 650
Cain, Mary B., 1, -, -, 80, 120, 200, -, 4000, 200, 825, 925
Davis, William R., -, -, 1, 75, 60, 25, -, 3000, 200, 500, 775
Reed, Peters, -, -, 1, 40, 50, -, -, 2000, 300, 300, 500
Maloney, Andrew, -, -, 1, 125, 200, 400, -, 8000, 300, 100, 1350
Hudson, James H., 1, -, -, 150, 250, 100, -, 6000, 400,700, 650
Maloney, John R., -, -, 110, 300, 150, -, 8000, 200, 900, 1300
Wadkins, Charles P., 1, -, -, 8, 8, -, -, 400, 50, 200, 150
Davis, Jehu, 1, -, -, 100, 300, 100, -, 4000, 250, 800, 750
Stokes, William, -, -, 1, 30, 40, -, -, 600, 20,75, 200
Wyatt, George, -, -, 1, 40, 100, 60, -, 3000, 60, 300, 295
Benson, John, -, -, 1, 35, -, 17, -, 1000, 15, 10, 275
Morris, Henneretta, 1, -, -, 80, -, 40, -, 2500, 20, 125, 200
Morgan, Samuel, -, 1, -, 42, -, -, -, 800, 75, 100, 350
Scott, Peter, -, -, 1, 60, -, 30, -, 2500, 25, 225, 400
Dickenson, William H., 1, -, -, 165, -, 75, -, 4000, 300, 150, 1200
Doy (Day), James H., -, -, 1, 80, -, 42, -, 1800, 45, 220, 550
Reece, John P., 1, -, -, 115, -, 10, -, 3000, 200, 300, 700
Griswold, Abner, 1, -, -, 80, -, 35, -, 3000, 60, 200, 500
Plum, Alexander, 1, -, -, 80, -, 82, -, 3000, 200, 400, 725
Reece, Evin, 1, -, -, 75, -, 30, -, 3000, 125, 400, 500
Quillin, A__, 1, -, -, 10, -, -, -, 100, -, -, -
Short, Robert H., 1, -, -, 46, -, 3, -, 2500, 106, 150, 600
Walker, James, 1, -, -, 72, -, 70, 1000, 25, 75, 175
Messick, Mines, -, -, 1, 100, -, 125, -, 3000, 150, 250, 500
Harrington, James, -, -, 1, 50, -, 25, -, 1600, 30, 250, 270
Barlow, Aaron, 1, -, -, 30, -, 15, -, 1500, 200, 700, 250
Barlow, Aaron, 1, -, -, 45, -, 10, -, 2000, 100, 200, 700
Litz, John W., -, -, 1, 25, -, 48, -, 750, 50, 150, 200
Harrington, Jonathan, 1, -, -, 100, -, 125, -, 4000, 200, 250, 375
Willis, Joseph W., -, -, 40, 35, -, 1000, 20, 100, 150
Tharp, George W., -, -, 1, 25, -, -, -, 500, 25, 100, 200
Lecompt, George, -, -, 1, 125, -, 75, -, 2000, 200, 300, 450
Harrington, Nathan, 1, -, -, 100, -, 130, -, 3000, 100, 300, 325
Sapp, Albert, 1, -, -, 100, -, 60, -, 3500, 75, 250, 325
Harrington, ____, 1, -, -, 12, -, 60, -, 2000, 200, 350, 425
Legats, Willard, -, -, 1, 5, -, 25, -, 2000, 100, 375, 225
Dorman, Solemon, 1, -, -, 15, -, 5, -, 900, 75, 150, 230
Norvel (Nowell), John S., 1, -, -, 70, -, 100, -, 2500, 200, 200, 500
Reese, William M., 1, -, -, 130, -, 84, -, 3000, 75, 200, 450
Winant, James W., -, 1, -, 40, -, 5, 5, 1500, 60, 60, 125
Tharp, Beniah, 1, -, -, 149, -, 50, -, 5000, 300, 800, 1800
Thistlewood, Benjamin, 1, -, -, 90, -, 25, -, 2000, 25, 400, 300
Nelson, Daniel P., -, -, 1, 15, -, -, -, 200, 5, 50, 175
Sapp, William, 1, -, -, 44, -, 40, -, 820, 100, 200, 225
Hands, Thomas, 1, -, -, 100, -, 40, -, 3000, 181, 500, 1100

Clark, William 1, -, -, 186, -, 80, -, 8000, 40, 300, 1190
Harrington, Danel, 1, -, -, 65, -, 31, -, 2000, 150, 200, 475
Short, John, -, -, 1, 80, -, -, -, 1000, 50, 125, 400
Tucker, David G., -, -, 1, 55, -, 52, -, 1000, 80, 100, 275
Peckaham, Joseph J., 1, -, -, 46, -, 58, -, 2500, 150, 50, 450
Jarrell, William B., 1, -, -, 75, -, 20, -, 1500, 75, 125, 300
Arnold, Robert M., 1, -, -, 200, -, 188, -, 1000, 500, 400, 925
Grier, John, -, -, 1, 350, -, 100, -, 1000, 300, 600, 600
Powell, Samuel, 1, -, -, 20, -, 5, -, 250, 20, 45, 100
Booth, Joseph, 1, -, -, 120, -, 89, -, 5000, 300, 500, 1050
Hill, Joshua, 1, -, -, 75, -, 25, -, 1000, 25, 200, 300
Hill, James, 1, -, -, 100, -, 200, -, 3000, 50, 200, 275
Culver, Isaac A. -, -, 1, 125, -, 390, -, 3000, 15, 25, 75
Abott, James, -, 1, -, 90, -, 90, -, 3500, 100, 75, 187
Abott, Warner, 1, -, -, 30, -, 5, 1000, 75, 100, 100
Case (Cox), Samuel, -, -, 1, 75, -, 40, -, 3000, 60, 100, 328
Slayton, John W., -, 1, -, 80, -, 50, -, 2000, 100, 200, 340
Case, John, -, -, 1, 20, -, 30, -, 4000, 100, 150, 850
Lyons, William, -, -, 1, 30, -, 217, -, 3000, 75, 250, 175
Harrington, Thomas M., 1, -, -, 100, -, 56, -, 4000, 200, 400, 900
Butler, William, 1, -, -, 100, -, 56, -, 4000, 200, 150, 1600
Short, Samuel, -, -, 1, 80, -, -, -, 2000, 200, 350, 250
Curtis, John R., 1, -, -, 50, -, 54, -, 1500, 5, 50, 150
Johnson, William J., 1, -, -, 75, -, 25, -, 2500, 250, 125, 300
Sampson, William W., 1, -, -, 50, -, 15, -, 1000, 100, 300, 600
Harrington, Clement C., 1, -, -, 125, -, 65, -, 7000, 150, 584, 1500
Weeks, William H., 1, -, -, 100, -, 80, -, 4000, 400, 800, 1700
Harrington, Nimrod, 1, -, -, 70, -, 670, -, 4000, 200, 340, 400
Thompson, William, 1, -, -, 18, -, 5, -, 1000, 50, 50, 350
Mastin, John, -, -, 1, 60, -, 30, -, 1500, 100, 75, 300
Legates, Robert B., -, -, 1, 20, -, 130, -, 1000, 20, 20, 75
Sampson, Alexander, 1, -, -, 140, -, 25, -, 4000, 300, 500, 1000
Bradly, Garret, -, 1, -, 100, -, 20, -, 2500, 50, 200, 750
Sherwood, Samuel, -, -, 1, 100, -, 60, -, 3500, 100, 450, 900
Neil, Robt. H., -, -, 1, 50, -, 30, -, 1000, 25, 250, 400
Tharp, Beniah, -, -, 1, 60, -, 130, -, 2000, 40, 190, 300
Fleetwood, Curtis, -, -, 1, 10, -, -, -, 200, 10, -, 110
Raughly, Thomas H., -, -, 1, 75, -, 39, -, 2000, 150, 300, 350
Harriss, Samuel, -, -, 1, 10, -, 150, -, 4000, 150, 700, 500
Powell, Nathanl. C., 1, -, -, 150, -, 160, -, 8000, 350, 550, 1500
Laws, George, 1, -, -, 50, -, 10, -, 1500, 125, 240, 325
Bradly, Thomas, -, -, 1,75, -, 27, -, 1500, 50, 100, 125
McCleavy, Mary, 1, -, -, 100, -, 100, 23, 4000, 200, 250, -
Lewis, William, -, 1, -, 200, -, 100, -, 3000, 20, 200, 450
Hatfield, James, -, -, 1, 200, -, 75, 25, 2500,75, 200, 300
Scrofield, Soleman R., 1, -, -, 130, -, 41, -, 1000, 150, 175, 1200

Stapleford, William J., -, -, 1, 140, -, 35, -, 3000, 200, 400, 800
Bradly, Solomon, -, -, 1, 48, -, 10, -, 1000, 75, 200, 345
Legates, William, -, -, 1, 75, -, 75, -, 2000, 50, 285, 250
Willis, William, 1, -, -, 150, -, 70, -, 3000, 200, 300, 50, 30, 200, 100, 375
Johnson, Nathaniel B., 1, -, -, 75, -, 75, -, 3000, 125, 500, 15, 100, 200, 72, 600
Bolden, George, -, -, 1, 100, -, 172, -, 1500, 20, 50, 5, -, 100, 50, 100
Collins, Levin, -, -, 1, 25, -, -, -, 250, 30, 70, -, -, 100, 50, 100
Marvel, William, 1, -, -, 100, -, 2, -, 3500, 100, 500, 15, 130, 175, 62, 956
Johnson, John W., -, -, 1, 30, -, 15, -, 1000, -, 40, -, -, -, -, -
Hammond, Theodore, -, -, 1, 45, -, 15, -, 1000, 5, 75, -, -, -, -, -
Dean, James, -, -, 1, 17, -, 10, -, 500, 100, 100, -, 4, -, -, 50
Dean, Henry, -, -, 1, 18, -, 6, -, 250, 25, 20, -, 6, 50, -, 37
Marvel, James R., -, -, 1, 40, -, 45, -, 1400, 75, 50, -, -, -, -, 220
Johnson, Zachariah, 1, -, -, 80, -, 29, -, 4000, 1800, 350, -, 100, 500, 80, 1200
Smith, Johnathan, 1, -, -, 45, -, 6, -, 2000, 30, 600, -, 40, 10, 2, 100
Shockley, Whetleton, -, -, 1, 30, -, -, -, 300, 20, 20, -, -, -, -, 100
Hendricks, John P., 1, -, -` 100, -, 36, -, 5000, 450, 650, 125, 180, 300, 150, 1150
McCauley, Clement S., 1, -, -, 130, -, 60, -, 2000, 150, 300, 40, 57, 180, 60, 700
Walls, William, 1, -, -, 20, -, 2, -, 600, 50, 150, 10, 18, 15, 3160
Fleming, Isaac, I, -, -, 6, -, 6, -, 300, 25, 50, 5, -, -, -, 50
Shockley, Marith, 1, -, -, 70, -, 5, -, 2000, 75, 250, -, 6, 200, 100, 460
Tatman, John M., -, -, 1, 40, -, 100, -, 1000, 8, 125, -, 32, 10, 3, 200
Collins, John, -, -, 1, 25, -, -, -, 25, 12, 50, -, -, -, -, 3
Tatman, James P., 1, -, -, 250, -, 250, -, 8000, 200, 200, 50, 136, 75, 12, 600
Johnson, David, 1, -, -, 6, -, -, -, 300, 30, 100, 10, 20, 20, 20, 100
Landis, William, 1, -, -, 94, -, 20, -, 2500, 250, 350, -, 60, 20, 3, 400
Deffadoffer, David, -, -, -, 40, -, 25, -, 1200, 75, 125, -, -, -, -, 75
Tatman, Collins, -, -, -, 150, -, 50, -, 2500, 25, 300, 10, -, -, -, 120
Hoey, James, 1, -, -, 150, -, 52, -, 5000, 300, 500, 10, 50, 500, 300, 127
Cordrey, Robert, -, -, 1, 80, -, 177, -, 3000, 100, 300, 40, 18, 125, 50, 460
Holfield, Joseph T., -, -, 1, 250, -, 150, -, 12000, 150, 400, 150, -, 200, 100, 1200
Prettyman, Remelton C., -, -, 1, 100, -, 20, -, 2500, 120, 450, -, 340, 150, 55, 500
Ralston, Robert, 1, -, -, 150, -, 150, -, 7000, 125, 600, 25, 25, 225, 100, 780
Scott, Seala, 1, -, -, 96, -, 4, -, 1500, 25, 100, 515, -, -, 350
Donovan, John W., -, -, 1, 100, -, 14, -, 1200, 10, 100, -, -, 100, 52, 225
Rickards, John W., -, -, 1, 122, -, 11, -, 3000, 300, 500, 20, 50, 200, 75, 8, 50
Alexander, Johnson, 1, -, -, 250, -, 59, -, 2000, 500, 1500, 300, 200, 600, 420, 2350
Johnson, Elisha, -, -, 1, 58, -, -, -, 1000, 24, 100, -, 20, -, -, 175
Duker (Dukes), Shadrach, -, -, 1, 40, -, 60, -, 800, 5, 100, -, 100, -, -, 75
Griffith, James H., -, -, 1, 60, -, 50, -, 500, 20, 50, -, -, 20, 8, 275
Tremble, Mary E., 1, -, -, 100, -, 100, -, 5000, 200, 300, -, 20, -, -, -
Nickols, John H., -, -, 1, 100, -, 112, -, 4000, 150, 700, 10, 18, 200, 62, 475
Prettyman, John, 1, -, -, 100, -, 50, -, 2500, 100, 500, -, 40, 175, 60, 625
Prettyman, Alfred, 1, -, -, 75, -, 35, -, 2000, 75, 300, -, 25, 250, 200, 450
Cordrey, Charles H., -, -, 1, 80, -, 26, -, 1000, 100, 500, 10, 25, 50, 12, 675

Sharp, Jesse, -, -, 1, 136, -, 100, -, 4000, 125, 200, 25, 16, 164, 35, 600
Dickerson, John T., 1, -, -, 4, -, -, -, 800, 10, 70, -, -, -, -, 150
Derickson, John W., 1, -, -, 40, -, -, -, 800, 60, 175, -, 10, 60, 24, 125
Long, Isaac J., 1, -, -, 30, -, 13, -, 1500, 50, 100, 10, 40, 100, 40, 180
Killen, John G., 1, -, -, 8, -, 4, -, 300, 50, 75, -, 21, 10, 2, 25
Sharp, William F., 1, -, -, -, -, 29, -, 500, 5, 70, -, 4, -, -, 50
Scott, Lydia, 1, -, -, 20, -, 29, -, 500, 10, 100, -, 4, -, -, 160
Scott, Hasty, 1, , -, 30, -, 30, -, 400, 5, 50, -, 7, -, -, 60
Cordray, Jacob, 1, -, -, 150, -, 75, -, 5000, 200, 400, 100, 33, 400, 125, 850
Dowed, James W., 1, -, -, 100, -, 70, -, 5000, 225, 500, -, 150, 175, 72, 105
Barker, Robert C., 1, -, -, 39, -, 1, -, 1000, 150, 125, -, 50, 32, 175,
Aldrich, Ralph W., 1, -, -, 20, -, 20, -, 800, -, -, -, -, -, -, 50
Thorp, Benjamin H., -, -, 1, 75, -, 25, -, 2000, 100, 300, 20, 50, 100, 52, 300
Thorp, Lewellen, 1, -, -, 130, -, 170, -, 6000, 50, 80, 40, 100, 150, 50, 400
Hill, Robert J., 1, -, -, 100, -, 70, -, 5000, 50, 400, -, 60, 5, -, 480
Fisher, Sherman, 1, -, -, 37, -, 10, -, 2400, 250, 230, 40, 30, 100, 40, 600
Prettyman, James B., 1, -, -, 68, -, 20, -, 4000, 400, 520, 25, 32, 100, 52, 150
Johnson, Purnell, -, -, 1, 16, -, -, -, -, 300, 400, -, -, -, -, 5
Harrington, Moses (?), 1, -, -, 120, -, 120, -, 7000, 100, 625, 50, 150, 300, 150, 100
Wyatt, Daniel, 1, -, -, 80, -, 75, -, 2500, 100, 200, 20, 20, 150, 52, 250
Ross, Reuben, 1, -, -, 60, -, 80, -, 4000, 200, 500, 50, 270, 150, 700
Smith, David, -, 1, -, 100, -, 64, -, 1500, 200, 500, 20, 100, 20, 100, 650
Brown, Major, -, -, 1, 60, -, 220, -, 3000, 125, 170, 15, 31, 100, 50, 580
Jones, George R., -, 1, -, 70, -, 35, -, 2500, 100, 500, 10, 125, 100, 35, 620
Porter, James, 1, -, -, 80, -, 90, -, 4000, 300, 600, 40, 100, 200, 100, 800
Anderson, William H., -, -, 1, 125, -, 50, -, 8000, 500, 600, 50, 125, 700, 150, 1380
Curtis, John R., 1, -, -, 110, -, 90, -, 4000, 300, 400, 25, 200, 300, 44, 1000
Burnham, Andrew R., 1, -, -, 40, -, 43, -, 4000, 200, 280, -, 100, 350, 110, 1000
Cooper, John R., -, -, 1, 100, -, 90, -, 2000, 150, 300, -, 16, 100, 50, 225
Martin, Thomas, -, -, 1, 75, -, 100, -, 3000, 50, 150, 15, -, 100, 50, 125
Booth, John, 1, -, -, 150, -, 84, -, 5500, 242, 6540, 100, 220, 500, 300, 1360
Sapp, Louder L., 1, -, -, 70, -, 35, -, 1500, 125, 170, 20, 100, 150, 50, 400
Colman, Phineas R., 1, -, -, 25, -, 84, -, 800, 75, 50, -, 50, 10, 2, 125
Longfellow, James, -, -, 1, 100, -, 60, -, 7000, 150, 200, 20, 100, 52, 500
Laramore, John, -, 1, -, 13, -, -, -, 400, 5, 75, -, -, -, -, -
Morris, James H., -, -, 1, 50, -, 70, -, 1000, 25, 75, -, -, 20, 30
Harrington, Samuel S., 1, -, -, 75, -, 75, -, 4000, 100, 300, 25, 50, 200, 100, 825
Pearson, Alexander, -, -, 1, 50, -, -, -, 600, 20, 100, -, -, -, -, 80
Woman, Nathan, -, -, 1, 1000, -, 165, -, 2000, 100, 200, 20, 50, 150, 75, 375
Morris, Alexander, -, -, 1, 80, -, 50, -, 2000, 50, 250, 30, -, 150, 60, 275
Minner, Jonathan, 1, -, -, 150, -, 60, -, 3000, 50, 250, 30, 90, 150, 5, 330
Warren, John W., 1, -, -, 250, -, 100, -, 4000, 175, 1000, -, 80, -, -, 120
Barcus, William, -, -, 1, 50, -, 10, -, 600, 200, 200, 10, -, 150, 100, 640
Masten, Anna, 1, -, -, 100, -, 30, -, 4000, 150, 450, 20, 112, 350, 120, 800

Sapp, William, -, -, 1, 15, -, -, -, 650, 95, 100, 150, -, 250, 10, 250
Morris, Hezekiah, -, 1, -, 22, -, 1, -, 1500, 12, 70, -, 26, -, -, -
Smith, Samuel W., -, 1, -, 113, -, 10, -, 1600, 30, 200, -, 21, 300, 10, 1350
Morris, Elisha, - , -, 1, 70, -, 20, -, 2000, 200, 400, 15, 42, 200, 75, 575
Minner, John, 1, -, -, 40, -, 8, -, 1000, 50, 100, -, 25, 5, 2, 180
Edwards, William D., -, -, 1, 200, -, 60, -, 3000, 100, 500, -, 32, 200, 100, 370
Foster, Benjamin, 1, -, -, 13, -, 1, -, 800, 25, 150, 5, 2, -, -, 100
Smith, Samuel, -, -, 1, 50, -, 50, -, 1000, 25, 50, 15, 34, 100, 52, 200
Abbler, Thomas E., -, -, 1, 10, -, 15, -, 300, 10, 25, 8, 10, 8, 4, 120
Minner, William M., -, -, 1, 70, -, 5, -, 2000, 75, 175, 25, 30, 75, 20, 300
Morris, John, -, -, 1, 101, -, 156, -, 1500, 30, 200, 5, -, -, -, 240
Dunham, John W., -, -, 1, 3, -, 10, -, 800, 10, 100, 5, 16, 15, 3, 150
Hill, John D., -, -, 1, 75, -, 50, -, 1500, 100, 300, -, 20, 100, 52, 360
Steel, Robert D., 1, -, -, 14, -, 3, -, 300, 20, 100, -, -, -, -, 55
Masten, Louden L., 1, -, -, 200, -, 156, -, 900, 600, 800, -, 85, 250, 75, 1200
Dyton, Samuel M., -, -, 1, 40, -, 75, -, 1500, 50, 350, 25, 25, 200, 200, 460
Killen, William E., 1, -, -, 85, -, 35, -, 2000, 250, 320, 20, 80, 200, 120, 600
Donophon, James, -, 1, -, 65, -, -, -, 1000, 100, 200, 5, 20, 150, 5, 425
Minner, Samuel, 1, -, -, 40, -, 18, -, 1200, 100, 175, 3, 24, 100, 52, 275
Minner, Samuel J., -, -, 1, 28, -, -, -, 250, 30, 100, -, -, -, -, 200
Tinley, Jonathan, -, -, 1, 30, -, 15, -, 500, 25, 51, 6, -, -, -, 100
Hopkins, Samuel, 1, -, -, 50, -, 32, -, 1500, 50, 200, 20, 17, 100, 32, 200
Jester, James G., 1, -, -, 120, -, 10, -, 4000, 300, 600, 10, 120, 410, 150, 1300
Clark, Walt__, 1, -, -, 50, -, 43, -, 2000, 100, 550, -, 24, 200, 75, 500
Jester, Albert, -, -, 70, -, 35, 1500, 250, 540, 25, 55, 200, 100, 700
Cooper, John G., -, -, 1, 175, -, 30, -, 3000, 100, 250, -, 23, 100, 7, 560
Reed, Susan, 1, -, -, 75, -, 20, -, 2000, 100, 30, 18, 40, 150, 52, 200
Hughes, Alexander W., 1, -, -, 175, 60, -, 5000, 100, 800, -, 66, 300, 100, 450
Hughes, Samuel W., -, 1, -, 60, -, 46, -, 1000, 75, 200, 10, 42, 10, 1, 150
Jester, William F., 1, -, -, 60, -, 40, -, 1500, 150, 300, 23, 100, 200, 100, 400
Catlin, William H., -, -, 1, 60, -, 100, -, 2000,125, 280, 10, 60, 200, 100, 590
Donophon, John E., 1, -, -, 60, -, 30, -, 1500, 100, 300, 10, 100, 150, 75, 400
Wilson, James, -, -, 1, 25, -, 75, -, 500, 20, 700, -, -, -, -, 205
Laramore, John T., -, -, 1, 150, -, 50, -, 3000, 75, 300, 25, 600, 250, 100, 600
Wyatt, John, 1, -, -, 64, -, 18, -, 1500, 50, 300, -, 50, 200, 75, 500
Sapp, William, 1, -, -, 150, -, 150, -, 6000, 300, 1560, 20, 600, 300, 100, 900
Hughes, Ebin, 1, -, -, 100, -, 40, -, 1800, 100, 350, 20, 60, 250, 100, 350
Morice, William, -, -, 1, 50, -, 50, -, 900, 40, 135, 1130, 6, 150
Pharo, William, -, -, 1, 70, -, 20, -, 800, 35, 200, -, -, 150, 50, 136
Draper, Thomas J., 1, -, -, 20, -, 14, -, 500, 40, 100, 10, 1120, 6, 182
Sapp, Walter, -, -, 1, 113, -, 56, -, 2000, 61, 175, 5, -, 64, 32, 200
Kenton, William, -, -, 1, 40, -, 40, -, 1000, 40, 75, -, -, -, -, 75
Benson, Evin, -, -, 1, 50, -, 50, -, 800, 15, 125, -, -, -, -, 150
Melvin, Nathaniel, -, 1, -, 30, -, 30, -, 700, 15, 75, 5, -, -, -, 77
Porter, Garrett, -, -, 1, 50, -, 14, -, 1000, 50, 200, 10, 48, 150, 30, 200
Sapp, John, -, -, 1, 15, -, 10, -, 3000, 360, 910, 24, 90, 250, 75, 900

Longrel, William, -, -, 1, 25, -, 10, -, 300, 10, 75, -, -, -, -, 115
Reed, George W., 1, -, -, 50, -, 26, -, 800, 25, 100, 10, 44, 25, 3, 200
Hammond, Oliver, -, -, 1, 50, -, 25, -, 700, 25, 100, 30, 28, 100, 50, 200
Slaughter, William, 1, -, -, 50, -, 25, -, 700, 40, 150, 8, -, 120, 50, 150
Cooper, John J., -, -, 1, 16, -, 85, -, 800, 24, 100, -, -, 25, 5, 300
Kalas, John, -, -, 1, 30, -, 70, -, 600, 25, 100, -, 12, 100, 50, 220
Hopkins, John, -, -, 1, 15, -, 35, -, 300, 30, 175, -, 1, -, -, 140
Reed, Edward P., 1, -, -, 30, -, 20, -, 500, 30, 200, -, -, -, -, 125
Vansant, John R., -, -, 1, 90, -, 15, -, 1000, 200, 400, -, 60, 200, 100, 250
Kates, William T., -, -, 1, 40, -, 66, -, 600, 20, 100, -, 12, 100, 40, 175
Hicks, Ignatius, -, -, 1, 50, -, 75, -, 1000, 40, 200, -, 20, 100, 50, 360
Hopkins, Robert, 1, -, -, 50, -, 90, -, 1200, 20, 250, -, 60, 50, 20, 225
Harrington, William _., 1, -, -, 9, -, -, -, 1600, 50, 130, -, 15, 25, 4, 125
Clark, Henry N., 1, -, -, 3, -, - -, 140, 180, 20, -, 17, 50, 30, 170
Hurd, David, -, -, 1, 12, -, 60, -, 1500, 50, 100, -, 12, -, -, 125
Shaw, Samuel L., -, -, 1, , 22, -, 8, -, 900, 25, 75, -, -, -, -, 75
Shaw, William, -, -, 1, 125, -, 91, -, 4000, 280, 400, 50, 209, 250, 70, 625
Wolcott, Henry C., -, -, 1, 80, -, 20, -, 5000, 150, 110, -, 100, 150, 50, 250
Jester, Isaac N., 1, -, -, 130, -, 30, -, 2500, 1000, 200, -, 25, 75, 11, 50
Pitcher, George W., -, -, 1, 20, -, -, -, 100, 150, 60, -, 20, 20, 3, -
Killen, William S., -, -, 1, 18, -, 12, -, 600, 40, 125, -, -, -, -, -
Sharp, William T., 1, -, -, 20, -, -, -, 5000, 50, 125, -, -, -, -, -
Brown, Isaac, 1, , -, 10, -, -, -, 3000, 21, -, -, 8, 25, 60, 550
Franklin, John, 1, -, -, 60, -, 30, -, 6000, 200, 150, -, 50, 200, 40, 550
Shockley, Levin, -, -, 1, 12, -, -, -, 600, 20, 50, -, 20, 50, 20, 325
Salmons, Minus, -, -, 1, 28, -, 2, -, 900, 50, 150, -, 10, 10, 2, 200
Sapp, Curtis, 1, -, -, 50, -, 6, -, 1000, 100, 125, -, 50, 100, 50, 300
Wix, Mariam, -, -, 1, 40, -, 51, -, 2500, 150, 175, -, 45, 60, 50, 200
Morris, James W., -, -, 1, 60, -, -, -, 1200, 40, 250, -, 66, 100, 50, 440
Hamilton, John W., -, -, 1, 40, -, 20, -, 2500, 25, 100, 5, -, 20, 2, 100
Cain, Noah, 1, -, -, 20, -, 8, -, 700, 90, 250, 20, 30, 200, 10, 340
Morgan, Nancy, 1, -, -, 70, -, 30, -, 2000, 50, 100, -, 23, 100, 50, 200
Barker, Joseph W., 1, -, -, 30, -, 18, -, 1000, 50, 100, - 80, 100, 50, 230
Billings, William, -, -, 1, 100, -, 52, -, 2000, 75, 225, 10, 56, 250, 100, 530
Callaway, James C., -, -, 1, 75, -, 20, -, 1500, 20, 200, -, 20, 100, 50, 385
Scott, Peter S., -, -, 1, 75, -, 25, -, 1500, 50, 150, -, -, 50, 50, 180
Derickson, Stephen H., 1, -, -, 26, -, 14, -, 800, 150, 325, 50, 12, 100, 40, 250
Wyatt, Wesley, -, -, 1, 120, -, 27, -, 2500, 125, 470, 10, 44, 400, 150, 870
Cain, William L., -, -, 1, 60, -, 40, -, 1500, 50, 150, -, 23, 125, 50, 560
Masten, Isaac, -, -, 1, 125, -, 25, -, 4000, 125, 400, 25, 46, 250, 100, 800
Baynard, Laureon, -, -, 1, 50, -, 70, -, 1000, 20, 100, 10, 20, -, -, 275
Jones, Walter, -, -, 1, 60, -, 60, -, 1500, 80, 200, 10, 42, 100, 50, 180
Tootle, Samuel, -, -, 1, 35, -, 50, -, 800, 20, 100, 5, 20, 125, 52, 200
Blade, James, 1, -, -, 11, -, 5, -, 500, 25, 75, 5, -, 10, 3, 175
Campher, Philemon, -, -, 1, 150, -, 550, -, 11000, 160, 400, 15, 25, 275, 100, 600
Harrington, Samuel, 1, -, -, 56, -, 70, -, 1500, 100, 250, 25, 23, 200, 56, 260

Porter, John, 1, -, -, 50, -, 20, -, 1500, 50, 350, 15, 22, 150, 75, 300
Hopkins, Wailman, 1, -, -, 100, -, 63, -, 2000, 150, 450, 10, 30, 200, 10, 350
Hopkins, James, 1, -, -, 80, -, 100, -, 4000, 500, 374, 50, 80 200, 95, 600
Hopkins, Philemon, -, -, 1, 60, -, 40, -, 2000, 150, 300, 10, 80, 150, 56, 600
Kimmey, Jacob L., - , -, 1, 40, -, 20, -, 600, 25, 75, - ,15, -, - , 150
Parus, William, -, -, 1, 75, -, 25, -, 1000, 50, 314, 10, 40, 150, 50, 575
Shellece, Phillip, -, -, 1, 50, -, 50, -, 1000, 25, 100, -, 30, 50, 30, 206
Neil, William C., -, -, 1, 75, -, 75, -, 4000, 20, 250, 10, 80, 100, 50, 500
Hockster, John, -, -, 1, 100, -, 5, -, 100, 25, 100, -, 20, 50, 20, 250
Dill, William C., -, -, 1, 150, -, 25, -, 4000, 50, 200, 25, 150, 151, 75, 400
Marine, David, 1, -, -, 80, -, 5, -, 4200, 200, 725, 10, 60, 210, 60, 75
Baynard, Asbery, -, -, 1, 90, -, 50, -, 2100, 20, 325, 15, 30, 150, 50, 390
Price, Nicholas B., -, -, 1, 100, -, 28, -, 4000, 300, 500, 50, 100, 250, 50, 650
Raughley, Alfred, -, -, 1, 56, -, 10, -, 2000, 150, 681, 15, 81, 175, 52, 460
Smith, Edmund B., 1, -, -, 100, -, 25, -, 4000, 200, 550, 20, 80, 175, 75, 625
Parris, Iseral, -, -, 1, 200, -, 200, -, 5000, 75, 700, 60, 60, 200, 75, 615
Benson, Robert J., -, -, 1, 60, -, 45, -, -, 1000, 15, 50, -, 10, 100, 50, 150
Melvin, Riley, 1, -, -, 140, -, 20, -, 5000, 300, 550, 37, 75, 200, 50, 800
Smith, Alexander, 1, -, -, 40, -, 20, -, 800, 75, 200, 10, 15, -, -, 275
Draper, Richard J., 1, -, -, 120, -, 28, -, 3500, 90, 800, 60, 86, 250, 100, 850
Laramore, Thomas, -, -, 1, 15, -, 15, -, 400, 10, 75, -, -, -, -, -
Bullock, James P., 1, -, -, 100, -, 30, -, 2000, 30, 800, 25, 46, 30, 170, 700
Hubbard, James P., 1, -, -, 100, -, 20, -, 1000, 30, 300, -, 17, 260, 70, 460
Harrington, Ann, 1, -, -, 60, -, 60, -, 1000, 20, 75, -, 20, 100, 50, 175
Cornish, Joseph D., -, -, 1, 60, -, 90, -, 1500, 56, 125, -, -, -, -, -
Porter, James, 1, -, -, 90, -, 6, -, 2000, 150, 400, 20, 25, 100, 30, 400
Swain, Samuel, -, -, 1, 70, -, 55, -, 2000, 60, 150, -, 60, 10, 3, 240
Hopkins, William, -, -, 1, 75, -, 50, -, 2000, 150, 300, 50, 75, 150, 50, 460
Travis, John, 1, -, -, 100, -, 40, -, 2500, 100, 350, -, 32, 150, 50, 400
Jewell, Richard, -, -, 1, 80, -, 60, -, 2000, 50, 350, 20, 72, 70, 50, 500
Wyatt, Thomas C., -, -, 1, 45, -, 10, -, 1500, 25, 75, -, -, -, -, -
Brown, Thomas, 1, -, -, 45, -, 10, -, 2000, 500, 300, 25, 100, 200, 75, 580
Dorman, Thomas J., -, -, 1, 40, -, 35, -, 700, 20, 50, -, 22, -, -, 260
Wyatt, Isaac, -, -, 1, 80, -, 56, -, 2000, 250, 400, 10, 40, 175, 50, 375
Brown, David, -, -, 1, 50, -, 50, -, 2500, 100, 250, 25, 20, 100, 25, 450
Blades, Alexander, -, -, 1, 40, -, 35, -, 2000, 80, 200, 20, 12, 50, 10, 450
McKnatt, George W, 1, -, -, 50, -, 25, -, 1500, 150, 200, 20, 40, 150, 50, 500
Anthony, Benjamin, -, -, 1, 56, -, 16, -, 1500, 75, 300, 25, -, 100, 50, 350
Anthony, Daniel, 1, -, -, 33, -, 16, -, 1200, 50, 200, 10, 3, 100, 50, 280
Anthony, Robert G., 1,-, -, 20, -, 17, -, 600, 75, 150, 10, 18, 50, 70, 125
Blades, Thomas, -, -, 1, 65, -, 30, -, 1200, 25, 250, 25, 45, 79, 32, 500
McKnatt, Nathan, 1, -, -, 25, -, 25, -, 500, 30, 150, 10, 20, -, -, 180
Brown, John, 1, -, -, 55, -, 55, -, 1500, 100, 300, 25, 17, 200, 75, 350
Brown, Tilghman, 1, -, -, 45, -, 17, -, 1600, 200, 300, 10, 33, 150, 52, 450
Baker, Alexander, -, -, 1, -, -, - -, -, 100, 250, 10, 50, 150, 52, 570

Raughley, Nathan F., -, -, 1, 180, -, 110, -, 6000, 325, 1000, 15, 140, 300, 100, 965

Cain, Sarah, 1, -, -, 38, -, -, -, 1500, 150, 300, -, 75, 150, 50, 480

Brown, Tilghman W., -, -, 1, 16, -, 23, -, 1200, 300, 270, 10, 105, 200, 75, 660

Morris, Curtis, 1, -, -, 55, -, 60, -, 2500, 150, 350, 10, 44, 100, 50, 600

Callaway, Eli, -, -, 8, 80, -, 94, -, 2500, 150, 200, 75, 150, 80, 52, 550

Smith, Charles, 1, -, -, 70, -, 38, -, 1200, 150, 200, 10, 30, 150, 50, 385

Price, Edward T., -, -, 1, 60, -, 20, -, 1400, 30, 200, 10, 26, -, -, 250

Harrington, James D., 1, -, -, 100, -, 45, -, 1500, 30, 175, 5, 22, 100, 52, 280

Draper, Benjamin, 1, -, -, 50, -, 35, -, 2500, 1000, 300, 50, 30, 300, 100, 730

Raughley, Robert J., 1, -, -, 60, -, 40, -, 1800, 100, 300, 5, 60, 30, 12, 328

Hopkins, Henry, 1, -, -, 170, -, 100, -, 5000, 425, 640, 50, 91, 385, 64, 1100

Sipple, Zado, 1, -, -, 90, -, 60, -, 3000, 200, 600, 20, 131, 300, 75, 820

Wootens, Edward S., 1, -, -, 81, -, 25, -, 1000, 25, 125, -, 50, -, -, 250

Baynard, Ferdinand, 1, -, -, 70, -, 30, -, 3000, 100, 300, 40, 61, 300, 50, 500

Thrawley, Henry, 1, -, -, 75, -, 25, -, 1500, 50, 200, 5, 60, 40, 12, 135

Trice (Price), Charles W., 1, -, -, 21, -, -, -, 300, 40, 100, -, -, -, -, 40

Callaway, Jonatha, 1, -, -, 75, -, 81, -, 2000, 75, 300, 10, 50, 100, 30, -

Harrington, Albert, -, -, 1, 100, -, 100, -, 3500, 300, 500, 50, 170, 200, 50, 1175

Jewell, Samuel M., 1, -, -, 50, -, -, -, 800, 50, 100, -, 30, -, -, 365

Smith, Robert H., 1, -, -, 140, -, 26, -, 4000, 500, 500, 50, 175, 400, 600, 1500

Blades, William H., -, -, 1, 192, -, 60, -, 5000, 200, 630, 25, 68, 218, 56, 1025

Stafford, Robert H., -, -, 1, 70, -, 14, -, 1200, 100, 300, 25, 40, 75, 50, 430

Walls, Lafayette, -, -, 1, 100, -, 50, -, 3000, 200, 300, 25, 200, 200, 75, 920

Thrawley, John F., 1, -, -, 65, -, 35, -, 2500, 200, 300, 20, 135, 200, 80, 500

Sipple, Thomas, -, -, 1, 100, -, 41, -, 3000, 100, 200, 15, 50, 140, 51, 350

Lynch, Noah, 1, -, -, 50, -, 55, -, 800, 50, 200, -, -, 40, 4, 280

Pond, Samuel, -, -, 1, 50, -, 50, -, 100, 125, 150, -, 50, 20, 5, 300

Anthony, Joseph, -, -, 1, 25, -, 10, -, 400, 20, 75, -, 10, -, -, 120

Cain, Loderick, 1, -, -, 150, -, -, -, 1500, 200, 500, 5, 51, -, -, 750

Jester, William F., 1, -, -, 40, -, 35, -, 1000, 30, 300, 5, -, 50, 10, 240

Derickson, Thomas, -, -, 1, 100, -, 70, -, 2000, 25, 100, 8, 42, 18, 50, 375

Cain, Daniel, 1, -, -, 80, -, 20, -, 1500, 50, 200, -, 8, 50, 10, 300

Brown, William M., 1, -, -, 80, -, 15, -, 2800, 175, 600, 6, 5, 300, 100, 760

Hemping, William, 1, -, -, 30, -, -, -, 300, 60, 30, -, -, -, -, -

Hemping, Simon, 1, -, -, 60, -, 19, -, 1000, 75, 150, 20, -, 150, 50, 350

Stubs, John H., -, -, 1, 130, -, 10, -, 1500, 25, 175, -, 16, 150, 52, 325

Travis, Samuel, -, -, 1, 100, -, 30, -, 2000, 50, 300, 8, 30, 200, 52, 370

Cain, John, 1, -, -, 50, -, 20, -, 1000, 50, 75, -, 12, 100, 50, 335

Cain, William, 1, -, -, 108, -, -, -, 1800, 50, 350, -, 15, 150, 52, 387

Bowen, Levi, -, -, 1, 108, -, 12, -, 2500, 125, 350, 10, -, 160, 50, 725

Bowen, William B., -, -, 1, 75, -, 20, -, 1000, 40, 125, -, 16, 25, 5, 290

Cain, Caleb, -, -, 1, 50, -, 20, -, 1000, 50, 200, -, 20, 150, 50, 387

Camper, Thomas, -, -, 1, 75, -, 25, -, 900, 25, 100, 10, 32, 40, 10, 320

Anthony, Robert, 1, -, -, 67, -, 10, 90, 50, 200, 5, 16, 100, 4, 225

Welch, Walter, -, -, 1, 100, -, 100, -, 3500, 350, 500, -, 37, 300, 100, 820

Eisenbrey, John M., -, -, 1, 106, -, 35, -, 1500, 75, 200, -, 45, 100, 50, 368

Raughley, Shadrach, 1, -, -, 240, -, 23, -, 18000, 500, 1200, 40, 200, 600, 150, 1650

Thomas, Robert H., -, -, 1, 100, -, 25, -, 2500, 350, 650, -, 64, 300, 100, 1400

Trice, James R., -, -, 1, 40, -, 20, -, 1500, 100, 550, -, 10, 30, 160, 52, 670

Merrkan, John H., 1, -, -, 60, -, 12, -, 2500, 175, 270, 10, 40, 200, 100, 430

Minner, Nimrod, -, -, 1, 12, -, -, -, 300, 52, 75, -, -, 50, 20, 215

Callaway, Henry, 1, -, -, 50, -, -, -, 3000, 35, 300, 10, 50, 200, 75, 550

Ross, Hooper, -, -, 1, 60, -, -, -, 2000, 400, 400, -, 35, 60, -, 566

Kelley, George L., -, 1, -, 40, -, 2, -, 800, 20, 50, -, -, -, -, -

Pierce, James, -, 1, -, 30, -, 25, -, 80, 20, 50, -, -, -, -, -

Hopkins, Elias P., 1, -, -, 95, -, 9, -, 4000, 450, 400, -, 104, 300, 100, 1000

Shelten, Charles, 1, -, -, 60, -, 5, -, 4000, 300, 250, -, 1000, 250, 100, 700

Waples, Robert, -, -, 1, 80, -, 40, -, 1500, 125, 200, 5, -, -, -, 260

Ross, James H., 1, - -, 100, -, 25, -, 2000, 150, 225, 10, 32, 80, 28, 400

Melven, Sydenham, 1, -, -, 160, -, 60, -, 3000, 300, 350, -, 100, 200, -, 670

Cleaves, John C., -, -, 1, 106, -, 17, -, 2000, 200, 400, 20, 30, 200, 52, 460

Cochran, John S., 1, -, -, 31, -, 20, -, 1500, 300, 200, -, -, -, -, 200

Ross, James, 1, -, -, 60, -, 35, -, 2000, 200, 300, -, 75, 100, 50, 450

Hopkins, James P., 1, -, -, 100, -, 35, -, 2500, 200, 200, -, 78, 150, 50, 410

Simpson, David, -, -, 1, 200, -, 50, -, 3000, 300, 200, -, 45, 75, -, 350

Harnon, Steuart, -, -, 1, 50, -, 50, -, 1000, 10, 30, -, -, -, -, 180

Baker, Benjamin, -, -, 1, 90, -, 10, -, 1500, 20, 100, -, 50, -, -, 270

Cahall, John A., 1, -, -, 120, -, 50, -, 3000, 150, 300, -, 180, 75, 25, 915

Williams, William T., 1, -, -, 32, -, 3, -, 800, 20, 25, -, 33, -, -, 250

Graham, Jacob S., 1, -, -, 60, -, 22, -, 2500, 250, 230, 10, 91, 200, 100, 1450

Burton, William, -, -, 1, 80, -, 20, -, 2500, 125, 250, -, 175, 250, 100, 680

Thomas, John R., -, -, 1, 80, -, 30, -, 1000, 50, 100, 5, 22, 100, 30, 187

Lynch, James H., -, -, 1, 75, -, 5, -, 1200, 350, 100, -, 36, 20, 3, 200

Lewis, Stephen C., 1, -, -, 16, -, 16, -, 700, 75, 200, 10, 46, 50, 10, 350

Lewis, Evan, -, -, 1, 40, -, 20, -, 1000, 200, 250, 5, 75, 50, 10, 170

Cahall, William H., -, -, 1, 100, -, 40, -, 3000, 100, 200, 25, 156, 200, 32, 660

Salmons, Isaac, -, -, 1, 50, -, 50, -, 1000, 25, 30, -, -, -, -, -

Outten, Charles, 1, -, -, 70, -, 35, -, 1000, 75, 130, 5, 20, 50, 12, 300

Pepper, John P., -, -, 1, 60, -, 40, -, 1000, 20, 100, -, 12, -, -, 130

Simpson, Joseph, 1, -, -, 15, -, 15, -, 400, 20, 75, -, -, -, -, -

Hardesty, George L., -, -, 1, 60, -, 20, -, 800, 50, 150, 8, 26, 7, 2, 200

Maloney, James B., 1, -, -, 100, -, 20, -, 2000, 300, 400, 10, 46, 100, 50, 450

Tucker, ___, 1, -, -, 30, -, -, - 520, -, -, -, 15, -, -, 80

Ferrens, Jesse, 1, -, -, 60, -, -, -, 600, 25, 80, -, 12, -, -, 75

Ferrens, Mary, 1, -, -, 70, -, 10, -, 800, 25, 50, -, 20, -, -, 165

Outten, John H., 1, -, -, 70, -, 10, -, 1000, 30, 150, 10, 35, -, -, 150

Hopkins, Sarah, 1, -, -, 50, -, 30, -, 800, 25, 50, -, -, -, -, 87

Layton, William, 1, -, -, 45, -, 35, -, 800, 75, 275, -, 45, -, -, 170

Jones, Edwards F., -, -, 1, 200, -, 100, -, 3000, 200, 300, -, 160, 200, 50, 270

Smith, Araminta, 1, -, -, 100, -, 30, -, 3000, 100, 500, -, 78, 280, 80, 500

Blaudan, Thomas, -, -, 1, -, -, 109, -, 6500, 100, 150, -, 50, 150, 50, 400
Smith, Marten B., 1, -, -, 36, -, 30, -, 2000, 50, 250, -, 56, -, -, 180
Jacobs, Isaac, 1, -, -, 36, -, -, -, 1000, 40, 20, -, 30, 180, 52, 326
Hardesty, George W., 1, -, -, 70, -, 30, -, 2000, 50, 150, -, 35, -, -, -
Layton, Tilghman, -, -, 1, 65, -, 60, -, 1200, 15, 150, -, 18, -, -, -
Hobs, Nathan, -, -, 1, 25, -, 30, -, 700, 15, 250, -, 28, -, -, -
Bredan, Mary, -, -, 1, 60, -, 27, -, 600, 10, 60, -, -, -, -, 76
Smith, John W., 1, -, -, 125, -, 57, -, 4000, 400, 600, 24, 105, 250, 100, 1225
Willey, Zackariah, -, -, 1, 66, -, 30, -, 1000, 25, 200, -, -, 100, 50, 450
Anderson, Ezekiel, 1, -, -, 40, -, 46, -, 800, 25, 60, -, 47, 150, 52, 280
Cain, George W., 1, -, -, 80, -, 47, -, 3000, 200, 385, 25, 68, 250, 75, 500
Fleming, Matthew, -, -, 1, 30, -, 20, -, 600, 25, 50, -, 10, -, -, 900
Simpson, James, -, -, 1, 75, -, 25, -, 2000, 100, 250, -, 30, 150, 52, 450
Meredith, Robert J., -, -, 1, 200, -, 60, -, 5000, 200, 600, 10, 36, 300, 88, 880
Harrington, Clement A., 1, -, -, 200, -, -, -, 8000, 1500, 800, 25, 250, 700, 200, 1800
Legates, Elijah, -, -, 1, 60, -, 50, -, 2000, 30, 10, -, 40, -, -, 460
Purnell, William, -, -, 1, 65, -, 25, -, 1500, 60, 300, -, 16, 50, -, 334
Cain, Benjamin F., 1, -, -, 100, -, 40, -, 5000, 200, 300, 20, 68, 150, 52, 830
Thorp, William, 1, -, -, 100, -, 60, -, 5000, 500, 850, 50, 250, 275, 100, 770
Callaway Peter, 1, -, -, 89, -, 19, -, 3500, 300, 575, 28, 50, 300, 8, 512
Knox, William H., 1, -, -, 115, -, 35, -, 4000, 250, 660, 25, 125, 400, 100, 997
Collins, George W., 1, -, -, 60, -, 62, -, 1900, 250, 200, 25, 100, 200, 50, 520
Tharp, Samuel, 1, -, -, 70, -, 211, -, 4000, 200, 500, -, 50, 325, 80, 800
Workman, Thomas, -, -, 1, 40, -, 100, -, -, 20, 75, -, 16, -, -, 160
Lighter, Jeremiah, -, -, 1, 30, -, 30, -, 500, 20, 90, -, -, -, -, 280
Duker, William H., -, -, 1, 70, -, 125, -, 1500, 20, 100, -, -, -, -, 187
Rash, Purnell, 1, -, -, 90, -, -, -, 2000, 200, 300, 20, 100, 250, 100, 580
Callaway, Eli, -, -, 1, 100, -, 16, -, 1200, 75, 300, 10, 32, 100, 50, 330
Johnson, Robert E., -, -, 1, 100, -, 69, -, 3500, 25, 200, 25, 30, 150, 52, 415
Colison, William W., 1, -, -, 85, -, 15, -, 2500, 100, 225, 30, 70, 225, 40, 460
Merrikan, Zachariah D., 1, -, -, 55, -, 10, -, 1000, 30, 150, -, 70, 120, 52, 300
Anderson, Benaiah T., 1, -, -, 110, -, 15, -, 2500, 383, 400, -, 48, 375, 120, 700
Swain, Gilbert R., -, -, 1, 20, -, 30, -, 600, 5, 50, -, -, -, -, 220
Bowman, George W., 1, -, -, 80, -, 18, -, 1500, 25, 100, -, -, -, -, 150
Andrews, James, -, -, 1, 50, -, 20, -, 1000, 60, 75, 10, 20, 100, 78, 400
Smith, Samuel J., -, -, 1, 85, -, -, -, 1000, 50, 150, 15, 57, 40, 4, 175
Spence, Luther E., -, -, 1, 5, -, 20, -, 500, 66, 175, -, 8, -, -, 110
Collison, Charles M., 1, -, -, 69, -, 3, -, 500, 200, 400, -, 25, -, -, 187
Jester, George, 1, -, -, 30, -, 16, -, 300, 10, 100, -, -, -, -, 170
Morgan, James H., 1, -, -, 100, -, 35, -, 3500, 225, 300, -, 80, 300, 100, 780
Merrikan, Rebecca A., 1, -, -, 150, -, 25, -, 4000, 200, 700, 24, 40, 300, 100, 925
Wells, Nathan, -, -, 1, 21, -, -, -, 400, 20, 30, -, -, -, -, 190
Butler, Willis W., -, -, 1, 140, -, 75, -, 2500, 200, 375, 15, 30, 250, 100, 500
Ryon, David W., -, -, 1, 180, -, 45, -, 6000, 200, 400, 25, 43, -, -, 260
Ward, Jesse, 1, -, -, 40, -, 22, -, 2000, 240, 635, 30, 63, 200, 54, 620

Collison, Benjamin, 1, -, -, 60, -, -, -, 1200, 100, 150, -, 28, -, -, -
Sharp, Nehemiah, -, -, 1, 75, -, 25, -, 1500, 20, 100, -, -, -, -, 175
Jones, Samuel L., -, -, 1, 75, -, 50, -, 1500, 10, 100, -, -, -, -, -
Wilson, Edward, -, -, 1, 100, -, 50, -, 2000, 90, 425, 5, 30, 150, 75, 400
Porter, John F., -, -, 1, 90, -, 20, -, 2000, 160, 400, 10, 60, 200, 75, 450
Hamilton, James B., -, -, 1, 50, -, 40, -, 1500, 25, 75, 5, 15, 100, 50, 325
Morton, Charles W., 1, -, -, 70, -, 17, -, 1000, 150, 400, 10, 35, 25, 52, 425
Adams, Garrison, -, -, 1, 70, -, 60, -, 800, 80, 140, -, 7, -, -, 200
Cannon, David, -, -, 1, 150, -, 60, -, 4000, 150, 450, 20, 60, 200, 52, 500
Hopkins, David J., 1, -, -, 100, -, 100, -, 6000, 300, 780, 20, 145, 300, 100, 780
Russel, Francis, -, -, 1, 100, -, 160, -, 2000, 150, 400, 10, -, 125, 52, 570
Jones, William, -, -, 1, 80, 50, -, -, 1500, 80, 175, 25, 85, 150, 50, 515
Fisher, James H., 1, -, -, 100, 50, -, -, 2000, 500, 600, 20, 100, 250, 100, 600
Wright, William H., -, -, 1, 50, -, 75, -, 1000, 30, 300, 10, 60, 200, 75, 365
Parris (Parvis), Matthew, 1, -, -, 130, -, 48, -, 1500, 20, 50, -, -, -, -, 125
Williamson, Charles, 1, -, -, 175, -, 75, -, 3000, 200, 250, 15, 42, 150, 50, 435
Messick, James B., -, -, 1, 80, -, 48, -, 1200, 220, 400, -, 32, 100, 20, 312
Saulsbery, Garrison, 1, -, -, 50, -, 34, -, 1000, 55, 50, -, 32, 80, -, 260
Layton, James, -, -, 1, 65, -, 10, -, 700, 15, 125, -, 25, 20, -, 275
Saulsbery, Henry, -, -, 1, 60, -, 40, -, 1000, 65, 300, -, -, -, -, -
Collins, Isaac, -, -, 1,16, -, -, -, 200, 10, 50, -, -, -, -, 100
Scott, Evan, -, -, 1, 48, -, -, -, 300, 20, 50, -, -, -, -, 15
Cahall, Alfred, 1, -, -, 58, -, 28, -, 1200, 250, 300, -, 52, -, -, 277
Oday, William, -, -, 1, 60, -, 40, -, 100, 25, 50, -, 10, -, -, 310
Peters, William R., 1, -, -, 75, -, 57, -, 2000, 150, 250, -, 75, 200, 40, 300
Adams, Charles M., 1, -, -, 80, -, -, -, 800, 60, 100, 10, -, -, -, -
Layton, Stephen, -, -, 1, 100, -, 100, -, 4000, 100, 500, 10, 128, 300, 100, 1500
Vincent, Collins, -, -, 1, 100, -, 150, -, 2500, 100, 250, 25, 14, 150, 50, 200
Loyed, Ephraim S., 1, -, -, 20, -, 16, -, 500, 20, 50, -, 10, -, -, 100
Smith, George F., 1, -, -, 25, -, 35, -, 750, 25, 150, 10, 12, -, -, 150
Goslin, James M., -, -, 1, 90, -, 60, -, 1500, 50, 200, 10, -, -, -, 300
Redden, William A., 1, -, -, 85, -, 30, -, 2000, 75, 500, -, 30, 250, 100, 530
Johnson, George W., -, -, 1, 75, -, 25, -, 2500, 80, 300, 10, 100, 200, 100, 565
Rickard, George W., -, -, 1, 50, -, 20, -, 1400, 60, 250, 5, 60, 150, 52, 250
Hopkins, Zebulon, 1, -, -, 100, -, 100, -, 6000, 800, 600, 10, 200, 300, 100, 1150
Williams, John, 1, -, -, 100, -, 50, -, 3000, 200, 500, 20, 140, 250, 68, 960
Cordray, Jeremiah, 1, -, -, 100, -, 127, -, 2000, 30, 250, -, 200, 100, 50, 560
Dean, Charles, -, -, 1, 100, -, 200, -, 2000, 50, 250, -, 50, 100, 40, 400
Anderson, David P., 1, -, -, 75, -, 30, -, 2000, 300, 300, 26, 50, 200, 100, 280
Harrington, Peter D., 1, -, -, 90, -, 40, -, 1500, 100, 300, 10, 40, -, -, 300
Redden, James D., -, -, 1, 110, -, 36, -, 3000, 200, 400, 10, -, 40, -, 490
Baynard, Charles H., 1,-, -, 58, -, 2, -, 1500, 200, 400, 10, 42, 150, 50, 418
Collison, George W., -, -, 1, 75, -, -, -, 1000, 200, 320, 25, 16, 125, 52, 450
Jones, Samuel O., 1, -, -, 100, -, 40, -, 1500, 120, 300, 20, 38, 200, 75, 531
Jones, Samuel P., -, -, 1, 60, -, 38, -, 1000, 15, 150, -, -, -, -, -
Raughley, George, -, -, 1, 110, -, 75, -, 2000, 200, 450, 20, 50, 150, 52, 560

Harrington, Jehu (John) P., -, -, 1, 100, -, 50, -, 2000, 300, 650, -, 150, -, -, 850
Rust, James, 1, -, -, 65, -, 30, -, 1800, 150, 400, 10, 35, 200, 70, 690
Collison, Robert P., -, -, 1, 75, -, 75, -, 2000, 1150, 225, 10, 85, 150, 52, 340
Murphy, Robert, -, -, 1, 60, -, -, -, 500, 20, 140, -, 40, -, -, 135
Hopkins, William N., 1, -, -, 150, -, 58, -, 2000, 100, 300, 10, 40, 150, 52, 470
Morris, John W., -, -, 1, 100, -, 50, -, 2000, 50, 100, -, -, 50, 10, 280
Evans, Auther J., -, -, 1, 40, -, 160, -, 2000, 10, 75, -, -, 55, 15, 100
Hopkins, Robert G., -, -, 1, 100, -, 200, -, 4000, 300, 700, 50, 100, 200, 60, 927
Jones, Anna M., 1, -, -, 90, -, 40, -, 2000, 125, 400, 50, 45, 250, 80, 475
Cannon, Hubert, 1, -, -, 60, -, 20, -, 600, 24, 200, -, -, -, -, -
Wroten, Eli, 1, -, -, 70, -, 64, -, 1000, 80, 300, 20, 35, 175, 52, 235
Smith, William, H., 1, -, -, 30, -, -, -, 300, 50, 100, -, -, -, -, -
Wroten, James H., 1, -, -, 30, -, -, -, 300, 50, 100, -, -, -, -, -
Spence, George, -, -, 1, 100, -, 40, -, 800, 15, 50, -, 8, -, -, 150
Paswaters, William, -, -, 1, 40, -, -, -, 400, 25, 60, -, -, -, -, 155
Rickards, James, -, -, 1, 50, -, 20, -, 800, 20, 50, -, -, -, -, 120
Morris, Thomas E., 1, -, -, 60, -, 8, -, 1500, 250, 350, 20, 100, 200, 80, 520
Sudars, William, 1, -, -, 200, -, 225, -, 4000, 200, 400, 20, 40, 250, 100, 571
Adkinson, Elizabeth, 1, -, -, 60, -, 15, -, 1000, 175, 275, -, -, -, -, 235
Booth, John, 1, -, -, 50, -, 69, -, 1000, 100, 75, -, 62, 200, 75, 635
Lewes, Nimrod, -, -, 1, 75, -, 15, -, 600, 20, 75, 1000, 200
Adams, David H., 1, -, -, 50, -, 25, -, 400, 25, 100, -, -, -, -, 120
Betts, Adryan, -, -, 1, 50, -, -, -, 500, 20, 100, -, -, 50, 52, 180
Raughley, John W., 1, -, -, 95, -, 45, -, 1000, 250, 250, -, 40, -, -, 265
Lane, Charles H., -, -, 1, 30, -, -, -, 500, 25, 50, -, 10, -, -, 600
Thorp, William H., -, -, 1, 40, -, -, -, 500, 20, 100, 5, 25, 100, 40, 275
Smith, Lewis, 1, -, -, 20, -, 61, -, 600, 50, 75, -, 24, -, -, 120
Hall, Zeborah, 1, -, -, 50, -, 50, -, 800, 10, 100, -, -, 50, 12, 90
Hall, Samuel F., 1, -, -, 50, -, 45, -, 800, 50, 100, -, 35, -, -, 250
Jones, Henry S., 1, -, -, 30, -, 8, -, 500, 50, 75, -, -, -, -, 150
Cain, James M., 1, -, -, 85, -, 15, -, 2000, 140, 400, 15, 80, -, -, 450
Reynolds, Thomas, -, -, 1, 75, -, 25, -, 3, 30, 350, 15, 46, 100, 52, 345
Harris, William B., 1, -, -, 11, -, -, 2, 3000, 1000, 125, -, 50, 25, -, 500
Stevens, J. W., 1, -, -, 65, -, -, -, 10000, 100, 425, -, 125, 325, -, 1000
Jarman, Charles B., 1, -, -, 26, -, -, -, 1500, 200, 100, -, 1156, 600, 150, 2500
Lowber, Jonathan H., -, -, 1, 16, -, -, -, 200, 50, 200, -, 40, 80, 25, 500
Postles, Stephen, 1,-, -, 46, -, -, -, 12000, 500, 1000, 200, 85, 700, 100, 1100
Haregrove, Thomas W., 1, -, -, 26, -, -, -, 6000, 20, 750, -, 30, 200, 78, 600
Van Burkalow, William, -, -, 1, 75, -, 15, -, 5000, 150, 450, -, 46, 200, 50, 800
Dager, Henry M., 1, -, -, 60, -, 10, -, 5000, 350, 400, -, 50, 40, 50, 100
Mifflin, Holliday J., -, -, 1, 75, -, 40, -, 5000, 200, 300, -, 100, 150, 40, 100
Jenkins, Jabez S., -, -, 1, 150, -, 100, -, 8000, 300, 800, -, 150, 300, 100, 1200
Dager, John, 1, -, -, 35, 65, -, -, 10000, 800, 500, -, 350, 100, 300, 5000
Mifflin, Samuel H., -, -, 1, 80, 5, -, -, 1000, 250, 300, -, 75, 300, 100, 500
Willcutts, Purnell, -, -, 1, 65, 65, -, -, 6500, 500, 300, -, 80, 150, 70, 1200
Wilson, Thos. H., 1, -, -, 30, 25, -, -, 7000, 800, 1500, -, 150, 800, 160, 3000

Shahan, John, -, 1, -, 90, 5, 40, -, 2500, 50, 140, -, -, 10, 5, 150
Blood, Simeon, 1, -, -, 90, 33, 46, -, 10000, 500, 500, 200, 175, 500, 100, 1450
Mifflin, Daniel, 1, -, -, 100, 50, -, 5000, 150, 300, -, 150, 300, 75, 1500
Satterfield, Phillip J., 1, -, -, 100, 15, -, -, 4000, 200, 400, -, 85, 150, 40, 1000
Montague, Wm. H., -, -, 1, 125, 10, 20, -, 5000, 150, 400, -, 37, 25, 6, 600
Harris, Luther & Willard, -, -, 1, 80, 20, 40, -, 7000, 500, 500, -, 125, 40, 10, 1020
Slay, George, -, -, 1, 100, 30, 40, -, 10000, 350, 800, -, 186, 500, 100, 350
Webb, Wm. R., 1, -, -, 5, 25, 35, -, 6000, 200, 700, -, 106, 300, 60, 2700
Caulk, John S., 1, -, -, 104, 8, 25, -, 6000, 500, 600, -, 125, 175, 35, 1000
Griffith, Asa, 1,-, -, 40, 20, 3, -, 6000, 1000, 800, 400, 100, 200, 45, 1500
Dawnham, Thomas, 1, -, -, 52, 10, 8, -, 4500, 150, 200, -, 60, 50, 15, 700
Kent, Phila A., 1, -, -, 17, 9 ½, -, -, 7000, 300, 550, -, -, 300, 65, 700
Lindale, William P., 1, -, -, 30, 50, -, -, 8000, 500, 400, 50, 100, 500, 50, 600
Fisher, M. J., 1, -, -, 50, 10, 12, -, 6000, 175, 400, -, -, -, -, 350
Connor, John S., 1, -, -, 55, 15, 15, -, 3500, 200, 300, -, 100, 200, 45, 850
Raush, Charles D., 1, -, -, 37, 25, -, -, 6000, 250, 400, 50, 75, 200, 45, 500
Sarde, Robert, 1, -, -, 21, 6, -, -, 3000, 300, 250, -, 20, 150, 10, 250
Lawrence, Alanson, 1, -, -, 16, 29, -, -, 6000, 450, 190, 5, 25, 60, 15, 200
Lingo, Samuel W., -, -, 1, 3, 33, -, -, 3000, 100, 125, -, 12, 90, -, 40, 200
Bancroft, E. H., 1, -, -, 100, 35, -, -, 1500, 1200, 1500, 25, 400, 1800, 450, 5300
Carey, A. H., 1, -, -, 100, 65, -, -, 1600, 1850, 1300, 50, 510, 1600, 420, 4800
Jackson, Edward, 1,-, -, 75, 10, 25, -, 5000, 300, 400, 50, 225, 500, 130
Postles, Henry W., 1, -, -, 24, 6, 5, -, 1750, 60, 200, -, 15, -, -, 600
Rickards, John M., 1, -, -, 45, 15, -, -, 7000, 200, 300, 30, 80, 300, 70, 200
Everett, Samuel J., 1, -, -, 35, 10, -, -, 10000, 1000, 700, 75, 180, 750, 200, 3000
Slaymaker, A. W., 1, -, -, 46, 14, -, -, 4500, 250, 485, 10, 90, 640, 170, 2350
Peters, John, 1,-, -, 130, 80, 15, -, 23000, 2500, 1000, 250, 450, 1980, -, 5500
Green, James, 1, -, -, 70, 6, 20, -, 7000, 150, 350, 10, 225, 225, -, 1150
Derborough, John C., 1, -, -, 70, 42, -, -, 10000, 300, 1100, 10, 300, 500, -, 6000
Charles, Sarah A., 1, -, -, 74, -, 6, -, -, 6000, 50, 200, -, 100, -, -, 500
Hanson, Thomas P., 1, -, -, 25, 112, -, -, 15000, 1350, 1000, 150, 610, 2500, -, 10870
Brown, Jacob G., 1, -, -, 92, 105, -, -, 30000, 1000, 800, 100, 800, 1800, -, 13000
Brown, C. Gundy, -, -, 1, 45, 60, 2, -, 10000, 1400, 500, 50, 295, 800, -, 4600
Brown, A. N., -, -, 1, 56, 45, 2, -, 10000, 1400, 500, 50, 367, 500, -, 2500
Maloney, William E., -, -, 1, 85, 25, 40, -, 7500, 500, 700, 15, 300, 300, -, 1500
Barnard, Daniel P., 1, -, -, 85, 20, 35, -, 12000, 1500, 1000, 15, 250, 1000, 200, 2000
Jackson Thomas, 1, -, -, 150, -, 70, -, 6000, 300, 600, -, 250, 200, 60, 2350
Allaband, William R., -, -, 1, 110, 30, 40, -, 4500, 800, 800, -, 120, 350, 115, 1360
Kinney, Edgar J. (Estate), 1, -, -, 55, 23, 5, -, 4500, 200, 350, -, 25, 470, 100, 1750
Johnson, Henry, -, -, 1, 90, 20, 10, -, 6000, 500, 600, -, 120, 200, 60, 100
Evans, William, 1, -, -, 70, 11, 25, -, 6000, 200, 500, -, 48, 125, 32, 650
Richards David P., 1, -, -, 24, 18, 18, -, 5000, 100, 250, -, 20, 200, 50, 500
Draper, Thos., 1, -, -, 453, 130, 40, -, 30000, 700, 1625, 400, 300, -, 3000
Lodge, John, 1, -, -, 60, 30, 10, -, 5000, -, 200, -, 100, 200, -, 600

Frear, Cornelius, 1, -, -, 175, 35, 25, -, 18000, 300, 400, 50, 300, 720, -, 3400
Husbands, Benj., 1, -, - ,165, 35, -, -, 7000, 100, 450, 15, 150, 350, -, 3000
Burton, Edwards, 1,-, -, 177, 20, 20, -, 12000, 200, 600, 50, 190, 500, -, 1500
Moore, Benjamin H., -, -, 1, 140, 85, 25, -, 8000, 400, 700, 20, 140, 275, -, 1800
Faucett, Samuel, -, 1, -, 88, 12, 14, -, 7000, 300, 600, -, 200, -, -, 1000
Vincent, Morris B., -, -, 1, 160, 58, 10, -, 6000, 50, 300, -, 100, 150, -, 2000
Lloyd, John A., 1, -, -, 29, 6, -, -, 3500, 250, 300, -, 125, 60, -, 600
Loper, Harvey, 1, -, -, 95, 20, 20, -, 400, 200, 600, -, 125, 100, -, 2000
Phillips, George V., -, -, 1, 180, 25, 30, -, 3500, 200, 700, -, 200, 250, -, 1400
Cooper, Samuel B., 1, -, -, 50, 25, 8, -, 3000, 350, 300, 10, 80, 260, 52, 2000
Cleaver, John, -, -, 1, 160, 40, 60, -, 10000, 300, 1200, 50, 270, 1000, 250, 4000
Lewis, Robert H., 1, -, -, 115, 12, 40, -, 4500, 350, 450, 50, 150, 450, 90, 1000
Gray, William A., 1, -, -, 43, 707, -, 7000, 400, 600, 100, 200, 500, 125, 3500
Draper, Henry R., 1, -, -, 225, 3, 55, -, 10000, 1000, 1000, 25, 360, 300, 75, 2000
Robinson, Clayton, -, -, 1, 97, 5, 30, -, 2500, 150, 400, -, 65, 75, 20, 575
Barnitt, Isaac, 1, -, -, 82, 14, 24, -, 5500, 400, 400, 25, 70, 160, 45, 700
Cooper, Thomas B., 1, -, -, 95, 5, 20, -, 2400, 400, 8500, -, 500, 50, 13, 900
Catlin, Edward, -, -, 1, -, -, -, -, 2000, 100, 350, -, 23, 25, 6, 600
Lewis, Annie E., 1, -, -, 88, 12, 12, -, 4000, 250, 400, 20, 1555, 135, 35, 1250
Kent County Almshouse, 1, -, -, 260, 13, 30, -, 15000, 500, 2000, 175, 360, 600, 155, 5530
Loose, Cornelius, -, 1, -, 75, 30, 40, -, 4000, 100, 360, -, 40, -, -, 780
Dill, Alexander C., 1, -, -, 100, 3, 55, -, 5000, 300, 600, -, 175, 200, 50, 700
Cooper, T. L., -, -, 1, 103, 1, 80, -, 2000, 60, 400, 30, 23, 100, -, 500
Ford, Christopher, -, -, 1, 120, 10, 75, -, -, 150, 600, -, -, -, -, 1000
Stubbs, William L., 1, -, -, 92, 8, 70, -, 3000, 100, 450, -, 50, 100, -, 600
Smith, J. Colby, -, 1, -, 150, -, -, -, 6000, 600, 600, 50 168, 250, -, 1200
Gooden, William, 1, -, -, 80, 5, 13, -, 1500, 100, 600, 20, 66, 250, -, 750
Gooden, William G., 1, -, -, 85, 1, 40, -, 3000, 105, 255, 10, 50, 250, -, 500
Clark, Richard, -, -, 1, 125, -, 40, -, 4000, 125, 450, 20, 85, 75, -, 800
Cook, Nathan, -, -, 1, 132, 9, 30, -, 8000, 300, 1200, 160, 225, 75, -, 1100
Cook, Thomas, -, -, 1, 220, 50, 50, -, 70000, 260, 900, 50, 335, 450, -, 1400
McGinnis & Reed, -, -, 1, 102, 12, 20, -, 5200, -, 340, 50, 175, 250, -, 1000
Cowgill, Ezekiel J. D., 1, -, -, 65, 10, -, -, 5000, 320, 800, 50, 100, 400, -, 1770
Evans, William K., 1, -, -, 87, 55, -, -, 11500, 800, 750, 75, 100, 500, -, 2800
Proud, Lewis, -, -, 1, 35, 130, -, -, 10000, 150, 500, -, 60, 800, -, 6000
Cooper, Henry C., 1, -, -, 133, 15, 15, -, 12000, 800, 850, 25, 200, 340
Evans, John W., 1, -, -, 80, 40, 20, -, 10000, 350, 600, 25, 150, 300, -, 1000
Derby, Henry W., 1, -, -, 88, 23, 95, -, 5700, 150, 500, -, 340, 408, -, 1400
Pfleegon, William, -, -, 1, 68, 60, 12, -, 10000, 650, 700, 15, 150, 300, -, 2500
Blood, Wiliam H., -, 1, -, 175, 16, 80, 40, 6000, 175, 400, 25, 55, 2240, -, 800
Johnson, Elisha, -, -, 1, 125, 45, 30, -, 7000, 750, 850, 15, 480, 800, -, 2200
Conwell, Charles, 1, -, -, 115, 30, 15, -, 8000, 400, 600, 50, 210, 400, -, 1800
Barcus, Greenbury, -, -, 1, 130, 60, 5, -, 12000, 600, 600, 20, 390, 500, -, 6000
Wells, Albert W., -, -, 1, 34, 5, 10, -, 2500, 150, 225, 25, 145, 276, -, 700
Brown, William, 1, -, -, 90, 10, 15, -, 5000, 225, 350, 25, 100, 50, -, 1200

Green, James, 1, -, -, 129, 71, 35, -, 5950, 500, 1000, 50, 420, 350, -, 2750
Jackson, M. & G.D., -, -, 1, 164, 96, -, -, 20000, 100, 1000, 40, 160, 1000, -, 10000
Mensch, Andrew C., 1, -, -, 57, 8, -, -, 300, 150, 425, 10, 200, 250, -, 1200
Jester, James H., 1, -, -, 115, 15, 5, -, 5400, 500, 500, 50, 150, 300, -, 1800
Pickery, Thomas, 1, -, -, 550, 60, 30, -, 25000, 1000, 2000, 100, 1400, 1400, -, 12000
Wharton, Samuel, 1, -, -, 200, 20, 40, -, 10400, 400, 650, 100, 350, 300, -, 4000
McBride, Daniel L., 1, -, -, 102, 60, 12, -, 6800, 200, 550, 30, 224, -, -, 4500
Thomas, Frank, 1, -, -, 65, 30, -, -, 3600, 150, 250, -, 50, -, -, 1400
Frear, Dewitt, 1, -, -, 160, 32, 15, -, 15000, 1000, 550, 100, 280, 400, -, 3200
Saxton, Cornelius, -, -, 1, 152, 20, 15, -, 5400, 250, 500, 20, 300, 360, -, 2200
Kelley, Joseph, -, -, 1, 103, 32, 15, -, 5000, 200, 700, 25, 200, 300, -, 1500
Townsend, Samuel, 1, -, -, 173, 70, 30, -, 13500, 250, 600, 20, 150, 550, -, 850
Barrett, Philip, -, -, 1, 100, 31, 15, -, 5000, 350, 600, 25, 300, 400, -, 1500
Darling, Thomas, -, -, 1, 50, 38, 250, -, 5000, 225, 400, 5, 200, 400, -, 2000
Davis, Jehu, 1, -, -, 41, 11, -, -, -, 2600, 150, 260, 5, 5, 25, -, 850
Anderson, James, 1,-, -, 75, 10, 15, -, 8000, 600, 900, 20, 250, 200, -, 1800
Graham, William, 1, -, -, 105, 12, 2, -, 6000, 300, 700, 25, 250, 250, -, 800
Graham, George H., -, -, 1, 135, -, 15, -, 8000, 300, 800, 30, 46, 200, -, 800
Burchenal, Joseph S., 1, -, -, 37, 9, 10, -, 5000, 250, 125,20, 100, 450, -, 1350
Jones, Clement, -, -, 1, 54, 15, 7, -, 2000, 150, 300, 15, 70, 25, -, 600
Patterson, John, 1, -, -, 95, 8, -, -, 8000, 150, 650, 20, 150, 300, -, 1500
Hargadine, Henry R. Sr., 1, -, -, 95, 15, 40, -, 6000, 250, 400, 25, 125, 300, -, 1250
Stutz, John, 1, -, -, 86, 8, 6, -, 2500, 300, 550, -, 290, -, -, 585
Hargadine, Henry R. Jr., -, -, 1, 13, 9, 8, 30, -, 2500, 100, 400, 100, 114, 75, -, 600
Caulk, Robert K., -, -, 1, 150, -, 5, 100, -, 7500, 500, 600, 100, 100, 200, -, 800
Tinley, Jacob S., -, -, 1, 165, 15, 20, -, 4000, 300, 1000, 20, 70, 80, -, 100
Walhearter, George W., 1,-, -, 83, 10, 60, -, 4600, 225, 1160, 25, 90, 100, -, 600
Cohee, Edward T., -, -, 1, 96, 4, -, -, 4000, 100, 250, 10, 22, -, -, 600
Gooden, William T., -, -, 1, 138, 12, 10, -, 4000, 50, 560, 25, 50, 200, -, 700
Cook, James, H., -, -, 1, 108, 11, 70, - 3000, 200, 400, 100, 70, -, -, 900
Gooden, Ambrose B., -, 1, -, 180, 3, 40, -, 6000, 75, 150, 50, 70, 200, -, 500
Price, James, -, -, 1, 100, 12, 100, -, 4000, -, -, -, -, -, -, 500
Gooden, Thomas, 1, -, -, 100, 5, 50, -, 3750, 600, 700, 50, 66, 275, -, 1100
Gooden, Isaac, 1, -, -, 120, 15, 75, -, 4200, 400, 725, 50, 110, 300, -, 1300
Moore, Nathan, -, -, 1, 120, 8, 50, -, 4000, 200, 500, 25, 45, 200, -, 600
Gruwell, Peter C., 1, -, -, 95, 5, 74, -, 2000, 150, 500, 40, 70, 100, -, 600
Frasher, William J., 1, -, -, 72, 5, 24, -, 2500, 150, 300, 50, 25, 90, -, 500
Cook, John H., 1, -, -, 118, 8, 15, -, 4000, 450, 600, 95, 90, 200, -, 1000
Frasher, Richard C., 1, -, -, 40, 7, 20, -, 2000, 150, 300, 50, 22, 150, -, 800
Frasher, A. C., 1, -, -, 32, 7, 16, -, 1500, 250, 275, 25, 20, 50, -, 500
Meredith, Peter, 1, -, -, 52, 16, 23, -, 5000, 500, 475, 50, 52, 300, -, 1700
Meredith, Alexander F., -, -, 1, 40, 10, 30, -, 200, 240, 175, 5, 30, 210, -, 550
Gruwell, John, 1, -, -, 110, 25, 40, -, 1500, 300, 560, 10, 83, 150, -, 500

Dill, Benjamin F., -, -, 1, 122, 12, 35, -, 2500, 250, 450, -, 110, 185, -, 700
Gruwell, Joseph D., 1, -, -, 142, 20, 50, -, 3150, 525, 640, 15, 88, 200, -, 1200
Melvin, John S., 1, -, -, 75, 25, -, 75, 2000, 102, 250, -, -, -, -, 500
Shockley, Thomas, 1, -, -, 40, 15, -, -, 3000, - 300, -, -, -, -, 800
Cubbage, Benjamin O., -, -, 1, 100, 20, 40, -, 3200, 200, 450, 5, 40, 8, -, 1000
McGinnis, John H., -, -, 1, 113, 25, 106, -, 5000, 350, 1000, 20, 44, -, -, 1500
Frazier, Ezekiel C., 1, -, -, 73, 10, 50, -, 2600, 600, 300, 50, 73, 110, -, 650
Cohee, James L., 1, -, -, 85, 4, 30, -, 1000, 150, 300, 25, 47, -, -, 600
Cheffin, Eli, -, -, 1, 100, 25, 75, -, 6000, 125, 525, -, 100, 200, -, 1000
Frashier, Robert 1, -, -, 115, 20, 15, -, 7200, 500, 415, 15, 300, 100, -, 1000
Jackson, Isaac R., -, -, 1, 185, 15, 55, -, 7000, 400, 600, -, 312, -, -, 1000
Steele, John W., -, -, 1, 125, 8, 15, -, 2500, 400, 600, 15, 60, 350, -, 800
Smith, Margarett, 1, -, -, 133, 45, 12, -, 3600, 100, 400, -, 125, 300, -, 1400
Rawley, Peter, 1, -, -, 105, 20, 90, -, 4000, 200, 400, 20, 85, 20, -, 725
Cohee, Benjamin, -, -, 1, 250, 15, 100, -, 10000, 500, 700, 25, 275, 300, -, 1500
Carter, Edward J., 1, -, -, 64, 100, 117, -, 30000, 800, 1340, 40, 162, 1000, -, 8830
Dill, Luther M., 1, -, -, 94, 6, 89, -, 4000, 250, 600, 25, 67, 2225, -, 500
Smith, Henry, -, -, 1, -, 8, -, -, 2500, 100, 400, -, 50, -, -, 550
Cooper, Peter L., 1, -, -, 240, 50, 40, -, 10000, 1000, 1000, 300, 400, 400, -, 2000
Evits, John, -, -, 1, 134, 16, 45, -, 3000, -, 1100, -, -, -, -, 1100
Steele, Nathan, 1, -, -, 70, 40, 6, -, 4000, 400, 550, 20, 128, 250, -, 1400
Murry, George N., 1, -, - , 30, 20, 20, -, 5000, 50, 100, -, 50, 75, -, 1200
Owens, George, -, -, 1, 125, -, -, 10, 11000, 1000, 700, -, 170, 137, 36, 4000
Raymond, Walter, -, 1, -, 4, -, -, -, 300, -, -, -, -, -, -, 50
Saxton, John, 1, -, -, 120, -, 10, -, 5000, 900, 1000, -, 300, 150, 36, 1500
Scanlon, Jeremiah, 1, -, -, 22, -, 4, -, 1200, 150, 175, -, 20, 40, -, 800
Sullivan, Joseph, -, -, 1, 50, 1, 20, -, 5500, 250, 700, -, 120, 115, 36, 1500
Abram, Robert 1, -, -, 7, -, 2, -, 400, 50, 100, -, 5, -, -, 200
Sullivan, William, -, -, 1, 90, 1, -, 20, 5500, 250, 700, -, 160, 150, 36, 1500
Coverdale, Samuel, -, -, 1, 55, -, 10, -, 2500, 200, 200, -, 50, 60, -, 700
Williams, Joseph W., -, -, 1, 65, 1, -, -, 5000, 200, 500, -, 120, 60, -, 800
Coverdale, John, -, -, 1, 200, -, 56, -, 15000, 275, 1000, -, 280, 200, 36, 2000
Warren, George R., 1, -, -, 80, -, -, -, 3000, 50, 300, -, 70, 100, 36, 500
King, William, -, 1, -, 10, -, -, -, 450, 30, 80, -, 10, -, -, 200
McQueen, James, 1, -, -, 3, -, -, -, 2000, 200, 450, -, 40, -, -, 600
Williams, Benjamin, 1, -, -, 50, 12, -, 2000, 200, 500, -, 50, 160, 12, 500
Scanlon, Jon, 1, -, -, 23, -, -, -, 1000, 50, 100, -, 20, -, -, 160
Lane, Enos, 1, -, -, 29, -, -, -, 900, 100, 250, -, 35, -, -, 400
Williams, Caleb, 1, -, -, 100, 2, 20, -, 4000, 300, 700, -, 140, 200, 36, 1000
Green, Eugene, 1, -, -, 120, 1, 10, -, 4000, 350, 700, 25, 114, 280, 72, 1100
Doody, Margaret, 1, -, -, 6, -, -, -, 1000, 50, 300, -, -, -, -, 200
Williams, Curtis, -, -, 1, 105, -, 10, -, 4000, 200, 700, -, 114, 90, 36, 1300
Cohee, Anthony, -, 1, -, 30, 1, 20, 50, 3000, 150, 275, -, 37, 35, -, 500
Faley, Timothy, 1, -, -, 15, -, -, -, 800, 100, 100, -, 22, 25, -, 250
Gray, William 1, -, -, 20, -, -, -, 600, 120, 250, -, 47, 21, -, 350
Morrel, Avery, -, -, 1, 300, 1, 30, -, 16000, 500, 1000, 100, 425, 200, 72, 2500

Binns, Benjamin, 1, -, -, 50, 1, -, -, 5000, 150, 250, -, 15, 216, 12, 700
Slaughter, David, -, -, 1, 41, -, -, -, 2500, 150, 150, -, -, 50, -, 600
Barnett, John, 1, -, -, 60, -, 1, 20, -, 5000, 200, 500, -, 100, 150, 32, 900
Barnett, Jacob, 1, -, -, 125, 2, -, -, 4500, 200, 1000, -, 150, 100, 20, 900
Wright, John W., 1, -, -, 150, 1, 10, -, 8000, 300, 1000, -, 150, 150, 36, 1000
Holt, Henry P., 1, -, -, 30, 1, -, -, 3000, 250, 200, -, 10, 75, 20, 1000
Aaron, Morris, 1, -, -, 4, -, -, -, 300, -, 100, -, 64, 34, -, 50
Wilcutts, William J., 1, -, -, 80, 2, 25, -, 3000, 250, 600, -, 88, 40, 10, 800
Postles, David, -, -, 1, 70, -, 50, -, 3000, 250, 275, -, 60, -, 36, 600
Lindale, William, -, -, 1, 80, -, 20, -, 3000, 150, 300, -, 2000, 150, 36, 500
Russel, Elias, 1, -, -, 120, -, 4, -, 6000, 500, 800, -, 250, 150, 36, 1350
Knight, Hewett L., 1, -, -, 130, 1, -, -, 6000, 200, 800, -, 60, 30, -, 1000
Roach, Robert, -, -, 1, 100, -, -, -, 5000, 100, 450, -, 100, 100, 8, 1000
Davis, George S., -, -, 1, 100, 1, 40, -, 8000, 250, 600, -, 200, 50, 26, 1500
Reed, James H., 1, -, -, 26, -, -, -, 2000, 50, 140, -, -, -, -, 100
Cox, Mathews, -, 1, -, 200, 1, 30, -, 7000, 250, 800, -, 100, 260, 39, 1000
Stevenson, Tindley, -, -, 1, 150, -, 40, -, 6000, 250, 40, -, 120, 125, 36, 1200
Grier, Kemp Robert, 1, -, -, 120, 1, 22, -, 6000, 600, 650, 75, 200, 234, 72, 1500
Robbins, David, -, -, 1, 180, 1, 35, -, 500, 150, 400, -, 150, 60, 12, 80
Stout, Edmund, 1, -, -, 120, -, 15, -, 12000, 300, 500, -, 100, 200, 72, 2000
Wilcutts, David, -, -, 1, 80, 1, 28, -, 5000, 150, 300, -, 75, 70, 20, 700
Conner, Jno. B., 1, -, -, 240, -, 10, -, 14000, 300, 700, -, 200, 110, 32, 2100
Conner, John J., 1, -, -, 175, -, 22, -, 10000, 250, 400, -, 175, 200, 32, 1250
Slaughter, James, -, -, 1, 150, 2, -, -, 7000, 250, 350, -, 120, 110, 36, 1000
Slaughter, John, -, -, 1, 300, -, 400, -, 18000, 300, 600, -, 270, 250, 72, 2500
Barnett, Mary, 1, -, -, 100, -, 5, -, 6000, 200, 400, -, 80, 75, 36, 300
More, Griffin, -, -, 1, 60, -, 10, 3, 2500, 125, 300, -, 50, 70, 10, 450
Davis, James H., -, -, 1, 50, -, 20, -, 2000, 150, 250, -, 70, -, -, 500
Conwell, Benjamin T., -, -, 1, 150, 1, -, -, 10000, 400, 800, -, 250, 200, 72, 2000
Vincent, Thomas W., 1,-, -, 18, 2, -, -, 1300, 75, 150, -, 70, 25, 10, 300
Smith, Peter H., 1, -, -, 70, 1, -, -, 2400, 200, 350, -, 75, 75, 10, 1200
Garton, Edmund H., -, -, 1, 50, 1, 25, -, 3000, 150, 350, -, 50, 50, 12, 400
Lodge, Thomas E., 1, -, -, 61, -, 10, -, 1100, 100, 400, -, 47, 105, 36, 450
Voshell, Daniel, -, -, 1, 87, -, 1, 1, 12, -, 3000, 200, 300, -, 60, 50, 70, 500
Prettyman, John C., 1, -, -, 10, -, -, -, 500, 50, 125, -, -, -, -, 200
Johnson, William, -, -, 1, 110, 1, 20, -, 6000, 200, 400, -, 75, 100, 36, 900
Meredith, William, -, -, 1, 70, 1, -, -, 7000, 150, 300, -, 60, 60, -, 1700
Edmunds, John A., 1, -, -, 32, -, -, -, 800, 50, 70, -, 25, 15, -, 200
Edmunds, David, 1, -, -, 38, 1, 12, -, 700, 50, -, -, -, -, -, 120
Layman, John, -, -, 1, 54, 1, -, -, 1200, 200, 200, -, 40, 75, 12, 300
Faulkner, James, -, -, 1, 175, 1, 150, -, 10000, 500, 700, -, 125, 150, 36, 1500
Jackson, Lewis, 1, -, -, 55, 1, 10, -, 4000, 200, 300, -, 60, 50, -, 700
Kirbin, James, W., -, -, 1, 75, -, 75, -, 6000, 250, 450, -, 170, 110, 36, 900
Burk, Henry L., 1, -, -, 10, 1, -, -, 700, 50, 75, -, 25, -, -, 200
Layton, Albert, -, -, 1, 80, -, 55, -, 4000, 100, 200, -, 100, 70, 20, 600
Richards, A. C., -, -, 1, 150, 1, 25, -, 10000, 500, 800, -, 200, 250, 46, 2100

Dickson, William, -, -, 1, 200, 1, 25, -, 6000, 200, 500, -, 150, 150, 36, 1000
Abbert, Gilbert, -, -, 1, 75, 1, 72, -, 5000, 200, 500, -, 56, 110, 36, 800
Frazier, Thomas E., -, -, 1, 100, 1, 50, -, 6000, 100, 400, -, 150, 150, 800
Hammond, William, -, -, 1, 26, 1, -, -, 1300, 175, 150, -, -, -, -, 400
Pratt, Amsa, -, -, 1, 210, 1, -, -, 11000, 500, 1000, -, 150, 157, 36, 1100
Satterfield, John, 1, -, -, 80, 1, 20, -, 5000, 250, 500, -, 150, 250, 36,1000
McGill, Ebenezer, 1, -, -, 150, 1, 20, -, 4000, 300, 1000, -, 160, 70, 36, 450
Lafferty, William, 1, -, -, 75, 1, -, -, 12000, 300, 300, -, 100, 300, 72, 2500
Bateman, John W., 1, -, -, 70, 1, 38, -, 5000, 200, 500, -, 170, 210, 72, 900
Kersey, Thomas C., 1, -, -, 70, 1, 30, -, 3000, 175, 500, -, 210, -, -, 1800
Lowber, Charles, 1, -, -, 90, 1, -, 12, 5500, 400, 600, -, 100, 230, 36, 1500
Lynch, John W., -, -, 1,15, 1, -, -, 700, 50, 200, -, -, -, -, 150
Wyatt, Thomas E., -, -, 1, 180, 1, -, 125, 80000, 300, 400, -, 170, 140, 32, 800
Horsey, Thomas S., -, -, 1, 66, -, 1, -, -, 3500, 200, 300, -, 50, -, -, 700
Raughley, James, 1, -, -, 75, 1, -, 5, 300, 200, 500, -, 75, 100, -, 300
McIntosh, John, 1, -, -, 80, 1, -, 20, 4000, 120, 325, -, 37, -, -, 400
Jackson, Edward, 1, -, -, 100, 1, -, 25, 2500, 125, 200, -, 150, 150, 32, 400
Taylor, John W., 1, -, -, 164, -, 1, -, 14, 4000, 500, 800, -, 360, 240, 32, 1500
Stevenson, Hewlet, 1, -, -, 8, -, -, -, 200, 10, 50, -, 20, -, -, 150
Massey, John W., 1, -, -, 60, 1, -, -, 6000, 150, 320, -, 100, 300, 36, 1500
Mark, Flavel C., 1, -, -, 2, 1, -, -, 4000, 100, 300, -, 15, 20, 2, 350
Mark, Flavel C., 1, -, -, -, -, 8, -, -, -, -, -, -, -, -, 20
Hockster, Benjamin, 1, -, -, 28, -, -, -, 300, 25, 50, -, -, -, -, -
Saulsbury, E. H., 1, -, -, 53, -, -, -, 5300, 400, 300, -, 100, 150, -, 1200
McGilligan, George, 1, -, -, 9, -, -, -, 900, 25, 150, -, -, -, - 20
Murphy, William, 1, -, -, 18, -, -, -, 2500, 50, 300, -, 45, -, -, 400
Boynton, Charlott, 1, -, -, 80, -, 26, -, 2000, 20, 75, -, -, -, -, 150
Harrington, Jonathan, 1, -, -, 17, -, -, -, 2500, 75, 150, -, 2500, -, -, 250
Hoover, Samuel, 1, -, -, 50, -, 10, -, 5000, 125, 60, -, -, -, -, 700
Hoover, Samuel, 1, -, -, 75, -, 10, -, 2500, -, -, -, -, -, -, 75
Alexander, James, -, -, 1, 54, 1, 6, -, 2000, 175, 400, -, 75, 75, 22, 700
Satterfield, Emory, -, -, 1, 25, 1, -, 10, 1500, 100, 75, -, -, -, -, 200
Spence, Pierson, 1, -, -, 5, -, -, -, 120, -, -, -, -, -, -, 25
Reynolds, Rebecca, 1, -, -, 200, 1, -, 120, 5000, 100, 100, -, -, 70, 20, 40
Carpenter, George, 1, -, -, 100, 1, 30, -, 3000, 100, 200, -, 29, -, -, 200
Murrey, Stansbury, 1, -, -, 63, 1, -, 10, 2000, 150, 200, -, 70, -, -, 450
Pritchell, Warner, -, -, 1, 160, 1, 78, -, 6000, 175, 400, -, 190, 190, 36, 1500
Anderson, John W., 1, -, -, 70, -, 1, 10, -, 3700, 100, 300, -, 50, -, -, 750
Anderson, George W., 1, -, -, 150, 1, 25, -, 8000, 400, 300, -, 120, 300, 72, 1300
Baker, Isabella, 1, -, -, 100, 1, 40, -, 4300, 150, 250, -, 125, 121, 36, 1200
Ells, Daniel S., 1, -, -, 27, 1, -, -, 3000, 200, 300, -, 35, 50, 10, 400
Perkins, Joseph, -, -, 1, 80, -, 20, -, 5000, 500, 750, -, 112, 150, 36, 1000
Driggis, John, -, -, 1, 80, -, 40, -, 1100, 50, 50, -, 50, 25, -, 200
Reed, Jehu M., 1,-, -, 160, 1, 10, -, 10000, 2000, 1000, -, 200, 3000, 36, 6000
Reed, Jehu M., 1, -, -, 100, 1, -, -, 3000, -, -, -, 50, 200, 36, 800
Wilson, James, -, -, 1, 20, -, -, -, 1200, 25, 5, -, -, -, -, 100

Emerson, John P., 1,-, -, 69, -, 20, -, 4000, 300, 200, -, 33, 100, -, 700
Black, James, -, -, 1, 95, -, 50, -, 4000, 220, 275, -, 70, 30, 10, 650
Downham, John W., 1,-, -, 100, 1, 10, -, 5500, 150, 400, -, 120, 175, 36, 1000
Davis, William, -, -, 1, 50, -, 20, -, 2500, 75, 150, -, 25, -, -, 100
Johnson, Thomas, -, -, 1, 100, 1, 20, -, 3000, 100, 200, -, 21, 50, 10, 300
McIlvaine, William S., 1, -, -, 95, 1, -, -, 8000, 150, 350, -, 100, 100, 20, 1500
Bailey, Edward, 1, -, -, 220, 1, 10, -, 16000, 400, 1650, -, 330, 250, 72, 2500
Bailey, Edward, 1, -, -, 9, -, -, -, 300, -, -, -, -, -, -, 66
Cullen, William, 1, -, -, 15, 1, -, -, 800, 50, 100, -, 20, -, -, 300
Cullen, William, 1, -, -, 105, 1, 35, -, 4000, 150, 150, -, 30, -, -, 250
Maxwell, James, 1, -, -, 20, 1,-, -, 2000, 200, 150, -, 40, 90, 36, 700
Edwards, Philemon, 1, -, -, 25, 1, -, -, 1500, -, -, -, 23, -, -, 50
Harrington, Samuel, 1, -, -, 40, -, 2, -, 2000, 30, 200, -, -, 15, -, 700
Satterfield, William, 1, -, -, 200, 1, -, 100, 8000, 125, 800, -, 100, 50, -, 1000
Hodson (Hudson), Robert, 1, -, -, 44, 1, -, -, 3000, 300, 300, -, 40, 10, -, 1000
Hargadine, Robert, 1, -, -, 6, -, -, -, 2000, 100, 300, -, 50, 235, 36, 700
Roe, Jonathan, -, -, 1, 100, -, 50, -, 5000, 200, 500, -, 90, 200, 36, 1000
Dehority, George, -, -, 1, 12, -, -, -, 300, 25, 700, -, 60, 150, -, 300
Cole, Manlove, -, -, 1, 25, 1, -, -, 1000, 50, 1000, -, 20, 25, 10, 300
Spencer, Joshua, -, -, 1, 300, -, 5, -, 15000, 300, 800, -, 288, 300, 72, 1850
Roe, Samuel D., 1, -, -, 300, 2, 50, -, 15000, 300,800, -, 150, 100, 20, 1500
Mitten, James C., -, -, 1, 100, -, 15, -, 7000, 200, 600, -, 95, 100, 36, 900
Jones, Robert H., -, -, 1, 300, -, 150, -, 15000, 300, 1000, -, 235, 125, 36, 1600
Clifton, Jehu, -, -, 1, 120, 1, 15, -, 8000, 300, 800, -, 125, 200, 36, 1500
McIlvaine, Smith, 1, -, -, 17, -, -, -, 500, -, -, -, -, -, -, 25
Jones, Williard, -, -, 1, 133, 1, 50, -, 10000, 400, 800, -, 67, 90, 36, 1500
Simpson, William, 1, -, -, 95, -, 1, 5, -, 5000, 200, 300, -, 105, 125, 36, 600
Heistand, Levi, 1, -, -, 87, 1, 3, -, 8000, 400, 600, -, 86, 200, 36, 1000
Bidding, Wayne, 1, -, -, 165, -, 1, 56, -, 10000, 800, 900, -, 180, 250, 72, 1600
Evans, James H., -, -, 1, 120, 1, 30, -, 6000, 500, 1000, -, 200, 100, 36, 1100
Kelley, James, -, -, 1, 225, -, 80, -, 11000, 300, 700, -, 225, 100, 36, 1800
Milburn, Samuel, -, -, 1, 120, 1, 30, -, 8000, 100, 250, -, 88, 250, 36, 1000
Miten, Emma, 1, -, -, 55, 1, 20, -, 3000, 75, 300, -, 65, 25, -, 750
Rice, Charles, 1, -, -, 5, 1, 10, -, 6000, 500, 800, -, 100, 700, 36, 4000
Knapp, John W., 1, -, -, 40, 1, -, -, 4000, 250, 250, -, -, -, 100, 950
Cradick, John, 1, -, -, 100, 1, 50, -, 3000, 200, 200, -, 100, 100, 36, 500
Cradick,William, 1, -, -, 200, 1, 70, -, 6000, 500, 500, -, 150, 150, 36, 1200
Bradley, William, 1, -, -, 25, -, 7, -, 1000, 75, 150, -, 40, 25, -, 300
Boone, Joseph H., 1, -, -, 90, 2, 110, -, 3000, 200, 350, -, 120, 150, 36, 800
Harrington, Henry, -, -, 1, 165, 1, 40, -, 3000, 300, 250, -, 150, 250, 72, 1500
Harrington, Nathl., 1, -, -, 35, -, -, -, 1500, 75, 100, -, 75, -, -, 200
Clark, Patrick, -, -, 1, 47, -, -, -, 1500, 25, 50, -, 16, -, -, 250
Thomas, Zebulon H., -, -, 1, 90, 1, -, -, 4000, 150, 500, -, 50, 150, 36, 1200
Case, Joseph, 1, -, -, 18, 1, 20, -, 1000, 50, 100, -, 15, 20, -, 175
Case, William T., 1, -, -, 20, 1, -, -, 1000, 100, 100, -, 140, -, -, 300
Killen, George W., 1, -, -, 115, 1, 13, -, 4000, 400, 600, -, 150, 250, 36, 1200

Melvin, Riley, -, -, 1, 100, 1, -, 9, 4000, 150, 500, -, 135, 120, 36, 1200
Dickerson, John, -, -, 1, 100, -, -, -, 2500, 350, 390, -, 37, 75, 200, 800
Turner, Amos G., -, -, 1, 140, 1, 12, -, 9000, 500, 700, -, 200, 250, -, 2100
Charles, Warren, 1, -, -, 4, -, -, -, 200, 10, 70, -, -, -, -, 100
Harrington, Caleb, 1, -, -, 8, 1, -, -, 400, 15, 50, -, -, -, -, 200
Hitch, John, -, -, 1, 150, -, 50, -, 2500, 150, 300, -, 180, 225, 36, 1020
Davis, William, -, -, 1, 75, 1, 20, -, 1200, 150, 300, -, 4, 410, -, 350
Palmer, James, -, -, 1, 54, -, -, -, 4000, 80, 150, -, 60, 100, 36, 600
Rogers, Hezekiah, 1, -, -, 10, 1, -, -, 3500, 300,300, -, 25, 25, -, 350
Prattis, James, 1, -, -, 24, -, -, -, 800, 40, 100, -, 15, -, -, 100
Coarsey, Thomas B., 1, -, -, 80, -, -, -, 8000, 100, 500, -, 90, 100, 36, 700
Russel, Elias, 1, -, -, 160, -, 40, -, 8000, 100, 400, -, 120, 150, 36, 1000
Stevenson, Cornelius, -, -, 1, 100, -, 100, -, 6000, 100, 500, -, 75, 100, 36, 800
Whitaker, Peter, 1, -, -, 55, 1, -, 5, 5000, 100, 50, -, 60, 25, -, 400
Lollis (Sollis), Thomas, -, -, 1, 90, 1, 25, -, 4500, 200, 500, -, 150, 140, 36, 1000
Emory, Abner, -, -, 1, 60, -, 11, -, 3000, 200, 450, -, 120, 50, 20, 600
Emory, Thomas, -, -, 1, 95, -, 13, -, 4000, 350, 1200, -, 200, 200, 36, 2500
Emory, James, -, -, 1, 72, -, 10, -, 2200, 170, -, 200, -, 75, -, -, 400
Case, Charles C., 1, -, -, 106, 1, 22, -, 6000, 300, 1000, -, 154, 50, -, 1200
Clark, Henry, -, -, 1, 145, 1, 15, -, 7000, 200, 425, -, 100, 250, 36, 1200
Rogers, Cyrus P., 1, -, -, 81, -, 40, -, 8000, 1000, 400, -, 200, 500, 72, 1400
Brown, Joseph, -, -, 1, 250, 1, 37, -, 7000, 500, 700, -, 294, 450, 72, 2500
Perry, William, 1, -, -, 55, 1, -, 11, 1200, 100, 125, -, 50, 125, -, 450
Godwin, Tarnza (Tamza), -, -, 1, 150, -, 30, -, 10000, 200, 500, -, 90, -, -, 900
Conyer, Henry, -, -, 1, 100, -, 250, -, 10000, 500, 300, -, 110, 10, -, 900
Moore, Thomas W., -, -, 1, 90, -, -, -, 4000, 200, 250, -, 150, 100, 36, 1100
Fountain, Stephen, 1, -, -, 20, -, 7, -, 500, 60, 100, -, 15, -, -, 150
Lowber, Benjamin, 1, -, -, 7, -, -, -, 200, 25, 75, -, 5, 5, -, 100
Hall, Jno. W., 1, -, -, 250, -, 9, -, 15000, 300, 1000, -, 400, 450, 32, 2000
Emory, John, -, -, 1, 140, 1, 8, -, 2500, 250, 400, -, 170, 157, 10, 1100
Jackson, James, 1, -, -, 28, 1, -, -, 1000, 100, 150, -, 50, -, -, 350
Jackson, William, 1, -, -, 30, 1, -, -, 1000, 70, 200, -, 75, -, -, 450
Stout, Edmund, 1, -, -, 25, -, -, -, 700, 100, 100, -, -, 25, -, 200
Williams Risdon, 1, -, -, 160, -, 15, -, 3000, 300, 400, -, 170, 75, 15, 700
Harmer, David, -, -, 1, 200, 1, 20, -, 4000, 250, 215, -, 75, 100, 20, 850
Johnson, Charles H., -, -, 1, 150, 1, 10, -, 3500, 100, 240, -, 70, 28, 15, 1500
Lewis, Reuben, 1, -, -, 32, -, -, - 700, 50, 50, -, -, -, -, 200
Lowber, Peter, 1, -, -, 18, -, -, -, 400, 20, 30, -, -, -, -, 150
McIlvain, Thomas, -, -, 1, 88, 1, 3, -, 4000, 155, 275, -, 150, 200, 32, 1100
McIlvain, McIlroy, 1, -, -, 175, 1, -, -, 17500, 400, 700, -, 98, 300, 32, 2450
Conner, Jno. J., 1, -, -, 200, 1, 20, -, 9000, 225, 600, -, 175, 170, 32, 1800
Melvin, Jno. G., 1, -, -, 50, 1, 25, -, 2000, 100, 200, -, 50, -, -, 450
Manlove, James, -, -, 1, 150, 1, 10, -, 6200, 175, 200, -, -, 110, -, 1250
Willis, Jonathan N., 1, -, -, 200, 2, 46, -, 12600, 500, 1400, -, 275, 2000, -, 5200
Hickey, William H., -, -, 1, 140, 2, 42, -, 11500, 450, 1100, -, 225, 1000, -, 2800
Green, James, 1, -, -, 100, -, 32, -, 6500, 250, 1100, -, 135, 200, 72, 1300

Sipple, James, 1, -, -, 60, -, 20, -, 2000, 50, 100, -, 60, -, -, 250
Cahall, Dr. Thomas, 1, -, -, 46, -, -, -, 1200, 50, 200, -, -, -, -, 400
Parkinson, Robert, 1, -, -, 80, 1, -, -, 4000, 100, 220, -, 45, 50, -, 850
Quillan, James, -, -, 1, 110, -, 10, -, 5500, 270, 425, -, 150, 95, 32, 1100
Harrington, Alexander, 1, -, -, 40, -, 45, -, 2200, 200, 75, -, 40, 25, -, 300
Lowber, Rachael A., 1, -, -, 28, -, 11, -, 5000, 100, 100, -, 30, -, -, 400
Anderson, James B., 1, -, -, 200, -, -, -, 1000, 70, 100, -, 75, 125, 20, 250
Burton, Thomas B., 1, -, -, 16, 1, -, -, 900, 150, 125, -, 115, 87, 10, 370
Carter, Henry L., 1, -, -, 40, -, -, -, 3500, 450, 300, -, 100, 150, 36, 500
Burcharel, Joseph, -, -, 1, 100, -, 60, -, 7300, 250, 300, -, 176, 600, -, 1475
Sullivan, William, -, -, 1, 200, 1, 45, -, 10000, 400, 700, -, 220, 150, 64, 1850
Brown, William, 1, -, -, 115, -, 12, -, 8000, 150, 400, -, 85, 120, 36, 1400
Darby, Samuel W., -, -, 1, 250, -, 250, -, 15000, 1500, 2730, -, 300, 1020, 144, 4044
Bradley, Josiah, 1, -, -, 25, -, 15, -, 1200, 30, 100, -, 30, 15, -, 150
Bamborough, Thomas, 1, -, -, 35, 15, 11, -, 1000, 10, 75, 150, -, -, 52, 125
Grennell, Myron, 1, -, -, 25, 5, 30, -, 1500, 35, -, 25, -, -, -, 400
Russ, Charles, 1, -, -, 5, -, 1 30, -, 800, 10, 150, -, -, -, -, 28
Holder (Holden), Noah, 1, -, -, 75, -, 80, -, 3000, 25, 300, -, 50, 150, 52, 413
Donaphon, William, -, 1, -, -, -, -, -, -, -, -, 4, -, -, -, -, -
Jarrel, James, -, 1, -, 70, -, 10, -, 3000, 100, 200, -, 130, 20, 4, 315
Peck, Charles, 1, -, -, 40, -, 60, -, 2000, 200, 450, 200, 150, 350, 42, 600
Peck, S. L., 1, -, -, 150, -, 250, -, 10500, 600, 1200, 300, 100, 500, 210, 2000
Boys, Mary, -, 1, -, 40, -, 10, 20, 1500, 20, 150, 25, 22, -, -, -
Townsend, William 1, -, -, 160, -, 40, -, 2000, 200, 500, -, 90, 40, 5, 1000
Townsend, Elias, 1, -, -, 125, -, 80, -, 2500, 225, 500, 100, 100, 80, 20, 1500
Murrey, William, -, -, 1, 10, -, -, -, 150, 5, 60, -, -, -, -, 46
Shillcuts, William, -, -, 1, 20, -, -, -, 400, 20, 100, -, -, -, -, 20
Minner, William, 1, -, -, 15, -, 3, -, 300, -, -, -, -, -, -, 25
Conner, A. B., 1, -, -, 6, -, -, 5, 6000, -, 300, -, -, 25, 52, 200
Holden, Alexander, -, -, 1, 40, -, 10, -, 3000, 30, 400, -, -, -, -, 620
Needles, David, 1, -, -, 45, -, 20, -, 1500, 100, 300, -, 50, 5, 1, 400
Heirs of N. Bostick, 1, -, -, 3, -, 5, -, 300, -, -, -, -, -, -, 15
Moore, Archibald, 1, -, -, 4, -, -, -, 400, 1, 20, 10, -, 10, -, 60
Coldwell, Anderson, 1, -, -, 200, 10, 70, -, 10000, -, 200, 500, -, 200, 350, 52, 1500
Mason, George, -, -, 1, 125, 15, 60, -, 8000, 100, 200, 50, 500, 200, 37, 500
Fredell, Jacob, 1, -, -, 150, 30, 100, -, 7000, 1000, 1000, -, 100, 400, -, 2000
Fredell, Charles, 1, -, -, 130, 8, -, -, 3000, -, 250, 40, -, -, -, 200
Mason, John C., -, -, 1, 100, -, 70, -, 6000, 225, 400, -, 100, -, -, 75
Bastin, George, 1, -, -, 40, 5, 100, -, 4000, 100, 125, 66, 56, 40, -, 495
Clump, Chdiles, 1, -, -, 100, -, 100, -, 4000, 200, 250, -, 30, 160, 52, 740
Heyd, John Jr., 1, -, -, 120, -, 80, -, 7000, 500, 1200, 250, 175, 400, 52, 1500
Martin, James, -, -, 1, 90, -, 15, -, 3000, 50, 150, -, 30, -, -, 100
Dill, Alfred, 1, -, -, 70, -, 30, -, 1500, 150, 250, 40, 125, 10, 5, 175
Sparklin, Saulsbery, -, -, 1, 120, -, 200, -, 8000, 125, 300, -, 63, 190, 52, 1000

Fredell, Frederic, -, -, 1, 85, 18, 30, -, 5000, 200, 800, -, 25, 200, 40, 800
Needles, James R., 1, -, -, 20, -, 80, -, 3000, 100, 200, 15, -, -, -, 200
Fleming, Samuel, -, -, 1, 70, -, 12, -, 2000, 100, 350, -, 60, 25, 8, 300
Harrass, Isaac, -, -, 1, 15, -, 45, -, 800, 10, 125, -, 30, -, -, 75
Carson, M., -, -, 1, 56, -, 56, -, 3000, 100, 150, -, 30, -, -, 350
Needles, Catharine, 1, -, -, 42, -, 42, -, 1600, 20, 40, -, 40, -, -, 130
Carson, George W., 1, -, -, 9, -, 1, -, 500, 3, 85, 90, -, -, -, 90
Downham, Thos., -, -, 1, 35, -, 140, -, 90, 10, 70, -, 21, -, -, 210
Davos (Davis), Leanau (Laurence), 1, -, -, 55, -, 64, -, 1150, -, 20, -, -, -, -, -
Lynch, Riley, -, -, 1, 90, -, 25, -, 3500, 100, 300, -, 80, -, -, 275
Graham, George, 1, -, -, 100, -, 50, -, 2000, 40, 140, -, 38, -, -, 200
Johnson, George S., -, -, 1, 150, -, 90, -, 1004, 50, 200, -, 30, -, -, 371
Marker, Ferry, 1, -, -, 25, -, 25, -, 1000, 25, 150, -, -, -, -, 175
Walker, Charin, -, -, 1, 40, -, 7, -, 1000, 25, 50, -, -, -, -, 175
Right, James, 1, -, -, 50, -, 85, -, 1000, 206, 100, -, 30, -, -, 150
Beddle, W. James, 1, -, -, 40, -, 200, -, 2000, -, 15, -, -, -, -, 50
Kemp, Isac, -, -, 1, 20, -, 56, -, 500, 15, 40, -, -, -, -, -
Marker, Isac, 1, -, -, 16, -, 24, -, 1000, 35, 300, -, -, -, -, 75
Lemans, Lewis, 1, -, -, 35, -, 20, -, 1700, 10, 140, -, 30, -, -, 210
Furbish (Furnisk), C., 1, -, -, 40, -, -, -, 1500, 100, 150, -, 88, -, -, 200
Dod, Siles, 1, -, -, 10, -, 6, -, 1000, 25, 180, -, -, -, -, 100
Fleming, William, 1, -, -, 14, -, 7, -, 600, 2, 15, -, -, -, -, 35
Franer, William, 1, -, -, 130, -, 76, -, 3000, 125, 550, -, 40, 200, 40, 600
Shutts, Samuel, 1, -, -, 95, -, 25, -, 2000, 125, 250, -, 65, -, -, 500
Cook, Gis___, -, -, 1, 160, -, 40, -, 4000, 300, 600, -, 90, 100, 36, 1000
Dill, Edward, -, -, 1, 110, -, 25, -, 2000, 200, 400, 100, 77, 150, 36, 1090
Gooden, Benjamin, 1, -, -, 125, -, 125, -, 4000, 500, 900, -, 132, 160, -, 800
Scott, Francis P., -, 1, -, 95, -, 100, -, 2000, 50, 50, -, 60, 2, -, 100
Bosket (Basket), James, 1, -, -, 49, -, 13, -, 1500, 25, 125, 6, 12, -, -, 200
Coldwell, E. S., 1, -, -, 80, -, 40, -, 4000, 250, 500, 50, 85, -, -, 1000
Spencer, Perry, -, 1, -, 30, -, -, -, 275, -, -, -, 37, -, -, 90
Lorres, Robert, -, 1, -, 45, -, 20, -, 240, 10, 250, -, -, -, -, 120
Spencer, Perry Jr., 1, -, -, 50, -, 70, -, 1000, 30, 100, -, 6, -, -, 250
Mohler, John G., -, -, 1, 150, -, 100, -, 4000, 500, 500, -, 130, -, -, 500
Greewell, John, 1, -, -, 250, -, 100, -, 3500, 150, 300, 50, 54, 216, 75, 500
Clough. W. L., 1, -, -, 100, -, 46, -, 3000, 250, 400, -, 90, 450, -, 600
Reynolds, J. Robert, 1, -, -, 110, -, 100, -, 4500, 400, 1000, 50, 110, 500, 150, 2200
Reynolds, R. Franklin, 1, -, -, 50, -, 15, -, 1300, 50, 150, 200, 20, -, -, 150
Grills(Grells), Thomas, 1, -, -, 60, -, 20, -, 1500, 50, 150, -, 20, -, -, 100
Merieth, Jacob, 1, -, -, 60, -, 59, -, 2000, 100, 250, 20, 30, -, -, 250
Killen, Edward, 1, -, -, 45, -, 23, -, 600, 30, 160, -, 44, -, -, 110
Jarrell, Joseph, 1, -, -, 44, -, 14, -, 1200, 100, 200, 25, -, 15, -, 150
Marker, James, 1, -, -, 20, -, 20, -, 60, 10, 200, -, -, -, -, 100
Anderson, Hester, 1, -, -, 53, -, 12, -, 2600, -, 25, -, -, -, -, 500
Roch, Robert, 1, -, -, 20, -, 9, -, 1000, -, 75, 50, 30, -, -, 200

Benson, Mathew, 1, -, -, 19, -, 1, -, 800, 20, 125, -, -, -, -, 200
Ridder, Hiram B., 1, -, -, 58, 1, 25, -, 4600, 50, 150, -, 40, 75, 10, 50
Dill, Abner, 1, -, -, 150, -, 125, -, 3000, 255, 625, 150, 75, 30, -, 420
Burres, George, 1, -, -, 14, -, 11, -, 250, 50, 80, -, -, -, -, 100
Minner, Thomas, 1, -, -, 12, -, 11, -, 1500, 50, 145, -, 20, -, -, 230
Minner, Samuel, 1, -, -, 18, -, -, -, 600, 225, 200, 10, 10, -, -, 185
Lator, Charles, 1, -, -, 30, -, 35, -, 1000, 50, 280, 120, 30, -, -, 250
Sipple, Uriah, 1, -, -, 75, -, 75, -, 4000, 200, 500, 50, 40, -, -, 225
Chemberg, Alexandria, 1, -, -, 44, -, 36, -, 1400, 75, 1100, 100, 66, -, -, 250
Kemp, Jacob, 1, -, -, 15, -, 40, -, 800, 40, 60, 400, 10, -, -, 52
Kemp, Mathew, 1, -, -, 55, -, 4, -, 1500, 100, 140, -, 22, -, -, 250
Holden, Andrew, 1, -, -, 125, -, 105, -, 3500, 150, 500, -, 120, 125, 34, 400
Warren, Sarah, 1, -, -, 30, -, 319, -, 8000, 175, 830, 40, 15, 250, 67, 900
Dill, Catherine, 1, -, -, 70, -, 3, -, 1300, 40, 175, 55, 64, -, -, 350
Sipple, R. J., 1, -, -, 30, -, 33, -, 1200, 20, 250, -, 22, -, -, 175
Kooper, James, 1, -, -, 11, -, 5, -, 250, 70, 30, -, -, -, -, 50
Merieth, Peter L., 1, -, -, 180, 10, 10, -, 7000, 400, 900, -, 130, 500, 70, 3700
Kee, Thomas, -, -, 1, 70, 25, 40, 5, 2000, 50, 250, 20, 30, 50, -, 250
Clark, Nathan, 1, -, -, 100, -, 40, -, 1500, 600, 400, -, 120, -, -, 500
Harrington, Thomas, 1, -, -, 100, -, 35, -, 3000, 375, 425, -, 86, 100, -, 400
Mill, Milard, -, -, 1, 120, -, 48, -, 1500, 150, 390, -, 44, -, -, 100
Rouss, Thomas H., 1, -, -, 29, -, 30, -, 1500, 35, 190, -, 33, -, -, 325
Sapp, James, -, -, 1, 140, -, 60, -, 7000, 500, 1000, -, 84, 100, 26, 680
Kemp, Mathew, 1, -, -, 55, -, 31, -, 1500, 75, 216, -, 66, -, -, 430
Lister, Samuel, 1, -, -, 80, -, 56, -, 1500, 100, 225, -, -, 75, 40, 350
Heard, John, 1, -, -, 75, -, 97, -, 1800, 200, 375, 25, 84, -, -, 275
Miller, Ezekiel, 1, -, -, 15, -, 2, -, 350, 35, 200, -, 11, -, -, 135
Clark, John, -, -, 1, 63, -, 4, -, 1200, 50, 265, -, 48, -, -, 387
Pierson, William, -, -, 1, 60, -, 20, -, 1200, 53, 200, -, -, -, -, 300
Mereith, John R., 1, -, -, 150, -, 100, -, 3000, 300, 600, -, 66, -, -, 650
Greenlley, William, 1, -, -, 66, -, 10, -, 1500, -, 125, -, 35, -, -, 210
Sipkle (Syskle), Robert K., -, -, 1, 120, -, 15, -, 3500, 100, 550, -, 115, 200, 40, 600
Longlellow(Longfellow), Thomas, 1, -, -, 100, -, 41, -, 2500, 150, 375, 25, 125, 150, 75, 575
Heard, Peter, 1, -, -, 45, -, 39, -, 500, 20, 300, -, 22, -, -, 130
Kemp, Thomas, 1, -, -, 35, -, 18, -, 450, 15, 60, -, -, -, -, 30
Moor, George, 1, -, -, 130, -, 30, -, 7000, 300, 600, 50, 200, 340, 37, 1000
Melvin, Hinson, 1, -, -, 38, -, 6, -, 700, 24, 150, 20, 46, -, -, 180
Clump, Charine, 1, -, -, 65, -, 5, -, 1500, 50, 200, -, -, -, -, 290
Reed, John C., 1, -, -, 90, -, 99, -, 5000, 200, 475, -, 45, -, -, 260
Dill, Samuel C., 1, -, -, 20, -, 1, -, 3000, 100, 200, -, 11, 50, -, 190
Jarves, Thomas, -, -, 1, 60, -, 15, -, 1500, 50, 150, -, 22, -, -, 75
Cohee, James, 1, -, -, 75, -, 25, -, 1600, 50, 75, -, -, -, -, 206
Smith, David, -, -, 1, 11, -, -, -, 400, 50, -, -, -, -, -, 250
Clark, Mary, 1, -, -, 40, 20, 10, -, 300, 10, 50, -, 6, -, -, 50

Tribit, Wesley, 1, -, -, 8, -, 84, -, 600, 175, 100, -, 8, -, -, 60
Cooper, Samuel, -, -, 1, 70, -, 120, -, 2000, 50, 200, -, 22, -, -, 250
Sylvester, Samuel, 1, -, -, 130, -, 18, -, 1400, -, 600, 1000, 88, -, -, 300
Jump, William C., 1, -, -, 75, -, 28, -, 2000, 50, 200, -, 56, -, -, 440
Edwards, William, 1, -, -, 150, -, 150, -, 3000, 177, 300, -, 60, -, -, 600
Draper, John L., 1, -, -, 12, -, 11, -, 600, 800, 150, -, 10, -, -, 91
Heard, James, 1, -, -, 100, -, 28, -, 1000, 245, 300, -, 32, -, -, 260
Voshel, William, 1, -, -, 50, -, 57, -, 1000, -, 25, -, 42, -, -, 275
Pinder, Edward, 1, -, -, 75, -, 85, -, 3000, 24, 150, -, -, -, -, 100
Reynolds, George, 1, -, -, 50, -, 51, -, 1000, 10, 100, -, -, -, -, 75
Warren, Andrew, 1, -, -, 45, -, 8, -, 1800, 24, 300, -, 23, -, -, 300
Antney, Samuel, 1, -, -, 15, -, 15, -, 150, 10, 30, -, -, -, -, 40
Draper, Henry, 1, -, -, 30, -, 10, -, 400, 10, 10, -, -, -, -, 100
Vickry, S. William, 1, -, -, 40, -, 10, -, 600, 20, 100, -, -, -, -, 100
Hughes, William, 1, -, -, 200, -, 116, -, 2000, 250, 200, -, -, -, -, 280
Cooper, John W., 1, -, -, 200, -, 100, -, 15000, 2000, 1200, 500, 300, 200, -, 3600
Heard, James K., -, -, 1, 260, -, 100, -, 1500, 125, 300, -, 55, -, -, 200
Dill, Peter, 1, -, -, 100, -, 39, -, 1500, 15, 250, -, 67, -, -, 350
Earvin, John, 1, -, -, 100, -, 40, -, 1400, 100, 300, -, 75, -, -, 190
Dill, Samuel, 1, -, -, 80, -, 45, -, 1000, 100, 250, -, 43, -, -, 225
Greer, Thomas, 1, -, -, 300, -, 25, -, 8000, 250, 1200, -, 325, -, -, 1800
Gullett, Thomas, -, -, 1, 150, -, 75, -, 1500, 30, 200, -, 22, -, -, 375
Hopkins, Bateman, 1, -, -, 140, -, 20, -, 1600, 150, 1150, -, 130, -, -, 850
Schalinger, John, 1, -, -, 160, 15, 40, -, 5000, 600, 1000, -, 200, 200, 45, 1200
Welch, Thomas, -, -, 1, 120, -, 140, -, 4000, 140, 700, -, 115, -, -, 500
Cooper, Samuel B., 1, -, -, 70, -, 20, -, 1800, 125, 400, -, 54, 75, -, 600
Shaurlot, ___, 1, -, -, 55, -, 25, -, 1200, 50, 250, -, -, -, -, 200
Heard, Benjamin, 1, -, -, 70, -, 25, -, 2000, 75, 250, -, 75, 75, -, 466
Kane, Ruben, 1, -, -, 85, -, 18, -, 1500, 10, 100, -, -, -, -, 200
Dill, Ander, 1, -, -, 75, -, 30, -, 1500, 40, -, 200, 33, -, -, 200
Wiatt, John, -, -, 1, 75, -, 60, -, 1500, 50, 300, -, -, -, -, 500
Moore, Benjamin, -, -, 1, 75, -, 25, -, 2250, 175, -, -, 45, -, -, -, 410
Scott, Levy, -, 1, -, 45, -, 35, -, 800, 40, -, -, 24, -, -, 135
Knox, Thomas, -, -, 1, 50, -, 33, -, 2000, 50, 200, -, 30, -, -, 250
Burnitt, James K., 1, -, -, 240, 40, 220, -, 16000, 600, 2000, -, 300, 1200, 2000, 2500
Godwin, John W., 1, -, -, 25, -, 12, -, 800, 5, 100, -, 30, -, -, 100
Leach, Dr. John, 1, -, -, 120, -, 25, -, 3000, 120, 300, -, 65, 100, 52, 1200
Shabinger, Louis, 1, -, -, 100, 20, 50, -, 6000, 200, 800, 50, 125, 200, 120, 4000
Wilds, David S., 1, -, -, 60, 40, -, -, 8000, 250, 400, -, 100, 200, 35, 500
Wilds, David S., 1, -, -, 151, -, -, -, 12000, 600, 1000, 100, 150, 1100, 6, 1940
Wilds, David S., -, -, 1, 151, 15, 26, 8, 7000, 200, 350, 50, 100, 250, -, 500
Temple, James D., -, -, 1, 140, 15, -, -, 11000, 500, 760, 50, 80, 500, -, 1400
Cooper, Wm. H., 1, -, -, 78, -, 7, -, 6000, 200, 400, 130, 170, 750, -, 775
Minner, Stephen, -, -, 1, 240, 30, 60, -, 18000, 1000, 1400, -, 185, 350, 8, 3470

Williams, James, 1, -, -, 130, 60, 20, 30, 20000, 400, 1700, 100, 100, 700, 80, 3500

Brayman, Theodore F., 1, -, -, 20, 15, -, -, 1000, 25, 100, -, -, 100, -, 200

Berry, Scott, -, -, 1, 30, -, 3, -, 500, 25, 100, -, -, 100, -, 100

Bailey, Thomas E., 1, -, -, 300, 70, 100, -, 30000, 1500, 1500, 100, 400, 1300, 160, 5000

Williams, Samuel, -, -, 1, 150, 25, 25, -, 7000, 300, 800, 25, 100, 180, 20, 920

Moore, James, 1, -, -, 75, 30, 25, -, 3000, 100, 500, 50, 112, 340, -, 1200

Morgan, John, -, -, 1, 40, 8, 12, -, 1200, 50, 200, -, -, -, -, 260

Durham, Robert J., 1, -, -, 15, -, -, -, 300, 75, -, -, -, -, -, 240

Moore, Andrew H., 1, -, -, 24, 5, -, -, 3500, 15, 150, 100, 25, 350, 65, 700

Ford, Martin B., 1, -, -, 5, -, 32, -, 740, 100, 100, -, -, -, -, 50

Ford, Martin B., 1, -, -, 75, 5, 1, -, 2000, 300, 600, 50, 50, 5, -, 400

Moore, Wm. H. & John, 1, -, -, 80, 20, 40, -, 7000, 150, 500, -, 63, -, -, 1000

Wilds, Michael D., 1, -, -, 39, 22, -, -, 2500, 100, 300, -, -, 75, 8, 300

Clark, George, 1, -, -, 15, -, 8, -, 300, 75, 100, -, -, -, -, 200

Moore, John, 1, -, -, 110, 25, -, -, 12000, 500, 1100, -, 200, 800, -, 2500

Cole, John, -, -, 1, 68, 15, 10, -, 1000, 100, 200, -, -, 38, 6, 600

Moore, Thomas, 1, -, -, 60, 11, 20, -, 2500, 100, 500, 20, 50, 300, -, 500

Carney, John, -, 1, -, 9, 3, -, -, 360, 20, 54, -, -, -, 2, 100

Cole, John, 1, -, -, 15, -, 2, -, 350, -, -, -, -, -, 2, 200

Roush, Charles, 1, -, -, 19, -, 3, -, 700, 100, 200, -, 5, -, 1, 250

Smith, John M. Dr., 1, -, -, 9, -, -, -, 3500, 25, 250, -, -, 125, -, 300

Davis, Lewelyn T., -, -, 1, 40, 3, 27, -, 4000, 50, 250, 3, 49, 100, -, 275

Sanders, George, 1, -, -, 6, 3, -, -, 270, 50, 100, -, -, -, -, 140

Stroud, John H., 1, -, -, 250, 50, 72, -, 12000, 500, 2000, 50, -, 400, -, 1000

Hardcastle, Peter, -, -, 1, 300, 50, 100, -, 30000, 1200, 2500, -, 500, 560, 60, 4500

Dear, John, 1, -, -, 5, -, 2, -, 280, 150, 100, -, -, -, -, 100

Griffith, Nathaniel H., 1, -, -, 100, 8, 80, -, 5000, 340, 500, 50, 140, 165, 10, 830

Hutchinson, Samuel Jr., 1, -, -, 109, 16, 4, -, 9000, 500, 200, 50, 120, 500, -, 1603

Wright, George D., -, -, 1, 125, 25, 30, -, 10000, 500, 800, 50, 200, 400, 20, 3000

Nowell, Sipple James, 1, -, -, 100, 25, 10, 17, 10000, 500, 500, -, 125, 350, 10, 1500

Farris, Calvin, -, -, 1, 80, 16, 75, -, 2000, 200, 225, 25, 150, 200, 16, 400

Boyd, John _., 1, -, -, 120, 31, 35, -, 9000, 400, 650, -, 225, 450, -, 1400

Morgan, William, 1, -, -, 18, 8, -, -, 1000, 100, 200, -, 15, -, -, 200

Bryan, Jacob, 1, -, -, 30, 7, -, -, 1200, 100, 200, 20, 3, 110, -, 400

Frasure, Alexander, -, -, 1, 125, 23, 25, -, 6000, 125, 500, 24, 90, 90, -, 850

Surgeon, William, 1, -, -, 113, 34, 20, -, 10000, 500, 1000, 20, 150, 900, 1, 2300

Price, David Jones, -, -, 1, 40, 60, 10, -, 8000, 300, 500, -, 42, 300, 10, 3700

Boyer, Wm. M., -, -, 1, 12, 14, 2, -, 1000, 50, 300, -, -, -, -, 300

Stewart, James, -, -, 1, 60, 45, 15, -, 3000, 75, 300, -, 90, 100, 10, 800

Phillips, Gardiner W., -, -, 1, 111, 9, 5, -, 4000, 50, 500, -, -, 175, -, 300

Rhine, James, -, -, 1, 5, -, 3, -, 200, 50, 75, -, -, -, -, 75

Webb, Isaac C., -, -, 1, 120, 20, 15, -, 4000, 75, 400, 20, 56, -, -, 525

Garrison, Jedeiah, -, -, 1, 185, 25, 10, -, 10000, 300, 800, 40, 150, 250, 8, 2000

Voshell, Obadie, -, -, 1, 186, 20, 7, -, 9000, 800, 1200, 50, 240, 400, 10, 1700
Hill, Charles, -, 1, -, 80, 3, 40, -, 3200, 100, 300, -, 50, -, -, 350
Downs, Nehemia C., 1, -, -, 154, 6, 17, -, 8000, 50, 500, -, 125, 200, 20, 1700
Evans, Richard, 1, -, -, 7, -, -, -, 500, 10, 100, -, -, -, -, 125
Sanders, Perry, -, 1, -, 12, -, -, -, 300, 10, 30, -, -, -, -, 75
Ennis, George R., -, 1, -, 54, -, -, -, 1500, 40, 200, 25, 33, 30, 6, 100
Attix, Thomas, 1, -, -, 120, 21, 110, -, 5000, 250, 700, -, 130, 250, 12, 1400
Attix, Thomas & Benjamin, -, -, 1, 50, 12, 18, -, 2000, 75, 325, -, 50, 30, 6, 575
Hawkins, Anderson, 1, -, -, 14, -, 19, -, 500, 20, 75, -, -, -, -, 175
Prior, John, 1, -, -, 8, -, 4, -, 200, 25, 75, -, -, -, -, 75
Jones, William, 1, -, -, 69, 17, 40, 30, 3000, 155, 350, 50, 100, 175, 10, 468
Jones, Loadaman Samuel, -, -, 1, 100, 2, 40, 5, 2000, 50, 240, -, -, -, -, 500
Jones, John H., -, -, 1, 20, 2, -, -, 1400, 75, 150, -, -, -, -, 500
Lafferty, Oliver J., -, -, 1, 100, 30, 110, -, 8000, 150, 750, 25, 142, 250, 10, 300
Warren, George H., -, -, 1, 50, 10, 11, -, 2000, 75, 150, 10, -, -, -, 500
Cook, James A., 1, -, -, 80, 15, 19, -, 2500, 100, 300, -, 100, -, -, 700
Downs, David O., 1, -, -, 118, 15, 40, 10, 8000, 150, 950, -, 125, 258, 10, 1280
Graham, Robert, 1, -, -, 32, 1, 2, -, 2500, 40, 125, 2, 28, 86, -, 250
Hazel, Wm. B., 1, -, -, 85, 2, 20, -, 5000, 300, 630, 25, 100, 175, 10, 625
Hazel, George W., 1, -, -, 94, -, 1, 22, -, 2500, 100, 200, -, 75, -, -, 475
Carney, William A., 1, -, -, 34, -, -, -, 1200, 150, 200, 50, 45, 180, 25, 350
Murphy, William, -, -, 1, 100, -, 60, -, 4500, 150, 800, 50, 99, 150, 5, 900
Lewis, Reese, 1, -, -, 60, 1, 3, -, 2500, 50, 180, -, 15, 50, 10, 150
Ward, George, 1, -, -, 10, 4, 16, -, 600, 50, 175, -, -, -, -, 130
Downs, Edwin B., 1, -, -, 96, 4, 47, -, 4000, 150, 500, -, 150, 250, 10, 835
Jacobs, George W., -, -, 1, 38, 15, 27, -, 2500, 100, 340, -, 28, -, 4, 450
Rash (Rush), John, 1, -, -, 55, 6, 12, -, 2000, 75, 180, -, 28, 130, -, 260
Jacobs, William H., 1, -, -, 30, 5, 51, -, 2900, 75, 310, -, 7, 20, 4, 375
Taylor, Samuel, -, -, 1, 110, 18, -, -, 3800, 100, 550, 6, 125, 150, 8, 950
Good, John M., -, -, 1, 85, 8, -, -, 2000, 50, 340, 350, 80, 450, 840
Craig, John, -, -, 1, 140, 10, 150, -, 5000, 250, 685, -, 126, 265, -, 906
Clark, David S., 1, -, -, 60, 20, 24, -, 5000, 300, 1200, -, 200, 500, 20, 2000
Clark, John H., 1, -, -, 90, 16, 25, -, 5000, 200, 600, 12, 112, 300, 20, 1500
Durham, Hewett, -, -, 1, 50, 10, 30, -, 1800, 75, 200, -, 30, 25, -, 225
Leverage, Nathaniel, 1, -, -, 95, 6, 35, -, 4000, 215, 386, -, 105, 150, 8, 600
Concealor, Edward, -, -, 1, 50, 4, 50, -, 2000, 100, 465, -, 63, -, -, 380
Taylor, Wm. W., -, -, 1, 147, 14, 14, -, 10000, 500, 1185, 36, 125, 525, 10, 2650
Guessford, James, 1, -, -, 80, 12, 40, -, 5000, 500, 900, 50, 162, 200, 10, 1200
Croker, Moses, -, -, 1, ½, 10, -, 1000, 75, 175, -, -, -, -, 300
Burrows, Francis M., 1, -, -, 125, 12, 38, -, 8750, 250, 645, 15, 212, 523, 8, 1260
McDonald, James F., 1, -, -, 59, 6, 60, -, 3000, 175, 400, -, 30, -, -, 300
Forum, John W., 1, -, -, 50, -, 14, -, 1200, 75, 175, -, -, -, -, 150
Bourke, Emlie P., 1, -, -, 30, 12, 25, -, 3000, 75, 372, 140, 15, 306, 20, 570
Hutton, John, 1, -, -, 100, 50, 60, -, 2500, 350, 600, -, 117, 50, 10, 600
Cook, George, -, -, 1, 500, 4, 194, -, 15000, 700, 1200, -, 260, 500, 20, 2500
Robinson, Robinson, -, -, 1, 30, -, 70, -, 1000, 40, 100, -, 20, 30, -, 150

Remley, John, 1, -, -, 40, -, 107, -, 2550, 75, 100, -, -, -, -, 175
Deats, George, -, -, 1, 50, -, 125, -, 2500, 25, 50, -, 22, -, -, 150
Underwood, Nathan T., 1, -, -, 163, 45, 15, -, 12150, 500, 98, 7, -, 180, 250, -, 2000
Hepburn, John P., -, -, 1, 80, 40, 180, -, 3000, 125, 400, -, 52, 140, -, 400
Sylvester, John T., -, -, 1, 100, -, 40, -, 2500, 100, 240, -, 30, 25, 5, 450
Burrows, Thomas, -, -, 1, 50, 10, -, -, 2000, 100, 300, -, 42, -, -, 261
Hutchinson, William, 1, -, -, 199, 8, 30, -, 10000, 400, 1480, 50, 262, 776 80, 2986
Grady, John, 1, -, -, 65, 10, 40, 20, 2800, 25, 250, 11, 33, 70, -, 520
Voshell, John, -, -, 1, 240, -, 40, -, 2500, 150, 600, -, 100, 150, -, 800
Knotts, Frederick, -, -, 1, 145, 65, 40, -, 10000, 350, 876, -, 100, 600, 15, 300
Shahan, Mary, 1, -, -, 66, 24, 15, -, 1800, 400, 600, -, 200, 162, -, 700
Gerker, Henry, -, -, 1, 68, 25, 30, -, 2500, 100, 230, -, 42, 200, -, 300
Numbers, Charles, 1, -, -, 46, 6, 31, -, 1600, 25, 260, -, 39, 150, -, 400
Numbers, John, 1, -, -, 26, 6, 6, -, 1000, 75, 265, -, 50, 25, 5, 275
Shorts, Miles T., 1, -, -, 30, 3, 3, -, 600, 50, 100, -, -, -, -, 175
Claytor (Clayton), William, 1, -, -, 63, -, 15, -, 650, 25, 75, -, -, -, -, 150
Husbands, Howell B., 1, -, -, 105, 20, 25, -, 6000, 150, 600, 80, 275, 10, 1030
Moore, Robt. P., -, -, 1, 120, 40, 40, -, 5000, 125, 450, -, 150, 250, 10, 1300
Benson, Elijah, -, -, 1, 100, 6, 96, -, 3000, 25, 400, -, 90, -, -, 460
Lofland, Asa, -, -, 1, 69, 6, 12, -, 1500, 50, 200, -, 56, 50, -, 200
Loose, Peter, 1, -, -, 355, 130, 20, -, 16000, 400, 1100, 50, -, 300, -, 5600
David, John T., -, -, 1, 100, 20, 20, -, 7000, 300, 600, 27, 75, 250, 10, 1400
Shorts, James T., -, -, 1, 75, -, 23, -, 1500, 60, 240, 100, 41, -, -, 150
David, Nemiah, -, -, 1, 130, 20, 100, -, 2000, 50, 200, -, 50, 108, 4, 425
Hutchinson, Samuel Sr., 1, -, -, 100, 14, 35, -, 5000, 450, 780, -, 79, 148, 3, 500
Shorts, William Henry, -, -, 1, 100, 2, 44, -, 2500, 75, 250, -, -, 24, -, 424
Shorts, James M., 1, -, -, 71, 15, 143, -, 4500, 100, 400, -, 54, 150, -, 2, 600
Shorts, Charles S., -, -, 1, 121, 4, 25, -, 4560, 100, 1000, 5, 32, 10, -, 633
Shorts, Henry, 1, -, -, 105, 10, 35, -, 3750, 150, 610, -, 40, 200, 5, 1355
Shorts, Thomas R., -, -, 1, 30, 6, 50, -, 1720, 75, 170, -, 20, -, -, 250
Clark, Walter, -, -, 1, 100, 2, 84, -, 3000, 100, 300, -, 85, 320, 10, 525
Clark, John W., -, -, 1, 120, 7, 53, -, 4000, 125, 800, -, 75, 350, 8, 720
Hazell, James P., 1, -, -, 82, 8, 10, -, 4000, 250, 500, 12, 80, 50, -, 882
David, Enoch, 1, -, -, 85, 16, 62, -, 4000, 50, 510, -, 56, 144, 3, 850
Shorts, Joshua A., 1, -, -, 80, -, 39, -, 5000, 150, 500, -, 102, 150, 12, 100
Martin, William, 1, -, -, 60, 4, 61, -, 2500, 100, 390, -, 90, 100, 2, 400
Taylor, Thomas R., 1, -, -, 127, 15, 12, -, 5000, 100, 610, -, 58, 100, 4, 1200
David, James F., -, -, 1, 100, -, 60, -, 1500, 150, 427, 25, 100, 110, 1, 325
Hazell, Matthew, 1, -, -, 230, 20, 40, -, 5000, 150, 800, 40, 120, 100, -, 1150
Jones, William S., 1, -, -, 143, 12, 10, 8, 5190, 175, 795, -, 40, -, -, 1273
Anthony, Henry, 1, -, -, 70, 30, -, -, 7500, 150, 500, -, 54, 360, 40, 1468
Stevens, Thomas, 1, -, -, 30, -, -, -, 1000, 50, 140, -, -, -, -, 150
Lapham, Isaac, 1, -, -, 119, 83, -, 28, 20000, 400, 800, 50, 225, 1400, 200, 7000
Wilds, James F., 1, -, -, 110, 39, 15, -, 8000, 250, 650, 15, 170, 275, 3, 600

Struts, Edward, 1, -, -, 100, 16, 3, -, 53, 55, 500, 470, -, 45, 216, -, 460
Reed, Austin, -, -, 1, 90, 30, 3, -, 2500, 100, 550, -, 30, -, -, 600
Spruance, David L., 1, -, -, 150, 28, 30, -, 10000, 250, 1165, -, 150, 488, 6, 1840
Goldsborough, John F., -, -, 1, 90, 30, -, -, 500, 370, 1725, -, 120, 162, 2, 1150
Reese, David, 1, -, -, 140, 30, -, -, 8800, 350, 1085, -, 220, 546, 52, 2172
Anthony, James W., 1, -, -, 90, 30, 15, -, 10800, 500, 1000, 25, 100, 300, 4, 1000
Middleton, Benjamin F., -, -, 1, 170, 30, 100, -, 18000, 700, 1200, -, 240, 500, 6, 2400
Bedwell, John M., -, -, 1, 90, 6, 25, -, 6000, 150, 300, -, 75, 162, 10, 600
Bedwell, John D., -, -, 1, 110, 48, 15, -, 8000, 175, 625, -, 230, 180, 8, 1470
Macy, Peter, -, -, 1, 97, 3, 20, -, 5000, 100, 635, 25, 75, 150, 12, 461
Hoffecker, William D., -, -, 1, 100, 70, 16, -, 10000, 500, 1035, -, 200, 7, 2, 1980
Jones, David, 1, -, -, 100, 10, 40, -, 3400, 100, 400, -, 50, -, -, 500
Buckmaster, William W., -, -, 1, 196, 23, 17, -, 10000, 300, 800, -, 156, 1500, 150, 2664
Stevens, William A., 1, -, -, 38, 3, -, -, 2400, 50, 415, -, 102, 275, 23, 630
Stevens, William, 1, -, -, 100, 25, 30, -, 12000, 175, 710, -, 78, 482, 4, 1400
David, William R., -, -, 1, 120, 40, 20, -, 20000, 200, 935, 6, 86, 496, 8, 2390
Maree, William W., -, -, 1, 110, 15, 12, -, 6500, 300, 300, -, 60, 300, -, 1000
Maberry, Thomas J., -, -, 1, 90, 6, 10, -, 2500, 125, 350, 50, 80, 130, 10, 800
Dawson, Josiah, -, -, 1, 75, 2, 100, -, 2200, 50, 250, -, 60, 8, 4, 250
Maberry, Henry S., -, -, 1, 100, -, 25, -, 2000, 60, 250, -, 96, 135, 2, 300
Clayton, William, -, -, 1, 100, 15, 55, -, 5000, 200, 250, -, 30, -, -, 400
Rheins, Charles L., -, -, 1, 60, 8, 12, 3, 1600, 50, 125, -, 35, -, 6, 375
Robinson, William, -, -, 1, 60, -, 40, -, 1000, 75, 150, -, 30, -, 2, 100
Ellis, John, 1, -, -, 18, -, -, 4, 500, 50, 125, -, -, 25, -, 100
Hillyard, William, 1, -, -, 70, 3, 27, -, 800, 150, 250, -, -, -, -, 250
Heinholt (Reinholt), George, -, -, 1, 92, 23, 15, -, 4000, 200, 1000, -, 56, 10, 4, 1200
Coe, Russel, 1, -, -, 25, 45, 20, -, 7000, 250, 700, -, 270, 486, 52, 2250
Myers, William, -, -, 1, 30, 4, 60, -, 1000, 25, 200, -, -, -, -, 100
Collins, John M., -, -, 1, 140, 15, -, -, 7000, 250, 700, -, 125, -, -, 1300
Hudson, John P., 1, -, -, 100, 50, -, -, 12000, 350, 725, -, 80, 320, 50, 2000
Casperson, William, -, -, 1, 20, -, -, 8, 3000, 100, 119, -, 14, 65, -, 100
Rees, Thomas A., -, -, 1, 172, 25, 45, -, 12000, 250, 800, 50, 300, 536, 13, 2550
Donovan, James A., -, -, 1, 150, 50, -, -, 12000, 50, 600, 20, 60, 500, 12, 3270
Hill, Robert, 1, -, -, 100, 42, 59, -, 7000, 100, 540, -, 100, 500, 20, 2000
Crawford, James S., 1, -, -, 335, -, 40, -, 15000, 500, 800, 60, 90, 375, 10, 1400
Ross, Charles G., 1, -, -, 100, 35, 40, -, 5000, 150, 1000, 50, 80, 350, 10, 1800
Stevens, Wesley, 1, -, -, 95, 5, 35, -, 5000, 150, 600, 52, 52, 175, 2, 800
Harris, George, -, -, 1, 144, 16, 110, 40, 5000, 200, 200, -, 40, 175, 8, 1150
Finn, John, 1, -, -, 140, 15, 42, 35, 8000, 300, 800, -, 43, 400, -, 1400
Ga___, Georgiana, -, -, 1, 120, 40, 40, -, 12000, 500, 1100, -, 155, 500, 10, 2700
Myers, John W., 1, -, -, 60, 10, -, -, 2500, 300, 400, -, 75, -, 3, 800
Pratt, Henry S., 1, -, -, 130, 20, 25, -, 10000, 300, 800, 50, 125, -, -, 1800
Burrows, Ebenezer, -, -, 1, 275, 25, 75, -, 10000, 400, 800, 50, 240, 400, 10, 1250

Pratt, John W., 1, -, -, 185, 16, 37, -, 9000, 150, 900, 100, 135, 250, 20, 2400

Savin, John A., 1, -, -, 80, 10, 36, -, 5000, 200, 500, 60, 30, 175, 5, 650

Maberry Thomas, 1, -, -, 30, -, 30, -, 1200, 50, 150, -, 32, -, 4, 250

Davis, George, -, -, 1, 330, 40, 230, -, 12000, 350, 1100, -, 130, 435, 6, 2400

Jackson, Andrew N., -, -, 1, 54, -, 26, 10, 1000, 75, 150, -, 30, -, 8, 220

Hurlock, Isaac P., -, -, 1, 60, -, 65, -, 1000, 80, 275, -, 30, 84, -, 200

Blackiston, Samuel, 1, -, -, 80, 1, 47, -, 1000, 50, 450, -, 60, 10, 4, 500

Yates, William E., 1, -, -, 50, 6, 44, -, 1700, 50, 200, 100, 32, 150, 2, 400

Whittington, John, -, -, 1, 44, 1, 116, -, 1500, 25, 100, -, -, 30, 10, 105

Rheins, John G., 1, -, -, 54, 6, 40, -, 2000, 25, 175, -, -, 8, 2, 300

Johnson, William, -, -, 1, 60, 30, 40, 10, 3000, 100, 450, -, 56, 100, 2, 715

Arnold, George G., 1, -, -, 100, 60, 14, 6, 16000, 300, 600, 120, 120, 484, 33, 1800

Dulin, Risdon, -, -, 1, 133, 12, 20, -, 6000, 300, 500, -, 150, 165, 4, 860

Numbers, Charles, 1, -, -, 92, 2, 2, -, 4000, 250, 600, -, 64, 135, 26, 800

Voshell, John W., -, -, 1, 140, -, 10, -, 1800, 75, 350, -, 150, 125, -, 150

Hurlock, Benjamin F., -, -, 1, 100, 96, 67, -, 12000, 500, 1300, -, 18, 500, 52, 310

West, Benjamin M., -, -, 1, 160, 20, 15, -, 800, 500, 1000, -, 160, 330, 2, 2100

Fluharty, William (overseer), 1, -, -, 40, 14, -, -, 1600, 100, 300, 35, 27, 100, -, 330

Bilderback, John L., 1, -, -, 155, 55, 15, -, 15000, 500, 1000, 128, 138, 500, 152, 2200

Massey, Henry P., 1, -, -, 110, 20, -, -, 8000, 200, 600, 10, 225, 500, 52, 1500

Bailey, John C., 1, -, -, 135, 50, 22, 20, 12000, 400, 800, 50, 200, 800, 52, 2450

Wright, Eben, -, -, 1, 90, 2, -, 3, 2500, 25, 250, -, 78, 75, 2, 600

Bell, John F., -, -, 1, 125, 20, 30, -, 5000, 300, 850, -, 20, 225, -, 1050

Downs, Joseph C., 1, -, -, 75, 8, 75, -, 3000, 200, 400, 40, 96, 150, 52, 925

Downs, Fletcher, -, -, 1, 50, 10, 45, -, 1200, 100, 300, -, 60, 50, 24, 340

Weller, Silas L., -, -, 1, 130, 30, 5, -, 3500, 75, 315, 60, 40, -, -, 350

Urian (Urias), Howard, T., 1, -, -, 122, 12, 10, -, 2200, 250, 375, -, 48, 150, 10, 775

Sammons, James, -, -, 1, 125, 55, 42, -, 5000, 250, 310, -, 37, 75, 23, 1000

Lecount, Charles, 1, -, -, 148, 12, 3, -, 3500, 250, 575, -, 70, 145, 2, 1000

Sinex, Isaac, -, -, 1, 80, 75, 10, -, 4000, 400, 400, -, 80, 148, 20, 1700

Carron, James, -, -, 1, 250, 2, 38, -, 2500, 50, 200, -, 60, 150, 5, 500

Graham, John W., 1, -, -, 100, 30, 50, -, 5000, 200, 500, 50, 60, 600, 7, 1400

Graham, John W., 1, -, -, 18, 30, 22, -, 1000, -, -, -, -, 150, -, 1000

Dickson, John, 1, -, -, 25, -, 10, -, 500, 75, 200, -, 25, -, -, 175

Tignor, John W., -, -, 1, 161, 30, 50, -, 10000, 450, 755, 100, 97, 310, 8, 1414

Mahan, Jacob B., -, -, 1, 90, 3, 60, -, 3500, 200, 900, 40, 90, 210, 10, 500

This agricultural census was filmed from original records in the Delaware State Archives in Wilmington Delaware by the Delaware State Archives Microfilm office.

There are some 100 columns of information on each individual. Only the head of household is addressed. I have chosen to use only the first 16 columns of the information because I feel that this information best illustrates the wealth of the individuals. In columns 2-4 the number 1 is used as an indicator only and represents no numeric value. These are shown below:

1. Name of Individual
2. Owns
3. Rents for Fixed Money Rental
4. Rents for Shares of Products
5. Improved Tilled Land
6. Improved Permanent Meadows, Pastures, Orchards, and Vineyards
7. Unimproved Woodland & Forest
8. Unimproved including Old Fields, and Not-Growing Wood
9. Farm Value of Farm including Land, Fences, and Buildings
10. Farm Value of Farming Implements and Machinery
11. Value of Livestock
12. Fences—Cost of Building and Repairing in 1879
13. Cost of Fertilizers Purchased in 1879
14. Labor—Amount Paid for Wages for Farm Labor during 1879, including Value of Board
15. Labor—Weeks paid labor in 1879 including Value of Board
16. Estimates Value of all Farm Productions (sold, consumed or on hand for   1879)

Thus, the numbers following the names represent columns 2, 3, 4, 5, 6, 7, 8, 9, 10,11, 12, 13, 14, 15, 16.

The following symbol is used to maintain spacing where information in a column is left blank (-). This symbol is used where letters, names or numbers are not legible (_).   This census was split between two reels of microfilm. The first reel I transcribed columns 1-11 & 16, the second I transcribed 1-16.

In Columns 2-4, the number 1 is used to indicate the column choice. The number 1 has no numeric value in these three columns—it is simple used as an indicator.

Talley, Elizabeth E., -, 1, -, 35, ½, -, 4, -, -, 230, -, -, -, -, -
Berry, John, 1, -, -, 46, 2, 2, 3, 6500, 1500, 470, -, 70, 84, 28, 1000
Teal, Benjamin, -, 1, -, 11, -, -, -, 1800, -, 90, -, -, -, -, 200
Talley, Thomas L. Jr., 1, -, -, 70, -, 17, -, 8000, 269, 440, -, 25, 220, -, 800

Talley, Henry, 1, -, -, 30, -, 10, -, 4000, 150, 294, -, 40, 40, 6, 600
Campbell, David, 1, -, -, 80, 7, -, -, 10000, 400, 275, 50, -, 338, 52, 1100
Bird, Lewis, 1, -, -, 65, 1, -, -, 6000, 250, 280, -, -, 220, 52, 600
Forwood, Valentine, 1, -, -, 40, -, 17, 11, 6500, 200, 200, 100, -, -, -, 684
Dougherty, George, 1, -, -, 60, -, 20, 13, 7440, 200, 1000, -, -, 100, 16, 1000
Hand, Rachel, 1, -, -, 31, -, -, -, 3000, 200, 825, 100, 100, -, -, 600
Haughey, Peter, -, 1, -, 33, -, -, 22, 3500, 75, 320, -, -, -, -, 220
Pierce, Brandley, -, 1, -, 25, -, -, 10, 35, -, 210, -, 10, -, -, 100
Squibb, Lewis, 1, -, -, 28, -, -, 12, 430, -, 370, 500, -, -, -, 400
Day, John, 1, -, -, 16, -, -, 4, 2500, 125, 100, -, 16, -, -, 85
Derickson, William R., 1, -, -, 120, -, 30, 5, 15000, 400, 1000, -, 38, 700, -, 2000
Webster, Clark, 1, -, -, 75, -, 20, 5, 10000, 1500, 1052, -, -, -, -, 2000
Webster, Mary & Martha, 1, -, -, 5, -, -, -, 1000, 100, 150, -, -, 8, -, 150
Webster, Isaac, 1, -, -, 40, -, 5, -, 8000, 200, 600, -, -, -, -, 800
Talley, Thos., 1, -, -, 81, -, 14, 5, 12000, 500, 400, -, 120, 220, 39, 1200
Pierce, Joseph M., 1, -, -, 63, -, 3, 10, 5000, 300, 800, -, -, 200, -, 1000
Talley, Lewis F., -, -, 1, 55, -, 6, 4, 5000, 300, 600, -, -, -, -, 800
Bonsall, George, -, 1, -, 55, -, -, 15, 7000, 150, 400, -, 190, 150, -, 1000
Talley, Norris W., -, 1, -, 23, -, 2, -, 5000, 300, 350, -, 100, 280, 52, 800
Talley, Charles, 1, -, -, 36, -, 4, -, 4000, 250, 450, -, 37, -, -, 800
Talley, Penrose R. (Est.), 1, -, -, 50, -, 10, 30, 7000, 200, 495, -, 50, 25, 4, 1000
Miller, Penn, 1, -, -, 17, -, 5, 5, 2500, 200, 280, -, 50, 50, 32, 500
Mousley, Edward, 1, -, -, 10, -, -, 5, 840, 70, 35, -, -, -, -, 80
Wiltbank, Charles, 1, -, -, 30, -, 10, 8, 3000, 25, 220, -, 42, 25, 4, 300
Clark, Humphrey, 1, -, -, 40, -, 5, 5, 5000, 300, 720, -, 58, -, -, 700
Barlow, Henry M., 1, -, -, 100, -, -, 8, 9000, 400, 1400, -, 50, 220, 52, 1200
Day, Thomas R., 1, -, -, 90, -, 14, -, 8000, 500, 1000, -, 100, 100, 52, 1500
Pierce, Ezra, 1, -, -, 42, -, -, -, 3000, 200, 300, -, -, 200, 52, 1200
Nicholson, Emory, 1, -, -, 16, -, -, -, 2500, 100, 300, -, -, -, -, 500
Talley, Baldwin, 1, -, -, 57, -, 3, -, 5500, 300, 600, -, -, 120, -, 700
Talley, John, -, 1, -, 111, 12, 7, -, 8000, 350, 1007, -, 45, 400, -, 1500
Ebright, Zachariah, 1, -, -, 50, -, 40, 10, 7500, 75, 3500, -, -, 220, -, 800
Weer, William, 1, -, -, 45, -, 24, -, 4000, 300, 380, -, 80, -, -, 1000
Blackwell, John, 1,-, -, 3, -, -, -, 600, 100, 200, -, -, 100, -, 200
Kirk, Lewis, 1, -, -, 33, -, 4, 3, 3000, 100, 200, -, 15, 3, -, 250
Pierce, James C., 1, -, -, 11, -, -, -, 1700, 100, 100, -, -, 5, -, 200
Nicholson, Rebecca, 1, -, -, 31, -, 8, 4, 3000, 170, 500, -, -, -, -, 500
Foulk, Nelson C., -, 1, -, 20, -, 2, -, 4000, 150, 300, -, -, 50, -, 500
Talley, Robert, 1, -, -, 5, -, -, -, 2000, -, 45, 60, -, 5, -, 60
Wilkinson, James, 1, -, -, 25, -, 15, -, 3000, 300, 200, 50, 60, 45, -, 400
Foulk, William R., -, -, 1, 43, -, 3, 16, 5000, 100, 565, 50, 63, -, -, 800
Pierce, Sue H., 1, -, -, 41, -, 11, 4, 7000, 200, 400, 100, -, -, -, 600
Talley, James W., -, 1, -, 49, -, 1, -, 7000, 500, 500, -, 20, 240, -, 1000
Walston, William, 1, -, -, 50, -, -, -, 6500, 300, 850, -, -, 175, -, 1000
Warren, Matthew, 1, -, -, 25, -, 5, -, 3000, 100, 300, -, -, -, -, 400
Talley, Rachel A., 1, -, -, 7, -, 1, -, 3000, 125, 100, -, 25, 120, -, 500

Phillips, Lewis, -, 1, -, 20, -, 13, -, 2500, -, 1000, -, -, -, -, 300
Wilson, Hannah F., 1, -, -, 52, -, 8, -, 6000, 300, 610, -, 30, -, -, 1000
Pierce, Alfred, 1, -, -, 36, -, 8, -, 6000, 150, 372, -, -, -, -, 800
Cassady, Peter, 1, -, -, 48, -, 12, -, 5500, 75, 658, -, -, -, -, 600
Talley, Peter, 1, -, -, 22, -, 5, -, 4000, 200, 300, -, 50, 100, 25, 600
Miller, Charles, -, 1, -, 73, -, 7, -, 10000, 2000, 910, -, 30, 220, -, 1200
Weldin, Stephen G., 1, -, -, 152, 6, 60, -, 21850, 1000, 1180, 100, 147, 200, -, 1000
Paschall, Charles, 1, -, -, 60, 15, 5, -, 7000, 300, 1040, 100, -, 300, -, 1000
Meggenson, John, -, -, 1, 300, -, 50, -, 52500, 1200, 2900, -, 168, 830, -, 3000
Webber, John W., 1, -, -, 5, -, -, -, 3000, 100, 150, -, -, 18, -, 200
Weldin, Joseph, 1, -, -, 13, -, -, 5, 2500, 100, 400, -, -, -, -, 500
Allmond, John T., 1, -, -, 36, -, 4, -, 6000, 200, 495, 40, 40, 60, -, 700
Null, William W., -, 1, -, 2, -, 2-, 2500, -, -, -, -, -, -, -
Forwood, Joseph, -, -, 1, 61, 1, -, -, 6000, 300, -, -, 30, 190, -, 1000
Forwood, John, 1, -, -, 100, 8, 25, -, 10000, 800, 1600, -, 300, 150, -, 2050
Hanby, Joseph, 1, -, -, 89, 33, 4, -, 17000, 500, 1000, 55, 100, 100, -, 2160
Beeson, Edward, 1, -, -, 35, 5, 50, -, 8000, 100, 236, -, -, 100, -, 500
Mahaffy, Henry C., -, -, 1, 40, -, 13, -, 5600, 400, 700, -, -, 150, -, 1000
Mahaffy, Henry C., 1, -, -, -, 5, -, -, 700, -, -, -, -, -, -, 80
McCall, Lewis, 1, -, -, 10, -, -, -, 6000, 150, 125, -, -, -, -, 600
Beeson, John S., 1, -, -, 92, -, 4, -, 8000, 100, 560, 23, 125, -, -, 1000
Weldin, Mary & Beeson, Jane, 1, -, -, 3, -, -, -, 2000, -, -, 50, -, -, -, -
Weldin, William R., 1, -, -, 34, 5, -, -, 3900, 100, 160, 20, -, 75, -, 200
Weldin, Monroe & Anna M., 1, -, -, 62, -, -, -, 12000, -, 500, -, -, -, -, -
Clair, Jesse, 1, -, -, 5, -, -, -, 700, -, -, -, -, 10, -, 100
Talley, Lewis, 1, -, -, 2, -, -, -, -, -, -, -, -, -, -, -
Veale, William H., 1, -, -, 60, -, 5, -, 7000, 600, 967, 104, 100, 500, -, 1200
Rambo, Thomas, 1, -, -, 60, -, 5, -, 6000, 100, 245, -, 36, 20, -, 600
Habbart, Joseph, -, 1, -, 5, -, -, -, 1500, 100, 160, -, -, -, -, 164
Pierce, William H., 1, -, -, 14, -, -, -, 2500, 100, 250, 50, 10, -, -, 300
Veale, John F., 1, -, -, 88, -, 12, -, 7000, 300, 485, 15, 54, 100, -, 1000
Bird, William 1, -, -, 91, -, 14, -, 10500, 500, 600, -, 46, 37, -, 1500
Veale, George J., 1, -, -, 75, -, 25, -, 12000, 600, 665, -, -, 523, -, 1800
Leach, James, 1, -, -, 6, -, -, -, 3000, 100, 120, 20, -, -, -, 200
Lodge, Isaac N., 1, -, -, 125, -, 25, -, 16000, -, 200, -, 100, -, -, -,
Loughead, John, -, 1, -, 125, -, 25-, 16000, 800, 1045, -, -, 456, -, 2000
Sharpe, Jesse, 1, -, -, 210, -, 15, -, 20000, 300, 1000, -, 45, -, -, 1600
Mouseley, Isiah D., 1, -, -, 4, -, 2, -, 1500, 150, 205, -, 9, -, -, 260
Foulk, Thomas H., -, 1, -, 20, -, 5, -, 4500, 100, 284, -, -, -, -, -
Perkins, Daniel B., 1, -, -, 20, -, -, 4, 5000, 150, 487, -, 180, 44, -, 650
Bird, Harry C., 1, -, -, 60, 5, -, 10000, 400, 800, -, -, 240, -, 1200
King, Ether, -, 1, -, 125, 4, 50, -, 8950, 700, 1063, 60, 175, 207, -, 1541
Harvey, Jeramiah, -, -, 1, 70, -, 30, -, 10000, 300, 815, 100, -, 200, -, 1258
Meggenson, John, 1, -, -, 50, 3, 7, -, 13000, 850, 495, -, 5, 100, -, 600
Cole, Humphrey, -, 1, -, 45, -, 5, -, 7000, -, 400, -, -, 100, -, 800

Lodge, William P., 1, -, -, 115, 10, 25, -, 22500, 100, 2115, 50, 175, 700, -, 2500
Lodge, William C., 1, -, -, 14, -, -, -, 7000, 200, -, -, -, 360, -, 400
Bigger, George, -, -, 1, 95, -, 5, -, 15000, 150, 1000, 100, 31, 260, -, 1500
Lodge, Isaac W., 1, -, -, 28, -, 4, -, 3200, 200, 377, -, 15, 135, -, 550
Lodge, Clark, -, 1, -, 35, -, 5, -, 4000, 250, 500, 60, -, 50, -, 900
Johnson, John W., -, -, 1, 39, 1, 3, -, 10000, 3000, 300, 8, -, 200, -, 1000
Landerman, Isaac, 1, -, -, 65, 8, 3, -, 7600, 300, 1000, 50, 300, 300, -, 1600
Weldin, Mary, -, 1, -, 101, 3, 40, -, 20000, 592, 1200, -, -, 364, -, 1800
Martine, William P., -, -, 1, 65, 5, -, -, 10000, 300, 700, -, -, 100, -, 1000
Guest, J. Henry, 1, -, -, 42, -, 3, -, 8000, 300, 650, 50, -, 100, -, 1000
Conrow, Joseph D., 1, -, -, 35, 5, 6, 14, 2500, 200, 400, 50, 2200, 500, -, 1750
Wilkinson, Thomas, -, 1, -, 7, -, -, -, 3000, 30, 140, -, -, 4, -, 100
Weld, Albert H., 1, -, -, 14, -, -, 6, 8000, 100, 100, -, 50, 400, -, 900
Beeson, Ann & Mary J., 1, -, -, 115, -, 25, -, 15000, 125, 515, 50, 20, 440, -, 1000
Lippinscott, James P., 1, -, -, 147, 20, 8, -, 20000, 300, 1731, 100, 100, 800, 50, 1500
Closs, Thomas K., -, 1, -, 51, 4, 25, -, 16000, 300, 800, 50, 45, 100, -, 943
Lower, Charles B., 1, -, -, 37, 10, 3, -, 12000, 600, 850, -, 50, 500, -, 630
Pierce, Clark, 1, -, -, 65, -, 20, 20, 6000, 200, 700, 10, -, 24, -, 600
Oskins, John W., 1, -, -, 38, -, 15, -, 3800, 200, 385, 62, 8, 132, -, 600
Grubb, Isaac N., 1, -, -, 116, -, 34, -, 16000, 400, 1345, 200, 100, 690, -, 2200
Talley, Curtis M., -, 1, -, 41, -, 20, -, 6000, 500, 512, 13, 35, 262, -, 1392
Booth, Isaac L., -, 1, -, 11, -, 1, -, 2500, 100, 150, -, 18, 23, -, 230
Clair, Lewis, -, -, 1, 20, -, 35, 20, -, -, 95, -, -, -, -, 400
Wood, Jeramiah, 1, -, -, 5, -, -, -, 2800, -, 193, -, -, -, -, -
Orr, Robert, 1, -, -, 78, -, 6, -, 8400, 400, 1000, -, 175, 300, -, 1300
Phillips, William, 1, -, -, 40, 10, -, -, 7500, 600, 650, 100, -, 175, -, 1000
Guest, Joseph, 1, -, -, 35, 5, -, -, 8000, 200, 497, -, -, -, -, 800
Mahony, Michael B., 1, -, -, 869, -, -, 16000, 700, 960, 90, 45, 1030, -, 2058
Perkins, Christine, 1, -, -, 37, 5, -, -, 12000, 400, 590, -, -, 73, -, 700
McSorley, Francis, 1, -, -, 25, -, -, -, 5000, 100, 475, 25, -, -, -, 700
Talley, William A., 1, -, -, 40, -, 13, -, 4500, 150, 300, 35, 40, 100, -, 500
Pierce, Stephen, 1, -, -, 11, -, 1, -, 2400, 50, 62, 20, -, -, -, 100
Hanby, Jacob K., 1, -, -, 52, -, 5, -, 4000, 200, 587, -, 57, 120, -, 1000
Hanby, Curtis C., 1, -, -, 22, -, 10, 3, 3500, 200, 3000, 12, 15, 1000, -, 525
Hanby, George W., 1, -, -, 62, -, 6, -, 6000, 200, 600, 45, 18, 25, -, 600
McAlaster, James, -, 1, -, 47, 3, 15, -, 6000, 100, 505, -, -, -, -, 1010
Chaytor (Clayton), Armina D., 1, -, -, 7, -, -, -, 3000, 50, 86, 18, -, 120, -, 340
Anderson, John H., -, 1, -, 15, -, 3, -, -, 50, 50, 1, -, 35, 243, -, 300
Taggart, John H., 1, -, -, 10, -, 2, -, 11025, -, 150, 100, 24, 600, -, 300
Bigger, Elwood, -, -, 1, 112, -, 55, -, 16700, 300, 1115, -, -, 24, -, 1500
Dixon, Taylor, 1, -, -, 36, 1, 2, -, 9000, 100, 300, -, 30, 300, -, 500
Grubb, James, 1, -, -, 57, -, -, -, 7000, 200, 3500, 23, -, 100, -, 500
Cloud, William, -, -, 1, 50, -, -, -, 10000, 300, 764, 75, 40, 200, -, 1000
Bird, Alfred, 1, -, -, 11, -, -, -, 6000, 10, 100, 20, -, -, -, 100
Edwards, Richard W., -, 1, -, 13, -, -, -, 3000, 200, 457, 7, -, -, -, 326

Forwood, Amor (Amos) G., 1, -, -, 81, -, 9, -, 18000, 300, 1000, 100, 16, -, -, 1500
Young, Thomas S., 1, -, -, 39, 8, 8, -, 27000, 200, 600, 200, 150, 720, 52, 500
Wistar, Isaac J., 1, -, -, 4, 3, -, -, 10000, 100, 600, -, 20, 400, 52, 200
Graff, Charles, 1, -, -, 3, 4, -, -, 25000, -, 120, 140, 6, 432, 52, -
Myers, William H. T., 1, -, -, 50, 5, 6, -, 12000, 150, 600, -, 200, 750, 52, 1500
Addicks, J. Edward, 1, -, -, 3, 4, -, -, 12000, 200, 150, 100, -, 300, 52, 200
Cresson, William P., 1, -, -, 9, 2, -, -, 6000, 10, -, -, -, -, -, -
Bush, Ellen, 1, -, -, 70, 10, 20, -, 11000, 500, 400, -, -, -, -, 2000
Sloan, Albert, 1, -, -, 3, 6, 3, -, 1500, 75, -, 30, 18, -, -, 200
Derley, Ellen E., 1, -, -, 21, -, 2, -, 15000, 150, 250, -, 40, 300, 52, 250
Cozzens, John, -, 1, -, 65, 30, 15, -, 33000, 400, 1500, 25, -, 240, 52, 2000
Casey, William H., -, 1, -, 155, -, 5, -, 25000, 400, 2000, 25, 30, 300, 52, 3000
Danzenbaker, Francis, -, -, 1, 55, 7, -, -, 3000, 200, 700, -, -, -, -, 1000
Vernor (Vernon), William G., -, 1, -, 40, -, 5, -, 10000, 400, 430, 50, 25, 200, 52, 1000
Ennis, Valentine R., -, 1, -, 50, -, -, -, 7000, 150, 723, 100, -, -, -, 1000
Ford, Franklin, 1, -, -, 115, 15, 30, -, 27000, 500, 1515, -, 35, 720, 52, 2500
Beauregard, Fred, -, 1, -, 65, 5, -, -, 10500, 200, 920, -, -, -, -, 800
Goodley, Thomas P. Jr., -, -, 1, 42, -, 3, -, 6600, 300, 350, -, 50, -, -, 900
Grubb, Richard, 1, -, -, 19, -, -, -, 3200, 200, 371, 150, -, -, -, 300
Prince, John M.C., 1, -, -, 65, -, 6, -, 8500, 200, 700, 50, 40, 258, 36, 1200
Primrose, Thomas L., -, 1, -, 160, 60, 60, -, 28000, 200, 945, 50, 200, -, 1205
Cloud, Lot, 1, -, -, 100, -, 50, -, 11250, 200, 1000, 100, 25, 450, 76, 1382
Petildemange, John S., 1, -, -, 95, -, 3, 22, 10000, 200, 880, 10, 72, -, -, 1405
Cloud, George, A., -, 1, -, 45, -, 28, -, 5000, 150, 500, 15, 40, 200, 52, 577
Pierce, Walter, 1, -, -, 73, -, 33, -, 7000, 125, 1475, -, 75, 475, -, 1257
Carpreseters (Goodpastur), Julia A. & Joseph, 1, -, -, 69, -, 25, -, 7000, 25, 50, 25, 30, -, -, 940
Hanby, Samuel, 1, -, -, 10, -, 14, 29, 8550, 200, 911, 100, 350, 400, -, 1000
Hanby, Samuel 1, -, -, 70, -, 20, -, 6750, -, 500, -, -, -, -, 800
Hanby, Alfred D., 1, -, -, 13, -, -, -, 4000, 50, 327, 50, 100, -, -, 203
Hanby, James G., 1, -, -, 75, -, 6, 19, 10000, 200, 683, 200, 166, 240, 52, -
Bunting, Joshua, 1, -, -, 40, 24, 6, -, 10000, 400, 736, 50, 50, 360, 52, 1000
Perkins, E. S., 1, -, -, 103, -, 4, -, 20400, 2500, 825, 100, 35, 577, 52, 2500
Mouseley, William R., -, -, 1, 93, 17, 70, -, 15000, 150, 989, 14, 10, 340, 52, 1000
Weldin, William P., 1, -, -, 86, 12, -, 4, 10200, 150, 640, 60, 40, -, -, 1000
Hamilton, John, -, -, 1, 60, -, 3, -, 9000, 75, 355, -, -, 264, 152, 800
Casey, Robert, 1, -, -, 68, -, 7, -, 7500, 350, 680, 25, 60, 20, -, 600
Casey, Robert, 1, -, -, 218, -, 7, -, 5000, -, -, -, -, -, -, 300
Del Beet Sugar Co., -, 1, -, 80, -, -, -, 16000, 300, 200, -, -, -, -, -
Hanby, Samuel W., 1, -, -, 35, -, -, 1, 4000, 150, 500, 25, 50, 203, 32, 500
Wilson, Hannah, 1, -, -, 11, -, 1, 2000, -, 60, -, -, -, -, 500
Day, John W., 1, -, -, 70, -, 40, -, 10000, 500, 800, 200, 100, 30, 2, 2200
Trimble, John, -, -, 1, 75, -, 6, -, 8000, 100, 695, 25, 50, 25, 2, 800
Furey, David G., 1, -, -, 8, -, -, -, 2000, -, 600, -, -, 40, 4, 200
Kellum, John, H., 1, -, -, 70, 4, 20, -, 10000, 100, 800, 70, 86, 200, 20, 1500

Graves, James, -, 1, -, 11, -, -, -, 2200, 25, 225, -, 20, 10, 1, 350
Hinkson, Minshall, 1, -, -, 55, 9, 24, -, 9000, 300, 500, 40, 75, 75, 6, 1500
Hunter, Joseph, -, -, 1, 70, -, 17, -, 6000, 200, 800, 50, 35, 35, 20, 1500
Chandler, John, 1, -, -, 9, -, -, -, 2000, -, 80, 25, -, -, -, 200
Ramsey, Hugh, 1, -, -, 110, -, 40, -, 10000, 400, 800, -, 125, 150, 20, 2000
Palmer, Moses, 1, -, -, 100, -, 35, -, 10000, 400, 536, 25, 80, 50, 8, 2000
Duffey, Neal, 1, -, -, 30, 10, 4, -, 4400, 75, 175, 10, 40, 100, 25, 600
Chandler, Jehu M., -, 1, -, 10, -, -, -, 1000, 25, 125, -, -, 15, 2, 200
Fraim, Benjamin, 1, -, -, 30, -, 7, 7, 7000, 150, 300, 5, 20, 300, 52, 600
Nickerson, Wm. T., -, -, -, 1, 100, -, 15, -, 17250, 350, 500, 35, 200, 400, 42, 1200
Rotthouse, William, 1, -, -, 50, -, 33, 10, 5000, 300, 1000, 100, 400, 1000, 150, 4000
Lynam, Atwood D., -, 1, -, 18, 3, 7, -, 4000, 100, 200, 15, 38, 20, 3, 1000
Talley, William, 1, -, -, 133, -, 7, -, 14000, 1000, 1500, 100, 190, 400, 80, 3308
Talley, Elihu D., -, -, 1, 100, 4, 30, -, 10000, 700, 1100, 125, 80, 455, 100, 2990
Leach, Charles, 1, -, -, 55, -, 10, 15, 8000, 175, 525, 35, 42, 6, 1, 800
Harkins, T. Smith, -, 1, -, 60, -, 36, 12, 5700, 50, 175, 20, 56, 200, 52, 600
Anderson, James, -, 1, -, 65, -, 35, 15, 10000, 275, 700, 25, 40, 6, 1, 900
Righter, Margaret, 1, -, -, 39, -, 33, -, 7000, 175, 500, 40, 45, 60, 20, 700
Hornby, James E., 1, -, -, 53, -, 17, -, 7000, 200, 540, 25, 55, 7, 1, 1391
Husbands, John, 1, -, -, 65, 10, 85, 10, 11000, 500, 815, 50, 100, 29, 4, 1200
Husbands, William A., -, -, 1, 95, 5, 7, 8, 20000, 500, 1065, 150, 50, 50, 7, 2800
Wilson, William, 1, -, -, 45, 16, 20, 4, 8500, 500, 800, 100, 150, 300, 75, 2000
Crockett, Robert, -, -, 1, 75, 20, 55, -, 7000, 100, 300, 50, 50, 150, 25, 700
Husbands, William E., -, 1, -, 47, -, 1, 2, 8000, 500, 700, 20, 50, 150, 40, 2000
Cloud, William M., 1, -, -, 55, -, -, -, 15000, 500, 1000, 100, 100, 125, 52, 1000
Naylor, Wm. P., -, 1, -, 15, 2, -, -, 3000, 150, 300, 10, 48, 85, 52, 600
Sharpley, John F., 1, -, -, 30, -, 10, -, 3200, 500, 700, 25, 50, 200, 60, 500
McKimm, James J., 1, -, -, 28, -, -, -, 2800, 50, 200, 40, 45, 12, 3, 300
Pierce, J. Bayard, -, 1, -, 28, -, 12, -, 3500, 250, 300, 50, 40, 15, 3, 400
Ewing, Henry J., 1, -, -, 10, -, 1, -, 3000, 200, 500, 25, 30, -, -, 300
Derrick, Samuel H., 1, -, -, 6, -, -, 2, 1200, 200, 175, 5, 41, 25, 4, 200
Viney, John W. Jr., -, -, 1, 12, -, 10, -, 2000, 100, 250, 10, 15, -, -, 200
Hand, James A., 1, -, -, 12, -, -, 3, 1500, 30, 250, -, 50, 50, 8, 300
Talley, Samuel M., 1, -, -, 55, 5, 3, 6300, 300, 880, 50, 65, 20, 3, 1025
Perkins, J. A. Bayard, 1, -, -, 25, -, 44, 10, 3950, 125, 525, 15, 24, 15, 3, 625
Talley, Jehu, 1, -, -, 7, -, 1, -, 2500, 125, 150, 5, 15, 20, 3, 300
Harvey, Samuel M., -, -, 1, 45, -, 15, -, 6500, 300, 450, 75, 32, 125, 28, 800
Parry (Perry), Oliver H., 1, -, -, 90, -, 20, -, 12000, 250, 500, 50, 150, 300, 52, 1000
Sharpley, Jacob A., 1, -, -, 17, -, -, 3, 2000, 100, 225, 20, 30, -, -, 400
Healy, Patrick, -, 1, -, 50, -, 3, 39, 10600, 300, 1000, 64, 75, 350, 52, 1300
Pattison, William, 1, -, -, 10, -, -, -, 3000, 300, 200, -, 20, 150, 25, 700
Williams, John H., -, 1, -, 15, -, -, -, 3000, 25, 150, 10, 25, -, -, 300
Banks, Robert, -, 1, -, 160, -, 140, -, 22500, 1000, 1000, 45, 100, 66, 144, 1620
Weldin, Jacob R., 1, -, -, 160, -, 21, -, 13575, 600, 1600, 75, 150, 936, 156, 2800

Wilson, David J., 1, -, -, 6, 6, 6, 4, 2500, 50, 200, 25, -, -, -, 300
Talley, John H., -, 1, -, 50, -, 2, -, 5200, 300, 450, 75, 50, 240, 36, 500
Doherty, George B., 1, -, -, 20, -, 5, 5, 4000, 75, 275, 30, 40, 6, 1, 600
Talley, Curtis, 1, -, -, 70, -, 30, -, 5000, 125, 600, 20, 100, 300, 50, 1000
Simon, Charles, 1, -, -, 25, -, -, -, 5000, 300, 310, 23, 200, 120, 40, 1600
Tomlin, George W., -, 1, -, 42, 10, 8, -, 6000, 150, 490, 150, 60, 224, 52, 800
Husbands, Thomas, 1, -, -, 45, -, 5, -, 7500, 500, 680, 50, -, 300, 52, 1390
Petildemange, Francis, 1, -, -, 70, -, 10, 24, 11500, 500, 700, 40, 30, 500, 90, 2000
Husbands, Abraham, 1, -, -, 20, -, 1, 6, 2000, 150, 330, -, 25, 100, 52, 600
Husbands, Wm. F., 1, -, -, 30, 12, 20, -, 5000, 85, 400, 10, 25, 12, 2, 600
Weldin, J. Atwood, -, 1, -, 25, 5, 10, -, 4000, -, 90, -, 30, 30, 5, -
Weldin, George W., 1, -, -, 50, -, 10, -, 10000, 400, 1000, 50, 9, 240, 52, 1400
Dougherty, John, 1, -, -, 30, 10, -, -, 3500, 100, 450, -, 100, 6, 1, 469
E. I. DuPont DeNemours & Co., 1, -, -, 49, -, 49, -, 8900, -, -, 50, -, 200, 33, 1100
McGilligan, James, -, 1, -, 80, -, 17, -, 9700, 800, 1000, -, 70, 300, 52, 1000
Anderson, John W., -, 1, -, 95, -, 35, 8, 13800, 800, 1175, 300, 125, 435, 78, 1815
Roland, Richard C., -, 1, -, 20, -, 62, -, 8000, 50, 140, 35, 34, 10, 2, 400
Talley, John R., 1, -, -, 58, -, 2, -, 6000, 250, 300, 40, 60, 150, 52, 2000
Morton, Samuel, -, -, 1, 54, 6, 6, -, 7000, 500, 450, -, 20, 152, 52, 1500
Talley, Charles B., -, 1, -, 45, -, 25, -, 6000, 100, 540, 5, 40, 128, 37, 600
Miller, George L., 1, -, -, 55, -, 15, -, 7000, 250, 500, 36, 41, 270, 40, 900
Wilson, Edmund, 1, -, -, 96, -, 2, -, 21000, 400, 2000, 20, -, 300, 52, 12850
McDonald, Alexander -, -, 1, 45, -, 5, -, 10000, 150, 500, 25, -, 200, 52, 1600
Talley, George W., 1, -, -, 75, -, 6, -, 12000, 600, 1225, 15, -, 400, 70, 1500
Webster, Henry G., 1, -, -, 70, -, 26, -, 12000, 250, 610, 25, 16, 300, 52, 700
Dixon, Samuel H., 1, -, -, 75, -, 20, -, 20000, 1000, 850, -, 40, 834, 156, 920
Shipley, Hannah (Est.), 1, -, -, 39, -, 109, -, 35000, 300, 500, -, 75, 520, 364, 900
Miller, Martin, 1, -, -, 60, -, 5, -, 8000, 200, 500, 20, 20, 80, 15, 900
Talley, John of Thos., 1, -, -, 40, -, 20, -, 6000, 300, 500, 25, -, 50, 10, 900
Carom, Joseph, -, -, 1, 106, -, -, -, 16000, 600, 925, 20, 32, 350, 58, 2675
McIntyre, James, -, 1, -, 40, -, -, -, 10000, 485, 628, 11, 52, 272, 39, 1315
Oakes, Francis L., -, -, 1, 15, -, -, -, 10000, 100, 100, -, 20, -, -, 520
Oakes, William, -, 1, -, 19, -, -, -, 8000, 100, 150, -, 9, 48, 8, 600
Talley, Joseph B., -, -, 1, 12, -, -, -, 10000, 200, 700, 10, -, 300, 40, 1455
McDonald, Alexander Jr., -, -, 1, 63, -, 7, -, 14210, 500, 1200, 80, 199, 300, 35, 1800
Miller, Robert S., -, 1, -, 8 ½, -, -, -, 12000, 75, 180, 5, 6, 186, 52, 400
Naylor, Isaac, 1, -, -, 60, -, 4, 15, 16000, 250, 770, 25, 28, 250, 40, 1600
Allmond, John G., -, -, 1, 40,70, -, -, 9000, 400, 735, -, -, 600, 104, 1000
Bringhurst, Edward, 1, -, -, 32, -, 15, -, 7000, 500, 935, 52, 40, 800, 104, 900
Edgemoor Iron Co., 1, -, -, 150, 60, 5, -, 30000, 2500, 1600, -, 1200, 1800, 364, 4200
Potter, William -, -, 1, 12, 5, 6, 12, 8000, 200, 300, -, 250, 275, 52, 1390
Elliott, Isaac S., 1, -, -, 200, 20, 35, -, 40000, 300, 1000, 100, 600, 800, 100, 4000
Talley, William W., -, 1, -, 40, 12, 4, -, 7000, 400, 550, 75, 23, 220, 52, 1485
Elliott, J. Cloud, 1, -, -, 60, 30, 8, -, 15000, 250, 500, 75, 75, 300, 52, 1000

Jessup & Moore Paper Co., 1, -, -, 30, 5, 15, -, 6000, 200, 200, 80, -, 468, 156, 775

Mouseley, Joseph & Alfred, 1, -, -, 70, -, 20, -, 9500, 150, 425, 10, -, -, -, 975

Stidham, Susan, 1, -, -, 55, -, 20, -, 7500, 100, 150, 80, 35, 300, 52, 630

Murphy, Alfred D., 1, -, -, 50, -, 7, -, 5700, 100, 415, -, 40, 240, 52, 700

Greenfield, Richard, -, 1, -, 12, -, 3, -, 3000, 75, 1170, 10, -, -, -, -

Jessup & Moore Paper Co., 1, -, -, 25, -, 15, 35, 10000, 100, 100, -, -, 30, 40, 430

Stirling, Hugh, 1, -, -, 80, -, 70, -, 15000, 100, 1500, 100, 200, 350, 52, 650

Dyson, George, -, 1, -, 55, 4, 12, -, 12000, 150, 600 -, 85, 250, 52, 1200

Twaddell, Charles & Jas., 1,-, -, 60, 14, 40, -, 11400, 400, 950, 50, 25, 300, 52, 1200

Malin, Charles, 1, -, -, 8, -, -, -, 2000, -, 225, -, 25, -, -, 275

Graves, John A., 1, -, -, 70, -, 9, -, 12000, 300, 700, 30, 250, 750, 104, 1600

Mouseley, Joseph M., 1, -, -, 10, -, -, 10, 4000, 75, 225, 50, -, -, -, 400

Starr, Moses, 1, -, -, 25, -, 5, 15, 5400, 75, 350, -, -, -, -, 700

Wilson, Norris, 1, -, -, 80, -, 6, -, 10000, 300, 600, 20, 50, 300, 52, 100

Betty, Robert, -, 1, -, 125, -, 40, -, 13000, 600, 1000, 100, 50, 25,4, 1000

Talley, Elihu, 1, -, -, 80, 8, 30, -, 12000, 500, 1200, 15, 150, 600, 100, 2300

McDermott, Patrick, 1, -, -, 50, -, 15, 10, 6000, 300, 500, 100, 25, 300, 52, 500

Cross, William, -, 1, -, 30, -, 20, -, 4000, 150, 400, 85, 50, -, -, 400

Carney, John, 1, -, -, 10, -, -, 20, 2000, 50, 220, -, -, -, -, 300

Robinson, Thomas S., 1, -, -, 140, -, -, -, 40000, 1200, 2300, 100, -, 1000, 190, 5200

Scott, William, 1, -, -, 11, -, -, -, 6000, -, 50, 30, 75, 50, 8, 125

Flicks, Ellis, -, 1, -, 20, -, -, -, 6500, 200, 200, 10, 250, 750, 52, 3000

Flicks, Ellis, -, 1, -, 17, -, -, -, 8000, 50, 125, 10, 50, 25, 40, 500

Harris, Thomas C., 1, -, -, 7, -, -, -, 5500, 100, 150, -, -, 150, 36, 600

Dupont, M. E., 1, -, -, 30, 3, -, -, 15000, 500, 800, 50, 150, 1500, 52, 1000

Griffith, Job, -, 1, -, 100, 1, 10, -, 20000, 500, 1000, 100, 200, 200, 36, 2000

Ely, Oliver P., 1, -, -, 30, -, -, -, 10000, 200, 400, 25, 50, 300, 32, 800

Clinton, George W., 1, -, -, 23, -, -, -, 10000, 400, 450, 200, 200, 100, 10, 1500

Gettz, Doniphan, -, 1, -, 10, -, 2, 2, 1500, 150, 200, 25, 200, 500, 52, 1000

Stidham, Gilpin P., 1, -, -, 12, -, -, -, 5000, 150, 200, 715, 500, 100, 10, 300

Cannon, James, 1, -, -, 34, -, 2, -, 4000, 200, 500, 20, 75, -, -, 600

Armor, James, 1, -, -, 60, 6, -, -, 7000, 600, 800, 20, 100, 50, 4, 1000

Journey, Margret, 1, -, -, 57, -, 3, 2, 7000, 400, 600, 10, 40, 200, 36, 500

Robinson, William O., 1, -, -, 90, -, 30, -, 8000, 300, 400, 100, 65, 300, 52, 1000

Hendricson, Charlotte, 1, -, -, 75, 12, 15, -, 25000, 600, 1500, 20, 200, 200, 52, 1600

Snyder, Jacob, -, -, 1, 80, 320, -, 20000, 300, 1200, 100, 100, 250, 52, 1500

West, Elizabeth C., 1, -, -, 20, 8, 4, 6, 4500, 50, 300, 25, 100, 150, 35, 500

Chandler, Samuel G., 1, -, -, 80, -, 20, 6, 5000, 400, 700, 25, 30, 50, 50, 600

Springer, Lewis C., 1, -, -, 50, 6, 22, 1, 8000, 250, 800, -, -, 50, 20, 500

Rolston, William, 1, -, -, 14, -, 5, 2, 1800, -, 25, -, 25, -, -, 100

Biderman, Herman F., 1, -, -, 13, 1, -, 3, 3000, 75, 150, 15, 75, -, -, 300

Traynor, Annie, 1, -, -, 5, 2, -, 2, 1500, 50, 175, 30, 10, 30, 3, 200

Talley, Isaac S., 1, -, -, 60, 1, 1, 10, 360, 150, 500, 75, 60, 50, 5, 1000
Palmer, Abraham, 1, -, -, 92, 4, 5, -, 10000, 500, 1200, 75, 60, 600, 52, 1500
Miller, Caleb, -, 1, -, 5, -, 10, -, 1500, 50, 50, 50, 50, -, -, 200
Ward, Patrick, 1, -, -, 40, -, 5, -, 5000, 450, 400, 25, 60, -, -, 400
McGrellis, George, 1, -, -, 40, 1, 6, 1, 10000, 200, 200, 100, 100, 175, 12, 700
McCallion, Michael, -, 1, -, 12, 3, 8, 2, 2200, 50, 100, 50, 26, 20, 3, 300
Dougherty, Joseph J., 1, -, -, 52, -, 5, -, 6000, 200, 500, 100, 200, 100, 4, 1000
Conley, James, 1, -, -, 720, 10, 6, -, 18000, 1200, 1500, 100, 50, 1100, 52, 2500
Morrow, Robert, 1, -, -, 73, -, 3, -, 20000, 400, -, 50, 50, 75, 10, 1000
McGrellis, James, 1, -, -, 10, 6, -, 3500, 100, 100, 10, 25, 25, 10, 500
Concannon, Martin, 1, -, -, 55, 1, 3, -, 6000, 400, 500, 25, 50, 25, 10, 600
Leach, James, 1, -, -, 90, 1, 5, -, 20000, 500, 800, 100, 150, 150, 25, 1000
Wilson, James, 1, -, -, 25, 1, 8, -, 5000, 500, 700, 15, 30, 400, 52, 800
Frederick, Peter, 1, -, -, 5, -, 37, 2000, 200, 150, 25, 40, 25, -, 500
Vernon, Otley, 1, -, -, 200, -, 25, -, 25000, 800, 2000, 200, 600, 700, 52, 4000
Morrow, Joseph, 1, -, -, 90, -, 10, -, 12000, 300, 1000, 10, 40, 50, 2, 1000
McCollom, Patrick, 1, -, -, 40, 7, 4, -, 6000, 300, 900, 10, 60, 50, 30, 900
Dixson, Coborn, 1, -, -, 60, 9, 37, -, 10000, 100, 500, -, 60, 110, 36, 600
Lowther, Thomas, 1, -, -, 90, -, 10, -, 10000, 500, 1000, -, 100, 300, 52, 1000
Lowther, John, -, -, 1, -, 40, -, 5, -, 4000, 400, 400, 25, 100, 500, 52, 800
Bartran, Benjamin, 1, -, -, 27, -, 1, -, 9000, 800, 500, 25, 10, 200, 4, 1000
Hoopes, Edward, 1, -, -, 27, 4, 2, -, 8000, 300, 400, 15, 150, 400, 52, 600
Barry, Thomas, -, 1, -, 65, 1, 2, -, 6000, 300, 600, 50, 120, 50, 4, 800
Cusick, John, 1, -, -, 6, 1, 3, -, 2400, 25, 150, 75, 25, 15, 2, 200
Graves, Lewis, 1, -, -, 110, 1, 25, -, 12000, 500, 1000, 50, 100, 400, 13, 1000
Jackson, Charles, 1, -, 1, -, 90, 7, 15, -, 8000, 400, 800, 75, 125, 250, 52, 1200
Harigan, James, 1, -, -, 25, -, 15, -, 3000, 200, 300, 50, 100, 100, 10, 300
Mackey, Daniel, 1, -, -, 40, -, 10, -, 4000, 200, 400, 20, 30, 25, 2, 500
Hughes, Austin, 1, -, -, 6, -, -, -, 2000, -, 120, 20, -, -, -, 200
Brannon, John, 1, -, -, 60, 1, 10, -, 5000, 200, 600, 20, 70, 105, 36, 800
Doughtery, Patrick, 1, -, -, 38, -, 4, 4, 4000, 200, 400, 20, 40, 50, 20, 500
Holmes, Jackson, 1, -, -, 65, 9, 7, -, 7500, 300, 500, 15, 55, 80, 10, 32
Graves, Hayes, 1, -, -, 10, 5, -, -, 3000, 100, 100, 10, 30, 10, 2, 300
Boyd, Andrew, 1, -, -, 14, 1, 5, -, 1500, 100, 200, 40, 42, -, -, 300
Conners, Thomas, 1, -, -, 50, 4, 16, 2, 6500, 400, 600, 20, 75, 25, 7, 1300
Klair, Fredrick, 1, -, -, 85, 3, 7, -, 12000, 500, 700, 40, 60, 10, 3, 1500
Dilworth, James, 1, -, -, 170, 6, 20, -, 15000, 1000, 1200, 50, 300, 600, 52, 4500
Dilworth, Franklin, 1, -, -, 60, 8, 2, 7, 8000, 600, 800, 50, 150, 250, 32, 1500
Cooney, Conard, 1, -, -, 16, 2, -, -, 2000, 25, 100, 30, 30, 20, 2, 250
Cloud, George D., -, 1, -, 100, 1, 20, -, 12000, 150, 1000, 100, 125, 600, 52, 1200
Gillice, Thomas, 1, -, -, 14, 4, 2, 5, 2000, 50, 150, -, -, 50, 6, 200
Klair, Bennett J., 1, -, -, 115, 2, 20, -, 14000, 500, 1200, 100, 50, 400, 52, 1500
Kane, Benery, 1, -, -, 35, 4, 6, -, 3500, 200, 500, 100, 150, 50, 15, 800
Davis, Jesse, -, -, 1, 30, 1, 3, 7, 5000, 200, 250, 25, 30, 25, 4, 500
Chandler, James, 1, -, -, 25, 1, 6, -, 4000, 200, 275, 50, 100, 50, 12, 500
Cloud, James P., -, 1, -, 70, 12, 20, -, 14000, 600, 800, 100, 125, 290, 36, 1600

Cloud, Jackson, -, 1, -, 45, 1, 9, -, 7000, 200, 300, 50, 40, 175, 30, 1000
Gamble, Samuel, 1, -, -, 55, 1, 5, -, 6000, 600, 600, 75, 40, 50, 8, 1000
Lynch, Patrick, -, 1, -, 80, 13, 11, -, 9000, 300, 800, 20, 40, 200, 52, 1000
May, Jacob M., -, -, 1, 90, 5, 10, -, 10000, 300, 700, 40, 75, 300, 40, 900
Springer, Jeremiah, 1, -, -, 35, 6, 8, 2, 4500, 300, 500, 30, 36, 120, 13, 200
Plankinton, Edwin, 1, -, -, 23, -, 4, -, 3500, 300, 800, 100, 70, 50, 15, 250
Trane, Michael, 1, -, -, 55, -, 5, -, 6000, 150, 300, 60, 25, 150, 52, 500
Muller and Brother, _____, 1, -, -, 55, 15, 9, 6, 10000, 500, 600, 50, 130, 75, 15, 1000
Murrey, Sylvester, 1, -, -, 16, -, -, -, 3500, 175, 450, -, 85, 20, 2, 500
Clarke, William, 1, -, -, 100, -, 10, -, 15000, 500, 1200, 20, 60, 600, 52, 1500
Williams, Thomas, -, 1, -, 165, 1, 28, -, 20000, 450, 1500, 130, 200, 800, 52, 2800
Ford, Abraham, -, 1, -, 125, 3, 50, -, 20000, 700, 1000, 100, 150, 400, 52, 1400
Woodward, Abner, 1, -, -, 70, 10, 14, -, 10000, 500, 500, 25, 100, 200, 36, 800
Lowther, William, 1, -, -, 65, 7, 25, -, 9500, 300, 1500, 150, 100, 200, 52, 1500
Smith, Richard S., -, 1, -, 15, 2, -, 2, 2500, 100, 175, 10, 25, 15, 2, 500
Alcorn, James, 1, -, -, 30, -, 3, 6, 3000, 100, 350, 40, 40, 30, 8, 300
Donoghoue, Susan, 1, -, -, 30, 2, 8, 3, 3500, -, 100, 21, 10, 10, 2, 250
Hobson, Lawrence, 1, -, -, 70, 2, 7, 10, 7000, 150, 500, 10, 20, 30, 5, 1800
White, James, 1, -, -, 36, 7, 3, -, 12000, 700, 500, 25, 50, 200, 36, 800
Morrow, Samuel D., -, 1, -, 35, -, 10, -, 10000, 200, 600, 10, 100, 250, 52, 1200
Hall, Thomas A., -, 1, -, 78, 2, 6, -, 12500, 800, 1000, 50, 200, 25, 4, 1600
Green, Charles, 1, -, -, 130, 11, 11, -, 40000, 1000, 1000, 50, 150, 1500, 52, 2000
Spence, Mathew, -, 1, -, 10, -, -, -, -, 20, 600, -, -, 25, 4, 300
Spence, Samuel, -, 1, -, 70, 1, 4, -, 9000, 150, 300, 25, 110, 225, 52, 400
Parkin, William M., 1, -, -, 30, 4, 3, 2, 4000, 50, 200, 48, 10, 50, 8, 400
Hipple, Jacob C., -, 1, -, 119, 2, 3, -, 1, 15000, 500, 1200, 30,75, 25, 4, 1000
Carpenter, James L., 1, -, -, 133, -, 10, -, 14300, 500, 1300, 100, 40, 750, 52, 1500
Nicholson, E. C. S., 1, -, -, 400, 3, 8, -, 10000, 100, 600, 25, 50, 200, 36, 600
Murphy, James W., 1, -, -, 32, 2, -, -, 6000, 250, 900, 25, 20, 250, 52, 800
Passmore, William P., 1, -, -, 145, -, 30, -, 20000, 800, 1500, 700, 100, 500, 52, 2000
Hendrickson, John, -, 1, -, 87, 7, 6, -, 10000, 250, 900, 25, 75, 50, 20, 1000
Clement, Franklin W., 1, -, -, 110, 13, 23, -, 7500, 700, 1000, 15, 75, 350, 20, 1000
Mace, George B., -, -, 1, 25, -, 15, 40, 8000, 5, 30, -, -, 10, 2, 200
Lynch, Michael, -, 1, -, 10, -, 6, 5, 8000, 100, 500, 40, 15, 50, 8, 300
Armor, William, 1, -, -, 50, 10, 5, -, 7500, 500, 600, 20, 50, 150, 40, 1000
Talley, Winfield S., -, 1, -, 47, 1, 64, -, 5600, 250, 600, 50, 100, 150, 36, 800
Ritchie, James, 1, -, -, 20, 1, 5, -, 5000, 100, 225, 40, 10, 5, 1, 500
Wood, James, 1, -, -, 32, -, -, 4, 3000, 250, 250, 50, 45, 10, 2, 400
Kelly, John, -, 1, -, 30, -, 6, -, 1700, 25, 150, 20, 30, 25, 18, 500
McDonald, Bridget, 1, -, -, 5, -, -, 2, 800, -, 100, 35, 10, 15, 2, 250
Barnes, Warfield, 1, -, -, 30, -, 13, -, 1800, -, 10, -, -, -, -, 100
Humphrys, John, 1, -, -, 70, -, 12, 14, 7000, 150, 400, 50, 25, -, -, 600
Fenn, Oscar _., -, -, 1, 150, 30, 20, 2, 20000, 500, 1500, 50, 25, 225, 36, 3000

Wilson, Robert, 1, -, -, 28, 1, 1, -, 5000, -, 200, 500, 10, 25, 50, 16, 500
Wilson, Thomas, 1, -, -, 80, -, 7, -, 10000, 300, 1000, 50, 12, 400, 30, 1200
Ely, George W., 1, -, -, 90, 35, 19, -, 18000, 500, 1800, 50, 150, 500, 52, 300
Woollens, Thomas E., 1, -, -, 20, -, 10, 3, 3000, 200, 350, 25, 60, 150, 24, 600
Bakey, Peter, 1, -, -, 8, -, -, 2, 3000, 50, 100, 15, 15, 15, 2, 350
Bankes, Edwin R., -, 1, -, 85, 5, 45, -, 12000, 300, 500, 200, 190, 200, 52, 1500
Hobson, George L., -, 1, -, 60, 20, 10, 3, 10000, 400, 1000, 40, -, 175, 32, 1000
Oakes, Albin, 1, -, -, 9, -, 2, -, 2500, 100, 200, -, 50, 200, 52, 500
Carpenter, Alfred D., -, 1, -, 32, -, -, 8, 4000, 350, 650, 25, -, 175, 52, 600
Swayne, Henery, 1, -, -, 68, -, 12, -, 10000, 200, 600, 10, 50, 100, 30, 1000
Coning, William, -, 1, -, 100, -, 20, -, 2000, 700, 1200, 150, 150, 275, 52, 2500
Ball, John, -, 1, -, 65, 2, 8, -, 15000, 150, 500, 50, 100, 200, 2, 1000
Ferris, William, 1, -, -, 20, -, -, -, 10000, 200, 450, 10, -, 240, 52, 1500
Pyle, Joseph, 1, -, -, 12, -, -, -, 25000, 75, 600, 40, 100, 325, 52, 200
Thompson, William H., 1, -, -, 30, -, 10, -, 5000, 50, 40, -, -, 50, 10, 400
Law, William, 1, -, -, 20, -, 5, -, 15000, 20, 500, 20, -, 50, 10, 500
Chandler, J. Poulson, 1, -, -, 130, 47, 80, -, 15750, 300, 1430, 200, 210, 1000, 50, 4026
Thompson, William, 1, -, -, 100, 33, 15, -, 14000, 1200, 700, 300, 300, 600, 52, 300
Husband, Adolphus, 1, -, -, 120, 35, 200, -, 15000, 400, 1000, 50, 200, 500, 52, 3500
McCullough, Joshua H., 1, -, -, 107, -, 10, 14050, 400, 1060, 150, 112, 500, 52, 300
Graves, Lewis J., 1, -, -, 27, -, 9, -, 5000, 250, 495, 20, 50, 175, 35, 650
Chaxelle, Prosper, 1, -, -, 8, 3, -, -, 3500, 50, 175, 30, 50, 40, 5, 200
Woodward, George H., -, 1, -, 110, -, 8, -, 20000, 400, 1200, 300, 200, 300, 2, 2500
Wood, George H., -, 1, -, 110, -, 8, -, 20000, 400, 1200, 300, 200, 300, 52, 2500
Mousley, George K., 1, -, -, 70, -, 35, -, 5600, 300, 600, 75, 175, 225, 52, 1000
Thompson, Isaac S., 1, -, -, 60, 7, 19, -, 6500, 250, 600, 5, 80, 150, 52, 1000
Haley, Daniel, 1, -, -, 36, -, 10, -, 4000, 300, 400, 50, 90, 50, 8, 1000
Murphy, Patrick, 1, -, -, 16, 5, -, -, 2500, 50, 500, 40, 50, 15, 3, 500
Kane, Thomas, 1, -, -, 10, -, -, -, 2000, 100, 100, 30, 24, 100, 6, 300
Golding, Daniel, 1, -, -, 17, 4, -, -, 2000, 100, 250, 50, 25, 50, 6, 500
Gregg, Samuel, 1, -, -, 50, 8, 13, -, 1200, 50, 800, 50, 200, 600, 52, 2000
Thompson, John B., -, 1, -, 170, -, 30, -, 22000, 400, 1500, 100, 300, 400, 52, 2500
Rycott, Mathew, -, 1, -, 100, 20, 25, -, 20000, 400, 1200, 100, 150, 300, 52, 2000
Seal, William P., 1, -, -, 26, 2, -, -, 8000, 300, 350, 50, 70, 175, 32, 1500
Nichols, Cathrine, 1, -, -, 16, -, 1, -, 5000, 50, 4000, 30, 20, 50, 10, 500
McCullough, John, 1, -, -, 14, 2, 4, -, 3000, 100, 300, 10, -, 20, 3, 250
Pugh, James R., 1, -, -, 7, -, -, -, 4500, -, 150, 20, -, 20, 4, 200
Moore, Cromwell, 1, -, -, 60, 8, 15, 10, 8000, 200, 500, 20, 50, 50, 8, 1000
Banning, John, 1, -, -, 80, -, 34, 6, 17000, 600, 600, 75, 300, 700, 52, 2500
Beatty, Benjamin, 1, -, -, 30, 20, 45, -, 8000, 100, 300, 15, -, 50, 8, 600

Negendank, Charles, -, -, 1, 16, -, -, 6000, 150, 250, 15, 40, 300, 52, 500
Dilworth, William Lewis, 1, -, -, 145, -, 13, -, 15000, 1000, 1500, 30, 300, 90, 52, 3000
Sharpless, A. & J. D., 1, -, -, 18, -, -, 12, 7000, 150, 650, 15, -, 150, 25, 300
Green, Elwood R., -, 1, -, 60, 3, 18, -, 8000, 350, 750, 50, 70, 275, 36, 1400
Grady, Michael, 1, -, -, 11, -, -, -, 2000, 125, 250, 10, 15, 50, 8, 350
Chandler, Marshal S., 1, -, -, 100, 15, 40, -, 20000, 400, 3000, 50, 75, 600, 52, 3000
Wright, Howard, -, -, 1, 50, 15, 15, 5, 6000, 100, 350, 20, 15, 50, 8, 800
Kent, Henry, 1, -, -, 10, -, -, -, 1800, 50, 100, 10, 10, 10, 2, 300
Carpenter, William, -, 1, -, 70, 6, 10, -, 10000, 300, 800, 10, 50, 300, 52, 1000
Chandler, Thomas M., -, 1, -, 56, 1, 5, -, 10000, 300, 600, 20, 25, 300, 52, 1200
Haley, Benjamin, -, -, 1, 175, 20, 16, -, 28000, 700, 2500, 150, 350, 850, 52, 5000
Dougherty, Charles B., 1, -, -, 12, -, -, -, 5000, 50, 200, 80, 60, 30, 4, 300
E. I. DuPont DeNemours & Co., 1, -, -, 500, -, 215, 258, 100000, 13191, 170, 250, 5000, 52, 13800
Riddle, Leander F., 1, -, -, 65, -, 50, -, 25000, 500, 150, 100, -, 200, 30, 650
Bancroft, Joseph & Sons, 1, -, -, 17, 5, 20, -, 4500, 300, 400, 25, 75, 350, 52, 600
Wilson, William Jr., 1, -, -, 148, 20, 30, -, 30000, 600, 2400, 250, 125, 550, 52, 3000
Brinckle, W. R., 1, -, -, 55, 20, -, -, 70000, 100, 400, 50, 25, 450, 52, 1000
LeCarpentier, Charles, 1, -, -, 50, 20, 10, 25, 8000, 200, 300, 100, 75, 600, 52, 1500
LeCarpentier, Edward Jr., -, 1, -, 40, 20, 10, 15, 7150, 200, 550, 100, 70, 800, 52, 1600
Toy, James, 1, -, -, 16, -, -, -, 12000, 200, 500, 15, 15, 35, 6, 400
McHugh, Bernard, -, 1, -, 22, -, -, -, 16000, 50, 200, 10, 20, 25, 4, 300
DuPont, Mary V., 1, -, -, 40, -, 18, -, 12000, 500, 400, 100, 100, 500, 52, 820

NEXT 64 NAMES ARE FIRST NAME FIRST
Isaac N. Flinn, -, 1, -, 42, -, -, -, 7000, 50, 400, 50, 100, 190, 300, 500
John R. Lynam, 1, -, -, 65, 5, -, -, 65, 260, 400, 20, 50, 300, 150, 500
Wm. R. Flinn, 1, -, -, 133, 20, -, -, 10000, 250, 1200, 100, 225, 506, 52, 2500
David M. Price, 1, -, -, 8, -, -, -, 2500, 20, 50, -, -, 5, 1, 500
Enos Walters, -, 1, -, 120, -, 20, -, 15000, 400, 1000, 130, -, 200, 25, 200
John Evans, 1, -, -, 103, -, 20, -, 12000, 500, 600, 50, 50, 150, 52, 1000
John Stuart, 1, -, -, 60, -, 6, -, 6000, 200, 300, 25, 90, 1000, 32, 900
John Jordan, 1, -, -, 110, -, 5, -, 10000, 200, 500, 25, 75, 100, 30, 1200
John P. Armstrong, 1, -, -, 120, -, 5, 20, -, 12000, 1000, 1200, 50, 100, 200, 26, 1200
Lousice Lynam, 1, -, -, 113, 3, 10, -, 15000, 200, 1000, 50, -, 200, 20, 1000
Joseph Killgore, 1, -, -, 4, -, -, -, 5000, 5, 50, -, -, 50, 10, 400
Rufus Gregg, -, -, 1, 70, 11, 30, -, 10500, 700, 600, 50, 100, 100, 10, 1000
T.F. Moore, -, 1, -, 90, 20, 10, -, 10000, 200, 400, 100, 50, 150, 20, 1000
Elizabeth McCollough, 1, -, -, 8, -, -, -, 40000, 100, 100, 200, -, 300, 52, 500
Benjamin Rothwell, 1, -, -, 53, 6, 2, -, 53000, 200, 3500, 50, 50, 300, 52, 500

Thomas P. Lynam, 1, -, -, 150, 50, -, -, 10000, 200, 500, 50, 150, 500, 52, 2000
Robt. Lynam, 1, -, -, 29, 3, -, -, 5000, 150, 500, 10, 60, 100, 10, 100
Charley Newlin, 1, -, -, 137, 25, -, -, 13000, 500, 1000, 50, 100, 300, 25, 1500
Samuel Cranston, -, 1, -, 146, -, 4, 12, -, 15000, 500, 1000, 60, 200, 400, 30, 32
A. W. Everson, 1, -, -, 100, 25, 10, -, 12000, 400, 800, 50, 120, 300, 100, 3000
Mrs. Joema, Flinn, 1, -, -, 12, -, -, -, 5000, 50, 100, 25, 50, 50, 25, 500
Howard E. Flinn, 1, -, -, 20, -, -, -, 5000, 25, 150, 25, 25, 25, -, 400
Robert F. Lynam, 1, -, -, 29, 2, -, -, 4000, 150, -, -, -, -, -, -
Wm. Southgate, -, 1, -, 80, -, -, -, 7000, 400, 600, 50, 50, 300, 40, 1000
Alexander Irons, 1, -, -, 17, -, -, -, 5000, 20, 175, 10, 5, -, -, 500
Vincent G. Flinn, 1, -, -, 153, 60, 5, -, 18880, 500, 1500, 75, 70, 500, 52, 4000
John Q. Flinn, 1, -, -, 79, 2, -, -, 6000, 100, 600, 50, 90, 500, 52, 1200
John Q. Flinn, 1, -, -, 120, -, 6, -, 8000, 100, 700, 50, 100, 100, 30, 1500
John Flinn & Catherine B. Snyder, 1, -, -, 100, 16, 3, -, 12000, 100, 1000, 60, 75, 200, 20, 2500
John W. Moore, -, 1, -, 115, -, 10, -, 20000, 400, 1000, 100, -, 250, 500, 1500
R. L. Armstrong, 1, -, -, 120, 10, 20, -, 12000, 500, 500, 150, 100, 500, 100, 3000
Philip Quigley, 1, -, -, 190, 35, 28, -, 20000, 2500, 800, -, -, 300, 90, 1410
Abner Hillingsworth, 1, -, -, 140, 15, 5, -, 14000, 800, 1000, 50, 100, 300, 60, 2500
Gilpin P., Stidham, 1, -, -, 100, -, 2, -, 12000, 400, 1500, 100, 120, 300, 36, 1500
Mary E. Armsttrong, 1, -, -, 113, 16, 12, 7, 8000, 1500, 800, 50, 50, 180, 52, 1000
William Armstrong, 1, -, -, 112, 20, 4, 5, 8000, 500, 500, 50, 60, 200, 52, 1500
William M. Brackin, 1, -, -, 116, 20, 18000, 300, 1000, 50, 100, 250, 52, 2000
Edward Woodward, 1, -, -, 154, 10, 12, 20, 18000, 300, 1000, 50, 100, 500, 52, 3000
John R. Lynam, 1, -, -, 44, -, -, -, 5000, 100, 150, 50, 50, 100, 51, 1000
B. F. Zebley, -, -, 1, 150, 18, -, -, 18500, 500, 1000, 100, 60, 600, 52, 2000
Jos. S. Richardson, 1, -, -, 40, 7, -, -, 10000, 50, 600, 25, 50, 1500, 52, 500
George W. Vandergrift, -, -, 1, 95, 20, 10, -, 10000, 250, 500, 50, 30, 250, -, 1500
Penrose S. Lynam, 1, -, -, 130, 27, 10, -, 10000, 200, 1000, 50, 30, 300, 32, 1500
Thomas D. Lynam, 1, -, -, 160, 8, 7, 5, 1500, 500, 1200, 100, 125, 800, 26, 3000
John McBride, -, -, 1, 104, 8, 25, -, 12000, 1000, 1000, 40, 26, 260, 25, 2500
Ellen B. Stidham, 1, -, -, 70, 9, -, -, 10000, 200, 500, 40, 50, 250, 30, 1200
George Bremefle, -, -, 1, 60, 12, 50, -, 10000, 150, 600, 50, -, 350, 25, 1000
Edwin Q. Cranston, 1, -, -, 150, 25, 8, -, 20000, 200, 1500, 40, 100, 400, 52, 3000
David R. Lynam, 1, -, -, 104, 4, -, -, 20000, 400, 800, 50, -, 100, 25, 3000
Joseph P. Richardson, 1, -, -, 110, 6, 30, 10, 16000, 300, 2000, 100, 50, 300, 40, 2000
John Richardson, 1, -, -, 90, 15, 10, -, 16000, 200, 550, 100, -, 250, 26, 1500
Wm. Jenks Fell, 1, -, -, 106, 65, -, 10, 6000, 150, 400, 25, 25, 150, 24, 1000
Henry G. Banning, 1, -, -, 25, 18, 7, -, 35000, 125, 500, 50, 10, 400, 52, 300
David Lynam, 1, -, -, 53, -, -, -, 7500, 100, 250, 25, 25, 150, 30, 1000
William P. Richardson, 1, -, -, 60, 14, 10, -, 12000, 300, 500, 50, -, 500, 10, 1000
Ashton Richardson, 1, -, -, 60, 11, 2, -, 12000, 400, 760, 70, 159, 500, 75, 1153
John B. Gooding, -, -, 1, 85, 10, 6, -, 8000, 200, 700, 25, 50, 400, 50, 1500

James Brown, 1, -, -, 85, 6, -, -, 85, 500, 1200, 50, 75, 400, 52, 4000
Wm. Z. Derrickson, 1, -, -, 80, 14, -, -, 8000, 400, 1000, 25, 130, 200, 30, 2000
Jos. S. Foreman, -, 1, -, 120, 15, -, -, 10000, 200, 400, 25, 200, 150, 15, 900
Henry White, 1, -, -, 111, -, 1 ½, -, 15500, 300, 800, 50, 200, 75, -, 2500
James H. Hoffecker, 1, -, -, 100, 5, 1, -, 15000, 400, 1000, 50, 25, 300, 75, 1600
John R. Tatum, 1, -, -, 168, -, 40, -, 30000, 400, 1500, 130, 200, 450, 50, 2000
Isaac P. Foreman, -, 1, -, 90, 8, 10, -, 11000, 200, 800, 50, 65, 400, 52, 1200
END OF FIRST NAME FIRST SECTION

Lewis, Thomas, -, 1, -, 8, -, -, 1, 5000, 50, 100, -, -, -, -, 1000
Sutton, William, -, -, 1, 12, -, -, -, 30000, 125, 100, -, -, -, -, 1000
Pugh, James B., -, 1, -, 8, -, -, 6, 12000, -, 225, -, -, -, -, 800
Brown, Jacob, 1, -, -, 50, 26, -, 24, 15000, 150, 250, -, -, 500, 52, 3000
Alexander, Henry T., -, 1, -, 14, 9, -, 5, 8500, 100, 250, -, -, 200, -, 2000
Miller, Peter, 1, -, -, 13, 3, -, -, 8500, 75, 250, -, -, 200, 1800
Lutz, Henry, 1, -, -, 13, -, -, -, 11600, -, 200, -, 300, 300, -, 2000
Peoples, John, -, -, -, -, -, -, -, -, -, 100, -, -, -, -, -
Simmons, Bandray, -, -, -, -, -, -, -, -, -, 100, -, -, -, -, -
Benson, Nathaniel, -, -, -, -, -, -, -, -, -, 200, -, -, -, -, -
Tatnall, Joseph, 1, -, -, 40, 10, 45, -, 60000, 850, 600, -, -, -, -, -
Smith, David R., -, -, 1, 200, 1000, -, -, 50000, 100, 2400, 100, 130, 100, 52, 300
Kerney, Thos. W., -, -, 1, 60, 9, -, -, 30000, 500, 1200, -, 100, 225, 40, 17
Stroup, Eugene, -, -, 1, 80, 25, -, -, 60000, 550, 1460, 45, -, 418, 40, 1300
Conner, Patrick, Geo. Bush, 1, -, 10, -, -, -, 12000, -, 400, -, -, -, -, 1300
Conner, Bridget, Geo. Bush, -, -, -, -, -, -, -, -, -, -, -, -, -, -, -
Conner, Annie, Geo. Bush, -, -, -, -, -, -, -, -, -, -, -, -, -, -, -
Conner, Wm., Geo. Bush, -, -, -, -, -, -, -, -, -, -, -, -, -, -, -
Conner, Francis, Geo. Bush, -, -, -, -, -, -, -, -, -, -, -, -, -, -, -
Conner, James, Geo. Bush, -, -, -, -, -, -, -, -, -, -, -, -, -, -, -
Conner, Margaret, Geo. Bush, -, -, -, -, -, -, -, -, -, -, -, -, -, -, -
Conner, John, Geo. Bush, -, -, -, -, -, -, -, -, -, -, -, -, -, -, -
Conner, Mark, Geo. Bush, -, -, -, -, -, -, -, -, -, -, -, -, -, -, -
Barlow, Malachi, 1, -, -, 10, 14, -, -, 100000, -, 900, 4000, -, -, -, 1369
Denison, John W., -, -, 1, 114, 2, 22, -, 10000, 500, 1000, 25, 250, 450, 90, 2000
Petitdemange, Joseph, -, -, 1, 110, 25, 10, -, 13000, 500, 1700, 25, 90, 250, 40, 2000
Shakespeare, Benjamin, -, -, 1, 165, 14, 40, -, 16400, 700, 2500, 20, 60, 350, 104, 3000
Brackin, John, -, -, 1, 35, -, 1, -, 3600, 150, 500, -, 100, 75, 29, 475
Gebhart, George F., -, -, 1, 55, 4, 60, -, 6000, 200, 275, 10, 183, 125, 1252, 800
Chamberlain, Isaac T., 1, -, -, 80, 320, -, 15000, 250, 300, 100, 125, 150, 32, 1000
Pierson, Jacob, -, 1, -, 82, 6, 12, -, 10000, 300, 800, 20, 75, 300, 36, 1750
Reardon, Patrick, 1, -, -, 60, -, 3, -, 8000, 300, 500, 200, 100, 400, 52, 1000
Ector, Thomas M., 1, -, -, 35, 4, 7, 5, 7000, 300, 400, 10, 61, 150, 30, 700
Springer, John, -, -, 1, 83, -, 10, -, 9000, 200, 250, 10, 360, 500, 32, 1037
Ahern, David, 1, -, -, 6, ¼, -, -, 1200, 10, 95, -, -, 20, 6, 275

Hoopes, Enos, -, 1, -, 70, 10, 7, 2, 75, 200, 550, -, 100, 250, 40, 1000
Ford, Henry, 1, -, -, 7, 1, -, 1, 900, 15, 100, 25, -, -, -, 200
Hyde, Hannah, 1, -, -, 23, -, 3, -, 3500, 75, 220, -, 28, -, -, 500
Hyde, Daniel, 1, -, -, 6, 2, -, -, 1500, 50, 125, 10, 10, -, -, 200
McGovern, John T., 1, -, -, 7, -, -, -, 1000, 75, 100, -, -, -, -, 200
McGovern, Edward, 1, -, -, 21, -, 3, 2, 3000, 175, 400, 80, 65, -, -, 500
Ovil, John, 1, -, -, 8, -, -, 2, 1000, -, 150, -, 20, -, -, 100
Springer, George, 1, -, - 50, 9, 10, -, 7000, 800, 900, 50, 100, 300, 52, 1200
Lumford, Francis E., -, -, 1, 70, 3, -, 6, 7000, 200, 300, -, 100, 275, 36, 1200
Smith, James, -, 1, -, 36, ½, 4, -, 3000, 150, 200, 35, 18, 25, 5, 600
Pierson, John, -, -, 1, 100, 3, 12, -, 8000, 550, 800, 60, 60, 200, 40, 1745
Derrickson, Joseph, 1, -, -, 75, 1, 28, 5, 8000, 500, 800, 150, 100, 200, 36, 1000
Clair, James S., 1, -, -, 100, 5, 12, 7, 8500, 275, 600, 25, 160, 250, 52, 2000
Jones, Lemuel, 1, -, -, 14, ½, -, -, 3500, 100, 200, -, 18, 150, 32, 500
Ball, John, 1, -, -, 91, 11, 3, -, 8000, 300, 700, 20, 220, 500, 52, 200
Ball, Reben, 1, -, -, 110, -, 10, 15, 7000, 400, 900, 100, 318, 200, 36, 2080
Hanna, John, 1, -, -, 90, 6, 6, -, 9000, 500, 960, 20, 130, 400, 60, 1900
Menden, John, -, -, 1, 19, 5, 2, -, 3000, 100, 300, 400, 15, 20, 4, 250
Moore, William, 1, -, -, 90, 5, 5, -, 6500, 500, 500, 50, 200, 400, 52, 1200
Sesterhenn, Christian, 1, -, -, 9, -, 3, -, 7000, 150, 250, 40, -, 100, 12, 500
Springer, Stephen, 1, -, -, 90, 3, 5, -, 9000, 400, 600, 20, 150, 250, 52, 1375
Gordon, Howard H., 1, -, -, 80, 6, 9, 3, 5000, 350, 500, 150, 168, 208, 36, 850
Rohuebencane, Jacob, 1, -, -, 11, -, -, -, 3000, 40, 250, 10, 25, 25, 4, 500
McClar, George W., 1, -, -, 9, -, -, -, 1500, 50, 200, -, 5, -, -, 400
Foote, James, 1, -, -, 55, 1, 25, -, 6000, 250, 400, 165, 100, 50, 5, 1500
Yearsley, Susanna, 1, -, - 18, 4, -, -, 4000, 150, 250, 40, -, 25, 3, 500
Rubencane, Charles, 1, -, -, 113, 1, 5, -, 12000, 500, 1066, 100, 200, 480, 52, 2200
Pierson, James C., -, -, 1, 4, -, -, -, 4000, -, 50, 20, -, 10, 1, 100
Haylett, Ann, 1, -, -, 35, 6, 3, -, 5000, 200, 500, 50, 50, 400, 52, 900
Bartlett, Henry, -, 1, -, 48, 12, -, 8, 3000, 250, 600, 20, -, 200, 24, 653
Taylor, Robert S. M., 1, -, -, 27, 3, -, -, 4500, 350, 450, 10, 100, 200, 52, 1000
Cranston, William B., -, 1, -, 85, -, 5, 10, 9000, 600, 1000, 25, 100, 500, 90, 2000
Newlin, Alonzo, -, 1, -, 83, 27, 16, 11, 12000, 500, 1000, 75, 212, 420, 56, 2000
Klair, Egbert Q., 1, -, -, 114, 2, 19, 25, 12000, 400, 900, 150, 161, 332, 43, 1800
Allcorn, Samuel M., 1, -, -, 22, -, -, 8, 3500, 150, 275, -, 70, 225, 40, 500
Derickson, Cornelias, 1, -, -, 80, 10, 6, -, 6725, 450, 900, 25, -, 225, 36, 1000
Welch, Eli, -, -, 1, 140, 3, 10, -, 10000, 300, 1000, 50, 160, 500, 75, 2000
Bowing, Eliza, 1, -, -, 6, -, -, -, 800, 50, 150, 25, -, -, -, 200
Minehart, William, -, 1, -, 10, -, -, 3, 3000, 50, 100, -, 14, -, 20, 200
Lynam, George M.D., -, 1, -, 185, 50, 1, 60, 14000, 500, 1000, 20, 200, 500, 90, 2000
Justice, Robert, -, 1, -, 92, 16, 4, 10, 12000, 600, 600, 20, 235, 360, 38, 1120
Duncan, B. F., 1, -, -, 105, 1, 8, 1, 10000, 1000, 1000, 50, 100, 300, 40, 1200
Springer, James, 1, -, -, 75, 2, 15, 1, 5000, -, -, 10, 28, 50, 8, 600
Highfield, Morris, 1, -, -, 29, -, 13, 4, 4000, 200, 400, 20, 17, 100, 25, 300
Mitchell, Thomas C., -, -, 1, 92, 30, 6, 9, 9000, 90, 900, 56, 40, 340, 52, 1800

Laffety, John, 1, -, -, 12, -, -, -, 2000, 10, 150, -, -, 50, 4, 250
Pierson, Philip T., -, 1, -, 65, 5, 8, -, 6000, 250, 620, 20, 80, 125, 32, 1000
Shakespeare, James M., -, -, 1, 110, -, 10, 12, 10000, 500, 700, 50, 110, 325, 52, 1400
Gregg, James E., 1, -, -, 50, 10, 6, -, 8000, 400, 700, 20, 155, 308, 52, 1500
Jester, William D., -, -, 1, 60, -, 60, -, 6000, 100, 300, -, 125, 40, 4, 700
Vance, George, -, 1, -, 60, -, 7, -, 4500, 125, 400, -, 100, -, -, 800
Crosser, Lemuel, 1, -, -, 21, 1, 7, -, 2500, 150, 300, 50, 35, 10, 1, 400
Stewart, Hugh, 1, -, -, 20, -, -, 6, 3800, 225, 300, -, 34, 200, 40, 510
Gregg, Benjamin, 1, -, -, 140, -, 30, 10, 14000, 600, 1500, -, 20, 600, 104, 2000
Oskins, John E., -, -, 1, 98, 12, 20, 2, 6000, 410, 800, 20, 90, 250, 40, 1000
Lynch, Patrick, 1, -, -, 5, -, -, -, 1500, 50, 200, 15, 5, 10, 5, 400
Wilson, Stephen, 1, -, -, 150, 22, 30, -, 12000, 500, 700, 125, 125, 250, 40, 200
Gebhart, Benjamin F., 1, -, -, 62, 16, 25, -, 6000, 500, 500, 25, 50, 100, 6, 1500
Pierson, Hannah _., 1, -, -, 113, 2, 30, 2, 8000, 500, 1000, 20, 10, 350, 52, 1500
Peoples, William T., 1, -, -, 64, 1, -, -, 7000, 500, 750, 50, 190, 550, 52, 1546
Graves, Rachel, 1, -, -, 56, 1,6, -, 6000, 200, 350, 20, 100, 340, 48, 1090
Armor, Abraham F., -, -, 1, 90, 1, 9, 2, 11000, 300, 400, 50, 200, 350,52, 1200
Graves, Benjamin, -, -, 1, 41, 1, -, 1, 5000, 200, 700, -, 75, 200, 40, 525
Pierce, Henry, 1, -, -, 43, -, -, -, 5000, 200, 500, 35, 40, 125, 36, 935
Pierce, William, 1, -, -, 44, 1, 5, -, 5000, 200, 450, 20, 40, 200, 36, 925
Clairnan, Henry, -, -, 1, 48, -, 20, -, 5000, 200, 350, 50,70, 175, 36, 950
Scanling, Peter, 1, -, -, 5, -, 2, -, 800, -, 200, 50, 40, 25, 2, 250
Jefferis, William, -, 1, -, 100, 30, 5, -, 12000, 250, 530, 20, 50, 275, 52, 1120
Grace, Walter, 1, -, -, 34, -, -, -, 6000, 300, 500, 15, 50, 25, 3, 800
Harigan, Jerry, 1, -, -, 18, -, -, -, 3000, 200, 600, 20, 50, 150, 20, 820
Garnett, Samuel, 1, -, -, 20, 2, -, -, 6000, 250, 400, 20, 20, 30, 3, 800
Heald, Caleb, 1, -, -, 10, -, -, -, 6000, 200, 100, 15, 20, 225, 52, 250
Flinn, Isaac W., 1, -, -, 190, 1, 20, 6, 19000, 400, 1275, 70, 100, 675, 104, 1625
Lamborn, Lewis H., 1, -, -, 15, -, 6, -, 3000, 200, 150, -, 7, 100, 15, 500
McCarty, Cornelias, 1, -, -, 58, 2, 6, -, 6000, 200, 800, 20, 52, 50, 5, 1000
Thompson, John, -, 1, -, 93, 2, 15, -, 6000, 600, 1000, 100, 90, 175, 30, 1200
Derickson, Lewis, 1, -, -, 102, -, 1 ½, 11, -, 11000, 500, 800, 40, 200, 300, 52, 1850
Swainey, Miles, 1, -, -, 18, -, 1, 6, 2000, 150, 250, 100, 20, 150, 20, 600
Mclewrie, Lewis, 1, -, -, 109, 7, 15, -, 8000, 600, 900, 20, 140, 200, 25, 2200
Derickson, William M., 1, -, -, 68, 1, 5, 5, 7000, 200, 700, 25, 150, 400, 52, 1100
Derickson, Joseph, 1, -, -, 120, 5, 10, 3, 1200, 300, 1000, 100, 105, 700, 104, 2200
Durross, Bernard, 1, -, -, 44, -, 6, -, 5000, 300, 700, 20, 50, 300, 45, 1000
Woodward, Irwin, -, 1, -, 70, -, 5, -, 4500, 200, 1000, 20, 80, 100, 12, 1000
Lynam, Anna M., -, 1, -, 76, -, 4, -, 6000, 200, 800, 25, 60, 300, 36, 1400
Margaugle, Ephraim, 1, -, -, 9, 1, -, -, 5000, 100, 350, 10, 40, 200, 25, 600
Philips, Isaac, 1, -, -, 14, 8, 3, 5, 5000, 700, 400, 20, -, 375, 60, 800
Logan, Samuel, 1,-, -, 23, -, 2, -, 2500, 200, 200, 35, 35, 200, 25, 600
Yearsley, McCoy, 1, -, -, 8, 1 ½, -, -, 3000, 150, 150, 30, -, 25, 3, 500
Brackin, Thomas, -, 1, -, 43, -, 7, -, 5000, 100, 300, 30, 60, 200, 75, -

Credin, Daniel, 1, -, -, 28, 1, -, -, 5000, 250, 450, 10, 30, 50, 6, 2000
Philips, John M., -, 1, -, 120, 10, 5, -, 10000, 400, 1000, 20, 100, 225, 40, 1500
Chandler, Spence, 1, -, -, 155, 23, 20, -, 160000, 600, 1200, 100, 200, 450, 88, 2200
Wilson, Stephen S., 1, -, -, 4, -, 3, -, 3000, 100, 200, 10, 10, 100, 12, 450
Ball, George, 1, -, -, 134, 2, 17, -, 15000, 600, 1200, 100, 140, 500, 90, 2300
Ferrie, T. Walters, 1, -, -, 110, 16, 8, 3, 10000, 500, 1000, 20, -, 300, 36, -, 2000
Bartholmew, Edwards, -, 1, 87, 10, 6, -, 8000, 200, 700, 20, 75, 50, 5, 1000
Cripps, Benjamin R., -, -, 1, 63, -, -, -, 5500, 250, 500, 20, 70, 350, 40, 1000
Moss, Lemuel, 1, -, -, 50, -, -, -, 6000, -, -, 20, -, 250, 28, 900
Klair, Frank, 1, -, -, 28, -, 3, -, 2000, 75, 100, 10, -, 75, 10, 500
Hoops, Thomas, 1, -, -, 52, -, 5, -, 5000, 250, 500, 20, 34, 250, 15, 750
Mclevee, Barton, -, -, 1, 48, -, 7, -, 5000, 200, 600, 10, 65, 70, 10, 1000
Highfield, Calvin, 1, -, -, 19, -, -, -, 2800, 100, 200, 25, 10, 75, 9, 400
Hollingsworth, Hannah, 1, -, -, 108, 2, 7, -, 11000, 500, 1000, 100, 120, 100, 12, 2000
Whiteman, Frank, 1,-, -, 45, 1, 4, -, 5000, 200, 450, 50, 75, 50, 6, 900
Margargargle, Orlands, -, -, 1, 50, 7, 3,-, 8000, 600, 850, 25, 150, 375, 52, 2900
Sowder, Edward, 1, -, -, 95, 20, 30, -, 12000, 500, 1300, 50, 105, 352, 52, 1000
Hulett, Josiah, -, -, 1, 175, 32, 15, -, 14000, 500, 2000, 25, 1000, 650, 95, 3340
Faust, John, 1, -, -, 18, -, 1, -, 4000, 200, 300, 30, 775, 8500
Mitchell, Dr., Taylor, 1, -, -, 36, -, -, -, 5000, 150, 200, -, 70, 190, 26, 920
Brackin, Berthia, 1, -, -, 47, -, 2, -, 5000, 400, 600, 10, 40, 300, 48, 945
Taylor, John, 1, -, -, 31, -, 2, -, 4400, 250, 500, 10, 15, 125, 15, 800
Mitchell, Stephen H., -, -, 1, 136, 16, 20, -, 13000, 350, 1000, 100, 175, 190, 75, 2200
Bailey, Samuel, 1, -, -, 66, 19, 4, -, 6300, 400, 900, 50, 200, 670, 88, 2108
Clark, Henry, 1, -, -, 46, -, 4, -, 5000, 300, 500, 10, 45, 325, 52, 900
Fergison, Margaret, 1, -, -, 9, -, -, -, 2800, 100, 150, -, 7, 30, 4, 400
Graves, David, 1, -, -, 73, 2,8, -, 8000, 500, 600, 40,70, 400, 60, 150
Bailey, John, 1, -, -, 56, -, 4, -, 5000, 400, 500, 20, 35, 200, 40, 1100
Worth, William H., 1, -, -, 65, 2, 8, -, 7300, 100, 700, 175, -, 175, 32, 1500
Graham, Hamilton, 1, -, -, 95, 15, 5, -, 10000, 400, 1000, 150, 270, 300, 52, 2175
Reece, William, 1, -, -, 34, -, -, 1, 6000, 100, 600, 40, 170, 150, 30, 760
Griffin, James, -, -, 1, 90, 15, 5, 10000, 300, 400, 100, 100, 5, 8, 1200
Knowles, William, -, 1, -, 14, -, -, -, 2000, 100, 150, -, 10, -, -, 300
Fisher, John, -, 1, -, 5, ½, -, -, 1500, 50, 150, 10, 16, 25, 3, 400
Hulett, Martha, 1, -, -, 68, -, 6, -, 6000, 500, 900, 20, 70, 200, 25, 1630
Ralston, William, 1, -, -, 27, -, -, -, 3000, 100, 200, 10, 80, 20, 3, 550
Schwenderman, Laurence, 1, -, -, 6, -, -, -, 1800, 100, 175, -, -, 75, 10, 400
Boughman, Joseph, -, 1, -, 50, -, 15, -, 5000, 300, 650, 10, -, 400, 52, 100
Leach, Joseph, 1, -, -, 100, 2, 24, 4, 8000, 400, 1000, 20, 70, 250, -, 40
Jordan, Marshall, -, 1, -, 44, 2, 3, -, 4400, 150, 300, 10, -, 250, 36, 900
Armstrong, John, 1, -, -, 65, 30, 30, -, 8000, 500, 1000, 20, 100, 100, 12, 1800
Miller, Lewis, 1, -, -, 98, -, 6, -, 9000, 300, 700, 20, 65, 100, 12, 1800
Little, William, 1, -, -, 100, 3, 10, -, 10000, 400, 500, 50, 56, 600, 12, 1000

Sharpless, Samuel, 1, -, -, 140, 1, 10, -, 12000, 500, 1750, 25, 75, 400, 95, 1425
Mitchell,William, -, -, 1, 130, -, 20, -, 8000, 300, 700, 200, 160, 350, 75, 1700
Sharpless, A. & J. D., 1, -, -, 130, 26, 10, -, 14000, 500, 1700, 50, 100, 800, 60, 2200
Mason, Jonathan, 1, -, -, 60, 10, 25, -, 5000, 800, 700, 20, -, 200, 125, 1200
Lamborn, Chandler W., 1, -, -, 5, 10, 50, -, 1100, -, 100, -, -, 200, 40, 60
Dixon, Maris, -, -, 1, 115, 15, 35, -, 12000, 400, 1450, 20, 108, 350, 52, 1800
Chandler, Jacob, 1, -, -, 76, 2, 35, -, 8500, 500, 800, 25, 24, 350, 52, 1800
Poole, Thomas H., 1, -, -, 73, 1 ½, 10, -, 7000, 500, 600, 26, 100, 300, 50, 1500
Fisher, Richard S., 1, -, -, 87, 3, 12, -, 10000, 500, 1000, 20, 250, 600, 75, 2000
Armstrong, Achabald, 1, -, -, 72, 2, 30, -, 8000, 250, 350, 50, 100, 300, 52, 1600
Vandever, Amos, -, -, 1, 83, 10, 15, -, 7500, 300, 600, 25, 60, 60, 8, 1500
Cloud, Clarkson, 1, -, -, 22, 4, -, -, 4500, 150, 250, 100, 7, 175, 25, 750
Barker, Joseph, 1, -, -, 57, 12, 8, -, 7800, 500, 800, 20, 50, 175, 25, 1500
Dewese, Christopher, 1, -, -, 53, -, 1, 6, -, 4800, 200, 300, 25, 17, 100, 15, 700
Marshall, Hannah, 1, -, -, 6, -, 4, -, 3000, 150, 275, 10, 54, 20, 3, 600
Hobson, William, 1, -, -, 87, 2, 10, -, 7000, 500, 700, 25, 70, 100, 12, 1800
Mitchell, John, 1, -, -, 140, 4, 10, -, 15000, 500, 1600, 25, 105, 400, 52, 2400
Hendrickson, William, -, 1, -, 25, 1, -, -, 3000, 300, 400, 50,70, 50, 6, 900
Connell, Peter, 1, -, -, 44, 5, 8, -, 4500, 300, 425, 20, 80, 75, 9, 1000
Bumgust, J. R., 1, -, -, 48, -, -, -, 8000, 550, 650, 20, 70, 400, 52, 1300
Fell, William J., 1, -, -, 100, -, 60, -, 15000, 400, 100, 50, 100, 600, 70, 2000
Chandler, Dr. Swethan, 1, -, -, 7, -, -, -, 6000, 50, 300, 15, 15, 200, 40, 400
Jackson, William B., 1, -, -, 50, -, -, -, 10000, 300, 350, 100, 19, 300, 35, 1130
Lamborn, Chandler, 1, -, -, 120, 2, 30, -, 12000, 2000, 750, 25, 150, 500, -, 70
Groves, Samuel, 1, -, -, 998, 2, 10, -, 12000, 400, 1000, 30, 70, 200, 40, 2200
McVaugh (McSough), Franklin, 1, -, -, 28, -, 4, -, -, 5000, 250, 600, 100, 15, 350, 50, 900
Thompson, Henry, -, 1, -, 80, 4, 20, 4, 10000, 400, 1000, 55, 120, 500, 80, 1934
Moore, Samuel L., 1, -, -, 40, -, -, -, 3600, 75, 318, 5, 54, 50, 6, 568
Sentman, Joseph, 1, -, -, 12, ¼, -, ½, 1000, 30, 177, -, -, -, -, 107
Collins, George W., -, 1, -, 67, 1 ½, 15, -, 7500, 200, 727, 8, 70, 150, 40, 1600
Bell, Nelson, 1, -, -, 100, -, 65, -, 6000, 250, 700, -, -, -, -, 300
Carlisle, Theodore, 1, -, -, 50, -, 10, -, 48, -, -, 50, 178, 200, 56, 817
McCormick, John, 1, -, -, 16/12, -, 1 ½, -, -, 1800, 50, 305, 6, 60, 100, 10, 263
Mitchell, Joseph, 1, -, -, 85, 1, 13, -, 9000, -, 1300, 10, 200, 600, 104, 2000
Mitchell, Joseph, 1, -, -, 80, -, 15, -, 7500, 500, -, 15, -, -, -, 2000
Woodward, Frederick, 1, -, -, 65, -, 1, 14, -, 8000, 200, 400, 60, 50, 200, 45, 1400
Mitchell, Stephen, 1, -, -, 82, 1, 18, -, 9000, 200, 300, 5, 100, 300, 64, 1100
Woodward, E. & A., 1, -, -, 100, 12, 16, -, 10000, 200, 600, 67, 50, 240, 44, 1700
Mclen, Samuel S., -, 1, -, 130, 2, 20, 10, 10000, 250, 1100, 50, 68, 235, 40, 1800
Hoopes (Hooper), Hard___, 1, -, -, 53, -, 1, 8, -, 4960, 200, 500, 25, 158, -, -, 1025
Thompson, George, 1, -, -, 100, 1 ½, 20, -, 10000, 200, 500, 10, 75, 300, 52, 1700
Seal, Thomas G., 1, -, -, 64, 4, 12, -, 6500, 300, 750, 50, 100, 300, 52, 2387
McCormick, Ann, 1, -, -, 59, 1, 6, -, 5000, 225, 460, 25, 60, 8, 1, 900
Cleaver, James M., -, -, 1, 99, 1, 8, -, 6000, 200, 275, -, -, -, -, 1066

Harras, Isabella, -, 1, -, 73, -, 2, 5, -, 6000, 150, 480, 5, 17, 160, 34, 800
Fell, Watson, 1, -, -, 80, 3, 15, -, 8000, 500, 1055, 15, 68, 144, 40, 1738
Quil, John, 1, -, -, 40, 1, 30, 20, 4000, 200, 500, 95, 100, 100, 26, 1100
Aiken, Samuel, 1, -, -, 16, ¼, -, -, 1200, -, 57, 25, 40, -, -, 110
Aiken, James, 1, -, -, 58 ½, ½, 4, -, 3150, 150, 209, 50, 100, 200, 84, 600
Hopkins, Abel J., 1, -, -, 195, 12, 15, -, 16000, 800, 2400, 10, 150, 560, 156, 3000
Eastburn, Oliver W., -, 1, -, 90, 1, 15, 2, 6500, 400, 700, 34, 100, 250, 52, 1340
McCannon, Joel, 1, -, -, 33, 2 ½, -, -, 500, 300, 500, 5, 75, 15, 2, 960
Moore, Levi G., 1, -, -, 110, 3, 10, -, 9000, 450, 775, 75, 200, 325, 800, 1839
Chambers, George B., -, 1, -, 36, ¼, 3, -, 2000, 80, 390, -, -, -, -, 5
Foote, George, -, 1, -, 15 ½, ½, 2, -, 800, 30, 40, 2, 8, -, -, 110
Barlow, Mary A., 1, -, -, 18 3/5, -, 1/8, 1, -, 2000, 30, 110, -, 18, -, -, 200
Morrison, William A., 1, -, -, 125, 4, 60, 25, 13000, 200, 1175, 200, 105, 430, 88, 2100
Warren, Harry, 1, -, -, 120, 25, 30, -, 9000, 350, 900, 50, 75, 700, 150, 1900
Reynolds, John A., 1, -, -, 95, -, 5, -, 8000, 650, 1000, -, 75, 250, 64, 1862
Klair, George T., 1, -, -, 109, 2, 9, -, 10000, 400, 685, 15, 75, 224, 40, 1532
Denison, Samuel, 1, -, -, 79, 2, 15, -, 9000, 300, 855, 50, 150, 50, 50, 1675
Whiteman, Charles, -, 1, -, 25 ½, ½, -, -, 1300, 100, 180, -, -, 140, 52, 118
Pennock, William H., -, 1, -, 11, -, -, -, -, 25, 100, 25, -, -, -, -, -
Hanna, Joseph, 1, -, -, 56, 1, 8, -, 5200, 300, 430, 5, 30, 38, 4, 856
Derickson, Calvin, -, -, 1, 85, 2, 18, -, 9000, 300, 480, 150, 150, 349, 70, 2100
Walker, Sarah D., 1, -, -, 95, 25, 25, -, 10640, 300, 1250, 75, -, 500, 100, 1873
Chambers, Newton _., 1, -, -, 89, 4, 2, 2, 75, 200, 550, 30, 75, 250, 52, 1600
Denison, James, 1, -, -, 70, 2, -, -, 5000, 200, 490, 70, 80, 900, 75, 1231
Hanna, William T., 1, -, -, 26, 1, -, -, 1500, 100, 60, 20, 33, -, -, 157
Taylor, Samuel, 1, -, -, 79, 13, -, -, 7900, 200, 1000, 50, 117, 275, 80, 1678
Walker James H., 1, -, -, 81, 30, 13, 1, 9375, 300, 1459, 100, 164, 350, 80, 2800
Walker, Alfred W., 1, -, -, 165, 3, 15, -, 8000, 300, 710, 300, 160, 300, 50, 1488
Tweed, William B., -, 1, -, 2/8, ¼, -, -, 1000, 25, 50, 5, -, -, -, 24
Little, James, 1, -, -, 103, 2, 12, -, 4680, 100, 450, 50, 60, 300, 80, 1128
Chambers, John K., -, 1, -, 72 ½, ½, 10, -, 4000, 100, 335, 5, 38, -, -, 450
Robinson, George C., -, 1, -, ½, 1 ½, 12, -, 4375, 250, 572, 50, 100, 185, 52, 700
Kelly, William H., 1, -, -, 290, -, 3, -, 1500, 30, 135, 35, 32, -, -, 225
Harkness, Samuel, 1, -, -, 45, 1, 4, -, 2250, 150, 380, 5, 53, -, -, 412
Chambers, Sarah A., 1, -, -, 8 ¾, ¼, 1, -, 800, 40, 35, 15, -, -, -, 60
Kearns, Edward B., 1, -, -, 9 ½, ½, -, -, 1000, 80, 108, 3, 8, -, -, 288
Whiteman, Gilbert, 1, -, -, 10, ½, -, -, 1000, 100, 80, 10, 15, -, -, 100
Russell, Christiana, 1, -, -, 3 ¾, -, -, -, 800, -, 50, 2, -, -, -, 8
Jacquott, James, 1, -, -, 16, ¼, -, -, 800, 30, 270, 15, 16, -, -, 176
Taylor, Robert, 1, -, -, 91, 1, 60, 10, 8000, 250, 550, 150, 60, 230, 52, 1200
Pennock, Franklin, 1, -, -, 92 ½, ½, 7, 1, 6500, 325, 755, -, -, -, -, 9000
Russell, Samuel, 1, -, -, 35, ½, 4 ½, -, 2500, 100, 170, 2, 30, 20, 6, 500
Rankin, Joseph, 1, -, -, 47, 3, 50, -, 11000, 200, 930, 90, 80, 220, 52, 1550
Counselor, James H., 1, -, -, 4 ½, -, 11, -, 800, 15, 20, 35, -, -, -, 40
Jacobs, Sarah, 1, -, -, 76, -, 17, -, 3000, 75, 260, 4, 33, 25, 4, 300

Mote, Sarah C., 1, -, -, 99, 1, 26, -, 5000, 150, 430, 2, 70, 200, 52, 750
Jackson, Peter, 1, -, -, 80, -, 20, -, 4000, 200, 1050, 25, 200, 400, 104, 1469
Peters, Randolph, -, 1, -, 75, 50, 15, -, 8000, 100, 265, 3, 95, 1000, 172, 4660
McCormick, Levi & Tamar, 1, -, -, 68 ½, 1 ½, 5, -, 3375, 200, 560, 60, 104, 300, 80, 932
Eastburn, William M., 1, -, -, 120, 20, 20, -, 7500, 150, 575, 5, 100, 400, 104, 1722
Eastburn, David, 1, -, -, 83, 2, 19, -, 8000, 350, 75, 5, 300, 475, 58, 1806
Buckingham, Richard G., 1, -, -, 75, 4, 10, -, 5500, 200, 590, 50, 75, 100, 25, 1912
Pyle, Jehu, 1, -, -, 93 ½, 1 ½, 8, -, 6000, 150, 417, -, -, -, -, 1318
Moore, Thomas J., 1, -, -, 67, 2, 7, -, 5320, 150, 770, 10, 50, 80, 10, 1129
Buckingham, Alban, 1, -, -, 58, 1 ½, 6, -, 5000, 200, 647, 10, 35, 108, 42, 1204
Armstrong, William R., 1, -, -, 88, 2, -, -, 6300, 300, 480, 100, 74, 225, 52, 1461
Temple, John K., -, -, 1, 86 ¼, ¾, 4, -, 4500, 75, 284, 15, 36, 100, 16, 895
Eastburn, Samuel, 1, -, -, 74, 3, 27, -, 4500, 150, 840, 10, 34, 460, 104, 2110
Leach, Emaline, 1, -, -, 104 ½, 1 ½, 10, -, 8000, 300, 1245, 60, 164, 500, 104, 2304
Walker, Elizabeth A., 1, -, -, 98, 1 ½, 20, -, 10000, 200, 665, 10, 85, 30, 4, 1749
Hanna, Lewis, -, 1, -, 55, 4, 5, 3, 3840, 200, 440, 10, 100, 250, 54, 1085
Bowman, Henry B., -, 1, -, 15, ½, -, -, 2000, 100, 230, -, 52, 80, 37, 692
Lynam, R. Thomas, 1, -, -, 131 ½, ½, 1, -, 10000, 500, 1020, 20, 200, 400, 150, 2193
McKee, Andrew, 1, -, -, 71, -, 7, -, 5000, 200, 630, 10, 100, 75, 25, 1434
Pennock, Pusey, 1, -, -, 87 ¾, 1 ¼, 1, 1, 8000, 200, 590, 100, 50, 175, 42, 1545
Pennock, Alpheus, 1, -, -, 116, 4, 21, -, 8120, 225, 450, 20, 100, 25, 5, 1639
Yerkes, Charles L., -, 1, -, 80, 12, 10, -, 9000, 250, 810, 10, 100, 275, 56, 2050
Eastburn, Joseph, 1, -, -, 175, 25, -, -, 14000, 300, 1659, 50, 150, 700, 220, 3000
Whiteman, John K., 1, -, -, 128, 22, 15, -, 7500, 300, 1100, 75, 150, 200, 134, 1775
Eater, Robert M., -, 1, -, 93, 5, 25, 7, 8000, 200, 641, 2, 17, 170, 34, 1438
Taylor, Robert S., 1, -, -, 34 ¾, ¼, 15, -, 3500, 100, 373, 40, 25, 200, 52, 595
Paxson, Alfred I., 1, -, -, 56, -, ¾, -, -, 7000, 400, -, 10, 70, 60, 7, 1010
Little, William F., 1, -, -, 21, 2, 2, -, 1500, 800, 160, 3, 75, 200, 40, 466
Moore, Jacob, -, 1, -, 114, 4, 10, 10, 7000, 250, 573, 100, 315, 400, 68, 1428
Wollaston, Edmund, -, 1, -, 85, 12, 8, -, 7500, 450, 941, 50, 78, 244, 43, 1700
Trinder, Joseph, 1, -, -, 48, 2, 8, -, 4300, 100, 250, 60, -, 100, 25, 610
Welsh, Susanna L., 1, -, -, 24, 2, -, -, 3000, 100, 150, 25, 15, 150, 40, 272
Negendank, Louis, 1, -, -, 54, 2, 2, -, 4500, 400, 478, 150, 50, 150, 43, 708
Richardson, James T., -, 1, -, 164, 16, 20, -, 16000, 300, -, 400, 70, 375, 100, 2813
Derickson, Bayard T., -, 1, -, 100, 15, 18, -, 10000, 200, 985, 200, 199, 350, 75, 2112
Foote, Benjamin, 1, -, -, 14, 4, -, -, 1600, 100, 184, 20, 12, 10, -, 260
Haman, Edmund, 1, -, -, 79, 2, -, -, 6075, 400, 550, 100, -, 450, 52, 1669
Morrison, John C., 1, -, -, 95, 1 ½, 1 ½, -, 7500, 250, 650, 75, 80, 450, 104, 1187
Brown, James Jr., -, -, 1, 175, 6, -, 10, 18000, 500, 1695, 100, 140, 375, 92, 3988

Brown, John, -, -, 1, 146, 8, 6, -, 12000, 400, 1545, 100, 68, 380, 86, 2295
Powell, William, 1, -, -, 112, 5, 10, 20, 6615, 500, 1586, 20, 75, 380, 80, 2411
McFarlan, Robert, 1, -, -, 5, 7, -, 10, 5, 5500, 250, 420, 30, 70, 180, 60, 1120
Greenwalt, Robert, 1, -, -, 48, -, ¼, -, 2900, 150, 535, 4, 50, 150, 40, 837
Worrell, Samuel, 1, -, -, 103, 1, 4, -, 7560, 375, 640, 55, 100, 225, 60, 1814
Walker, Robert, 1, -, -, 70, 4, 15, -, 7000, 300, 565, 35, 125, 225, 40, 1200
Mitchell, Abner, 1, -, -, 153 ½, 18, 30, -, 13875, 500, 1548, 75, 300, 350, 208, 2807
Robinson, John, 1, -, -, 14, 1 ½, -, -, 900, 40, 138, 3, 10, -, -, 110
Whiteman, Henry, 1, -, -, 90, 4, 3, -, 7000, 200, 612, 20, 150, 275, 60, 1439
Harkness, James, -, 1, -, 34, 1, 3, 2, 2000, 50, 370, 20, 72, 45, -, 780
Mier, David E., 1, -, -, 120, 2, 25, 3, 7000, 100, 550, 125, 350, 300, 75, 1283
Campbell, William J., 1, -, -, 23, -, 4, -, 1500, 50, -, 35, -, -, -, 150
Rankin, Sarah, 1, -, -, 110, 8 ½, 26 ½, -, 10000, 550, 1145, 90, 150, 350, 60, 2200
Collins, Edward B., 1, -, -, 70, 1, 5, -, 4500, 250, 360, 20, 100, 250, 60, 1219
Davis, Eli, 1,-, -, 90, 6, 10, -, 5000, 300, 1050, 75, 116, 325, 60, 2291
Banks, Jane, -, -, 1, 203, 7, 7, -, -, 18000, 600, 1680, 25, 195, 600, 135, 4213
Foote, James R., 1, -, -, 51 ½, 1 ½, 5, -, 4000, 350, 450, 3, 70, 220, 40, 713
Boughman, George W., 1, -, -, 52, 4, -, -, -, 50, -, -, 87, 100, 90, 300
Holland, John, 1, -, -, 105, 3, -, -, 6480, 300, 610, 50, 150, 390, 80, 1391
Rankin, Robert T., 1, -, -, 118, 2, 40, -, 11000, 300, 950, 25, 180, 380, 90, 1877
Stinson, William P., 1, -, -, 20, ½, -, -, 2000, 100, 280, 10, 140, 100, 18, 685
Whiteman, Israel, 1, -, -, 173, 4, 10, -, 8255, 300, 1025, 125, 450, 80, 2006
Worrall, John W., 1, -, -, 50, 1 ½, -, -, 3500, 300, 335, 75, 150, 200, 60, 907
Morrison, Catherine, 1, -, -, 9, -, -, -, 900, 30, 60, 10, 10, 40, 14, 337
Foracre, William, -, 1, -, 5, ½, -, -, 700, -, -, 10, -, -, -, 5
Crockett, John E., -, 1, -, 56 ½, ½, 10, -, 2500, -, 125, 50, 32, -, -, 842
Lockard, Mary A., 1, -, -, 12, -, -, -, 1300, -, 30, -, -, -, -, 60
Morrison, Douglas, -, -, 1, 222, 3, 15, -, 16000, 600, 1800, 20, 231, 430, 90, 3357
Vought, Joseph H., 1, -, -, 137, 13, 13, -, 9000, 125, 525, 100, 114, -, -, 1798
Sayle, James, 1, -, -, 59, 7, 6, -, 4000, 250, 642, 184, 122, 417, 80, 858
Egee, Elmer, -, 1, -, 76, 3, 16, -, 3880, 100, 400, -, -, -, -, 600
Johnson, John C., 1, -, -, 95, 10, 20, -, 7000, 250, 630, 100, 200, 175, 52, 1754
Alexander, Guthrie, 1, -, -, 60, -, 15, -, 5600, 300, 300, 50, 50, 150, 6, 989
Cranston, Edward, 1, -, -, 52, -, 1, -, -, 150, 550, 50, 75, 235, 52, 1221
Satterthwait, Reuben, 1, -, -, 72 ½, 20 ½, 5, -, 10800, 300, 1550, 50, 100, 550, 160, 2105
Taylor, Annie, 1, -, -, 10, -, -, -, 1000, -, -, -, -, -, -, 100
Negendank, William, 1, -, -, 13, ¼, -, -, 3000, -, 300, 40, 35, 10, 1, 431
Dashper, John, -, -, 1, 125, 20, 5, -, 5500, 250, 890, 50, 115, 140, 40, 500
Wollaston, Albert, 1, -, -, 105, 5, 2, -, 10000, 500, 1417, 10, 70, 300, 50, 26, 78
Currinder, Charles, -, 1, -, 91, 7, -, 3, 8000, 300, 797, 50, 100, 250, 52, 1486
Murrey, Levi, -, -, 1, -, 2, 25, 25, 16500, 500, -, 150, 284, 600, 145, 4769
Stafford, James, -, 1, -, 186, -, 20, -, 6500, 300, 1255, 25, 65, 300, 52, 840
Greenwalt, John L., 1, -, -, 107, 2, 25, -, 6500, 1100, 875, 200, 94, 300, 75, 1532
Richards, Dutten, 1, -, -, 55, 2, 10, -, 2500, 300, 480, 80, 100, 50, 4, 1180

Ware, James R., 1, -, -, 17 ½, ½, -, -, 1500,70, 70, 40, 100, -, -, 179
MacDonald, Edward, 1, -, -, 28, 2, -, -, 1800, 50, 135, 10, 35, -, -, 400
Chillas, Estate of Arthur, 1, -, -, 125, 1, 40, 10, 10350, 300, 759, 5, 100, 150, 52, 1892
Yarnall, Margaret, 1, -, -, 8, ½, 1, -, 1500,80, 90, 3, 12,20, 4, 158
Brown, Joseph, -, 1, -, 37, 1, 7, 7, 2500, 400, 300, 35, 40, -, -, 600
Fitzsimmons, Robert, 1, -, -, 46, 1, 30, 2, 3200, 200, 396, -, 40, 30, 80, 555
Whiteman, Henry M., -, 1, -, 8, -, 8, -, 1200, 100, 111, 5, -, -, -, 94
VandenLehe, Jacob, -, -, 1, 66, 1, 4, -, 3500, 100, 340, 20, 100, 50, 5, 528
Dean, John W., -, 1, -, 12, ¼, -, -, 1600, 50, -, 2, -, -, -, 225
Higgins, Thomas, 1, -, -, 119, 5, 6, -, -, 500, 1280, 100, 200, 500, 104, 1837
Brown, Ann E., 1, -, -, 8, ½, -, -, 1000, 140, -, -, 35, 50, -, 172
Guthrie, Joseph, 1, -, -, 41, ½, 3, -, 3000, 50, 320, 30, 50, 150, 40, 502
Davis, Robert K., 1, -, -, 87, 4, 9, -, 7500, 300, 750, 25, 200, 400, 80, 1866
Whiteman, Jackson A., 1, -, -, 64, 4, 8, -, 4500, 150, 495, 65, 50, 150, 36, 1242
Harkness, William Estate of, 1, -, -, 139, 6, 30, 25, 12000, 300, 1175, 75, 80, 500, 104, 2000
Parker, John, 1, -, -, 23, ½, ½, -, 2000, 50, 205, 4, 34, 100, 36, 430
Minnie, Patrick, 1, -, -, 15, -, -, -, -, 50, 130, -, 35, -, -, 312
Lynch, Humphrey, 1, -, -, 3 ½, -, -, -, -, -, 70, -, -, -, -, 100
Taylor, David W., -, 1, -, 100, -, 70, -, -, 400, 1035, 7, 119, 450, 8, 1642
Dean Brothers, -, 1, -, 8, -, -, -, 800, -, 450, -, -, 175, 25, 40
Walton, William H., 1, -, -, 80, 2 ½, 15, -, 7500, 300, 975, 25, 150, 300, 104, 2087
Meyers, Taylor J., -, 1, -, 18, 5, 2, 7, 2000, 300, 340, 13, 12, -, 15, 505
Baldwin, Thomas L. J., 1, -, -, 60, 3, 4, -, 5600, 100, 500, 40, 50, 225, 40, 958
Walker, Thomas M., 1, -, -, 78, 3, 20, -, 7500, 300, 465, 40, 100, 220, 40, 1450
Flannery, Thomas, 1, -, -, 15, -, -, -, 900, 18, 149, 1, -, -, -, 230
Sutton, Robert, -, 1, -, 10, -, -, -, 3000, 500, 500, -, 12, 12, 3,1500
Everson, John, -, -, 1, 98, -, -, -, 10000, 500, 800, 30, 100, 150, 30, 2000
Ayars, Jeptha, -, -, 1, 140, -, -, -, 15000, 800, 800, 50, -, 200, 25, 3500
Wheatley, Ezekiel, -, 1, -, 147, -, -, -, 15000, 800, 1000, 10, 25, 200, 20, 2000
Calhoun, John W., -, 1, -, 108, -, -, -, 7000, 600, 500, 100, 75, 150, 30, 1200
Banks, William, -, -, 1, 150, -, -, -, 16000, 800, 1000, 100, 100, 800, 39, 2500
Biggs, Alexander, -, 1, -, 175, -, -, -, 17000, 1000, 2500, 100, 100, 600, 40, 2500
Newlon, Salomen, -, -, 1, 208, -, -, -, 16000, 500, 1000, 1000, -, 200, 52, 3000
Fols, Ezekiel, -, 1, -, 90, -, -, -, 8000, 300, 600, 50, 2, 400, 26, 1500
Moore, Robert, -, -, 1, 160, 1, -, -, 10000, 500, 1000, 100, 90, 600, 40, 2000
McMullin, John, -, -, 1, 165, 12, 4, -, 14000, 500, 800, 200, 125, 450, 32, 1800
Rogers, Eugene, -, -, 1, 210, -, 60, 16000, 500, 800, -, 50, 400, 36, 1500
Eckles, John S., -, -, 1, 150, -, -, -, 14000, 500, 500, -, 100, 250, 36, 1200
Huggins, Joseph, -, 1, -, 75, -, -, -, 7500, 200, 300, -, -, 2000, 30, 950
Edwards, George, 1, -, -, 10, -, -, -, 5000, 200, 200, -, 50, 400, 36, 1500
Janvier, Julian D., 1, -, -, 180, -, 10, -, 15000, 500, 500, -, 35, 300, 236, 2500
Clark, Elmer W., -, -, 1, 150, -, -, -, 20000, 1000, 1000, 100, -, 1000, 3, 6, 3500
Stoops, Edward A., -, -, 1, 170, 2, -, -, 12000, 500, 200, 40, -, 300, 36, 1800
Steele, Hugh E., -, 1, -, 375, 75, 4, -, 25000, 1200, 2500, 300, 600, 940, 52, 4500

Halcomb, Thomas, 1, -, -, 18, 2, -, -, 12000, 50, 350, -, 10, 400, 40, 500
Nivin, David G., 1, -, -, 150, 28, 10, -, 11280, 500, 2000, -, 100, 300, 52, 1500
Smith, Charles H., -, -, 1, 80, 60, -, 24, 7000, 125, 400, 330, 125, 100, 40, 840
Smith, Azariah F., -, -, 1, 250, -, 30, 20, 22500, 500, 2800, -, 118, 800, 36, 3000
Davidson, Alexander H., -, -, 1, 218, 32, -, 12, 20000, 800, 3000, 500, 80, 700, 52, 4000
Weldin, John, -, 1, -, 24, 1, -, 3, 8000, 500, 600, 150, 32, 400, 52, 560
Righter, Charles B., -, 1, -, 60, 20, -, 20, 6000, 200, 300, 1000, 200, 200, 40, 1000
Newkirk, William A., -, -, 1, 200, -, -, -, 16000, 500, 600, 125, 88, 300, 36, 2000
Jordan, Charles M., -, 1, -, 27, -, -, 3, 1800, 200, 325, -, -, 200, 36, 1100
McCoy, William B., -, -, 1, 131, -, 9, -, 8400, 450, 1200, 100, 150, 530, 52, 2000
Silver, Henry M., 1, -, -, 126, 4, -, -, 9800, 300, 600, -, 200, 200, 36, 900
Waters, Thomas, -, -, 1, 150, -, 50, -, 16000, 800, 1500, 100, 200, 300, 52, 2500
Richards, Sarah, 1, -, -, 90, -, 10, 5, 8000, 300, 600, 50, 62, 2500, 36, 600
Gray, Henry, 1, -, -, 10, -, 2, -, 1200, 150, 300, 50, 25, 100, 30, 200
Poole, Thomas, 1, -, -, 27, -, -, -, 2000, 50, 150, -, -, -, -, 600
Eastburn, Lloyd, -, -, 1, 50, -, 50, -, 4000, 50, 100, -, -, 100, 36, 400
Biddle, Deborah, 1, -, -, 100, 1, 25, -, 7500, 300, 800, -, 90, 150, 32, 1500
Biddle, John, -, -, 1, 160, -, 15, -, 17500, 400, 700, -, 180, 300, 36, 2450
Miles, Mary A., -, 1, -, 90, -, -, -, 10000, 150, 900, -, -, -, -, 1000
Hurst, James M., 1, -, -, 80, 2, -, 7, 9000, 600, 1000, 50, -, 600, 52, 1500
Estlin, Theodore, 1, -, -, 8, -, -, -, 3000, 150, 125, -, 50, 75, 12, 300
Peck, Isaac H., 1, -, -, 6, -, -, -, 1800, 50, 100, -, -, -, -, 900
Lentz, Joseph, -, 1, -, 10, -, -, -, 10000, 1000, 1000, -, 200, 500, 52, 3000
Lentz, David, -, 1, -, 13, -, -, -, 13000, 200, 300, -, 150, 500, 53, 2500
Sutton, William, -, 1, -, 15, -, -, -, 15000, 200, 250, -, 75, 200, 26, 3500
Flinn, James, 1, -, -, 11, 10, -, -, 6000, 150, 100, -, 50, -, -, 7300
Smith, James overseer, -, -, 1, 12, -, -, -, 6000, 100, 190, -, 50, 150, 24, 800
Greeman, Joel V., 1, -, -, 14, 3, -, -, 8000, 200, 300, -, 25, 200, 36, 200
Pugh, John, 1, -, -, 47, -, -, -, 9000, 300, 400, 20, 50, 300, 52, 600
Smith, William, -, -, 1, 196, -, 4, -, -, 30000, 1000, 1500, 125, 75, 458, 52, 4000
Kirk, LeRoy R., -, -, 1, 140, -, -, 150, 15000, 300, 800, -, 60, 500, 36, 1800
White, Henry M., -, -, 1, 180, -, 20, -, 16000, 250, 1200, 20, 60, 600, 52, 1800
Revis, Milburn, -, 1, -, 12, -, 15, 75, 3600, 500, 200, -, 60, 200, 36, 1100
Kelley, Abraham, -, 1, -, 45, -, 15, 75, 10000, 300, 500, -, -, 206, 52, 600
McGrier, William J., -, 1, -, 82, 60, -, -, 69000, 100, 400, -, -, 200, 36, 1200
McMahan, James, -, 1, -, 32, 116, -, -, 28000, 150, 1500, -, -, 600, 52, 1200
Burris, James W., -, 1, -, 14, -, -, 4, 4000, 125, 100, -, 3, 200, 36, 1000
Jackson, Samuel A., -, -, 1, 88, 2, -, -, 13500, 300, 600, -, 60, 300, 36, 200
Willis, John, 1, -, -, 7, 1 ½, -, -, 5000, 50, 120, -, 55, 130, 36, 700
Alrich, Lucas, 1, -, -, 58, 60, -, -, 35000, 500, 1500, -, -, 450, 52, 1500
Cross, Robert B., -, 1, -, 13, -, -, -, 3900, 150, 125, -, 9, 180, 52, 600
Rose, David C., -, 1, -, 14, 4, -, 2, 5000, 200, 300, -, 70, 150, 36, 1000
Fox, John, 1, -, -, 10, 7, -, -, 6000, 400, 500, -, 100, -, -, 1500
Mirely, Edward N., 1, -, -, 6, -, -, -, 12000, 800, 200, -, -, 500, 52, 800
Chalk, James, -, 1, -, 2, -, -, -, 800, 200, 300, -, 25, 189, 36, 800

Landers, John & Sons, 1, -, -, 10, -, -, -, 3000, 300, 250, -, 15, -, -, 2600
Russell, Sarah, -, 1, -, 9, -, -, -, 5000, 100, -, -, -, 250, 52, 600
Jordan, John J., -, -, 1, 70, -, -, -, 3500, 200, 800, -, -, 400, 52, 1000
Peach, William 1, -, -, 100, -, 20, 10, 12000, 500, 800, 25, 72, 750, 52, 1800
Stroup, William W., -, -, 1, 127, 2, 8, 7, 16000, 700, 1200, 50, 100, 500, 52, 2000
Tatnall, Ashton R., -, -, 1, 215, 90, 10, -, 40000, 500, 2000, -, 200, 10, 36, 4000
White, William, -, -, 1, 100, 15, -, 2, 17000, 500, 900, -, 75, -, -, 1600
Peters, Randolph, 1, -, -, 59, 16, -, -, 30000, 1000, 1500, -, 1000, 5000, 52, 65000
Moore, Hannah, -, -, 1, 100, 70, -, -, 17000, 1000, 1500, 100, -, 500, 52, 1600
Powell, John W., -, -, 1, 230, -, -, -, 18400, 400, 1600, -, 250, 450, 36, 3200
McKee, John P., -, -, 1, 450, -, 50, 25000, 600, 2000, -, 60, 650, 36, 1500
Barnaby, Joseph W., -, -, 1, 108, -, -, 2, 5000, 100, 300, 35, 66, 225, 32, 400
McFarland, Manlove, -, -, 1, 90, 8, -, -, 7840, 200, 800, -, 90, 300, 36, 1500
Stafford, John, -, -, 1, 283, -, -, -, 40000, 1000, 4000, 50, 305, 720, 36, 6500
Knotts, William T., -, -, 1, 117, 3, -, -, 11000, 450, 1500, 250, 90, 300, 36, 1100
Armstrong, Spencer P., -, -, 1, 100, -, -, -, 8000, 400, 900, -, 105, 450, 52, 1500
Vane, William S., -, -, 1, 90, 30, 15, -, 8000, 250, 800, -, -, 270, 36, 1100
Frail, John, 1, -, -, 1, 2, -, -, 700, 10, 300, -, -, -, -, 150
Roberts, Samuel P., -, -, 1, 50, -, -, -, 4000, 250, 700, 100, 25, 100, 12, 600
Fow, John, 1, -, -, 235, 2, 12, 15, 10000, 900, 700, -, 200, 300, 36, 2700
Tybout (Tybont), George Z., 1, -, -, 175, 190, -, -, 36500, 800, 8000, -, 400, 1600, 52, 3300
Forsythe, Elisha, -, 1, -, 200, 6, -, -, 25000, 400, 2000, -, 100, 600, 52, 1800
Porter, Samuel M., 1, -, -, 100, -, -, 20, 5000, 200, 500, 25, 75, 180, 36, 1100
Powell, William, -, 1, -, 140, 6, 1,3, 7000, 700, 800, -, 75, 400, 36, 1200
Burris, John W., -, -, 1, 35, 2, 20, 8, 5000, 200, 800, -, 50, 250, 36, 1300
Crossan, Eli, 1, -, -, 120, -, 60, -, 14400, 500, 900, -, 100, 200, 36, 2000
Ross, Robert S., -, -, 1, 130, 1, 7, 10, 6000, 300, 1000, -, 75, 300, 36, 1800
Walther, Frederick, 1, -, -, 82, 20, -, 10, 3000, 200, 600, -, 50, 300, 36, 1000
Diehl, John, 1, -, -, 240, 3, 10, -, 25000, 800, 1500, -, 250, 900, 36, 3500
Lank (Louk), William J., -, 1, -, 150, -, -, 50, 10000, 200, 1000, -, 125, 350, 36, 2000
Morrison, Thomas, -, -, 1, 120, 2, 15, -, 8000, 200, 500, 115, 160, 500, 36, 2000
Appleby, Rachel Anna, 1, -, -, 140, -, 12, 18, 8000, 300, 800, -, 188, 200, 40, 1500
Pearce, Benjamin C., -, 1, -, 100, 18, -, -, 25000, 300, 500, 250, 150, 250,36, 1000
Clark, Philip R., -, -, 1, 260, 20, 20, -, 25000, 500, 1000, 100, 120, 250, 52, 2500
Morrison, George W., -, 1, -, 101, -, -, -, 15000, 500, 11000, 1000, 100, 700, 52, 2000
Mitchell, Charles, -, -, 1, 110, -, 20, -, 6500, 150, 600, 10, 50, 200, 36, 1400
Taylor, Henry, 1, -, -, 85, -, 8, -, 8000, 400, 1200, -, 110, 400, 52, 1600
Allen, George, 1, -, -, 50, -, 10, 10, 5000, 200, 300, -, 50, 50, 16, 600
Cewden (Sowdon), Joseph & John, 1, -, -, 75, 20, 60, 120, 15000, 600, 560, -, 250, 400, 36, 1800
Dougherty, William, -, -, 1, 40, -, 60, -, 20000, 800, 900, -, -, 600, 36, 2800
Davidson, John W., -, -, 1, 192, 8, -, -, 12000, 600, 2000, -, 180, 400, 36, 2200
Whitfield, George, 1, -, -, 100, 2, -, -, 10000, 600, 1200, 100, 162, 800, 52, 2000

Brown, Joseph T., -, 1, -, 194, -, -, 31, 18000, 500, 3000, -, 250, 1200, 36, 7000
Atwood, Elnathan, 1, -, -, 33, -, -, -, 4500, 300, 400, 33, 13, 175, 36, 800
Downey, Edward, -, -, 1, 175, -, -, -, 12000, 300, 500, 100, 210, 450, 36, 3500
Holcomb, Bankson T., -, 1, -, 3, 15, -, -, 8000, 100, 200, -, -, -, -, 250
Morrison, John, -, -, 1, 146, 24, 4, -, 16000, 500, -, 150, 100, 540, 36, 2600
Jackson, John J., -, 1, -, 100, -, -, -, 10000, 500, 500, 150, 90, 450, 36, 1500
White, Thomas, 1, -, -, 115, 10, -, -, 900, 500, 800, 100, 10, 250, 52, 1500
Slack, Thomas, -, -, 1, 200, -, 5, 1, 20000, 400, 1500, 130, 132, 600, 52, 300
Sheldon, Albert D., -, -, 1, 115, -, 10, -, 6000, 300, 800, 250, 78, 700, 52, 2000
Edmundson, William, -, -, 1, 198, 11, -, 3, 15000, 400, 1000, 700, 300, 100, 10, 2800
Selletos, William W., -, -, 1, 100, -, -, 40, 8000, 400, 100, -, 100, 400, 36, 1800
Lofland, Alfred, 1, -, -, 120, -, -, -, 10000, 300, 800, -, 100, 400, 32, 1300
Davis, Nehemiah, 1, -, -, 72, 10, 14, -, 6000, 400, 700, -, 64, 450, 52, 1000
Bolton, James L., -, -, 1, 104, 6, -, -, 6000, 200, 700, 25, 180, 400, 40, 1600
Appleby, William L., -, -, 1, 107, 3, -, 1, 10000, 500, 800, -, 137, 400, 52, 1800
Hayes, John, -, -, 1, 140, 3, 60, -, 10000, 200, 800, -, 135, 450, 52, 1500
Stoops, William T., 1, -, -, -, -, -, -, -, -, -, -, -, -, -, -
Grose, James M., -, -, 1, 270, 20, -, 20, 25000, 500, 1000, 50, 280, 500, 36, 5000
Diehl, John C., -, -, 1, 120, -, -, -, 7000, 300, 400, -, 90, 400, 36, 1800
Thompson, William F., 1, -, -, 120, -, 5, 20, 10000, 700, 1000, 100, 250, 500, 36, 2000
George, James H., -, -, 1, 125, -, 25, -, 15000, 500, 800, -, 40, 500, 52, 1500
Rittenhouse, Ephraim, 1, -, -, 10, -, -, -, 4000, 100, 400, -, -, 400, 52, 950
Emory, John H., -, -, 1, 170, 2, 30, -, 20000, 600, 1000, -, 125, 360, 36, 3200
Lofland, Elias, -, 1, -, 160, 40, -, -, 15000, 600, 2000, -, 24, 700, 52, 3000
Morrison, Robert, -, 1, -, 87, -, -, -, 10000, 800, 1500, -, 100, 500, 52, 2300
Bartholomew, George, 1, -, -, 10, 8, -, -, 1000, 500, 100, -, -, 250, 52, 600
McKee, Elwood, -, -, 1, 172, -, 14, -, 15000, 800, 1600, -, 150, 600, 52, 3500
Jackson, Richard, 1, -, -, 140, 28, 10, -, 20000, 800, 1500, 100, 120, 600, 52, 2500
McCoy, James, 1, -, -, 112, 35, 30, -, 13000, 300, 2000, -, 200, 600, 52, 2800
Grubb, Isaac, 1, -, -, 215, 5, 20, 30, 15000, 800, 1500, 600, 450, 700, 36, 1600
Weldin, William Atwood, 1, -, -, 285, -, -, 15, 15000, 1000, 1600, 250, 250, 1000, 35, 2500
Stapleton, Martin, -, -, 1, 130, 3, 3, 15, 15000, 400, 1200, 150, 150, 400, 36, 1800
Stafford, Henry, -, -, 1, 188, 19, 25, 62, 20000, 400, 100, 60, -, 600, -, 3000
Simpson, George, 1, -, -, 135, 155, -, -, 25000, 300, 2000, -, 50, 700, 36, 3200
McFarland, John H., 1, -, -, 90, 60, 12, 10, 16000, 400, 1500, 150, -, 800, 52, 1800
Davis, George E., -, 1, -, 80, -, 20, -, 6000, 150, 2500, 100, 30, 700, 52, 1000
Moore, Stuney (Henry) H., -, -, 1, 290, -, 50, 10, 28000, 800, 2400, 100, 225, 1000, 52, 4000
Miller, Adam, 1, -, -, 7, 3, -, -, 2000, 200, 200, -, 30, 50, -, 600
Mousely, Elwood H., -, -, 1, 100, -, -, -, 10000, 700, 1500, -, -, 500, 52, 2000
McCoy, David, -, 1, -, 132, -, -, -, 13000, 500, 1500, 100, 210, 500, 52, 2800
Morgan, David T., -, 1, -, 195, 2, -, -, 30000, 503, 2000, 100, 100, 900, 36, 3500
Davis, Jason, 1, -, -, 200, 54, 10, -, 25000, 800, 1800, -, 500, 1000, 52, 3750

Guthrie, Jones, 1, -, -, 4/4, -, -, -, 12000, 100, 1500, -, -, 300, 52, 350
Rogers, Mary _., 1, -, -, 20, 40, 15, -, 50000, 200, 3400, 40, -, 1200, 52, 1800
McGowan, Charles, -, -, 1, 130, 30, -, -, 40000, 500, 2700, -, 84, 800, 52, 3800
Frismuth, Charles, -, 1, -, 8, -, -, ¾, 2500, 50, 250, -, -, 200, 36, 450
Simon, George D., 1, -, -, 30, -, -, -, 15000, 600, 1000, 400, 300, 800, 35, 3500
Gow, Euphemia, 1, -, -, 15, -, -, -, 60000, 60, 300, -, -, 300, 36, 800
Eggleton, Alexander, -, 1, -, 16, 24, -, -, 6000, 100, 600, -, -, 300, 52, 1000
Valentine, William, 1, -, -, ¾, -, -, -, 1000, 25, 150, -, -, -, -, 400
Taylor, John E., 1, -, -, 7 ½, -, -, -, 4000, 50, 150, -, -, -, -, 600
Sutton, Ephrain, 1, -, -, 5, -, -, -, 5000, 200, 400, -, -, 50, 8, 1500
Lentz, Joseph L., -, 1, -, 4, -, -, -, 1000, 150, 150, -, -, 50, 8, 800
Stedman, Joseph K., -, -, 1, 20, -, -, -, 6000, 100, 250, -, 300, 200, 36, 2000
LeFevre, Joseph, 1, -, -, 50, -, -, -, 8000, 150, 350, -, 100, 150, 36, 900
Hanson, George C., 1, -, -, 33, 4, -, -, 8000, 300, 600, -, 10, 500, 52, 100
White, George, -, -, 1, 300, -, -, -, 40000, 1000, 2000, 100, 500, 1200, 52, 6000
Keegan, James, -, -, 1, 87, -, 12, -, 10000, 300, 600, -, -, 400, 35, 1500
Reynolds, William N., 1, -, -, 14, -, -, -, 10000, 288, 350, -, -, 300, 52, 1500
Morris, James, -, -, 1, 10, -, -, -, 3000, 300, 350, 30, 200, 100, -, 1000
Lambson, Giles, -, -, 1, 157, 40, -, -, 25000, 300, 450, -, 100, 500, 36, 2500
Deputy, Samuel, -, -, 1, 58, -, -, -, 6000, 300, 430, 50, 90, 322, 64, 1857
Rambo, Andrew, -, -, 1, 143, -, 7, -, 18000, 453, 1705, -, 68, 660, 130, 4212
Tweed, Mansil, 1, -, -, 56, 20, 12, -, 15000, 100, 1130, 30, -, 300, 91, 1298
Mote, Jones, 1, -, -, 80, 8, 10, -, 6000, 200, 646, 125, 138, 360, 78, 1811
Pyle, Lamborn, 1, -, -, 20, 4, 11, -, 5000, 100, 225, -, -, 100, 30, 653
Chambers, Charles, 1, -, -, 20, -, -, -, 5000, 100, 200, 30, 30, 300, 54, 342
Chambers, Mary Jane, 1, -, -, 100, -, 5, -, 12000, 600, 1500, 100, 140, 444, 80, 2550
Johnston, John T., 1, -, -, 75, 11, 8, -, 10000, 200, 800, 25, 100, 400, 78, 2083
Wright, John, 1, -, -, 10, 4, -, -, 1200, 60, 200, 10, 10, 15, 2, -, 244
Ogram (Ograne), Thomas, -, 1, -, 20, -, -, -, 2800, 100, 100, -, 36, -, -, 243
Smith, William, 1, -, -, 96, -, 25, -, 10000, 500, 800, 75, 200, 400, 40, 2300
Mackey, James H., 1, -, -, 15, 8, 1, -, 5000, 50, 200, 25, 20, 50, 8, 806
Oliver, William, 1, -, -, 12, -, -, -, 2000, 100, 200, -, 25, -, -, 216
Rankin, William, 1, -, -, 80, -, 6, -, 6000, 150, 638, 20, 30, 200, 18, 1002
Rankin, Sarah A., 1, -, -, 30, 3, 10, -, 4000, 100, 240, 20, 30, 100, 14, 574
Steward, Andrew, 1, -, -, 32, -, 40, 4, 4000, 100, 200, 25,38, 200, 42, 950
Teaf, Ellen M., 1, -, -, 50, -, 4, -, 5000, 400, 600, 200, 80, 325, 82, 1420
Blackwell, Ephraim, 1, -, -, 5, -, -, -, 3000, 10, 150, -, 20, -, -, 248
McKeowon, John, 1, -, -, 38, -, 3, -, 4500, 100, 275, -, 30, 12, 2, 844
Evans, Owen, 1, -, -, 100, -, 20, -, 15000, 1000, 10000, 50, 50, 1100, 115, 2122
Conkey, Georgiana, 1, -, -, 8, -, -, -, 800, 50, 70, -, 10, 50, 6, 198
Wilson, Robert, -, 1, -, 18, -, 3, -, 800, 15, 80, -, 24, 12, 2, 201
Gregg, William, 1, -, -, 39, -, 5, -, 3000, 150, 325, 10, 100, 200, 37, 1004
Baker, Aaron, 1, -, -, 98, -, 15, -, 10000, 250, 100, 10, 160, 542, 81, 2228
Crossan, James L., -, -, 1, 100, 15, 15, -, 10000, 400, 1200, -, 175, 400, 80, 3534
Blandy, Charles W., -, -, 1, 30, 10, 4, -, 9000, 150, 300, 25, 31, -, -, 300

Bonnell, Samuel, 1, -, -, 100, 18, 25, -, 16000, 1500, 1500, 200, 225, 800, 145, 3039

Murphy, David J., 1, -, -, 115, 15, 5, -, 25000, 800, 1250, -, 60, 546, 109, 2990

Steel, John T., 1, -, -, 10, 3, 10, -, 9000, 1000, 1200, 250, 160, 400, 60, 2117

McLaughlan, Constantine, 1, -, -, -, 13, -, -, 8000, 50, 300, 50, -, 50, 10, 150

Steel, George, 1, -, -, 88, 16, 5, -, 8000, 600, 600, 50, 10, 450, 80, 1796

Steel, Addie, 1, -, -, 22, 4, -, 1, 4000, 200, 215, 10, 30, 90, 15, 598

Croes, James, -, 1, -, 60, -, -, -, 5000, 800, 500, 150, 80, 200, 38, 1155

Miller, William, -, 1, -, 30, -, -, 5, 1500, 10, 125, -, 25, -, -, 424

Morrison, Thomas, 1, -, -, 130, 45, 75, -, 18000, 1000, 2000, -, 300, 50, 104, 2321

Casho, George A., 1, -, -, 92, 1, 2, 5000, 500, 600, 100, 165, 300, 45, 2063

Casho Machine Co., 1, -, -, 20, 4, -, -, 13000, 200, 500, -, 40, 250, 52,725

Kerr, George G., -, -, 1, 90, 45, 50, -, 15000, 1000, 2000, 100, 200, 400, 100, 3050

Fisher, Levi, 1, -, -, 9, -, -, -, 2500, 40, 200, -, 10, -, -, 461

Miller, Samuel, 1, -, -, 160, 22, 25, -, 20000, 369, 1127, 100, 317, 675, 132, 2583

Frazier, Job, -, 1, -, 80, 2, -, -, 6000, 100, 580, 5, 93, 400, 85, 1078

Lewis, Albert G., 1, -, -, 85, 65, -, -, 10000, 365, 618, 30, 100, 200, 42, 1098

Homewood, William, 1, -, -, 185, 60, 24, -, 20000, 1000, 269, 100, 60, 530, 100, 3966

Lum, Thomas, 1, -, -, 100, -, 14, -, 9000, 700, 748, 40, 40, 380, 80, 2131

Mitchell, Edward, -, -, 1, 140, 5, 5, -, 8000, 500, 640, 10, 50, 270, 60, 2061

Hauthorn, William, 1, -, -, 98, 20, 45, -, 12500, 300, 700, 20, 60, 268, 60, 2198

Moore, Joseph A., -, -, 1, 70, -, 15, -, 4500, 200, 300, 10, 60, 160, 38, 1165

Sayres, Robert, -, 1, -, 60, -, 8, -, 3000, 200, 450, 10, 70, 80, 16, 901

Shepperd, Casper, 1, -, -, 50, 12, 2, -, 5000, 500, 500, 25, 30, 180, 30, 1043

Morrow, James, 1, -, -, 285, 5, 10, -, 30000, 1200, 2500, 40, 300, 636, 112, 5558

Thompson, Joel, 1, -, -, 100, 35, 20, -, 15000, 800, 2000, 100, 90, 700, 140, 2755

McKeowan, William, 1, -, -, 38, -, -, -, 3000, 200, 300, 10, 80, 30, 4, 809

Coyle, W. George, 1, -, -, 10, -, -, -, 5000, 100, 160, 50, 36, 40, 4, 250

Coyle, William M., 1, -, -, 85, -, 10, 7000, 200, 500, 30, 58, 250, 40, 978

Jones, Thomas W., 1, -, -, 60, 5, 20, 3, 6000, 250, 540, 10, 50, -, -, 1036

Robinson, Joshua, 1, -, -, 67, 8, 2, -, 8000, 700, 100, 10, 80, 305, -, 56, 1847

Lewis, Evan, -, 1, -, 60, 20, 20, -, 8000, 665, 1008, 210, 1418, 400, 75, 1450

Draper, Daniel F., -, -, 1, 143, 12, -, -, 9000, 650, 1060, -, 215, 573, 100, 3193

Kyle, William, 1, -, -, 50, -, -, 30, 3000, 100, 500, 100, 100, 320, 60, 632

Johnson, George, 1, -, -, 35, -, 18, 5, 4000, 100, 125, -, -, -, -, 556

Eastburn, Franklin, 1, -, -, 73, -, -, 6, 6000, 200, 430, 55, 100, 100, 16, 1763

Oldham, Alexander, -, 1, -, 5, -, -, 8, 800, 50, 50, 20, -, 200, 36, 418

Singer, John, 1, -, -, 65, -, 15, -, 5000, 150, 315, 75, 100, 325, 60, 916

Morton, George, 1, -, -, 20, 5, 3, -, 2000, 100, 430, -, 7, 130, 138, 430

Cannon, Abram, -, -, 1, 135, -, 25, -, 7000, 655, 870, 200, -, -, 150, 2497

Lea, James A., 1, -, -, 60, -, 8, 5, 4000, 300, 310, -, 12, 125, 15, 661

Pogue, Joseph, 1, -, -, 142, -, 20, 15, 10000, 600, 1060, 125, 70, 250, 50, 2443

Ruth, Alfred, 1, -, -, 85, -, 10, -, 9000, 1000, 5000, 65, 90, 600, 140, 2242

Ruth, Theodore, -, -, 1, 60, -, 12, 25, 8000, 300, 300, -, 75, 250, 52,1294

Finton, Mary, 1, -, -, 24, -, 2, -, 1000, 80, 150, -, 10, -, -, 400

Starr, Gilbert, 1, -, -, 22, -, 4, -, 800, 50, 175, -, 15, -, -, 425
Donnell, James, -, -, 1, 40, -, 5, -, 3000, 200, 375, 10, 25, 130, 30, 950
Rimenter, Isaac, -, -, 1, 18, -, -, -, 3000, 200, 215, -, -, 264, 49, 800
Elliott, John, 1, -, -, 50, -, -, -, 5000, 1000, 800, 200, 50, 250, 76, 1496
Armstrong, Robert, 1, -, -, 115, -, 25, 5, 10000, 500, 605, -, 60, 400, 65, 2934
Lynam, William, 1, -, -, 120, 10, 1, 10, 10000, 500, 1000, 10, 90, 377, 80, 2724
Brooks, George T., -, -, 1, 96, -, 30, -, 5000, 150, 675, 52, 129, 260, 40, 2412
Whitten, Thomas, 1, -, -, 296, -, 60, -, 16000, 500, 1500, 50, 300, 795, 120, 3565
Pritchard, Joseph, 1, -, -, 103, -, -, -, 10000, 500, 1200, 30, 100, 337, 60, 2282
Brooks, Alfred G., 1, -, -, 116, 15, 8, -, 7000, 200, 562, 70, 75, 452, 80, 2213
Palmer, George W., 1, -, -, 89, -, 40, 20, 8000, 250, 800, 58, 108, 598, 104, 2400
Hawthorn, Thomas, 1, -, -, 70, 20, 30, -, 75000, 400, 690, 118, 100, 340, 52, 1703
Hamilton, William, -, -, 1, 60, -, 20, -, 2500, 325, 200, 58, 76, 15, 2, 518
Duggan, Joseph, -, -, 1, 70, -, 8, 8, 3000, 150, 650, 160, 175, 299, 50, 812
Smally, William, 1, -, -, 70, -, -, -, 10000, 500, 1000, 20, 100, 260, 40, 1398
Peters, Benjamin, 1, -, -, 200, -, -, -, 18000, 400, 1200, 325, 175, 361, 62, 3051
Morrison, William, 1, -, -, 60, 25, 10, -, 10000, 500, 655, 60, 250, 375, 70, 1960
McElwee, Allan C., -, -, 1, 130, 7, 50, 10, 19000, 400, 800, 20, 275, 264, 40, 2142
Greenwalt, William, 1, -, -, 48, -, 8, 2, 4800, 150, 175, 10, 75, 180, 30, 1040
Rothwell, Abraham, 1, -, -, 100, -, 4, 4, 7500, 200, 700, 30, 132, 310, 60, 1920
Jones, Margaret, -, -, 1, 150, -, -, -, 12000, 350, 700, -, -, 320, 52, 4246
Maree, Margaret, 1, -, -, 70, 4, 8, -, 8000, 100, 400, 10, 18, 208, 32, 907
Stroud, William, 1, -, -, 98, 16, 2, -, 9000, 600, 1020, 100, 120, 518, 90, 2657
Morrison, James, 1, -, -, 100, -, 5, 6, 9000, 1000, 1000, 100, 125, 440, 72, 2074
Appleby, David, -, 1, -, 17, 3, -, -, 4000, 200, 600, -, 48, 360, 60, 581
Clay, William, 1, -, -, 12, 6, -, -, 4000, 40, 140, 30, 12, 255, 40, 480
Campbell, William J., 1, -, -, 40, -, -, -, 4000, 100, 250, 20, 40, 300, 48, 508
Cannon, Abram, 1, -, -, 155, -, 12, -, 8000, 300, 910, 200, 137, 352, 60, 2170
Grove, Jonathan, 1, -, -, 80, -, 15, -, 7000, 200, 596, 20, 35, 238, 45, 1265
Alrich, Samuel P., 1, -, -, 108, 14, -, -, 6000, 250, 610, 75, 50, 321, 60, 1869
Shannon, Abram P. Estate, 1, -, -, 68, -, 122,-, 5100, -, -, 50, 91, -, -, 693
Jester, Isaac, -, -, 1, 150, -, -, -, 6000, 250, 656, 25, 50, 432, 80, 1245
Ballen, Price H., -, -, 1, 130, -, -, -, 8000, 360, 500, 50, 100, 510, 90, 1950
Cluff, ___, 1, -, -, 30, -, -, -, 1500, 100, 200, -, -, 100, 12, 496
Clark, J. Curtis, 1, -, -, 425, 50, 25, -, 15000, 200, 1400, -, 80, 500, 100, 3750
Runner, Henry P., 1, -, -, 5, -, -, -, 1000, 50, 200, -, -, -, -, 383
Carlisle, Samuel, -, -, 1, 150, -, -, -, 7000, 406, 1500, 200, 130, 500, 120, 2210
Fisher, John G., -, -, 130, -, 10, -, 5400, 375, 1100, 20, 150, 220, 40, 1590
Churchman, Henry L., 1, -, -, 420, -, 3, -, 35000, 100, 6960, 250, 13, 1510, 200, 7444
Donohoue, Joseph, -, -, 1, 190, -, 20, -, 12000, 350, 930, -, 92, 400, 70, 2670
Naudain, Arnold Jr., 1, -, -, 110, -, -, -, 5000, 500, 500, -, 80, 300, 52, 1897
Naudain, Arnold Sr., 1, -, -, 65, -, 25, 20, 6000, 150, 640, 50, 125, 240, 40, 1972
McBride, William, 1, -, -, 140, -, 10, 11, 13500, 1000, 1200, 100, 120, 504, 104, 3272
Wright, James, -, -, 1, 200, -, 90, -, 18000, 600, 2209, 150, 122, 596, 100, 3572

Wilson, Edward, 1, -, -, 24, -, -, -, 1500, 100, 400, -, 25, 100, 12, 509
Morrison, Robert, 1, -, -, 120, 2, 10, -, 12000, 650, 6435, 20, 140, 425, -, 3072
Jones, John, 1, -, -, 80, 15, -, -, 9000, 400, 601, 24, 80, 298, -, 1950
Morrison, Samuel, -, -, 1, 200, -, 20, -, 9000, 400, 725, 10, 800, 500, ½, 2589
Dean, William, 1, -, -, 179, 6, -, 30, 12000, 600, 3480, 180, 300, 504, -, 4320
Lindsey, Samuel, 1, -, -, 235, 20, 40, -, 23500, 600, 2000, -, 200, 1000, -, 3054
Wilson, Edward R., -, 1, -, 90, -, 5, 8, 10000, -, 800, 100, -, 480, ½, 842
Worral, Nimrod, 1, -, -, 36, -, -, -, 4500, 500, 500, 48, 17, 300, -, 824
Hossinger, Andrew J., 1, -, -, 130, 15, 14, -, 11500, 500, 1000, 100, 500, 364, -, 2984
Haines, Eri (Evi) W., 1, -, -, 50, -, -, -, 8000, 400, 538, 10, 25, 500, -, 820
Woodrow, William A., -, 1, -, 60, -, -, -, 10000, 800, 949, 25, 50, 400, -, 1555
Porter, Edward D., 1, -, -, 36, -, -, -, 10000, 800, 600, 75, 175, 597, 142, 1624
Ellison, Jonathan L., 1, -, -, 250, 40, 10, -, 2100, 900, 1654, 50, 100, 70, 36, 3400
Davis, Solomon P., -, -, 1, 111, 17, 10, -, 8000, 450, 500, 25, 200, 375, 36, 1100
Davis, Nehemiah, - -, 1, 390, -, 65, -, 25000, 1100, 1300, 125, 255, 950, 36, 3700
Russell, Henry, -, -, 1, 106, -, 5, -, 5250, 300, 400, 25, 125, 250, 36, 800
Eliason, Andrew S., -, -, 1, 340, 50, 16, -, 2200, 1000, 1760, 100, 300, 1000, 36, 4000
McIntire, Thomas, 1, -, -, 150, 32, 20, -, 11000, 650, 1100, 140, 150, 450, 36, 1800
Price, Thomas L., -, -, 1, 200, 20, -, -, 12000, 600, 950, 25, 225, 550, 36, 1600
Eliason, John D., -, -, 1, 400, 27, 20, -, 26000, 1100, 1750, 75, 375, 1200, 36, 3100
Biggs, Abram B., -, -, 1, 180, 25, 20, -, 12000, 700, 200, 50, 280, 450, 36, 3000
Stevens, Samuel, -, -, 1, 175, -, 10, -, 11000, -, 450, 300, 175, 456, 36, 2100
Biggs, Abram B. -, -, 1, 240, 50, -, -, 13200, 600, 700, 25, 100, 300, 36, 2600
Biggs, William P., 1, -, -, 216, 50, -, -, 13000, 800, 800, 150, 250, 1000, 36, 3000
Dempsey, Thomas, 1, -, -, 150, 40, 5, -, 9000, 500, 600, 100, 75, 200, 36, 1200
Dempsey, John H., 1, -, -, 18, -, -, -, 800, 60, 175, 75, 18, 12, 3, 200
Wiley, Jonathan, 1, -, -, 4, -, -, -, 160, 25, 35, -, -, -, -, 40
Montgomery, Robert, 1, -, -, 24, -, 16, -, 2500, 150, 400, 60, 56, -, 20, 200
Seeney, David L., 1, -, -, 5, -, -, -, 175, 25, 50, -, -, -, -, 60
Gay (Guy), Elijah, -, -, 1, 40, -, -, -, 800, 30, 80, -, 30, -, 125, 200
Wright, Stephen, -, -, 1, 27, -, -, -, 600, 100, 75, -, 13, 10, 3, 220
Wiley, John, 1, -, -, 8, -, -, -, 120, 10, 50, -, -, -, -, 50
Racine, George M., -, -, 1, 100, -, -, -, 3000, 75, 125, 20, 50, 125, 32, 500
Benson, James A., 1, -, -, 130, -, 24, -, 1000, 450, 1000, 10, 75, 180, 36, 1200
Shockley, David, -, 1, -, 90, -, 68, -, 5800, 350, 550, -, 100, 75, 36, 400
Russell, John, 1, -, -, 90, -, 23, -, 4500, 400, 600, 100, 50, 100, 36, 400
Boys, Jacob, 1, -, -, 53, -, -, -, 460, 25, 170, 5, 30, 25, 2, 125
Cann, Edgar R., -, 1, -, 165, 20, -, 15, 14000, 700, 1000, 100, 200, 400, 36, 1600
Stradly, William, -, -, 1, 142, 22, 90, -, 4900, 300, 1500, 25, 40, 400, 40, 1300
LeFevre, John B. Jr., 1, -, -, 100, -, 62, -, 2000, 400, 800, 75, 250, 500, 36, 200
Ford, W. B., 1, -, -, 10, -, -, -, 4000, 40, 450, 10, 8, 100, 36, 200
Shetzline, Adam, -, 1, -, 7, -, -, -, 800, 200, 103, -, 168, 650, 32, 1000

Cleaves, Benjamin F., -, -, 1, 90, -, -, 31, 2300, 200, 200, 10, -, 10, 2, 328
Dawson, William H., -, -, 1, 140, -, 20, -, 10000, 500, 1000, 225, 200, 625, 50, 2000
Rust, George E., -, -, 1, 100, -, 40, -, 6000, 400, 1200, 50, 50, 400, 36, 1500
McMullen, James, -, -, 1, 110, -, 20, 5000, 200, 530, -, 50, 500, 36, 1500
Boys, Henry, -, -, 1, 130, -, 170, -, 8000, 251, 800, 175, 25, 450, 40, 1200
Thompson, Charles, -, -, 1, 126, 4, 9, -, 7500, 400, 600, 175, 80, 325, 32, 1400
Racine, John George, 1, -, -, 100, 2, 5, -, 7000, 300, 450, 20, 75, 300, 36, 800
Myers, Joseph C., -1, -, 40, -, -, 10, 2000, 75, 135, -, 10, 200, 30, 300
Porter, Thomas G., 1, -, -, 49, -, -, 6, 2700, 300, 360, -, 10, 50, 52, 500
Grimes, Robert, 1, -, -, 30, -, 3, 47, 2500, 75, 200, 10, 25, 25, -, 200
Moore, Abram S., 1, -, -, 80, 2, 20, -, 6000, 1150, 450, 10, 50, 165, 36, 500
Reynolds, George, -, -, 1, 47, -, -, 100, 3000, 100, 500, 30, 50, 150, -, 400
Hollett, Eli, -, -, 1, 75, -, -, 75, 1500, 50, 200, 100, 25, 50, -, 400
Reed, Robert, -, -, 1, 100, -, -, 50, 2000, 100, 300, 35, -, 50, -, 30
Dayett, Adam, -, 1, -, -, 42, -, 10, 25, 4000, 50, 400, 10, 25, 200, -, 400
George, Henry, -, -, 1, 90, -, 20, -, 7000, 50, 325, -, 50, 140, 32, 800
Batton, James, -, -, 1, 110, -, 40, -, 6000, 350, 700, 290, 100, 225, 36, 1400
Racine, Fred P., -, -, 1, 118, -, 30, -, 5000, 300, 700, 25, 150, 200, 36, 1000
Thompson, Nathaniel, -, -, 1, 90, -, 10, -, 6000, 300, 500, 25, 50, 175, 36, 800
Calhoun, William, -, -, 1, 110, -, 40, -, 5000, 100, 500, 150, 75, 200, 36, 500
Armstrong, Alexander, -, 1, -, 16, -, -, -, 900, 25, 95, -, -, 20, -, 100
Moody, Isaac, 1, -, -, 20, 3, -, -, 1500, 45, 150, 5, -, 15, -, 300
Stewart, William W., 1, -, -, 100, -, 25, 25, 8000, 100, 510, 260, -, 75, 36, 800
Janvier, Ferdinand, 1, -, -, 150, 15, 15, 10, 14000, 500, 1100, 75, 130, 700, 39, 1700
Titter, Mary E., 1, -, -, 100, -, -, 200, 7200, 400, 600, -, 37, 500, 36, 1400
Palmer, Benson E., 1, -, -, 8, -, -, -, 600, 50, 100, 40, -, -, -, 100
Vandergrift, William E., -, -, 1, 100, -, -, 76, 4000, 25, 200, 10, 50, 100, 34, 550
Murray, William E., -, -, 1, 100, -, -, 75, 4000, 25, 200, 10, 50, 100, 30, 250
King, Washington, -, -, 1, 75, -, -, 65, 4500, 70, 300, 15, 200, -, 30, 800
Jones, Andrew, 1, -, -, 12, -, -, -, 400, 25, 150, 5, -, 10, -, 150
Ferris, Brainard D., 1, -, -, 130, -, 43, 45, 15000, 300, 700, 25, 67, 300, 36, 1800
Ford, David, -, -, 1, 175, -, 2, 16, 1200, 350, 900, 25, 125, 400, 36, 2500
Boys, Henry, 1, -, -, 10, -, -, -, 600, 40, 150, 5, 5, 20, 20, 140
Sapp, Joseph, -, -, 1, 220, -, -, 96, 10000, 400, 900, 25, -, 600, 36, 2164
Ward, William, 1, -, -, 100, -, -, 90, 3000, 75, 175, 5, 50, 125, 32, 250
Cooch, J. Wilkins, -, -, 1, 225, 75, 100, -, 14000, 300, 600, 75, 100, 600, 36, 1300
Frazer, Richard S., -, -, 1, 60, 2, 30, 14, 4500, 300, 300, 125, 150, 75, 36, 450
Cornly (Conly), Samuel, 1, -, -, 100, 40, 20, -, 9000, 450, 900, 125, 100, 200, 36, 1600
Groves, George W., 1, -, -, 60, 12, -, 4, 4762, 300, 500, 50, 40, 50, 20, 1000
Brooks, Joseph & Mary P., 1, -, -, 120, 26, 5, -, 6000, 150, 375, 20, 84, 400, 36, 1000
Champion, Benjamin, 1, -, -, 26, -, -, -, 800, 10, 40, -, -, 10, -, 100
Miller, William, 1, -, -, 5, -, -, -, 200, 5, -, -, -, -, -, 20

Reese, J. R., 1, -, -, 113, -, 20, 33, 11046, 500, 1100, 50, 130, 550, 50, 1900

Powell, George W., -, 1, -, 56, -, -, 50, 3500, 25, 300, 30, 33, 180, 36, 400

Hall, John, -, -, 1, 120, -, 30, 30, 9000, 500, 900, 25, 200, 300, 36, 1500

Silcox, William, -, 1, -, 20, -, -, -, 1200, 50, 225, -, 25, 100, 19, 200

Jones, Calvin, 1, -, -, 120, -, -, 30, 2000, 200, 200, 125, 45, 225, 36, 825

Mcdill (Medill), George D., 1, -, -, 100, 39, -, -, 10500, 500, 1050, 75, 200, 375, 36, 2700

Willis, James R., -, -, 1, 130, -, -, 35, 13000, 400, 1150, 200, 60, 246, 36, 1254

Brooks, James I., -, -, 1, 180, -, -, 32, 5000, 450, 700, 120, 180, 400, 36, 2600

Morrison, Charles A., 1, -, -, 60, -, -, 15, 5000, 150, 375, 65, 50, 200, 36, 900

Lee, Benjamin, -, 1, -, 30, 29, 1, -, 1500, 20, 350, 20, 12, 18, 30, 351

Porter, Isaac, 1, -, -, 15, -, -, -, 500, 27, 75, 5, 7, 11, 30, 92

Powell, William G., 1, -, -, 80, -, -, 4, 4000, 150, 240, 125, 100, 400, 36, 360

Faulkner, William H., -, -, 1, 180, -, 15, 15, 13000, 500, 700, 20, 210, 200, 36, 1700

Bowen, Joseph, -, 1, -, 214, -, -, 20, 17000, 1300, 2000, -, -, -, -, 2000

Lark, Charles, -, 1, -, 40, -, -, -, 2200, 40, 150, 20, 40, 25, 40, 250

Caskey, David W., 1, -, -, 42, -, -, -, 5000, 250, 207, 5, -, 30, 20, 300

Cunningham, Stephen, -, -, 1, 120, -, -, 5, 10000, 240, 850, 15, -, 210, 36, 1000

Moody, John, 1, -, -, 90, -, -, 25, 7500, 150, 225, 120, 124, 200, 31, 900

Miller, John, -, -, 1, 200, -, 46, 50, 18000, 500, 600, 100, 400, 400, 36, 2600

Murray, James, 1, -, -, 75, -, -, 9, 5000, 150, 225, 25, 60, 500, 31, 500

Riddle, Margaret, -, 1, -, 16, -, -, 10, 1000, 40, 150, -, 5, 10, -, 75

Woodward, Joel, 1, -, -, 30, 6, -, 50, 4300, 200, 400, 15, 20, 50, 26, 550

Keeley, Meachel, 1, -, -, 34, -, -, 6, 2000, 25, 200, -, -, 50, 21, 210

Mahon, James, 1, -, -, 7, -, -, -, 200, -, 40, -, -, 5, -, 50

Cole, Thomas, 1, -, -, 18, -, -, 3, 540, 50, 150, 20, 25, 20, -, 100

Powell, Charles A., -, 1, -, 30, -, -, 30, 1000, 10, 120, -, 48, 10, -, 100

McCloskey, Authur, 1, -, -, 80, -, -, 50, 9000, 400, 850, 50,125, 400, 40, 700

Megget, Peter, 1, -, -, 64, -, -, 2, 6600, 400, 450, 30, 100, 500, 40, 1000

Barber, John, -, 1, -, 30, -, -, 5, 1960, 150, 225, 10, 50, 360, 36, 800

Slack, William, 1, -, -, 100, -, 10, 40, 6000, 200, 350, -, 25, 300, 36, 700

Rambo, Enos, -, -, 1, 60, -, 20, 20, 5000, 200, 350, 75, 100, 150, 40, 852

Stroud, Edward, 1, -, -, 148, -, 18, -, 6530, 400, 425, 20, 100, 250, 38, 1300

McConaughey, William, 1, -, -, 300, -, 200, -, 24000, 600, 1500, 150, 700, 1100, 40, 4500

Surrat, John, 1, -, -, 56, -, -, -, 2000, 75, 300, 10, 56, 30, 31, 300

O'Rourke, Timothy, 1, -, -, 27, -, -, 5, 1000, 50, 200, 10, 20, -, -, 300

O'Rourke, Bartholomew, 1, -, -, 50, -, -, 10, 1200, 50, 250, -, -, -, -, 400

Loughead, John, -, 1, -, 30, -, -, 40, 2500, 40, 100, 7, 25, 10, -, 276

Clendenin, Samuel, 1, -, -, 25, 10, 15, -, 1500, 100, 250, 5, 20, 40, 6, 370

Sullivan, Patrick, 1, -, -, 13, -, -, -, 600, 20, 175, -, 20, -, -, 160

Sapp, Benjamin, -, -, 1, 80, -, 50, -, 5220, 125, 325, 10, 30, 200, 41, 1200

Ash, Catherine, 1, -, -, 120, -, 20, -, 7500, 300, 475, 15, 195, 195, 38, 1700

Clarkson, Alex. R., 1, -, -, 75, -, 27, -, 6000, 300, 1400, 200, 125, 200, 40, 1200

Walton, Charles, 1, -, -, 90, 12, -, -, 8000, 300, 1200, 25, 100, 400, 52, 1700

Johnson, Robert S., 1, -, -, 40, -, -, -, 4000, 200, 300, -, 75, 200, 46, 1000
Shepperd, Edgar, -, -, 1, 45, -, -, 15, 3500, 150, 300, 15, 50, 45, 41, 600
Crow, James H., -, -, 1, 145, -, -, -, 12000, 250, 760, 10, 100, 150, 50, 1700
Murphey, Hugh, 1, -, -, 40, -, -, 31, 100, 400, 25, 37, 20, 46, 250
O'Rourke, James, 1, - -, 60, -, 20, 2, 3000, 25, 150, 10, 50, 60, -, 400
Walton, Rosana, 1, -, -, 25, -, 25, -, 3000, 95, 150, 20, 100, 125, 49, 300
O'Rourke, Timothy, 1, -, -, 18, -, 6, -, 1100, 40, 150, 40, 25, 50, 52, 200
Sullivan, James, 1, -, -, 40, -, -, 14, 3000, 50, 100, 80, 100, 200, 36, 400
James, Joseph, 1, -, -, 10, -, -, -, 600, 30, 100, 10, 6, 100, 32, 200
Lum, Elias, 1, -, -, 25, -, 15, -, 2000, 250, 400, 35, 35, 80, 10, 400
Harkstein, Godfrey, 1, -, -, 40, -, -, -, 2500, 25, 200, 125, 150, 150, 36, 310
Stewart, Samuel, 1, -, -, 10, -, -, 15, 1500, 25, 100, -, -, -, -, 100
Green, Levi, 1, -, -, 30, 10, -, -, 2000, 100, 195, 25, -, 125, 50, 475
Trusty, George, 1, -, -, 75, -, 30, -, 4500, 125, 200, 100, 50, 25, 52, 385
Wilson, Washington, 1, -, -, 10, -, -, 20, 900, 50, 75, -, -, 15, -, 100
Congo, Nash, -, 1, -, 70, -, 47, -, 2750, 50, 300, 10, 45, 30, 42, 450
Irons, William, 1, -, -, 22, -, -, -, 100, 25, 50, -, 10, 20, -, 100
Keniether, Valentine, 1, -, -, 38, -, -, 30, 2500, 30, 200, 20, 40, 30, -, 200
Keniether, Martin, 1, -, -, 65, -, -, 15, 2500, 200, 450, 72, 12, 100, 56, 700
Kendell, Henry, 1, -, -, 65, -, 24, -, 3200, 250, 650, 15, 45, 500, 40, 600
Lim, John, 1, -, -, 50, -, 47, -, 2000, 200, 150, -, 25, -, -, 150
Brown, John, 1, -, -, 70, -, 36, -, 3000, 100, 400, 200, 100, -, -, -
Catts, Levin, -, -, 1, 160, -, 20, -, 9500, 550, 950, 25, 110, 200, 40, 1670
Newman, Nathaniel, -, -, 1, 290, 30, 80, -, 20000, 1000, 1750, 320, 530, 1200, 40, 1600
McCoy, Thomas, -, -, 1, 210, -, -, 15, 13000, 500, 1000, 125, 230, 400, 36, 3500
Davidson, Thomas, -, -, 1, 196, -, -, 8, 12500, 500, 550, 25, 200, 450, 36, 1700
Duckey, Richard, 1, -, -, 16, -, -, -, 800, 25, 75, -, 15, 10, -, 100
Money, George W., 1, -, -, 12, -, -, -, 600, 40, 150, 20, 14, 15, 20, 275
Catts, Peter R., -, -, 1, 217, -, 50, -, 10000, 150, 400, 25, 135, 250, 40, 1300
Reynolds, John W., 1, -, -, 296, -, -, -, 15000, 500, 1500, 75, 325, 800, 36, 3000
Walters, James, -, -, 1, 200, -, -, -, 12000, 400, 300, 60, 150, 200, 36, 1400
McCoy, James A., 1, -, -, 125, -, -, -, 7200, 50, 200, 25, 60, -, 20, 800
Roy, John, -, -, 1, 20, -, -, -, 1000, 100, 200, 5, 20, 60, 10, 130
Wilson, Abram, 1, -, -, 16, -, -, -, 800, 50, 150, 5, 20, 25, 10, 200
Davidson, John, -, -, 1, 300, -, 30, 10, 14000, 600, 1000, 75, 336, 725, 36, 3000
Nichols, Isaac, 1, -, -, 30, -, -, 11, 500, 50, 100, 20, 20, 20, 36, 400
Veazey, James I., -, -, 1, 100, -, 26, -, 3800, 150, 500, 20, 50, 200, 36, 900
Huber, Simon, 1, -, -, 40, -, -, 20, 3000, 25, 250, 100, 15, 120, 30, 400
Boulden, George, 1, -, -, 160, -, 40, -, 12000, 400, 700, 75, 75, 500, 36, 1400
Murray, William, 1, -, -, 300, -, -, -, 18000, 650, 1100, 350, 225, 1000, 34, 3500
Spencer, Nathaniel, -, 1, -, 200, -, 100, -, 9000, 100, 150, 15, 50, -, -, 400
Barrow, Benson, 1, -, -, 55, -, -, 11, 1700, 100, 300, 50, 50, 50, 8, 600
Davis, John W., 1, -, -, 35, -, -, 6, 2600, 35, 200, 195, 40, 200, 32, 500
Wright, J. Thomas, -, -, 1, 100, -, 30, 18, 6500, 400, 750, 130, 100, 450, 36, 1200

Dayett, William Thomas, 1, -, -, 85, -, 14, -, 6500, 450, 550, 25, 150, 100, 10, 1200

Norris, William, -, -, 1, 140, -, 44, -, 4000, 50, 500, 200, 60, -, -, 500

Laws, Joshua, -, -, 1, 150, -, 100, 50, 10000, 450, 900, 30, 72, 75, 8, 1200

Ellison, Lewis G., -, -, 1, 250, -, 50, -, 14000, 500, 1300, 20, 50, 700, 36, 1352

Boulden, Jesse, 1, -, -, 15, -, -, -, 3000, 75, 100, 25, 25, 20, -, 225

Clark, Delaware, 1, -, -, 200, -, -, -, 12000, 500, 1500, 150, 2225, 750, 36, 1500

Bregator, James R., 1, -, -, 91, -, -, -, 7500, 400, 700, -, 50, 60, 20, 975

Kuley, Miachel B., -, 1, -, 76, -, 30, 30, 6500, 75, 400, 50, 40, 225, 52, 700

Batton, Mahlon, -, -, 1, 60, -, 14, 26, 2500, 350, 600, 80, 94, 350, 5, 500

Stewart, Charles B., 1, -, -, 150, -, 50, -, 9000, 200, 750, 25, 75, 275, 48, 700

Droolinger, Charles F., -, -, 1, 80, -, -, 11, 2700, 60, 150, -, 48, 100, 20, 350

Price, John, 1, -, -, 1, -, -, -, 1000, 30, 150, 10, -, 25, -, 300

Reed, Robert, 1, -, -, 87, -, -, -, 3600, 75, 200, 75, 50, 200, 131, 600

Adair, William, 1, -, -, 45, -, -, -, 1200, 20, 175, -, -, -, -, 175

Frazer, John, 1, -, -, 110, -, 15, -, 7000, 175, 275, 25, 95, 250, -, 936

Ward, John I., -, -, 1, 180, -, 60, 18, 8000, 250, 300, -, -, 100, 52, 700

Mahan, Wilson, -, -, 1, 170, -, 10, 20, 9000, 500, 1000, 125, 100, 40, 50, 2000

Sheldon, George, -, -, 1, 120, -, 20, 27, 5600, 200, 650, 75, 50, 25, 52, 713

Cann, I. Alibone, -, -, 1, 200, -, -, -, 12000, 600, 1400, 130, 120, 1800, 45, 2700

Cann, Robert M., -, -, 1, 152, -, 4, 4, 8000, 200, 900, 50, 50, 400, 36, 2046

Thornton, John H., -, -, 1, 12, -, -, -, 2000, 25, 250, 5, 25, -, -, 150

Lindell, Thomas, 1, -, -, 110, -, 40, -, 7500, 500, 550, 25, 100, 275, 36, 1000

McIntyre, Samuel, 1, -, -, 120, -, 50, -, 19000, 600, 850, 30, 50, 125, 52, 1100

McIntyre, James, 1, -, -, 100, -, 25, 25, 9000, 700, 700, 30, 50, 350, 36, 1000

Pordham (Fordham), Harry, -, -, 1, 160, -, 20, -, 10000, 500, 550, 100, 70, 300, 36, 1500

Brown, Thomas, 1, -, -, 25, -, -, -, 1300, 50, 150, 50, 10, 10, -, 195

Williams, James, -, -, 1, 20, -, -, -, 800, 25, 130, 10, 15, -, -, 162

Cunningham, Samuel, -, -, 1, 11, -, -, -, 500, 30, 175, 5, 20, -, -, 100

Lindell, George H., 1, -, -, 5, -, -, -, 510, 10, 100, 5, 20, -, -, 75

Crossland, Thomas, -, -, 1, 115, -, 25, -, 6000, 250, 325, 10, 125, 500, 40, 900

Boys, John B., 1, -, -, 120, -, 10, -, 6500, 225, 500, 25, 50, 300, 46, 1000

Hogg, John R., 1, -, -, 60, -, 11, -, 2300, 150, 600, 25, 15, 75, -, 765

Vansant, Mitchell, -, -, 1, 60, -, -, 8, 4000, 200, 400, 15, 75, 100, 20, 400

Cavender, Josephus, -, -, 1, 230, -, 30, 7, 13000, 500, 650, 25, 490, 600, 40, 2400

Ellison, Curtis B., 1, -, -, 106, -, -, -, 9000, 500, 1275, 100, 200, 500, -, 1600

Ellison, Thomas B., 1, -, -, 92, -, 20, -, 3000, -, -, -, -, -, -, 600

Parson, Aaron K., 1, -, -, 68, -, -, 30, 2700, 20, 250, -, -, -, -, 325

Gould, Thomas H., -, -, 400, -, 100, -, 90000, -, 200, 1200, 175, 300, 100, 49, 4000

Veazey, D. James L., 1, -, -, 100, -, -, -, 5000, 400, 500, 50, 75, 260, 40, 200

McCracken, Thomas W. (1), 1, -, -, 143, -, -, 10000, 200, 1000, 25, 175, 400, 36, 2300

McCracken, Thomas W. (2), -, -, 1, 200, -, 53, -, 16000, 200, 900, 25, 180, 250, 36, 1100

Paxon, Merritt H., 1, -, -, 150, -, 47, -, 10000, 400, 900, 125, 200, 300, 38, 1500

McCoy, James, -, -, 1, 70, -, 10, -, 4000, 130, 400, 25, 25, 75, 36, 400

Cunningham, William, 1, -, -, 40, -, 40, 7, 2300, 100, 175, 75, -, 50, 10, 235

Mathews, Wingate, 1, -, -, 2, 25, -, 50, 50, 13000, 500, 1700, 25, 175, 630, 37, 2375

Cavender, William, -, -, 1, 180, -, 20, -, 10000, 500, 1000, 25, 125, 600, 36, 1500

Boys (Bogs), Henry, 1, -, -, 19, -, -, -, 1500, -, -, -, -, -, -, 269

Beck, William Sr., -, -, 1, 200, 75, -, -, 20000, 800, 3340, 300, -, 1000, 142, 550

Beck, William Sr., 1, -, -, 150, 100, -, -, 7000, 200, 150, 100, -, 300, 100, 600

Frye, Frederic L., -, -, 1, 40, -, -, -, 300, 50, 150, 68, -, -, 16, 50

Jester, James M., -, -, 1, 165, 100, -, 25, 10000, 50, 640, 75, -, 250, 45, 1045

Reeves, Clement, 1, -, -, 200, 60, -, 60, 12000, 600, 1200, 50, 50,700, 175, 2500

Colburn, Arthur, 1, -, -, 1000, 100, -, -, 100000, 3500, 6000, 1000, -, 6000, 1300, 17000

Higgins, John C., -, 1, -, 200, 150, -, -, 10000, 600, 900, 50, 200, 1000, 300, 2300

Swan, John, -, -, 1, 250, 80, -, -, 10000, 2800, 1000, -, 87, 1200, 200, 2500

Grimes, George W., -, -, 1, 140, 40, 10, 10, 5000, 150, 555, -, -, 200, 50, 1000

Swan, Thomas H., -, -, 1, 140, 33, 5, -, 6000, 250, 600, -, 100, 290, 54, 1400

Bryan, Michael _., 1,-, -, 242, -, -, 3000, 200, 200, 45, 28, 400, 80, 728

Robson, John, 1, -, -, 75, 4, 3, -, 8000, 400, 500, 80, 150, 250, 44, 1250

McWhorter, Francis F., -, -, 1, 357, 50, -, -, 30000, 500, 1500, 40, -, 1000, 180, 3000

Garman, James, -, -, 1, 50, 1, -, -, 5000, 400, 400, 11, 48, 420, 75, 400

Hopkins, George E., -, 1, -, 200, 20, -, -, 10000, 300, 675, -, 125, 550, 108, 1100

Bird, James W., -, -, 1, 160, 20, 20, -, 12000, 500, 1200, 100, 175, 500, 145, 2000

Reybold, John F., -, -, 1, 240, 40, -, -, 20000, 1000, 2000, 140, 260, 1100, 190, 4000

Davis, Cornelius W., -, -, 1, 350, 75, 5, 10, 25500, 800, 2000, 80, 250, 600, 150, 5400

Lofland, John C., -, -, 1, 190, 20, -, -, 25000, 300, 750, 55, 50, 400, 150, 500

Gray, Francis S., -, -, 1, 190, 70, -, -, 25000, 500, 885, 50, 125, 500, 145, 5200

Stuckert, William M., 1, -, -, 192, 34, -, -, 25000, 1000, 2000, 60, 180, 800, 94, 4000

Sester (Jester), Edward, -, -, 1, 175, 27, -, 6, 14000, 700, 1000, -, 95, 600, 82, 3000

Ashcroft, William -, -, 1, 108, 22, 5, 30, 6500, 200, 300, 30, 150, 400, 118, 1525

Cleacie (Cleaves), John A., -, -, 1, 200, 42, 2, 2, 14000, 600, 1015, 47, 154, 1000, 212, 3000

Alston, Joab, -, 1, -, 11, 30, 3, 2, 6500, 100, 575, 25, 150, 300, 118, 1600

Colescott, William, -, 1, -, 300, 100, 5, 10, 25000, -, 6000, 50, 150, 1000, 135, 5000

Clark, John C., 1, -, -, 276, 90, 3, 10, 20000, 2500, 1850, 175, 400, 900, 160, 4500

Cleaves, Peter, -, -, 1, 180, 68, -, -, 14000, 800, 800, 30, 90, 800, 150, 1600

Jance (Hance,Vance), Edward, -, -, 1, 200, 160, -, -, 14000, 300, 1600, 20, -, 450, 150, 2000

Clark, Miles, -, -, 1, 228, 67, 10, 10, 12000, 500, 1500, 50, 200, 650, 118, 2500

Sterling, Ephraim, -, -, 1, 303, 70, 15, 4, 100000, 300, 100, 250, 437, 1000, 375, 1600

Compton, Charles, -, -, 1, 220, 80, -, 4, 8000, 300, 1000, 20, 210, 375, 110, 1500

Marcey, Thomas W., -, -, 1, 98, 22, 8, -, 5000, 600, 400, 50, 88, 150, 50, 900

Marcey, John, -, -, 1, 130, 60, 4, -, -, 9000, 600, 850, 50, 120, 190, 75, 1800

McCall, John, 1, -, -, 110, 25, -, 25, 5000, 300, 1000, 100, 60, 500, 200, 800

McCall, Samuel W., -, -, 1, 120, 40, -, -, 7000, 200, 1000, 100, -, 300, 150, 1500

Atwell, William, W., -, -, 1, 100, 40, -, -, 10000, 200, 600, -, 60, 400, 160, 1400

Records, Atwood, -, -, 1, 145, 60, 5, 10, 1000, 500, 400, 350, 90, 475, 160, 2000

Fisher, George W., -, -, 1, 275, 100, -, 30, 11000, 400, 19000, 400, 375, 1200, 150, 2500

Hill, Thomas H., -, 1, -, 132, 47, -, 23, 10000, 400, 1000, 20, 125, 800, 145, 2100

Hutchison, Benedict W., -, 1, -, 195, 50, -, -, 12000, 200, 1000, 15, 300, 700, 144, 2000

Hance, Edward, -, -, 1, 200, 160, -, -, 14000, 300, 1600, 20, -, 450, 150, 2000

Reybold, Anthony, -, -, 1, 350, 100, -, -, 25000, 1000, 3175, 100, -, 650, 175, 300

Pordham, James, -, -, 1, 310, 100, -, -, 20000, 800, 850, 255, 160, 650, 155, 3000

Clark, Theodore F., 1, -, -, 345, 170, 15, 40, 25000, 700, 2400, 100, 150, 1000, 224, 5000

Clark, Edward L., -, -, 1, 265, 70, 30, -, 18000, 700, 1000, 25, 190, 1000, 175, 3500

Janvier, Charles, -, -, 1, 290, 130, -, -, 25000, 700, 1500, 25, 200, 1000, 210, 4000

Dickinson, John W., -, -, 1, 236, 56, 10, 36, 16000, 400, 800, 30, 180, 540, 150, 3000

Kelley, George C., -, -, 1, 50, 22, -, -, 9000, 200, 450, 15, 45, 350, 88, 600

Brady, George F., 1, -, -, 226, 55, -, 60, 18000, 600, 1000, 50, -, 600, 135, 3000

Reybold, Barney, 1, -, -, 67, 30, -, -, 17000, 1550, 1200, 20, 300, 600, 144, 1000

Jamison, Clarence, 1, -, -, 208, 45, -, 5, 16000, 600, 1500, 100, 80, 700, 120, 1200

Belville, Thomas, 1, -, -, 120, 29, 1, -, 12000, 350, 540, 75, 140, 400, 104, 1660

Jones, Theodore, -, -, 1, 72, 27, 7, -, 5000, 2000, 500, 50, 49, 300, 100, 1000

Smith, Lydia, 1, -, -, 24, 10, -, 5, 4000, 400, 225, 35, 5, 150, 60, 550

Woodward, John H., -, -, 1, 98, 38, -, 10, 6000, 300, 60, 50, 95, 400, 72, 800

Corbet, Charles, 1, -, -, 200, 60, -, -, 15000, 1000, 3000, 25, 150, 1200, 182, 2827

Deputy, Anna M., -, -, 1, 250, 100, 15, -, 14000, 700, 1200, 25, 120, 650, 150, 3000

Reybold, Edwin C., -, -, 1, 390, 150, -, 40, 40000, 3000, 840, 400, 240, 200, 270, 10000

Clark, James H., 1, -, -, 330, 210, -, 40, 30000, 800, 3000, 400, 250, 1800, 225, 5500

Clark, Levi C., 1, -, -, 280, 115, -, 20, 20000, 600, 1500, 60, 300, 1000, 154, 2400

Clark, William C., 1, -, -, 265, 150, -, 50, 50000, 2000, 4000, 150, 50, 1600, 500, 5000

Cleaves, Charles, -, -, 1, 250, 34, -, 100, 50000, 200, 325, 50, 45, 570, 115, 2000

Clark, Thomas Jefferson, 1, -, -, 40, 13, -, -, 10000, 300, 300, 15, -, 150, 20, 150

Dutton, Joshua, -, -, 1, 20, -, -, 12, 2500, 150, 250,-, 25, 180, 75, 675

Cann, Richard T., 1, -, -, 380, 75, 6, 7, 25000, 500, 1500, 50, 200, 1500, 288, 5000

Pennington, Frank J., -, -, 1, 250, 120, 50, -, 25000, 500, 1500, 300, 116, 550, 40, 2000

Cochran, Wm. A., 1, -, -, 240, 50, 4, 40, 20000, 400, 1400, -, 250, 500, 40, 2000

Cochran, Richard W., 1, -, -, 185, 16, -, -, 20000, 600, 1200, -, 300, 400, 32, 1700

Herrick, Alfred, -, -, 1, 150, 21, -, 12, 15000, 200, 401, -, 150, 300, 35, 1000

Fenimore, Joshua B., 1, -, -, 18, 13, -, 2, 6000, 100, 650, -, 100, 100, 25, 150

Pharo, Horatio W., 1, -, -, 160, 45, -, 26, 12000, 1000, 500, 100, 100, 600, 104, 850

Dickerson, Samuel, -, -, 1, 460, 25, 40, -, 31000, 800, 2000, 50, 270, 1000, 180, 3500

Shanof, James T. -, -, 1, 230, 20, 2, 3, 20000, 800, 2000, 50, 500, 1000, 224, 2000

Jarrell, James, -, -, 1, 144, -, -, 15, 12000, 300, 800, 20, 90, 600, 108, 1250

Cochran, Robert T., 1, -, -, 120, -, -, -, 15000, -, 600, 100, 250, 500, 52, 1300

Cochran, Dan M., 1, -, -, 150, -, 25, -, 20000, 500, 900, 100, 125, 700, 72, 1200

Shanof, Senick F., 1, -, -, 700, 200, -, -, 70000, 4000, 3000, 500, 1500, 2000, 360, 8000

Dickinson, Eli, -, -, 1, 78, -, -, -, 5000, 250, 125, 92, 81, 300, 104, 1000

Nowland, Henry A., -, -, 1, 340, -, -, 40, 25000, 1000, 2000, 100, 300, 1200, 165, 4100

Jones, Wilson, -, -, 1, 45, 5, -, -, 1600, 50, 200, 25, 60, 125, 52, 110

Templeman, Eliza, 1, -, -, 90, -, -, -, 9000, 1000, 300, 50, 125, 350, 108, -

Hopkins, Thomas R., -, 1, -, 150, -, -, 27, 10000, 400, 200, 150, 150, 500, 120, 1500

Carpenter, William E., -, -, 1, 80, -, -, 5, 4000, 150, 300, -, 60, 120, 40, 300

Houston, Thomas J., 1, -, -, 175, -, -, 25, 10000, 150, 700, 25, 100, 290, 72, 800

Houston, John, 1, -, -, 54, -, -, -, 5400, -, 125, 50, 40, 175, 52, 300

Houston, George H., -, -, 1, 140, 10, 25, -, 10000, 500, 900, 50, 175, 900, 160, 1800

Barnett, William, -, -, 1, 200, -, -, -, 8000, 300, 1000, -, 100, 400, 80, 1200

Woods, Isaac, 1, -, -, 180, -, 20, -, 16000, 500, 2500, 100, 100, 600, 120, 2000

Jones, William, -, -, 1, 170, -, -, -, 17000, 400, 950, 50, 450, 350, 80, 1510

King, William E., -, 1, -, 150, -, -, -, 15000, 300, 500, 100, 100, 500, 200, 1500

Whitlock, Alonzo, -, -, 1, 238, 52, -, 5, 20000, 650, 1100, 314, 270, 759, 176, 2500

Cavender, Lewis A., -, -, 1, 100, -, 3, -, 6000, 100, 351, -, 50, 300, 144, 400

Joster, John, -, -, 1, 83, -, -, -, 2500, 10, 100, -, 30, 125, 52, 250

Green, Lewis, -, -, 1, 90, -, -, -, 3500, 100, 475, 25, 53, 300, 80, 350

Gold, Thomas, -, -, 1, 100, -, -, 25, 2500, 125, 330, 20, 40, 225, 75, 500

Cleaves, William S., -, -, 1, 400, -, 60, 25, 20000, 1150, 1200, 100, 200, 1000, 150, 2800

LeCompt, James, 1, -, -, 230, 30, -, -, 15000, 500, 2000, 50, 125, 500, 108, 1500

Galleger, Andrew, -, -, 1, 160, -, -, -, 10000, 300, 5000, 100, 150, 500, 100, 1000

Lore, William, 1, -, -, 160, 7,-, -, 15000, 200, 2000, 50, -, 400, 75, 1900

Vail, David C., -, -, 1, 300, -, -, 10, 18000, 300, 1500, 50, 200, 600, 108, 2500

Aspril, John A., -, -, 1, 300, -, -, 30, 18000, 400, 1500, 50, 120, 600, 120, 2500
Hopkins, David H., -, -, 1, 125, -, -, -, 10000, 500, 15, 10, 75, 100, 650, 125, 1000
Hudson, John P., 1, -, -, 175, -, -, -, 10000, 300, 1000, 50, -, 600, 100, 600
Jamison, Oliver V., 1, -, -, 235, 35, -, -, 12000, 500, 1200, 50, 190, 600, 110, 2500
Cannon, Greensbery T., -, -, 1, 140, -, 20, -, 12000, 400, 700, 100, 150, 300, 100, 1800
Cleaves, Julius G., -, -, 1, 260, -, 45, -, 12000, 500, 1400, 300, 150, 720, 114,1650
Vail, William, -, -, 1, 250, -, -, 50, 20000, 600, 1000, 25, 50, 540, 108, 1700
Riley, Sallie, A., -, -, 1, 200, -, -, -, 10000, 500, 1000, 20, 100, 500, 104, 1500
Reynolds, Frank D., -, -, 1, 200, -, -, -, 10000, 175, 1200, 200, 300, 500, 146, 1600
Swain, George H., -, -, 1, 180, -, -, -, 14000, 100, 500, 25, 200, 1200, 144, 2500
Plummer, Silas, 1, -, -, 175, -, -, -, 10000, 150, 200, -, 75, 400, 100, 1000
McMullen, James, 1, -, -, 180, -, 14, -, 12000, 500, 1000, 50, 70, 500, 140, 2000
Townsend, George W., 1, -, -, 80, -, -, 7, 8000, 400, 300, 50, 75, 300, 100, 1200
Carrow, John W., -, -, 1, 250, -, -, 13, 20000, 500, 2000, 25, 140, 800, 200, 1800
McWorter, Leontine N., 1, -, -, 140, -, -, -, 12000, 800, 1500, 25, 162, 400, 52, 1200
McWhorter, Thomas S., 1, -, -, 160, -, -, 4, 16000, 400, 1200, 25, 50, 450, 200, 1150
Sparks, William L., -, -, 1, 140, -, 11, -, 12000, 150, 600, 10, 75, 360, 84, 1000
Hopkins, Levin, -, -, 1, 340, -, 10, -, 20000, 600, 1500, 50, 100,800, 200, 2000
Vail, Jehu V., -, -, 1, 200, -, -, -, 18000, 800, 1000, 25, 150, 700, 108, 1300
McMullen, William, -, -, 1, 112, -, -, -, 8000, 200, 700, 25, 40, 350, 100, 800
Bennett, John R., -, -, 1, 250, -, 6, -, 20000, 800, 3000, 1000, 60, 650, 200, 3000
McCalister, Edward, -, -, 1, 180, -, 80, -, 18000, 600, 2000, 25, -, 600, 100, 2000
Vail, Samuel C., -, -, 1, 200, -, 52, -, 20000, 800, 1000, 25, 205, 600, 108, 2200
Ellison, William, -, 1, -, 186, -, 58, -, 15000, 500, 1500, 25, 400, 400, 108, 1000
Keyes, Arthur, -, -, 1, 60, 50, -, -, 8000, 200, 1200, 500, -, 700, 120, 1200
Nelson, John B., 1, -, -, 180, -, -, -, 15000, 1000, -, -, -, -, -, -
Moody, Alex, -, -, 1, 200, -, 50, -, 16000, 400, 1200, 50, 175, 700, 160, 2400
Cleaves (Cleaver), Isaac, 1, -, -, 182, -, 8, -, 18000, 300, 600, 25, 82, 200, 62, 1000
Vanhekil, Fredus, 1, -, -, 120, -, 16, -, 7500, 500, 500, 25, 75, 500, 72, 1000
Vandergrift, Andrew J., 1, -, -, 220, -, 20, -, 22000, 1200, 1800, 50, 100, 1200, 160, 2000
Rickards, Ezekiel, -, -, 1, 100, -, 30, -, 8000, 250, 700, 70, 35, 500,70, 1000
Price, William K., -, -, 1, 70, 9, -, -, 4000, 300, 210, 20, 15, 150, 50, 500
Cleaver, John, 1, -, -, 135, 42, 6, -, 5000, 300, 500, 25, 300, 250, 108, 600
McMullin, William, 1, -, -, 200, 5, 15, -, 12000, 500, 1000, 50, 200, 500, 100, 2000
Dilworth, Thomas T., 1, -, -, 450, -, -, -, 30000, 1500, 2310, 200, 400, 600, 480, 5000
Ludlow, Joseph N., -, -, 1, 112, 8, -, -, 9000, 300, 500, 25, 140, 500, 120, 1000
McClean, Hugh, -, -, 1, 4, -, -, -, 1000, 20, 130, 20, -, 200, 52, 35
Price, Henry, -, -, 1, 200, -, 85, -, 15000, 300, 1000, 50, 50, 1000, 130, 1500

Laurance, William S., 1, -, -, 416, -, -, -, 30000, 1000, 2500, 300, 150, 1000, 500, 600

Longland, Zenas P., 1, -, -, 70, 100, -, -, 8000, 300, 800, 100, -, 500, 150, 1500

Binge, William, -, -, 1, 120, 250, -, -, 10000, 250, 300, 100, -, 400, 150, 2001

Carpenter, James T., 1, -, -, 156,75, -, -, 14000, 600, 3500, 50, -, 600, 260, 3800

Carpenter, Robert S., -, -, 1, 21, -, -, -, 3000, 50, 100, 15, -, 150, 152, 500

Green, Wilson T., 1, -, -, 12, -, -, -, 900, 50, 500, 10, -, -, -, 175

Green, Wilson T. Jr., 1, -, -, 5, -, -, -, 600, 100, 75, 10, -, -, -, 75

Vanhekle, Fredus P., -, -, 60, 40, -, -, -, 4000, 300, 300, 100, 120, 100, 52, 700

Denney, Benjamin, 1, -, -, 3, -, -, -, 20000, 25, 100, -, -, 200, -, 70

Cleaves, Isaac S., -, -, 1, 80, 20, -, -, 6000, 500, 500, 50, -, 300, -, 1000

Bennett, William H., -, -, 1, 100, -, 5, -, 5000, 100, 300, 50, -, 400, -, 1000

Gott, John H., -, -, 1, 180, -, -, 30, 12000, 300, 500, 40, -, 500, 100, 2000

Pleasanton, Benjamin, -, -, 1, 200, -, -, 25, 10000, 200, 500, 50, -, 300, 108, 1000

Whitaker, Conrad M., -, -, 1, 100, -, -, -, 2500, 100, 50, -, -, -, -, 200

Dempsey, Archabald G., -, 1, -, 115, -, -, -, 7000, 200, 300, -, -, 300, 52, 900

Jones, Abram, -, -, 1, 100, -, -, -, 1700, 100, 200, 25, -, 100, 52, 700

Burgess, George O., 1, -, -, 120, -, -, 75, 6000, 100, 300, 25, -, 100, 50, 1000

Croft, Israel F., 1, -, -, 3, -, -, -, 1000, -, 50, 10, -, -, -, -

Foard, Richard H., -, -, 1, 125, -, -, 66, 4000, 75, 400, 50, 147, 200, 52, 625

Pool, Zadock, 1, -, -, 284, -, -, 40, 28000, -, 400, 250, 400, 800, 4500, 4500

Stevens, Edmund S., -, -, 1, 34, -, -, -, 5000, 300, 410, 10, -, -, -, 200

Moore, Richard, -, -, 1, 20, -, -, -, 800, 50, 200, -, -, -, -, 180

Higgins, Samuel, 1, -, -, 142, -, -, 10, 8000, 100, 300, 40, 30, -, -, 1100

Vandergrift, Leonard G. Jr., 1, -, -, 175, -, -, 18, 14000, 350, 500, 25, 125, 550, 104, 2000

Dunning, Hugh, 1, -, -, 4, -, -, -, 700, 20, 100, 5, -, -, -, 1550

Vandergrift, C. John, 1, -, -, 130, -, -, 48, 13000, 500, 1200, 25, 150, 600, 100, 2000

Graham, Philip S., 1, -, -, 80, -, 40, 60, 4000, 200, 150, 30, 13, -, -, 250

Vandergrift, Wilson E., 1, -, -, 125, -, -, 25, 6000, 250, 1000, 25, 116, 800, 180, 900

Vandergrift, Wilson, 1, -, -, 80, -, -, 64, 3000, -, -, 150, -, -, -, -

Segark(Legark), David, 1, -, -, 100,-, -, -, 500, 20, 50, -, -, -, -, -

Diehl, William B., 1, -, -, 120, -, 20, 22, 7000, 500, 1000, 50, 150, 500, 100, 1500

Vandergrift, George L., 1, -, -, 200, -, -, 50, 20000, 1000, 2000, 50, 100, 900, 140, 1500

Walters Henry C., 1, -, -, 140, -, -, 60, 10000, 300, 800, 25, 175, 500, 108, 1200

Alrichs, Rachel, 1, -, -, 60, 12, -, 10, 6000, 100, 200, 25, -, 150, 52, 700

Corbit, John C., 1, -, -, 12, -, -, 1, 10000, -, 400, -, 20, 25, 60, 200

Hukill, William A., -, -, 1, 90, -, -, 6, 10000, 100, 225, 25, 120, 400, 124, 1300

Walkins, Columbus, 1, -, -, 14, -, -, -, 2100, 50, 200, 50, 29, 2000, 52, 300

Corbit, Daniel W., 1, -, -, 50, -, -, -, 15000, 300, 400, 20, -, 200, 52, 610

Corbit, Daniel W., 1, -, -, 20, -, -, 3, 10000, -, 250, 25, 56, 50, 15, 350

Lord, Simeon, 1, -, -, 80, -, -, 30, 8000, 100, 500, 25, -, 150, 40, 400

Lachus, John, 1, -, -, 20, -, -, 5, 2500, 100, 200, 10, 25, 200, 50, 400

289

Wilson, James, -, -, 1, 100, -, 30, 60, 8000, 100, 500, -, 50, 250, 52, 500
Cleaver (Cleaves), William A., 1, -, -, 120, -, 6, 60, 8000, -, 500, -, 50, 200, 50, 500
Cleaves, Henry, 1, -, -, 5, 5, -, -, 5000, -, 200, -, -, 200, 52, 150
Vandergrift, Abram F., -, -, 1, 230, -, -, 20, 9200, 150, 1000, 25, -, 600, 125, 1150
Fleming, John, -, 1, -, 150, -, 25, 25, 6000, 100, 200, 30, 50, 400, 104, 700, 10
Nancy, Edward R., 1, -, -, 200, 150, -, 18, 12000, 500, 1200, 150, -, 700, 160, 2400
Gordon, James, 1, -, -, 10, 250, -, -, 8000, 150, 500, 100, -, 175, 52, 500
Riley, Thomas, -, -, 1, 200, -, -, 100, 12000, 400, 1300, 50, 56, 600, 108, 1900
Eaton, Richard, -, -, 1, 260, -, -, 28, 18000, 250, 700, 25, 100, 500, 118, 2800
Webb, George, -, -, 1, 85, -, -, -, 5000, 200, 350, 25, -, 100, 25, 250
Williams, William B., -, -, 1, 125, -, -, 6, 9000, 150, 300, 15, 75, 350, 80, 800
Vandergrift, James M., 1, -, -, 197, -, -, -, 16000, 500, 1600, 50, 152, 800, 175, 1900
Janvier, James J., -, -, 1, 208, -, 8, -, 18000, 500, 1200, 10, 100, 800, 170, 1750
Vail, John, -, -, 1, 250, -, -, 30, 15000, 400, 800, 25, 155, 500, 118, 1850
Croft, Francis J., 1, -, -, 4, -, -, -, 1000, -, 60, 5, -, -, -, 125
Polk & Hatt, 1, -, -, 32, -, -, -, 5000, 175, 200, 10, 150, 2500, 412, 5000
Toulson, James A., -, -, 1, 140, -, -, 50, 7000, 200, 300, 10, 100, 300, 80, 1000
Vandergrift, Isaac W., 1, -, -, 7, -, -, -, 700, 10, 300, 15, 100, 300,75, 970
Knotts, William, -, -, 1, 400, -, -, -, 30000, 400, 1500, 50, 160, 1000, 216, 3200
Stevens, Daniel, 1, -, -, 4, -, -, -, 800, 50, 100, 5, -, -, -, 70
Mailley, Richard L., 1, -, -, 5, -, -, -, 900, 25, 200, 5, 75, 100, 20, 100
Mailley, Charles E. A., 1, -, -, 5, -, -, -, 900, -, 50, 5, -, -, -, 100
Williams, Jonathan K., 1, -, -, 300, -, -, 20, 25000, 1200, 3250, 100, 270, 1200, 236, 4906
Appleton, John, 1, -, -, 1, -, -, -, 2500, 50, 100, 5, 25, 350, 52, 800
Lord, Victor, 1, -, -, 4, -, -, -, 800, 40, 100, -, -, -, -, 430
Tatman, Charles, 1, -, -, 4, -, -, -, 800, 50, 100, 5, -, 100, -, 100
Polk, Sarah B., 1, -, -, 7, -, -, -, 1000, -, 100, 15, -, 250, 60, 110
Polk, William, 1, -, -, 4, -, -, -, 1200, -, -, -, -, -, -, 300
Polk, William, 1, -, -, 10, -, -, -, 1500, -, 400, 12, -, -, -, 300
Polk, William, 1, -, -, 12, -, -, -, 1200, -, -, -, -, -, -, -
Hamilton, William N., -, 1, -, 5, -, -, -, 5000, -, 200, 10, -, 150, 52, 200
Evans, John, 1, -, -, 100, -, -, 100, 6000, 100, 400, 100, 200, 250, 52, 400
Cleaves, Joseph, 1, -, -, 235, -, 30, -, 16000, 600, 1250, 50, 230, 900, 160, 2500
Cochran, William R., 1, -, -, 350, 62, -, -, 35000, 1500, 3437, 100, 490, 1600, 280, 6400
Pennington, Samuel}, 1, -, -, 80, -, -, -, 10000, 450, 690, 10, 50, 428, 125, 1742
Pennington, Samuel}, 1, -, -, 249, 78, 5, -, 20000, 560, 1097, 100, 240, 600, 180, 1586
Boulden, Lorenzo, -, -, 1, 150, ½, 25, -, 10500, 500, 860, 50, 175, 516, 150, 2243
Simkins, William, -, -, 1, 204, 37, 12, -, 10000, 500, 860, 50, 275, 872, 242, 2215
Degan, Federick J., 1, -, -, 4, ½, -, -, 1000, 10, 50, 2, 5, 15, 2, 18
Boulden, Edward, -, -, 1, 100, -, -, -, 8000, 403, 880, 20, 148, 535, 150, 1372

Foard, James, -, -, 1, 155, 26, 10, -, 10000, 700, 1160, 25, 162, 467, 140, 2003
Rothwell, Samuel, -, -, 1, 255, -, 1, -, -, 20000, 100, 370, 25, 550, 700, 225, 1761
Beaston, Ephraim, -, -, 1, 325, -, 25, -, 20000, 1100, 1400, 50, 240, 768, 225, 3472
Eliason, Andrew, 1, -, -, 230, 41, 5, -, 23500, 1000, 1431, 50, 300, 1094, 300, 3994
Houston, William H., 1, -, -, 370, -, 50, -, 30000, 1000, 1745, 500, 392, 1250, 325, 3585
Gould, John B., -, -, 1, 220, 31, 12, -, 10000, 300, 720, 25, 150, 600, 160, 981
Cochran, Julian, 1, -, -, 200, 25, -, -, 20000, 800, 2649, 25, 270, 874, 224, 3152
Murphey, Thomas C., 1, -, -, 200, 20, 20, -, 20000, 500, 1675, 50, 150, 965, 280, 2980
Eliason, James D., -, -, 1, 140, -, 10, -, 8000, 300, 425, 10, 125, 150, 120, 1200
Whitlock, Robert W., -, -, 1, 225, -, -, -, 23000, 1800, 1544, 25, 268, 790, 230, 4342
Warren, David, -, -, 1, 350, -, 30, -, 25000, 800, 1465, 50, 515, 1040, 300, 3596
Dale, William, 1, -, -, 7, -, -, -, 500, 5, 100, 5, 1, 30, 10, 20
Linch, Purnell, 1, -, -, 145, 40, 20, -, 740, 400, 600, 20, 100, 250, 150, 700
Ladd, William, -, -, 1, 170, 10, 40, -, 12200, 500, 600, 115, -, 250, 60, 1426
Cochran, Thomas, -, -, 1, 300, 31, 25, -, 3000, 20000, 1945, 50, 620, 1300, 360, 5963
Gray, James, -, -, 1, 425, 3, 15, -, 35000, 1000, 2620, 50, 340, 1100, 320, 5482
Elison, Charles, -, -, 1, 190, 15, -, -, 19000, 1000, 1480, 50, 240, 750, 200, 3274
Clayton, Richard, 1, -, -, 201, -, 2, -, 14000, 500, 1500, 24, 240, 590, 141, 2814
Clayton, Thomas, 1, -, -, 200, -, 13, -, 15000, 800, 1175, 50, 450, 610, 151, 2569
Crosland, John B., -, -, 1, 130, 32, 30, -, 9800, 200, 401, 41, 65, 150, 75, 865
Walker, Morton E., 1, -, -, 246, 40, 10, -, 22500, 1000, 1905, 175, 250, 594, 150, 2502
Green, William 1, -, -, 270, 70, 15, -, 35600, 1000, 1716, 50, 250, 950, 200, 429
Hushebick, Andrew H., 1, -, -, 4, -, -, -, 3500, 200, 65, 24, 5, -, -, 100
Hoffecker, James R., 1, -, -, 143, 16, 15, -, 15000, 1000, 1300, 25, 97, 689, 180, 2063
Jones, Samuel F., -, -, 1, 225, 30, 5, -, 10000, 400, 1140, 25, 192, 375, 90, 2228
Armstrong, Benjamin, -, -, 1, 60, -, -, -, 2500, -, 150, 25, 78, 75, 20, 280
Armstrong, Benjamin, -, -, 1, 140, 25, 15, 4120, 400, 790, 25, 175, 550, 150, 1850
Sellers, C. Cadwolleder, 1, -, -, 180, 60, 30, -, 20000, 800, 603, 25, 80, 590, 156, 1179
Clayton, Joshua of Thomas, 1, -, -, 14, -, -, -, 15000, 200, 605, 50, 100, 350, 104, 224
Brady, William, 1, -, -, 200, 40, 15, -, 25000, 300, 1110, 75, 200, 830, 210, 2370
Jones, Mary A., 1, -, -, 4, -, -, -, 4000, -, 25, 25, 11, 30, 5, 62
Jones, Charles E., -, -, 1, 7, 3, -, -, 4000, 100, 50, 20 60, 180, 150, 1500
Hushebick, Clarance, -, -, 1, 5, -, -, -, 1000, -, 100, 10, 6, 7, 10, 280
Williams, J. R., 1, -, -, 225, 8, 17, -, 15000, -, -, 25, 250, 949, 187, 2120
Polk, Cyrus, 1, -, -, 300, 80, 5, -, 22750, 10000, 2600, 100, 72, 700, 200, 4124
Crockett, Alfred P., 1, -, -, 174, -, -, -, 12500, 400, 600, 25, 125, 400, 70, 1990

Price, Richard L., 1, -, -, 175, 29, 24, -, 15500, 1000, 1900, 125, 300, 550, 150, 2684

Mitchell, James W., -, -, 1, 157, -, -, -, 12000, 800, 1600, 20, 66, 418, 144, 1462

Polk, Wm., 1, -, -, 257, -, 50, -, 15000, 600, 925, 100, 168, 600, 125, 4445

Walker, Henry R., -, -, 1, 200, -, -, -, 20000, 1010, 2840, -, 412, 766, 120, 3289

Webb, Jon P., -, -, 1, 130, -, 30, -, 10600, 300, 1025, 5, 45, 210, 60, 1317

Cochran, Charles P., -, -, 1, 400, -, 16, -, 30000, 1000, 2890, 25, 550, 1644, 320, 3835

Wood, William, -, -, 1, 300, -, 60, -, 28000, 800, 1721, 25, 722, -, 190, 3860

Clayton, Richard, 1, -, -, 203, -, 2, -, 1400, 500, 1500, 24, 240, 590, 41, 2814

Clayton, Tomas of Joshua, 1, -, -, 200, -, 25, -, 1500, 800, 1175, 50, 450, 610, 151, 2569

Crosland, John D., -, -, 1, 170, 38, 30, -, 7500, 200, 701, 40, 65, 150, 75, 865

Walker, Morton E., 1, -, -, 246, 40, 10, -, 25000, 1000, 1905, 175, 250, 599, 150, 2502

Cochran, Edward R., 1, -, -, 265, 120, 15, -, 30000, 1000, 1905, 50, 200, 2500, 550, 5679

Cochran, Edwin R., 1, -, -, 228, 160, -, -, 21000, 600, 975, 50, 100, 1000, 230, 1331

Vail, Alexandria, 1, -, -, 41, -, -, -, 6000, 200, 350, 18, 16, 118, 22, 372

Herrick, Alfred, 1, -, -, 130, 20, 30, -, 13000, 500, 550, 27, 170, 250, 75, 1600

Wilson, Robert B., 1, -, -, 150, 35, 7, -, 15000, 200, 845, 45, 260, 520, 160, 1684

Pharo, Horatio W., 1, -, -, 187, 16, 25, -, 12000, 500, 495, -, 50, 600, 165, 4448

Davis, Mark H., -, -, 1, 240, 81, 10, -, 15000, 700, 1750, 25, 350, 761, 224, 2881

Clothier, John L., -, -, 1, 153, 70, 3, -, 10700, 300, 900, 10, 75, 125, 50, 1212

Wilson, George F., -, -, 1, 295, 76, 5, -, 21000, 600, 1686, 450, 125, 555, 150, 2576

Houlton, Jesse J., 1, -, -, 158, 26, 7, -, 10000, 400, 625, 20, 100, 250, 100, 1237

Ratledge, Robert, -, -, 1, 180, 35, 60, -, 15000, 500, 710, 100, 257, 500, 110, 1541

Cavender, Thomas, -, -, 1, 452, 40, 60, -, 30000, 1100, 2916, 100, 493, 1357, 272, 4629

Jones, John A., 1, -, -, 350, 46, -, -, 30000, 200, 1681, 100, 400, 1200, 250, 4810

Clayton, Joshua Jr., -, -, 1, 248, 100, 20, -, 21000, 800, 220, 331, 200, 600, 155, 2681

Clayton, Henry, 1, -, -, 216, 2, 5, -, 20160, 800, 2140, 300, 400, 900, 164, 3847

Derickson, Charles, 1, -, -, 470, 100, 35, -, 52000, 1100, 2070, 50, 620, 1610, 225, 5695

Lockwood, John J., 1, -, -, 60, -, -, -, 10000, 300, 29, 420, 75, 195, 93, 475

Willits, Merritt N., 1, -, -, 333, -, 14, -, 53000, 1000, 2000, 300, 150, 1094, 326, 2695

McWhorter, John F., -, 1, -, 8, -, -, -, 1500, 100, 200, -, 25, 4, 103

Burris, Nemiah, 1, -, -, 357, -, 60, -, 30000, 800, 1555, 300, 440, 1045, 250, 3999

Brown, James L., -, -, 1, 176, -, 20, -, 9000, 1000, 1332, 25, 225, 691, 165, 1614

Brown, James L., -, -, 1, 100, -, 40, -, 5500, 125, 275, 25, 80, 300, 70, 312

James, Samuel, 1, -, -, 3, -, -, -, 500, 5, 25, -, -, -, 10, 100

Daniels, William, 1, -, -, 100, 7, 10, -, 5100, 100, 400, 50, 60, 100, 300, 500

Moore, James, -, -, 1, 100, -, 11, -, 3000, 100, 400, 50, 75, 300, 80, 600
Atwell, Edward, -, -, 1, 130, -, 11, -, 3000, 500, 800, 20, 104, 800, 100, 1000
Greer, David, -, -, 1, 104, 1, 2, -, 1000, 50, 200, 20, 88, 100, 50, 300
Pearce, William W., -, -, 1, 130, -, 10, -, 7000, 100, 300, 50, 80, 200, 30, 800
Dohaney, Danuel, 1, -, -, 100, -, 30, -, 300, 103, -, -, -, -, 80, 100
Warran, Samuel, -, -, 1, 280, -, 20, -, 13000, 1000, 1000, 500, 200, 800, 120, 2000
Dalson, Sarah, -, -, 1, 180, -, 10, -, 3000, 100, 1000, 50, 60, 600, 20, 2000
Armstrong, John, -, -, 1, 144, -, 80, -, 3000, 100, 500, 40, 60, 500, 100, 600
Lewis, John, -, -, 1, 230, -, 15, -, 4000, 100, 300, 60, 40, 100, 20, 500
Sherdon, John, -, -, 1, 20, -, 2, -, 700, 10, 200, 20, -, 50, 20, 43
Jones, Miles, -, -, 1, 55, -, 48, -, 1000, 100, 1600, 25, 20, 100, 40, 300
Mahan, Pal___, -, -, 1, 175, -, 60, -, 6000, 200, 600, 30, 30, 200, 120, 500
Lamb, Thorn, 1, -, -, 100, -, 15, -, 2000, 200, 500, 50, 60, 100, 100, 300
Claten, Danul, 1, -, -, -, -, 1, -, 500, -, -, 10, 60, -, -, -
Lee, William, -, -, 1, 230, -, 15, -, 4500, 100, 800, 100, 20, 200, 100, 500
Marten, Alexander, 1, -, -, 50, -, 3, -, 1000, 100, 400, 60, 40, 50, 80, 300
McCarter, Jane, 1, -, -, 65, -, 25, -, 3000, 100, 300, 30, 50, 36, 80, 300
Dickeson, James, -, 1, -, 105, -, 21, -, 3000, 100, 300, 40, 30, 50, 80, 500
Daniels, Benjamin, -, -, 1, 55, -, 11, -, 1000, 50, 100, 10, 20, 100, 80, 400
Isaacs, John, -, -, 1, 50, -, 42, -, 600, 10, 200, 30, 20, 100, 66, 300
Landis, Cal__, 1, -, -, 100, -, 100, -, 5000, 100, 500, 100, 90, 600, 160, 1200
Skaggs, John, 1, -, -, 100, -, 28, -, 3000, 100, 600, 100, 30, 100, 60, 300
Skaggs, William, 1, -, -, 100, -, 11, -, 1600, 200, 500, 100, 30, 110, 40, 300
Gatt, George, -, -, 1, 55, -, 10, -, 700, 100, 100, 70, 20, 150, 30, 330
Harris, Hanson W., 1, -, -, 200, -, 50, -, 33,00, 100, 202, 20, 10, 156, 40, 400
Gattis, James, 1, -, -, 40, -, 1, -, 600, 50, 60, 20, -, 20, 20, 500
Tuck, Martha, 1, -, -, 100, -, 3, -, 1000, 100, 300, 60, 60, 100, 120, 400
Tatman, Richard, 1, -, -, 40, -, 4, -, 1000, 100, 100, 10, 10, 50, 40, 400
Caulk, Jacob, 1, -, -, 20, -, 30, -, 500, 50, 60, 40, -, 40, 40, 100
Ganrice, James, -, -, 1, 150, -, 20, -, 3000, 100, 200, 20, 20, 100, 10, 300
Maloney, Daniel, -, -, 1, 115, -, 10, -, 200, 150, 100, 90, 10, 20, 120, 500
Chears, William, 1, -, -, 100, -, 10, -, 1000, 50, 50, 30, 60, 200, 80, 400
Caulk, John, 1, -, -, 20, -, 20, -, 100, 5, 20, -, -, 50, 20, 100
Onel, Elizabeth, 1, -, -, 20, -, 1, -, 600, 5, -, 10, -, 60, 60, 70
King, William, 1, -, -, 40, -, 15, -, 1000, 50, 200, 20, 30, 100, 80, 500
Vandyke, George, -, -, 1, 190, -, 5, -, 3000, 100, 300, 100, 180, 200, 130, 1000
Morgin, Edwin, -, -, 1, 125, -, 151, -, 6000, 200, 500, 100, 150, 300, 140, 2000
McKay, Benjamin, 1, -, -, 32, -, 33, -, 960, 50, 170, 10, 80, 50, 100, 500
McKay, Benjamin, 1, -, -, 67, -, -, -, 900, -, -, -, -, -, -, 50
Barlow, Gedion, 1, -, -, 108, 10, 6, -, 1590, 100, 150, 100, -, 100, -, 700
Warran, Sarah A., -, -, 1, 80, 20, 9, -, 500, -, 80, -, -, -, -, 60
Riley, William E., 1, -, -, 260, 10, 11, -, 8000, 500, 2000, -, 300, 500, 100, 2000
Ginn, Benjamin, 1, -, -, 175, 9, 10, -, 4000, 100, 700, 100, 200, 100, 10, 1000
Bailey, Wathen, -, -, 1, 150, 8, 11, -, 6000, 100, 500, 150, 100, 200, 10, 100
Caulk, Alexander, 1, -, -, 40, 7, 25, -, 1000, 100, 200, 50, 20, 50, 30, 300
Caulk, Wesley, 1, -, -, 40, 13, 17, -, 1000, 100, 200, 40, 20, 50, 10, 920

Caulk, Lenard, 1, -, -, 40, 15, 19, -, 1000, 100, 200, 30, 10, 30, 10, 150
Caulk, Isaac, 1, -, -, 40, 10, 5, -, 1000, 100, 210, 40, 20, 15, 5, 100
Steskey, Newbary, 1, -, -, 27, 9, 5, -, 600, 10, 10, 20, -, 2, 110, 100
McVay, Samuel, -, -, 1, 20, 10, 20, -, 10000, 100, 500, 20, 200, 400, 100, 1500
Jones, Robert M., -, -, 1, 100, 10, 12, -, 3000, 90, 50, 11, 90, 100, 20, 300
Ahern, Dennis, 1, -, -, 20, 9, 5, -, 500, -, 80, 16, -, -, -, 200
Reed, Richard, -, -, 1, 100, 9, 5, -, 2000, -, 100, 17, 30, -, -, 200
Guinn, William, -, -, 1, 190, 6, 6, -, 4000, 100, 100, 18, 90, -, -, 1000
Burchart, Pealer, -, -, 1, 300, 3, 12, -, 4000, 100, 300, 21, 90, 200, 10, 600
Scott, William, -, -, 1, 100, 3, 6, -, 3800, 50, 200, 1, 160, 340, -, 300
Turner, Martin, -, -, 1, 10, 6, 5, -, 3000, 50, 150, 21, -, 100, 10, 230
Henry, Susan, 1, -, -, 10, 6, 3, -, 500, -, -, 22, -, -, -, 200
Hill, James, 1, -, -, 50, 3, 11, -, 600, -, -, 66, -, -, -, 100
Seny, James, -, -, 1, 20, 10, 10, -, 600, -, 165, -, -, 50, 50, -
Edgelle, William F., -, -, 1, 70, 11, 12, -, 8700, 300, 1100, 100, 200, 500, 100, 1000
Fenimore, Samuel, -, -, 1, 141, 20, 15, -, 8000, 300, 300, 100, 60, 500, 140, 1500
Green, Isaac, -, -, 1, 100, 10, 16, -, 9000, 200, 500, 66, -, 20, 30, 800
Macy, Isaac, -, -, 1, 110, 8, 9, -, 4000, 200, 500, 40, 60, 200, 32, 600
Kapple, Corner, -, -, 1, 50, 10, 8, -, 600, 50, 160, 30, -, 10, 20, 100
Romer, John, -, -, 1, 90, 11, 3, -, 500, 50, 100, 40, 20, 50, 60, 100
Thomas, Samuel C., 1, -, -, 174, 20, 10, -, 10000, 300, 800, 200, 150, 500, 30, 1200
Roberts, John W., 1, -, -, 155, 20, 10, -, 9000, 400, 800, 100, 100, 520, 70, 1200
Clayton, John, 1, -, -, 550, 10, 5, -, 600, 50, 50, 40, 20, 20, 50, 20
Budd, William, 1, -, -, 520, 12, 3, -, 400, 50, 40, 30, -, 10, 7, 40
Poor, George, 1, -, -, 510, 11, 5, -, 300, 50, 50, 50, 10, 12, 9, 10
Bead, George, 1, -, -, 600, 5, 100, -, 5000, 10, 20, 10, -, 25, 12, 50
Seims, John, -, -, 1, 600, 8, 12, -, 1500, 12, 50, 10, -, 25, 10, 50
Sweatman, Alford T., -, -, 1, 125, 8, 11, -, 600, 10, 40, 11, -, 16, 9, 60
Goldstone, George, -, -, 1, 100, 5, 6, -, 3000, 100, 30, 12, -, 18, 8, 300
Anderson, William, 1, -, -, 70, 4, 5, 600, -, 30, 22, -, 12, 20, 105
Hopkins, George, -, -, 1, 140, 3, 4, -, 3000, 70, 300, 12, -, 2, 10, 300
Gesford, Nathaniel, -, -, 1, 140, 3, 20, -, 2500, 100, 600, 11, -, 11, 10, 300
Moore, Elias V., -, -, 1, 222, 10, 15, -, 8470, 300, 900, 100, 102, 900, 170, 2250
Townsend, Wesley, 1, -, -, 170, 6, 3, -, 8200, -, 500, -, 100, 500, 120, 1000
Vandergrift, William M., 1, -, -, 150, 5, 6, -, 10500, 300, 420, 300, 100, 600, 80, 200
Walker, Zekiel, -, -, 1, 100, -, 7, -, 7200, 100, 400, -, 30, 150,70, 450
Appleton, John M., -, -, 1, 286, 10, 2, -, 10870, 300, 780, 200, 240, 1000, 180, 7000
Corbit, Daniel, 1, -, -, 100, 13, 1, -, 6000, 500, 1000, -, 50, 300, 45, 9050
Ellis, John, -, -, 1, 135, 5, 6, -, 7000, 400, 600, 200, 210, 1500, 10, 1706
Taylor, James T., 1, -, -, 140, -, 30, -, 8280, 10,75, 50, 60, 80, 20, 1480
Budd, James, -, -, 1, 20, -, -, -, 1200, 300, 200, 10, -, 175, 20, 275
Whitlock, H. G., -, -, 1, 260, 320, 20, -, 200, 500, 780, 80, 200, 500, 100, 1200

Weldin, John W., -, -, 1, 10, -, 180, -, 3000, 50, 290, 50, 73, 325, 20, 800
Weldin, William, -, -, 1, 300, -, 5, -, 13000, 100, 775, -, 100, 300, -, 1590
Townsend, Joshua, 1, -, -, 5, -, -, -, 500, 20, 50, 10, 10, -, -, 100
Keller, John B., 1, -, -, 50, 30, 125, -, 2375, 100, 100, 75, 30, 537, 537, 100
White, Thomas, 1, -, -, 12, -, -, -, 400, -, 2, -, -, 10, -, 10
Ray, James, 1, -, -, 6, -, -, -, 200, -, 2, -, -, 10, 1, 10
Aurton, Isaac, -, -, 1, 13, -, 2, -, 500, 100, 200, 20, -, 50, 40, 30
Ray, James, 1, -, -, 5, -, -, -, 500, -, 2, -, -, -, -, 50
Naylor, Robert, 1, -, -, 6, -, 5, -, 200, 5, 50, -, -, 20, 4, 50
Aurton, Samuel, -, -, 1, 130, -, 20, -, 100, 200, 300, 30, -, 100, 80, 700
Brockett, Horce, 1, - -, 15, -, -, -, 1000, 15, 100, 20, -, 100, 90, 40
Maddis, Henry, 1, -, -, 75, -, 49, -, 1500, 200, 325, 100, 30, 300, 79, 820
Webb, Charles, -, -, 1, 200, 10, -, -, 9000, 200, 323, 50, 150, 300, 99, 1000
Richards, Joshua, -, -, 1, 260, -, 30, -, 9100, 400, 50, 60, 100, 300, 100, 1600
Welch, John, -, -, 1, 120, 5, 15, -, 2800, 200, 400, 50, 20, 300, 70, 100
Vandyke, Lydia B., 1, -, -, 250, 45, 125, -, 15000, 500, 500, 25, 140, 500, 146, 1200
Williams, David, 1, -, -, 4, 1, 1, -, 200, 10, 50, 10, 5, 10, 5, 40
Cannon, Hannah, 1, -, -, 4, 2, 1, -, 250, -, 21, 10, -, 10, 1, 20
Jones, William, 1, -, -, 5, 3, 2, -, 300, -, -, -, -, -, -, 50
Baynolds (Reynolds), Aaron, -, -, 1, 175, 55, 50, -, 7875, 200, 500, 50, 80, 200, 120, 1000
Reynolds, Aaron, -, -, 1, 100, 20, 25, -, 9000, 300, 700, 2525, 90, 430, 160, 1300
Tinley, Stringer, 1, -, -, 100, 20, 20, -, 6000, 200, 600, 20, 60, 400, 120, 1150
Derma, John, -, -, 1, 280, 20, 20, -, 14000, 200, 600, 3000, 60, 500, 142, 1500
Collins, Jackson, -, -, 1, 260, 20, 10, -, 14000, 200, 600, 300, 1400, 140, 1900
Sallomus, Benjamin, -, -, 1, 130, 5, 10, -, 3540, 100, 300, 10, 60, 200, 120, 900
Walson, Mary, -, -, 1, 100, 3, 5, -, 2000, 50, 300, 90, 40, 100, 50, 200
Auston (Aurton), John, -, -, 1, 45, -, 10, -, 760, 100, 200, 20, 30, 100, 50, 300
Ingram, Abraham, 1, -, -, 5, -, -, -, 400, 60, 100, 50, 20, 50, 42, 75
Hobson, John, -, -, 1, 100, -, 20, -, 3000, 100, 300, 50, 40, 100, 80, 500
Collins, James R., -, -, 1, 200, 20, 30, -, 13000, 500, 800, 100, 100, 500, 200, 2000
Gurton, James, -, -, 1, 200, 5, 20, -, 6000, 300, 600, 60, 60, 500, 80, 900
Jones, Purnal T., -, -, 1, 175, -, 12, -, 8750, 300, 600, 100, 100, 500, 200, 2100
Skaggs, William, 1, -, -, 148, -, -, -, 2000, 100, 300, 50, -, 310, 120, 160
Skaggs, John, 1, -, -, 38, -, -, -, 1000, 25, 50, 10, -, 20, 40, 70
Wiggin, George, 1, -, -, 60, 10, 30, -, 1600, 100, 300, 100, 30, 200, 80, 270
Brothers, John, -, -, 1, 60, -, 2, -, 720, 10, 100, 30, -, 50, 40, 200
Cochran, Robert A., -, -, 1, 180, -, 5, -, 10800, 300, 500, 90, 100, 500, 120, 1900
Lockwood, Richard T., 1, -, -, 148, -, 5, -, 8980, 1000, 1010, 100, 350, 500, 120, 1235
Dodson, James, -, -, 1, 50, -, 12, -, 3600, 300, 500, 100, 60, 500, 120, 412
Appleton, Edward W., 1, -, -, 100, 10, 20, -, 9000, 300, 470, 100, 100, 500, 100, 600
Vanpelt, Joseph, -, -, 1, 127, -, 5, -, 4620, 300, 300, 50, 60, 500, 120, 600
McCoy, John, -, -, 1, 250, 20, 20, -, 16800, 300, 440, 300, 300, 800, 200, 2420

Appleton, Henry H., -, -, 1, 200, -, 1, -, 13400, 300, 1000, 100, 600, 1000, 240, 3000

Francis, William, -, -, 1, 40, -, 35, -, 3500, 200, 300, 60, 60, 100, 60, 300

Staats, John F., 1, -, -, 150, 10, 60, -, 5250, 200, 500, 50, 90, -, 400, 80, 600

Deakyne, John, -, -, 1, 120, 10, 10, -, 4500, 300, 497, 20, 100, 200, 80, 600

Lester, Henry S., -, -, 1, 180, -, 10, -, 6545, 200, 400, 100, 90, 400, 120, 1400

Wilson, Manlove, 1, -, -, 180, -, 20, -, 9420, 500, 500, 100, 100, 800, 120, 1000

Silcox, Edward, 1, -, -, 80, -, 10, -, 3000, 100, 150, 20, 75, 300,80, 280

Numering, Sevel, -, -, 1, 200, 12, 12, -, 12000, 200, 600, 100, 60, 600, 100, 1200

Smyth, James C., 1, -, -, 90, -, 3, -, 3800, 100, 120, 100, 90, 200, 80, 900

Davis, Manlove, 1, -, -, 157, -, 1, 1, -, 12000, 300, 500, 2000, 60, 500, 120, 900

Gibbs, Benjamin, 1, -, -, 200, -, 3, -, 12000, 300, 580, 80, 180, 600, 120, 1000

Harman, Israel, 1, -, -, 10, 1, 2, -, 500, 10, 100, 10, -, 20, 10, 20

Lattemous, Levi, 1, -, -, 80, 5, 25, -, -, 250, 800, 100, 100, 200, 100, 200

Sanders, T. J. H., -, -, 120, 2, 80, -, 9000, 300, 360, 100, 60, 600, 120, 1500

McCoy, John, -, -, 1, 150, -, 20, -, 7950, 300, 470, 100, 60, 400, 100, 880

Townsend, George L., -, -, 1, 120, -, 7, -, 7225, 300, 250, 90, 60, 600, 100, 950

Townsend, John Jr., 1, -, -, 110, -, 5, -, 4440, 300, 350, 100, 100, 500, 600, 700

Pearce, John, 1, -, -, 12, -, 1, -, 240, 20, 60, 10, 20, 20, 10, 110

Townsend, Su__, 1, -, -, 145, -, 12, -, 500, 300, 1000, 200, 300, 500, 120, 1000

Rothwel, John M., 1, -, -, 180, -, 16, -, 10000, 1200, 13000, 130, 150, 600, 120, 1000

Oneal, Patrick, -, -, 1, 100, -, 11, -, 900, 300, 500, 60, 60, 300, 80, 600

Clayton, Macomb, 1, -, -, 136, -, 12, -, 9000, 400, 600, 50, 90, 500, 120, 1300

Crawford, Dr. James V., 1, -, -, 242, -, -, -, 15000, 600, 840, 100, 220, 700, 125, 1550

Tatman, Cyrus, 1, -, -, 135, -, -, -, 9000, 500, 1000, 50, 200, 600, -, 1100

Nandain, Richard L., 1, -, -, **230**, -, 10, -, 13900, 500, 700, 200, 400, 1000, 160, 1700

Ginn, Samuel B., 1, -, -, 250, -, 50, -, 10000, 360, 400, 90, 60, 500, 120, 1200

Nandain, A. S., 1, -, -, 140, -, 10,-, 7550, 300, 500, 100, 130, 400, 120, 1500

Cochran, Robert A., -, -, 1, 300, -, 5, -, 8000, 500, 1200, 100, 180, 1000, 200, 2300

Davis, Isaac M., -, -, 1, 310, -, 10, -, 8200, 300, 500, 100, 80, 700, 100, 900

Deakins, W. W., -, -, 1, 100, -, -, -, 3000, 200, 200, 20, 30, 100, 80, 200

Ginn, Samuel, 1, -, -, 200,-, 80, -, 10000, 500, 6000, 90, 90, 500, 120, 1000

Hanson, Richard F., -, -, 1, 125, -, 13, -, 12000, 400, 500, 75, 120, 600, 120, 1500

Milbrone, George W. C., -, -, 1, 300, -, 45, -, 12000, 400, 1000, 100, 150, 800, 160, 1200

Matts, William N., -, -, 1, 173, -, 3, -, 5250, 500, 800, 100, 100, 480, 120, 1000

King, John, 1, -, -, 200, -, 12, -, 1000, 600, 700, 100, 120, 600, 126, 1500

Vandyke, Jacob C., 1, -, -, 100, -, 10, -, 6000, 100, 200, 100, 100, 200, 30, 1000

Carter, James, 1, -, -, 40, -, 1, -, 2000, 200, 300, 2000, 60, 100, 50, 600

Gibbs, Doval, 1, -, -, 250, -, 5, -, 15000, 700, 1000, 100, 120, 800, 100, 2000

Kanely, Ben F., -, -, 1, 200, -, 10, -, 12000, 300, 1000, 120, 120, 500, 120, 1000

Fenimore Edward, 1, -, -, 250, -, 250, -, 15000, 500, 900, 250, 20, 1800, 240, 1900

Davis, Harry, -, -, 1, -, 10, -, 12750, 800, 1000, 200, 150, 1000, 200, 1800

Finter, William A., -, -, 1, 30, -, 12, -, 1000, 100, 400, 60, 60, 200, 60, 200

Roberts, Joseph, 1, -, -, 176, -, 10, -, 11180, 600, 800, 100, 120, 800, 120, 1600

Townsend, Richard, 1, -, -, 150, -, 20, -, 9300, 500, 600, 100, 100, 500, 120, 1000

Perry, Theoden, T., 1, -, -, 170, -, -, -, 12600, 800, 1000, 200, 120, 1000, 140, 1400

Brises, Charlotte, 1, -, -, 9, -, -, -, 500, 10, 50, 10, -, 50, 40, 100

Matthews, James T., -, -, 1, 175, 25, 75, -, 8000, 200, 350, 100, 120, 200, 75, 1500

Reed, Milard F., -, 1, -, 80, -, 130, -, 3000, 75, 200, -, -, 50, 25, 400

Watson, Levi M. (B), -, -, 1, 80, -, 80, -, 5600, 50, 200, 30, 45, 100, 50, 500

Campbel, William J., 1, -, -, 150, 30, 10, -, 5000, 150, 250, 75, 60, 325, 75, 925

Woodkeeper, Henry H., -, 1, -, 200, -, 75, -, 7000, 200, 530, -, 60, 243, 60, 600

Ginng, James, 1, -, -, 100, 11, 68, -, 3000, 50, 375, -, 78, 10, 2, 275

Vansant, Ira, -, -, 1, 150, -, 40, -, 3000, 50, 265, -, -, 400, 75, 400

Leatherm(an), John R., -, -, 1, 100, -, 60, -, 2800, 40, 170, 50, -, 20, 4, 140

Powell, Robert, -, -, 1, 105, 20, 195, -, 6000, 50, 525, 2, 32, 52, 100, 25, 720

Barnett, John W., 1, -, -, 56, 12, 20, -, 3000, 50, 195, -, 45, 100, 25, 320

Durham, John J., -, 1, -, 5, -, 2, -, 300, 10, 50, -, -, -, -, 50

Parvis, Dr. John H., 1, -, -, 31, 47, 6, -, 3500, 100, 620, -, 45, 50, 25, 455

Richardson, Jacob, 1, -, -, 87, 2, -, -, 3000, 75, 415, 150, -, 250, 70, 200

Brockson, Richard C., 1, -, -, 10, 1, -, -, 1000, 20, 150, -, 30, 50, 15, 90

Fie, John, 1, -, -, 40, 2, 8, -, 3000, 100, 170, -, -, 20, -, 100

Derrickson, John A., 1, -, -, 144, 2, 6, -, 4000, 200, 635, 40, -, 200, 80, 496

Derrickson, John A., 1, -, -, 12, 6, 60, -, 1500, -, 100, -, -, 50, 20, 200

Hayden, Abram, 1, -, -, 92, 22, 50, -, 6000, 200, 555, -, 112, 200,75, 700

Staats, Abram, 1, -, -, 22, -, -, -, 2000, 100, 100, -, 21, 100, 50, 80

McLane, Lewis, 1, -, -, 20, 5, -, -, 2000, 50, 100, -, -, -, -, 200

Hayden, Abram Jr., -, 1, -, 20, 10, -, -, 2000, 25, 100, -, -, 75, 20, 200

Carpenter, John R., -, -, 1, 300, 40, 15, -, 15000, 300, 1235, 25, 125, 300, 100, 1655

Staats, James H., 1, -, -, 90, -, 7, -, 3000, 250, 600, -, 90, 200, 40, 750

Buckson, James, -, -, 1, 150, 30, 15, -, 8000, 300, 1000, -, 45, 300, 100, 700

Maloney, Thomas, -, 1, -, 125, 20, -, 5, 4000, 100, 285, -, -, 50, 15, 200

Farrell, Francia J., 1, -, -, 20, -, 3, -, 1000, 50, 156, -, -, -, -, 175

Wadsley, Joseph, H., -, 1, -, 19, 9, -, -, 1000, 50, 50, -, -, -, -, 50

Latta, John A., -, 1, -, 80, 5, 12, -, 3000, 100, 350, -, 30, 80, 36, 500

Walker, George D., -, -, 1, 145, 25, 25, -, 8000, 100, 490, -, 112, 80, 40, 620

Staats, David, 1, -, -, 30, -, -, -, 600, 50, 50, -, -, -, -, 100

Collins, Morris & Saml. A., 1, -, -, 92, 8, -, -, 2000, 100, 800, 50, 30, 100, 50, 250

Armstrong, Richardson H., 1, -, -, 75, -, 80, -, 5000, 300, 300, 25, -, 150, 30, 500

Armstrong, George D., -, -, 1, 150, 2, 40, -, 6000, 100, 1000, 25, -, 420, 120, 1100

Bice, William, 1, -, -, 130, -, 20, -, 4000, 225, 520, -, -, -, -, 500

Donoho, William R., -, -, 1, 80, 3, 3, -, 2000, 75, 125, -, 105, -, 343, 343

Kirkley, William 1, -, -, 100, 1, 48, -, 3000, 75, 250, Farm Not Occupied Last Year (covers columns 12-16 and beyond)

Cooper, John F., 1, -, -, 75, 5, -, -, 2500, 100, 250, -, 90, 100, 33, 450

Greer, William J., -, -, 1, 194, 2, -, -, 8000, 400, 975, -, 270, 400, 132, 1525

Staats, Isaac R., 1, -, -, 85, 18, -, -, 3000, 200, 435, -, 70, 200, 80, 775

Ennis, Annias, 1, -, -, 80, 10, 10, -, 5000, 1700, 2500, 40, 110, 800, 150, 2000

Ennis, Annias, 1, -, -, 114, 46, 20, -, 9000, 1000, 2000, 40, 100, 700, 130, 2500

Ennis, Annias, 1, -, -, 96, 1, 10, -, 4000, 800, 1500, 50, 100, 500, 110, 1800

Ennis, Annias, 1, -, -, 125, 13, 10, -, 6600, 200, 1000, 50, 150, 500, 110, 2000

Staats, Isaac, 1, -, -, 91, 4, -, -, 4000, 200, 600, 30, 28, 175, 60, 700

Nailor, Levi Scott, -, -, 1, 23, -, 4, -, 1500, 100, 75, -, 40, 75, 25, 350

Lockerman, John, -, -, 1, 100, -, 20, -, 3000, 250, 456, 15, 120, 50, 20, 700

Warren, Charles B., -, -, 1, 100,1 5, -, -, 4500, 150, 295, -, 60, 250, 82, 475

Sevil, Able, 1, -, -, 85, 15, 18, -, 3500, 300, 500, -, 80, 200, 70, 310

Parker, Joseph D., -, 1, -, 40, 3, -, -, 1800, 50, 100, -, 30, -, -, 200

Keen, Henry, 1, -, -, 14, -, -, -, 1000, 50, 150, -, 15, -, -, 150

Alfree, James, -, 1, -, 200, 25, -, -, 7500, 150, 510, -, -, 200, 70, 1000

Boggs, Nathaniel P., -, -, 1, 250, -, 30, -, 5000, 125, 250, -, 90, 2000, 40, 700

Stayton, David C., -, -, 1, 110, 10, 40, -, 5000, 100, 412, -, 60, -, -, 432

Carpenter, Alfred, -, -, 1, 140, 25, 10, -, 4000, 100, 300, -, 75, 100, 35, 500

Walker, Charles W., -, -, 1, 100, -, 30, -, 1900, 50, 150, -, 30, -, -, 125

Lewis, Caleb B., -, -, 1, 150, 3, 40, -, 6000, 100, 400, -, 90, 100, 35, 400

Smeed, William, -, -, 1, 210, -, -, -, 9000, 300, 1150, 160, 90, 550, 200, 1300

Jones, Edward F., -, -, 1, 140, 4, 20, -, 2500, 500, 600, 40, 400, 200, 66, 1000

Bennet, Samuel, -, -, 1, 135, 15, 25, -, 4000, 200, 1000, -, 150, 200, 66, 700

Blindt, Michael, 1, -, -, 65, 5, 50, -, 3000, 200, 600, -, 45, -, -, 700

Staats, George, -, -, 1, 140, 200, -, -, 7000, 200, 720, -, 90, 250, 85, 650

Philips, Benjamin, -, -, 1, 100, -, 40, -, 4000, 100, 335, -, 60, 50, 18, 500

David, James L., -, -, 1, 36, 4, 60, -, 1200, 50, 400, 20, -, -, -, 225

Derrickson, Robert, 1, -, -, 135, 36, 34, -, 8000, 300, 1110, -, 90 675, 225, 1475

David, Benjamin, 1, -, -, 85, 100, -, -, 5000, 200, 550, 75, 30, 225, 100, 400

Warren, George, W., -, -, 1, 240, 60, -, -, 15000, 1000, 2000, -, 90, 700, 250, 2275

Grieves, William V., -, 1, -, 40, 70, -, -, 5000, 100, 75, -, -, 50, 20, 200

Collins, John P., 1, -, -, 140, 50, 10, -, 7000, 300, 800, -, 60, 350, 120, 1000

Warren, Robert M., -, -, 1, 150, 100, 15, -, 5000, 200, 400, -, 110, 20, 66, 970

Deakyne, Thomas, 1, -, -, 180, 10, 10, -, 5000, 200, 400, -, 30, 150, 60, 300

Daley, James, 1, -, -, 45, -, 6, -, 1000, Farm Not Occupied This Year (covers columns 10-16 and beyond)

Spearman, Pen B., -, -, 1, 100, -, 10, -, 2500, 100, 315, -, 90, 50, 20, 555

Carttrell, John, 1, -, -, 25, -, 8, -, 600, 25, 50, -, -, -, - 50

Reynolds, William H., 1, -, -, 75, 7, 6, -, 1600, 50, 145, -, 64, -, -, 263

Deakyne, Alexander, 1, -, -, 88, 12, 4, -, 5000, 500, 600, -, 180, -, -, 1055

Gardner, John M., -, 1, -, 35, -, 5, -, -, 100, 150, -, -, -, -, 500

Shivler, Charles E., -, -, 1, -, -, -, -, -, -, 190, -, -, -, -, -

Cooper, James S., -, -, 1, 55, -, -, -, 2000, 100, 250, -, -, 50, 18, 500

Collins, Frank, 1, -, -, 35, -, -, -, 3500, 100, 300, -, 30, -, -, 200

Deakyne, William C., -, -, 1, 100, 86, 24, -, 3500, 200, 415, -, 200, 300, 100, 1566
Deakyne, A. C. & Jos. B., 1, -, -, 160, 8, 23, -, 3500, 300, 665, 75, 200, 350, -, 835
Jerrell, John, 1, -, -, 9, -, -, -, 1000, 25, 100, -, -, -, -, 300
Deakyne, George A., -, -, 1, 150, 25, 20, -, 6000, 200, 450, -, 160, 220, 77, 1055
Fennimore, Jesse T., 1, -, -, 20, -, 6, -, 800, -, 250, -, -, -, -, 500
Downs, Oliver C., -, -, 1, 104, -, 20, -, 4000, 150, 350, 150, 180, 150, 50, 400
Hartup, Mary W., 1, -, -, 120, 16, -, -, 6000, 300, 515, -, 90, 355, 120, 950
Staats, James T., -, -, 1, 60,7, -, -, 2500, 200, 430, -, -, 25, 8, 600
Lofland, George R., -, -, 1, 160, 18, -, -, 6500, 500, 500, 30, 195, 300, 100, 1100
Biles, Jacob K., -, -, 1, 90, 35, -, -, 8000, 500, 500, 175, 150, 500, 150, 1500
McNamee, Charles, -, -, 1, 178, 52, -, -, 5000, 300, 1200, 240, 100, -, 300, 100, 800
Deakyne, William, -, 1, -, 45, -, -, -, 1000, 30, 150, -, -, 50, 20, 150
Sammons, Nehemiah, -, -, 1, 80, 3, 117, -, 2000, 50, 100, -, -, 50, 18, 200
Allston, John, 1, -, -, 160, 14, 26, -, 5000, 250, 500, 80, 150, 200, 70, 500
Hill, Jacob, -, -, 1, 105, 20, 25, -, 5000, 200, 1300, -, 100, 250, 100, 700
Reynolds, William L., 1, -, -, 75, -, 120, -, 3000, 100, 325, -, 30, 100, 33, 300
Truax, B. Frank, 1, -, -, 160, 20, 120, -, 5000, 300, 900, 320, 196, 550, 210, 1400
Carey, Philip J., -, -, 1, 65, 20, 2, -, 3000, 200, 350, -, 45, 50, 18, 600
Middleton, Thomas J., 1, -, -, 150, -, 86, -, 4200, 250, 60, -, -, 400, 160, 630
Noland, S. James, 1, -, -, 50, 10, 9, -, 1000, 50, 100, -, 30, -, -, 300
Hollitt, Samuel, -, -, 1, 20, 14, 7, -, 1200, 25, 100, -, 180, -, -, 150
Jones, George W., -, 1, -, 20, -, -, -, 1000, 50, 100, -, -, -, -, 200
Dill, Philomen, 1, -, -, 35, -, 5, -, 500, 20, 25, -, -, -, -, 150
Perry, Absalom, 1, -, -, 30, -, 29, -, 1000, 50, 50, -, -, -, -, 250
Lightcop, Joseph W., -, -, 1, 165, 64, -, -, 9600, 400, 100, -, 125, 450, -, 2000
Jones, Thomas, -, 1, -, 100, -, 35, -, 2500, 50, 150, -, 180, 100, 35, 860
Records, Mortimer, 1, -, -, 50, -, 15, -, 7000, 200, 400, -, 45, 150, 50, 400
Monro, Alexander, 1, -, -, 30, -, 10, -, 1200, 50, 250, -, -, -, -, 200
Moore, Gilbert, 1, -, -, 13, 1, 1, -, 800, 50, 150, -, -, -, -, 200
Hill, Samuel J., 1, -, -, 150, -, 100, -, 4000, 200, 500, 100, 100, 520, 160, 800
Money, William H., 1, -, -, 100, 20, -, -, 4800, 300, 600, -, 120, 50, -, 2100
Nailor, William, 1, -, -, 75, -, 5, -, 4800, 500, 800, 80, 120, 300, 100, 1100
Fennemore, Lewis, -, -, 1, 85, 15, -, -, 3500, 50, 250, -, -, 100, 33, 500
Jones, William J., -, -, 1, 105, 20, 200, -, 7000, 150, 450, 50, 30, 250, 100, 800
Walker, Harry C., -, 1, -, 20, -, 130, -, 3000, 25, 20, -, -, -, -, 60
Lattamus, Alexander, 1, -, -, 78, 12, 1, -, 4000, 200, 400, -, 68, 150, 50, 500
Chadwick, Thomas, -, -, 1, 74, -, 4, -, 2300, 50, 150, -, 55, -, -, 200
Reynolds, William F, -, -, 1, 56, 2, 6, -, 1500, 100, 200, -, 30, 75, 25, 250
Reynolds, James, 1, -, -, 35, 6, 100, -, 1000, 100, 200, -, 25, 50, 17, 300
Reynolds, James, 1, -, -, 70, -, 20, -, 900, 20, 110, -, -, -, -, 125
Alfree, William, 1, -, -, 75, -, 1, -, -, 2500, 150, 800, 25, 60, -, -, 400
Hamilton, William, 1, -, -, 6, 1, 71, -, 700, 20, 75, -, -, -, -, 100
Allen, William R., 1, -, -, 70, -, 25, -, 2000, 50, 150, -, 30, 30, 10, 300
Nailor, William B., -, -, 1, 51, 44, 8, -, 2500, 100, 500, 40, 30, 100, 33, 600

Ford, John W., 1, -, -, 90, 10, 44, -, 3000, 400, 600, 70, 130, 20, 67, 600
Neff, Henry L., 1, -, -, 10, 4, 20, -, 300, 50, 160, -, -, -, -, 100
Greenley, Robert, 1, -, -, 20, -, -, -, 900, 50, 150, -, -, -, -, 200
Prior, James R., 1, -, -, 1, 62, 2, 10, -, 1400, 100, 300, 50, 45, 150, 50, 450
Boyd, Thomas R., -, -, 1, 65, -, 50, -, 1000, 75, 250, -, 30, 25, 8, 300
Lauterwasser, Henry, -, 1, -, 21, -, -, -, 800, 30, -, 85, -, -, -, -, 150
Stradly, John B., -, -, 1, 80, -, 80, 20, 2200, 75, 110, -, -, -, -, 250
Hevern, John B., 1, -, -, 13, -, -, -, 300, -, 30, -, -, -, -, 100
Frances, William, -, -, 1, 76, -, 2, -, 3500, 50, 150, -, 45, 75, 25, 500
Nailor, Goldsmith C., -, -, 1, 88, 12, 12, -, 3500, 100, 250, 75, 60, 50, 17, 325
Collins, Benjamin C., -, -, 1, 90, 30, 20, -, 5600, 250, 420, -, 75, 260, 88, 1000
Stephenson, Joseph, -, -, 1, 206, 41, 150, -, 10000, 500, 750, 200, 150, 300, 100, 1200
Young, John, 1, -, -, 122, 28, 50, -, 7000, 600, 700, -, 150, 300, 100, 1200
Fennemore, William, -, 1, -, 35, -, -, -, 1200, 100, 150, -, -, -, -, 350
Gray, William, -, -, 1, 115, 25, 15, -, 3500, 100, 450, 150, 60, 150, 50, 500
Bennett, John, -, -, 1, 64, 8, 12, -, 3000, 100, 150, -, 30, -, -, 600
Reed, George W., -, -, 1, 25, -, 15, -, -, 25, 75, -, 15, -, -, 125
Marrim, Richard S., 1, -, -, 85, 3, 100, -, 2500, 150, 250, -, 60, 150, 50, 600
Beck, Samuel, 1, -, -, 70, -, 30, -, 2500, 175, 150, 150, 90, 200, 75, 600
Beck, George W., -, -, 1, 25, -, -, -, 500, 20, 50, -, 30, 80, -, 300
Donley, Barney, 1, -, -, 80, 3, 250, -, 600, 200, 335, -, 60, -, -, 500
Reed, John T., -, 1, -, 70, -, 128, -, 3000, 50, 100, -, -, -, -, 150
Ellingsworth, Robenson, -, -, 1, 75, -, 100, -, 2000, 50, 260, -, 30, -, -, 300
Seemons, John L., 1, -, -, 27, -, 25, -, 1800, 75, 250, 15, 25, -, -, 200
Powell, James, 1, -, -, 60, -, 20, -, 2500, 1000, 350, 50, 120, 300, 100, 800
Powell, James, 1, -, -, 60, -, 20, -, 1200, 200, 300, -, -, 100, 33, 350
Sparks, Amanda, -, -, 1, 108, 291, 5, -, 3500, 300, 400, 50, 60, 50, 17, 1125
Wright, Robert W., 1, -, -, 90, -, 90, -, 3000, 150, 400, 25, 60,-, -, 800
Thomas, William C., 1, -, -, 45, -, 58, -, 2200, 100, 300, -, 30, 50, 17, 300
Keiffer, Nicholas, 1, -, -, 40, -, 10, -, 1100, 50, 75, -, -, -, -, 250
Moffat, William, -, -, 1, 60, 15, 165, -, 3000, 50, 175, -, -, -, -, -
Price, Ann M., 1, -, -, 5, 8, 20, -, 2000, 50, 150, -, 15, 100, 25, 200
Donovan, William W., -, -, 1, 190, 40, 10, -, 11000, 1000, 1500, -, 195, 125, 31, 1580
Davis, Daniel T., -, -, 1, 200, 40, 40, -, 11000, 600, 800, -, 200, 500, 125, 1400
Deakyne, Napoleon B., -, -, 1, 125, 4, 75, -, 4000, 200, 350, -, 105, 150, 38, 1000
Money, Benjamin, -, -, 1, 120, 20, 2, -, 4000, 200, 400, -, 135, 200, 50, 900
Webster, Andrew W., 1, -, -, 50, -, 27, -, 2200, 150, 200, -, 60, 150, 45, 440
Mattiford, Charles H., -, -, 1, 78, 22, 22, -, 3500, 100, 150, 120, 90, 200, 50, 800
Moore, James B., 1, -, -, 13, 2, -, -, 650, 25, 100, -, -, 50, 13, 225
Brockson, James, -, -, 1, 200, 44, 40, -, 11000, 400, 1100, 30, 150, 500, 125, 2500
Lee, David, 1, -, -, 10, -, -, -, 400, 20, 50, -, -, -, -, 100
Roberts, Samuel, 1, -, -, 120, 10, 45, -, 7500, 300, 830, 25, 120, 450, 110, 2000
Bedwell, George D., -, -, 1, 80, 20, 10, -, 3300, 50, 250, -, 60, 110, 30,778
Hannafee, Patrick, -, -, 1, 148, 3, -, -, 4500, 300, 500, 50, 90, 200, 50, 1100

Jones, James, -, 1, -, 69, 1, 15, -, 2200, 200, 275, -, 60, 150, 40, 600
Deakyne, Charles B., -, -, 1, 55, 25, 90, -, 6500, 125, 310, -, 25, 130, 33, 500
Gardner, Mary A., 1, -, -, 80, 3, 200, -, 4000, 175, 300, 50, 120, 100, 25, 500
Huggins, Robert N., 1, -, -, 32, -, -, -, 1600, 100, 200, -, -, 100, 25, 2500
Pierson, Edwin, -, -, 1, 22, -, 8, -, 1000, 25, 75, 25, -, 50, 12, 220
Donovan, George R., -, -, 1, 172, 53, 5, -, 9000, 600, 825, -, 180, 800, 200, 2000
Buchanan, George W., 1, -, -, 45, -, -, -, 2020, 175, 250, -, 30, 150, 35, 350
Blackiston, Richard (B), -, 1, -, 15, -, 5, -, 600, 20, 75, -, 22, -, 200
Brodaway, William M., -, -, 1, 75, -, 25, -, 3000, 50, 260, -, 90, 75, 20, 600
Willard, Johnson, 1, -, -, 145, -, 15, -, 5400, 200, 300, 50, 150, 350, 85, 1200
Riggs, William E., 1, -, -, 160, 20, -, -, 13000, 350, 650, 200, 210, 500, 125, 2200
Riggs, William E., 1, -, -, 100, -, -, -, 5500, 100, 400, 50, 90, 250, 63, 1200
Thompson, Silas, -, -, 1, 125, 25, 8, -, 6000, 300, 600, 30, 120, 350, 85, 1000
Truax, Benjamin, -, -, 1, 175, 25, 5, -, 6000, 200, 550, 35, 100, 250, 62, 1000
Armstrong, Samuel A., 1, -, -, 243, 32, 91, -, 13300, 450, 1300, 175, 100, 70, 175, 3000
Wells, Daniel, 1, -, -, 45, 12, -, -, 4000, 150, 200, 50, 20, 160, 40, 400
Cavender, John A., 1, -, -, 44, 16, -, -, 3000, 100, 250, -, 75, 250, 62, 700
Pryor, William, -, -, 1, 40, -, -, -, 2500, 100, 200, -, 30, 100, 25, 400
Wells, Henry H., 1, -, -, 129, 21,35, -, 7500, 200, 1000, 100, 45, 400, 100, 1000
Hazard, Thomas B., 1, -, -, 108, 28, 4, -, 6000, 200, 400, -, 56, 275, 72, 650
Saxton, James T., -, -, 1, 150, 50, 75, -, 10000, 500, 1000, 100, 110, 350, 100, 1250
Rothwell, William, -, -, 1, 164, 27, -, -, 8000, 400, 700, 75, 290, 350, 100, 725
Deakyne, John, -, -, 1, 160, 1, 50, -, 8000, 75, 460, -, 105, 225,56, 550
Kennedy, Joseph, -, -, 1, 68, 12, 20, -, 5000, 150, 375, -, 105, 200, 50, 600
Cummons, Alexander G., 1, -, -, 30, 28, -, -, 20000, 500, 1000, 50, 150, 600, 150, 2000
Bennett, Jacob C., -, 1, -, 62, -, -, -, 2500, 25, 150, -, -, 1000, 30, 150
Johnson, James & Bros., -, -, 1, 180, 20, -, -, 10000, 500, 1200, -, 150, 700, 180, 1500
Carrow, James, -, 1, -, 50, -, -, -, 4000, 50, 350, -, 30, 10, 28, 300
Rothwell, Benjamin F. C. & Bros., -, -, 1, 145, 25, 20, -, 6000, 400, 1000, 50, 300, 700, 180, 1750
Wells, Frank, 1, -, -, 125, 9, -, -, 8500, 400, 600, -, 135, 360, -, 1000
Middleton, Thomas, 1, -, -, 110, 40, 10, -, 10000, 100, 1000, -, 100, 550, 140, 2700
Perry, William, -, -, 1, 72, 5, 150, -, 4000, 75, 500, -, 15, 100, 25, 600
Briggs, William, -, -, 1, 120, 30, -, -, 8000, 200, 650, -, 90, 200, 50, 1000
Stephenson, Jonathan, 1, -, -, 75, 20, -, -, 10000, 300, 400, 50, 90, 200, 50, 800
Cavender, Theodore W., -, -, 1, 91, 53, -, -, 10000, 500,700, -, 120, 600, 150, 2500
Donovan, Thomas B., -, -, 1, 155, 35, -, -, 8000, 350, 600, -, 90, 300, 76, 1100
Keen, David, 1, -, -, 107, 8, 25, -, 3500, 300, 200, -, 90, 100, 25, 500
Alfree, William A., -, -, 150, 50, 50, -, 10000, 350, 900, 100, 120, 400, 100, 1500
Lockerman, William, -, -, 1, 65, -, 51, -, 3500, 100, 250, 50, 30, 175, 44, 450

Clayton, Philip (B), 1, -, -, 5, 2, -, -, 1450, 20, 50, -, -, -, -, 150
Hill, Vincent O., 1, -, -, 200, 40, -, -, 9600, 500, 900, -, 120, 700, 175, 1600
Ferguson, Richard 1, -, -, 40, 20, 15, -, 4000, 100, 335, 50, 30, 150, 38, 350
Johnson, Samuel (Mu), 1, -, -, 50, -, 75, -, 1500, 50, 100, -, -, 50, 13, 200

This agricultural census was filmed from original records in the Delaware State Archives in Wilmington Delaware by the Delaware State Archives Microfilm Office.

There are some 100 columns of information on each individual. Only the head of household is addressed. I have chosen to use only the first 16 columns of the information because I feel that this information best illustrates the wealth of the individuals. In columns 2-4 the number 1 is used as an indicator only and represents no numeric value. These are shown below:

1. Name of Individual
2. Owns
3. Rents for Fixed Money Rental
4. Rents for Shares of Products
5. Improved Tilled Land
6. Improved Permanent Meadows, Pastures, Orchards, and Vineyards
7. Unimproved Woodland & Forest
8. Unimproved including Old Fields, and Not-Growing Wood
9. Farm Value of Farm including Land, Fences, and Buildings
10. Farm Value of Farming Implements and Machinery
11. Value of Livestock
12. Fences—Cost of Building and Repairing in 1879
13. Cost of Fertilizers Purchased in 1879
14. Labor—Amount Paid for Wages for Farm Labor during 1879, including Value of Board
15. Labor—Weeks paid labor in 1879 including Value of Board
16. Estimates Value of all Farm Productions (sold, consumed or on hand for   1879)

Thus, the numbers following the names represent columns 2, 3, 4, 5, 6, 7, 8, 9, 10,11, 12, 13, 14, 15, 16.

The following symbol is used to maintain spacing where information in a column is left blank (-). This symbol is used where letters, names or numbers are not legible (_). This census was split between two reels of microfilm. The first reel I transcribed columns 1-11 & 16, the second I transcribed 1-16.

Steel, John, R., -, -, 1, 60, -, 20, -, 2500, 150, 300, 10, -, 100, 57, 300
McGee, Mary J., -, -, 1, 40, -, 10, -, 500, 20, 35, 5, -, -, -, 100
Tunnell, Daniel, -, -, 1, 80, -, 50, -, 3500, 150, 250, 15, 13, 50, 5, 50
Tunnell, William W., -, 1, -, 9, -, -, -, 200, -, 30, 3, -, -, -, 100
Tunnell, George E., -, -, 1, 45, -, 25, -, 2000, 20, 100, 6, 6, 10, 2, 300
Rickards, Joseph, -, -, 1, 54, -, 30, -, 2500, 50, 75, 25, 10, 100, 57, 350
Hall, Henry, -, -, 1, 46, -, 30, -, 2500, 20, 100, 10, 13, 10, 2, 300

Murray, James, R., -, -, 1, 20, -, -, -, -, -, -, -, -, -, -, -
Murray, Henry B., 1, -, -, 100, -, 90, -, 4500, 100, 600, 50, 20, 200, 40, 900
Rickards, William H., 1, -, -, 27, -, -, -, 1000, 85, 325, 5, -, -, -, 250
Godwin, David, 1, -, -, 75, -, 100, -, 3000, 100, 350, 50, 30, -, -, 250
Townsend, James H., -, -, 1, 40, -, -, -, 800, 35, 200, 7, 6, -, -, 200
Evans, Jenkins H., -, -, 1, 36, -, -, -, 600, 10, 20, 5, 6, -, -, 125
Hall, Charles H., 1, -, -, 60-, -, 30, -, 2000, 50, 275, 20, -, -, -, 150
Walter, Caleb J., -, -, 1, 32, -, 30, -, 1200, 25, 150, 5, 5, -, -, 150
Mitchell, John M., 1, -, -, 4, -, 9, -, 250, -, -, 2, -, -, -, 50
Littleton, Isaac S., -, -, 1, 22, -, 3, -, 800, 25, -, -, -, -, -, 151
Rickards, Kendal, 1, -, -, 48, -, 20, -, 3500, 260, 600, 50, 5, 50, 10, 500
Rickards, Charles S., -, -, 1, 50, -, 15, -, 1800, 150, 400, 10, 15, 60, 12, 400
Evans, Edmund J., -, -, 1, 40, -, 6, -, 700, 60, 150, 6, -, -, -, 100
Godwin, David C., -, -, 1, 40, -, 10, -, 1000, 40, 200, 10, 15, -, -, 200
Lynch, Jacob W., 1, -, -, 12, -, 13, -, 1040, 75, 150, 5, 4, 15, 3, 100
Hudson, George T., 1, -, -, 30, -, 10, -, 1500, 150, 100, 10, 10, -, -, 300
Evans, Stephen R., 1, -, -, 60, -, 30, -, 2300, 150, 230, 25, 10, 30, 6, 225
Lynch, George F., -, -, 1, 20, -, 5, -, 600, 75, 160, 5, -, -, -, 100
Lynch, James H., 1, -, -, 50, -, 10, -, 1500, 175, 300, 10, 10, -, -, 350
Hudson, Joseph G., 1, -, -, 40, -, 12, -, 1400, 125, 100, 5, 20, 100, 40, 300
Rickards, Stephen, 1, -, -, 37, -, -, -, 2300, 300, 300, 10, -, -, -, 300
Long, Mark C., 1, -, -, 55, -, 14, -, 2000, 100, 225, 10, -, -, -, 250
Derickson, Sarah, 1, -, -, 35, -, 20, -, 1800, 250, 300, 15, 6, 100, 45, 225
Lynch, Reuben, 1, -, -, 75, -, 35, -, 1600, 150, 200, 30, 10, 15, 3, 250
Bennett, Leven H., 1, -, -, 100, -, 50, -, 3000, 200, 300, 35, 15, 150, 50, 350
Helm, William L., -, -, 1, 40, -, -, -, 1800, 15, 25, -, 18, 25, 5, 200
Rogers, Joshua J., -, -, 1, 70, -, 5, -, 1800, 35, 250, 10, -, -, -, 500
Layton, Caleb, -, -, 1, 100, -, -, -, 2000, 800, 175, 15, -, 150, 51, 475
Derickson, George J., 1, -, -, 8, -, -, -, 300, 20, 100, 4, -, -, -, 65
McGee, John E., -, -, 1, 25, -, 25, -, 2000, 20, 20, 15, 10, 20, 4, 250
Townsend, John S., 1, -, -, 20, -, 4, -, 1200, 50, 225, 5, 6, -, -, 190
Derickson, Benj. G., 1, -, -, 15, -, 25, -, 500, 10, 50, 5, -, -, -, 50
Steel, Miers B., 1, -, -, 45, -, 9, -, 3300, 400, 400, 5, 17, 50, 8, 700
Williams, Milby, -, -, 1, 35, -, 20, -, 2000, 25, 75, 5, -, -, -, 250
Williams, York R., -, -, 1, 45, -, 10, -, 1400, 10, 50, 5, -, -, -, 195
Williams, Nathaniel J., -, -, 1, 4, -, 7, -, 200, -, -, 1, -, -, -, 25
McGee, Nathaniel, -, -, 1, 16, -, -, -, 600, 20, 60, 5, -, -, -, 50
Steel, Thomas N., 1, -, -, 20, -, 10, -, 1500, 35, 300, 3, -, 25, 5, 225
Quillin, George J., -, -, 1, 10, -, -, -, 300, 50, 150, -, 7, -, -, 125
Pusey, Isaac W., -, -, 1, 40, -, -, -, 2000, 75, 125, 10, -, -, -, 375
Furman, Lemuel H., 1, -, -, 24, -, 40, -, 1000, 100, 255, 10, -, -, -, 125
Furman, Mary, 1, -, -, 60, -, 40, -, 1500, 95, 150, 10, 7, -, -, 165
Evans, John W., 1, -, -, 50, -, 25, -, 1800, 40, 100, 10, -, -, -, 50
Messick, Mary A., 1, -, -, 70, -, 35, -, 1200, -, 35, 5, 12, -, -, 75
Williams, Daniel J., -, -, 1, 40, -, 20, -, 1000, 15, 40, 10, -, -, -, 75
Walker, Quinby, 1, -, -, 73, -, -, -, 1500, -, -, -, -, -, -, 100

Holloway, Peter W., -, 1, -, 40, -, 40, -, 1000, 40, 40, 20, -, -, -, 100
Walker, Daniel G., 1, -, -, 30, -, 15, -, 800, 40, 125, 10, -, -, -, 75
Steel, John, 1, -, -, 40, -, 10, -, 1200, 50, 200, 15, -, -, -, 150
Torbert, William E., 1, -, -, 40, -, 15, -, 1000, 50,2 00, 15, -, 1, 7, 150
Clark, Giddeon W., 1, -, -, 5, -, 30, -, 700, 10, 15, -, -, -, -, 25
Moore, Peter E., 1, -, -, 5, -, 7, -, 500, 10, 110, 5, -, -, -, 25
Townsend, Charles H., 1, -, -, 40, -, 10, -, 1800, 30, 125, 10, -, -, -, 160
West, George E., 1, -, -, 25, -, 30, -, 1100, 50, 200, 5, -, -, -, 155
Evans, George W., -, -, 1, 20, -, 60, -, 800, -, 5, 5, -, -, -, 75
Bishop, John H., 1, -, -, 9, -, 5, -, 200, 20, 70, -, -, -, -, 10
Layton, John W., 1, -, -, 14, -, 9, -, 600, 30, 60, -, -, -, -, 65
Derickson, Charles W., 1, -, -, 12, -, 8, -, 500, -, 20, -, -, -, -, 25
Godfrey, George W., 1, -, -, 12, -, 8, -, 500, -, 20, -, -, -, -, 25
Carey, Cornelius P., -, 1, -, 20, -, 75, -, 800, -, -, 5, -, -, -, 50
Marvel, Kendal M., 1, -, -, 16, -, 4, -, 700, 40, -, -, -, -, -, 150
Jestace, John H., 1, -, -, 5, -, -, -, -, -, -, -, -, -, -, -
Cobb, Joseph G., -, -, 1, 45, 10, 100, -, 1000, 50, 50, 10, -, -, -, 100
Murray, George P., 1, -, -, 5, -, -, -, 570, 10, 75, -, -, -, -, 10
Bennett, Joshua R., 1, -, -, 50, 10, 60, -, 2500, 150, 300, 15, -, -, -, 200
Lynch, Gilbert, -, -, 1, 56, -, 40, -, 1500, 60, 175, 25, -, -, -, 150
Townsend, Isaac _., -, -, 1, 50, 50, 50, -, 1500, 200, 250, 10, -, -, -, 275
Clark, Nathaniel C., 1, -, -, 7, -, 3, -, 200, 20, -, 2, -, -, -, 25
Lynch, Edward J., 1, -, -, 23, -, 29, -, 1000, 10, 100, 40, -, 110, 20, 75
Calhoon, Hetty B., 1, -, -, 30, -, -, -, 500, 10, -, 5, -, -, -, 125
Holt, Henry, 1, -, -, 50, 50, 70, -, 1275, 100, 250, 10, -, -, -, 200
Murray, Caleb W., 1, -, -, 40, 5, 35, -, 2000, 80, 150, 10, -, -, -, 125
Philips, George, -, 1, -, 40, -, 50, -, 1000, -, 25, 5, -, -, -, 25
Ellis, Isaiah, 1, -, -, 40, 25, 150, -, 3000, 150, 125, 12, -, -, -, 330
West, Nancy W., 1, -, -, 20, -, 10, -, 1000, -, 35, 5, -, -, -, 100
McGee, Josiah, -, 1, -, -, 10, -, -, -, 6, -, -, -, -, -
Robinson, George M., -, -, 1, 30, 6, 50, -, 800, -, 50, 15, -, -, -, 100
Derickson, Benjamin B., -, 1, -, 100, 100, 50, -, 3500, 50, 500, 25, -, -, -, 550
Cannon, Joseph H., 1, -, -, 16, -, 40, -, 1000, 50, 125, -, -, -, -, 100
Calhoon, George W., 1, -, -, 44, 25, 18, -, 2300, 200, 300, 10, -, 20, 5, 250
Dasey, James T., 1, -, -, 45, 15, 50, -, 10, 25, 300, 10, -, -, -, 275
Dasey, John M., 1, -, -, 45, 15, 50, -, 10, 25, 100, 10, -, -, -, 100
Johnson, George D., 1, -, -, 20, 9, 25, -, 1500, 50, 200, 40, -, 50, 10, 200
Chericks, Elizabeth, -, -, 1, 25, 20, 35, -, 1000, 40, 125, 10, -, 50, 10, 125
Johnson, George P., 1, -, -, 20, -, 50, -, 800, 20, 100, 5, -, -, -, 75
Dale(Dole), Peter R., 1,-, -, 45, 7, 7, -, 850, 40, 100, 10, -, -, -, 100
Steel, Joshua, 1, -, -, 7, -, 20, -, 300, -, -, -, 5, -, -, 50
Townsend, Joshua C., 1, -, -, 20, -, 60, -, 2000, 200, 300, 10, 14, 150, -, 200
Taylor, John E., -, 1, -, 4, -, 4, -, 300, -, 30, 3, -, -, -, 40
Steel, Isaac M., 1, -, -, 6, -, 4, -, 300, 10, 75, 5, -, -, -, 100
Dickerson, Hester, 1, -, -, 40, -, 30, -, 800, -, 35, 5, -, -, -, 400
Evans, Mary A., -, -, 1, 18, -, 4, -, 500, -, -, 5, -, -, -, 100

Collins, Jehu B., -, 1, -, 18, -, 20, -, 500, -, 5, -, -, -, -, 100
Collins, Stephen A., -, -, 1, 50, 50, 50, -, 1500, 20, 50, 10, -, -, -, 125
Mitchell, Nathan J., 1, -, -, 35, -, 59, -, 1100, 50, 150, 10, -, -, -, 100
Wells, James C., 1, -, -, 10, -, 12, -, 450, 30, 60, 5, -, -, -, 75
Wilbourn, Hettie A., 1, -, -, 30, 6, 12, -, 700, 25, 70, -, -, -, -, 90
Banks, James H., -, -, 1, 50, 20, 40, -, 1000, 50, 75, 5, -, -, -, 125
Parsons, John B., -, -, 1, 30, 5, 90, -, 1000, 50, 125, 5, -, -, -, 100
Frances, Alfred L., -, -, 1, 30, -, 100, -, 1000, 20, -, 50, -, -, -, 125
Evans, Isaac H., 1, -, -, 80, 60, 100, -, 2000, 100, 300, 50, 30, 50, 10, 700
Dasey, David G., -, -, 1, 35, -, 10, -, 1000, 10, 70, 5, -, -, -, 200
Banks, Henry, 1, -, -, 85, -, -, -, 800, 25, 100, 5, -, -, -, 75
Aydelotte, David, 1, -, -, 30, -, 40, -, 1500, 40, 200, -, -, -, -, 125
Rickards, John W., 1, -, -, 20, 7, -, -, 500, -, 20, 40, -, -, -, 30
Quillin, Robert W., 1, -, -, 40, 5, 10, -, 1500, 50, 300, 10, -, -, -, 150
Marvel, Benj. P., -, -, 1, 10, -, 8, -, 200, -, -, 5, -, -, -, 75
Derickson, James L., -, -, 1, 45, -, 25, -, 2000, 50, 200, 10, 12, -, -, 350
Pusey, Jacob P., 1, -, -, 20, -, -, -, 1000, 50, 150, 5, 10, 25, 5, 255
Melson, John, 1, -, -, 40, -, 6, -, 2000, 30, 60, 5, -, -, -, 350
Barnett, Luke J., 1, -, -, 27, -, -, -, 1500, 50, 1500, 5, 20, 50, 10, 325
Barnett, George L., 1, -, -, 50, 75, 40, -, 2000, 50, 200, 10, -, -, -, 150
Eashum, Jacob, -, -, 1, 75, -, 20, -, 3000, 40, 175, 10, 10, -, -, 450
Melson, James, 1, -, - 8, -, -, -, 500, -, -, -, -, -, -, 100
Short, Thomas, -, -, 1, 35, -, 10, -, 1800, -, 175, -, -, -, -, 300
Mitchell, Leonard B., -, -, 1, 30, -, -, -, 1200, -, 40, 5, -, -, -, 225
Andrews, Leben, -, -, 1, 35, -, 5, -, 1500, 25, -, 5, -, -, -, 350
Williams, Henry, -, -, 1, 20, -, -, -, 800, -, -, -, -, -, -, 200
Hall, Charles, -, -, 1, 45, -, 50, -, 3000, -, 70, 10, -, -, -, 495
Hall, Isaac F., -, -, 1, 5, -, 40, -, 2000, -, -, 5, -, -, 80
Hudson, William M., 1, -, -, 20, -, -, -, 1000, 10, 85, 5, -, -, -, 175
Short, John B., 1, -, -, 30, -, 10, -, 1800, 150, 500, 25, -, -, -, 200
Morris, William W., 1, -, -, 40, -, 5, -, 2000, 50, 450, 10, -, -, -, 200
Rust, John B. W., -, -, 1, 64, -, 26, -, 2000, 200, 200, 95, 28, 12, 2, 400
Knox. George W., 1, -, -, 16, -, -, -, 400, 5, 70, 8, -, -, -, 20
Dodd, Rufus W., 1, -, -, 15, -, 4, -, 400, -, -, 5, -, -, -, 75
Tunnell, Isaac, 1, -, -, 11, -, 5, -, 400, 10, 10, -, -, -, -, 50
Helm, Ebe W., 1, -, -, 5, -, 7, -, 1000, 35, 300, 5, -, 40, 40, 40
Evans, Charles D., -, 1, -, 4, -, 8, -, 600, 40, 200, -, -, -, -, 40
Tunnell, Henry M., 1, -, -, 50, -, 40, -, 1500, 25, 225, -, -, -, -, 175
Derickson, Benjamin E., 1, -, -, 16, -, 11, -, 600, 10, 50, -, -, -, -, 105
Murray, Joshua S., 1, -, -, 3, -, -, -, 300, 10, 100, -, -, -, -, 40
Jones, George E., -, -, 1, 2, 3, 2, -, 600, 20, 125, -, -, -, -, 20
Holt, Cyrus, 1, -, -, 5, -, 3, -, 600, 20, 150, 10, -, -, -, 70
Williams, Benjamin R., -, 1, -, 4, -, -, -, 300, -, 40, -, -, -, -, 50
Hacker, Jacob, 1, -, -, 120, -, 130, -, 5000, 300, 2000, 300, 78, 500, -, 400
Dasey, Johnathan C., -, -, 1, 15, -, -, -, 600, 20, 30, 5, -, -, -, 40
Davis, James G., -, 1, -, 20, -, 10, -, 800, -, -, -, -, -, -, -, 50

Townsend, Peter, 1, -, -, 24, -, 60, -, 1500, 250, 260, 25, -, -, - 125
Derickson, George J., 1, -, -, 20, -, 4, 4, 700, 50, -, 8, -, -, -, 100
Betts, John, -, -, 1, -, -, 30, -, 38, -, 1500, 20, 360, 30, 120, 250, 50, 450
West, Philip, 1, -, -, 20, -, 55, -, 700, 50, 225, 25, -, -, -, 175
Steel, Thomas R., 1, -, -, 45, 30, 75, -, 2000, 150, 500, 30, 19, 50, 30, 200
Townsend, Luke, 1, -, -, 25, -, 60, -, 1500, 125, 120, 12, -, -, -, 200
Rickards, Mitchel, -, -, 1, 60, -, 40, -, 2000, 50, 200, 10, -, -, -, 300
Jones, Isaac W., 1, -, -, 132, -, 100, -, 6000, 500, 750, 50, 50, 480, 101, 1078
Townsend, Ebe, 1, -, -, 30, 30, 70, -, 3500, 300, 1000, 50, 80, 200, 50, 600
Jones, Isaac W., 1, -, -, 132, -, 100, -, 6000, 500, 750, 50, 50, 480, 101, 1078
Rickards, Thomas S., -, -, 1, 30, 10, -, -, 600, 25, 200, 15, -, -, -, 135
Tunnell, George W., 1, -, -, 40, 10, 5, -, 1200, -, 40, -, -, -, -, 150
Burton, George E., 1, -, -, 40, 10, 5, -, 1200, -, 40, -, -, -, -, 150
Hall, John C., 1, -, -, 28, 28, 3, -, 1500, 40, 160, 10, -, -, -, 150
Bennett, James D., 1, -, -, 8, 10, 20 -, 1200, 10, -, -, -, -, -, 80
Williams, William L. M., 1, -, -, 20, -, 15, -, 1500, 10, 100, 5, -, -, -, 60
Littleton, Hiram D., -, -, 1, 56, 75, 20, -, 3000, 50, 250, 40, -, -, -, 200
Williams, William S. H., 1, -, -, 5, -, 3, -, 700, 50, 250, 10, -, -, -, 75
Banks, Lemuel S., 1, -, -, 2, -, 10, -, 150, -, 50, -, -, -, -, 15
Dasey, Joshua J., 1, -, -, 4, -, 6, -, 200, -, 30, 10, -, -, -, 20
Poole, Benjamin, 1, -, -, 3, -, 4, -, 1500, -, 125, -, -, -, -, 25
Rickards, John M., 1, -, -, 25, 15, 13, -, 1200, -, 15, -, -, -, -, -, 85
Derickson, Lemuel, 1, -, -, 12, -, -, -, 500, 25, 25, 5, -, -, -, 15
Williams, Lemuel W., 1, -, -, 4, -, -, -, 1000, -, 5, -, -, -, -, 200
Hall, John T., 1, -, -, 20, -, -, -, 1000, 25, 20, -, -, -, -, 225
Dasey, Thomas R., 1, -, -, 12, -, 18, -, 1000, -, -, -, -, -, -, 75
Evans, David W., 1, -, -, 7, -, 8, -, 800, 25, 75, 5, -, -, -, 100
Betts, Thomas W., 1, -, -, 5, -, -, -, 600, -, -, -, -, -, -, 50
Holt, James F., 1, -, -, 8, -, 5, -, 1000, 50, 175, 5, -, -, -, 100
Bennett, James N., 1, -, -, 4, -, -, -, 500, -, 5, -, -, -, -, 25
Bennett, Jehu, 1, -, -, 60, -, 20, -, 2000, 50, 200, 10, -, -, -, 175
Bennett, Peter W., 1, -, -, 30, -, 20, -, 1500, 150, 225, 10, 5, -, -, 135
Gray, James B., -, -, 1, 90, -, 8, -, 2000, 50, 400, 15, 38, -, -, 750
Rickards, Charles C., 1, -, -, 8, 10, 2, -, 600, 10, 200, 10, -, -, -, 150
Rickards, Isaac D., 1, -, -, 25, -, 35, -, 800, 20, 70, 10, -, -, -, 60
Rickards, Sarah, -, -, 1, 12, -, -, -, 500, 10, 35, 5, -, -, -, 125
Lynch, Isaiah, 1, -, -, 50, -, 30, -, 1500, 50, 180, 10, 15, -, -, 250
West, James D., 1, -, -, 3, -, -, -, 1000, -, 200, -, -, -, -, 30
Watson, Henry H., 1, -, -, 11, -, -, -, 2000, 125, 385, 10, -, -, -, 50
Brasure, Wm. T., 1, -, -, 10, -, 6, -, 2000, 25, 200, 5, -, -, -, 100
McCabe, Edward H., 1, -, -, 9, -, 10, -, 700, 10, 100, -, -, -, -, 50
Simsler, Caleb M., 1, -, -, 7, -, 4, -, 700, 60, 120, 5, -, -, -, 60
Evans, Nathaniel W., 1, -, -, 15, -, 8, -, 700, 25, 160, 10, -, -, -, 185
Wilgus, James L., -, -, 1, 25, -, 5, -, 1000, 50, 150, 10, -, -, -, 140
Tubbs, William R., 1, -, -, 5, -, -, -, 1250, 10, 75, 50, -, -, -, 50
Coffin, John L., 1, -, -, 3, -, -, -, 500, 10, 65, -, -, -, -, 25

Lynch, Jacob E., 1, -, -, 10, -, -, -, 1500, 60, 225, 5, 7, -, -, 100
Lynch, Joshua J., 1, -, -, 7, -, 1, -, 600, 25, 50, -, -, -, -, 100
Wilgus, John, 1, -, -, 47, -, -, -, 2000, 75, 150, 30, 60, -, -, 300
Wilgus, Jacob A., 1, -, -, 10, -, -, -, 1000, 100, 500, 25, 21, -, -, 100
Wilgus, Robert, 1, -, -, 15, -, 10, -, 1500, 50, 200, -, 14, 120, 24, 230
Derickson, Edward, -, -, 1, 25, 15, -, -, 500, 25, 150, 5, -, -, -, 150
Burton, John C., 1, -, -, 3, -, -, -, 250, 10, 10, -, -, -, -, 25
Littleton Minus, -, -, 1, 12, -, 20, -, 300, -, 25, -, -, -, -, 100
Hall, Charles W., 1, -, -, 12, 12, 6, -, 800, -, 20, -, -, -, -, 75
Rickards, Archibol E., 1, -, -, 18, -, 18, -, 1000, -, 35, -, -, -, -, 25
Burbage, Ananias T., -, -, 1, 60, -, 5, 185, 2500, 40, 135, 15, -, -, -, 225
West, George H., 1, -, -, 40, 8, -, 40, 3000, 100, 200, 25, -, -, -, 500
Gray, Milby, -, -, 1, 35, -, 8, -, 1000, 10, 110, 5, -, -, -, 150
Wharton, John, 1, -, -, 20, -, 25, 8, 900, 10, 150, 5, -, -, -, 75
Williams, Zadoc L., -, -, 1, 20, -, 10, -, 1200, 50, 175, 15, -, -, -, 110
Jones, Joshua R., -, -, 1, 60, -, 10, 20, 2500, 75, 370, 15, 10, 60, 30, 420
Quillin, Joseph, H., -, -, 1, 65, -, -, 30, 1000, 50, 225, 10, -, -, -, 250
Derickson, Leven H., 1, -, -, 32, -, 8, -, 1200, 100, 225, -, -, 40, 8, 148
Lewis, William S., -, -, 1, 20, -, 8, -, 1000, -, 125, 10, -, -, -, 100
Lynch, Aaron, 1, -, -, 40, -, 16, -, 2500, 75, -, -, -, -, -, - 150
Warrington, David M., 1, -, -, 8, -, 8, -, 800, 25, 175, 10, -, -, -, 150
Lynch, Joshua T., 1, -, -, 20, -, 10, -, 1000, 25, 100, 5, 1000, 125
Wilgus, William, 1, -, -, 14, -, -, -, 500, 25, 40, 5, -, -, -, 125
Lynch, Samuel, 1, -, -, 16, -, -, -, 500, -, -, -, -, -, -, 125
Lynch, Lemuel, -, -, 1, 30, -, 70, -, 2500, 115, -, 75, -, -, -, 265
Lynch, Stephen, -, -, 1, 50, -, 85, -, 3000, 75, 250, 50, -, -, -, 450
Hudson, Irvin, 1, -, -, 15, -, -, -, 500, 20, 100, 5, -, -, -, 125
Dickerson, Ezekiel Jr., -, -, 1, 50, -, 12, -, 2500, 100, 275, 5, -, -, -, 300
Bennett, George L., -, -, 1, 40, -, 35, -, 3000, 50, 150, 10, -, -, -, 200
Stephenson, John M., -, -, 1, 14, -, 7, -, 1000, 50, 135, -, -, -, -, 225
Derickson, Hetty E., 1, -, -, 40, -, 25, -, 2000, 30, 175, 5, -, -, -, 225
Derickson, Henry, 1, -, -, 40, -, 20, -, 2000, 25, 125, 10, -, -, -, 300
Derickson, Isaiah, 1, -, -, 58, -, 23, -, 2000, 25, 300, 10, -, -, -, 375
Lynch, Joseph J., 1, -, -, 50, -, 33, -, 2000, 50, 250, 10, -, -, -, 250
Rickards, Geo. W., -, -, 1, 80,-, -, -, 3000, 20, 200, 10, -, -, -, 500
Lynch, Allard, 1, -, -, 30, -, 10, -, 1000, 25, 100, 10, -, -, -, 200
Betts, Silas J., 1, -, -, 25, 10, 15, 15, 1803, 100, 700, 15, -, -, -, 100
Evans, Elijah, 1, -, -, 25, -, -, 20, 1000, 25, 125, 10, -, -, -, 100
Evans, Selby H., 1, -, -, 25, -, 20, 40, 1500, 40, 190, 50, -, -, -, 160
Dickerson, William W., 1, -, -, 10, -, 3, 5, 600, 25, 50, 30, -, -, -, 75
Dolby, Peter, -, -, 1, 40, -, 15, 5, 2000, 25, 290, 15, -, -, -, 300
Pusey, Nathaniel S., -, -, 1, 25, -, 5, -, 1500, 20, 100, 5, 14, -, -, 165
Hearn, Alfred W., 1, -, -, 50, -, 4, -, 1500, 75, 225, 40, -, 20, 4, 375
Collier, Mary J., 1, -, -, 30, -, 20, -, 1500, 30, 150, 10, -, 20, 4, 90
Evans, Elisha, 1, -, -, 45, -, 30, 15, 2500, 75, 305, 20, -, -, -, 375
Evans, Stephen W., 1, -, -, 35, -, 4, 84, 1000, 80, 275, -, -, -, -, 175

Turner, William D., 1, -, -, 2, -, 9, -, 600, -, 3, -, -, -, -, 25
Evans, John W., 1, -, -, 4, -, -, -, 225, -, -, -, -, -, -, 20
Evans, William W., 1, -, -, 30, -, 4, -, 700, 15, 75, 10, -, -, -, 175
Lewis, Johnathan W., 1, -, -, 17, -, -, -, 600, 40, 150, 20, -, -, -, 100
Lynch, John, -, -, 1, 30, -, 25, -, 1000, 70, 150, 10, -, -, -, 150
Truitt, L. M., -, 1, -, 50, -, 20, -, 900, 20, 36, 10, -, -, -, 155
Quillin, Nathaniel T., 1, -, -, 39, -, 10, -, 1000, 40, 225, 5, -, -, -, 130
Evans, Asher T., -, -, 1, 10, -, -, 5, 300, -, 125, 5, -, -, -, 125
Dasey, Thomas, 1, -, -, 50, -, 50, -, 2000, 100, 175, 25, -, -, -, 325
Jefferson, Mary, 1, -, -, 4, -, 6, -, 200, -, -, -, -, -, -, 25
Showell, Robert, -, -, 1, 6, -, -, -, 200, 15, 50, -, -, -, -, 40
Evans, Sallie H., 1, -, -, 18, -, 2, -, 600, -, 4, 10, -, -, -, 75
Evans, Arlando B., 1, -, -, 20, -, 5, -, 700, -, 25, 5, -, -, -, 50
Turner, Isaac P., 1, -, -, 15, -, 3, -, 700, 20, 100, 10, -, -, -, 100
Hudson, Henry, 1, -, -, 50, -, 50, -, 2500, 100, 300, 25, 33, 75, 22, 500
Jones, John W., 1, -, -, 180, -, 100, -, 8000, 200, 600, 50, -, 350, 116, 1125
Layton, William D., 1, -, -, 108, -, 22, -, 4000, 130, 700, 20, 55, 25, 5, 700
Derickson, Robert, 1, -, - 6, -, 7, -, 200, -, 18, -, 5, -, -, -, 65
Furman, John P., 1, -, -, 20, -, 35, -, 1200, 25, 100, 10, -, -, -, 100
Lathbery, Edward T., 1, -, -, 5, -, -, -, 200, -, 10, 3, -, -, -, 60
Calhoon, Ephraim, -, -, 1, 80, -, 10, 30, 3000, 30, 300, 15, 6, 40, 24, 350
Calhoon, John B., 1, -, -, 75, -, 50, 30, 2000, 75, 330, 15, -, 25, 5, 300
Tunnell, Joshua C., -, 1, -, 20, -, 40, -, 800, 50, 100, -, 30, -, -, 100
Johnson, John A., 1, -, -, 25, -, 35, -, 1200, 30, 250, 10, -, -, -, 110
Gray, William T., 1, -, -, 20, -, 10, -, 1200, 40, 130, 15, -, -, -, 200
Melson, Stephen, -, 1, -, 4, -, -, -, 400, 25, 80, 5, -, -, -, 60
Rickards, Samuel D., 1, -, -, 40, -, 20, -, 900, -, 40, 10, -, -, -, 100
Poole, Alfred P., 1, -, -, 40, -, 25, -, 900, 50, 125, 10, -, -, -, 170
Chamberlain, Benj. D., 1, -, -, 45, -, 20, -, 1200, 50, 225, 10, -, -, -, 125
Johnson, Hettie A., 1, -, -, 25, -, 8, -, 1200, 25, 100, 5, -, -, -, 225
Sockwriter, Samuel C., -, -, 1, 30, 24, -, 1000, -, 100, 5, -, -, -, 175
Polyte, Caleb, -, -, 1, 65, -, 3, -, 1000, 40, 80, 10, -, -, -, 225
Miller, John H., -, -, 1, 12, -, 20, -, 350, 10, 60, 20, -, -, -, 125
Banks, John R., 1, -, -, 30, -, 12, -, 850, 25, 275, 20, -, -, -, 100
Aydelotte, Stephen, -, -, 1, 60, -, 180, -, 3000, 50, 200, 15, -, 100, 52, -, 252
Jefferson, Richard, -, 1, -, 10, -, 20, -, 600, -, -, -, -, -, -, 60
Blizzard, Pruda, -, -, 1, 20, -, 20, -, 800, 15, 50, 5, -, -, -, 100
Banks, Ebe, -, -, 1, 30, -, 10, -, 500, 10, 50, 10, -, -, -, 90
Rickards, William C., 1, -, -, 12, -, 28, -, 900, 70, 225, -, -, -, -, 100
Simpler, Curtis, -, 1, -, 10, -, 8, -, 600, -, -, -, -, -, -, 55
Rickards, William A., 1, -, -, 6, -, 14, -, 5000, 50, 125, 5, -, -, -, 100
Mitchell, Elizabeth, - 1, -, 8, -, 2, -, 400, -, -, -, -, -, -, 60
Tracy, Robert, 1, -, -, 20, -, 10, -, 600, 20, 100, 10, -, -, -, 80
Tracy, James, 1, -, -, 4, -, 6, -, 300, -, -, -, -, -, -, -
Johnson, Joshua H., 1, -, -, 36, -, 6, -, 700, 80, 165, 10, -, -, -, 225
Tracy, John, 1, -, -, 10, -, 17, -, 600, -, 20, 5, -, -, -, 100

Chandler, Henry J., -, -, 1, 50, -, 10, -, 1200, 50, 200, -, -, -, -, 200
Lynch, Caleb, 1, -, -, 40, -, 20, -, 1800, 175, 275, 10, -, -, -, 250
Sockwriter, Isaac H., -, -, 1, 35, -, 30, -, 800, 25, 100, 25, -, -, -, 200
Moore, Robert D., -, -, 1, 36, -, 24, -, 1800, 40, 70, 15, -, -, -, 260
Johnson, Stephen H., 1, -, -, 25, -, 12, -, 1800, 175, 225, -, -, -, -, 144
Collins, George, -, -, 1, 36, -, 20, -, 1500, 15, 110, 20, 25, 25, 5, 225
Christopher, Edward, 1, -, -, 4, -, -, -, 200, -, 10, 2, -, -, -, 40
Gray, William H., 1, -, -, 55, -, 34, -, 2500, 150, 330, -, -, -, -, 225
Derickson, James W., 1, -, -, 14, -, 5, -, 800, -, -, 5, -, -, -, 60
Hitchens, Jacob, 1, -, -, 3, -, -, -, 500, -, -, -, -, -, -, 50
Howard, George A., 1, -, -, 20, -, 25, -, 1700, 100, 35, 30, -, -, -, 100
Lynch, Edward, 1, -, -, 50, -, 20, -, 1500, 150, 235, 15, -, -, -, 250
West, Sarah E., 1, -, -, 30, -, 10, -, 500, 10, 20, 5, -, -, -, 75
Tunnell, Charles E., 1, -, -, 70, -, 30, -, 2500, 75, 380, 25, -, -, -, 250
Tunnell, Daniel Jr., -, -, 1, 20, -, 60, -, 1050, 30, 100, 5, -, -, -, 75
Chamberlain, Geo. W., 1, -, -, 35, -, 15, -, 1000, 40, -, -, -, -, -, 190
Turner, Elizabeth, 1, -, -, 20, -, -, -, 300, -, -, 10, -, -, -, -, 25
Polite, William P., 1, -, -, 100, -, 25, -, 3500, 75, 475, 6, 47, 25, 5, 775
Rickards, Ananias H., 1, -, -, 30, -, 10 -, 1000, 30, 150, 10, -, -, -, 150
Derickson, Handy L., 1, -, -, 8, -, 3, -, 800, 50, 75, 15, -, -, -, 75
Derickson, Nathaniel M., 1, -, -, 2, -, 6, -, 800, -, -, -, -, -, -, 20
Howard, William D., 1, -, -, 3, -, 3, -, 800, 40, 180, 5, -, -, -, 35
McCabe, Arthur J., 1, -, -, 30, -, 130, -, 2500, 140, 300, 20, 40, 250, 52, 500
Hickman, Henry W., 1, -, -, 50, -, 40, -, 3000, 200, 550, 20, 30, 200, 52, 600
Derickson, Isaac E., -, 1, -, 72, -, 136, -, 3000, 200, 500, 25, 28, 150, 50, 350
Derickson, Ananias, -, 1, -, 30, -, 50, -, 1600, 100, 200, 20, 15, -, -, 200
Derickson, Edward, -, 1, -, 55, -, 40, -, 2000, 125, 200, -, -, -, -, 200
Derickson, Jehu F., 1, -, -, 70, -, 30, -, 4000, 400, 450, 50, 21, 150, 56, 456
Miller, John H., -, -, 1, 72, -, 8, -, 2500, 50, 130, 10, 20, -, -, 400
Short, James I., -, -, 1, 50, -, 15, -, 2500, 50, 165, 10, 20, -, -, 400
Hickman, Mary H., 1, -, -, 40, -, 20, -, 3000, 200, 200, 15, -, 125, 40, 325
Viccurs, Elisha, -, -, 1, 60, -, 20, -, 1800, 150, 300, 15, -, -, -, 300
Derickson, John L., 1, -, -, 74, -, 15, -, 3000, 140, 350, 100, -, 125, 52, -, 300
Tingle, John, 1, -, -, 50, -, 50, -, 2500, 150, 575, 25, 31, -, -, 425
Clogg, John A., -, -, 1, 44, -, 30, -, 1800, 75, 325, 15, -, -, -, 350
Howard, B. F., 1, -, -, 40, -, 8, -, 1200, 125, 340, 8, -, -, -, 339
Godwin, Ebe W., 1, -, -, 12, -, 6, -, 800, 50, 125, 10, -, -, -, 100
Hastings, Joshua, 1, -, -, 24, -, 6, -, 1000, 100, 350, 15, -, -, -, 300
Davidson, Joseph B., -, -, 1, 65, -, 10, -, 2000, 100, 275, 10, 10, -, -, 275
Hastings, Catherine, -, -, 1, 25, -, 10, -, 500, 20, 175, 10, -, -, -, 175
Bishop, Joshua W., 1, -, -, 65, -, 40, -, 2500, 200, 380, 20, 60, -, -, 475
McComerick, John, -, -, 1, 18, -, 2, -, 500, -, 8, 10, -, -, -, 125
Lynch, William A., 1, -, -, 70, -, 10, -, 2500, 125, 335, 25, 15, -, -, 350
Lynch, Burton W., 1, -, -, 40, -, 3, -, 800, 25, 90, 10, -, -, -, 150
Lynch, David S., 1, -, -, 18, -, 4, -, 600, 40, 225, 2, -, -, -, 150
Lynch, James A., -, -, 1, 38, -, 5, -, 1200, 75, 175, 5, -, -, -, 125

Harriss, Joseph G., 1, -, -, 75, -, 10, -, 2800, 150, 390, 10, 20, 100, 45, 425
Bunting, Charles, -, -, 1, 56, -, 10, -, 1500, 40, 200, 15, 12, -, -, 375
Holloway, Armwell L., 1, -, -, 1, 45, -, 5, -, 1500, 100, 175, 10, -, -, 300
Coffin, Major, -, -, 1, 60, -, 20, -, 2000, 50, 215, 15, -, -, -, 317
McCabe, Levin T., 1, -, -, 8, -, 50, -, 600, 150, 260, 5, -, -, -, 40
McCabe, William O. Jr., -, 1, -, 6, -, -, -, 300, 100, 125, -, -, -, -, 35
Hudson, Ananias J., -, -, 1, 28, -, -, -, 700, 30, 200, 10, -, -, -, 130
McCabe, Curtis J., -, -, 1, 12, -, 30, -, 700, 100, 20, 10, -, -, -, 100
Long, Joseph S., -, -, 1, 15, -, -, -, 500, 40, 70, 5, -, -, -, 110
Morris, Isaac H., 1, -, -, 32, -, 6, -, 1000, 75, 235, 5, 16, -, -, -, 330
Derickson, Irena, 1, -, -, 3, -, -, -, 400, -, 40, 5, -, -, -, 30
Wilgus, Thomas L., 1, -, -, 18, -, 3, -, 800, 150, 125, 10, -, -, -, 250
Evans, Hettie E., 1, -, -, 18, -, 70, -, 1500, -, 40, 15, -, -, -, 75
Holloway, George C., -, -, 1, 50, -, 5, -, 800, 20, 50, 15, -, -, -, 100
Furman, Goettce, 1, -, -, 50, -, 25, -, 1200, 125, 225, 20, 6, 150, 5, 2200
Hickman, Nahaniel W., -, -, 1, 40, -, 24, -, 800, 30, 125, 10, -, -, -, 100
Lynch, Ezekiel, 1, -, -, 56, -, 25, -, 1500, 100, 300, 15, -, -, -, 250
Lynch, Lambert T., -, -, 1, 22, -, 7, -, 600, 60, 140, 5, -, -, -, 150
Furman, George P., 1, -, -, 40, -, 64, -, 2000, 200, 325, 15, -, -, -, 250
Furman, Edward J., 1, -, -, 40, -, 443, -, 1800, 200, 325, 15, 25, 10, 2, 250
Rickards, Robert, 1, -, -, 80, -, 60, -, 4000, 200, -, -, -, -, -, 492
McCabe, Garrison, 1, -, -, 45, -, 25, -, 2000, 159, 155, 20, 715, 3, 275
Hudson, John H., 1, -, -, 30, -, 5, -, 1000, 75, 180, 10, -, -, -, 200
Davis, Samuel J., -, -, 1, 6, -, 20, -, 2500, 150, 260, 5, 20, 75, 15, 450
Bunting, Edward D., 1, -, -, 30, -, 30, -, 2000, 350, 300, 15, 30, 35, 12, 480
Warrington, George, 1, -, -, 50, -, 30, -, 2000, 100, 350, 25, -, -, -, 300
Morris, Lemuel M., 1, -, -, 19, -, -, -, 800, 50, 175, 5, -, 25, 5, 325
Morris, Leven, 1, -, -, 50, -, 20, -, 1200, 30, 125, 20, -, -, -, 300
Jones, Thomas J., -, -, 1, 50, -, 30, -, 2000, 30, 200, 15, 18, -, -, 425
Evans, Zadoc Jr., 1, -, -, 30, -, 26, -, 1000, 50, 150, 12, -, -, -, 165
McCabe, Pemberton, -, -, 1,14, -, 10, -, 800, 30, 216, 5, 3, -, -, 225
McCabe, Mary D., 1, -, -, 50, -, -, -, 1200, 50, 125, 10, -, -, -, 225
Bennett, Catherine P., 1, -, -, 43, -, 4, -, 800, 20, 50, 10, -, -, -, 150
Tyre, James M., 1, -, -, 80, -, 20, -, 2000, 100, 300, 20, 5, -, -, 325
Evans, Joshua T., 1, -, -, 28, -, 10, -, 800, 50, 150, 10, -, -, -, 25
Holloway, Ebenezar, -, -, 1, 44, -, 45, 1600, 75, 360, 15, -, -, -, 225
Lynch, Elijah, 1, -, -, 40, -, 8, -, 1000, 100, 100, 10, -, -, -, 125
Anderson, James, 1, -, -, 37, -, 20, -, 1200, 75, 215, 10, 9, -, -, 225
Law, James H., 1, -, -, 50, -, 30, -, 1775, 225, 800, 25, 28, -, -, 250
Stephens, Joshua J., -, -, 1, 100, -, 60, -, 3500, 200, 350, 20, 56, 150, -, 50
Morris, Armwell, 1, -, -, 45, -, 25, -, 1000, 60, 200, 10, -, -, -, 275
Morris, Charlie O., 1, -, -, 20, -, 15, -, 550, -, -, -, -, -, -, 25
Lynch, John B., 1, -, -, 25, -, 28, -, 800, 50, 150, 5, 10, -, -, 225
McCabe, George, -, -, 1, 88, -, 30, -, 1500, 25, 125, 15, 8, -, -, 280
Long, Charles H., -, -, 1, 48, -, 7, -, 3000, 125, 340, 10, 18, 50, 12, 450
McCabe, William B., 1, -, -, 2, -, 10, -, 400, 30, 85, 4, -, -, -, 50

McCabe, Lebo, 1, -, -, 50, -, 20, -, 2000, 150, 400, 10, -, -, -, 250
Bunting, Merrell, 1, -, -, 8, -, 12, -, 300, 30, 130, 5, -, -, -, 100
Bunting, Peter B., -, -, 1, 44, -, 20, -, 1200, 75, 200, 12, 8, -, -, 225
Stephenson, Thomas, 1, -, -, 40, -, 38, -, 1200, 100, 200, 5, 7, -, -, 250
Ennis, John J., -, -, 1, 190, -, 10, -, 1000, -, 100, 10, -, -, -, 150
Timmons, Josiah, 1, -, - 5, -, 35, -, 800, 50, 175, 5, -, -, -, 50
Hickman, Angaline, 1, -, -, 75, -, 100, -, 3000, 300, 800, 10, 50, 45, 8, 250
Carey, James A., 1, -, -, 8, -, 7, -, 1200, 75, 134, -, 10, -, -, -, 175
Holloway, Elisha, 1, -, 60, -, 195, -, 3500, 250, 30, 25, 30, 35, 7, 450
Tingle, William H., -, -, 1, 45, -, 15, -, 1000, 40, 160, 15, 10, -, -, 350
Hudson, Ananias, -, 1, -, 4, -, -, -, 300, -, -, -, -, -, -, 50
Holloway, Thomas, 1, -, -, 25, -, 85, -, 2000, 40, 150, 15, 12, -, -, 225
Jacobs, John, 1, -, -, 40, -, 37, -, 2000, 40, 100, 10, -, -, -, 300
Weldin, Abbert, -, -, 1, 100, -, 80, -, 2000, 25, 300, 50, 600, 30, 12, 200
Bunting, George, 1, , -, 9, -, 4, -, 700, 75, 225, 5, -, -, -, 150
Bunting, Milby, -, -, 1, 76, -, 100, -, 3000, 50, 100, 15, -, 25, 5, 450
McCabe, Thomas, -, -, 1, 75, -, -, -, 2000, 40, 160, 15, -, -, -, 425
Hudson, David C., 1, -, -, 30, -, 10, -, 1000, 125, 150, 10, 6, 15, 3, 150
Derickson, George T., -, -, 1, 30, -, 10, -, 1400, 50, 140, 20, 15, -, -, 250
Lynch, Henry W., 1, -, -, 20, -, 16, -, 400, 50, 175, 10, -, -, -, 250
Dasey, Virginia E., 1, -, -, 36, -, 24, -, 620, 24, 188, 10, 8, -, -, 125
Dasey, Peter R., 1, -, -, 20, -, 8, -, 400, 100, 210, 10, -, 5, 1, 110
Hickman, Caleb L., 1, -, -, 11, -, 1, -, 400, -, -, 3, -, -, -, 100
Hickman, Matilda, 1, -, -, 8, -, 2, -, 350, -, -, -, -, -, -, 60
Evans, Joshua B., -, -, 1, 24, -, 10, -, 1000, 50, 185, 2, -, 5, 1, 225
Anderson, Joshua W., 1, -, -, 50, -, 19, -, 1500, 125, 125, 15, -, -, -, 300
Dukes, Thomas, 1, -, -, 80, -, 20, -, 2000, 275, 325, 30, -, 60, 24, 275
Evans, Jacob W., 1, -, -, 34, -, 55, -, 1800, 150, 170, 10, -, -, -, 175
Wilbourn, George, 1, -, -, 30, -, 10, -, 800, 20, 175, 10, -, -, -, 75
Taylor, John T., 1, -, -, 48, -, 48, -, 1500, 30, 200, 15, 15, -, -, 200
Turner, John M., -, -, 1, 36, -, 40, -, 1200, 40, 85, 10, -, -, -, 175
Hudson, Jacob H., 1, -, -, 95, -, 170, -, 2000, 150, 450, 25, 30, 75, 29, 350
Wilbourn, Roberson, -, -, 1, 16, -, -, -, 100, 25, 25, 5, -, -, -, 115
McGee, William H., -, -, 1, 42, -, 40, 32, 400, 30, 100, 10, -, -, -, 80
Walter, George, -, -, 1, 62, -, 75, 50, 2000, 40, 250, 25, 10, -, -, 375
Taylor, John M., 1, -, -, 45, -, 25, -, 1200, 150, 150, 10, -, -, -, 225
Furman, Mary E., 1, -, -, 16, -, 24, -, 600, 100, 200, 10, -, -, -, 100
McDowell, Joshua, -, -, 1, 120, -, 40, -, 2000, 20, 125, 25, 10, -, -, 325
Wharton, John B., 1, -, -, 64, -, 20, -, 2000, 40, 100, 20, -, -, -, 450
Davis, James, 1, -, -, 40, -, 20, -, 1000, -, 75, 10, -, -, -, 175
Wharton, Isaac R., 1, -, -, 40, -, 15, -, 800, -, 50, 15, -, -, -, 175
Quillin, Ebe D., 1, -, -, 100, -, 75, -, 3300, 50, 500, 40, -, 150, 52, 652
Townsend, Zadoc P., -, 1, -, 68, -, 30, 25, 2000, 125, 350, 15, -, -, -, 475
Dasey, Isaac B., -, -, 1, 45, -, 5, -, 2000, 100, 200, 10, 7, -, -, 400
Quillin, William P., -, -, 1, 100, -, 70, 30, 2500, 50, 200, 15, -, -, -, 350
Dasey, Samuel H., 1, -, -, 35, -, 35, 30, 1500, 125, 325, 10, -, -, -, 300

Taylor, John, 1, -, -, 5, -, 4, -, 250, 15, 50, 5, -, -, -, 60
Hall, George H., 1, -, -, 6, -, 10, -, 450, 5, 115, -, -, -, -, 40
Davis, Josiah, -, -, 1, 40, -, 200, 40, 1800, 100, 250, 15, -, -, -, 150
Surman, Samuel L., -, -, 1, 100, -, 30, -, 1800, 30, 100, 30, -, -, -, 300
Murray, Lambert C., -, -, 1, 60, -, 5, -, 1400, 50, 375, 25, 18, -, -, 350
Dukes, Elisha C., 1, -, -, 4, -, -, -, 900, 180, 280, 15, -, 5, 1, 70
Lockwood, Benjamin B., -, -, 1, 40, -, 25, -, 1500, 65, 150, 15, 10, -, -, 250
Hudson, Nancy W., 1, -, -, 12, -, 12, -, 1000, 100, 190, 40, -, 10, 2, 128
Gray, Michael, 1, -, -, 68, -, 30, -, 1800, 150, 600, 25, 25, -, -, 350
Hickman, James H., -, -, 1, 28, -, -, -, 500, 15, 60, 10, -, -, -, 150
Derickson, Lemuel L., -, -, 1, 20, -, 12, -, 1000, 15, 230, -, -, 10, 8, 150
Evans, William T., 1, -, -, 40, -, 10, -, 1000, 40, 125, 15, -, -, -, 125
Evans, Lemuel H., -, -, 1, 45, -, 45, -, 1500, 125, 250, 10, 14, -, -, 275
Carey, James M., 1, -, -, 24, -, 10, -, 800, 50, 75, 15, -, -, -, 130
Hudson, Joshua C., -, -, 1, 65, -, 36, -, 1250, 50, 60, 10, -, -, -, 25
Hickman, Richard, -, -, 1, 55, -, 145, -, 1800, 50, 150, 18, -, 12, 2, 325
Johnson, Henrietta, J., 1, -, -, 30, -, 70, -, 1200, 75, 260, 15, -, 8, 4, 150
Bunting, Ezekiel W., 1, -, -, 45, -, 55, -, 3000, 450, 300, 20, 100, 50, 10, 600
Rogers, Jacob, -, 1, -, 10, -, 25, -, 500, -, 30, 5, -, -, -, 100
McCabe, Edward, 1, -, -, 50, -, 40, -, 1000, 50, 200, 18, -, -, -, 225
Williams, Lemuel S., 1, -, -, 50, -, 30, 40, 1500, 100, 400, 20, -, -, -, 300
Green, David, -, -, 1, 11, -, -, -, 700, -, 20, 5, 10, -, -, 175
McKenney, Edward, -, -, 1, 48, -, 3, -, 1500, 30, 125, 10, -, -, -, 225
Hudson, Alexandria, 1, -, -, 25, -, 20, -, 1000, 90, 180, 10, -, 288, 150
Tingle, Charles C., 1, -, -, 15, -, 11, -, 1000, 35, 280, 30, -, -, -, 100
McGee, John, -, -, 1, 80, -, 50, -, 2000, -, 30, 20, -, -, -, 250
Viccurs, Burton, 1, -, -, 60, -, 188, -, 1500, 100, 200, 20, 15, -, -, 225
Adkins, Thomas A., 1, -, -, 52, -, 16, -, 1500, 75, 250, 15, -, -, -, 250
Adkins, Rufus M. of T., -, -, 1, 16, -, 2, -, 700, 10, -, -, -, -, -, 125
Holloway, David, -, -, 1, 56, -, 44, -, 1000, 30, 115, 10, -, -, -, 200
Hudson, William, 1, -, -, 75, -, 25, -, 1600, 125, 400, 25, 15, 40, 40, 400
Townsend, Jacob B., 1, -, -, 20, -, 125, -, 600, 75, 200, -, -, -, -, 100
Collins, James B., 1, -, -, 12, -, 4, -, 150, 20, 70, 5, -, -, -, 60
Collins, Stephen, -, -, 1, 6, -, 90, -, 600, 15, 40, 5, -, -, -, 40
Derickson, Jos. W. H., 1, -, -, 64, -, 166, 15, 1800, 100, 240, 20, 28, 75, 50, 250
Derickson, Ezekiel, 1, -, -, 6, -, -, -, 140, 10, 65, 2, -, -, -, 100
Hudson, Samuel W., 1, -, -, 50, -, 12, 25, 1200, 75, 250, 15, 22, 60, 15, 350
Collins, Josiah, -, -, 1, 58, -, 50, -, 1200,70, 175, 30, -, -, -, 225
Holloway, Margaret, 1, -, -, 1, -, -, -, 300, 10, 30, 5, -, -, -, 100
Truitt, James I., 1, -, -, 4, -, -, -, 200, -, 60, 3, -, -, -, 55
Truitt, James J., 1, -, -, 25, -, 5, -, 400, 15, 200, 15, -, -, -, 140
Rogers, Ananias, 1, -, -, 30, -, 8, -, 800, 25, 45, 10, -, -, -, 150
Adkins, Joshua B., 1, -, -, 56, -, 16, 6, 900, 100, 300, 15, -, 60, 50, 375
Bennett, Jehu D., 1, -, -, 35, -, 27, -, 1800, 200, 160, 3, 10, 35, 7, 355
Pepper, William E., 1, -, -, 5, -, -, 5, -, 125, 140, -, -, -, -, - 75
Williams, John W., -, -, 1, 25, -, -, 25, 1100, 40, 250, 15, -, -, -, 225

Baker, John H., 1, -, -, 50, -, 40, 20, 1200, 20, 200, 15, -, -, -, 300
Walter, Elias J., -, -, 1, 30, -, 5, 20, 400, 15, 40, 20, -, -, -, 125
Hudson, John H., 1, -, -, 44, -, 25, -, 2000, 250, 300, 20, 22, 50, 10, 250
McCabe, Joseph, -, -, 1, 35, -, -, -, 600, 75, 200, 10, -, -, -, 195
Shockley, Gillis M., -, -, 1, 60, -, 25, -, 1500, 25, 100, 20, -, -, -, 200
Brasure, James, 1, -, -, 6, -, -, 5, 1200, -, 200, 5, 5, -, -, 100
Brasure, James L., 1, -, -, 34, -, 55, -, 300, 50, 170, 10, 10, -, -, 225
Brasure, Catharine, -, -, 1, 25, -, 15, -, 500, 8, 50, 8, -, -, -, 100
Baker, William C., -, -, 1, 128, -, 30, 100, 3000, 50, 180, 30, -, -, -, 300
Hudson, Ananias, 1, -, -, 20, -, -, -, 500, 20, 80, 5, -, -, -, 100
Hudson, Marian, -, -, 1, 25, -, 10, 20, 800, 40, 150, 10, -, -, -, 100
Fisher, Nathaniel, L., 1, -, -, 8, -, -, -, 200, 12, 75, -, -, -, -, 20
Dasey, Joseph N., 1, -, -, 40, -, 25, -, 1800, 125, 250, 20, 25, -, -, 450
Bowton, James S., 1, -, -, 25, -, 30, -, 1000, 50, 200, 15, -, -, -, 200
Savage, James, -, -, 1, 6, -, -, -, 550, -, 40, 5, -, -, -, 75
McGee, Arthur L., 1, -, -, 36, -, 20, -, 1200, 75, 225, 25, -, -, -, 200
Betts, William H., 1, -, -, 28, -, 57, -, 1300, 125, 200, -, -, -, -, 400
Rickards, Stephen E., -, -, 1, 40, -, 26, -, 1200, 100, 160, 15, -, -, -, 150
Evans, Leven, 1, -, -, 3, -, -, -, 150, 20, 100, 5, -, -, -, 60
Oliver, William 1, -, -, 60, -, 30, 12, 1400, 40, 300, 20, -, -, -, 325
Freeman, Charles, 1, -, -, 2, -, 13, -, 350, 10, 25, 15, -, -, -, 25
Tomas, Rebecca, 1, -, -, 8, -, -, -, 100, -, -, -, -, -, -, 15
Aydelotte, Nancy, -, 1, -, 4, -, -, -, 150, -, -, 2, -, -, -, -, 15
Dickerson, Sarah M., 1, -, -, 18, -, -, -, 400, -, 75, -, -, -, -, 25
Hudson, James W. T., 1, -, -, 12, -, 15, -, 600, 25, 140, 5, -, -, -, 100
Hudson, Stephen H., 1, -, -, 4, -, -, -, 110, -, 18, -, -, -, -, 15
Walter, George, -, -, 1, 40, -, 30, -, 1000, 50, 125, 15, -, -, -, 325
Dickerson, Ezekiel, 1, -, -, 50, -, 120, -, 1500, 75, 125, 12, -, -, -, 200
Hudson, Ebe W., 1, -, -, 20, -, 5, -, 500, 25, 140, 10, -, -, -, 50
Bradford, Samuel J., 1, -, -, 12, -, 13, -, 600, 65, 250, 10, -, -, -, 50
Rogers, Henreitta, 1, -, -, 36, -, 20, -, 1000, 20, 125, 10, -, -, -, 100
Carey, John L. B. R., 1, -, -, 75, -, 40, -, 3500, 200, 250, 10, 40, 100, 325, 100
Hudson, Leven, 1, -, -, 48, -, 32, -, 3000, 250, 300, 15, 40, 30, 6, 525
Coffin, Kendal, -, -, 1, 42, -, 30, -, 4000, 25, 125, 20, 42, -, -, 600
Lekites, John, 1, -, -, 35, -, 25, -, 2000, -, 5, 10, -, -, -, 300
Ryon, David D., -, -, 1, 65, -, 30, -, 2500, 130, 460, 15, 25, 115, 56, 800
Lynch, Joshua G., 1, -, -, 55, -, 45, -, 2000, 100, 190, 15, -, -, -, 350
Lynch, Eligah C., -, -, 1, 60, -, 54, -, 1800, 150, 260, 15, -, 20, 4, 250
Lynch, Alfred, 1, -, -, 32, -, 37, -, 1500, 225, 225, 15, -, -, -, 225
Lynch, Manane, F., 1, -, -, 30, -, 39, -, 1800, 75, 325, 5, -, -, -, 325
Evans, Isaiah E., -, -, 1, 40, -, 29, -, 1500, 40, 710, 12, 10, 5, 1, 275
Murray, Laben H., 1, -, -, 65, -, 59, -, 2500, 80, 140, 15, -, -, -, 150
Morris, Byard, 1, -, -, 20, -, 40, -, 1500, 100, 275, -, 15, 20, 4, 300
Derickson, Peter, 1, -, -, 50, -, 45, -, 2500, 100, 180, 15, -, -, -, 200
Bunting, Catharine, 1, -, -, 60, -, 15, -, 1500, 100, 125, 10, -, -, -, 275
Bunting, Hester, 1, -, -, 2, -, 2, -, 300, -, 5, 3, -, -, -, 5

Murray, Peter, -, -, 1, 72, -, 150, -, 3500, 40, 125, 15, -, -, -, 300
Timmons, Henry, 1, -, -, 42, -, 8, -, 1500, 75, 250, 8, -, -, -, 325
Hickman, Joshua B., 1, -, -, 15, -, 6, -, 700, 45, 50, 20, 10, -, -, 200
Tubbs, James, 1, -, -, 40, -, 20, -, 1000, 75, 195, 10, -, -, -, 175
Johnson, Isaac R., 1, -, -, 30, -, 20, -, 1000, 100, 150, 25, 15, -, -, 100
Evans, Jedadiah D., 1, -, -, 10, -, 6, -, 600, -, 3, -, -, -, -, 50
Rickards, James K., -, -, 1, 48, -, 15, -, 2500, 200, 340, 20, 15, 85, 52, 325
Robert, James S., 1, -, -, 90, -, 10, -, 2000, 150, 380, 25, -, -, -, 450
Carey, Sarah, 1, -, -, 6, -, 6, -, 300, -, 10, -, -, -, -, 100
Long, Stephen L., -, -, 1, 176, -, 25, -, 6000, 125, 450, 30, 60, 50, 16, 1125
Hudson, Nahaniel P., -, -, 1, 125, -, 50, -, 4000, 75, 450, 25, 35, -, -, 700
Lofton, William, -, -, 1, 45, -, -, -, 1000, 50, 125, 10, -, -, -, 160
McCabe, Elijah W., 1, -, -, 40, -, 23, -, 1400, 30, 300, 15, 33, -, -, 158
Bunting, William B., 1, -, -, 4, -, 5, -, 200, -, 32, -, -, -, 30
Hudson, Elijah C., -, -, 1, 16, -, 40, -, 1000, 20, 100, 13, -, -, -, 100
Murray, L. Washington, 1, -, -, 35, -, 20, -, 1000, 60, 135, 10, -, -, -, 150
McCabe, Elisha, 1, -, -, 25, -, 14, -, 1000, 65, 85, 10, -, -, -, 200
Hickman, James A., 1, -, -, 15, -, 10, -, 1000, 70, 108, 10, -, -, -, 100
Moore, Elias T., 1, -, -, 3, -, 9, -, 600, 125, 100, 10, -, -, -, 60
Moor, William, -, -, 1, 58, -, 66, -, 3000, 150, 175, -, -, 25, 5, 450
Murray, Asher B., -, -, 1, 150, -, 50, -, 3600, 100, 425, 10, -, 90, 52, 500
Miller, John, 1, -, -, 18, -, -, -, 400, 75, 80, 5, -, -, -, 125
Baker, Charles J., 1, -, -, 3, -, 2, -, 400, -, 15, 15, 5, -, -, 10
Holloway, Thomas J., 1, -, -, 40, -, 15, -, 1200, 100, 300, 10, -, 30, 6, 200
Evans, George W., 1, -, -, 10, -, 14, -, 1000, -, 5, 10, -, -, -, 75
Hudson, George L., 1, -, -, 10, -, -, -, 500, 50, 115, 10, -, -, -, 100
Long, Isaiah, 1, -, -, 25, -, 55, -, 1500, 50, 125, 10, -, -, -, 125
Campbell, Sarah, 1, -, -, 4, -, 8, -, 100, -, -, -, -, -, -, 5
McCabe, William S., 1, -, -, 12, -, 25, -, 6000, 235, 275, -, -, -, -, 100
Conaway, William P., -, -, 1, 16, -, 10, -, 1000, 75, 50, -, -, -, -, 150
Williams Henry H. I., 1, -, -, 40, -, 10, -, 1800, 100, 60, 20, 10, -, -, 100
Handy, George, -, -, 1, 40, -, 8, -, 1800, 35, 75, 15, 12, -, -, 125
McCabe, John D., 1, -, -, 75, -, 50, -, 3000, 125, 170, -, -, -, -, 300
McCabe, John Q., 1, -, -, 4, -, -, -, 1000, 80, 135, 12, -, -, -, 25
Bishop, James R., 1, -, -, 8, -, -, -, 1000, 150, 440, 50, -, -, -, 150
Harrison, Joseph G., 1, -, -, 18, -, -, -, 2000, 150, 150, 20, -, -, -, 100
Andrew, Benjamin B., 1, -, -, 5, -, -, -, 300, 25, 45, -, -, -, -, 100
Hudson, William L., 1, -, -, 5, -, 20, -, 1000, 20, 350, -, -, -, -, 60
Eashum, William, -, -, 1, 64, -, 15, -, 1800, 40, 125, 15, -, -, -, 325
Murray, Joshua B., 1, -, -, 60, -, 20, -, 2000, 75, 400, 20, 18, -, -, 400
Stephens, William H., 1, -, -, 45, -, 55, -, 2000, 40, 150, 10, -, -, -, 125
Hudson, William H., -, -, 1, 50, -, 56, -, 2000, 40, 300, 15, 10, -, -, 325
McNeal, Joseph G., -, -, 1, 200, -, 200, -, 6000, 300, 460, 50, 90, 300,75, 825
Gray, Peter S., -, -, 1, 50, -, 15, -, 1500, 75, 225, 25, 15, 60, 30, 175
Holloway, George W., 1, -, -, 30, -, 5, -, 1200, 25, 45, 10, 8, -, -, 115
McCabe, Ebe L., 1, -, -, 44, -, 6, -, 2500, 150, 365, 15, 6, -, -, 600

Murray, Joseph G., -, -, 1, 88, -, 30, -, 3000, 115, 225, 20, -, 15, 4, 475
Stephenson, Charles, 1, -, -, 60, -, 40, -, 3000, 200, 260, 25, 25, 50, 15, 425
Stephenson, George M., -, -, 1, 70, -, -, -, 1800, 45, 140, -, -, 25, 8, 225
Tingle, Elias H., -, -, 1, 50, -, -, -, 1200, 50, 275, 15, 10, 40, 16, 400
Polite, George R., -, -, 1, 60, -, 15, -, 1500, 30, 300, 5, 10, 50, 17, 300
Johnson, William J., 1, -, -, 35, -, 45, -, 1200, 20, 20, 25, -, -, -, 175
Hancock, William H., -, -, 1, 4, -, 78, -, 600, -, 5, 10, -, -, -, 40
Bunting, Joseph B., 1, -, -, 25, -, 4, -, 800, 125, 130, 8, 10, -, -, 200
Wells, Garrison T., 1, -, -, 45, -, 22, -, 1000, -, -, -, -, -, -, -
Hudson, Peter R., 1, -, -, 120, -, 80, -, 6000, 200, 500, 25, 40, 100, 52, 825
Hudson, Elizabeth, 1, -, -, 3, -, -, -, 500, -, 50, 5, -, -, -, 50
Hudson, Renattis, -, -, 1, 26, -, 4, -, 1500, -, 150, 20, -, -, -, 40
Lockwood, Julia A., 1, -, -, 40, -, 25, -, 2000, 100, 115, 15, -, 10, 2, 325
Stephens, Robert H., -, -, 1, 75, -, 25, -, 2500, 65, 275, 15, 20, 65, 30, 425
Long, Armwell L., 1, -, -, 40, -, 12, -, 2300,70, 425, 10, -, 30, 7, 325
Long, Isaiah C., -, -, 1, 52, -, 15, -, 2300, 65, 130, 15, -, 30, 10, 350
Long, Zeno P., -, -, 1, 120, -, 10, -, 4000, 200, 600, 25, 40, -, -, 600
Campbell, George, 1, -, -, 75, -, 25, -, 2500, 175, 420, 20, 12, 35, 10, 400
Hastings, Charlie H., -, -, 1, 93, -, 25, -, 3500, 200, 350, 15, -, 150, 64, 500
Hudson, Elisha L., 1, -, -, 50, -, 50, -, 1800, 75, 215, -, 10, -, -, 425
Hickman, John, -, -, 1, 30, -, 80, 3, 500, 40, 50, -, -, -, -, 140
Stephens, Joseph, -, -, 1, 50, -, 40, -, 1800, 50, 135, 15, -, -, -, 240
Hudson, Isaiah, -, -, 1, 60, -, 40, -, 1800, 100, 245, 15, -, -, -, 370
Long, Nathaniel, -, -, 1, 140, -, 160, -, 4000, 110, 280, 20, 60, -, -, 700
Taylor, Leven L., -, -, 1, 52, -, 118, -, 1800, 15, 95, 15, 6, -, -, 300
Baker, Elijah S., -, -, 1, 35, -, 255, -, 1200, -, 4, -, -, -, -, 8-
Baker, Johnathan F., 1, -, -, 95, -, 27, -, 3000, 130, 370, 25, 12, 125, 52, 550
Hudson, Jerry, -, -, 1, 80, -, 76, -, 1800, 75, 175, 10, -, -, -, 225
Mumford, Charlie, 1, -, -, 25, -, 220, -, 1500, 175, 350, -, -, 20, 6, 300
Hudson, Charles W., -, -, 1, 10, -, 140, -, 1200, 50, 100, 10, -, -, -, 100
Mitchell, John E., -, -, 1, 4, -, 175, -, 1000, -, -, -, -, -, -, -
Fisher, Lambert D., -, -, 1, 30, -, 475, -, 1500, 15, 50, 10, -, -, -, 13
McCabe, Asher H., -, -, 1, 40, -, 53, -, 1750, 60, 230, 8, -, 10, 2, 140
Campbell, Henry, -, -, 1, 32, -, -, -, 500, 15, 100, 10, 20, -, -, 100
Godfrey, John B., -, -, 1, 80, -, 40, -, 3050, 50, 300, -, -, -, -, 500
Hudson, Isaiah, -, -, 1, 80, -, -, 50, -, 4000, 60, 260, 15, -, -, -, 425
Ryon, Elisha, -, -, 1, 108, -, 20, -, 2500, 100, 750, 25, 12, 100, 52, 840
Murray, John, -, -, 1, 56, -, 25, -, 1500, 25, 300, 6, -, -, -, 275
Campbell, James W., -, -, 1, 112, -, -, -, -, 25, 180, 30, 12, 45, 29, 275
Murray, Curtis W., -, -, 1, 50, -, 12, -, 1100, 125, 200, 15, -, 100, 50, 100
Stephens, Curtis, -, -, 1, 40, -, 25, -, 1000, 50, 160, -, -, -, - 325
Baker, Salathael, -, -, 1, 40, -, 25, -, 1800, 50, 245, -, -, 75, 16, 425
Harrison, Leven J., -, -, 1, 12, -, 33, -, 1200, 100, 130, -, -, -, -, 125
Murray, Milbourn, 1, -, -, 50, -, 50, -, 1500, 50, 140, 168, 15, -, 10, 2, 240
Long, Joshua P., 1, -, -, 8, -, -, -, 400, 5, 5, 5, -, -, -, 35
Long, Lydia, -, -, 1, 40, -, 10, -, 1500, 65, 220, 10, -, -, -, 125

316

Rogers, John L. J., 1, -, -, 92, -, 68, -, 3000, -, 160, 24, -, 10, 3, 375
Holloway, Jacob A., 1, -, -, 200, -, 153, -, 5000, 150, 850, 50, 18, 75, 26, 1125
Rogers, Solomon, -, -, 1, 5, -, -, -, 500, 50, 300, -, -, -, -, 550
Bishop, Gideon W., -, -, 1, 76, -, 30, -, 3000, 110, 250, -, 17, 40, 12, 500
McGee, Milliom J., -, -, 1, 36, -, 2, -, 1000, 30, 165, 10, -, 5, 36, 325
Lynch, William H., 1, -, -, 30, -, 10, -, 1000, -, 100, 195, 20, -, -, -, 225
Hudson, Hannah, 1, -, -, 2, -, 10, -, 60, -, -, -, -, -, -, -, 12
Evans, Henry J., -, -, 1, 50, -, 28, -, 1600, 30, 75, 25, -, -, -, 300
Simpler, Manane, S., -, -, 1, 150, -, 100, -, 3500, 75, 375, 40, -, 100, 40, 275
Townsend, Henry, -, -, 1, 280, -, 150, 200, 4000, 25, 550, 20, -, -, -, 300
Bailio, Clemeth, -, -, 1, 25, -, 50, 25, 600, 20, 75, 10, -, -, -, 175
Burton, Elizabeth L., 1, -, -, 100, -, 60, 15, 1500, 50, 125, 15, -, -, -, 400
Evans, John M., -, -, 1, 25, -, -, 5, 500, 80, 130, 10, -, -, -, 125
Moore, Kendal, -, -, 1, 100, 4, 50, -, 2000, 15, 150, -, 13, 75, 28, 250
Hearn, W. H., 1, -, -, 54, -, -, -, 1000, 50, 200, -, 15, 150, 25, 500
Hearn, Isaac T., 1, -, -, 5, -, -, -, 500, -, -, -, -, -, -, 300
Moore, Johnathan, 1, -, -, 54, 24, 2, -, 5000, 40, 40, 100, 100, 300, 60, 52
Moore, Wilson, 1, -, -, 17, -, -, -, 1500, 50, 250, -, -, -, -, 75
Moore, Francis G., 1, -, -, 4, -, -, -, 200, -, -, -, -, -, -, 120
Moore, John M. C., 1, -, -, 14, 4, -, -, 1200, 50, 100, 500, 50, 100, 25, 300
Quillen, John B. & Co., 1, -, -, 25, 14, -, -, 1000, 50, 100, -, 120, 200, 52, 600
Moore, David H., 1, -, -, 5, -, -, -, 1000, 2, 10, -, -, 20, 8, 30
Hopkins, Mary A., 1, -, -, 5, -, -, -, 1000, -, 3, 100, 6, -, -, 75
Wright, Jeremiah M., -, -, 1, 37, 25, 10, -, 1000, 30, 300, 200, -, 100, 20, 300
Coulborn, James A., 1, -, -, 12, -, -, -, 900, 50, 80, -, 3, 25, 8, 30
Spicer, Robert T., -, -, 1, 150, 2, 20, 15, 2000, 25, 400, 15, -, 15, 4, 300
Massey, Lorenzo M., 1, -, -, 5, 1, -, -, 1000, 100, 150, 50, -, 50, 12, 50
Wilson, George, -, -, 1, 80, 15, 20, -, 1800, 50, 300, -, -, -, -, 300
Dreden, Benjamin, -, -, 1, 3, -, 8, -, 150, -, -, -, -, -, -, 235
Plummer, William H., -, -, 1, 65, 20, 20, -, 1500, 25, 80, -, 32, -, -, 175
Prettyman, John, 1, -, -, 100, 5, 25, -, 3500, 50, 400, -, 64, -, -, 500
Baker, Smith, 1, -, -, 50, 3, 2, -, 1000, 100, 350, -, -, 40, 8, 300
Hitch, Elijah, 1, -, -, 80, 10, 36, 5, 1400, 250, 200, 50, 21, 100, 25, 100
Oneal, James G., -, -, 1, 20, -, 10, -, 250, 45, 275, 10, 6, -, -, 40
Mills, Edward W., -, -, 1, 60, -, 25, -, 1500, 25, 100, -, -, -, -, 150
Spicer, Curtis, -, -, 1, 20, -, 7, -, 800, 25, 80, 10, -, -, -, 100
Massey, Geo. W., -, -, 1, 90, -, 40, 5, 1800, 15, 150, 15, 50, -, 500
Allen, John W., -, -, 1, 40, -, 40, -, 900, -, 25, 30, 24, -, -, 125
Burris, Edmund H., 1, -, -, 90, -, 20, -, 1500, 40, 150, 25, -, -, -, 350
Moore, John, 1, -, -, 140, -, 50, -, 3000, 75, 350, 10, -, -, 300
Swain, James, -, -, 1, 100, -, 25, -, 1600, 24, 200, 16, -, -, -, 300
Hastings, Kendall F., -, 1, -, 5, -, 3, -, 200, 6, 50, -, -, -, -, 50
Plummer, Hudson D., -, -, 1, 120, -, 25, -, 1000, 45, 225, 40, 40, -, -, 300
Short, Neima, 1, -, -, 260, 10, 30, -, 3000, 200, 600, 25, 25, 100, 40, 800
Culver, Hardy, -, -, 1, 130, -, 120, -, 3000, 100, 400, -, -, -, -, 1800
Richards, A. W., -, -, 1, 175, -, 85, -, 2600, 100, 325, -, 64, 100, 52, 300

Messick, William, 1, -, -, 100, -, 65, -, 900, 75, 150, 15, 15, -, -, 200
Hearn, Joseph B., 1, -, -, 70, -, 10, -, 800, 30, 100, 25, -, -, -, 100
Messick, Clayton M., -, -, 1, 100, -, 50, -, 1500, 30, 100, 10, 15, -, -, 300
Messick, Nathan C., 1, -, -, 45, 20, 18, -, 2500, 100, 500, 18, -, 10, -, 100
Cormer, Henry, -, -, 1, 200, -, 200, -, 3000, 100, 250, 22, -, 100, -, 400
Spicer, H. John, 1, -, -, 40, 5, 17, -, 1500, 100, 300, 5, 15, 125, 52, 300
Mills, John C., -, -, 1, 100, 3, 1, 100, 20, 3000, 50, 250, 20, -, 100, 45, 100
Hitchens, John of S, 1, -, -, 75, 5, 15, 5, 1000, 15, 130, 10, -, -, -, 200
Lingo, William E., 1, -, -, 130, 15, 130, 10, 3000, 150, 350, 25, 32, 25, 7, 600
Gray, William P., 1, -, -, 100, 10, 25, -, 1500, -, -, 10, -, -, -, 200
Oneal, George 1, -, -, 75, -, 50, 10, 2000, 125, 300, 20, 20, 150, 40, 400
Vaughn, Sallie A., 1, -, -, 70, -, 50, 20, 300, 20, -, -, -, -, -, 30
Window, Joseph, 1, -, -, 25, 3, 2, -, 700, 75, 50, 10, 17, -, -, 75
Hastings, Michael, -, -, 1, 100, -, 100, -, 2000, 75, -, 75, -, -, -, 75
Oneal, William 1, -, -, 100, 8, 100, -, 3000, 200, 700, 20, 90, 75, -, 1000
Collins, John M., 1, -, -, 100, 20, 80, -, 8000, 75, 350, 25, -, 50, -, 300
Moore, William S., 1, -, -, 80, 25, 10, -, 13000, 150, 350, -, 250, 1600, 400, 3150
Morgan, James M., 1, -, -, 50, 10, 5, -, 700, 15, 60, 10, -, -, -, 150
Webb, Alfred A., -, 1, -, 4, -, -, -, 100, -, -, -, -, -, -, 10
Chase, James W., 1, -, -, 5, 1, -, -, 500, 75, 200, -, -, -, -, 200
Batson, Aaron, 1, -, -, 4, -, -, -, 2785, -, 3, -, -, 5, -, 80
Andrew, James M., -, 1, -, 31, 4, -, -, 600, 25, 100, -, -, -, -, 50
Morgan, Jacob, 1, -, -, 40, -, 25, 10, 700, 20, 40, -, -, -, -, 100
Lank, Nancy J., 1, -, -, 50, -, 75, 25, 2000, 40, 100, 20, 20, -, -, 200
Lank, Robert J., 1, -, -, 35, 5, 50, 10, 2000, 100, 200, -, -, -, -, 125
Kinny, William S., 1, -, -, 56, 2, 80, -, 1000, 75, 178, 10, -, 25, 6, 125
Oneal, Joseph, -, -, 1, 86, 2, 30, -, 600, 10, 60, -, -, -, -, 165
White, Thomas E., -, -, 1, 100, 2, 200, 100, 1500, 35, 125, -, -, 83, 20, 250
Culver, Asa R., 1, -, -, 100, 2, 125, -, 2000, 50, 300, 20, -, 100, 25, 900
Neuson, Joseph, -, -, 1, 14, -, -, -, 300, 10, 30, -, -, -, -, 50
Clifton, Mary L., 1, -, -, 5, -, 10, 15, 400, 25,75, -, -, -, -, 25
Waller, George W., 1, -, -, 5, -, -, -, 200, -, -, 5, -, -, -, 400
Clifton, Jesse D., 1, -, -, 5, -, -, -, 200, -, -, 5, -, -, -, 30
Hastings, William H., -, -, 1, 35, -, 15, -, 600, 25, 100, -, -, -, -, 200
Coulborn, George F., 1, -, -, 105, 3, 10, -, 1000, 110, 75, -, 17, -, -, 100
English, James H., -, -, 1, 50, -, 10, -, 300, -, -, -, -, -, -, 200
Hastings, Johnathan, -, -, 1, 100, -, 90, 10, 2000, 40, 200, -, 30, -, -, 300
Hastings, John H., -, -, 1, 50, 8, 50, -, 1000, 15, 125, -, -, -, -, 300
Truitt, Henry C., 1, -, -, 80, 10, 40, -, 1000, 30, 100, -, -, -, -, 200
Boyce, George, -, -, 1, 10, 4, -, -, 150, -, 10, -, -, -, -, 50
Nicholson, George H., 1, -, -, 40, 10, 80, 70, 1500, 80, 140, -, -, -, -, 200
Perkins, Wesley, -, -, 1, 50, 4, 90, 40, 1500, 20, 70, -, 12, -, -, 100
James, John E., 1, -, -, 10, -, 50, 200, 900, 100, 300, -, -, -, -, -
Murry, William F., 1, -, -, 50, 3, 85, 70, 1200, 100, 400, -, -, -, -, 200
Phillips, Henry W., 1, -, -, 65, 3, 25, 6, 1500, 40, 175, -, 16, 50, 10, 250
Lowe, John, 1, -, -, 115, 18, 15, -, 3000, 100, 300, 10, 50, -, -, 400

Hitchens, Gillis S., 1, -, -, 90, 12, -, 40, 3000, 400, 400, 15, 25, 30, 10, 60
Messick, Charles M., 1, -, -, 160, 12, 15, 20, 2500, 100, 440, -, 50, 150, 52, 400
Dukes, John H., 1, -, -, 5, -, -, -, 300, 75, 250, 10, 15, -, -, 75
Hastings, Gillis, 1, -, -, 4, 7, -, -, 250, 7,50, -, -, -, -, 88
Dukes, Thomas P., 1, -, -, 78, 2, 10, -, 600, -, 150, -, 110, -, -, 391
Cannon, Johnson, 1, -, -, 90, 10, 100, -, 1800, 75, 200, 10, -, -, -, 150
West, Burton, 1, -, -, 125, 20, 150, -, 1800, 75, 350, 14, 40, -, -, 200
Ricard, Andrew, -, -, 1, 75, -, 30, 10, 800, 75, 300, 10, -, -, -, 200
Parmer, Absolom, -, -, 1, 15, 3, 15, -, 250, 15, 50, -, -, -, -, 75
Hastings, Philip, -, -, 1, 75, -, 10, -, 600, 10, 40, 15, -, -, -, 125
Messick, Jewell H., 1, -, -, 80, 10, 60, -, 2500, 150, 400, 20, 70, 150, 52, 300
Dolby, John W., 1, -, -, 100, -, 20, -, 1200, 15, 250, 10, 25, -, -, 150
Dolby, Isaac, 1, -, -, 145, -, 20, 15, 1500, 50, 200, -, 3, 30, 10, 125
Spicer, William G., 1, -, -, 50, 10, 30, -, 1200, 150, 400, 10, 40, 10, 4, 500
Spicer, Nelly, 1, -, -, 36, 3, 30, -, 700, -, 30, -, -, -, -, 100
James, Cowell W., -, -, 1, 100, 25, -, 5, 1000, 70, 200, -, 30, -, -, 150
Jones, Branson D., 1, -, -, 160, 20, 50, -, 5000, 500, 1200, 25, 100, 300, 80, 2000
James, Charles, -, -, 1, 100, -, 100, -, 3000, 25, 400, 10, 25, -, -, 400
James, Noah W., 1, -, -, 75, 10, 40, -, 2500, 150, 500, 10, -, 100, 40, 800
Conaway, James C., 1, -, -, 265, 15, 50, -, 4000, 125, 350, 10, -, 100, 40, 1000
Bryan, Burton, -, -, 1, 300, 15, 125, -, 20000, 100, 250, 25, 100, -, -, 1200
Bryan, Thomas, -, -, 1, 75,-, 25, -, 1000, 20, 100, -, -, -, -, 200
Truitt, George T., -, -, 1, 50, 25, 40, -, 1000, 25, 250, -, -, -, -, 300
Collins, Kenny, -, -, 1, 50, 2, 25, -, 600, 30, 75, -, -, -, -, 125
Hopkins, ___ J., 1, -, -, 65, -, 20, 30, 1200, 40, 200, 10, 35, -, -, 200
Melson, John P., 1, -, -, 80, -, 20, -, 1000, 30, 300, 10, -, -, -, 100
Holister, Mary, -, 1, -, 27, -, 27, -, 500, -, 5, -, -, -, -, 50
Phillips, Robert H., -, -, 1, 75, -, 25, -, 1000, 25, 75, 10, 18, -, -, 150
Records, Willard S., -, -, 1, 55, 6, 20, -, 1000, 150, 125, 10, 20, -, -, 150
Torbert, Hamilton, 1, -, -, 66, -, 50, -, 800, 25, 130, -, 20, -, -, 100
Evans, Nathan, 1, -, -, 80, -, 40, -, 1500, 100, 200, 10, -, -, -, 150
Griffith, John, -, -, 1, 50, -, 50, -, 1200, 20, 40, 10, -, -, -, 250
Elliott, Benjamin, 1, -, -, 142, -, 50, -, 2500, 150, 1000, -, -, -, -, 700
Hastings, Peter, 1, -, -, 90, 5, 5, -, 1000, 100, 150, 10, -, -, -, 150
Thompson, Samuel W., 1, -, -, 148, 10, 60, -, 2500, 150, 450, 50, 50, -, -, 800
Thompson, William, -, 1, -, 65, -, 35, -, 600, 10, 50, -, -, -, -, 75
Adamno, Batson, -, -, 1, 55, 5, 15, 10, 600, 10, 50, 15, -, -, -, 100
Freeny, Elijah, -, -, 1, 75, 5, 10, 20, 900, 15, 40, -, -, -, -, 100
Eller, Thomas A., 1, -, -, 50, -, 20, -, 700, 15, 200, 10, 25, -, -, 100
Johnson, Thomas, -, -, 1, 20, -, 5, -, 200, 15, 100, -, -, -, -, 50
Boyce, Noah, 1, -, -, 4, -, -, -, 200, -, 10, -, -, -, -, 75
Phillips, Robert S., 1, -, -, 80, -, 70, -, 2000, 75, 200, -, 1, -, -, 350
Lank (Lauk), Louisa J., 1, -, -, 56, 4, 10, -, 1500, 30, 200, -, 10, 100, 25, 200
Whatly, James B., 1, -, -, 120, 30, 100, -, 4000, 200, 600, 100, 125, 250, 50, 700
Outter, Perry, -, -, 1, 60, -, 100, -, 3000, 40, 100, 20, 20, -, -, 150
Morris, Isaac, -, -, 1, 70, -, 60, -, 2200, 30, 100, 10, 25, -, -, 250

Morgan, Elijah A., 1, -, -, 8, -, -, -, 600, 25, 150, -, -, -, -, 36
Morgan, George W., 1, -, -, 75, -, 25, -, 1000, 15, 60, -, 21, -, -, 75
Morgan, Kendall, 1, -, -, 150, -, 50, -, 2000, 30, 200, -, -, -, -, 300
Boyce, John D., 1, -, -, 40, -, 88, 60, 1000, 30, 100, 10, -, -, -, 150
Murphy, William M., 1, -, -, 50, -, 15, 10, 600, 30, 60, 10, -, -, -, 150
Waller, John W., 1, -, -, 40, -, -, -, 1500, 100, 300, 25, -, 50, 12, 800
Moore, Henry S., -, -, 1, 120, -, 80, -, 3000, 75, 250, 10, 25, -, -, 400
Moore, Henry S., 1, -, -, 23, -, 24, -, 400, -, -, -, -, -, -, 155
Moore, George W., 1, -, -, 115, -, 15, -, 1500, -, -, -, -, -, -, 350
Waller, William, -, -, 1, 75, -, -, -, 800, 25, 30, 10, -, -, -, 150
Chipman, Irwin M., -, -, 1, 145, 10, 25, -, 3000, 75, 300, 10, 30, 200, 45, 800
Chipman, Thomas H., -, -, 1, 88, 8, 60, -, 2500, 40, 250, -, 27, -, -, 400
Knowles, William, -, -, 1, 8, -, -, -, 150, -, -, -, -, -, -, 30
Knowles, Thomas, 1, -, -, 25, -, 30, 15, 600, 15, 50, 10, -, -, -, 75
Riggin, Henry C., 1, -, -, 35, -, 55, 10, 600, 15, 75, 10, -, -, -, 100
Eskridge, Jeremiah, 1, -, -, 50, 36, -, 25, 2000, 75, 300, -, -, 100, 20, 100
Baker, Asbury, -, -, 1, 120, 36, -, 75, 3000, 75, 210, -, -, -, -, 600
Spicer, Tilghman, 1, -, -, 110, 5, -, 50, 2500, 200, 400, -, -, -, -, 400
Owens, Elijah, -, -, 1, 70, 10, 75, -, 2500, 15, 90, -, 25, 100, 25, 300
Moore, Luther T., 1, -, -, 82, 8, 40, -, 2500, 50, 400, -, 30, 150, 40, 350
Massey, Elijah, -, -, 1, 120, -, 10, -, 3000, 50, 300, -, -, -, -, 300
Knowles, Wilson, 1, -, -, 10, -, -, -, 300, 5, 10, -, -, 20, 4, 100
Marvil, James A., 1, -, -, 110, -, 20, 20, 2000, 40, 200, -, 40, 200, 50, 300
Spicer, George W., -, -, 1, 100, 30, 10, 20, 1800, 50, 80, -, -, 40, 10, 200
Moore, Samuel, -, -, 1, 80, -, 20, 40, 900, 25, 40, 8, -, -, -, 75
Dickerson, Elisha, -, -, 1, 125, -, 50, 10, 2000, 75, 250, -, 90, -, -, 358
Cannon, George, -, -, 1, 5, -, -, -, 200, -, 10, -, -, -, -, 25
Moore, Wm. G., 1, -, -, 85, 16, 40, -, 2500, 30, 200, 10, 50, 100, 25, 300
Phillips, Nathaniel G., 1, -, -, 140, -, 75, -, 1600, 50, 250, -, 35, 50, 15, 400
Loyd, Geo. S., 1, -, -, 200, 8, 200, -, 4000, 35, 250, -, 18, 100, 25, 500
Hill, Joseph, -, -, 1, 68, -, 4, -, 500, 10, 100, -, -, -, -, 150
Shiles, David, -, -, 1, 20, -, 10, -, 300, -, 10, -, -, -, -, 60
Huston, Zarah C., -, -, 1, 110, -, 40, -, 1800, 30, 175, 10, 13, -, -, 225
Hastings, Ebing, 1, -, -, 150, 20, 100, -, 5000, 100, 300, 20, -, 125, 30, 600
Conaway, Isaac F., -, -, 1, 50, -, 50, 50, 1200, 20, 75, -, -, -, -, 150
English, Wm. T., 1, -, -, 75, 10, 15, -, 800, 40, 100, 10, -, -, -, 200
Phillips, William F., 1, -, -, 2, -, -, -, -, 50, 200, -, -, -, -, 45
Knowles, James L., 1, -, -, 50, -, 50, -, 700, 10, 25, -, -, -, -, 45
Warrington, William W., -, -, 1, 50, -, 50, -, 700, 30, 100, -, -, -, -, 175
Watson, Joshua M., -, -, 1, 50, -, 50, -, 800, 50, 100, -, -, -, -, 150
Moore, Joseph A., 1, -, -, 65, 65, 65, -, 1500, 30, 50, -, -, -, -, 200
Hastings, Jel___ J., -, -, 1, 25, -, 5, -, 600, 35, 175, 25, -, 50, 12, 150
Williams, Townsend, -, -, 1, 70, -, -, -, 1000, 35, 125, -, -, -, -, 200
Knowles, Thomas J., 1, -, -, 10, -, 15, 6, 1000, -, 25, 10, -, -, -, 75
Baker, John W., 1, -, -, 13, -, -, -, 500, 20, 125, 15, 7, -, -, 100
Parker, Joseph H., -, -, 1, 100, -, 40, -, 3000, 40, 200, 10, -, -, -, 300

Pusey, John S., -, -, 1, 100, 20, 15, -, 6000, 100, 400, 25, 100, 300, 75, 1200
Oneal, Thomas, 1, -, -, 146, -, 20, -, 4000, -, 270, -, -, -, -, 700
Mills, John, 1, -, -, 100, -, 14, -, 4000, 15, 40, -, -, 100, 25, 350
Holt, Miles S., -, -, 1, 67, 3, 10, -, 2000, 100, 250, -, 45, -, -, 250
Messick, Samuel, -, -, 1,80, 20, -, -, 1800, 40, 175, 10, -, -, -, 250
Waller, John W., 1, -, -, 30, -, -, -, 350, 50, 275, -, -, -, -, 75
Burris, George, -, -, 1, 75, -, 20, -, 2000, 30, 125, 40, 100, -, -, 325
Morgan, John W., -, -, 1, 75, -, 20, -, 1900, 30, 100, -, 95, 72, 24, 300
Wheatly, Stansbury J., 1, -, -, 180, 3, 20, -, 4000, 150, 500, -, 95, 200, 65, 700
Benson, John T., -, -, 1, 8, -, -, -, 60, 5, 75, -, -, -, -, 100
Benson, Marand E., 1, -, -, 66, -, 30, -, 1000, 30, 90, -, -, -, -, 400
Boyce, Robert H., -, -, 1, 140, 13, 20, -, 3000, 100, 150, -, 32, 125, 30, 600
Boyce, David H., 1, -, -, 60, 5, 4, -, 1100, 125, 150, 20, 32, 150, 40, 400
Boyce, David Harley, -, -, 1, 55, -, -, -, 550, 25, 175, 13, 13, -, -, 200
Moore, Elenor, 1, -, -, 70, 7, 100, 50, 7000, 100, 200, 10, -, 130, 40, 400
Hastings, James, 1, -, -, 40, -, 10, -, 600, 15, 50, -, -, -, -, 50
Ingram, John, -, -, 1, 90, 10, 20, -, 3000, 50, 75, -, -, -, -, 300
Hastings, Levin, 1, -, -, 45, 40, 20, -, 2200, 10, 40, -, 20, 100, 25, 201
Swain, Joshua, -, -, 1, 60, 10, 15, 10, 2000, 10, 60, -, -, -, -, 300
Moore, Mathias T., -, -, 1, 31, 15, -, -, 800, 40, 100, -, -, 75, 23, 1400
Wiley, Geo. E., -, -, 1, 100, 60, 75, 10, 4000, 250, 600, 10, 10, 250, 62, 500
Wiley, Geo. E., -, -, 1, 14, -, -, -, 500, -, -, -, -, -, -, 265
Waller, Irma, -, -, 1, 125, 10, 35, 40, 1800, 50, 200, -, -, -, -, 350
Oneal, William H., 1, -, -, 25, -, 50, 30, 1000, 100, 300, -, -, 50, 12, 150
Taylor, Elias, 1, -, -, 16, -, -, -, 1000, 50, 75, -, -, -, -, 78
Hastings, Thomas, -, -, 1, 80, -, 40, -, 1000, 20, 60, -, -, -, -, 75
Taylor, Alexander C., -, -, 1, 32, -, 10, -, 500, 5, 100, -, -, -, -, 175
Hitchins, John, 1, -, -, 35, -, 10, -, 1000, 25, 50, -, -, -, -, 150
Messick, Joshua J., 1, -, -, 6, -, 30, -, 800, 50, 150, -, -, -, - 100
Chipman, John A., 1, -, -, 70, 10, 60, 10, 1500, 30, 200, 10, 30, 40, 10, 150
Oneal, David, -, -, 1, 100, 30, 125, -, 4000, 150, 200, 10, 19, 300, 75, 300
Chipman, William H., 1, -, -, 220, 20, 40, 25, 2500, 100, 125, 100, -, 130, 15, 200
Whaly, William, 1, -, -, 130, 10, 20, 20, 1900, 200, 300, 25, 50, 100, 25, 500
Hitch, Levin S., 1, -, -, 100, -, 60, -, 2000, 175, 700, -, 60, 100, 25, 200
Chipman, Charles, 1, -, -, 43, -, -, -, 800, 20, 100, -, -, -, -, 100
Bullock, Richard M., -, -, 1, 70, -, 225, 150, 7500, 25, 200, 10, -, 400, 10, 200
Wiley, James, 1, -, -, 80, -, 20, 10, 1200, -, 30, -, -, -, -, 200
Wiley, James, 1, -, -, 60, -, 20, 10, 1200, 90, 125, 15, -, 75, 15, 300
Graham, James R., 1, -, -, 60, 15, 25, -, 2000, 50, 250, 10, 32, 150, 40,400
Waller, William S., -, -, 1, 175, 6, 10, 10, 7000, 100, 400, 10, 100, 150, 40, 800
Vaughn, Joseph, 1, -, -, 100, -, 125, 50, 250, 25, 785, 200, -, 60, 12, 300
Windson, Samuel D., 1, -, -, 11, -, 4, 15, 700, 35, 140, 5, -, 15, 5, 175
Lynch, Lynch, 1, -, -, 58, -, 10, 5, 1000, 30, 100, 30, -, 20, 5, 200
Matthews, George, 1, -, -, 50, -, 20, 30, 1000, 40, 200, 30, -, 40, 20, 175
Elliott, John, -, -, 1, 60, -, 40, 100, 1500, 25, 75, -, 30, 10, 3, 300
Workman, Thomas, -, -, 1, 150, 7, 220, -, 5000, 25, 500, -, 100, 100, 25, 700

Workman, Edward, -, -, 1, 50, 5, 20, -, 1200, 30, 125, -, 25, 20, 5, 225
Burton, Joseph B., -, -, 1, 70, -, 80, 20, 1000, 25, 250, -, -, -, -, 175
Lambden, Robert, 1, -, -, 150, 30, 10, -, 5000, 250, 1200, -, 35, 350, 90, 500
Morris, John R., -, -, 1, 150, -, 58, -, 4000, 50, 200, -, -, -, -, 600
Whaley, James, -, -, 1, 50, -, 75, -, 1300, 40, 100, 10, -, -, -, 250
Hitchens, Levin S., 1, -, -, 75, -, 50, 25, 1200, 125, 300, 25, -, 25, 6, 250
Baily, Sandy, 1, -, -, 30, 35, 300, -, 3500, 10, 30, -, -, -, -, 150
Hearn, Michael, -, -, 1, 200, 3, 75, -, 5000, 80, 40, 25, -, -, -, 800
Morris, George T., -, -, 1, 70, -, 10, -, 1000, 40, 75, 15, 100, -, -, 200
Hitchens, Noah, -, -, 1, 75, -, 40, -, 1200, 50, 150, 10, 35, 25, 6, 300
Giles, Isaac, -, -, 1, 80, 6, 25, -, 1100, 40, 125, 25, -, 20, 5, 300
Jerman, Philip, 1, -, -, 70, -, 30, -, 1200, 100, 500, 10, 300, 75, 400
Giles, William, 1, -, -, 100, -, 60, 40, 3000, 100, 300, 20, -, 200, 60, 500
Giles, Jacob B., 1, -, -, 25, 35, 15, -, 1000, -, -, -, -, 200, -, 300
McGee, George W., 1, -, -, 90, 20, 60, -, 2000, 200, 400, 25, 300,75, 800
Moore, John T, 1, -, -, 200, 10, 40, -, 8000, 100, 500, -, 400, 250, 70, 1000
Morgan, William W., 1, -, -, 125, 10, 115, -, 5000, 400, 700, -, 125, 200, 60, 1000
Morgan, Isaac J., 1, -, -, 60, 2, 10, -, 1400, 10, 250, -, 70, -, -, 350
Wilson, John R., 1, -, -, 60, 20, 20, -, 7000, 100, 200, -, 100, 500, 125, 1500
Hastings, James W., -, -, 1, 75, -, 75, -, 2500, 50, 125, 10, 30, 25, 6, 300
Vincent, Joseph W., 1, -, -, 50, 15, 15, -, 1200, 40, 150, -, 21, 150, 40, 600
Jones, Samuel, -, -, 1, 3, -, -, -, 400, 25, 150, -, -, -, -, 24
Hastings, James H., 1, -, -, 65, -, 25, -, 1000, 100, 400, -, -, 175, 40, 350
Hitch, George P., 1, -, -, 150, 31, 10, -, 3500, 60, 400, 25, -, 200, 50, 400
Lowe, James W., 1, -, -, 70, -, 30, -, 1000, 50, 75, 100, 17, 30, -, 250
Matthews, Philip C., 1, -, -, 125, 15, 100, -, 3000, 250, 400, 100, 100, 600, 52, 1050
Gordy, William, -, -, 1, 62, 2, 60, -, 1500, 50, 250, -, -, -, -, 400
Conaway, James E., -, -, 1, 50, -, 100, -, 1800, -, 50, -, -, -, -, 125
Hitchens, Edmund, 1, -, -, 100, 5, 50, -, 1200, 60, 200, -, -, 25, 6, 300
Burris, Sovereign H., -, -, 1, 60, -, 40, -, 800, 25, 50, 20, -, -, -, 200
Shiles, John W., 1, -, -, 70, 5, 35, -, 1000, 25, 100, 20, 20, -, -, 200
Gundy (Gunly), Joseph, 1, -, -, 3, -, -, -, 75, -, -, -, -, -, -, 15
Scott, Thomas, 1, -, -, 45, -, 100, -, 2300, 175, 550, 10, -, -, -, 500
Scott, Thomas, -, -, 1, 115, -, 60, -, 1750, -, -, -, -, 30, 4, 125
Rodney, John D., 1, -, -, 75, -, 125, -, 2000, 150, 800, -, 35, -, -, 200
Harman, George, -, -, -, -, -, -, -, -, -, -, -, -, -, -, -
Betts, Handy, 1, -, -, 75, -, 25, -, 800, 20, 100, 10, -, -, -, 200
Short, Isaac, 1, -, -, 50, -, 20, -, 600, 25, 125, 20, -, -, -, 125
Warrington, William S., 1, -, -, 80, -, 20, -, 900, 20, 175, 10, -, 50, 5, 250
Mesius, West, -, -, 1, 60, -, 30, -, 720, 20, 100, -, -, -, -, 75
Truitt, John, -, -, 1, 70, -, 30, -, 700, 20, 80, -, -, -, -, 90
Oneal, Joseph, 1, -, -, 16, -, 25, -, 450, 30, 130, 10, 10, 25, 6, 228
Parker, John H., -, -, 1, 70, -, -, -, 560, 30, 125, -, -, -, -, 175
Betts, Handy, -, -, 1, 100, -, -, -, 800, 25, 100, -, -, 25, -, 200
Harmin, George, -, -, 1, 150, -, -, -, 1200, 30, 150, -, -, -, -, 300

Vincent, Marian, -, -, 1, 100, -, -, -, 800, 20, 75, -, -, -, -, 150
Dolby, Hamilton, 1, -, -, 100, -, 100, -, 1600, 30, 100, -, -, -, -, 225
Sanders, Anna, 1, -, -, 7, -, -, -, 100, -, -, -, -, -, -, 20
Hitchens, Edmond, -, -, 1, 90, -, 60, -, 1500, 20, 100, -, -, -, -, 150
James, Reuben, 1, -, -, 100, -, 40, -, 1200, 80, 150, -, -, -, -, 300
Hastings, Asbury, 1, -, -, 50, -, 200, -, 2000, 50, 75, -, -, -, -, 50
Hitch, Levin, 1, -, -, 250, -, 270, -, 10000, 200, 800, 100, -, 2000, 500, 150
Messick, George, 1, -, -, 30, -, -, -, 300, 25, 75, -, -, -, -, 50
Whaley, Isaac T., 1, -, -, 118, -, -, -, 1200, 100, 300, -, 35, 100, 28, 500
Whaley, Isaac N., 1, -, -, 85, -, 10, -, 1000, 50, 200, -, 17, 50, 10, 225
Whaley, William S., -, -, 1, 75, -, 25, -, 1000, 75, 350, -, -, 200, 50, 500
Mitchell, William, 1, -, -, 100, 5, 55, -, 1600, 125, 350, -, 35, 100, 25, 600
Burton, William L., 1, -, -, 70, -, 30, -, 1200, 40, 150, -, -, 25, 6, 200
Lewis, George, -, -, 1, 10, 5, -, -, 1000, 30, 75, -, -, 50, 12, 75
Hastings, Zedekiah, 1, -, -, 140, -, 60, -, 2000, 100, 300, 10, 40, 100, 25, 300
Cormean, Elias G., -, -, 100, -, 100, -, 1400, 25, 100, -, -, 25, 6, 200
Cannon, Jacob G., 1, -, -, 90, -, 10, -, 1500, 100, 200, 16, 16, 100, 25, 300
Cannon, Wingate, -, -, 1, 50, -, 30, -, 800, 40, 100, 10, -, -, -, 125
Walker, Henry, -, -, 1, 70, -, 30, -, 1000, 25, 150, -, -, -, -, 200
Niblet, George, -, -, 1, 40, -, 40, -, 300, -, -, -, -, -, -, 10
Mathews, Isaac J., -, -, 1, 40, -, 20, -, 400, 20, 50, -, -, -, -, 100
Matthews, George, -, -, 1, 23, -, -, -, 400, 30, 200, -, -, -, -, 150
Elliott, Josiah, -, -, 1, 70, -, 50, -, 1000, 40, 150, -, -, -, -, 175
Truitt, Greensbury, -, -, 1, 80, -, 50, -, 1000, 40, 100, -, -, -, -, 150
Elliott, Elisha, 1, -, -, 80, -, 40, 20, 1400, 75, 175, -, -, -, -, 200
Hudson, David H., 1, -, -, 10, 10, 10, -, 1000, 300, 400, -, 20, 50, 12, 150
Hudson, David, 1, -, -, 70, -, 55, -, 1500, -, -, -, -, -, -, 300
Matthews, Henry C., 1, -, -, 100, -, 75, -, 1500, 200, 350, -, 60, 100, 25, 450
West, William J., 1, -, -, 70, -, 130, -, 2000, 350, 400, -, 80, 100, -, 300
Harman, Zayinah, -, -, 1, 15, -, -, -, 200, 20, 20, -, -, -, -, 50
Conaway, Daniel, -, -, 1, 40, -, 40, -, 500, 20, 50, -, -, -, -, 100
Matthews, Stansbury C., 1, -, -, 180, -, 50, -, 1200, 200, 300, -, -, 100, 25, 300
Dickerson, Richard, -, -, 1, 75, -, 25, -, 700, 25, 50, -, -, -, -, 100
Truitt, James, 1, -, -, 125, -, 75, -, 1400, 200, 400, -, -, -, -, 350
Outwell, William, 1, -, -, 75, -, 25, -, 800, 150, 200, -, -, 100, 25, 300
Powell, Hiram, -, -, 1, 80, -, 20, -, 800, 30, 100, 10, -, 125, 25, 150
English, James W., 1, -, -, 80, -, 30, -, 800, 30, 50, 12, -, -, -, 100
Brittingham, John, 1, -, -, 50, -, 29, -, 500, 30, 200, -, -, -, -, 150
Gordy, John, 1, -, -, 100, -, 45, -, 1200, 150, 500, 13, -, 100, 25, 600
Rodney, William H., 1, -, -, 85, 8, 30, -, 2000, 150, 300, 15, -, 250, 60, 500
Hall, James, -, -, 1, 40, 5, 40, -, 800, 50, 150, -, -, 50, 12, 150
Dennis, William H., -, -, 1, 50, 5, 40, -, 900, 40, 125, 18, 30, -, -, 125
Hudson, Stephen, -, -, 1, 50, -, 50, -, 900, 75, 75, 20, -, 50, 12, 200
Wharton, Wm. H., -, -, 1, 80, -, 90, -, 1000, 40, 100, 10, -, 25, 6,125
Wryatt, William, 1, -, -, 18, -, -, -, 400, 25, 75, -, -, -, -, 100
Cannon, J. Gibson, 1, -, -, 160, -, 236, -, 4000, 200, 1000, 30, 150, 300, 75, 1000

Dickson, George W., -, -, 1, 120, 40, 20, -, 5000, 100, 300, 10, -, 300, -, 300
Boyce, William M., -, -, 1, 140, -, 60, -, 2000, 75, 250, -, -, 100, 20, 500
Lewis, Henry C., 1, -, -, 240, 10, 150, -, 10000, 400, 1200, 70, 70, 400, 100, 1000
Pusey, George W., -, -, 1, 80, 25, 10, -, 3000, 50, 100, -, -, 100, 20, 800
Hood, Isaac, 1, -, -, 5, -, -, -, 300, 25, 80, -, -, -, 10, 100
Long, Jesse D., -, -, 1, 5, -, -, -, 300, 50, -, -, -, 75, 15, 200
Truitt, M. G. 1, -, -, 218, -, 250, -, 4500, 150, 400, -, -, 200, 50, 300
Moore, Mary, 1, -, -, 25, -, 10, -, 1000, -, -, -, -, 40, 10, 80
Truitt, Wm. John, 1, -, -, 115, -, 115, -, 2000, 100, 150, -, -, 75, 25, 150
Johnson, Benton, H., 1, -, -, 75, 15, 15, -, 5000, 300, 250, 50, 100, 100, 20, 1200
Johnson, James S., 1, -, -, 15, -, 10, -, 600, 50, 50, -, -, -, -, 200
Robinson, Burton M., 1, -, -, 50, -, 10, -, 1500, 50, 150, 50, 50, 100, 10, 300
Collins, Eli, 1, -, -, 60, -, 40, -, 2000, 50, 100, 25, -, 125, 30, 250
Clendaniel, Luke, -, 1, -, 50, -, 300, -, 4000, 75, 100, 40, -, -, -, 150
Argo, David H., -, 1, -, 40, -, 60, -, 1000, 50, 35, 20, 20, 20, 10, 100
Jones, Elizabeth D., 1, -, -, 50, -, 6, -, 1000, 100, 200, 50, 50, 100, -, 250
Rust, George A., 1, -, -, 40, -, 60, -, 1500, 150, 500, 200, 40, 100, 10, 200
Jefferson, James R., 1, -, -, 45, -, 15, -, 1000, 100, 150, 30, 75, 15, 2, 300
Jester, Thomas, -, 1, -, 30, -, 30, -, 800, 25, 75, 25, -, 10, 2, 50
Reed, Curtus C., 1, -, -, 50, -, 100, -, 1000, 50, 75, 25, 50, 50, -, 400
Ellingsworth, Rufus, -, -, -, 1, 60, -, 54, -, 1000, 100, 200, 100, 20, 10, 2, 400
Morris, William H., -, -, 1, 125, -, 125, -, 5000, 75, 250, 100, 50, 15, 2, 500
Morris, William H., -, -, 1, 60, -, 66, -, 2000, 100, 200, 25, -, 2, -, 300
Ellingsworth, Noble C., 1, -, -, 75, -, 25, -, 2500, 125, 400, 50, 25, 10, 3, 500
Ellingsworth, Frederic F., -, -, 1, 75, -, 25, -, 1000, 20, 120, -, 10, -, -, 300
Robinson, William W., 1, -, -, 30, -, 10, -, 400, 20, 75, 25, -, -, -, 75
Morris, Letitia, -, 1, -, -, 30, -, 17, -, 400, 20, 80, 10, -, -, -, 75
Warren, George H., 1, -, -, 75, -, 15, -, 1400, 100, 500, 20, 64, 10, -, 500
Clifton, William W., -, -, 1, 70, -, 100, -, 3400, -, -, -, -, -, -, -
Abbott, Mary E., -, -, 1, 70, -, 100, -, 3400, 50, 150, 50, -, -, -, 400
Wolfe, David E., 1, -, -, 125, -, 240, -, 7000, 800, 500, 150, 570, 400, 52, 1200
Dodd, Pettyjohn, -, -, 1, 60, -, 140, -, 3000, -, 50, 40, -, -, -, 60
Burton, Charles H., -, -, 1, 200, -, 300, -, 1000, 100, 150, 100, -, -, -, 200
Salmons, Robert, -, 1, -, 75, -, 25, -, 2000, 40, 250, 20, -, 20, 3, 250
Brittingham, Moses M., 1, -, -, 40, -, 18, -, 1000, 50, 250, 25, 3,10, 1, 200
Brittingham, Smith S., 1, -, -, 15, -, 25, 500, 25, 250, 20, -, -, -, 100
Betts, David H., -, -, 1, 10, -, 10, -, 800, 50, 250, 50, -, -, -, 100
Donovan, Alfred, 1, -, -, 100, -, 300, -, 4000, 100, 200, 50, 50, 6, 300
Lofland, James H., 1, -, -, 75, -, 75, -, 2000, 200, 600, 50, 50, 50, 10, 600
Warren, Stephen, 1, -, -, 90, -, 40, -, 1000, 100, 200, 50, 30, 20, 4, 300
Johnson, Purnell R., -, -, 1, 100, -, 35, -, 3000, 200, 800, 50, 100, 130, 40, 500
Carpenter, George, -, -, 1, 30, -, 70, -, 500, 50, 150, 50, 75, 10, 2, 200
Carey, Robert F., -, -, 1, 60, -, 15, -, 1000, -, -, -, -, -, -, -
Abbott, William, 1, -, -, 46, -, -, -, 1000, 75, 500, 50, 50, -, -, 500
Dodd, Jessee, 1, -, -, 40, -, 50, -, 1000, 50, 400, 50, 20, 10, 2, 400
Milbey, Peter H., -, -, 1, 5, -, 20, -, 200, 20, 50, 10, -, -, -, 40

Fisher, Joseph, 1, -, -, 20, -, 24, -, 500, 50, 30, 10, -, 15, -, 150
Lindle, Joshua, 1, -, -, 30, -, 44, -, 600, 100, 200, 20, 20, 30, 6, 300
Pettyjohn, James, -, -, 1, 30, -, 75, -, 600, 20, 150, 10, -, -, -, 188
Abbott, Alfred, -, -, 1, 100, -, 250, -, 1500, 50, 200, 25, -, 10, 2, 800
Martin, Henry L., -, -, 1, 100, -, 250, -, 1000, 25, 200, 26, -, -, -, 400
Donovan, William H., 1, -, -, 100, -, 170, -, 2500, 40, 200, 25, -, 15, 3, 300
Walls, Nehemiah E., 1, -, -, 40, -, 22, -, 700, 20, 100, 25, -, -, -, 100
Pettyjohn, Robert, 1, -, -, 60, -, 60, -, 700, 40, 200, 100, -, -, -, 200
Walls, Charles H., -, -, 1, 60, -, 40, -, 1000, 40, 200, 30, -, 10, 2, 200
Warren, Asa F., 1, -, -, 35, -, 20, -, 500, 25, 150, 10, -, 20, 4, 200
Workman, James, -, -, 1, 100, -, 150, -, 1300, 150, 250, 60, -, 30, 6, 400
Workman, Jacob, -, -, 1, 75, -, 75, -, 1000, 50, 150, 25, -, 10, 2, 250
Messick, George H., 1, -, -, 100, 50, 150, -, 2500, 75, 400, 50, -, -, -, 500
Messick, Levi Decd., 1, -, -, 45, -, 45, -, 900, -, -, 50, -, -, -, 70
Messick, Elizabeth, 1, -, -, 65, -, 60, -, 700, 25, 60, -, -, -, -, 60
Jester, Isaac, -, -, 1, 40, -, 20, -, 700, 100, 150, -, -, -, -, 500
Jester, William, -, -, 1, 40, -, 20, -, 700, 50, 60, -, -, -, -, 200
Smith, Truford, -, -, 1, 40, -, 20, -, 600, 50, 75, -, -, -, -, 200
Dickerson, James, -, -, 1, 80, 50, 250, -, 3000, 50, 200, 50, -, -, -, 300
Pettyjohn, Benjamin T., -, -, 1, 100, -, 100, -, 1500, 100, 300, 50, 10, 25, 8, 300
Walker, Harriett, 1, -, -, 25, -, 50, -, 500, 40, 100, 10, -, -, -, 100
Pride, James, -, -, 1, 47, -, 100, -, 100, 50, 100, 40, 75, 20, 4, 200
Moseley, Elsey H., -, -, 1, 75, -, 110, -, 1000, 20, 100, 20, -, -, -, 150
Prettyman, Nathan J., -, -, 1, 30, -, 200, -, 2000, 20, 75, 25, -, -, -, 625
Reed, Peter, -, -, 1, 50, -, 50, -, 1000, 50, 200, 20, -, 15, 3, 200
Ellingsworth, Jones, -, -, 1, 100, -, 150, -, 5000, 50, 150, 40, 50, 50, 6, 300
Fisher, Myers R., 1, -, -, 60, 20, 20, -, 2500, 100, 400, 100, 100, 25, 4, 600
Lindle, John H., -, -, 1, 150, -, 150, -, 6000, 100, 250, 50, 25, 10, 2, 600
Johnson, Minos P., 1, -, -, 40, -, 20, -, 700, 40, 200, 50, -, 75, 30, 300
Wilson, Nehemiah J., 1, -, -, 75, -, 35, -, 1500, 100, 300, 50, 11, 50, 10, 300
Spicer, Theodore J., -, -, 1, 25, -, -, -, -, 50, 200, 50, 80, -, -, 400
Reed, Jones T., -, -, 1, 50, -, 25, -, 1500, 100, 300, 75, 20, 30, 4, 400
Reed, Somerset Sr., 1, -, -, 150, -, 150, -, 5000, 300, 600, 200, 120, 100, 20, 700
Reed, Somerset Sr., 1, -, -, -, -, 100, -, 1000, -, -, -, -, -, -, -
Reed, Somerset Jr., 1, -, -, -, -, 85, -, 500, -, -, -, -, -, -, -
Reed, John W., -, -, 1, 60, -, 40, -, 1500, 150, 300, 100, -, -, -, 300
Magee, George W., -, -, 1, 70, -, 80, -, 2000, 50, 150, 100, 150, -, -, 300
Dodd, Edward E., 1, -, -, 40, -, 27, -, 1000, 200, 250, 50, 10, 30, 6, 300
Wilson, William W., 1, -, -, 30, -, 8, -, 1000, 100, 500, 50, 20, 20, 4, 300
Mosley, Elsey H., 1, -, -, 10, -, 20, -, 200, -, -, 30, 10, -, -, 50
Coulter, James, 1, -, -, 35, -, 30, -, 1500, 50, 200, 50, 50, 30, 6, 500
Moseley, Levi, 1, -, -, 50, -, 90, -, 2000, 60, 300, 50, 50, 50, 10, 500
Donovan, Theodore W., 1, -, -, 15, -, 50, -, 700, -, -, -, 25, 10, 2, 120
Donovan, Nancy, 1, -, -, 50, -, 160, -, 2500, 100, 500, 100, 15, 25, 4, 500
Dutton, Jonathan, -, -, 1, 75, -, 150, -, 2000, 50, 250, 50, -, 40, 6, 200
Blizzard, John T., 1, -, -, 75, -, 40, -, 1500, 100, -, 300, -, -, -, 400

Ingrahm, Nathaniel, 1, -, -, 10, -, 10, -, 600, 10, -, 25, 20, 20, 4, 100
Wilson, Peter J., -, -, 1, 65, -, 80, -, 2000, 50, 200, 50, 20, 40, 6, 200
Wilson, Thomas P., 1, -, -, 40, -, 25, -, 1000, -, -, 40, 25, 20, 5, 100
Wilson, Thomas P. of W, 1, -, -, -, -, -, 28, -, 400, -, -, -, -, -, -, -
Macklin, Virden, 1, -, -, 50, -, 160, -, 1500, 100, 200, 100, 50, 75, 20, 300
Wilson, Thomas of E., 1, -, -, 100, -, 50, -, 2000, 100, 300, 100, 30, 30, 6, 600
Wilson, Reuben P., 1, -, -, 35, -, 15, -, 500, 40, 100, 10, -, 30, 8, 100
Hood, Nathaniel, -, -, 1, 80, -, 170, -, 2000, 30, 200, 40, -, -, 20, 200
Morris, James D., 1, -, -, 60, -, 50, -, 1000, 200, 500, 60, -, 150, 50, 300
Sharp, Absolom R., -, -, 1, 100, -, 75, -, 2000, 100, 400, 100, -, 10, 2, 400
Martin, William S., -, -, 1,100, -, 200, -, 5000, 100, 300, 100, 100, 40, 6, 600
Hill, Sarah, 1, -, -, 25, -, 20, -, 800, 40, 80, 10, -, 30, 6, 75
West, James B., -, -, 1, 13, -, 4, -, 300, 20, 100, 10, -, -, -, 100
Walker, Thomas, 1, -, -, 100, -, 88, -, 2000, 100, 300, 50, 30, 100, 40, 500
Walker, Thomas, 1, -, -, -, -, 200, -, 2500, -, -, -, -, -, -, 100
Vent, William S., 1, -, -, 60, -, 20, -, 1000, 100, 500, 50, 20, 10, 2, 400
Atkins, James P., 1, -, -, 50, -, 50, -, 1000, 50, 100, 20, -, 10, 2, 300
Coulter, Andrew J., -, -, 1, 40, -, 50, -, 1000, 40, 200, 40, -, 20, 4, 300
Prettyman, William S., 1, -, -, 70, -, 30, -, 1500, 100, 300, 50, 10, 100, 20, 400
Prettyman, William, 1, -, -, 60, -, 50, -, 800, -, -, 30, 40, 60, 10, 200
Marvel, Hiram, 1, -, -, 50, -, 40, -, 1500, 50, 250, 50, 30, -, -, 300
Blizzard, Stephen E., 1, -, -, 50, -, 63, -, 1000, -, -, 20, -, -, -, 180
Magee, Moses, -, -, 1, 100, -, 100, -, 2000, 50, 400, 50, 120, 10, 2, 500
Johnson, Greenbury P., 1, -, -, 50, -, 28, -, 1500, 100, 300, 40, 50, 20, 3, 300
Virden, Joseph B., 1, -, -, 100, -, 40, -, 3000, 300, 800, 50, 150, 100, 40, 1000
Veasey, Nathaniel T., -, 1, -, 10, -, 5, -, 600, 100, 200, 20, 20, -, -, 377
Hopkins, Willliam, 1, -, -, 25, -, 10, -, 80, 50, 100, 20, -, 10, 2, 200
Dickinson, Samuel J., 1, -, -, 30, -, 10, -, 800, 40, 100, 25, 20, 15, 3, 150
Martin, John D., 1, -, -, 80, -, 14, -, 2000, 100, 500, 60, 50, 100, 40, 800
Martin John, 1, -, -, 30, -, 60, -, 1000, -, -, 40, -, 40, 8, 200
Warrington, Peter R., -, -, 1, 80, -, 100, -, 1500, 100, 300, 60, -, 60, 12, 400
Veasey, Josiah M., 1, -, -, 70, -, 70, -, 2000, 200, 400, 60, 50, 10, 40, 600
Veasey, Josiah M., 1, -, -, 40, -, 50, -, 1500, -, -, 40, 40, 20, 4, 200
Veasey, Joseph W., 1, -, -, 50, -, 25, -, 1000, -, -, 30, 30, 30, 5, 250
Ennis, David, 1, -, -, 140, -, 30, -, 3000, 150, 250, 100, -, 200, 40, 500
Warrington, Nathaniel R., 1, -, -, 15, -, 5, -, 600, 40, 150, 10, -, -, -, 100
Hood, Philip H., -, -, 1, 60, -, 26, -, 600, 30, 150, 40, -, -, -, 200
Bailey, Joshua W., -, -, 1, 100, -, 10, -, 1500, 50, 300, 50, 10, -, -, 300
Holland, Andrew, J., 1, -, -, 100, -, 130, -, 3000, 100, 300, 100, 50, 100, 40, 600
Walls, Thomas W., -, -, 1, 40, -, 100, -, 2000, 100, 250, 50, -, -, -, 200
Conwell, Asa F., 1, -, -, 100, -, 70, -, 2500, 150, 300, 80, 20, 40, 10, 400
Mirch, William T., 1, -, -, 70, -, 10, -, 2000, 100, 300, 60, 20, 30, 4, 300
Carpenter, John B., -, -, 1, 40, -, 60, -, 1000, 30, 100, 30, -, -, -, 100
Mirch, Mathew J., -, -, 1, 100, -, 30, -, 2000, 50, 300, 80, 50, 20, 4, 300
Sherman, Thomas & George, 1, -, -, 90, -, 50, -, 1500, 50, 200, 50, 40, -, -, 300
Carpenter, Benton H., 1, -, -, 60, -, 40, -, 2000, 100, 400, 60, -, 100, 40, 500

Sherman, John, 1, -, -, 100, -, 170, -, 3000, 40, 300, 60, -, 100, 30, 400
Sherman, John, 1, -, -, -, -, 142, -, 1000, -, -, -, -, -, -, 200
Watson, Robert S., -, -, 1, 60, -, 100, -, 2000, 50, 150, 100, 100, 200, 40, 600
Atkins, Allen R., -, -, 1, 60, -, 20, -, 1500, 50, 100, 50, 70, 40, 5, 500
Coffin, Elisha J., -, -, 1, 200, -, 100, -, 10000, 200, 300, 100, 100, 100, 8, 800
Walker, Curtis, -, -, 1, 50, -, 100, -, 1500, 25, 100, 40, -, -, -, 200
Carpenter, Benjamin, -, -, 1, 100, -, 75, -, 2000, 100, 300, 80, 60, 60, 10, 500
Stephenson, Sarah H., 1, -, -, 50, -, 50, -, 1000, 30, 60, 20, 50, 6, 200
Douglass, Charles M., -, -, 1, 150, -, 100, -, 4000, 400, 600, 100, 60, 200, 40, 1000
Holland, Elisha, 1, -, -, 20, -, 125, -, 1500, -, -, -, -, -, -, 145
Holland, Joseph C., -, -, 1, 90, -, 10, -, 1500, 200, 300, 20, 50, 80, 6, 400
Brown, George W., -, -, 1, 52, -, -, -, -, -, -, -, -, -, -, 569
Lank, Peter C., 1, -, -, 50, -, 20, -, 2000, 100, 300, 60, 50, 30, 6, 400
Reynolds, Burton E., 1, -, -, 70, -, 30, -, 1500, 150, 300, 100, 50, 20, 3, 800
Hudson, Wollsey W., 1, -, -, 75, -, 18, -, 2000, 100, 200, 50, -, 10, 2, 500
Perry, Thomas J., 1, -, -, 130, -, 100, -, 5000, 300,700, 200, 150, 200, 40, 1200
King, Charles H., 1, -, -, 115, -, 20, -, 2500, 50, 400, 100, 20, 40, 6, 600
White, William N., 1, -, -, 50, -, 30, -, 1500, 100, 300, 100, 60, 150, 40, 600
White, William W., 1, -, -, 40, -, 90, -, 2500, -, -, 30, -, -, -, 276
Sharp, Nathaniel, -, -, 1, 62, -, -, -, -, 60, 300, -, -, -, -, 502
White, Benjamin, 1, -, -, 80, -, 20, -, 3000, 300, 500, 100, 100, 170, 40, 800
White, Benjamin, 1, -, -, 60, -, 60, -, 2500, -, -, -, 30, 50, -, -, 334
Harmon, Eli, 1, -, -, 10, -, -, -, 3000, 40, 100, 10, -, -, -, 100
Baynum, John D., 1, -, -, 25, -, 20, -, 600, 20, 200, 20, -, -, -, 200
Bruton, Lott, 1, -, -, 17, -, -, -, 500, -, -, 30, -, -, -, 100
Warrington, Alfred C., 1, -, -, 60, -, 12, -, 2000, 200, 3000, 50, 110, 60, 10, 800
Fisher, John, 1, -, -, 50, 40, 44, -, 3000, 150, 400, 40, 40, 100, 30, 80
White, Henry H., 1, -, -, 30, -, 10, -, 1000, 200, 700, 50, 70, 100, 40, 800
Fisher, Theodore, H., -, -, 1, 60, -, 10, -, 1500, 70, 200, 50, 50, 50, 10, 400
Fisher, James, 1, -, -, 60, 150, 15, -, 3000, 200, 500, 50, 50, 150, 40, 800
Fisher, James, 1, -, -, 20, -, 10, -, 600, -, -, 30, 20, 40, 8, 200
Davis, James, 1, -, -, 25, -, -, -, 100, 100, 200, 30, 20, 10, 2, 200
Holland, John P., 1, -, -, 30, 40, 70, -, 2000, 100, 400, 50, 30, -, -, 500
Lindle, Jones, -, -, 1, 30, -, 30, -, 1000, 30, 70, 300, -, 80, 20, 250
Lindle, William, -, -, 1, 50, -, 40, -, 1500, 30, 125, 40, 50, 20, 4, 300
Joseph, John W., -, -, 200, 70, 50, 50, 8000, 300, 800, 100, 150, 300, 40, 800
Trewitt, William A., -, -, 1, 50, 80, 50, -, 3000, 100, 700, 50, 30, 100, 20, 400
Tingle, William, -, -, 1, 20, -, -, -, 600, 50, 100, 20, 30, 40, 6, 150
Palmer, Sylvester, -, -, 1, 30, -, 40, -, 1000, 50, 200, 20, -, -, -, 100
Welch, James, -, -, 1, 30, -, 40, -, 1000, 50, 300, -, -, -, -, 150
Rust, Absolom, 1, -, -, 80, 9, 17, -, 3000, 200, 250, 100, 50, 200, 40, 800
Rust, William T., 1, -, -, 25, 9, 35, -, 1000, 80, 200, 50, 30, 50, 10, 300
Palmer, John W., -, -, 1, 320, -, 60, -, 2500, 200, 200, 50, 30, 150, 30, 500
Craig, James L., -, -, 1, 80, -, 40, -, 2000, 100, 200, 50, 30, 100, 10, 400
Carey, John P., -, -, 1, 75, -, 300, -, 3000, 50, 200, 50, -, 50, 10, 300
Bryan, William R., -, -, 1, 100, -, 130, -, 2000, 80, 200, 50, 20, 20, 3, 300

Pepper, Henry N., -, -, 1, 150, -, 150, -, 3000, 200, 300, 80, 120, 200, 52, 700
Martin, James M., 1, -, -, 50, -, 25, -, 2500, 50, 120, 40, 50, -, -, 300
Roach, William, -, 1, -, 9, -, -, -, 400, 10, 50, 5, -, -, -, 100
Burton, Benjamin H., 1, -, -, 80, -, 70, -, 2000, 100, 300, 50, 70, 40, 6, 500
Simples, John R., 1, -, -, 50, -, 90, -, 1000, 25, 100, 25, -, 10, 2, 200
Abbott, George F., -, -, 1, 60, 10, 40, -, 2500, 200, 600, 50, -, 100, 50, 600
Atkins, Edward C., 1, -, -, 60, 15, 40, -, 1500, -, -, 40, -, 60, 20, 200
Maull, Purnell J., 1, -, -, 40, -, 10, -, 1000, 80, 200, 30, -, 10, 2, 300
Wilson, Hiram, -, 1, -, 50, -, -, -, -, 70, 100, 60, 40, -, -, 300
Rust, Absolum, -, -, 1, 90, -, 20, 35, 1800, 50, 100, 50, 20, 20, 3, 300
Carey, James T., 1, -, -, 60, -, 40, -, 1000, 100, 300, 50, -, 50, 6, 400
Maull, Henry G., -, -, 1, 60, -, 40, -, 1000, 25, 150, 50, -, 10, 2, 400
Conaway, Nathaniel, 1, -, -, 60, -, 40, -, 700, 30, 100, 30, -, -, -, 263
Reynolds, Richard, -, -, 1, 24, -, -, -, -, 50, 200, 50, 70, 300, 4, 300
Carpenter, John -., -, -, 1, 60, 50, 10, -, 1000, 50, 200, 40, -, -, -, 293
Sharp, Henry, 1, -, -, 100, 5, 40, -, 1200, 100, 200, 50, 20, -, -, 400
Hudson, Henry C., 1, -, -, 80, -, 80, -, 3000, 100, 300, 80, 75, 60, 10, 600
Bartlett, George I., -, -, 1, 40, 10, 40, -, 700, 50, 100, 30, 20, -, -, 200
Maull, James E., -, -, 1, 100, 20, 6, -, 1000, 80, 500, 50, -, 10, 2, 400
Dutton, Harvey G., -, -, 1, 120, 50, 200, -, 5000, 100, 700, 100, 50, 180, 40, 800
Dutton, William O., -, -, 1, 20, -, 25, -, 400, 30, 200, 10, -, -, -, 100
King, Cornelius H., 1, -, -, 80, 10, 10, -, 1200, 100, 400, 50, 50, 10, 2, 300
Moore, Joseph R., -, -, 1, 100, 20, 40, -, 1200, 100, 200, 40, 150, 100, 20, 500
King, Joseph S., -, -, 1, 60, -, 30, -, 600, 100, 300, 60, 100, 4, 90, 300
Richards, Rachel C., 1, -, -, 20, -, -, -, -, -, -, -, -, -, -, -, 292
Dutton, Peter W., -, -, 1, 50, -, 20, -, 550, 100, 150, 20, -, -, -, 200
Warrington, Samuel C., 1, -, -, 150, -, 15, 25, 10000, 200, 800, 50, 100, 150, 40, 1000
Warrington, Samuel C., 1, -, -, 80, -, -, -, 2000, -, -, 10, 30, 50, 8, 400
Donovan, Jno. B., -, 1, -, 10, -, -, -, 5000, 100, 100, 30, -, 50, 10, 200
Sharp, Nathaniel, -, -, 1, 30, -, 20, -, 500, 50, 100, 20, -, -, -, 100
Burton & Dorman, 1, -, -, 130, -, 75, -, 3000, 50, 100, 25, -, -, -, 400
Holland, Jeremiah, 1, -, -, 25, 3, -, -, 400, 300, 200, 20, -, -, -, 150
Holland, Cyrus, 1, -, -, 23, 3, -, -, 400, 40, 100, 10, -, -, -, 150
Holland, Elsey, 1, -, -, 24, 3, -, -, 500, 60, 150, 10, -, -, -, 200
Atkins, Samuel, -, -, 1, 80, -, 20, -, 1500, 50, 200, 40, -, -, -, 400
Wilson, William R., 1, -, -, 20, -, 10, -, 2000, 50, 150, 20, -, 70, 10, 250
Clifton, John, -, -, 1, 80, -, 20, -, 2000, 50, 200, 40, 30, 20, 3, 400
Clendaniel, George, 1, -, -, 13, -, -, -, 200, 10, 40, 10, -, -, -, 50
Donovan, Thomas, 1, -, -, -, -, 100, -, 1000, -, -, -, -, -, -, -
Ponder, James, 1, -, -, -, -, 300, -, 7500, -, -, -, -, -, -, -
Ponder, James, 1, -, -, 50, -, 50, -, 2000, -, -, 50, -, -, -, 150
Ponder, James, 1, -, -, 12, -, -, -, 600, -, -, -, -, -, -, 150
Martin, Samuel J., 1, -, -, 100, -, 45, -, 1600, 50, 320, 60, -, 30, 4, 400
Martin, Samuel J., 1, -, -, 10, -, -, -, 1200, 20, 60, 30, -, 40, 6, 100
Black, J. L. & Brother, 1, -, -, 38, -, 15, 600, -, -, 40, -, 50, 10, 150

Black, J L., & Brother, 1, -, -, 56, 14, -, -, 700, -, -, 40, -, 30, 6, 100
Burton, Daniel R., 1, -, -, 7, -, -, -, 400, 30, 100, 20, 30, 40, 6, 150
Betts, Robert W., 1, -, -, 66, -, -, -, 1600, 200, 300, 50, 65, 100, 40, 500
Betts, Robert W., 1, -, -, -, -, 147, -, 2000, -, -, -, -, -, -, 200
Workman, J. Wesley, -, -, 1, 30, -, -, -, -, 100, 300, 50, 20, 50, 8, 500
Naylor, David H., 1, -, -, 100, 20, 25, -, 3000, 100, 400, 60, 110, 200, 40, 700
Vaughn, Charles, 1, -, -, 50, 20, 15, -, 2500, 300, 400, 60, 40, 200, 40, 700
Davidson, James C., -, -, 1, 80, -, -, 300, 3800, 100, 300, 60, -, 100, 40, 400
Roach, Talbert J., -, -, 1, 100, -, -, 500, 10000, 100, 500, 100, - 100, 20, 1000
Wiltbank, David A., -, -, 180, 150, 80, -, 13000, 500, 1200, 100, 150, 300, 50, 1200
Robinson, Thomas E., 1, -, -, 100, 70, 20, -, 6000, 1500, 600, 100, 150, 200, 40, 1200
Robinson, John S., 1, -, -, 100, 70, 25, 70, 3000, 200, 500, 80, 30, 40, 6, 1000
Robbins, James C., 1, -, -, 120, 36, 30, -, 3000, 200, 500, 80, 20, 100, 40, 700
Morris, John J., 1, -, -, 100, 200, 40, -, 5000, 600, 700, 100, 100, 200, 40, 1000
Reed, William L., -, -, 1, 60, 40, 30, -, 2500, 100, 400, 80, 30, 20, 4, 500
Dutton, Robert, -, -, 1, 75, 200, 22, -, 3000, 150, 500, 50, -, 20, 4, 500
Reed, James C., -, -, 1, 100, -, 18, 30, 1500, 100, 1400, 80, 60, 30, 6, 500
Roach, William W., -, -, 1, 80, 25, 10, -, 2000, 100, 200, 50, 50, 20, 4, 400
Pepper, Thomas B. -, -, 1, 75, -, 25, -, 1000, 80, 200, 250, -, 40, 6, 400
Loffland, Samuel M., 1, -, -, 120, -, 50, 60, 2000, 100, 800, 100, 125, 200, 40, 800
Heaveloe, Joshua, -, -, -, -, -, -, -, -, -, -, -, -, -, -, 15
Reed, Abraham, 1, -, -, 80, 60, 24, -, 2000, 200, 500, 80, 40, 100, 20, 600
Reed, Abraham, 1, -, -, 7, 111, -, 1000, -, -, -, -, -, -, 200
Cropper, Warner, -, -, 1, 20, -, -, -, 500, 25, 100, 20, -, -, -, 160
Collins, William T., 1, -, -, 36, -, 12, -, 2000, 150, 250, 40, 24, 150, 40, 500
Reynolds, James A., -, -, 1, 60, -, 8, -, 1500, 50, 100, 40, -, 50, 10, 400
Reed, James E., 1, -, -, 60, -, 104, -, -, 100, 300, 50, -, 50, 10, 500
Houston, Jacob, -, -, 1, 10, -, -, -, 400, 10, 40, 10, 6, -, -, 95
Serwithia, Allen, -, -, 1, 100, -, 75, 55, 2000, 100, 100, 25, -, -, -, 400
Heather, Horatio N., 1, -, -, 50, 10, -, -, 1000, 20, 100, 10, -, 30, 6, 250
Heaveloe, Joseph, -, -, 1, 25, -, -, -, -, 80, 200, 50, -, 50, 8, 400
Lindle, David P., -, -, 1, 20, -, -, 6, 500, 40, 60, 10, -, -, -, 100
Johnson, David M., 1, -, -, 40, -, -, -, 650, 60, 120, 20, -, 10, 2, 150
Fields, Joseph H., -, -, 1, 21, -, -, -, -, 100, 150, 30, 70, 20, 2, 400
Johnson, Henry W., 1, -, -, 100, -, 15, -, 1500, 100, 150, 40, -, 100, 20, 400
Conwell, Nancy, -, -, 1, 130, -, 20, -, 1500, 50, 150, 50, -, 100, 40, 400
Heavelow(Heaveloe), James, 1, -, -, 60, -, 40, -, 1500, 150, 200, 30, -, 10, 2, 300
Williams, Harry, -, 1, -, -, 20, -, -, -, 500, 20, 150, 40, -, 80, 30, 300
Betts, Isaac S., 1, -, -, 60, -, 20, -, 2000, 100, 500, 40, 40, 100, 30, 600
Johnson, Abraham W., -, -, 1, 30, -, -, -, -, 100, 250, 50, -, 100, 20, 400
Roach, John T., -, -, 1, 120, -, 50, 30, 3000, -, 300, 80, 10, -, -, 500
Russell, David, 1, -, -, 90, 12, 50, -, 400, -, -, -, -, -, -, 100
Conwell, Hester D., 1, -, -, 60, 30, 10, -, 2000, 50, 300, 60, 60, 180, 43, 600
White, Frederic, W., 1, -, -, 80, 40, 20, -, 1600, 40, 200, 50, 60, 20, 3, 400

Jones, Kensey B., -, -, 1, 90, 30, -, -, 3000, 100, 300, 60, 50, 40, 4, 600
Pettyjohn, Ebenezer & Levin, -, -, 1, 80, 50, 50, -, 3000, 100, 300, 60, -, 40, 3, 500
Dutton, Jessee, -, -, 1, 60, 20, 50, -, 1500, 50, 250, 50, -, 60, 10, 400
Milbey, William C., -, -, 1, 50, 35, 25, -, 2000, 50, 250, 40, -, 40, 6, 400
Craig, William, -, -, 1, 75, 100, 40, -, 3000, 100, 300, 60, -, 50, 4, 600
Holland, Joseph, 1, -, -, 140, -, 17, 170, 3600, 400, 600, 100, 40, 200, 40, 1000
Holland, Henry, -, -, 1, 10, -, 5, 20, 400, 30, 100, 10, -, -, -, 100
Warrington, Roland P., 1, -, -, 150, -, 30, 40, 2000, 100, 300, 60, 80, 80, 10, 600
Reed, Burton A., -, -, 1, 50, -, 10, -, 1000, 50, 200, 40, -, -, -, 300
Lofland, Samuel M., 1, -, -, 16, -, 10, -, 30, -, -, 20, -, -, -, 100
Hazzard, John A., 1, -, -, 130, 100, 20, -, 8000, 100, 200, 50, -, 150, 30, 800
Jones, Henry K., 1, -, -, 15, -, 85, -, 1000, -, -, 20, -, -, -, 80
Reed, Philip R., 1, -, -, 50, -, -, 20, 1500, 100, 400, 30, -, -, -, 400
Fox, J. W. & Brothers, 1, -, -, 80, -, 50, -, 1300, 150, 200, 60, 400, 200, 50, 500
Robbins, David, 1, -, -, 90, -, 60, -, 4000, 200, 300, 70, 110, 150, 40, 700
Morris, Robert R., 1, -, -, 120, -, 40, 50, 5000, 400, 350, 70, 110, 150, 40, 800
Conwell, David M., 1, -, -, 110, -, 15, -, 2500, 100, 400, 50, 80, 100, 30, 700
Russell, Alfred, 1, -, -, 40, -, 20, 40, 2000, 100, 500, 60, 20, 140, 40, 400
Russell, Alfred, 1, -, -, 85, -, -, 64, 2500, -, -, -, -, 100, 40, 500
Russell, Alfred, 1, -, -, -, -, 40, -, 1500, -, -, -, -, -, -, -
Pettyjohn, Able, -, -, 1, 80, -, 40, 200, 4000, 150, 400, 40, -, 20, 4, 400
Conwell, John T., 1, -, -, 120, -, 10, 10, 4000, 100, 550, 60, 200, 150, 40, 800
Conwell, John T., 1, -, -, -, -, 21, -, 1000, -, -, -, -, -, -, -
Prettyman, George, 1, -, -, 12, -, 88, -, 700, 25, 80, 40, 10, -, -, 100
Manship, William E., 1, -, -, 8, -, -, -, 1200, 30, 100, 20, 10, 30, 6, 150
Atkins, John W., 1, -, -, 17, -, 8, 36, 2500, 100, 200, 30, 20, 20, 4, 300
Atkins, John W., 1, -, -, -, -, 100, -, 1500, -, -, -, -, -, -, 125
Atkins, Edwin C., 1, -, -, 78, -, 50, 36, 2000, -, -, 50, -, 150, 40, 500
Loffland, Elias, 1, -, -, 10, -, -, -, 1500, 20, 100, 20, -, 50, 8, 200
Carey, Cornelius J., 1, -, -, 80, -, 20, -, 1000, 100, 200, 40, 20, 150, 40, 300
Jackson, Peter B., 1, -, -, 6, -, -, 60, 2000, 50, 150, 40, -, 150, 40, 300
Carey, Robert H., 1, -, -, 6, -, 14, -, 500, 20, 200, 20, -, 100, 12, 150
Collins, E. Lamdon, 1, -, -, 8, -, 4, -, 1200, 10, 150, 10, 10, 20, 3, 100
Oliver, Joseph A., 1, -, -, 13, -, -, -, 100, 40, 120, 15, -, 20, 4, 100
Russell, Mrs. Molly, 1, -, -, 21, -, -, -, 400, -, -, 20, -, 50, 8, 100
Betts, Joseph J., 1, -, -, 60, -, -, -, 3000, 100, 300, 60, 50, 150, 40, 600
Chandler, Lewis B., 1, -, -, 12, 2, -, -, 800, 20, 100, 35, 45, 800, 10, 150
Martin, Sarah, -, -, -, -, -, -, -, -, -, -, -, -, -, -, -
Martin, Sarah, 1, -, -, 12, -, -, -, 600, -, 50, -, -, -, -, 125
Lank (Lauk), William D., 1, -, -, 20, -, 32, -, 600, -, 35, 20, -, -, -, 200
Simpler (Simples), David J., 1, -, -, 12, -, -, -, 600, 40, 80, 10, -, 20, 3, 200
Marshall, Mitchell A., 1, -, -, -, -, 350, -, 2500, -, -, -, -, -, -, 150
Atkins, Joseph C., 1, -, -, 5, -, 10, -, 900, 25, 200, 10, 10, 50, 8, 100
Polk, John C., 1, -, -, 20, -, 10, -, 700, -, -, -, -, -, -, 250
Fisher, Eliza, 1, -, -, -, -, 200, -, 3000, -, -, -, -, -, -, 150
Parker, Pricilla, 1, -, -, 10, -, -, -, 500, -, -, -, -, -, -, 100

Lingo, Henry B., 1, -, -, -, -, 40, -, 40, -, -, -, -, -, -, -
Collins, John A., 1, -, -, 30, -, 20, -, 1000, -, -, -, -, -, -, 100
Hazzard, John C., 1, -, -, 45, -, -, -, 2000, 50, 100, 10, -, 50, 10, 200
Welch, Nehemiah D., 1, -, -, 9, -, -, -, 200, 40, 120, 10, 10, -, -, 200
Jefferson, Asa W., -, 1, -, 10, -, -, -, 500, 100, 125, 20, 100, 120, 40, 500
Manship, Alfred H., 1, -, -, 20, -, 2, 2000, 50, 100, 20, 20, 20, 3, 200
Mason, Charles H., 1, -, -, 80, -, -, 20, 2500, 120, 400, 50, 30, 80, 20, 1000
Donovan, John C., 1, -, -, 40, -, 90, -, 1200, 100, 200, 50, 10, 50, 8, 500
Short, Burton, -, -, 1, 30, -, 140, -, 1250, 50, 100, 20, -, -, -, 200
Hazzard, William A., 1, -, -, 100, -, 70, -, 9000, 200, 800, 100, 100, 300, 50, 1000
Hazzard, William A., 1, -, -, -, -, 100, -, 3000, -, -, -, -, -, -, 120
Carey, James R., 1, -, -, 60, -, 30, 10, 1500, 30, 75, 50, -, 50, 10, 300
Pepper, Josiah M., -, -, 1, 100, -, 70, -, 2000, 50, 150, 60, 80, 100, 30, 700
Wiltbank, John H., 1, -, -, 100, -, 20, 17, 5000, 500, 400, 50, 160, 300, 52, 1500
Wiltbank, John, 1, -, -, 23, -, -, -, 2300, 40, 100, 25, 15, 30, 4, 400
Betts, James G., 1, -, -, 16, -, -, -, 500, 50, 60, 15, -, 40, 8, 100
Bryan, Robert B., 1, -, -, 80, -, 65, 23, 1800, 150, 400, 50, 75, 60, 6, 600
Simpler Brothers, -, -, 1, 80, -, 80, -, 1500, 100, 300, 50, 10, -, -, 500
Palmer, Greensbury W., 1, -, -, 50, -, 50, -, 1200, 100, 200, 40, -, 20, 4, 400
Jackson, Peter B., 1, -, -, -, -, 100, -, 1000, -, -, -, -, -, -, 200
Walker, Thomas, 1, -, -, -, -, 150, -, 1500, - -, -, -, -, -, 100
Prettyman, William C., 1, -, -, 60, -, 10, -, 2000, 75, 300, 40, 60, 100, 10, 500
VanKirk, Wm. 1, -, -, 70, 3, 45, -, 1800, 75, 175, -, 45, 40, 10, 350
Wiswell, Frederic C., 1, -, -, 55, 10, -, 5, 2500, 75, 250, 50, 100, 200, 50, 600
Carlisle, Manlove R., 1, -, -, 12, -, -, -, 1500, 100, -, 15, 50, 150, 50, 200
Gilchrist, J. B., 1, -, -, 58, 23, -, -, 10000, 200, 250, 100, 100, 550, 150, 2500
CuyKendall, E. C., 1, -, -, 67, 15, -, -, 10000, 200, 150, -, 100, 1000, 200, 2600
Woodall, B. F. B., -, -, 1, 30, 12, -, -, 2000, 100, 60, -, 75, 150, 36, 400
Fiddaman, H. B., 1, -, -, 40, -, -, -, 1200, -, -, -, -, -, -, -
Smith, R. G., 1, -, -, 25, 15, -, -, 1500, 150, 400, 30, 25, 150, 40, 300
Hudson, George, -, 1, -, 60, 2, -, -, 2000, 50, -, -, -, 50, 15, 300
Boyce, J. W., -, -, 1, 160, -, 40, -, 5000, 75, 125, -, 62, 50, 25, 400
Clendaniel, Jacob, -, -, 1, 150, 10, 20, -, 5000, 50, 290, -, -, -, -, 800
Benston, William, -, -, 1, 100, 10, 40, -, 3000, 75, 400, -, 65, -, -, 1000
Vreland, Henry, 1, -, -, 100, 20, 25, -, 3000, 200, 350, 50, 60, 50, 10, 800
Vreland, Garret, 1, -, -, 20, 15, 10, -, 800, -, -, -, 15, -, -, 200
Hopkins, W. E., 1, -, -, 29, -, -, -, 4000, 304, 225, -, 125, 300, 100, 1200
Toll, Charles, 1, -, -, 18, 17, -, 5, 1500, -, -, -, -, 100, 25, 550
Hall, A. K., 1, -, -, 3, -, -, -, 300, -, 75, 10, 15, 40, 15, 75
Smith, J. J., 1, -, -, 3, -, -, -, 300, -, 75, -, 10, 40, 12, 80
Fox, Lyman, 1, -, -, 60, 10, 4, -, 1200, -, 150, 50, 32, 120, 30, 1000
Pierce, W. N., 1, -, -, 100, 8, 47, -, 4000, 300, 500, 10, 142, 300, 80, 1400
Shockley, Wilson, 1, -, -, 55, 5, 4, -, 1500, 200, 300, 20, 60, 10, -, 600
Webb, Abner, 1, -, -, 3, 10, 4, -, 900, -, 75, -, 15, -, -, 450
Gray, William, 1, -, -, 9, -, -, -, 900, -, 200, -, -, 30, 8, 100

Warrington, Stephen H., -, -, 1, 150, 42, 140, 268, 13000, 300, 1000, 25, 56, 300, 80, 2885

Deputy, J. H., 1, -, -, 95, 15, 50, 40, 4000, 350, 650, 50, 100, 250, 70, 2000

Davis, P. T., 1, -, -, 90, 10, 5, 12, 3000, 300, 1000, -, 35, 200, 60, 1030

Shockley, Elias of W, 1, -, -, 100, 1, 5, 50, 2500, 100, 900, -, 105, -, -, 630

Prettyman, William, 1, -, -, 18, 12, 17, -, 1000, 250, 250, -, 30, -, -, 250

Davis, Thomas M., -, 1, -, 150, 4, 50, 25, 5000, 400, 400, -, 200, 300, 80, 1250

Collins, Samuel P., 1, -, -, 42, -, 10, 60, 2500, 150, 500, -, 50, -, -, 400

Davis, R. H., 1, -, -, 38, -, -, -, 2850, 150, 200, -, 40, 75, 15, 150

Blair, Charles A., -, 1, -, 12, -, -, 3, 1000, 50, 300, -, 25, 50, 15, 200

Davis, Thos. J., 1, -, -, 4, -, -, -, 400, 150, 400, -, -, 20, 5,70

Daniel, E. M., 1, -, -, 104, -, 10, -, 3000, 160, 425, -, 40, -, -, 350

Prettyman J. T., 1, -, -, 6, -, -, -, 1000, -, 200, -, 10, -, -, 50

Deputy, B. B., -, 1, -, 4, -, -, -, 1200, 50, 70, -, -, -, -, 50

Deputy, B. B., 1, -, -, 3, -, -, -, 500, -, -, 5, -, -, -, -, -

McColly, H. W., 1, -, -, 120, -, 6, -, 12000, 200, 500, 50, 75, 500, 12, 5, 1675

Hudson, Sarah A., 1, -, -, 9, -, -, -, 900, -, -, -, -, -, -, 150

Fisher, Eliza A., 1, -, -, 6, -, -, -, 1000, -, -, -, -, -, -, 50

Cirwithen, Isaac, -, -, 1, 150, 25, 25, 200, 10000, 250, 800, -, 150, 300, 80, 1452

Bennett, Purnel S., -, -, 1, 118, -, 12, 5, 3000, 100, 200, -, 60, -, -, 520

Daniel, Molten R., 1, -, -, 70, -, 10, 30, 1500, 50, 175, 15, 62, 60, 16, 475

Deputy, Zacariah, 1, -, -, 70, 12, 30, 23, 1500, 150, 200, -, 62, 50, 15, 950

Sharp, Roulen (Rueben) P., 1, -, -, 92, 8, 10, -, 2500, 200, 300, 100, 150, 140, 30, 1050

Kerson, Peter, -, -, 1, 140, -, 10, 20, 3000, 200, 550, -, 16, -, -, 500

Higman, Elijah T., -, -, 1, 155, 45, 20, 200, 7000, 150, 600, -, 50, 50, 16, 1700

Morgan, Wm. -, -, 1, 75, 10, 25, 60, 2000, 80, 200, -, 100, 20, 4, 550

Watson, Thos. A., 1, -, -, 60, -, 92, 70, 1200, 200, 325, -, -, 150, 40, 550

Burton, Daniel, -, -, 1, -, -, 10, -, -, -, -, -, -, -, -, -

Scott, James, -, -, 1, 80, -, 10, -, 800, 20, 30, -, -, -, -, 125

Morgan, Hiram, -, -, 1, 100, -, 10, 100, 1800, 50, 150, -, 50, 25, 6, 500

Marshall, William, 1, -, -, 15, -, -, -, 2000, -, 100, -, -, -, -, 255

Gray, William, 1, -, -, 9, -, -, -, 900, -, 250, 10, -, 5, -, 40

Daniel, Walter, 1, -, -, 5, -, -, -, 250, -, 75, -, -, -, -, 25

Truitt, William E., 1, -, -, 5, -, -, -, 300, -, 50, -, -, -, -, 30

Humphreys, Thomas, 1, -, -, 60, 5, 3, 10, 5000, -, 250, 50, 80, 200, 50, 605

Deputy, Thomas H., -, -, 1, 80, 6, 14, -, 1500, 75, 200, -, 20, -, -, 300

Lindall, Jessee, -, -, 1, 100, -, 25, 8, 4000, 25, 350, -, 60, -, -, 400

Bozman, Wm. J., -, -, 1, 100, 15, 4, 10, 2000, 50, 200, -, 40, -, -, 500

Cannon, Lucrecia, 1, -, -, 6, -, 3, -, -, -, -, -, -, -, -, -

Carpenter, Barton D., -, -, 1, 60, 20, 5, 80, 2000, 120, 200, -, 18, -, -, 300

Spencer, Joseph S., 1, -, -, 40, -, 14, 70, 1800, 300, 350, 50, 35, 10, -, 350

Deputy, James G., -, -, 1, 60, -, 12, 90, 2000, 100, 250, -, 40, -, -, 350

Hanniford, Eli, -, -, 1, 30, -, 30, 10, 400, -, -, -, -, -, -, 40

Clendaniel, Thomas, -, -, 1, 110, -, 10, 150, -, 250, 1000, 20, 180, 350, 90, 1500

(Kilgore), Wilson, -, -, 1, 35, -, 25, -, -, -, -, -, 10, -, -, 80

Wilkins, William, -, -, 1, 60, -, -, 80, 1200, 50, 300, -, 50, -, -, 150

Hammond, James H., -, -, 1, 100, 12, 10, 100, 5000, 150, 700, 20, 150, 200, 50, 1000

Draper, Samuel, -, -, 1, 18, -, -, -, 400, -, -, -, -, -, -, -

Townsend, George H., -, -, 1, 100, 12, 10, 50, 4000, 500, 100, 120, 100, 200, 40, 1035

Townsend, George H., 1, -, -, 71, -, 12, 25, 2000, -, -, 20, 65, 50, 15, 450

Watson, Wm. P., 1, -, -, 70, -, 10, 20, 2000, 100, 300, -, 35, 30, 7, 400

Brittingham, Smith, -, -, 1, 20, 6, -, 5, 1000, 50, 150, -, 10, -, -, 150

Victor, John W., -, -, 1, 110, -, 12, -, 2000, 50, 75, 100, 110, 60, 16, 500

Swaine, Alfred E., -, -, 1, 60, -, 10, 80, 1500, 150, 150, -, 20, 10, 2, 500

Potter, Jno. H., 1, -, -, 70, 10, 58, 18, 3500, 250, 350, 365, 65, 45, 15, 775

Brinkley, Robert, 1, -, -, 6, -, -, -, -, -, -, -, -, -, -, -

Cannon, Wesley, 1, -, -, 6, -, 4, -, -, -, -, -, -, -, -, -

Cannon, Isaiah, 1, -, -, 6, -, 4, -, -, -, -, -, -, -, -, -

Donovan, Thomas, 1, -, -, 40, -, 15, 5, 100, 124, 250, -, 25, 10, -, 200

Shockley, George, 1, -, -, 60, -, 5, 30, 2500, -, -, -, -, -, -, 325

Hill, John, 1, -, -, 6, -, 5, 1, 300, 20, 100, -, -, -, -, 70

Wilkins, William, -, -, 1, 12, -, 5, -, 400, 25, 150, -, -, -, -, 50

Hill, Perlina, 1, -, -, 20, -, 30, 20, 500, 10, 75, -, -, -, -, 100

Jones, Edward, -, -, 1, 90, 5, 10, 20, 2000, 100, 150, 10, 75, 75, 25, 500

Potter, Benjamin E., 1, -, -, 145, -, 42, 172, 5000, 250, 500, -, 95, 150, 40, 750\

Corsdon, John, 1, -, -, 40, -, 10, 50, 1500, 10, 100, -, 35, -, -, 180

Mills, Miles T., 1, -, -, 90, 2, 12, -, 3000, 250, 500, 24, 64, 100, 30, 650

Mills, David W., 1, -, -, 40, 2, 13, -, 2000, 200, 200, 15, 65, 75, 25, 480

Fitzgerald, David, -, -, 1, 100, 13, 20, 70, 3000, 150, 200, 25, 80, 60, 15, 750

McColley, T. P., -, -, 1, 60, -, 10, 15, 1300, 20, 50, 40, 5, -, -, 150

Boyce, Jas H., 1, -, -, 50, 2, 13, -, 1000, 50, 100, -, 16, 15, 3, 250

Fitzgerald, Elizabeth, -, -, 1, 100, 15, 10, 20, 3500, 50, 300, -, 60, 10, 2, 550

Ennis, Stephen R., 1, -, -, 65, 3, 20, -, 2000, 100, 150, 50, 70, 125, 35, 600

Watson, William, 1, -, -, 160, -, 70, -, 5000, 200, 500, -, 80, 250, 80, 900

Laws, Harry I., 1, -, -, 85, 15, 20, -, 2500, 200, 220, 10, 50, 75, 26, 600

Walls, Burton, -, 1, -, 4, -, -, -, 100, -, 40, -, -, -, -, 30

Rogers, Daniel, -, -, 1, 50, 3, 14, 15, 1000, 50, 75, -, 40, -, -, 300

Warren, David, -, -, 1, 45, -, -, 35, -, 500, 10, 10, -, -, -, -, -

Short, Charles, 1, -, -, 70, -, 9, -, 2000, -, 160, -, -, -, -, 9

Prettyman, Elizabeth, -, -, 1, 80, 4, 5, 70, -, -, 430, -, -, -, -, 772

Parcks, William, 1, -, -, 42, -, 15, -, 2000, 50, 150, -, 40, 10, 3, 250

Wells, James V., 1, -, -, 75, 15, 5, -, 1500, 75, 150, -, 20, 50, 18, 350

Tease, John, 1, -, -, 10, 15, 20, 33, 2000, 100, 200, -, -, -, -, 250

Marshall, Joseph, 1, -, -, 40, 3, 16, -, 1000, 50, 200, 700, -, -, -, -, 50

Burton, Jacob, 1, -, -, 9, -, -, -, 300, 20, 70, -, -, -, -, 100

Watson, Henry S., 1, -, -, 120, 10, 7, 30, 2000, 50, 75, -, -, -, -, 350

Morgan, Uriah, 1, -, -, 5, -, 4, -, 200, -, 75, -, -, -, -, 41

Burton, Peter, 1, -, -, 5, -, 75, -, 500, -, 40, -, -, -, -, 6

Ingram, Nehemiah R., -, -, 1, 50, -, 50, -, 800, -, 30, -, -, -, -, 54

Coulter, Thomas, 1, -, -, 100, -, 580, -, 4000, 200, 300, -, 125, 200, 40, 400
Groves, Soloman, -, -, 1, 100, -, 20, -, 2000, -, 150, -, -, -, -, 152
Shockley, Elias, 1, -, -, 1, -, -, -, 200, -, 20, -, -, -, -, 91
Weatherly, Samuel, -, -, 1, 40, -, -, -, 500, -, -, -, -, -, -, 10
Hill, Sarah, 1, -, -, 60, -, -, -, 500, -, -, -, -, -, -, 92
Calhoon, Joseph, 1, -, -, 80, 7, 22, -, 1600, 150, 250, 20, 32, -, -, 460
Calhoon, Elizabeth, 1, -, -, 48, 2, 30, -, 800, -, 20, 250, 32, -, -, 60
Beideman, Oliver, 1, -, -, 10, -, 15, -, 800, 25, 200, 700, 16, 30, 6, 50
Benston, Cassa, -, -, 1, 125, -, 65, 10, 2000, 75, 150, -, 32, -, -, 300
Reynolds, William, -, -, 1, 80, -, 40, 15, 1800, 10, 55, -, -, -, -, 125
Johnson, R. S., 1, -, -, 5, 300, 10, 2, 1000, -, -, -, -, -, -, 300
Leverage, Robt. _., -, -, 1, 65, -, 60, 10, 1200, 40, 150, -, -, -, -, 200
Grove, Nehemiah, 1, -, -, 5, -, 5, -, 250, -, -, -, -, -, -, 20
Watson, Jacob, 1, -, -, 50, 5, 4, -, 600, 25, 150, -, -, -, -, 150
Evans, George, 1, -, -, 4, -, -, -, 100, -, -, -, -, -, -, 30
Shockley, Robenson, 1, -, -, 55, 4, 6, -, 1500, 50, 150, -, 60, -, -, 200
Cirwithen, Solomon, 1, -, -, 4, -, -, -, 80, -, -, -, -, -, -, 25
Young, Charles, 1, -, -, 30, -, 12, -, 1000, 50, 270, -, 38, -, -, 200
Young, Nathan, 1, -, -, 27, -, 4, -, 300, -, 50, -, -, -, -, 80
Young, Joseph E., -, -, 1, 22, -, 6, -, 600, 50, 50, -, -, -, -, 40
Shockley, James M., 1, -, -, 50, -, 15, -, 500, 25, 200, -, 30, 10, 2, 250
Young, Jacobs, 1, -, -, 15, -, 3, -, 400, 25, 100, -, -, -, -, 200
Bennett, John W., 1, -, -, 80, 5, 20, -, 2000, 200, 450, -, 100, 200, 40, 600
Bennett, James D., 1, -, -, 15, -, 10, -, 1000, 150, 150, 300, -, 25, 5, 200
Shockley, Chas. M., 1, -, -, 40, 2, 3, -, 1000, 50, 250, -, 50, 50, -, 400
Shockley, Antony R., 1, -, -, 54, -, 4, -, 1000, 50, 200, 40, 65, 50, 10, 250
Shockley, Solomon, 1, -, -, 40, -, 8, -, 1000, 75, 250, 20, 40, 20, 5, 250
Cirwithen, Thos., -, 1, -, 40, -, 30, -, 400, -, 150, -, -, -, -, 100
Davis, N. H., 1, -, -, 15, -, 5, -, 200, -, 6, -, -, -, -, 50
Crapper, Sarah J., -, -, 1, 10, -, 5, -, 100, -, -, -, -, -, -, 25
Draper, Sarah, -, -, 1, 10, -, 5, -, 150, -, -, -, -, -, -, 25
Vincent, William, 1, -, -, 8, -, 10, -, 160, -, -, -, -, -, -, 60
Young, Joseph, 1, -, -, 20, -, -, -, 200, -, 56, -, -, -, -, 75
Groves, Joseph, 1, -, -, 7, -, 2, -, 100, -, -, -, -, -, -, 50
Davis, John, 1, -, -, 60, -, 3, -, 600, 185, -, -, -, -, 50
Argoe, Joseph, 1, -, -, 35, -, 50, -, 1000, 50, 100, 20, 18, -, -, 200
Carn, Be__ M., -, -, 1, 180, 70, 15, 120, 5000, 500, 500, 50, 100, 118, 30, 2000
Shockley, David, -, -, 1, 76, 4, 20, -, 1500, 50, 100, 15, 30, -, -, 350
Titus, Jos. V., 1, -, -, 90, 40, 30, -, 5000, 500, 600, 1500, 75, 400, 100, 2500
Titus, David, -, -, 1, 100, 40, 60, 12, 4000, 200, 500, -, 75, 150, 30, 1200
McColley, Stephen H., 1, -, -, 50, -, 6, 50, 2000, 175, 225, 35, -, -, 350
WynKoop, Cornelius P., 1, -, -, 180, 80, 20, -, 10000, 500, 400, -, 35, 200, 60, 1400
Roach, David S., 1, -, -, 35, -, 8, -, 600, 40, 200, -, 18, -, -, 200
Coffen, David B., 1, -, -, 4, -, 5, -, 100, -, 50, -, -, -, -, 12
Clifton, William, -, -, 1, 50, -, 4, 24, -, -, 265, -, -, -, -, 200

Roach, Robert, 1, -, -, 75, -, 5, 50, 1800, 40, 225, -, 40, -, -, 320
West, Thomas, -, -, 1, 90, -, 10, 30, 1000, 50, 260, -, -, -, -, 200
Walls, John H., -, -, 1, 100, -, 20, 20, 2500, 100, 300, -, -, 100, 40, 650
Jefferson, Wm., -, -, 1, 120, -, 10, -, -, -, 615, -, -, -, -, 688
Argoe, Albert -, -, 1, 65, -, -, 40, 800, 50, 50, -, -, -, -, 136
Roberts, Sarah A., 1, -, -, 10, -, -, -, 300, -, -, -, -, -, -, 15
Coffin, David, -, -, 1, 36, -, -, -, 400, -, 75, -, -, -, -, 15
Jones, Theodore, -, -, 1, 100, -, -, 10, 2000, 40, 150, -, -, 20, 6, 300
Russell, William -, -, 1, 70, -, -, 20, 2000, 40, 300, -, 30, -, -, 350
Roach, Edward, 1, -, -, 50, -, 12, 80, 1000, 50, 200, -, -, -, -, 250
Roach, Thomas, -, -, 1, 90, 50, 25, 50, 5000, 100, 300, -, 50, -, -, 800
Draper Brothers, 1, -, -, 200, 12, 20, 40, 14000, 800, 1100, 25, 140, 600, 150, 2500
Milman, Samuel, 1, -, -, 33, 5, -, -, 500, -, 40, -, -, -, -, 300
Davis, Jos. M., 1, -, -, 65, 38, -, 150, 3000, 400, 100, -, 70, 250, 80, 1225
Stevenson, Peter R., -, -, 1, 55, 5, -, 30, 1500, 100, 350, -, -, -, -, 200
Roach, George, 1, -, -, 60, -, 3, 30, 1200, 40, 300, -, -, -, -, 150
Wilson, George H., 1, -, -, 100, -, 10, 50, 2500, 150, 300, -, 35, -, -, 500
Draper, Susan R., 1, -, -, 125, 3, 125, 50, 5500, 500, 1000, -, 35, 200, 60, 1000
Messick, N. J., 1, -, -, 60, -, 16, -, 1500, 50, 1000, -, 50, 100, 30, 500
Conner, David, 1, -, -, 50, -, -, -, 900, 50, 200, -, 35, -, -, 100
Reynolds, David, 1, -, -, 150, -, -, 50, 2500, 150, 400, -, 70, 25, 6, 650
Wilson, Riley, 1, -, -, 100, 5, 40, 20, 2500, 50, 250, -, 80, -, -, 400
Shockley, Jessee, 1, -, -, 32, -, -, -, 600, 25, 100, -, 11, -, -, 200
Shockley, Charles, 1, -, -, 40, -, 5, -, 2500, 100, 200, -, 65, 100, 30, 600
Shockley, Jeremiah, -, -, 1, 40, -, -, -, 1000, 100, 200, -, 32, 40, 12, 200
Truitt, Eliza, 1, -, -, 100, -, 60, 40, 2000, 50, 100, -, -, -, -, 150
Scagg, Thomas, -, -, 1, 60, -, -, 600, -, -, -, -, -, -, 25
Warren, William, -, -, 1, 50, -, -, -, 2000, -, 460, -, 50, -, -, 150
Walls, Jesse, -, -, 1, 60, -, 10, 50, 800, -, 100, -, -, -, -, 175
Roach, James, -, -, 1, 75, -, 20, 50, 2000, 150, 300, -, -, 20, 6, 300
Roach, Thomas J., -, -, 1, 50, -, 25, 40, 2000, 125, 250, -, 25, 50, 15, 350
Bennett, Riley W., 1, -, -, 100, -, 20, 40, 3000, 150, 400, 200, 60, 250, 100, 450
Davis, Thomas J., 1, -, -, -, -, 33, -, 300, -, -, -, -, -, -, -
Davis, Robert H., 1, -, -, -, -, 25, 10, 300, -, -, -, -, -, -, -
Draper, George H., 1, -, -, 100, 35, 30, 97, 3000, 500, 800, -, 275 450, 150, 2200
Fountain, Wm. H., -, -, 1, 50, -, 25, -, 1500, 100, 250, -, -, 50, 25, 300
Higman, Jas, 1, -, -, 100, 9, 50, 20, 2000, 150, 450, 10, 40, -, -, 450
Pierce, H. J., 1, -, -, 80, 18, 15, 200, 3000, 250, 500, -, 80, 50, 12, 800
Davis, Lot W., -, -, 1, 120, -, 40, 140, 2500, 100, 400, -, 50, -, -, 500
Williams, James, -, -, 1, 12, -, -, -, 1000, 40, 75, -, 20, -, -, 100
Piper, Mitchel, -, -, 1, 60, -, 40, -, 1500, 100, 250, -, -, -, -, 250
Jackson, Wesley, 1, -, -, 60, 4, 20, 10, 800, 25, 125, -, -, -, -, 100
Shockley, Bethuel, 1, -, -, 6, -, -, -, 100, -, 40, -, -, -, -, 50
Shockley, Lemuel, 1, -, -, 7, -, -, -, 100, 20, 70, -, -, -, -, 80
Young, Charles, 1, -, -, 10, -, -, 5, 600, -, 100, -, -, -, -, 150

Ingram, Jno. B., 1,-, -, 80, -, -, 20, 1000, -, 50, 200, -, 18, 30, 8, 350
Hudson, Robert, -, -, 1, 85, -, -, 10, 500, 10, 50, -, -, -, -, 200
Lindall, D. R., -, -, 1, 60, 25, 400, 3000, 50, 200, -, -, 50, 12, 200
Hazzard, David, 1, -, -, 6, -, -, -, 100, -, 40, -, -, -, -, 50
Pettyjohn, Wm., -, -, 1, 30, -, -, -, -, -, 100, -, -, -, -, 150
Mason, James, -, -, 1, 50, -, 8, -, 1000, 15, 50, -, -, -, -, 225
Davis, John S., 1, -, -, 110, -, 30, 150, 2000, 100, 250, -, 20, 10, -, 300
Bennett, W. H., 1, -, -, 14, -, -, -, 1000, 40, 100, -, -, -, -, 200
Shepherd, Jas. B., 1, -, -, 40, 10, 50, -, 1500, 100, 200, -, 20, -, -, 250
Pettyjohn, George, -, -, 1,, 90, 10, 5, 5, 2000, 25, 150, -, -, -, -, 250
Darby, William, 1, -, -, 60, -, 5, -, 600, -, 150, -, -, -, -, 225
Burton, Jacob, -, -, 1, 4, -, -, -, 100, -, 50, -, -, -, -, 150
Collins, Morris, 1, -, -, 40, -, -, -, 1000, -, 250, -, 20, -, -, 20
Todd, William, 1, -, -, 20, -, -, -, 400, -, 100, -, -, -, -, 125
Roach, David, -, -, 1, 80, -, 20, 40, 3000, 100, 300, -, 50, -, -, 700
Messick, John D., -, -, 1, 20, 20, 10, -, -, -, 150, -, -, -, -, 150
Hazell, Robert, -, -, 1, 50, 3, 15, 30, 1500, -, 100, -, -, -, -, 150
Morgan, Wm. B., 1, -, -, 60, -, 2, -, 600, 25, 75, -, -, -, -, 150
Argoe, John, -, -, 1, 100, -, 20, 40, 2500, 150, 2000, -, 50, -, -, -
Milman, David, -, -, 1, 60, -, 20, 15, 1500, 100, 300, -, 38, -, -, 350
Clendaniel, William, -, -, 1, 70, -, 10, 10, 2000, 100, 200, -, 38, 30, 10, 450
Warren, Bennet, 1, -, -, 60, 4, -, -, 2000, - 420, -, 18, -, -, 300
Gillace, James, 1, -, -, 30, 5, -, -, 300, -, -, -, -, -, -, 100
Carpenter, Robt. S., -, -, 1, 35, -, 15, -, 800, -, 300, -, -, -, -, 150
Bennett, Elizabeth, 1, -, -, 60, -, 8, -, 1500, 100, 200, -, 25, -, -, 450
Bennett, David R., -, -, 1, 80, -, 137, -, 2000, 100, 300, -, -, -, -, 400
Hickman, George, -, -, 1, 62, -, 30, -, 800, 50, 100, -, 18, -, -, 303
Male, Geo. W., 1, -, -, 8, -, 60, -, 2000, 150, 250, -, -, -, -, 350
Milman, David, 1, -, -, 60, -, 40, -, 900, -, -, -, -, -, -, 200
Macklin, George, -, -, 1, 75, -, 15, -, 800, -, 75, -, 20, -, -, 150
Young, David, -, -, 1, 40, -, -, -, 500, -, 100, -, -, -, -, 150
Watson, John, -, -, 1, 50, -, -, 50, 500, -, 100, -, -, -, -, 150
Roach, Theodore, -, -, 1, 100, -, 10, -, 800, -, 250, -, -, -, -, 300
Hazell, Kinsey, -, -, 1, 40, -, 15, 10, 800, -, 200, -, 40, -, -, 200
Bennett, Jos. R., 1, -, -, 20, -, 5, -, 500, 50, 100, -, -, -, -, 500
Jones, James, -, -, 1, 80, -, 10, -, 1000, 50, 200, -, 40, -, -, 300
Waples, Frank, 1, -, -, 120, -, -, 125, 1000, 75, 375, -, 25, 150, 50, 250
Draper, Henry C., 1, -, -, 130, -, 4, 30, 4000, 500, 400, -, 80, 150, 50, 600
Truitt, John S., 1, -, -, 28, -, -, -, 800, -, 200, -, 20, -, -, 200
Hickman, Wm. _, 1, -, -, 42, -, 10, -, 700, 40, 200, -, 30, -, -, 200
Young, Phillip, 1, -, -, 7, -, -, -, 100, -, 186, -, -, -, -, 50
Young, Robert, 1, -, -, 20, -, -, -, 300, -, 100, -, -, - -, 150
Warren, Frances A., 1, -, -, 75, -, 10, 15, 1000, 100, 400, -, 30, -, -, 300
Prettyman, John, 1, -, -, 40, -, 5, -, 1000, -, 250, 50, 30, 40, 12, 300
Davis, Henry, 1, -, -, 40, -, 4, -, 800, -, 100, -, -, -, -, 300
Wilkins, Wesly, -, -, 1, 50, -, 30, 20, 600, -, 50, -, -, -, -, 150

Jones, Wm. H., -, -, 100, -, 20, 20, 2500, 100, 300, -, 40, 50, 20, 600
Shockley, Jeremiah, -, -, 1, 50, -, 10, -, 600, 25, 150, -, 18, -, -, 250
Sceine, Elizabeth, -, -, 1, 100, -, 25, 25, 2500, -, 150, -, -, -, -, 200
Harrington, James H., -, -, 70, -, 150, 50, 3000, 100, 400, 140, 100, -, -, 450
Jefferson, Nathaniel, -, -, 1, 53, -, 14, -, 600, 40, 150, -, -, -, -, 150
Jefferson, Thomas W., 1, -, -, 75, 2, 25, -, 800, 100, 300, -, 100, 10, -, 275
Donovan, Anna R., 1, -, -, 75, -, 100, -, 3000, 100, 300, -, 50, -, -, 175
Donovan, Gibson, -, -, 1, 70, -, 75, -, 1000, 40, 225, -, 30, -, -, 185
Betts, Robert, 1, -, -, -, -, 40, 60, 800, -, -, -, -, -, -, 70
Walls, Robert J., -, -, 1, 30, -, 60, 10, 1000, 100, 200, -, -, -, -, 350
Purnell, Eliza J., 1, -, -, 30, -, 15, -, 700, 25, 100, -, -, -, -, 100
Abbot, Lydia E., 1, -, -, 100, -, 25, 25, 1000, 25, 50, -, -, -, -, 200
Cary, Burton, 1, -, -, 60, -, 8, 10, 800, 40, 80, -, -, -, -, 100
Abbot, Henry, 1, -, -, 30, -, 10, 20, 300, -, -, -, -, -, -, 38
Warren, Peter, -, -, 1, 5, -, -, -, 100, -, -, -, -, -, -, -
Clifton, John, -, -, 1, 80, -, 320, -, 3000, 50, 150, -, -, -, -, 300
Morris, James H., -, -, 1, 50, 25, 25, 800, 40, 200, -, 25, -, -, 300
Morris, Isaac C., -, -, 1, 27, -, 3, -, 500, 40, 200, -, 50, -, -, 250
Morris, Josiah, 1, -, - 30, -, 40, -, 600, -, -, -, -, -, -, 150
Thurston, Jonathan H., 1, -, -, 50, 60, 60, 80, 4000, 100, 350, 7, -, -, 700
Jones, Erasmus, 1, -, -, 70, -, 100, 30, 2000, 300, 400, -, 60, -, -, 550
Hudson, Bashaba, 1, -, -, 80, 2, 25, -, 1500, 75, 100, 25, 50, 75, -, 250
Stewart, Eliza, 1, -, -, -, -, 23, 7, 400, -, -, -, -, -, -, 125
Clifton, Annie, 1, -, -, -, -, 40, 10, 400, -, -, -, -, -, -, 75
Hudson, Mary, 1, -, -, -, -, 30, -, -, -, -, -, -, -, -, -
Prettyman, Lemuel, -, -, 1, 20, -, 80, -, 400, -, -, -, -, -, -, 50
Reston, Lyman, 1, -, -, 16, 4, -, -, 3000, 15, 150, -, 10, -, 30, 120
Risler, George W., 1, -, -, 60, 20, 115, -, 2500, -, 50, -, 30, -, 20, 120
Risler, George W., 1, -, -, 62, 15, 10, -, 3000, -, -, -, 30, -, 10, 150
Smith, Thompson, 1, -, -, 10, 1, 30, -, 700, -, 30, 10, -, -, -, 100
Risler (Rider), Laura, 1, -, -, 28, 10, 7, -, 2000, -, -, 300, -, -, 10, 225
Lynch, James, -, -, 1, 120, 10, 45, -, 2500, 50, 100, 25, 50, -, 25, 300
Wilkinson, John W., 1, -, -, 60, -, -, -, 500, 150, 250, -, 30, -, -, 200
Norcross, John D., 1, -, -, 75, 10, 50, 65, 2000, 23, 75, -, -, -, -, 150
Morris, Frank, -, -, 1, 50, -, 300, 25, 2000, 50, 100, -, -, -, -, 125
Greenly, David, -, -, 1, 50, 5, -, -, 5, 700, 100, 200, -, 20, 30, -, 150
Douthard, Eweene, 1, -, -, 12, 3, -, -, 800, 125, 223, -, 40, -, 25, 125
Simmons, Bawdy, 1, -, -, 10, 5, -, -, 1500, -, -, -, -, -, -, 50
Clendaniel, Isaac, -, -, 1, 550, 3, 51, -, 2500, 250, 450, 30, 40, 100, -, 500
Stewart, James, 1, -, -, 75, 4, 18, -, 1000, 100, 125, 10, 50, 100, 25, 250
Banta, Edwin, 1, -, -, 125, 8, 7, -, 2000, 250, 100, 25, -, 100, 25, 200
Small, Alfronso, 1, -, -, 30, -, 30, -, 1000, -, -, -, 20, -, -, 100
Johnson, George H., -, 1, -, 20, 4, -, -, 1000, 50, 100, -, 25, 100, -, 150
Carlisle, Manlove, 1, -, -, 5, 2, 40, -, 1000, 20, 20, -, -, -, -, 50
Sprague, Barney, 1, -, -, 48, 10, -, -, 500, -, -, -, -, -, -, 100
Smith, Richard, 1, -, -, -, 20, 14, -, 2000, -, -, -, 75, 120, -, 300

Macklin, John S., 1, -, -, 20, -, -, -, 400, 150, 250, -, 25, -, 25, 150
Lord, Luther, -, -, 1, 90, 3, 70, -, 1500, 75, 75, -, 20, -, -, 175
Wilkinson, David, 1, -, -, 25, -, 15, 5, 300, 50, 50, -, 30, -, -, 150
Dickerson, David, -, 1, -, 50, 8, -, -, 500, 50, 75, -, 20, -, -, 225
Conner, James, -, -, 1, 70, -, 80, 10, 1000, 25, 50, -, -, -, -, 100
Morrison, Senge, 1, -, -, 60, 10,70, -, 2500, 50, 200, 100,70, -, 130, 250
Stout, David, 1, -, -, 7, 3, -, -, 500, 25, -, -, -, -, -, 100
Clifton, Garrett, 1, -, -, 12, -, -, -, 1540, 50, 50, -, -, -, -, 320
Holt, Henry, 1, -, -, 30, 16, -, -, 5000, 50, 125, -, 60, -, 350, 270
Shubert, Benniah, 1, -, -, 23, 10, -, -, 3000, 50, 175, 50, -, -, -, 300
Stevenson, Jesse, -, -, 1, 90, 7, 3, -, 2000, 15, 225, -, 10, -, -, 230
Fitzgerald, John, -, -, 1, 75, 5, 150, 20, 2500, 75, 150, -, 40, -, -, 250
Walton, Edwin, 1, -, -, 18, -, -, -, 1000, 115, 200, -, 10, 75, -, 200
Walton, Maggie, 1, -, -, 35, 14, 23, -, 1000, -, -, -, -, -, -, 170
Walton, Harry (Walter), 1, -, -, 36, -, -, -, 1000, -, -, -, -, -, -, 200
Smith, James, -, -, 1, 30, -, 100, 20, 800, 100, 125, -, -, -, -, -
Webb, Purnell, -, -, 1, 60, -, 100, 10, 800, 125, 200, -, -, -, -, 250
Truitt, Sarah, 1, -, -, 30, -, 50, 10, 400, 25, -, -, -, -, -, 100
Truitt, Benjamin, 1, -, -, 40, -, 50, 10, 600, 75, 100, 50, -, -, -, 125
Shockley, Purnell, -, -, 1, 35, 4, 50, -, 500, 50, 75, -, -, -, -, 150
Pettyjohn, Desolene, -, -, 1, 30, -, 34, -, 300, 100, 400, 50, 20, -, -, 250
Warren, David O., 1, -, -, 20, 2, 63, -, 800, 100, 200, -, 20, -, 10, 250
Warren, Sallie, 1, -, -, 75, -, 75, -, 1000, -, -, -, -, -, -, -
Warren, Fannie, 1, -, -, 4, -, 82, -, 1000, -, -, -, -, -, -, -
Warren, Emma, 1, -, -, -, -, 46, -, 1000, -, -, -, -, -, -, -
Carey, James, -, -, 1, 50, 20, 10, -, 100, 30, 75, -, 30, -, -, 500
Johnson, Alexander, -, -, 1, -, -, 143, -, 2300, -, -, -, -, -, -, -
Deputy, Nancey, 1, -, -, 75, 15, 20, -, 1300, 100, 250, 30, 50, -, -, 200
Deputy, William W., 1, -, -, 30, 3, 28, -, 800, 100, 230, 300, 40, -, -, 260
Deputy, John H., 1, -, -, 50, 3, 10, -, 800, 125, 200, 50, 60, -, -, 250
Deputy, Emma, 1, -, -, -, 15, 70, -, -, -, -, -, -, -, -, 250
Deputy, Joseph, 1, -, -, -, -, -, 47, -, 500, -, -, -, -, -, -, -
Deputy, Charles, 1, -, -, -, -, 40, -, 500, -, -, -, -, -, -, -
Deputy, Ida, 1, -, -, -, -, 45, -, 500, -, -, -, -, -, -, -
Deputy, James, 1, -, -, 30, 5, 40, -, 1500, 200, 300, 200, 40, 100, 30, 300
Griffith, Lide, 1, -, -, 10, -, -, -, 150, -, 10, -, -, -, -, -
Dawson, Isaac, 1, -, -, 50, 5, 50, -, 1000, 100, 125, -, 60, -, -, 300
Carpenter, Joseph, -, -, 1, 60, 2, 30, -, 1000, 75, 60, 25, 40, -, -, 200
Johnson, William, 1, -, -, 40, 6, 17, -, 800, 30, 300, -, 20, 125, 75, 200
Fenner,Owen, 1,-, -, 40, 7, 8, -, 700, 50, 50, -, 10, -, -, 200
Dawson, John O., 1, -, -, 100, 4, 60, 12, 1000, 125, 300 -, 30, -, -, 300
Belknap, John M., 1, -, -, 12, -, 50, 3, 200, -, -, -, -, -, -, 100
Ellensworth, Salathiel, -, -, 1, 60, 2, 50, 10, 500, 50, 175, -, -, -, -, 250
Miller, William, 1, -, -, 75, 3, 20, 10, 1200, 100, 350, 400, 40, -, -, 300
Austin, John, 1, -, -, 50, 10, 70, -, 1200, 100, 200, -, 70, -, -, 300
Watson, William, 1, -, -, -, -, 80, 70, 1000, -, -, -, -, -, -, -

Clifton, James W., 1, -, -, 40, -, 30, -, 800, 100, 200, 50, 40, 150, 100, 200
Austin, Henry W., 1, -, -, 80, 10, 100, -, 3000, 200, 400, -, 140, 75, 25, 500
Austin, William H., 1, -, -, 78, -, 5, 70, 9, 1000, 150, 175, 25, 80, 50, -, 350
Tatman, James, -, -, 1, 60, 8, 70, 5, 1500, 125, 200, -, 60, 40, -, 300
Fleming Harry, 1, -, -, 4, 2, -, -, 200, -, -, -, -, -, -, 30
Carpenter, Charles, -, -, 1, 90, 3, 50, 10, 1200, 100, 175, 25, 20, -, -, -
Burris, William -, -, 1, 20, -, 50, -, 800, 30, 75, -, 20, -, -, 150
Hevaloe, William, -, -, 1, 75, -, 150, 25, 2000, 75, 150, 28, 40, -, -, 300
Felton, Elias R., -, -, 1, 260, 28, 240, -, 10000, 150, 325, 100, 160, 480, 32, 100
Lecompt, John H., 1, -, -, 6, -, -, -, 600, 25, 100, 300, 20, 25, 25, 150
Houston, Curtis, 1, -, -, 100, 5, 100, -, 2500, 100, 175, -, 30, 75, -, 250
Clendaniel, Joshua, -, -, 1, 150, 5, 150, -, 4000, 300, 400, -, -, -, -, 400
Truitt, William, 1, -, -, 40, 5, 20, -, 1500, 75, 100, -, 20, -, -, 200
Houston, James, 1, -, -, 60, 14, 80, -, 8000, 200, 250, -, 120, 144, -, 600
Campbell, James L., 1, -, -, 10, 40, -, -, 3000, 150, 280, 50,70, 102, 75, 800
Dickerson, James B., -, -, 1, 75, 5, 20, -, 2000, 75, 100, -, 30, -, -, 200
Beardsley, Robert, 1, -, -, 14, -, 14, -, 800, 50, 75, -, 40, 72, -, 150
Beardsley, Trueman, 1, -, -, 25, -, 5, -, 1000, 50, 100, 20, 30, -, -, 175
Beardsley, Emer, 1, -, -, 13, -, 5, -, 800, -, -, -, 24, -, -, 100
Corse, William P., 1, -, -, 50, 17, -, -, 2500, 50, 170, -, 125, 113, 39, 367
Bogart, John A., -, -, 1, 15, 10, -, -, 2000, 50, 150, -, -, 112, -, 400
Reynolds, Zachariah, -, -, 1, 168, 60, 80, -, 10000, -, 400, -, 140, 220, -, 1200
Wroten, Robert, -, 1, -, 20, 4, -, -, 2000, 50, 75, -, 20, -, -, 250
Layton, David, 1, -, -, 20, 5, -, -, 2000, 50, 125, -, 20, -, -, 450
Lofland, Mark, 1, -, -, 53, 15, -, -, 3000, 200, 290, 2000, 75, 36, 25, 575
Rhodes, Margaret, 1, -, -, 70, 30, 100, -, 2000, 25, 75, -, -, -, -, -, 300
Dickerson, Andrew, -, -, 1, 100, 25, 86, -, 2000, 100, 200, -, -, -, -, -, 300
Davis, Mark, 1, -, -, -, -, 20, -, 200, -, -, -, -, -, -, -
Shockley, Stephen, 1, -, -, 8, -, -, -, 80, 25, 100, -, 20, 75, 25, 126
Small, Abel S., 1, -, -, 400, -, 300, -, 20000, 2000, 3000, 100, 1000, 200, 1000, 5000
Stevens, George L., 1, -, -, 25, 50, 25, -, 2000, 100, 100, -, 50, -, 75, 400
Cooper, Hamilton, -, -, 1, 46, 6, 10, -, 1500, 75, 150, -, -, -, -, 200
Transul, William, 1, -, -, 96, 40, -, -, 3000, 50, 100, -, 50, -, 75, 700
Collins, Winlock, 1, -, -, 18, -, 36, -, 500, -, 100, -, 20, -, 60, 200
Hill, William, 1, -, -, 60, -, 10, -, 800, 50, 100, -, 25, -, -, 400
Donovan, Burton, -, -, 1, 150, -, 10, -, 5000, 100, 300, -, 150, 100, -, 150
Lodge, Frank, -, -, 1, 25, -, 28, -, 1700, -, 200, -, 40, -, -, 100
Swain, George, -, -, 1, 150, 20, 150, -, 5000, 100, 400, -, 130, -, 100, 400
Carlisle, Manlove, 1, -, -, 100, 15, 100, -, 3000, 150, 300, -, 75, -, 100, 17
Hall, Silas, 1, -, -, 3, 17, -, -, 200, 25, -, -, 20, -, -, 150
Causey, John W., 1, -, -, 150, 50, 10, -, 15000, 500, 500, 50, 400, 400, 500, 1600
Causey, John W., 1, -, -, 70, 20, 10, -, 7000, 150, 200, -, 150, 100, 300, 1600
Vauls, William, 1, -, -, 13, -, -, -, 1300, -, 60, 30, 25, -, 30, 160
Causey, William F., 1, -, -, 40, -, -, -, 20000, -, -, 150, 150, -, -, 600
Davis, Robert H., 1, -, -, 12, -, -, -, 1200, 100, 300, -, 75, -, 75, 250

Gray, William, 1, -, -, 5, -, -, -, 500, 100, 100, -, 40, -, 40, 90
Deputy, James, 1, -, -, 60, 10, 130, -, 2000, 300, 400, -, 120, -, -, 600
Deputy, Solomon, -, -, 1, 70, 3, 130, -, 2000, 30, 190, 500, 20, -, -, 150
Hammond, Samuel, 1, -, -, 95, 6, 50, -, 4000, 150, 300, 50, 100, -, 75, 800
Fleetwood, Nancey, -, 1, -, 35, -, 100, -, 500, -, 50, -, -, -, -, -
Deputy, Henry, 1, -, -, 100, 6, 175, -, 3000, 300, 400, -, -, -, -, 900
Deputy, William, 1, -, -, 70, -, 35, -, 1500, 200, 300, -, 20, 75, -, 350
Hemmons, Joshua, 1, -, -, 20, -, 22, -, 400, 100, 125, -, 20, -, -, 200
Hemmons, William, 1, -, -, 15, -, 25, -, 300, 75, 75, -, -, -, -, 100
Hemmons, James, 1, -, -, 21, -, 21, -, 300, 50, 75, -, -, -, -, 125
Vauls, Sallie, 1, -, -, 15, -, 75, -, 300, 25, 20, -, -, -, -, 100
Wharton, Jams, 1,-, -, 20, -, 5, -, 200, 50, 50, -, 20, -, -, 150
Medhurst, Edmond, 1, -, -, 70, 9, 44, -, 2000, 100, 300, 25, 60, -, 25, 550
Russel, Charles, 1, -, -, 30, 2, 15, -, 1500, 100, 250, -, 6, -, 25, 400
Betts, Emma, 1, -, -, 50, -, 10, -, 2000, 150, 100, 100, -, 50, -, 300
Sockrider, William, -, 1, -, 50, 10, 25, -, 600, 50, 50, 25, -, -, -, 130
Ingraham, Manship, -, 1, -, 15, 3, 20, -, 400, 50, 50, 25, -, -, -, 130
Horton, Seymour, 1, -, -, 45, 7, -, 5, 400, -, -, -, -, -, -, 100
Williams, Whittington, 1, -, -, 100, 10, 100, -, 2000, 150, 200, 200,70, -, -, 600
Welsh, Nathaniel, -, 1, -, 100, -, 165, -, 2000, 150, 350, -, 70, -, -, 300
Hudson, Houston, 1, -, -, 75, -, 100, -, 2000, 250, 400, 75, 200, 100, 75, 400
Betts, Edward, 1, -, -, 80, 10, 35, -, 1500, 200, 400, 200, 75, -, -, 300
Stewart, Henry, 1, -, -, 94, 10, 52, 6, 2000, 150, 225, -, 20, 60, 30, 400
Hudson, Samuel, 1, -, -, 45, -, 15, -, 700, 175, 125, 100, 60, 30, -, 275
Webb, Joshua, 1, -, -, 130, 20, 40, -, 3000, 300, 500, 600, 80, -, 30, 1200
Walls, John, 1, -, -, 70, 12, 73, -, 2500, -, -, -, 20, -, -, 275
Griffith, Cyrus, -, 1, -, 30, -, 100, 12, 600, 25, 25, -, -, -, -, 30
Tatman, John, 1, -, -, 5, -, 95, -, 200, -, -, -, -, -, -, 25
Bailey, William C., 1, -, -, 40, -, 10, -, 400, 25, 100, -, 20, -, -, 125
Bailey, Zachariah T., -, 1, -, 50, 6, 10, -, 300, 200, 75, -, 20, -, -, 200
Truitt, William E., 1, -, -, 7, 7, 80, -, 600, -, -, -, -, -, 50, 300
Johnson, Catharine J., 1, -, -, 40, 2, 12, -, 500, -, 200, 25, 20, -, -, 175
Johnson, John H., 1, -, -, 60, 10, -, -, 1500, 250, 200, -, 60, -, 25, 100
Coverdale, John, 1, -, -, 30, 4, 39, -, 700, 75, 100, 25, 20, -, -, 200
Smith, William, 1, -, -, 40, -, 15, -, 800, 100, 120, -, -, -, -, 200
Clendaniel, Samuel, 1, -, -, 200, 10, 300, -, 3000, 400, 1000, 30, 60, 300, -, 1650
Clendaniel, Samuel E., 1, -, -, 50, 2, 100, -, 800, 100, 100, -, 20, -, -, 275
Rion, Thoms, -, 1, -, 120, 2, 120, -, 2000, 100, 200, -, -, -, -, 350
Donovan, James, -, 1, -, 100, 2, 30, -, 1500, 100, 130, -, 30, 75, -, 625
Donovan, Russel, -, 1, -, 30, -, 50, -, 1000, 30, 100, -, 25, -, -, 300
Hellems, Alexander, 1, -, -, 50, -, 119, -, 1000, 100, 200, -, 30, 75, -, 850
Banning, Rinkey, 1, -, -, 30, 2, 50, -, 500, 75, 25, -, 40, -, -, 120
Wharton, William, -, 1, -, 35, 5, 95, -, 600, 100, 130, 300, -, -, -, 225
Williams, Morgan, -, 1, -, 25, -, 35, -, 300, 30, -, -, -, -, -, 100
Macklin, William B., -, -, 7, 75, 25, 50, -, 2000, 100, 100, -, 50, -, -, 300
Waston (Watson), James, -, -, 1, 25, 2, 60, -, 800, 40, 125, -, 25, -, -, 250

Reed, Elias, 1, -, -, 70, 8, 80, -, 2000, 75, 150, -, 50, -, -, 300
Fay, Cyrus, 1, -, -, 40, 12, 30, -, 1000, 50, 50, 30, 30, -, -, 850
Macklin, Henry D., -, -, 1, 38, -, 3, -, 800, 200, 600, -, -, -, -, 400
Short, William, 1, -, -, 50, 15, 30, 100, 2000, 75, 300, -, 60, -, 25, 450
Jester, James, 1, -, -, 8, -, 3, -, 200, -, 90, -, 10, 10, 20, 50
Veasey, William 1, -, -, 3, -, 3, -, 150, -, 50, -, 3, -, 10, 75
Macklin, Salathiel -., -, -, 1, 30, 40, 60, -, 1000, 100, 300, -, 30, -, -, 400
Warren, David, 1, -, -, 30, -, 30, -, 500, 80, 75, -, 30, -, -, 200
Short, John W., -, 1, -, 80, 12, 30, -, 2000, 75, 300, -, 20, -, -, 500
Derickson, Nathaniel, -, -, 1, 75, 10, 10, 30, 2000, 150, 300, 30, 100, 130, -, 300
Abbott, Nemiah, -, -, 1, 90, -, 25, -, 2500, 100, 75, 80, 45, -, 25, 300
Clendaniel, Samuel 1, -, -, 60, -, 150, 90, 3000, 100, 400, 100, 30, 75, 100, 300
Clendaniel, Jehu, 1, -, -, 50, 2, 75, -, 2000, 100, 600, 30, 100, 230, 30, 650
Warren, Isaac, 1, -, -, 35, 16, 6, 41, 100, 80, 100, 40, 16, -, 15, 375
Milman, Jesse, 1, -, -, 35, -, 20, 20, 500, 100, 75, 30, 30, -, -, 150
Prettyman, Ralph, 1, -, -, 70, -, 25, 18, 800, 75, 140, 60, 54, -, -, 150
Milman, Michael, 1, -, -, 30, -, 33, -, 500, 100, 150, -, 10, -, -, 200
Milman, James, -, -, 1, 25, -, 25, 10, 300, 70, 60, -, -, -, -, 150
Warren, David M., 1, -, -, 60, -, 40, -, 1000, 100, 300, 50, 30, -, -, 300
Warren, Stephen, 1, -, -, 50, -, 25, 60, 300, 60, 150, 30, -, -, -, 200
Shepherd, James, -, -, 1, 30, -, -, 10, 300, 50, 100, 20, -, -, -, 100
Murphy, James, -, -, 1, 50, -, 50, -, 500, 75, 300, 25, -, -, -, 150
Shew, John, 1, -, -, 75, 28, 30, -, 1500, 60, 125, 50, 50, 100, 30, 400
Baily, John S., 1, -, -, 40, 10, 16, -, 2500, 100, 150, 250, 40, -, 30, 425
Lynch, Robert, -, -, 1, 75, -, 20, -, 1000, 100, 300, 20, 50, 25, 25, 400
Bebee, Elmore, -, 1, -, 30, -, 30, -, 5000, 100, 200, 30, 40, -, -, 200
Gilman, Robert, 1, -, -, 30, 35, -, -, 1000, 200, 200, 20, 110, 250, 104, 450
Paige, George, 1, -, -, 20, 30, 15, -, -, 75, 75, 200, 10, 50, 125, 250
Clendaniel, John H., 1, -, -, 32, 10, 15, -, 2000, 75, 375, 50, 20, -, 2, 225
Clendaniel, John, 1, -, -, 86, 10, 100, -, 2500, 300, 1000, 75, 150, 216, 25, 1125
Reed, Greenbury, -, -, 1, 33, -, 50, 12, 1000, 75, 53, 500, 34, 100, -, 200
Dickerson, Willard, 1, -, -, 40, 10, 100, -, 2000, 100, 150, 60, 75, 100, 50, 250
Arnold, Erasmus, 1, -, -, 20, 10, 28, -, 1000, 40, 150, 100, -, -, 1, 100
Arnold, Gustavius, -, -, 1, 15, 3, 75, -, 1000, -, -, 25, -, -, 25, 130
Clendaniel, John, 1, -, -, 22, -, 60, 40, 800, 50, 200, 50, -, -, 5, 150
Bannon, Mark, -, -, 1, 30, -, -, -, 500, -, -, 25, -, -, -, 120
Walls, Henry, -, -, 1, 10, -, 100, -, 500, 50, 50, 20, -, -, -, 100
Morris, Bivins, 1, -, -, 12, -, 8, 15, 500, 25, 135, 40, -, 1, -, 125
Williams, Richard, 1,-, -, 40, 6, 100, 4, 400, 30, -, -, -, -, -, 225
Lofland, Littleton, 1, -, -, 40, 6, 75, -, 500, 125, 150, -, 20, -, 25, 240
Morris, Josiah, -, -, 1, 30, -, 20, 10, 300, 50, 100, -, -, -, -, 150
Warren, James, 1, -, -, 75, 2, 37, 5, 500, 125, 125, 300, 20, -, -, 320
Milman, John W., -, -, 1, 20, -, -, 20, 200, -, -, -, -, -, -, -
Warren, Henry, 1, -, -, 100, 10, 90, -, 2000, 200, 200, -, 40, -, 25, 4000
Milman, Jonathan, 1, -, -, 20, -, -, 10, 200, 100, 75, -, -, -, -, 150
Milman, John, 1, -, -, 25, -, 20, 18, 500, -, -, -, 40, -, -, 125

Warren, Isaac F., 1, -, -, 160, 10, 150, 10, 4700, 100, 800, 50, 175, 280, 40, 760
Lofland, Alfred, -, -, 1, 25, -, 10, 10, 700, 10, 150, -, 40, -, -, 300
Stayton, Nicholas, -, -, 1 82, 2, 60, -, 2000, 150, 400, 90, 154, 150, 50, 820
Simpson, Isaac, 1, -, -, 5, 2, 10, -, 2500, 50, 100, -, 150, -, 12, 1010
Griffith, Charles, -, -, 1, 40, 2, 40, 120, 1000, 100, 150, -, 20, -, -, 180
Rion, James, -, -, 1, 5, 4, 50, 12, 1000, 100, 150, -, -, -, -, 225
Gleeson, Charles H., 1, -, -, 75, 12, 75, 25, 1500, 75, 200, -, 10, -, -, 300
Samons, George, -, -, 1, 15, -, 3, 12, 200, -, -, -, -, -, -, 51
Hammond, Mathias, -, -, 1, 15, -, 25, -, 1000, 50, 200, -, 20, -, -, 250
Sayton (Saxton), Andrew, 1, -, -, 15, 4, 40, -, 800, 100, 150, -, 20, -, -, 240
Johnson, Benjamin, -, -, 1, 20, -, 20, 65, 700, 150, -, 100, 70, -, -, 100
Clifton, Mary E., 1, -, -, 60, 12, 10, 10, 1000, 100, 200, -, 30, -, 25, 400
Welsh, George, 1, -, -, 40, 6, -, 70, 800, 50, 125, -, 40, -, -, 300
Dickerson, Charles, -, -, 1, 60, -, 30, 10, 500, 100, 175, -, -, -, -, 200
Milby, James, 1, -, -, 55, 4, 17, -, 900, 125, 125, -, 20, -, -, 278
Shew, Putman, 1, -, -, 30, 20, 31, -, 1500, 100, 150, 20, 25, -, -, 375
 Lion, Davis, 1, -, -, 15, 10, 17, -, 700, -, -, -, -, -, -, 350
Baker, Elvas B., 1, -, -, 44, -, 60, -, 1200, 100, 50, 300, 24, -, -, 200
Morgan, Bradsbery, -, -, 1, 100, 10, 40, -, 1500, 150, 350, -, -, 100, 25, 500
Griffith, James, -, -, 1, 50, -, 50, 35, 1000, 50, 25, -, 100, -, -, -, 150
Loffland, Parker, 1, -, -, 50, 5, 30, -, 1500, 150, 100, -, 20, -, 50, 250
Lofland, James, -, -, 1, 50, -, 30, -, 1000, 50, 100, 315, -, -, 200
Calhoun, William, -, -, 1, 60, 9, 60, 10, 700, 100, 100, 100, 20, -, -, 300
Biediman, Wileby, -, -, 1, 75, 12, 25, 30, 700, 150, 175, -, -, -, -, 300
Clendaniel, George, 1, -, -, 160, 20, 140, -, 4000, 400, 650, 100, 140, 175, 25, 1200
Paswaters, B__y, -, -, 1, 70, 10, 60, 10, 1500, 110, 125, -, 40, 50, -, 325
Paswaters, Nancy, 1, -, -, 50, 2, 100, 20, 1000, 100, 150, 20, 40, -, -, 240
Biediman, William, 1, -, -, 40, 7, 60, -, 1500, 250, 400, 500, 50, -, -, 450
Burris, William, -, -, 1, 50, -, 40, 10, 800, 100, 100, -, -, -, -, 175
Warren, Asbury, 1, -, -, 37, -, -, -, 500, 100, 75, -, -, -, -, 150
Shockley, William, 1, -, -, 60, 3, 60, 10, 800, 200, 100, 20, 20, -, -, 150
Shockley, James, 1, -, -, 30, 20, 620, 60, 800, 175, 150, -, 40, -, 25, 200
Milman, Rush, 1, -, -, 30, -, 39, -, 400, 50, 50, -, -, -, -, 100
Truitt, Andrew, -, -, 1, 30, 15, 9, -, 500, 150, 200, -, 10, -, -, 275
Russel, George, 1, -, -, -, 3, -, 125, 500, -, -, -, -, -, -, -
Watson, David, 1, -, -, 45, 3, -, 69, -, 600, 50, 150, -, -, -, 75, 170
Watson, Heirs, 1, -, -, 30, -, 70, 10, 500, -, -, -, -, -, -, 100
Hudson, Henry, -, -, 1, 30, 2, 90, 25, 700, 50, 50, -, -, -, -, 110
Adhern, Elzey, -, -, 1, 10, 40, 10, 300, -, -, -, -, -, -, 100
Truitt, Catharine, 1, -, -, -, 10, 50, 25, 300, -, -, -, -, -, -, 200
Warren, Frank, 1, -, -, 25, 3, 10, 10, 300, 75, 150, 20, -, -, -, 150
Jester, William, -, -, 1, 10, -, -, 50, 500, 50, 100, 25, -, -, -, 225
Truitt, Joshua, -, -, 1, 70, -, -, -, 2000, 250, 350, -, 40, -, -, 400
Rion, William, -, -, 1, 100, 8, 40, -, 1000, 100, 375, 1, 40, -, -, 375
Clendaniel, Avery, 1, -, -, 75, 10, 80, 20, 1500, 300, 500, -, 20, -, 40, 575

Clendaniel, James, -, -, 1, 25, 2, 23, -, 1500, 200, 400, -, 70, 70, 25, 430
Conaway, Albert, 1, -, -, 20, -, 20, -, 500, 100, 200, -, -, -, -, 200
Milman, William, -, -, 1, 40, -, 100, -, 500, 100, 100, 1, -, 50, -, -, 175
Macklin, Samuel, -, -, 1, 40, -, 100, -, 1500, 150, 50, 75, -, -, -, 300
Shockley, Elias, -, -, 1, 40, 20, 4, 10, 1500, 125, 175, -, -, -, -, 400
Wroten, James, 1, -, -, 25, 10, 3, -, 700, 125, 100, -, 50, 75, -, 425
Webb, John, 1, -, -, 100, 20, 40, -, 2000, 200, 400, -, 100, 25, 25, 925
Masten, Clement, 1, -, -, 80, 8, 7, 15, 1500, 150, 350, -, 80, -, -, 600
Readie, John, 1, -, -, 75, 6, 50, 50, 1000, 150, 175, -, -, -, -, 450
Harden, Ezra, 1, -, -, 30, -, -, 40, 1000, 100, 200, 25, 20, 50, 25, 150
Houston, Clement, 1, -, -, 100, 10, 112, -, 3000, 250, 240, 100, 150, -, 25, 630
Short, Henry, -, -, 1, 100, 15, 100, -, 3000, 200, 125, 80, 80, 50, -, 600
Daniel, Brinkley, 1, -, -, 75, -, 23, -, 2500, 200, 200, 1, 60, 150, 10, 300
Walls, Asa B., 1, -, -, 100, -, 2, 100, 12, 2000, 30, 250, -, 30, 125, 50, 700
Hill, John, 1, -, -, 50, -, 12, 10, 1000, 100, 200, -, 20, 25, -, 200
Betts, Solomon, 1, -, -, 65, -, 16, -, 2500, 300, 350, 60, 60, -, 10, 500
Miller, Nemiah, -, -, 1, 50, -, 50, -, 500, 50, 50, -, -, -, -, -, 150
Carpenter, Burton, -, 1, -, 20, 6, 15, 15, 1000, 50, 100, 15, 10, -, 20, 250
Gates & Stewart, 1, -, -, 15, 10, 40, 10, 800, -, -, -, -, -, -, 250
Dubois, John, 1, -, -, 30, 10, -, -, 1000, 100, 100, 100, 30, -, -, 200
Sackett, George, 1, -, -, 85, 55, 40, -, 4000, 200, 300, 100, 300, -, 1, 800, 3000
Truitt, Catharine, 1, -, -, -, -, 50, 30, 800, -, -, -, -, -, -, -
Lynch, George M., -, -, 1, 40, 8, 10, -, 1000, 50, 75, 50, 75, -, 25, 350
Fisher, George, 1, -, -, 50, 4, 20, -, 800, 100, 150, 25, 20, -, -, 300
Workman, Philip, -, -, 1, 100, 16, 13, -, 1200, 150, 300, -, 40, -, 25, 400
Mason, Joseph, -, -, 1, 100, -, 40, -, 3000, 200, 400, -, 60, -, 25, 500
Holstein, James, -, -, 1, 75, -, 100, 25, 3000, 50, 100, -, -, -, -, 200

FIRST NAME FIRST FOLLOWS
Harry Ingram, -, -, 1, 60, 65, 795, -, 10000, 100, 300, 25, -, 50, -, 800
Isaac Burton, 1, -, -, 15, -, 20, 5, 500, 15, 250, 25, 7, 150, -, 400
Jonathan Dickerson, -, -, 1, 65, -, 300, -, 2500, 50, 250, 75, -, 10, -, 250
Wallace B. Thompson, 1, -, -, 30, -, 200, -, 3300, 150, 200, 40, 36, 40, -, 300
Joshua S. Morris, 1, -, -, 100, -, 100, -, 5000, 150, 400, 50, 25, 25, -, 800
Hirah M. Johnson, -, -, 1, 60, -, 70, 5, 600, 75, 100, 10, -, 10, -, 250
Robert G. Hudson, 1, -, -, 40, -, 50, -, 1200, 100, 150, 10, 16, 15, -, 200
William Wilson, 1, -, -, 40, -, 60, -, 1200, 100, 200, 40, 25, 15, -, 300
Joshua S. Fisher, -, -, 1, 40, -, 70, 10, 1000, 20, -, 20, 20, 10, -, 200
William C. Wilson, -, -, 1, 90, -, 100, -, 1000, 20, 150, 20, 50, -, -, 300
Adam M. Ennis, 1, -, -, 40, -, 10, -, 700, 50, 350, 20, -, 15, -, 300
John Pride, -, -, 1, 100, -, 50, -, 1000, 40, 75, 40, -, 20, -, 200
Harry B. Truitt, -, -, 1, 30, -, 10, -, 800, 30, 100, 10, -, 10, -, 150
Ramsford S. Johnston, 1, -, -, 155, -, 157, -, 3500, 250, 500, 50, 350, 100, 75, 300
Gardner W. Marvill, 1, -, -, 40, -, 90, -, 500, 25, 225, 10, -, -, -, 250
Jacob Prettyman, -, -, 1, 50, -, 100, -, 1000, 30, 150, 10, 25, -, -, 250
George W. Clark, -, -, 1, 60, -, 140, 10, 700, 25, 150, 30, -, 5, -, 250

Robert W. Joseph, -, -, 1, 80, -, 75, -, 1000, 15, 150, 10, 12, 10, -, 250
Martin B. Russel, -, -, 1, 80, -, 40, -, 800, 15, 150, 25, -, 20, -, 260
Arteamus W. Betts, -, -, 1, 60, -, 120, -, 600, 10, 60, 10, -, -, -, 200
Henry W. Simpson, -, -, 1, 40, -, 43, -, 700, 10, 200, 25, -, 25, -, 500
Philip R. Marvel, 1, -, -, 150, -, 150, -, 3000, 25, 150, 10, 12, 40, -, 250
George W. Marvel, 1, -, -, 60, -, 5, -, 600, 10, 30, 5, -, 5, -, 200
Louisa W. Johnson, -, -, 1, 100, -, 600, -, 3000, 30, 200, 10, -, 15, -, 530
George A. Johnson, -, -, 1, 75, -, 50, -, 1000, 25, 100, 10, -, -, -, 400
William Mariner, -, -, 1, 20, -, -, -, 300, 10, 100, 20, -, -, -, 200
Aaron B. Marvel, 1, -, -, 100, -, 200, -, 5000, 100, 300, 50, 90, 90, 25, 500
Joseph Marvel, 1, -, -, 100, -, 100, 20, 1600, 75, 300, 10, 30, 60, 10, 500
Isaac C. Moore, 1, -, -, 50, -, 50, -, 1000, 50, 150, 5, -, 54, 10, 400
Robert J. Davidson, 1, -, -, 45, 8, 105, -, 1800, 20, 250, 10, 3, 60, 20, 300
William Jefferson, -, -, 1, 30, 20, 25, -, 800, 35, 25, 10, -, 10, -, 400
Hirahm S. Smith, -, -, 1, 20, 20, 10, -, 800, 20, 75, 5, -, 10, -, 350
Harrison Rogers, 1, -, -, 35, -, 30, -, 600, 75, 150, 10, 30, -, -, 400
Letty P. Rogers, 1, -, -, 60, -, 20, -, 600, 10, 150, 10, 5, 50, -, 350
John P. Johnson, -, -, 1, 60, -, 60, -, 1500, 5, 75, 5, -, -, -, 300
Joseph Prettyman, 1, -, -, 60, -, 60, -, 1000, 15, 300, 10, -, 25, -, 300
George W. Rogers, 1, -, -, 50, -, 50, -, 700, 50, 300, 10, 10, -, -, 400
John T. Rogers, 1, -, -, 70, 5, -, -, 600, 25, 250, 20, 12, -, -, 350
Daniel Rogers, 1, -, -, 35, -, 5, -, 500, 15, 200, 10, 15, -, -, 300
Edward E. Lofland, -, 1, -, 20, -, 5, -, 350, 10, 150, 5, -, -, -, 150
Andrew M. Johnson, -, -, 1, 40, -, 47, -, 1500, 60, 125, 10, 100, -, -, 300
John P. Ennis, 1, -, -, 40, -, 10, -, 500, 60, 200, 10, -, -, -, 300
Nuter Marvel, 1, -, -, 65, -, 15, -, 2000, 100, 200, 20, 30, 60, -, 300
George Stoeckel, 1, -, -, 35, -, 80, -, 1000, 100, 200, 10, 40, 60, 50, 800
Charles J. Stoeckel, 1, -, -, 50, -, 5, -, 500, 50, 100, 10, 10, -, -, 400
George P. Stoeckel, 1, -, -, 50, -, 5, -, 500, 50, 100, 10, 10, -, -, 400
Benjamin Roach, -, -, 1, 50, -, 20, -, 800, 150, 200, 10, -, 20, -, 500
William M. Joseph, -, -, 1, 50, -, 200, -, 1500, 20, 200, 10, -, 7, -, 300
Thomas W. Johnson, 1, -, -, 75, 25, -, 1000, 25, 100, 10, 16, 5, -, 400
Joseph Burris, -, -, 1, 100, -, 25, -, 1500, 25, 200, 10, 20, -, -, 350
Asa Johnson, 1, -, -, 50, -, 60, -, 1500, 50, 250, 20, -, -, -, 400
Levi Johnson, 1, -, -, 60, -, 60, -, 600, 30, 200, 10, -, 10, -, 150
Joel W. Rogers, -, -, 1, 70, -, 20, -, 600, 25, 175, 10, -, 5, -, 400
John L. Short, 1, -, -, 100, -, 50, -, 3000, 150, 500, 25, 100, 150, 50, 110
Ananias Mears, -, -, 1, 50, -, 25, -, 800, 30, 150, 10, 10, 8, -, 500
William Conaway, -, -, 1, 40, -, 30, -, 800, 50, 100, 20, -, 10, -, 300
Daniel Short, 1,-, -, 140,-, 20, -, 1600, 150, 200, 20, 50, 75, -, 700
George Cannon, -, -, 1, 40, -, 10, -, 500, 50, 150, 10, -, -, -, 400
James T. West, -, -, 1, 50, -, 29, -, 790, 20, 75, 10, -, -, -, 200
Minos Rogers, 1, -, -, 30, -, 25, -, 500, 50, 250, 25, 50, 15, -, 500
Wilmer M. Rogers, -, -, 1, 30, -, 5, -, 400, 100, 150, 10, -, -, -, 400
William M. Williams, -, -, 1, 50, -, 50, -, 1000, 100, 200, 40, 15, -, -, 500
John C. Prettyman, -, -, 1, 40, 30, -, 600, 50, 150, 20, -, 12, -, 500

Lenard Short, -, -, 1, 60, -, 30, -, 1000, 100, 200, 20, -, -, -, 600
Cornealus Prettyman, 1, -, -, 70, -, 10, -, 2000, 75, 300, 25, 50, -, -, 600
Branson Dolby (Dalby), -1, -, -, 70, -, 20, -, 1000, 40, 200, 20, 5, -, -, 300
Robert H. Mumford, -, -, 1, 40, -, 10, -, 800, 25, 200, 25, -, -, -, 450
Curtis Rogers, 1, -, -, 25, -, -, -, 300, 30, 150, 10, -, 10, -, 400
Theoplin S. Rogers, 1, -, -, 50, -, 10, -, 500, 50, 20, 25, 20, 5, -, 300
Elizabeth Rogers, 1, -, -, 25, -, 2, -, 450, 10, 25, 5, -, -, -, 200
George N. Rogers, 1, -, -, 60, -, 15, -, 450, 10, 75, 5, 5, -, -, 200
Joseph Williams, -, -, 1, 50, -, 30, -, 500, 30, 100, 10, -, -, -, 300
Curtis W. Rogers, 1, -, -, 30, -, 10, -, 600, 40, 200, 5, -, -, -, 400
John W. Rogers, 1, -, -, 40, -, 70, -, 1000, 40, 250, 20, 9, 25, -, 800
Margaret A. Short, 1, -, -, 75, -, 60, -, 1000, 50, 300, 40, -, 25, -, 600
Levin Stean Jr., -, -, 1, 20, -, 30, -, 500, 10, 100, 5, -, 35, -, 150
Elijah Adkins of E, 1, -, -, 15, -, 49, -, 1000, 50, 125, 10, 26, 30, -, 400
Edward U. Dill, 1, -, -, 50, -, 200, -, 1300, 50, 150, 100, -, -, -, 400
William T. Burton, 1, -, -, 55, -, 5, -, 700, 10, 100, 20, 850, -, 500
James H. Evans, 1, -, -, 60, -, 65, -, 700, 10, 75, 50, 5, 25, -, 300
Levin Stean Sr., -, -, 1, 60, -, 200, -, 2000, 150, 700, 75, 35, 200, -, 600
Daniel S. Short, -, -, 1, 60, -, 10, - 1000, 20, 100, 10, 14, -, -, 350
George Butler, -, -, 1, 60, -, 40, -, 150, 25, 100, 10, -, -, -, 400
James E. Rogers, -, -, 1, 75, -, 75, -, 804, 30, 75, 10, -, 10, -, 400
Curtis W. Scott, -, -, 1, 55, -, 8, -, 1000, 20, 100, 5, -, -, -, 400
William O. Short, 1, -, -, 200, 100, -, -, 6000, 300, 1100, 100, 100, 200, -, 1200
William Monroe, 1, -, -, 80, -, 60, -, 2000, 200, 400, 25, 24, 50, -, 500
James H. Toomey, -, -, 1, 50, -, 25, -, 800, 25, 100, 10, -, -, -, 400
Nancy Short, 1, -, -, 60, -, 60, -, 1000, 15, 150, 10, -, 10, -, 500
Benjamin J. Williams, -, -, 1, 60, -, 20, -, 1200, 14, 75, 10, 22, -, -, 500
John W. Short, 1, -, -, 60, -, 80, -, 3000, 200, 1100, 50, 20, 100, -, 600
Thomas W. Short, 1, -, -, 150, -, 40, -, 2000, 200, 1000, 25, 50, 25, -, 1000
Thomas W. Stean, -, -, 1, 90, -, 50, -, 2000, 25, 200, 20, 10, -, -, 800
Eli S. Short, 1, -, -, 50, -, 30, -, 1500, 150, 250, 10, 30, 30, -, 600
George W. Bryan, -, -, 1, 90, -, 10, -, 1000, 50, 200, 10, -, -, -, 600
Joshua Bryan, -, -, 1, 100, -, 300, -, 3000, 100, 1200, 20, 50, 10, -, 800
William D. Records, 1, -, -, 100, -, 50, -, 3000, 100, 900, 30, 20, 100, -, 800
Walter J. Melson, 1, -, -, 40, -, 20, -, 900, 75, 250, 10, 12, -, -, 300
Wilson U. Black, 1, -, -, 20, -, 10, -, 350, 20, 125, 20, -, 15, -, 50
Charles R. Messick, 1, -, -, 30, -, 20, -, -, 50, 20, 100, 15, -, -, -, 200
Jason M. Adkins, 1, -, -, 30, -, 20, -, 500, 20, 100, 10, 10, 5, -, 150
Isaac T. Adkins, 1, -, -, 25, -, 6, -, 400, 6, 10, 20, -, 5, -, 75
Clayton M. Adkins, 1, -, -, 20, -, 22, -, 400, 50, 100, 10, 10, 5, -, 150
Joseph Parker, 1, -, -, 45, -, 48, -, 1000, 50, 600, 20, 80, 100, -, 600
Joshua Phillips, 1, -, -, 100, -, 75, -, 2500, 100, 700, 50, 50, 150, -, 1200
Peter P. Parker, 1, -, -, 30, -, 30, -, 500, 25, 75, 10, 12, 25, -, 190
John W. Phillips, 1, -, -, 25, -, 10, -, 4000, 1000, 1000, 25, 60, 200, -, 1200
William A. Messick, 1, -, -, 30, -, 12, -, 1500, 175, 100, -, 25, 50, -, 150
Isaac E. Kinekin, 1, -, -, 20, -, 32, -, 300, 10, 175, 10, -, 20, -, 150

George W. Parker, -, 1, -, 40, -, 100, -, 800, 8, 150, -, -, 50, -, 100
Erasmus M. Massey, 1, -, -, 75, -, 25, -, 1000, 100, 250, 10, -, -, -, 400
Burton Messick, 1, -, -, 50, -, 30, -, 700, 50, 100, 10, 10, 6, -, 300
Nathan C. Messick, 1, -, -, 60, -, 50, -, 1000, 200, 300, 15, 10, 15, -, 500
Henry W. Short, 1, -, -, 60, -, 100, -, 1000, 30, 75, 25, -, 25, -, 400
Josiah J. Mathis, -, -, 1, 80, -, 100, -, 1000, 50, 200, 15, 20, 10, -, 400
George Scott, -, 1, -, 25, -, 5, -, 500, 125, 150, 5, 10, 5, -, 50
Robinson Webb, -, -, 1, 50, -, 100, -, 1000, 10, 125, 10, -, 20, -, 30
Philip Short of S, 1, -, -, 100, -, 100, -, 2000, 100, 400, 20, 18, 100, -, 300
John C. Phillips, 1, -, -, 157, -, 50, -, 3000, 75, 500, -, 35, 100, -, 900
Joseph Phillips, 1, -, -, 200, -, 200, -, 4000, 150, 400, 50, -, 100, -, 1200
Eliza E. Phillips, 1, -, -, 35, -, 4, -, 500, 30, 150, 5, 11, -, -, 300
Burton Phillips, 1, -, -, 100, -, 75, -, 1500, 75, 250, 10, -, 40, -, 500
Timmon Hibithia, -, -, 1, 50, -, 25, -, 500, 25, 100, 10, -, -, -, 300
John M. Rogers, -, -, 1, 30, -, 10, -, 500, 10, 50, 10, -, 5, -, 300
Miles W. Cannon, -, -, 1, 50, -, 40, -, 600, 30, 150, 8, -, -, -, 300
Jacob W. Bryan, 1, -, -, 75, -, 35, -, 1000, 30, 100, 20, -, 50, -, 225
James Rogers, -, -, 1, 50, -, 25, -, 500, 50, 125, 10, -, -, -, 250
Fisher W. Toomey, -, -, 1, 100, -, 100, -, 1000, 50, 200, 20, -, -, -, 300
Robert J. Conoway, -, -, 1, 50, -, 35, -, 800, 50, 160, 10, -, 15, -, 150
Alfred R. Layton, 1, -, -, 75, -, 50, -, 1300, 50, 100, 30, 48, 15, -, 350
James H. Layton, 1, -, -, 60, -, 40, -, 800, 25, 125, 20, -, -, -, 150
David C. Truitt, -, -, -, 1, 50, -, 150, -, 2000, 15, 150, 10, -, -, -, 200
Theodore Walls, -, -, 1, 75, -, 125, -, 2000, 25, 75, 5, -, 18, -, 300
John M. Hopkins, -, -, 1, 90, -, 10, -, 1800, 30, 175, 10, -, -, -, 300
Clayton M. Adkins, 1, -, -, 30, -, 6, -, 500, 50, 150, 10, -, 10, -, 250
William Parker, 1, -, -, 20, -, 8, -, 400, 10, 75, 5, -, 10, -, 150
Beverly W. Adkins, 1, -, -, 25, -, 12, 50, 700, 30, 75, 10, -, 10, -, 250
Samuel Coffin, -, -, 1, 30, -, 20, -, 1000, 100, 425, 40, -, 40, -, 400
William S. Rogers, 1, -, -, 20, -, 70, -, 1200, 50, 300, -, -, -, -, 300
William S. West, -, -, 1, 125, -, 75, -, 2500, 50, 400, 10, -, 40, -, 550
James S. Parsons, -, -, 1, 100, -, 200, -, 1200, 40, 300, 10, -, 25, -, 400
Miles M. Johnson, -, -, 1, 80, -, 220, -, 1200, 30, 75, 10, -, -, -, 300
Benson W. Thompson, -, -, 1, 100, -, 100, -, 800, 10, 100, 10, -, 5, -, 250
Isaac T. Bowden, -, -, 1, 25, -, 30, -, 400, 25, 60, -, -, 10, -, 200
Handy J. Littleton, -, -, 1, 60, -, 65, -, 2000, 50, 200, 10, -, 10, -, 500
Elijah N. Carey, 1, -, -, 80, -, 100, -, 2000, 100, 300, 30, -, 50, -, 500
Shelley Shockley, 1, -, -, 40, -, 50, -, 1000, 100, 125, 10, -, -, -, 300
John C. West, -, -, 1, 60, -, 100, -, 800, 15, 100, -, -, -, -, 300
Thomas R. Shockley, 1, -, -, 100, -, 30, -, 100, 100, 400, 10, -, 20, -, 400
Joshua A. Parkes, -, -, 1,30, -, -, -, 500, 125, 70, 5, -, -, -, 200
William M. Mears, -, -, 1, 30, -, -, -, 600, 30, 150, -, -, -, -, 300
Nathaniel Warrington, -, -, 1, 100, -, 30, -, 3000, 40, 500, 15, -, 50, -, 1000
Eben Taylor, -, -, 1, 75, -, 10, -, 1200, 40, 400, 15, 20, -, -, 500
William E. Ottwell, -, -, 1, 35, -, 10, -, 700, 30, 100, 10, -, -, -, 200
Hamilton B. Truitt, -, -, 1, 120, -, 75, -, 2500, 200, 400, -, 30, 250, -, 500

Ira Roberts, -, -, 1, 50, -, 10, -, 500, 30, 75, 10, -, -, -, 200
Cyrus Moore, -, -, 1, 150, -, 50, -, 2500, 225, 550, 25, 60, 30, -, 950
Caleb Carey, -, -, 1, 40, -, 30, -, 800, 30, 175, 10, -, -, -, 300
John H. Roberts, -, -, 1, 25, -, 10, -, 500, 25, 75, 10, -, 5, -, 250
William H. Short, -, -, 1, 100, -, 12, -, 300, 100, 50, 25, -, 100, -, 800
Eli P. West, -, -, 1, 75, -, 13, -, 3000, 75, 250, 15, -, 35, -, 700
William B. Fosque, -, -, 1, 100, -, 50, -, 3000, 50, 150, -, -, 40, -, 400
John H. Parsons, -, -, 1, 40, -, 65, -, 1000, 10, 60, 5, -, -, -, 100
Joseph S. Hudson, 1, -, -, 40, -, 10, -, 400, 15, 100, 40, -, -, -, 400
Benjamin Olophant, -, -, 1, 125, -, 75, -, 900, 15, 200, 20, -, -, -, 400
George T. Timmons, -, -, 1, 60, -, 15, -, 1000, 50, 300, 10, -, 25, -, 500
William Moore, 1, -, -, 60, -, 22, -, 700, 30, 60, 20, -, -, -, 300
Isaac M. Brittingham, -, -, 1, 100, -, 60, -, 3500, 150, 500, 55, -, 50, -, 1200
William H. Hitchens, -, -, 1, 50, -, 50, -, 2000, 50, 600, 20, -, 75, -, 900
James W. Timmons, 1, -, -, 111, -, 15, -, 3500, 150, 500, 25, -, -, -, 1200
Johnathan C. Timmons, -, -, 1, 9, -, -, -, 500, 6, 5, -, -, -, -, 250
Samuel A. Mitchell, -, -, 1, 60, -, 20, -, 700, 75, 150, 30, 10, -, -, 300
Samuel P. West, -, -, 1, 12, -, 5, -, 800, 10, 175, -, -, -, -, 500
Riley Hosey, -, -, 1, 50, -, 25, -, 1200, 15, 100, 10, -, -, -, 700
William Hickman, 1, -, -, 13, -, 2, -, 700, 15, 150, -, -, -, -, 400
Mary E. Parsons, -, -, 1, 100, -, 80, -, 4000, 50, 400, 10, -, 8, -, 800
Henry B. Watson, -, -, 1, 40, -, 10, -, 1000, 75, 225, 25, -, 15, -, 800
Caleb _. Brimer, -, -, 1, 100, -, 100, -, 4000, 75, 400, -, -, -, -, 1000
John L. Littleton, -, -, 1, 33, -, 10, -, 1000, 25, 125, 8, -, -, -, 400
Peter Mitchell, -, -, 1, 50, -, 30, -, 1200, 50, 250, 10, -, 10, -, 600
Isaac B. Mitchell, -, -, 1, 8, -, 3, -, -, 10, 150, -, -, -, -, 600
Joseph Murry of M, 1, -, -, 42, -, 3, -, 1500, 10, 100, 20, -, -, -, 500
Elisha T. Wilkinson, -, -, 1, 30, -, 20, -, 900, 10, 150, 10, -, -, -, 500
Robert H. Houston, 1, -, -, 40, -, 100, -, 2300, 150, 800, 200, 35, 75, -, 600
James J. Williams, 1, -, -, 100, -, 113, -, 5000, 300, 800, 25, -, 100, -, 900
Samuel P. Hitchens, -, -, 1, 75, -, 10, -, 1000, 25, 300, 10, -, -, -, 700
Joseph A. Kallock, 1, -, -, 100, -, 22, -, 3000, 100, 300, 25, -, -, 25, 900
William S. Kollock, 1, -, -, 100, -, 22, -, 3000, 100, 400, 10, -, -, 40, 900
Joshua T. Carey, -, -, 1, 30, -, -, -, 600, 15, 200, 10, -, -, 25, 600
Peter H. Carey, -, -, 1, 100, -, 30, -, 3000, 100, 300, 10, -, 45, -, 800
Milby Adkins, -, -, 1, 34, -, 2, -, 1800, 60, 300, -, -, 11, -, 60
George W. Hudson, -, -, 1, 30, -, 46, -, 2500, 60, 125, 10, -, -, -, 600
Isaac S. Hudson, -, -, 1, 100, -, 25, -, 3000, 75, 300, -, -, 20, -, 1300
Henry B. Mitchell, 1, -, -, 50, -, 27, -, 2500, 75, 300, 10, -, 25, -, 800
Ephraim W. Low, -, -, 1, 100, -, 60, -, 4000, 25, 350, -, -, 60, -, 1200
James T. Wilkinson, -, -, 1, 30, -, 100, -, 1500, 20, 150, 8, 20, 20, -, 500
Washington B. Moore, -, -, 1, 100, -, 40, -, 4000, 60, 200, -, -, 5, -, 600
Marshall Smith, -, -, 1, 30, -, -, -, 100, 40, 300, 10, -, -, -, 400
Daniel S. Mitchell, -, -, 1, 25, -, -, -, 1000, 20, 175, 10, 55, -, 600
Edward H. Littleton, -, -, 1, 40, -, 27, -, 3000, 75, 300, 40, -, -, -, 500
Lenard W. Moore, -, -, 1, 30, -, 30, -, 1500, 25, 250, -, -, -, -, 350

John W. Hitchens, -, -, 1, 25, -, -, -, 1000, 30, 100, -, -, -, -, 350
Thomas R. Toomey, -, -, 1, 20, -, -, -, 500, 10, 40, 5, -, -, -, 300
Benjamin M. Ellingsworth, -, -, 1, 12, -, -, -, 300, -, 100, -, -, -, -, 300
Woolsey Toomey, -, -, 1, 40, -, 5, -, 800, 20, 250, 5, -, -, -, 300
John M. Hitchens, -, -, 1, 100, -, -, -, 2000, 25, 225, -, -, 12, -, 550
Minas W. Mitchell, -, -, 1, 100, -, 75, -, 1500, 30, 200, -, -, 5, -, 550
Samuel P. Winbrow, -, -, 1, 110, -, 75, -, 2500, 100, 450, -, 25, -, -, 800
Robert Parsons, -, -, 1, 70, -, 15, -, 1800, 30, 150, -, -, -, -, 450
Thomas Ingram, -, -, 1, 75, -, 15, -, 1500, 50, 2100, 10, 30, -, -, 600
Samuel H. Layton of M, -, -, 1, 80, -, 75, -, 4000, 75, 450, 15, -, -, -, 1000
Lemuel Parsons, -, -, 1, 125, -, 20, -, 2000, 75, 400, 10, -, 75, -, 800
James Wilkinson, -, -, 1, 60, -, -, -, 2000, 50, 300, -, -, 10, -, 600
John A. Moore, -, -, 1, 100, -, 10, -, 2000, 30, 250, -, -, -, -, 550
Alexander Wimbrow, -, -, 1, 50, -, 5, -, 1500, 25, 125, -, -, -, -, 500
Levin Hudson, -, -, 1, 110, -, 8, -, 3000, 100, 200, -, -, -, -, 700
Edward B. Toomey, -, -, 1, 60, -, -, -, 1500, 20, 200, 10, -, -, -, 450
George M. Hudson, -, -, 1, 80, -, 4, -, 1800, 35, 150, -, -, -, -, 450
Thomas G. Wells, -, -, 1, 30, -, 50, -, 1800, 60, 400, -, -, -, -, 400
Edward G. Wells, -, -, 1, 45, -, 55, -, 2000, 50, 300, -, -, -, -, 650
Miras B. Moore, -, -, 1, 100, -, 50, -, 3000, 50, 400, -, -, -, -, 750
Richard W. Long, -, -, 1, 100, -, 50, -, 3000, 50, 300, -, -, -, -, 500
David B. Tingle, -, -, 1, 30, -, 4, -, 700, 40, 300, -, -, -, -, 400
William E. Hudson, -, -, 1, 60, -, 50, -, 2000, 150, 300, -, -, -, -, 600
Saylathul Baker, -, -, 1, 100, -, 10, -, 2500, 150, 400, -, -, -, -, 700
Hirahm Dukes, -, -, 1, 150, -, 50, -, 3000, 150, 500, -, -, -, -, 700
James B. Middleton, -, -, 1, 75, -, -, -, 1500, 40, 200, -, -, -, -, 300
George H. Burton, -, -, 1, 50, -, 50, -, 800, 15, 100, -, -, -, -, 250
James H. Ingram, -, -, 1, 50, -, 25, -, 800, 10, 75, -, -, -, -, 225
Mary Phillips, -, -, 1, 50, -, 50, -, 700, 30, 75, -, -, -, -, 250
Alfred B. Marvel, 1, -, -, 60, -, 140, -, 2000, 100, 400, -, -, -, -, 500
George Ingram, -, -, 1, 60, -, 150, -, 1500, 50, 60, -, -, -, -, 250
David H. Frame, 1, -, -, 10, -, 5, -, 300, 18, 60, -, -, -, -, 250
Hazell Thompson, 1, -, -, 30, -, 25, -, 800, 20, -, 40, -, -, -, -, 200
Joshua T. Short, -, -, 1, 45, -, 85, -, 2500, 25, 125, -, -, -, -, 400
John Burton, -, -, 1, 95, -, 5, -, 2000, 100, 400, 20, -, -, 10, 700
Benjaman R. Waggamon, -, -, 1, 200, -, 109, -, 4000, 300, 500, 14, 52, -, -, 1000
Henry C. Long, -, -, 1, 65, -, 10, -, 1500, 35, 2509, -, -, -, 500
Clayton M. Hudson, 1, -, -, 22, -, 1, -, 800, 40, 400, 40, -, -, -, 400
Asher W. McCabe, 1, -, -, 40, -, 30, -, 1000, 75, 300, 15, -, 15, -, 500
Levin J. Parker, -, -, 1, 60, -, 25, -, 1000, 75, 250, -, -, -, -, 500
Sedynum C. Hudson, -, -, 1, 75, -, 40, -, 1000, 75, 430, 10, -, -, -, 700
Isaac J. Williams, 1, -, -, 50, -, 71, -, 1000, 75, 250, -, -, -, -, 400
George F. Woolford, -, -, 1, 119, -, -, -, 2500, 150, 400, -, -, 5, -, 700
William C. Long, -, -, 1, 50, -, 20, -, 650, 30, 200, -, -, 25, -, 150
Alfred Hudson, -, -, 1, 125, -, 20, -, 2500, 50, 300, -, -, -, -, 500
William J. Daisey, 1, -, -, 50, -, 20, -, 2000, 150, 450, -, -, -, -, 400

Lemuel J. Hudson, -, -, 1, 50, -, 150, -, 1500, 20, 100, -, -, -, -, 400
Eli Bunting, 1, -, -, 30, -, -, -, 500, 10, 40, -, -, -, -, 200
John C. Hazzard, 1, -, -, 80, -, 50, -, 2500, 30, 125, -, -, -, -, 350
George L. Davidson, 1, -, -, 70, -, 63, -, 2500, 40, 300, 15, -, -, -, 800
George R. Davidson, 1, -, -, 10, -, 12, -, 400, 10, 20, -, -, -, -, 200
Wilson Campbell, 1, -, -, 40, -, 20, -, 500, 30, 200, -, -, -, -, 300
Isaac W. Timmons, -, -, 1, 65, -, 40, -, 2000, 65, 250, -, -, -, -, 600
Isaac C. McCabe, 1, -, -, 50, -, 17, -, 2000, 50, 250, 10, -, 50, -, 500
Caschius C. Brasure, -, -, 1, 25, -, 10, -, 500, 15, 60, -, -, 5, -, 200
Perry McN. Brasure, 1, -, -, 50, -, 150, -, 1200, 25, -, -, -, -, -, 500
Henry S. Cady, 1, -, -, 120, -, 50, -, 3400, 50, 345, 25, 33, -, -, 600
Charles E. Dingle, -, -, 1, 50, -, 90, -, 1000, 10, 30, -, -, -, -, 200
Charles N. Collins, 1, -, -, 7, -, 16, -, 1500, 5, 100, -, -, -, -, 200
Alfred Tingle, -, -, 1, 60, -, 20, -, 1500, 35, 225, 5, -, -, -, 300
Absalom R. Walls, 1, -, -, 60, -, 25, -, 1400, 50, 200, 100, -, -, -, 350
John E. Davis, 1, -, -, 100, -, 100, -, 8000, 125, 400, 80, 45, 150, -, 375
Jonathan Casey, 1, -, -, 40, -, 100, -, 1500, 150, 500, 50, -, -, -, 600
John Musgrove, -, -, 1, 40, -, 110, -, 75, 20, 75, -, -, -, -, 200
Catharine Timmons, -, -, 1, 50, -, 100, -, 1500, 15, 150, -, -, -, -, 350
Aaron Timmons, -, -, 1, 50, -, 100, -, 1500, 50, 400, -, -, -, -, 404
Joshua Williams, 1, -, -, 25, -, 4, -, 300, 20, 75, 10, -, -, -, 250
Ever Long, 1, -, -, 40, -, 75, -, 1800, 50, 4000, -, -, -, -, 400
Eli Campbell, -, -, 1, 150, -, -, -, 3400, 150, 500, -, -, -, -, 550
Wm. M. Hastings, -, -, 1, 90, -, 75, -, 2000, 50, 300, 5, -, -, -, 375
Jessey McGee, -, -, 1, 100, -, 100, -, 3000, 50, 350, -, -, -, -, 425
John S. Brasure, -, -, 1, 60, -, 40, -, 2200, 125, 350, -, -, -, -, 400
Burton C. Hudson, -, -, 1, 100, -, 60, -, 2000, 50, 200, 10, -, -, -, 500
Miras B. Hudson, -, -, 1, 60, -, 20, -, 1000, 30, 175, 5, -, -, -, 375
Joshua B. Chandler, -, -, 1, 100, -, 100, -, 1200, 20, 100, -, -, -, -, 250
Hiram J. Thompson, 1, -, -, 25, -, 15, -, 500, 15, 100, -, -, -, -, 300
John C. Tingle, 1, -, -, 40, -, 85, -, 800, 25, 150, 10, 10, -, -, 225
George Ingram of C, -, -, 1, 75, -, 150, -, 1000, 25, 60, -, -, 6, -, 475
Sheppard P. Moore, 1, -, -, 14, -, 11, -, 500, 10, 30, -, -, -, -, 200
Wingett E. Mathews, 1, -, -, 30, -, 17, -, 500, 50, 150, -, -, 8, -, 20
George W. Moore, -, -, 1, 100, -, 300, -, 3500, 100, 125, -, -, 25, -, 500
Nathaniel W. Vickers, -, -, 1, 75, -, 125, -, 1000, 30, 160, -, -, -, -, 350
George M. Rickets, -, -, 1, 50, -, 10, -, 700, 10, 100, -, -, -, -, 225
George Tingle, -, -, 1, 50, -, 100, -, 1000, 25, 60, -, -, -, -, 250
Joseph H. Thomas, 1, -, -, 40, -, 10, -, 400, 10, 60, -, -, -, -, 200
James H. Adkins, 1, -, -, 5, -, 25, -, 200, 10, 35, -, -, -, -, 125
Ephraim West, -, -, 1, 85, 20, 15, -, 500, 50, 150, -, -, -, -, 250
William Timmons, 1, -, -, 20, 7, 3, -, 300, 5, 95, -, -, -, -, 150
Benjaman, C. Wingett, 1, -, -, 25, -, 31, -, 700, -, 75, -, -, -, -, 200
Isaac H. Bishop, 1, -, -, 42, -, 8, -, 600, 30, 175, 10, -, 10, -, 400
Hezekiah Wingett, 1, -, -, 10, -, 37, -, 450, 100, 150, 5, -, -, -, 300
Isaac Hall, -, -, 1, 30, 25, 30, -, 1000, 30, 75, -, -, -, -, 300

Cannon Wingett, 1, -, -, 40, 3, 150, -, 1500, 250, 325, 20, -, 25, -, 400
John Timmons, -, -, 1, 50, 7, 187, -, 2000, 75, 150, 10, -, -, 10, 150
Eli W. Hitchens, -, -, 1, 40, -, 200, -, 2800, 20, 75, 6, -, 5, -, 175
John T. Phillips, 1, -, -, 50, -, 50, -, 700, 75, 200, 17, -, 75, -, 250
John H. McCray, -, -, 1, 50, -, 30, -, 600, 10, 60, 5, -, -, -, 200
Stean Hazzard of C, -, -, 1, 30, -, 70, -, 800, 15, 100, 5, -, -, -, 200
Elijah Holtson, -, -, 1, 10, -, -, -, 200, 15, 50, 5, -, -, -, 175
Joshua Hitchens, -, -, -, 61, 60, -, 140, -, 1000, 15, 60, -, -, -, -, 125
Edward F. Morris, -, -, 1, 100, -, 200, -, 6000, 25, 100, 10, -, -, 5, 200
John L. Mumford, 1, -, -, 100, -, 200, -, 6000, 300, 700, 100, -, 200, -, 900
Elisha M. Lynch, 1, -, -, 50, -, 160, -, 800, 25, 125, 125, -, -, -, 125
Elijah Davis, -, -, 1, 12, -, -, -, -, 30, 60, -, -, -, -, 100
George Truitt, 1, -, -, 6, -, 3, -, 250, 10, 65, -, -, -, 200
Robert Joines, 1, -, -, 25, -, 13, -, 1500, 30, 250, 10, -, -, -, 250
Jerry M. Waggomon, 1, -, -, 15, -, 28, -, 600, 10, 150, 10, -, -, -, 250
Isaac J. Thompson, 1, -, -, 50, -, 50, -, 800, 30, 175, 10, -, 18, -, 400
Charles H. McCray, -, -, 1, 40, -, 40, -, 1000, 30, 125, 10, -, -, -, 400
Elisha W. & George W. Cannon, 1, -, -, 100, -, 200, -, 5000, 500, 700, 50, 50, 150, -, 400
John H. Dukes, -, -, 1, 25, -, 5, -, 400, 20, 60, -, -, -, -, 150
Joseph McCray, -, -, 1, 80, -, 10, -, 1000, 30, 150, 10, -, -, -, 300
Mitchell, Bunting, -, -, 1, 80, -, 10, -, 1000, 30, 225, -, -, -, -, 450
David R. Hopkins, 1, -, -, 4, -, -, -, 500, 5, 100, -, -, -, -, 100
James Andrews, -, -, 1, 50, -, 100, -, 2000, 20, 125, 10, 35, -, -, 400
Charles H. Hopkins, -, -, 1, 50, -, 100, -, 1000, 20,75, -, -, -, -, 150
Robert J. Houston, 1, -, -, 85, -, 93, -, 4000, 250, 450, 30, 125, 110, 25, 400
Samuel T. Walls, -, -, 1, 90, -, 90, -, 1500, 30, 100, 10, -, -, -, 300
Joseph E. Johnson, -, -, 1, 30, -, 30, -, 2000, 25, 250, -, -, 47, 15, -, 100
James Hood, 1, -, -, 65, -, 40, -, 600, 35, 95, -, -, -, -, 200
Joseph P. Morris, 1, -, -, 60, -, 40, -, 3000, 50, 125, 10, -, 25, -, 350
Joseph Betts, 1, -, -, 11, -, -, -, 700, 10, 50, -, -, 30, -, 150
Abraham Ingram, -, -, 1, 60, -, 30, -, 2500, 100, 300, 25, -, -, 10, 500
Peter Parker, -, -, 1, 125, 25, 850, -, 12000, 100, 300, 40, 50, 50, 20, 800
David H. Johnson, 1, -, -, 60, -, 60, -, 1504, 20, 125, -, -, 40, -, 540
Peter Frame, 1, -, -, 40, -, 90, -, 1100, 25, 75, -, -, 40, -, 250
David H. Clogg, 1, -, -, 30, -, 43, -, 500, 30, 200, -, -, -, -, 200
Alfred L. Gray, 1,-, -, 40, -, 70, -, 2000, 150, 175, -, -, -, -, 400
William Parsons, 1, -, -, 18, -, 15, -, 600, 25, 100, 5, 42, -, -, 250
Janett Fosque, 1, -, -, 20, -, 15, -, 500, 5, 25, 20, -, -, -, 150
Robert W. Poole, 1, -, -, 20, -, 32, -, 1200, 100, 200, 10, 10, 70, -, 300
Rhody Daldvan, 1, -, -, 20, -, 25, -, 1500, 10, 20, 10, -, -, -, 200
John J. Moore, 1, -, -, 8, -, 9, -, 700, 30, 90, 15, -, -, -, 200
John H. Long, -, -, 1, 30, -, 45, -, 2500, 250, 400, 15, -, 150, -, 500
Joshua J. Derickson, 1, -, -, 100, -, 150, -, 5000, 200, 750, -, -, 100, 40, 800
Henry C. Rickets, -, -, 1, 50, -, 50, -, 800, 10, 100, 10, -, -, -, 250
Wingett Short, -, -, 1, 75, -, 50, -, 1000, 20, 75, -, -, -, -, 400

Thomas Johnson, -, -, 1, 75, -, 50, -, 75, -, 740, 30, 50, -, -, -, -, 200
Wesly W. Savage, -, -, 1, 50, -, 100, -, 2500, 50, 200, -, -, -, -, 500
Robert B. Marvill, -, -, 1, 50, 100, -, 700, 10, 50, -, -, -, -, 450
William L.Morris, 1, -, -, 10, -, 30, -, 500, 10, 25, -, -, -, -, 150
Peter N. B. Helem, 1, -, -, 150, -, 100, -, 2500, 50, 400, -, -, -, -, 350
Armwell L. Christopher, 1, -, -, 50, -, 38, -, 800, 40, 75, -, -, -, -, 150
Peter D. Showell, 1, -, -, 60, -, 55, -, 1000, 40, 65, -, -, -, -, 200
Samuel D. Ward, -, -, 1, 60, -, 80, -, 704, 20, 75, 10, -, -, -, 150
Harry Burton, -, -, 1, 50, -, 60, -, 800, 25, 75, 5, -, -, -, 175
Theodore Houston, -, -, 1, -, 10, 10, -, 600, 15, 705, -, -, -, -, 150
John Balis, -, -, 1, 75, -, 100, -, 900, 20, 75, 5, -, 10, -, 300
William N. Thouroughgood, 1, -, -, 100, -, 100, -, 3000, 30, 225, 10, 10, 25, -, 300
Mitchell Scott, -, -, 1, 50, -, 30, -, 700, 40, 100, 5, -, -, -, 200
John R. Wilkinson, -, -, 1, 40, -, 50, -, 800, 15, 100, 5, -, -, -, 200
William Toomey, -, -, 1, 60, -, 100, -, 900, 175, 600, -, -, -, -, 400
John T. Davis, -, -, 1, 50, -, 60, -, 800, 20, 100, 10, -, -, -, 125
William R. Taylor, 1, -, -, 6, -, 1, -, 600, 40, 130, -, -, -, -, 300
Lambert Campball, 1, -, -, 11, -, 5, -, 2500, 20, 125, 10, 10, -, 5, 400
Charles F. Gum, 1, -, -, 16, -, -, -, 3000, 15, 225, -, -, -, -, 450
Manain Gum, 1, -, -, 20, 10, 15, -, 6000, 100, 175, 10, -, 60, 15, 300
Francis M. Gum, 1, -, -, 5, -, -, -, 2000, 50, 115, -, -, 40, -, 150
Alfred Long, -, -, 1, 100, -, 50, -, 3000, 50, 250, 10, -, -, -, 250
Robert Conquest, -, -, 1, 40, -, 75, -, 1000, 15, 75, 5, -, -, -, 200
Ebe D. Gray, -, -, 1, 6, -, 11, -, 1000, 5, 50, -, -, -, -, 100
John H. Long, 1, -, -, 10, -, 18, -, 1200, 40, 125, 10, -, 5, -, 150
George L. Gray, 1, -, -, 7, -, 7, -, 1200, 40, 75, 5, -, 35, -, 175
Thomas B. Windson, 1, -, -, 20, -, -, -, 1500, 50, 200, 10, 25, -, 15, 200
Robert M. White, 1, -, -, 20, -, 71, -, 2000, 30, 200, 10, -, -, -, 250
Arnold Bethards, -, -, 1, 50, -, 100, -, 3000, 20, 150, 10, -, -, -, 300
William P. Rogers, -, -, 1, 50, -, 100, -, 2000, 40, 225, 10, -, -, -, 350
William H. Stean, 1, -, -, 15, 5, -, -, 1000, 30, 75, 5, -, -, -, 175
William C. Parkhurst, 1, -, -, 4, 20, -, -, 1500, 40, 425, 10, -, 15, -, 150
Thomas Z. Barker, -, -, 1, 100, 20, 240, -, 5000, 60, 200, 20, -, 30, -, 350
George P. Morris, 1, -, -, 40, -, 20, -, 2500, 40, 175, 15, 20, 20, -, 325
Lavenia E. Burton, 1, -, -, 40, -, 20, -, 3000, -, 50, 10, -, 10, -, 200
Tilghman S. Johnson, 1, -, -, 20, -, -, -, 2000, 200, -, -, -, -, -, 400
William R. Ellis, 1, -, -, 45, -, 59, -, 3000, 40, 250, 50, 50, -, 75, 350
Edward W. Houston, 1, -, -, 50, -, 5, -, 3000, 25, 300, 75, 10, -, 15, 175
John M. Houston, 1, -, -, 100, -, 25, -, 4000, 500, 1500, 200, 100, 150, -, 1200
END FIRST NAME FIRST

Downing, Hiram T., 1, -, -, 50, -, -, -, 6000, 400, 400, 150, 100, 25, 3, 550
Tam, William, 1, -, -, 180, 60, 23, -, 18000, 500, 800, 200, 50, 200, 60, 1400
Pepper, Walter W., 1, -, -, 90, 10, 85, -, 5000, 250, 390, 10, 42, 14, 8, 500
Campbell, Wm. H., -, -, 1, 20, -, -, -, 700, -, -, -, -, -, -, 150
Joseph, John M., -, -, 1, 40, -, 40, 50, 2000, 100, 300, 20, -, 100, 24, 400

Day, Mary, 1, -, -, 100, 30, 30, 10, 5000, 100, 600, -, 4, 5, 1, 600

Wilson, Theodore J., -, -, 1, 90, 1, 4, -, 900, 20, 121, -, -, -, -, 300

Harmon, Robert K., -, -, 1, 60, 40, 120, -, 3000, 25, 100, 30, -, -, -, 247

Elliott, Wingate E., -, -, 1, 50, 10, 100, -, 3000, 15, 140, 40, -, -, -, 331

Martin, James, 1, -, -, 125, 10, 75, -, 5000, 100, 500, 75, -, 200, 40, 1200

Short, Isaac L., -, -, 1, 40, -, 10, -, 1000, 125, 250, -, -, 19, 4, 200

Hearn, Benj. B., -, -, 1, 150, -, 100, -, 3000, 50, 300, 5, -, -, -, 400

Smith, Prettyman D., 1, -, -, 75, 4, 30, -, 1000, 75, 200, 5, 8, 75, 50, 400

Messick, George P., -, -, 1, 80, -, 40, -, 4000, 25, 75, 10, -, -, -, 300

Pepper, David M., 1, -, -, 175, 12, 37, -, 4000, 100, 350, 75, 25, 100, 20, 600

Pepper, Catharine, 1, -, -, 100, 20, 80, -, 6000, 75, 400, 50, 40, 300, 50, 1600

Pepper, Catharine, 1, -, -, 30, 20, 70, -, 2000, -, -, 20, -, 100, 16, 300

Littleton, John S., -, -, 1, 28, 2, 26, -, 850, 10, 50, 5, -, -, -, 300

Pepper, Charles T., 1, -, -, 160, 20, 80, -, 9000, 500, 1200, 20, 150, 5, 76, 156, 2000

Pepper, David, 1, -, -, 200, 25, 100, -, 10000, 300, 1000, 30, 75, 300, 80, 2500

Walls, James G. & Henry C., -, -, 1, 125, 10, 75, -, 2500, 20, 150, 5, -, -, -, 300

Calhoun, Gideon W., -, -, 1, 70, -, 35, -, 2500, 40, 350, 20, 18, -, -, 500

Smith, Mitchell, -, -, 1, 100, -, 15, -, 3500, 100, 500, 25, 72, 144, 35, 600

Messick, Daniel H., -, -, 1, 39, 2, -, -, 1000, 25, 50, 5, -, -, -, 300

Derrickson, Wm. E., -, -, 1, 70, 1, 15, -, 1600, 25, 50, 23, -, -, -, 200

Dickerson, Russell, 1, -, -, 37, -, -, -, 650, 20, 200, 5, -, -, -, 200

Marvel, Dagworthy D., 1, -, -, 80, 8, 35, -, 1500, 25, 60, 10, -, -, -, 250

Warren, John, 1, -, -, 25, 2, 12, -, 2000, 40, 200, -, 4, 12, 2, 200

Lynch, Greensbury, 1, -, -, 5, 3, 14, -, 4000, 25, 200, 15, -, -, -, 250

Pepper, Alfred L., 1, -, -, 30, 4, 15, -, 1200, 20, 80, 10, -, 24, 5, 250

Pepper, Greensbury H., 1, -, -, 70, 5, 30, -, 1600, 15, 200, 20, -, -, -, 250

Walls, George, 1, -, -, 100, 6, 60, -, 2000, 25, 300, 20, -, -, -, 250

Donovan, Asbury, 1, -, -, 50, 16, 30, -, 1000, 100, 600, 100, -, -, -, 400

Hudson, Mason (Major) W., 1, -, -, 65, 10, 45, -, 1600, 25, 100, 20, 20, 20, 4, 500

Adams, Horace, -, -, 1, 10, -, 3, -, 500, 10, 100, 10, -, -, -, 20

Wood, Mather, 1, -, -, 50, 6, 50, 6, 1500, 50, 150, 10, 17, 40, 10, 500

Coffin, Thomas P., -, -, 1, 25, -, 160, -, 3000, 20, 80, 10, -, -, -, 100

Marvel, Thomas R. & Joshua B., -, -, 1, 75, 10, 150, -, 3000, 25, 150, 20, 60, -, -, 560

Derickson, Joseph M., -, -, 1, 25, -, 15, 5, 500, 25, 65, 10, -, -, -, 100

Calhoun, Thomas L., -, -, 1, 180, 35, 200, -, 4000, 150, 500, 15, -, -, -, 500

Wilson, Jonathan, -, -, 1, 50, 1, 15, 35, 1000, 20, 75, 5, 5, -, -, 100

Wilson, Elias R., 1, -, -, 50, -, 75, -, 3000, 24, 200, 5, 19, -, -, 250

Wilson, George F., 1, -, -, 100, 2, 50, -, 3000, 25, 250, 5, 20, -, -, 500

Wilson, George of E, 1, -, -, 60, 4, 20, -, 1500, 24, 150, 10, -, -, -, 500

Morris, Robert, -, -, 1, 60, -, 25, -, 2000, 25, 100, 10, -, -, -, 75

Salmons, Joseph P., -, -, 1, 60, 6, 70, -, 5000, 70, 250, 10, 175, 120, 36, 700

Wilkins, John P. & Samuel R., -, -, 1, 120, 10, 50, -, 3000, 100, 350, 5, 18, -, -, 550

Warren, James H., 1, -, -, 40, 5, 40, -, 1500, 25, 250, 15, -, -, -, 250

Willin, Samuel, -, -, 1, 60, 3, 10, -, 1500, 10, 50, 10, -, -, -, 200
Charnean (Charnece), Alfred J., -, -, 1, 50, 10, 75, -, 2000, 40, 150, 25, -, -, -, 500
Jacobs, Eugene, 1, -, -, 60, 2, 28, -, 1200, 75, 125, 30, 75, 60, 15, 400
Wilson, Nathaniel H., 1, -, -, 70, 38, -, 3000, 100, 300, 10, 44, -, -, 400
Carmean, Henry T., 1, -, -, 20, -, 155, -, 1000, -, -, -, -, -, -, 2000
Messick, Sallie T., 1, -, -, 60, -, 30, -, 800, -, 5, -, -, -, -, 100
Saulesbury, Francis M., -, -, 1, 185, 50, 100, -, 5000, 300, 400, 200, 600, 320, 80, 1200
Roach, Daniel, -, -, 1, 80, 4, 20, -, 1000, 30, 150, 10, -, 16, 4, 200
Calhoun, George W., 1, -, -, 12, 1, 33, -, 525, 25, 100, 5, -, -, -, 150
Stuart, David A., 1, -, -, 9, -, -, -, 250, 15, 120, -, -, -, -, 75
Stuart, Ebie T., 1, -, -, 4, -, 25, -, 250, -, 35, 25, -, -, -, 20
Hopkins, G. Robert, 1, -, -, 19, 2, 30, -, 550, -, -, -, -, -, -, 30
Bennum, Nehemiah W., -, 1, -, 19, -, 12, -, 1000, 50, 100, 25, 30, -, -, 75
Bennum, Henry O., 1, -, -, 50, 6, 65, -, 1500, 10, 50, 20, 5, 10, 3, 150
Warrington, Wm. A., -, 1, -, 15, 1, -, 36, -, 800, 60, 175, -, -, -, -, 100
Dodd, Peter A., -, -, 1, 100, -, 100, -, 4000, 50, 200, 10, 18, 40, 10, 500
Atkins, Kendal, 1, -, -, 75, 1, 65, 10, 1500, 20, 500, 10, 17, -, -, 400
Dodd, John H., 1, -, -, 60, 1, 65, -, 1000, 50, 200, 10, 17, 16, 4, 300
Hudson, Thomas E., -, -, 1, 55, 10, 50, -, 1000, 25, 50, 10, -, -, -, 300
Dodd, Peter P., 1, -, -, 100, 2, 250, -, 2500, 50, 300, 25, -, 160, 40, 600
Lyons, Laban L., 1, -, -, 15, -, 195, -, 2000, -, -, 2, -, 15, 4, 20
Martin, Emily O., 1, -, -, -, -, 200, 15, 2000, -, -, -, -, -, -, 6
Barker, Joseph R., 1, -, -, 100, 20, 100, -, 4000, 75, 580, 25, 50, 160, 40, 1000
Rust, Peter, 1, -, -, 75, 10, 75, -, 2500, 25, 250, 25, 36, 160, 40, 500
Wilkins, James B., -, -, 1, 100, 3, 12, -, 4000, 75, 400, 25, 110, 26, 6, 1000
Wilkins, John H., -, -, 1, 80, 18, 80, -, 6000, 100, 100, 30, 20, 60, 15, 1100
Chase, Mary R., 1, -, -, 20, -, 2, -, 800, -, -, -, -, 60, 15 100
Pettyjohn, Lupinkster, 1, -, -, 35, 20, 80, -, 1200, -, 25, 10, -, -, -, 200
Sharp, Josiah, 1, -, -, 50, -, 100, -, 1500, 50, 125, 10, -, -, -, 300
Hancock, Wm., 1, -, -, 50, 2, 63, -, 1000, 25, 250, 10, -, 10, 2, 250
Watson, John F., -, -, 1, 40, -, 40, -, 1000, 5, 25, -, -, -, -, -
Wilson, Wm. H., -, -, 1, 50, 2, 50, -, 1000, 25, 90, 20, 30, 14, 3, 200
Collins, James C., 1, -, -, 40, -, 50, -, 1000, 40, 175, 10, -, -, -, 250
Morris, George F., -, -, 1, 160, 20, 70, -, 3500, 25, 100, 5, -, 48, 12,300
Millman, George W., 1, -, -, 20, 2, 125, -, 2000, 15, 50, 10, -, -, -, 100
Roach, Eli J., -, -, 1, 40, 2, 90, -, 800, 25, 75, 10, -, -, -, 75
Dickerson, James, 1, -, -, 20, 3, 30, -, 1500, 25, 200, 20, -, 2, 46, 300
Morris, John, -, -, 1, 10, -, 4, -, 400, -, -, -, -, -, -, 60
Millman, Elisha, -, -, 1, 50, 340, -, 2000, 40, 350, 20, 36, 160, 40, 600
Green, George, -, -, 1, 50, 3, 50, -, 2000, 25, 150, 10, 108, -, -, 300
Short, Isaac W., 1, -, -, 20, 4, 88, -, 2000, 20, 200, 30, -, 20, 5, 250
Workman, John W., 1, -, -, 80, 6, 45, -, 2000, 50, 400, 20, 10, -, -, 600
Pettyjohn, Wm., 1, -, -, 80, 7, 165, -, 1500, 100, 200, 25, -, -, -, 250
Pettyjohn, George W., -, -, 1, 35, -, -, 75, -, 1000, 15, 75, 5, -, -, -, 200
Pettyjohn, William, -, -, 1, 40, 12, 60, -, -, 1440, -, 10, -, 10, -, -, -, 100

Warren, Robert, 1, -, -, 4, 1, 55, 5, 1200, 30, 80, 20, -, 16, 4, 250
Calloway, Wm., -, -, 1, 50, -, 5, -, 1000, 20, 60, 8, -, -, -, 75
Pettyjohn, George W., 1, -, -, 50, 4, 50, -, 1000, 50, 200, 20, -, -, -, 500
Spocer(Spicer), Wm. J., 1, -, -, 70, 10, 50, -, 2000, 100, 400, 20, 30, -, -, 600
Greenly, Robert, 1, -, -, 75, 3, 180, 25, 3000, 25, 400, 20, 36, 40, 10, 500
Pepper, Greensbury H., 1, -, -, 15, -, 65, 10, 600, 20, 125, 25, -, -, -, 75
Messick, Robert H., -, -, 1, 60, -, 35, 30, 700, 35, 75, 10, -, -, -, 125
Wilson, John W., 1, -, -, 1, -, 40, -, 300, 25, 100, -, -, -, -, 139
Megee, Levin J., 1, -, -, 30, -, 70, -, 1000, 25, 275, 10, -, 30, 6, 200
Calhoun, Wm., -, -, 1, 45, 15, 55, -, 700, 15, 100, 5, -, -, -, 225
Sport, John W., -, -, 1, 50, -, 200, -, 2500, 30, 75, 10, 15, -, -, 175
Spicer, Charles, -, -, 1, 75, -, 90, 25, 1400, 10, 100, 35, 20, -, -, 150
Megee, George E., 1, -, -, 9, -, 3, -, 500, -, -, -, -, -, -, 50
Sharp, Asa J., 1, -, -, 40, -, 110, -, 2500, 25, 200, 15, -, -, -, 250
Sharp, Kearsey, 1, -, -, 75, -, 325, -, 4800, 200, 350, 75, -, -, -, 800
Massey, Mathias, -, -, 1, 20, 2, 75, -, 900, 8, 25, -, -, -, -, 100
Green, Jesse, -, -, 1,15, -, 1, 300, 30, 60, 5, -, -, -, 75
Williams, Joseph, -, -, 1, 40, 40, 50, -, 1200, 30, 125, 15, -, -, -, 150
Conaway, Isaac A., -, -, 1, 10, 45, 5, -, 500, 20, 65, 50, -, -, -, 10
Sharp, John, 1, -, -, 100, -, 215, 10, 7000, 30, 150, -, -, 10, 2,150
Short, John of J, -, -, 1, 50, 1, 1, -, 750, 15, 100, 6, 25, -, -, 150
Sharp, Jacob H., 1, -, -, 15, -, 35, 10, 1000, 20, 150, 12, 10, -, -, 150
Donovan, Peter, -, -, 1, 15, -, 10, 50, 400, 20, 50, 20, -, -, -, 50
Peck, Isaac A., 1, -, -, 100, 15, -, 50, 8000, 100, 400, 10, 16, 100, 20,700
Swain, Spencer A., -, -, 1, 50, 2, 150, 50, 1500, 25, 100, 25, -, -, -, 150
Donovan, Rhuben, 1, -, -, 75, 5, 40, -, 2500, 50, 600, 50, 35, 200, 50, 600
Evans, James N., 1, -, -, 100, 12, 50, -, 5000, 30, 475, 20, 30, 200, 50, 650
Torbert, George, 1, -, -, 60, -, 36, -, 2500, 25, 250, 10, 10, -, -, 450
Steel, James M., -, -, 1, 60, 6, 45, 50, 2000, 20, 250, 10, 10, -, -, 250
Donovan, Riley, 1, -, -, 10, 4, 2000, 15, 4000, 25, 400, 25, 25, 50, 12, 500
Marvel, John, -, -, 1, 10, -, 30, 10, 600, 15, 40, 15, -, -, 25
Donovan, Eli W., -, -, 1, 50, -, 40, -, 800, 15, 300, 15, -, -, -, 250
Donovan, Zachariah, 1, -, -, 50, -, 250, 30, 3000, 30, 200, 25, -, -, -, 250
Workman, Robert B., 1, -, -, 40, 1, 40, 5, 1000, 25, 200, 20, -, -, -, 400
Abbott, James, 1, -, -, 35, 1, 42, 5, 1000, 40, 150, 15, 20, -, -, 500
Marker, John W., -, 1, -, 25, 1, 100, 90, 1000, 25, 100, 10, -, -, -, 250
Jones, Zachariah, 1, -, -, 30, -, 90, 10, 1000, 100, 200, 20, -, -, -, 200
Donovan, Robert, 1, -, -, 35, 2, 65, -, 100, 50, 200, 10, -, 50, 12, 250
Dutton, Levin P., 1, -, -, 40, 5, 100, -, 600, 50, 400, 10, 10, -, -, 500
Messick, George, 1, -, -, 30, -, 100, 18, 1000, 50, 400, 5, -, 40, 10, 300
Messick, Henry T., 1, -, -, 30, 4, 35, -, 1000, 25, 200, 20, -, 40, 10, 500
Donovan, Zachariah, -, -, 1, 15, -, 5, 5, 600, 20, 200, 10, -, -, -, 200
Burton, Benj. D., 1, -, -, 50, 4, 60, -, 1500, 50, 250, 15, -, -, -, 300
Lynch, Joshua A., 1, -, -, 100, 12, 50, -, 2000, 100, 400, 20, 25, -, -, 1000
Pettyjohn, Robert, -, -, 1, 75, -, 55, -, 1000, 20, 250, 10, -, -, -, 250
Donovan, James G., -, -, 1, 20, -, 400, 5, 4500, 15, 60, -, -, -, -, 225

Pettyjohn, Wm. H., -, -, 1, 60, 2, 140, -, 1500, 25, 200, 10, -, -, -, 250
Abbott, Riley W., -, -, 1, 60, 4, 135, 5, 1500, 20, 75, 10, -, -, -, 275
Donovan, John P. & Wm., 1, -, -, 135, 3, 103, 35, 1500, 50, 400, 20, -, -, -, -, 600
Coughter, Charles E., 1, -, -, 80, 30, 100, -, 6000, 50, 225, 150, 55, 300, 60, 1500
Stergus, Wm. N., -, -, 1, 40, 2, 40, -, 1000, 50, 400, 10, 16, 70, 14, 300
Robbins, Joseph L., 1, -, -, 60, 2, 40, -, 2000, 30, 500, 5, -, 84, 24, 500
Macklin, John, 1, -, -, 140, 9, -, 100, -, 4800, 130, 350, 75, 45, 200, 50, 800
Middleton, Charles W., 1, -, -, 100, 15, 150, 123, 3730, 54, 110, 25, -, 100, 20, 250
Sweeney, James, -, -, 1, 35, 5, 25, -, 1500, 40, 60, 10, -, -, -, 200
Brown, Alfred L., 1, -, -, 50, 5, 50, -, 800, 25, 100, 30, -, -, -, 200
Davis, Solomon & Brinkly, 1, -, -, 75, 2, 50, -, 1200, 75, 500, 25, 12, -, -, 500
Butabee (Butcher), Noah, 1, -, -, 50, 2, 50, -, 1202, 25, 100, 25, 5, 40, 10, 450
Lofland, John W., -, -, 1, 50, -, 250, 20, 3500, 30, 75, 15, 10, -, -, 300
Donovan, James C., -, -, 1, 75, 6, 220, 25, 5000, 25, 200, 25, -, -, -, 350
Scott, Robert, -, -, 1, 60, 6, 50, 1200, 50, 250, 25, -, -, -, 300
Lecates, Job, -, -, 1, 25, 3, 10, -, 1000, 25, 150, 5, -, -, -, 200
Conaway, Dixon, -, -, 1, 40, -, 400, 5, 4450, 25, 125, 10, -, -, -, 300
Donovan, Enoch W., 1, -, -, 4, 1, -, -, 1200, 50, 600, 25, 18, -, -, 300
Reynolds, David M., 1, -, -, 50, 2, 112, 14, 3000, 100, 400, 10, -, -, -, 300
Torbert, George, 1, -, -, 100, 12, 300, -, 7000, 150, 450, 25, -, 400, 80, 1200
Dickerson, George, -, -, 1, 100, 6, 40, 20, 1000, 20, 50, 20, 40,-, -, 300
Ward, Wm., -, -, 1, 80, 1, 60, 20, 1000, 25, 100, 5, -, -, -, 300
King, John, 1, -, -, 15, -, 20, 3, 300, 10, 90, 5, -, -, -, 150
West, Stockly, -, -, 1, 75, -, 200, 35, 3000, 50, 250, 20, -, -, -, 400
King, Wingate, 1, -, -, 80, 5, 80, 10, 2000, 100, 600, 10, -, 300, 60, 650
Cooper, Theodore E., 1, -, -, 30, 1, 20, -, 500, 10, 110, 5, -, -, -, 100
King, Robert, -, -, 1, 30, -, 160, -, 1800, 25, 125, 10, -, -, -, 150
Donovan, Margaret, 1, -, -, 25, -, 75, 1000, 10, 60, 10, -, -, -, 125
Salmons, Wingate, 1, -, -, 25, -, 25, -, 60, 30, 100, 5, -, -, -, 100
Donovan, Peter, -, -, 1, 60, -, 50, 50, 1200, 20, 200, 25, -, -, -, 200
Shoot, Thomas W., 1, -, -, 50, 3, 25, -, 800, 30, 225, 10, -, 20, 5, 300
Day, John, H., 1, -, -, 70, 2, 54, -, 2000, 100, 300, 20, 38, 120, 30, 500
Jefferson, James S., -, -, 1, 25, -, 25, -, 500, 25, 100, 5, -, -, -, 125
Pride, George E., -, -, 1, 20, -, -, -, 350, 25, 75, 5, -, -, -, 175
Macklin, Bayard, 1, -, -, 75, -, 75, -, 3000, 200, 300, 10, 30, 200, 40,1200
Conaway, Isaac P., -, -, 1, 35, -, 15, -, 1000, 25, 250, 6, -, 61, 245
Swain, Theophilus, -, -, 1, 110, 5, 22, -, 400, 50, 300, 15, 54, 120, 24, 850
Conaway, Thomas Sr. & Thomas W., -, -, 1, 80, 10, 90, -, 4500, 25, 260, 15, -, 24, 6, 550
Wingate, John M. C., -, -, 1, 30, -, 54, -, 1000, 75, 300, 10, 10, -, -, 200
Willey, William, -, -, 1, 60, 8, 20, -, 1000, 25, 150, 10, 27, -, -, 400
Smith, Lemuel, -, -, 1, 60, 10, 25, -, 4000, 50, 300, 5, 225, -, -, 1000
Brion, Lewis, -, -, 1, 150, -, -, -, 4000, 50, 250, 25, -, 25, 6, 600
Russell, James N., 1, -, -, 15, 2, -, -, 1000, 10, 100, 10, 14, -, -, 150
Russell, Isaac, 1, -, -, 15, 4, -, -, 1100, 40, 260, 35, 22, -, -, 125

Russell, Isaac, -, -, 1, 13, -, -, -, 1004, -, -, 10, 55, -, -, 40
Day, John R., -, -, 1, 180, 20, -, -, 10000, 60, 700, 10, 66, 12, 1000
Fleetwood, Cyrus, 1, -, -, 12, 1, -, -, 900, 10, -, 5, -, -, -, 50
Conaway, Curtis A., 1, -, -, 113, 12, 5, -, 7000, 225, 1200, 50, 100, 300, 60, 1425
Harris, George, 1, -, -, 13, 8, -, -, 4000, 25, 100, 25, -, 100, 20, 650, 20
Hatfield, Thomas W., 1, -, -, 6, 12, -, -, 1000, 20, 160, 10, -, 125, 25, 300
Layton, Caleb S., 1, -, -, 16, 4, -, -, 8000, 50, 25, 20, -, 150, 35, 150
Lynde, James W., 1, -, -, 25, 6, -, -, 2000, 30, 200, 10, 32, 15, 3, 300
Ewinge, Adolphus P., 1, -, -, 40, -, 100, 1, 4000, 50, 500, 50, -, 50, 10, 225
Stockley, Charles C., 1, -, -, 210, 100, 275, 50, 11000, 300, 750, 100, 100, 600, 150, 2000
Wootten, Hon. Edward, 1, -, -, 72, 12, -, -, 8000, 100, 425, 100, 60, 250, 65, 350
Pepper, Thomas, 1, -, -, 80, -, 180, -, 5000, 50, 150, 20, -, -, -, 250
Faucett, Wm. A., 1, -, -, 120, -, 33, -, 3000, 50, 200, 20, 25, -, -, 400
Long, Henry W., 1, -, -, 5, -, -, -, 1500, -, 200, -, -, 12, -, 100
Willin, Isaac, 1, -, -, 12, -, -, -, 1200, 50, 150, 10, -, -, -, 100
Atkins, John S., 1, -, -, 70, 30, 59, 9, 4000, 50, 100, 50, 200, 200, 50, 150
Wright, Gardner H., 1, -, -, 60, 17, 20, 17, 10000, 200, 350, 100, 150, 200, 50, 675
Cullen, Charles M., 1, -, -, 48, -, -, 15, 10000, 100, 100, 50, -, 50, 10, 200
Tunnell, Wilber F., 1, -, -, 66, 8, -, 10, 1000, 100, 400, -, 100, 160, 32, 650
Albruy, Robert W. D., 1, -, -, 100, 30, 77, -, 15000, 250, 500, 200, 135, 350, 125, 7000
Waples, Joseph B., 1, -, -, 6, -, 52, -, 9000, 25, 200, 20, 2, 30, 10, 100
Stockley, John, -, 1, -, 360, -, 5, 30000, 500, 2000, 100, 300, 1496, 156, 3500
Conaway, Gilley J., -, -, 1, 75, -, 27, -, 1500, 50, 200, 20, 6, -, -, 300
Richards, Charles H., 1, -, -, 18, -, -, -, 8500, 200, 400, 50, 100, 300, -, 350
Conaway, Isaac, 1, -, -, 10, -, -, -, 2250, 50, 75, 5, -, -, -, 125
McKin, John L., 1, -, -, 5, -, -, -, 2000, -, -, 10, -, 25, -, 75
Catlin, Lamerson, 1, -, -, 20, -, 8, -, 1500, 40, 215, 15, -, 130, 52, 151
Jones, Joseph S., 1, -, -, 49, 1, 1, -, 1500, 60, 240, 15, 100, 10, -, 310
Smith, Sallie, 1, -, -, 40, 1, 51, -, 750, 18, 135, 10, -, 20, 4, 75
Truitt, Joseph S., -, -, 1, 45, 1, 60, -, 1200, 18, 217, -, -, -, -, 120
Jones, Edward L., -, -, 1, 60, -, 40, -, 600, 15, 45, -, -, -, -, 70
Truitt, Noah S., -, -, 1, 30, 1, 50, -, 560, 7, 85, -, -, -, -, 40
Roberts, George W., -, -, 1, 75, 12, 12, -, 1200, 60, 300, 15, 40, 20, 5, 425
Gordy, Aaron K., -, -, 1, 40, -, 10, -, 500, 25, 333, -, -, -, -, 65
Phillips, Nathaniel H., -, -, 1, 20, 340, -, 400, 21, 350, -, -, -, -, 30
Truitt, Greensbury M., 1, -, -, 100, 7, 40, -, 1500, 75, 470, 20, -, 65, -, 950
Truitt, Burton P., 1, -, -, 70, 2, 18, -, 1000, 28, 267, 10, -, -, -, 208
Lingo, Minos B., 1, -, -, 70, 1, 17, -, 1600, 25, 308, 15, 24, 25, -, 225
Littleton, Loury J., 1, -, -, 8, 1, -, -, 150, -, 20, 12, -, 5, -, 33
West, Joseph P., -, -, 1, 50, -, 20, -, 700, 18, 125, 15, -, -, -, 190
West, Benjamin B., -, -, 1, 65, 5, 45, -, 700, 15, 150, 12, -, -, -, 135
Beachum, James E., -, -, 1, 35, 2, 15, -, 600, 8, 146, 5, -, -, -, 154
Cannon, William E., 1, -, -, 5, -, -, -, 3000, 150, 300, 5, 5, -, -, 25
Tyree, Ebenezer H., -, -, 1, 60, 2, 75, -, 1200, 10, 175, 40, 30, 20, 7, 140

Grinby, Stephen P., -, -, 1, 75, 1, 25, -, 1800, 50, 325, 50, 16, 25, 7, 250
Collins, Isaac, -, -, 1, 35, 1, -, -, 500, 40, 80, 20, 6, -, -, 140
Jones, Elijah W., 1, -, -, 40, 1, 40, -, 1000, 25, 130, 10, -, -, -, 100
Parsons, Peter J., -, -, 1, 10, -, -, -, 200, 12, 55, 5, -, 25, -, 200
Truitt, James B., -, -, 1, 40, 1, 30, -, 1500, 8, 230, 15, -, -, -, 390
Short, Elijah C., 1, -, -, 60, 4, 35, -, 1500, 50, 435, -, 5, 10, 100, 52, 555
Wootten, Loucine, -, -, 1, 25, 1, 25, -, 400, 3, 10, 3, -, -, -, 25
Collins, Josiah W., 1, -, -, 60, -, 6, 40, -, 1000, 35, 170, 7, 12, 100, 30, 270
Collins, Ephraim, 1, -, -, 80, 2, 12, -, 800, 1, 75, 10, -, -, -, 180
Collins, Joseph S., 1, -, -, 36, -, 3, 10, -, 1000, 75, 440, 10, 15, 30, 5, 285
Moore, Wilson T., -, -, 1, 45, 1, 10, -, 400, 15, 85, 5, -, -, -, 50
Lowe, Benjamin B., -, -, 1, 40, -, 27 -, 600, 7, 170, 15, 96, 276
Warington, John Q., -, -, 1, 45, 1, 50, -, 500, 10, 50, 5, -, -, -, 63
Adkins, Samuel J., -, -, 1, 50, 3, 30, -, 800, 25, 160, 15, 8, -, -, 187
Collins, Wm. J., -, -, 1, 20, 2, 4, -, 300, 15, 30, -, -, -, -, 95
West, Cornelius D., 1, -, -, 70, 1, 10, -, 2000, 90, 490, 10, -, 175, 59, 390
West, Joseph B., 1, -, -, 60, 1, 15, -, 800, 35, 270, 15, -, 50, 18, 190
Hudson, Elijah H., -, -, 1, 48, -, 22, -, 1000, 25, 162, 15, -, -, -, 188
Parker, James B., -, -, 1, 25, ½, 25, -, 800, 16, 150, 20, -, 70, 24, 120
Turner, John, -, 1, -, 8, ¼, 40, -, 500, 7, 22, -, -, 18, 6, 5
Legates, S. Edward, -, -, 1, 40, -, 15, -, 600, 7, 145, 10, -, -, -, 135
Jones, Joseph B., 1, -, -, 25, 2, 20, -, 850, 80, 200, 5,10, 16, 5, 145
West, Handy B., -, -, 1, 45, ½, 20, -, 1200, 80, 455, 10, -, 160, 52, 525
West, Joshua G., 1, -, -, 30, ½, 75, -, 1500, 60, 340, 10, 15, 25, 8, 240
West, Joseph B. of J, -, -, 1, 42, 3,60, -, 2000, 30, 150, 20, -, 10, 5, 178
Hudson, James, -, -, 1, 15, -, 1000, -, 225, 8, 100, -, -, -, -, 140
Brumbly, Joseph S., -, -, 1, 15, -, 50, -, 1500, 20, 272, 10, -, 5, 5, 445
Mitchell, Lowder W., -, -, 1, 20, ½, 20, -, 900, 12, 155, 7, -, 5, 2, 360
Short, Isaac B., 1, -, -, 35, 1, 165, -, 3000, 85, 575, 25, 15, 175, 52, 415
Pusey, Edward C., -, -, 1, 30, ½, 20, -, 800, 15, 29, 10, 10, -, -, -, 270
Johnson, Abbert H., -, -, 1, 10, -, -, -, 200, -, 8, 1, -, 8, 4, 200
Jones, George W. of I., -, -, 1, 20, ½, 25, -, 650, 20, 200, 12, -, 70, 28, 395
Wells, William B., -, -, 1, 50, -, 15, -, 700, 8, 145, 12, -, 60, 35, 215
Brumbly, John, -, -, 1, 20, -, 20, -, 40, 15, 165, 20, -, 15, 5, 230
Hitchens, John T., -, -, 1, 20, -, 20, -, 500, 10, 40, -, -, -, -, 100
Brasure, William, -, -, 1, 85, -, 30, -, 1500, 20, 250, 100, -, 50, 18, 550
Hastings, Wm. B. -, -, 1, 50, 1, 40, -, 1500, 15, 185, 80, -, 35, 10, 450
Hosea, George, -, -, 1, 25, -, 11, -, 400, 15, 115, 5, -, -, -, 215
Mitchell, Minos W. of Lea, -, -, 1, 20, -, 80, -, 1500, 6, 200, -, -, -, -, 158
Mitchell, John R. of Lem, -, -, 1, 30, -, 70, -, 1500, 25, 240, -, -, -, -, 185
Moore, John B., -, -, 1, 55, -, 10, -, 900, 20, 180, -, -, -, -, 350
Moore, Levin T., -, -, 1, 70, -, -, -, 1000, 30, 330, 4, -, -, -, 420
Downs, William H., -, -, 1, 25, 2, 2, -, 350, 5, 65, 10, -, -, -, 125
Collins, Elijah W., 1, -, -, 65, 3, 40, -, 2000, 60, 460, 100, 55, 175, 70, 375
Collins, Jacob P., 1, -, -, 60, -, 15, -, 800, 10, 90, 5, -, -, -, 160
Corden, Jacob B. -, -, 1, 45, -, 20, -, 750, 5, 80, 50, -, -, -, 160

Wooten, John H., -, -, 1, 8, 4, -, -, 125, 4, 3, -, 1, -, -, 15
Ward, Gordy, -, -, 1, 60, 2, 15, -, 800, 12, 115, 20, -, 15, 4, 250
Timmons, Ezekiel, 1, -, -, 70, 2, 50, -, 900, 15, 295, 5, -, 35, 10, 160
Jones, Jacob S., 1, -, -, 65, -, 6, 65, -, 850, 25, 150, 8, 20, -, -, 150
Pennel, William J., -, -, 1, 100, 1, 10, -, 2000, 35, 250, 5, 3, 30, 17, 550
Short, John H., 1, -, -, 47, ½, -, -, 700, 50, 183, -, -, 90, 30, 255
Short, Urias, 1, -, -, 75, 1, 8, -, 1400, 100, 480, -, 15, -, -, 195
Campbell, Charles H., -, -, 1, 11, -, -, - 700, 8, 5, -, -, -, -, 110
Carey, Philip P., 1, -, -, 68, 4, 3, -, 700, 7, 212, -, -, -, -, 185
Wootten, Edward S. -, -, 1, 22, -, -, -, 500, 14, 105, 5, -, -, -, 225
Mitchell, Elijah J., -, -, 1, 22, -, 1, -, 500, 6, 122, 4, -, -, -, 197
Short, Robert, 1, 50, 4, 15, -, 1800, 97, 425, 10, -, 200, 60, 395
Hickman, Levin J., -, -, 1, 29, -, 25, -, 800, 23, 145, 12, -, 200, 6, 128
Parsons, James B., -, -, 1, 35, -, -, -, 450, 20, 21, 15, -, -, -, 270
Downs, John, -, -, 1, 45, -, -, -, 600, 5, 100, -, -, -, -, 160
Matthews, John, -, -, 1, 45, -, -, -, 600, 10, 130, 1, -, -, -, 160
Beachum, Elish J., -, -, 1, 16, -, -, -, 300, 15, 75, -, -, -, -, 130
Mitchell, Isaac S., -, -, 1, 28, -, 6, -, 300, 18, 125, 5, -, -, -, 178
Baker, James M., -, -, 1, 80, 1, 10, -, 1500, 60, 364, 15, -, -, -, 378
Pennel, Silas J., 1, -, -, 100, 1, 40, -, 1200, 20, 300, 25, -, -, -, 327
Moore, Joseph H., -, -, 1, 45, ½, 21, -, 700, 12, 732, -, -, -, -, 240
McClain, Levin L. W., 1, -, -, 50, 3, 20, -, 1000, 37, 210, -, -, 7, 3, 310
Collins, Elias T., 1, -, -, 42, 4, 2, -, 800, 30, 125, 5, -, -, -, 265
Workman, James P., -, -, 1, 15, -, -, -, 150, 10,35, -, -, -, -, 25
West, Wingate, 1, -, -, 25, 1, 10, -, 400, -, 85, -, -, -, -, 65
Jones, Isaac S., 1, -, -, 50, 3, 25, -, 1500, -, 520, 10, -, 40, 20, 280
Jones, George W., 1, -, -, 65, 6, 43, -, 1200, -, 340, 18, -, 80, 20, 220
Workman, William H., -, -, 1, 54, 12, -, 1000, 18, 134, 5, -, -, -, 310
Daisey, James, -, -, 1, 26, -, 7, -, 600, 15, 120, -, -, -, -, 105
Betts, Minos W., -, -, 1, 54, -, 3, -, 700, 5, 60, 10, -, -, -, 50
Townsend, Asher H., -, -, 1, 26, -, 7, -, 900, 17, 135, -, -, -, -, 135
Sampson, Charles H., -, -, 1, 40, -, -, -, 650, 7, 80, -, -, -, -, 178
Downs, Thomas H., -, -, 1, 20, -, 80, -, 1000, 10, 130, 20, -, -, -, 200
Moore, Henry C., -, -, 1, 30, -, 80, -, 1200, 15, 90, 20, -, -, -, 375
Short, Henry C., 1, -, -, 36, -, 19, -, 1300, 15, 185, 6, -, -, -, 360
Baker, Jonathan, -, -, 1, 50, -, 80, -, 1000, -, 157, -, -, -, -, 180
Short, Shadrach, 1, -, -, 120, 1, 20, -, 2800, 100, 765, 15, 19, 520, 144, 1180
Dennis, Jonathan, -, -, 1, 60, -, 80, -, 1100, 25, 25, 5, -, -, -, 225
Baker, Seth W., -, -, 1, 50, 1, 8, -, 950, 20, 32, 10, -, -, -, 295
Mitchell, Marshall, -, -, 1, 30, -, -, -, 600, 8, 80, -, -, -, -, 180
Smith, Robert W., -, -, 1, 30, 1, 20, -, 900, 17, 237, 40, -, 12, 3, 280
Short, Mias B., 1, -, -, 44, 12, 60, -, 1200, 30, 520, -, -, 100, 40, 225
Brumbly, Isaac, -, -, 1, 28, -, 6, -, 400, 12, 140, 10, -, -, -, 227
Savage, John, -, -, 1, 44, -, -, -, -, 8, 46, 3, -, -, -, 114
Crampfield, Wm. P. -, -, 1, 40, -, -, -, -, 7, 48, -, -, 8, 4, 84
Gravner, James E., -, -, 1, 50, -, -, -, -, 32, 165, -, -, -, -, 110

Moore, Wolsey B., -, -, 1, 50, -, -, -, -, 15, 160, 5, -, -, -, 325
Short, Roben J., 1, -, -, 60, 1, 20, -, 960, 230, -, -, 10, 2, 230
Mitchell, Milbourn D., -, -, 1, 15, -, 5, -, 300, 5, 115, -, -, -, -, 88
Savage, William T., -, - 1, 60, -, 20, -, -, 49, 325, 7, -, -, -, 415
Olophant, Elijah J., -, -, 1, 30, -, -, -, -, 8, 42, -, -, -, -, 225
Hickman, Daniel H., -, -, 1, 12, -, 5, -, 350, 10, 85, -, -, 12, 2, 63
Hickman, John W., 1, -, -, 100, -, 10, -, 1500, 80, 630, -, -, 40, 18, 245
Donaway, Edward M., -, -, 1, 75, -, 63, -, 2000, 17, 240, -, -, -, -, 320
Baker, Jonathan, -, -, 1, 25, 1, 21, -, 700, -, 31, -, -, -, -, 135
Wootten, Isaac N., -, -, 1, 72, -, 60, -, 1900, -, 162, 5, -, -, -, 190
Daisey, Wm. F., -, -, 1, 25, 1, -, -, 250, 8, 13, -, 1, -, -, -, 62
Hastings, Wm. S., -, -, 1, 90, 2, 55, -, 1000, 30, 135, 10, -, -, -, 65
Brittingham, Benj. B., -, -, 1, 100, 20, 20, -, 1200, 40, 320, 10, -, -, -, 200
Parker, Henry J., -, -, 1, 54, 1, 35, -, 400, 18, 152, -, -, -, -, 290
Pusey, Nathan K., -, -, 1, 46, 1, -, -, 600, 8, 152, -, -, -, -, 255
Gray, Joshua M., -, -, 1, 68, -, 90, -, 1800, 15, 200, 7, -, -, -, 200
Gunby, Jacob M., 1, -, -, 50, 1, 22, -, 1400, 15, 179, 10, -, -, -, 270
Gunby, Lourena E., 1, -, -, 40, 5, 20, -, 2000, 25, 235, 20, -, -, -, 315
Donaway, Henry, 1, -, -, 56, 1, 69, -, 1000, 85, 290, 75, -, -, -, 278
Jermen, Leonard, 1, -, -, 38, 1, 12, -, 450, 18, 129, -, -, -, -, 23
Moore, John W., -, -, 1, 40, 2, 8, -, 300, 20, 157, 10, -, -, -, 248
Truitt, James H., -, -, 1, 29, -, 54, -, 700, 15, 205, 5, -, -, -, 50
Hitchens, John E., -, -, 1, 40, -, 89, -, 1000, 15, 174, 5, -, -, -, 60
Baker, Joseph E., 1, -, -, 70, -, 40, -, 1300, 65, 390, 7, -, -, -, 370
Donaway, Thomas H., -, -, 1, 80, -, 60, -, 2000, 35, 317, -, -, 18, 6, 195
Wells, William H., 1, -, -, 60, 1, 40, -, 1000, 125, 312, -, -, -, -, 162
Gray, Henry W., -, -, 1, 85, -, 22, -, 750, 22, 248, -, -, -, -, 82
Baker, Levin H., -, -, 1, 90, -, 60, -, 1800, 50, 365, -, -, 150, 52, 255
Mitchell, Rufus W., 1, -, -, 88, -, 12, -, 1500, 70, 345, -, -, -, -, 375
Jestur, Isaac, 1, -, -, 58, 1, 17, -, 1000, 33, 340, 10, -, 16, 8, 200
Lewis, Silas J., 1, -, -, 50, 1, 30, -, 1100, 27, 285, 80, -, -, -, 200
Donaway, Wm. P., 1, -, -, 65, -, 39, -, 1500, 12, 325, 5, -, -, -
Baker, Abisha, -, -, 1, 53, -, 54, -, 2000, 12, 185, -, -, 75, 52, 245
Lewis, John W., -, 1, -, -, 50, -, 75, -, 600, 25, 120, 3, -, 15, 4, 110
Donaway, Thomas O., 1, -, -, 75, 2, 25, -, 1200, 30, 364, 20, -, -, -, 190
Morris, William B., -, -, 1, 33, -, 10, -, 430, 15, 109, 10, -, -, -, 75
Hudson, Joseph W., 1, -, -, 72, -, 85, -, 1400, 115, 380, 80, 78, 150, 52, 392
Tingle, Mathias, -, -, 1, 83, -, 17, -, 1200, 50, 350, -, -, -, -, 250
Gordy, Stansbury, -, -, 1, 65, -, 6, -, 800, 20, 170, 30, -, -, -, 230
Donaway, John G., 1, -, -, 75, -, 20, -, 1200, 40, 315, -, -, 10, 2, 202
Baker, Samuel, -, -, 1, 90, -, 2, -, 1000, 125, 360, -, -, 40, 14, 297
Baker, Lambert A., -, -, 1, 50, -, 3, 20, 1000, 30, 182, -, -, -, -, 155
Bouden (Bowden), Joseph J., -, -, 1, 18, -, 54, -, 600, 12, 140, -, -, -, -, 105
Collins, Noah J., -, -, 1, 12, -, 50, -, 150, 8, 30, -, -, -, -, 38
Bowden, John H., -, -, 1, 8, -, 75, -, 175, 5, 85, 25, -, -, -, 35
Downs, James M., -, -, 1, 6, -, 75, -, 175, 4, 15, 12, -, -, -, 15

Ake, Thomas J., -, -, 1, 16, -, 40, -, 300, 15, 50, 6, -, -, -, 185
Donaway, Peter C., -, -, 1, 30, -, -, -, 450, 8, 140, 5, -, -, -, -
Macollister, William N., -, -, 1, 80, -, 100, -, 1620, 20, 230, 5, -, 1, 94, 340
Sampson, James, -, -, 1, 45, -, 41, -, 700, 10, 174, -, -, -, -, 241
Truitt, Edwards C., -, -, 1, 40, -, -, -, 600, 5, 150, -, -, -, -, 183
Truitt, Martha, 1, -, -, 85, -, 75, -, 1300, 75, 257, -, -, -, -, 277
Wells, John V., -, 1, -, 20, -, 3, -, 275, 20, 125, -, -, -, -, 70
Short, James W., 1, -, -, 50, -, 10, -, 1400, 75, 900, -, 45, 125, 52, 35
Wyatt, Elijah L., -, -, 1, 40, 1, -, -, 800, 12, 163, -, -, -, -, 250
Wootten, Isaac, 1, -, -, 26, -, -, -, 2000, 45, 300, -, -, 100, 26, 96
West, James B., -, -, 1, 72, -, 68, -, 2000, 10, 116, -, -, -, -, 435
Ake, John S., 1, -, -, 40, 2, 9, -, 1000, 20, 134, 5, -, -, -, 217
Ake, Hiram J., 1, -, -, 40, 1, 12, -, 1000, 20, 115, -, -, -, -, 194
Ake, Thomas H., 1, -, -, 45, -, 25, -, 700, 17, 216, 5, -, -, -, 140
Morris, D. B., -, -, 1, 35, -, 15, -, 500, 20, 112, -, -, -, -, 220
Jerman, James E., 1, -, -, 28, -, 20, -, 500, 12, 79, 3, -, -, -, 77
Baker, Joseph D., 1, -, -, 60, 2, 40, -, 1600, 29, 470, 15, -, -, -, 290
Downs, Elijah R., -, -, 1, 60, -, 35, -, 600, 9, 67, 8, -, -, -, 150
Melson, Benjamin G., -, -, 1, 38, -, 25, -, 500, 6, 7, 73, -, -, -, 785
Matthews, Catharine, 1, -, -, 66, -, 1, 50, -, 1050, 77, 165, 5, 20, -, -, 258
Cormean, Jacob L., 1, -, -, 20, 2, 15, -, 600, 2, 45, 10, -, -, -, 65
Selby, William H., 1, -, -, 2, -, 6, -, 200, 25, 165, 5, -, -, -, 17
Lany, James, 1, -, -, 5, -, 2, -, 125, 1, 1, 3, -, -, -, 5
Gordy, Peter B., -, 1, -, 130, 10, 72, -, 2800, 100, 758, 20, 30, 75, 70, 320
West, Albert, -, -, 1, 75, -, 24, -, 800, 3, 48, 10, -, -, -, 130
Hudson, Elijah, 1, -, -, 85, 6, 30, -, 1600, 37, 275, 15, 32, -, -, 315
Downs, Mary R., 1, -, -, 64, 4, 10, -, 700, 42, 258, -, -, -, -, 220
Hearn, Isaac T., 1, -, -, 50, 2, 50, -, 1000, 200, 686, 30, 30, -, -, 285
Phillips, John H., 1, -, -, 44, 1, 44, -, 800, 50, 322, 5, -, -, -, 173
West, John H., 1, -, -, 100, 6, 129, -, 2000, 50, 396, 25, 50, 50, 30, 370
Watson, James B., -, 1, -, 60, 3, 30, -, 500, 10, 175, 10, 30, 10, 4, 145
Gordy, Benjamin F., 1, -, -, 50, 2, 15, -, 1400, 25, 388, -, -, 15, 4, 267
White, William B., -, -, 1, 50, -, 314, -, 3000, 50, 167, -, -, 50, 15, 140
Foskey, Elijah W., -, -, 1, 80, 1, 102, -, 1800, 25, 152, 10, -, 20, 10, 203
Niblett, Levin H., -, -, 1, 32, -, 30, -, 650, 10, 100, 5, -, -, -, 205
Pusey, George W., -, -, 1, 70, -, 50, -, 1200, 52, 177, 10, -, -, -, 235
Hearn, Amanda, 1, -, -, 70, 5, 50, -, 1600, 80, 242, 16, -, -, -, 180
Brittingham, Hiram B., 1, -, -, 40, 1, 20, -, 500, 15, 83, 20, 6, -, -, 160
White, William S., 1, -, -, 60, 1, 30, -, 800, 18, 28, -, -, 45, 12, 149
Cordry, William T., -, -, 1, 60, -, 70, -, 1600, 50, 153, -, -, -, -, 255
Hearn, Clement C., 1, -, -, 120, 5, 50, -, 2500, 118, 671, 20, 38, -, -, 318
Niblett, George, -, -, 1, 16, -, 20, -, 450, 7, 66, 5, -, -, -, 97
Cordry, Martin H., -, -, 1, 56, -, 56, -, 1200, 45, 230, 10, -, -, -, 270
Condry, Jobe S., -, -, 1, 30, -, -, -, 600, 12, 75, 15, -, -, -, 105
Dukes, Paynter, -, -, 1, 65, -, 110, -, 1750, 40, 225, 15, 28, -, -, 310
Littleton, Minos S., -, -, 1, 25, -, 66, -, 725, 8, 63, 20, -, -, -, 100

Cannon, Joseph H. B., 1, -, -, 30, 2, 54, -, 1250, 400, 210, 812, -, 75, 14, 150
Brittingham, John H., -, -, 1, 84, 10, 50, -, 2000, 25, 250, 8, 28, -, -, 470
Hastings, George W., -, -, 1, 50, -, 108, -, 1000, 30, 200, 10, -, -, -, 100
King, Benjamin S., 1, -, -, 30, 4, 6, -, 900, 20, 60, 3, -, -, -, 84
Gordy, John L., -, -, 1, 100, -, 10, -, 1200, 50, 290, -, -, -, -, 612
Pennel, Lemuel, 1, -, -, 75, -, 50, -, 1000, 20, 240, 10, -, -, -, 295
Mitchell, Benjamin S., -, -, 1, 24, -, 15, -, 400, 5, 75, -, -, -, -, 190
Hearn, James E., 1, -, -, 55, 3, 10, -, 1200, 25, 617, 20, 25, -, -, 296
White, William J., -, -, 1, 20, -, 25, -, 600, 10, 100, 3, 10, -, -, 112
Heall(Neall), Nathaniel, 1, -, -, 40, 1, 23, -, 700, 10, 175, -, -, -, -, 148
Brittingham, Kendall P., -, -, 1, 80, -, 20, -, 1200, 35, 230, 10, 14, -, -, 283
Hearn, Joseph W., 1, -, -, 40, 3, 30, -, 1400, 75, 394, 12, 27, 50, 10, 345
Jones, Jacob R., 1, -, -, 35, 2, 40, -, 600, 75, 101, 25, -, -, -, 205
Niblett, Purnel J., -, -, 1, 40, -, 20, -, 400, 6, 80, 30, -, -, -, 153
McFaden, James, P., 1, -, -, 20, 1, 10, -, 800, 8, 225, 5, -, -, -, 74
West, Byard I., -, -, 1, 52, 4, 60, -, 1000, 20, 164, -, -, 10, 4, 354
White, Robert C., 1, -, -, 10, -, 93, -, 900, -, 9, -, -, -, -, 50
Warrington, Edward D., -, -, 1, 85, 6, 85, -, 1800, 15, 85, 10, -, -, -, 90
Mitchell, Joshua, -, -, 1,50, 2, 6, -, 500, 20, 101, 5, -, -, -, 200
Nicholson, Elijah J., -, -, 1, 12, 2, -, -, -, 6, 92, 1, -, -, -, 35
King, Mathias R., 1, -, -, 100, -, 25, -, 1500, 100, 200, 10, -, -, -, 270
King, William C., -, -, 1, 20,8, 30, -, 400, 5, 75, 2, -, -, -, 40
Foskey, James, -, -, 1, 70, -, 22, -, 560, 8, 110, -, -, -, -, 125
Downs, Benjamin M., 1, -, -, 86, 1, 18, -, 1100, 15, 137, 5, -, -, -, 150
Hastings, James W., -, -, 1, 48, 1, 22, -, 900, 15, 157, 5, -, -, -, 161
Palmer, Viletus H., -, -, 1, 80, -, 50, -, 1200, 25, 185, 8, 15, -, -, 240
Hastings, Benjamin B., 1, -, -, 115, 8, 40, -, 1800, 58, 425, 20, 30, 200, 52, 565
Phillips, William H., 1, -, -, 72, 2, 66, -, 1500, 40, 287, 12, 3, -, -, 180
Mitchell, George W., -, -, 1, 40, 1, 8, -, 500, 5, 120, -, -, 6, 3, 72
Ward, James H., 1, -, -, 50, -, 25, -, 1200, -, 280, -, -, -, -, 215
Mitchell, Nancy, 1, -, -, 24, -, 23, -, 600, 12, 225, 10, -, 20, 8, 247
Mitchell, Richard H. J., 1, -, -, 24, 1, 23, -, 800, 20, 135, 20, -, -, -, 237
Hitchens, Joseph H., -, -, 1, -, 1, -, -, -, 15, 121, -, -, -, -, -
Niblett, Charles H., -, -, 1, -, 3, -, -, -, 10, 195, 15, 6, -, -, 197
Betts, Charles R., 1, -, -, 100, 2, 100, -, 1800, 45, 486, 26, 6, -, -, 462
Downs, Jessie S., -, -, 1, 34, -, 3, -, 700, 18, 112, 5, -, -, -, 83
Evans, Lemuel, 1, -, -, 25, -, 25, -, 500, 10, 140, -, -, -, -, 60
Truitt, Philip T., 1, -, -, 52, 4, 40, -, 1900, 15, 235, 10, -, -, -, 309
Condry, Joseph W., -, -, 1, 40, 1, 20, -, 700, 10, 110, 6, -, -, -, 136
Phillips, Benjamin B., -, -, 1, 48, -, 54, -, 1000, 75, 298, 20, -, 20, 3, 245
Evans, Samuel C., 1, -, -, 80, -, 50, -, 1100, 37, 253, 12, 12, -, -, 189
Benson, Samuel E., -, -, 1, 100, 2, 25, -, 1200, 17, 285, -, -, -, -, 246
Workman, Parson T., 1, -, -, 100, -, 100, -, 1800,, 30, 326, 15, -, -, -, 390
Ennis, Stephen R., -, -, 1, 4, -, -, -, 125, 8, 115, 3, -, -, -, 15
Sturgis, Elijah, -, -, 1, 32, -, -, -, 550, 8, 227, 90, -, -, -, 175
LeKites, James H., -, -, 1, 60, -, 100, -, 1280, 140, 215, 15, -, -, -, 210

Prettyman, Burton C., 1, -, -, 75, -, 50, -, 4000, 225, 425, 50, 25, 75, 30, 300
Hurdle, William T., 1, -, -, 66, -, 75, 25, 1500, 125, 655, 50, 9, 150, 40, 140
Warrington, William T., 1, -, -, 250, 35, 148, -, 8660, 300, 1226, 100, 100, -, -, 1000
Carpenter, Wingate S., -, -, 1, 50, -, -, -, -, 25, 120, -, -, -, -, 253
Drain, Gardner H., -, -, 1, 50, -, 93, -, 1430, 25, 142, 15, -, -, -, 117
Wright, Nicholas, -, -, 1, 40, -, 160, -, 1600, 25, 55, 15, -, -, -, 125
Steel, Thomas, -, -, 1, 90, -, 110, -, 2000, 35, 475, 25, -, -, -, 190
Veasey, William -, -, 1, 160, -, 70, -, 5000, 165, 350, 50, 34, 4, 92, 502
Perry, John M., 1, -, -, 40, -, 140, -, 2000, 105, 270, 40, 68, -, -, 513
Mustard, John, 1, -, -, 60, -, 44, -, 2000, 110, 390, 25, 40, -, -, 280
Marsh, Mary A., -, -, 1, 120, -, 77, -, 1600, 25, 180, 20, -, -, -, 233
Mustard, David, 1, -, -, 52, -, 52, -, 2000, 20, 169, 100, -, -, -, 157
Walls, Peter S. 1, -, -, 44, -, 18, -, 1000, 80, 245, 20, 50, 100, -, -, 475
Bennum, Henry O., -, -, 1, 120, -, 80, -, 2000, 500, 620, 50, 100, -, -, 475
Norwood, Samuel B., 1, -, -, 75, 12, 105, -, 1500, 150, 595, 25, 18, -, -, 325
Joseph, Theodore, -, -, 1, 40, -, 100, 108, 1700, 25, 64, 25, -, -, -, 95
McIlvane, Wrixham M., 1, -, -, 25, -, 75, -, 1500, 55, 240, 10, 33, -, -, 50
Prettyman, Wm. F., -, -, 1, 100, -, 65, -, 2000, 30, 210, 25, -, -, -, 218
Joseph, Jesse E., 1, -, -, 15, -, 58, 35, 1000, 130, 310, 25, -, 150, 36, 648
Wilson, Wingate, -, -, 1, 50, -, 50, 117, 1500, 20, 117, 30, -, -, -, 120
Johnson, William A., -, -, 1, 20, -, 10, -, 400, 30, 355, 20, -, -, -, 125
Atkins, Clayton, -, 1, -, 25, -, 25, 50, 500, 150, 555, 20, 36, 75, 25, 164
Johnson, Whittington, 1, -, -, 75, -, 25, -, 2500, 65, 270, 20,33, -, -, 369
Hunter, Joseph, 1, -, -, 80, -, 100, -, 3000, 235, 375, 30, 44, 125, 40, 470
Lingo, Joseph B., 1, -, -, 100, 43, 24, -, 3000, 325, 731, 30, 66, 30, 8, 768
Warrington, Silas M. Jr., 1, -, -, 90, -, 90, -, 3000, 150, 272, 25, -, -, -, 257
Joseph, Hezekiah, -, -, 1, 50, -, 27, -, 1000, 30, 135, 20, -, -, -, 141
Joseph, David H., -, -, 1, 45, -, 14, -, 600, 50, 182, 20, -, -, -, 125
Vaughn, Samuel M., 1, -, -, 70, -, 70, 26, 1500, 120, 411, 25, -, -, -, 200
Joseph, Alfred, -, -, 1, 70, -, 85, 70, 3000, 25, 103, 28, -, -, -, 90
Marvel, Manaen B., -, -, 1, 125, -, 115, -, 2000, 75, 415, 25, -, -, -, 577
Marvel, David, -, -, 1, 100, -, 100, -, 2000, 40, 135, 20, -, -, -, 220
Marvel, Thomas, 1, -, -, 40, -, 20, -, 800, 20, 94, 15, -, -, -, 110
Carey, William S., -, -, 1, 75, -, 69, 75, 2000, 70, 317, 30, -, -, -, 220
Burton, Peter W., 1, -, -, 60, -, 100, 230, 2730, 75, 140, 40, 22, -, -, 125
Walls, Asa, -, -, 1, 75, -, 55, -, 1500, 35, 205, -, 70, -, -, 130
Steel, James, -, -, 1, 50, -, 50, 75, 1200, 15, 100, 20, -, -, -, 160
Joseph, Zachariah S., 1, -, -, 78, -, 60, -, 1600, 115, 455, 30, 35, -, -, 275
McIlvane, John S., 1, -, -, 48, -, 35, 37, 600, 25, 81, 30, 35, -, -, 138
Wells, Josiah D., 1, -, -, 65, -, 40, -, 75, 75, 220, 30, 18, -, -, 250
Wright, Selina, 1, -, -, 40, -, 20, -, 600, -, 25, 20, -, -, -, 60
Drain, Daniel, 1, -, -, 53, -, 52, -, 1500, 180, 347, 25, 14, -, -, 210
Stockley, John M, 1, -, -, 37, -, 10, -, 800, 110, 203, 20, 35, -, -, 95
Wright, Phillip, -, -, 1, 40, -, 10, -, 300, 30, 105, 15, -, -, -, 40
Hazzard, Anne, 1, -, -, 40, -, 40, 42, 900, 50, 130, 15, -, -, -, 175

McColley, Edward A., -, -, 1, 60, -, 40, -, 1500, 45, 294, 30, -, -, -, 372
Hart, Samuel, 1, -, -, 61, -, 76, -, 1200, 120, 185, 15, 9, 50, -, 210
Hazzard, John E., 1, -, -, 80, -, 50, -, 2000, 155, 533, 40, 25, 25, -, 240
Hazzard, Alphonso, 1, -, -, 100, -, 50, 40, 3000, 175, 605, 30, 35, -, -, 230
Johnson, Perry M., 1, -, -, 35, -, 35, -, 700, 50, 135, 25, -, -, -, 75
Harris, William, -, -, 1, 170, -, 130, -, 3000, 135, 615, 50, 66, -, -, 450
Harris, George, -, -, 1, 70, -, -, -, 700, 50, 210, 15, -, -, -, 160
Wilson, Daniel B., -, -, 146, -, 71, -, 2000, 230, 662, 15, 19, 150, 75, 430
Johnson, William A., -, -, 1, 30, -, -, -, 1300, 60, 225, 15, -, -, -, 60
Steelman, John, -, 1, -, 75, -, 50, 50, 1600, 100, 130, 20, 40, 45, 25, 228
Burton, Lemuel P., 1, -, -, 139, -, 100, -, 3000, 405, 515, 30, 36, -, -, 275
Webb, Charles, -, -, 1, 84, -, 43, -, 2000, 50, 308, 20, -, -, -, 200
Webb, George M., 1, -, -, 75, -, 35, 40, 1500, 90, 205, 20, 11, -, -, 175
Robinson, Thomas, 1, -, -, 75, -, 50, 147, 2500, 85, 245, 20, 82, -, -, 220
Vessels, Miars B., 1, -, -, 100, -, 100, 80, 3800, 190, 537, 40, 22, -, -, 260
Joseph, John H., 1, -, -, 26, -, 26, -, 550, 75, 145, 15, -, -, -, 125
Messick, William E., -, -, 1, 15, -, -, -, 600, 40, 170, 30, -, -, -, 135
Marsh, Samuel P., 1, -, -, 60, 20, 20, 59, 2000, 145, 260, 25, 11, -, -, 290
Marsh, John A. Jr., 1, -, -, 22, -, -, -, 800, 60, 90, 20, 10, -, -, 110
Marsh, John A. Sr., 1, -, -, 70, 20, 70, -, 1600, 190, 310, 25, 18, -, -, 278
Marsh, James P. W., 1, -, -, 100, -, 50, 50, 2000, 195, 618, 30, 44, -, -, 578
Robinson, Thomas, 1, -, -, 40, 100, -, 60, 2000, 75, 5, 40, 40, -, -, -, 290
Robinson, George, 1, -, -, 100, 50, 100, 79, 2500, 53, 305, 60, 10, -, -, 325
Ennis, Peter R., -, -, 1, 90, -, 50, 57, 1500, 40, 165, -, -, -, 130
Robinson, Tilghman L., 1, -, -, 30, -, 20, -, 500, 150, 335, 15, 22, -, -, 140
Hazzard, William A., 1, -, -, 50, -, 50, 50, 1000, 65, 165, 30, -, 65, 32, 155
Robinson, William S., 1, -, -, 70, 30, 50, 10, 1600, 315, 420, 30, 22, 35, 15, 350
Simpler, Amos, 1, -, -, 40, 10, -, 10, 200, 30, 225, 20, -, -, -, 225
Hazzard, Robert, 1, -, -, 30, 30, 50, 49, 1000, 20, 87, 25, -, -, -, 95
Rickards, James, -, -, 1, 40, 50, 10, -, 800, 15, 150, 25, -, -, -, 98
Burton, Sarah P., 1, -, -, 20, 8, 5, -, 500, 27, 155, 25, -, -, -, 190
Miller, Samuel R., -, -, 1, 28, 28, 19, -, 500, 15, 240, 20, 6, -, -, 125
Johnson, Arthur H., -, -, 1, 40, 40, 50, 168, 1500, 15, 195, 30, -, -, -, 85
Waples, David M., -, -, 1, 90, -, 55, 55, 2500, 80, 215, 25, -, -, -, 250
Lawson, Radock, 1, -, -, 40, -, 30, 30, 1000, 95, 3000, 30, 18, -, -, 300
Scott, Mitchell, 1, -, -, 35, -, 10, -, 600, 45, 152, 20, -, -, -, 100
Brereton, James L., 1, -, -, 25, -, -, -, 300, 40, 113, 15, -, -, -, 150
Davidson, John W., -, -, 1, 150, -, 50, 100, 3200, 270, 443, 40, -, -, -, 380
Davidson, George P., -, -, 1, 85, -, 50, 165, 2500, 45, 285, 30, -, -, -, 125
Davidson, John, -, -, 1, 30, -, 30, -, 600, 15, 180, 15, -, -, -, 75
Mustard, David, 1, -, -, 45, 10, 65, -, 1200, 45, 110, 25, 7, -, -, 175
Mustard, Jams S., -, -, 1, 50, 5, 25, -, 1800, 55, 190, 20, 11, -, -, 210
Burton, Walter D., -, -, 1, 50, -, 15, -, 1000, 80, 405, 15, -, -, -, 125
Burton, Peter R., 1, -, -, 200, -, 100, -, 6000, 650, 1050, 25, 50, -, -, 440
Joseph, Elisha J., -, -, 1, 35, -, 50, 15, 1000, 30, 165, -, -, -, -, 60
Brereton, James A., 1, -, -, 75, -, 100, 38, 2500, 390, -, 30, 18, -, -, 290

Massey, John S., 1, -, -, 60, 12, 100, 25, 2000, 195, 520, 35, -, -, -, 210
Johnson, Walter S., -, -, 1, 20, -, -, 30, 500, 85, 165, 20, -, -, -, 95
Robinson, Peter, -, -, 1, 50, -, 50, 100, 2000, 25, 125, 20, -, -, -, 180
Joseph, Thomas A., 1, -, -, 70, -, 58, -, 2000, 225, 550, 30, 18, -, -, 175
Lingo, Salathiel, 1, -, -, 70, -, 180, -, 3000, 320, 495, 40, 36, 160, 40, 300
Reynolds, Thomas R., 1, -, -, 30, -, 35, -, 700, 30, 158, 20, -, -, -, 100
Wells, Joseph, -, -, 1, 25, -, -, 25, 500, 20, 60, 25, -, -, -, 75
Burton, John S., 1, -, -, 40, -, 35, -, 800, 65, 200, 20, -, -, -, 225
Copes, Joseph, -, -, 1, 40, -, 160, -, 1500, 30, 50, 30, -, -, -, 80
Lynch, William, 1, -, -, 24, -, 48, -, 1000, 100, 130, 10, -, -, -, 100
Lynch, Robert, 1, -, -, 55, -, 25, 25, 1000, 60, 215, 25, 5, -, -, 95
Green, John O., 1, -, -, 50, 15, 5, 10, 1000, 95, 157, 25, -, 145, 50, 175
Massey, John B., 1, -, -, 35, -, 65, 50, 1100, 25, 142, 20, -, -, -, 100
Joseph, Zachariah, -, -, 1, 65, -, 96, 100, 1600, 70, 450, 25, -, -, -, 250
Hood, James, 1, -, -, 75, 60, 115, -, 2500, 172, 250, 20, 18, -, -, 260
Lynch, Peter R., 1, -, -, 75, 50, 73, 7, -, 2500, 205, 600, 30, 27, -, -, 225
Clendaniel, Joseph E., -, -, 1, 50, -, 150, -, 2000, 20, 135, 25, -, -, -, 160
Collins, Joseph L., 1, -, -, 20, -, 30, 15, 700, 20, 190, 20, -, -, -, 150
Lingo, William, 1, -, -, 15, -, 2, -, 500, 65, 310, 5, -, -, -, 375
Lingo, McClain, 1, -, -, 70, -, 30, -, 800, 30, 90, 25, 66, -, -, 150
Warrington, John, 1, -, -, 25, -, 5, -, 300, 70, 175, 15, 18, -, -, 150
Murray, Edward, 1, -, -, 66, -, 20, -, 100, 155, 490, 15, -, -, -, 320
Steel, John, 1, -, -, 25, -, 25, -, 700, 35, 145, 10, -, -, -, 88
Lank, Samuel J., -, -, 1, 100, 30, 100, 47, 3000, 85, 285, 15, -, -, -, 335
Burton, Thomas W., 1, -, -, 140, 40, 200, -, 5000, 310, 600, 40, -, -, -, 425
Wilson, George P., 1, -, -, 70, 25, 33, 66, -, 2000, 150, 160, 10, -, -, -, 130
Wilson, Jacob, 1, -, -, 50, 25, 17, -, 1500, 35, 185, 25, -, -, -, 135
Lingo, Alfred B., -, -, 90, 15, 65, -, 2500, 35, 430, 25, 18, -, -, 270
Hart, Arthur, -, -, 1, -, -, -, -, -, -, 225, 40, -, -, -, 350
Hobbs, John R., -, -, 1, 60, 120, -, 60, 2400, 30, 450, 10, -, -, -, 260
Warrington, Silas T., -, -, 1, 100, 130, 10, -, 2500, 70, 256, 15, 48, -, -, 300
Burton, Lewis, -, -, 1, 125, 63, 15, -, 3000, 300, 430, 20, 1110, 16, 8, 290
Steel, Thomas W., -, -, 1, 95, 140, 5, -, 2500, 110, 275, 15, 27, -, -, 490
Massey, Peter J., 1, -, -, 35, 70, -, 15, 1500, 170, 315, 25, -, -, -, 150
Wilson, Major H., -, -, 1, 30, 20, -, -, 500, 20, 175, 10, -, -, -, 95
Goslee, Salathiel B., 1, -, -, 37, 38, -, -, 1000, 90, 200, 5, 6, -, -, 179
Goslee, Edward S., 1, -, -, 115, 115, -, -, 2500, 175, 559, 10, -, 180, 52, 190
Miller, William F., -, -, 1, 47, 30, 10, -, 1500, 35, 190, 15, -, -, -, 200
Burton, John E. M., 1, -, -, 200, 40, 325,70, 5000, 615, 1500, 75, 120, 576, 156, 1000
Harman, Theodore P., -, -, 1, 50, -, 30, -, 1000, 25, 210, 15, 18, -, -, 150
Baylis, Benjaman M., 1, -, -, 130, -, 110, -, 2000, 150, 690, 25, 18, -, -, 750
Baker, William S., 1, -, -, 75, -, 34, -, 1500, 125, 390, 15, 9, -, -, 240
Lingo, Daniel _., 1, -, -, 200, 50, 150, -, 5000, 420, 558, 20, 27, -, -, 550
Burton, Elisha, -, -, 1, 50, 50, 90, -, 2000, 15, 176, 20, -, -, -, 140
Barton, Gideon W., 1, -, -, 60, 20, 73, -, 1300, 40, 239, 15, 27, -, -, 160

Street, Isaac, 1, -, -, 50, 12, 85, -, 1500, 20, 145, 25, -, -, -, 165
Lingo, Charles H., 1, -, -, 60, 59, 6, -, 2000, 1 80, 40, 10, 40, 100, 42, 270
Warrington, William F., -, -, 1, 25, -, 27, 25, 600, 25, 65, 10, 30, -, -, 95
Rust, Robert, -, -, 1, 58, 5, 231, 150, 3000, 242, 375, 12, -, -, -, 216
Marvel, William C., -, -, 1, 115, 12, 100, -, 2500, 75, 244, 30, 18, -, -, 270
Hansor, John C., -, -, 1, 65, -, 34, 10, 800, 10, 65, 15, -, -, -, 140
Collins, Charles E., 1, -, -, 120, -, 43, 43, 1400, 50, 268, 20, -, -, -, 190
Lingo, John A., 1, -, -, 80, -, 45, -, 3000, 190, 750, 40, 80, -, -, 400
Thompson, John W., 1, -, -, 40, -, 62, -, 800, 50, 176, 8, -, -, -, 170
Goslee, William W., 1, -, -, 60, -, 37, -, 1200, 173, 376, 6, 11, -, -, 185
Harmon, Isaac, 1, -, -, 200, -, 191, 10, 6000, 98, 457, -, 25, 75, 100, 24, 1210
McGee, David C., -, -, 1, 15, -, 22, -, 400, 25, 80, 5, -, -, -, 95
Atkins, Noah, 1, -, -, 51, -, 25, -, 1900, 80, 257, 10, -, -, -, 190
Street, Wingate, -, -, 1, 90, -, 7, -, 1500, 65, 180, 12, -, 36, 12, 200
Johnson, John B., 1, -, -, 30, -, 34, -, 800, 30, 170, 15, -, -, -, 195
Harmon, John, -, -, 1, 50, -, 25, -, 1000, 25, 85, 20, -, -, -, 150
Atkins, Edward P., -, -, 1, 80, -, 96, -, 1500, 26, 200, 25, -, -, -, 80
Warrington, Kendal, 1, -, -, 75, -, 38, 38, 1500, 125, 270, 20, -, -, -, 320
Lingo, Paynter E., 1, -, -, 110, -, 62, -, 1600, 90, 582, 25, -, -, -, 280
Clark, James H., 1, -, -, 140, -, 57, -, 1000, 45, 250, 20, -, -, -, 156
Carmean, Nathaniel, 1, -, -, 40, -, 10, 10, 1100, -, 115, 15, -, -, -, 80
Burton, William C., 1, -, -, 297, 53, 50, -, 8000, 655, 2568, 50, 110, -, -, 1837
McGee, Isaac W., -, -, 1, 100, -, 50, 50, 2000, 50, 200, 20, -, -, -, 155
Johnson, Robert B., -, -, 1, 40, -, 60, -, 900, 30, 193, 20, -, -, -, 220
Johnson, Sylvester H., 1, -, -, 13, -, 2, -, 250, 37, 120, 10, 6, -, -, 100
Rust, Charles H., 1, -, -, 40, -, 90, -, 700, 15, 65, 20, -, -, -, 50
Burton, Robert H., -, -, 1, 100, -, 60, -, 2000, 30, 235, 10, -, -, -, 282
Burton, William C. Jr., -, -, 1, 150, -, 50, -, 3000, 152, 480, 15, -, -, -, 112
Joseph,Willard, 1, -, -, 40, -, 24, -, 1000, 100, 240, 14, 10, -, -, 170
Miller, Nehemiah, -, -, 1, 30, -, 32, 10, 500, 45, 195, 10, -, -, -, 160
Joseph, Paynter, 1, -, -, 25, -, 8, -, 500, 20, 90, 5, -, -, -, 95
Townsend, Daniel C., 1, -, -, 90, -, 94, 54, 3700, 190, 665, 15, 5, 165, -, 170
Hudson, Joseph W., -, -, 1, 150, -, 52, -, 1000, 25, 180, 15, -, -, -, 108
Stevenson, Kendal R., 1, -, -, 75, -, 90, 50, 2500, 75, 169, 20, 22, -, -, 170
Stevenson, Robert D., 1, -, -, 75, -, 90, 50, 2500, 20, 262, 15, -, -, -, 160
Burton, Joshua R., 1, -, -, 100, -, 95, -, 2000, 90, 695, 25, 18, -, -, 285
Thoroughgood, Simon W., -, -, 1, 110, -, 21, -, 2500, 260, 360, 5, 30, -, -, 410
Prettyman, William H., -, -, 1, 83, -, 87, -, 1500, 90, 490, 25, 22, 170, 55, 290
Allen, Edward, -, -, 1, 128, -, 60, 62, 2500, 40, 370, 20, 39, -, -, 185
Plummer, John, -, -, 1, 34, -, 33, 33, 1000, 20, 80, 10, -, -, -, 165
Burton, John B., 1, -, -, 84, -, 42, 42, 1500, 155, 141, 16, -, -, -, 155
Walls, John H., -, -, 1, 30, -, 30, 25, 600, 35, -, 20, 150, -, -, 110
Virden, James H., 1, -, -, 65, -, 30, 31, 1000, -, 120, 20, -, -, -, 115
Frame, Paynter, 1, -, -, 225, 25, 131, 20, 5000, 255, 755, 40, 118, 175, 120, 613
Warrington, Silas J., -, -, 1, 60, -, 40, -, 700, 75, 195, 10, 30, -, -, 98
Wilson, Josiah, 1, -, -, 30, -, 50, -, 800, 35, 180, 20, -, -, -, 85

Simpler, Josiah, 1, -, -, 85, -, 50, 135, 1500, 100, 25, 50, 16, -, -, 208
Blizzard, Stephen E., 1, -, -, 60, -, 136, -, 2500, 155, 185, 45, 7, -, -, 165
Harmon, Nehemiah, -, -, 1, 60, -, 6, -, 600, 24, 80, 60, 10, -, 135
Rust, Peter W., 1, -, -, 60, -, 115, -, 2000, 150, 310, 20, 20, -, -, 385
Johnson, Charles R., 1, -, -, 25, -, 55, -, 500, 15, 90, 10, -, -, -, 75
Cooper, James J., -, -, 1, 26, -, 13, -, 500, -, 100, 12, 13, -, -, 170
Joseph, James S., -, -, 1, 52, -, 30, 30, 100, 25, 180, 15, -, -, -, 110
Walls, Gideon, 1, -, -, 65, -, 35, -, 1000, 30, 350, 10, -, -, -, 145
Walls, John, 1, -, -, 50, -, 7, -, 1000, 35, 80, 12, -, -, -, 150
Wilson, George W., -, 1, -, 12, -, 28, -, 1000, 52, 140, 10, -, -, -, 95
Norwood, Steven M., -, -, 1, 40, -, 10, 30, 400, 15, 84, 15, -, -, -, 85
Harmon, John, -, -, 1, 95, -, 45, 50, 1500, 16, 115, 12, -, -, -, 170
Johnson, James P., 1, -, -, 35, -, 53, -, 1000, 40, 140, 5, -, -, -, 240
Clark, Robert, -, -, 1, 75, -, 56, -, 1200, 85, 428, 10, -, -, -, 160
Harmon, Mitchel, -, -, 1, 50, -, 40, -, 800, 20, 105, 10, -, -, -, 160
Collins, Thomas P., 1, -, -, 80, -, 55, 25, 1500, 70, 224, 30, 5, -, -, 170
Lingo, George F., 1, -, -, 40, -, 40, 29, 1200, 25, 69, 15, -, -, -, 145
Dorman, Peter W., 1, -, -, 80, -, 80, -, 1800, 60, 193, 20, -, -, -, 170
Dorman, Abraham H., 1, -, -, 60, -, 73, -, 1500, 95, 259, 25, -, -, -, 380
Lingo, Eliza A., -, -, 1, 120, -, 136, -, 2000, 35, 305, 35, 44, -, -, 250
Street, David P., 1, -, -, 100, -, 144, -, 2500, 50, 345, 20, 10, -, -, 235
Thoroughgood, Edward, -, -, 1, 140, 30, 214, -, 3500, 85, 305 20, 34, 25, 10, 350
Morris, John W., -, -, 1, 50, -, 75, 25, 1800, 35, 135, 15, -, -, -, 140
Thoroughgood, John C., 1, -, -, 150, -, 100, 81, 3000, 325, 638, 30, -, -, -, 345
Rust, Peter A., 1, -, -, 45, -, 60, 15, 1000, 82, 173, -, 18, -, -, -, 115
Wright, David, -, -, 1, 42, -, 59, 15, 1000, 20, 85, 10, -, -, -, 120
Walls, Jonathan W., -, -, 1, 45, -, 20, 79, -, 600, 25, 33, 20, -, -, -, 45
Lawson, Selby, -, -, 1, 137, -, 83, -, 1500, 45, 355, 25, 20, -, -, 305
Wilson, Charles W., 1, -, -, 25, -, 58, -, 600, 12, 95, 10, -, -, -, 140
Thoroughgood, Lemuel D., 1, -, -, 130, -, 125, -, 2500, 215, 335, 20, 30, -, -, 245
Davis, Henry, -, -, 1, 85, 15, 40, 45, 1500, 35, 155, 18, -, -, -, 185
Rust, William T., -, -, 1, 60, -, 40, 25, 1000, 15, 80, 5, -, -, -, 150
Marvel, Andrew J., -, -, 1, 140, -, 110, -, 2500, 30, 210, 15, 105, -, -, 190
Thoroughgood, James F., 1, -, -, 60, -, 74, -, 1500, 110, 180, 20, -, -, -, 178
Waples, Cornelius, 1, -, -, 63, -, 63, -, 1000, 100, 235, 30, -, -, -, 130
Shockley, Peter S., -, -, 1, 100, -, 100, 50, 2000, 20, 60, 20, -, -, -, 85
Vickers, Obed, -, -, 1, 50, -, 50, -, 1000, 25, 160, 14, -, -, -, 65
Dary, George W., -, -, 1, 40, -, 260, -, 3000, 40, 140, 20, -, -, -, 218
Davidson, James J., -, -, 1, 50, -, 28, -, 1000, 75, 265, 10, 35, -, -, 95
Thompson, Jeremiah, 1, -, -, 105, -, 89, -, 2000, 15, 165, 5, -, 12, 5, 225
Collins, Horatio P., -, -, 1, 100, -, 250, 66, 3000, 45, 160, 30, -, 40, 13, 185
Gibbs, Thomas, -, 1, -, 30, -, 38, 30, 600, 45, 85, 10, -, -, -, 115
Milver(Miller), Major, -, -, 1, 65, -, 85, 100, 2000, 55, 150, 15, -, -, -, 130
Coffin, James B., 1, -, -, 60, -, 30, 30, 1000, 60, 277, 16, -, -, -, 225
Coffin, David H., 1, -, -, 25, -, 20, 13, 800, 65, 130, 15, -, -, -, 140
Frame, Henry C., 1, -, -, 100, -, 152, -, 2520, 180, 375, 20, 25, 136, 35, 250

McGee, John V., -, -, 1, 30, -, 70, 50, 1800, 16, 72, 20, -, -, -, 60
Barker, Thomas S., 1, -, -, 45, -, 55, -, 1200, 155, 240, 15, 22, -, -, 215
Barker, Henry L., 1, -, -, 150, -, 50, -, 3000, 127, 520, 30, 40, 150, 50, 450
Joseph, William C., -, -, 1, 75, -, 125, 100, 2500, 90, 415, 24, 11, -, -, 170
Joseph, Charles H., -, -, 1, 20, -, 10, 10, 500, 40, 175, 10, -, -, -, 75
Cannon, William D., 1, -, -, 30, -, 30, -, 600, 15, 140, 20, -, -, -, 95
Morris, George, -, -, 1, 50, -, 100, 50, 1200, 15, 80, 16, -, -, -, 90
Messick, Lydia A., 1, -, -, 30, -, 60, -, 1000,80, 153, 20, -, -, -, 100
Hobbs, Mathias B., -, -, 1, 50, -, 25, 25, 900, 15, 95, 18, -, -, -, 85
Rogers, Nathaniel I., 1, -, -, 40, -, 60, -, 700, 24, 100, 10, -, -, -, 90
McGee, William F., 1, -, -, 50, -, 35, 35, 1200, 25, 115, 20, -, -, -, 160
Simpler, James B., 1, -, -, 40, -, 95, 90, 1500, 65, 228, 25, -, -, -, 90
McGee, John W., -, -, 1, 200, -, 100, 200, 3500, 85, 290, 40, 18, -, -, 340
Rust, Thomas B., -, -, 1, 60, -, 34, 42, 800, 25, 155, 15, -, -, -, 40
Hurdle, Joseph C., 1, -, -, 35, -, 27, 28, 700, 50, 235, 15, -, -, -, 110
Davidson, Samuel, -, 1, -, -, 100, -, 85, 100, 1795, 40, 75, 20, -, -, -, 115
Rust, Thomas, 1, -, -, 75, -, 50, -, 1500, 55, 240,15, 20, -, -, 175
Walls, Gilley S., 1, -, -, 70, -, 80, -, 1600, 120, 200, 20, 11, -, -, 275
Green, George P., -, -, 1, 50, -, 45, -, 800, 77, 190, 15, -, -, -, 120
Johnson, Albert J., 1, -, -, 100, -, 72, -, 1500, 80, 265, 25, 6, -, -, 275
Johnson, David, 1, -, -, 50, -, 30, -, 700, 30, 215, 15, -, -, -, 120
Johnson, Erasmus C., 1, -, -, 30, -, 50, 47, 650, 15, 75, 10, -, -, -, 85
Johnson, Buchanan, -, -, 1, 35, -, 75, 89, 1500, 25, 75, 15, -, -, -, 90
Wilson, Kendal R., -, -, 1, 50, -, 50, -, 1500, 105, 190, 20, 11, -, -, 165
Pettyjohn, Truitt, 1, -, -, 150, -, 215, 100, 4000, 100, 455, 25, 36, -, -, 280
Rust, James, 1, -, -, 125, -, 34, -, 1500, 115, 265, 20, -, 12, 4, 250
Rust, William B., 1, -, -, 30, -, 70, 50, 1000, 40, 160, 15, -, -, -, 65
Griffith, McElroy M., -, -, 1, 55, -, 25, 40, 800, 25, 76, 15, -, -, -, 700
Lawson, George W., 1, -, -, 60, -, 10, -, 700, 20, 75, 25, -, -, -, 105
Bryan, George T., -, -, 1, 60, -, 30, 10, 1000, 25, 180, 15, -, -, -, 325
Lawson, David C., -, -, 1, 40, -, -, 10, 700, 30, 100, 5, 35, -, -, 210
Lawson, Jehue, 1, -, -, 65, -, 22, 35, 800, 75, 323, 20, -, -, -, 130
Hurdle, Jacob F., 1, -, -, 55, -, 25, 45, 900, 100, 150, 15, 11, -, -, 135
Lawson, Robert P., -, -, 1, 60, -, 28, 40, 1100, 75, 100, 15, 35, -, -, 145
Hurdle, William W., 1, -, -, 100, -, 41, 41, 2500, 200, 255, 30, -, -, -, 280
Hickman, Harbison, Walker Farm, 1, -, -, 100, -, 5, -, 6000, 300, 1550, 150, 200, 300, 60, 1250
Hickman, Harbison, Hickman Farm, 1, -, -, 50, -, -, -, 2000, 100, 250, -, -, 90, 12, 600
Hickman, Harbison, Long Farm, 1, -, -, 90, -, -, -, 4000, 200, 300, -, 100, 140, 28, 1000
Maull, Samuel R., Rowland Farm, -, -, 1, 40, -, -, -, 1000, 60, 150, -, -, -, -, 400
Maul,Thomas S., 1, -, - ,40, -, -, -, 2000, 175, 250, -, -, -, - ,600
Maull, James R., 1, -, -, 30, -, -, -, 1000, 60, 200, -, -, -, -, 400
Russel, Thomas C., -, -, 1, 60, -, -, -, 1500, 175, 200, -, -, -, -, 500
Parker, John, -, -, 1, 10, -, -, -, 800, 45, 225, -, -, -, -, 300

Lyons, Charles, -, -, 1, 40, -, -, -, 2500, 100, 300, -, -, 100, 52, 600
Henell, Robert, 1, -, -, 60, -, -, -, 4040, 200, 375, -, 100, 100, 52, 700
Metcalf, John, 1, -, -, 60, -, -, -, 1500, 150, 250, 10, -, 100, 20, 600
Lyons, Labon L., 1, -, -, 50, -, -, -, 2500, 240, 300, 10, 120, 200, 44, 750
Richardson, Ephraim R., 1, -, -, 20, -, -, -, 1400, 50, 240, -, -, 75, 15, 300
Waples, Michael M., -, -, 1, 60, -, -, -, 2600, 100, 300, -, 140, 100, 20, 600
McIlvain, Comfort, 1, -, -, 100, -, -, -, 2500, 100, 300, 10, -, 100, 20, 500
Morris, Elihu J., 1, -, -, 110, -, -, -, 2750, 150, 250, 15, 150, 100, 20,750
Salmond, George, -, -, 1, 60, -, -, -, 1600, 75, 200, 10, -, -, -, 400
Hitchens, Edward D., 1, -, -, 120, -, -, -, 3200, 200, 400, 20, 140, 50, 30, 1000
Waples, Henry, -, -, 1, 40, -, -, -, 1400, 35, 200, 5, -, -, -, 300
Houston, Shepard P., -, -, 1, 90, -, -, -, 2400, 75, 200, 5, -, 100, 20, 400
Maull, David R., -, -, 1, 40, -, -, -, 1200, 75, 275, 10, -, -, -, 428
Lyons, Henry, -, -, 1, 100, -, -, -, 2500, 75, 275, 5, 30, -, -, 728
Lodge, John K., 1, -, -, 40, -, -, -, 200, 75, 175, 5, 50, 100, 20, 850
Lodge, George W., 1, -, -, 20, -, -, -, 1200, 50, 150, -, 20, -, -, 300
Lodge, Samuel J., 1, -, -, 27, -, -, -, 1500, 50, 150, -, 40, 20, 4, 375
Hickman, Harbison, N. W. Hickman Farm, 1, -, -, 112, -, -, -, 6000, 150, 500, 40, 300, 250, 50, 1350
Hickman, Harbison, Clayton Farm, 1, -, -, 144, -, -, -, 5000, 140, 500, 20, 350, 240, 48, 1900
West, William A., 1, -, -, 60, -, -, -, 2400, 75, 275, 10, 30, 100, 20, 2000
Chase, George, -, -, 1, 100, -, -, -, 2000, 75, 375, -, -, 30, 6, 460
Harrison, William J., -, -, 1, 50, -, -, -, 1500, 75, 300, 5, 40, -, -, 625
Wolfe, Henry, 1, -, -, 200, 100, -, -, 4000, 150, 350, 100, 150, 100, 20, 950
Stokely, William, -, -, 1, 120, 20, -, -, 3000, 75, 150, -, -, 50, 10, 525
Stokeley, Henry, -, -, 1, 140, -, 20, -, 3500, 100, 175, -, -, 50, 10, 554
Williard, James, -, -, 1, 240, 120, -, -, 7000, 300, 500, 200, 150, 150, 30, 850
Marshall, John P., 1, -, -, 125, -, 20, -, 4000, 100, 300, -, -, 100, 20, 800
Thomas, Groom, -, 1, -, 200, -, 30, -, 6000, 200, 500, 150, 150, 160, 32, 1125
Drane, David, -, -, 1, 100, -, 25, -, 2000, 75, 75, -, -, -, -, 300
Hickman, Harbison, Houston Farm, 1, -, -, 288, -, 40, -, 4000, 300, 600, 350, 320, 300, 60, 1940
Walker, Thomas, 1, -, -, 200, -, 30, -, 4000, 100, 400, 75, 100, 100, 20, 150
Carey, Joseph, -, -, 1, 100, -, -, -, 2000, 100, 300, -, 25, -, -, 275
Hausee, Robert, 1, -, -, 100, 20, 10, -, 1500, 50,75, -, -, -, -, 700
Hood, John N., 1, -, -, 240, 40, 20, -, 4000, 150, 50, 100, 100, 100, 20, 1350
Wolfe, William P., 1, -, -, 210, 30, 20, -, 2500, 75, 150, -, 30, 50, 50, 1100
Green,William -, -, 1, 120, 25, 10, -, 1200, 50, 125, -, 6, -, 20, 450
Wolfe, Daniel, 1, -, -, 105, 15, 15, -, 2300, 100, 275, -, 60, 50, 10, 900
Paynter, Richard G., 1, -, -, 40, -, -, -, 1000, 50,75, -, -, -, -, 300
Hart, Joseph, -, -, 1, 100, 20, 10, -, 3000, 100, 250, -, 600, 100, 20, 100
Frazier, Henry, -, -, 1, 80, 10, -, -, 1600, 75, 150, -, -, -, -, 500
Metcalf, Stephenson, -, -, 1, 20, -, -, -, 600, 30, 75, -, -, -, -, 225
Dodd, Joseph H., 1, -, -, 700, 30, 50, -, 7000, 340, 250, 200, 300, 200, 40, 2500

Warrington, Charles K., 1, -, -, 160, -, 40, -, 4000, 150, 600, 10, 100, 100, 20, 1200

Wiltbanks, David, -, -, 1, 40, -, -, -, 800, 50, 100, -, -, -, -, 200

Welch, Joseph, 1, -, -, 244, -, 30, -, 3000, 100, 300, -, 30, 120, 24, 800

Harmon, Pernell, -, -, 1, 40, -, -, -, 1000, 50, 200, -, -, -, -, 200

Harmon, Thomas, -, -, 1, 110, -, 20, -, 2500, 75, 200, 5, -, -, -, 350

Fisher, Hiram C., 1, -, -, 160, -, 30, -, 3200, 100, 275, -, 60, -, -, 700

Dodd, James A., 1, -, -, 280, -, 20, -, 8500, 250, 950, 100, 150, 200, 40, 2140

Groom, William T., -, -, 1, 210, -, 39, -, 4000, 100, 400, -, 30, 75, 15, 600

Howard, Samuel J., 1, -, -, 177, -, 28, -, 2500, 75, 250, 3, -, -, -, 575

Drane, Jacob, -, -, 1, 129, -, 20, -, 2000, 75, 225, 5, -, -, -, 450

Hazzard, David, -, 1, -, 140, -, -, -, 3000, 75, 240, 60, 30, 100, 20, 100

Sommers, James, -, 1, -, 300, 25, 30, -, 4000, 100, 250, 60, 60, 120, 24, 1950

Baker, William -, -, 1, 100, -, -, -, 1500, 75, 300, 60, 30, 60, 12, 650

Huelser (Huelson), James C., 1, -, -, 200, 25, 20, -, 3500, 150, 350, 100, 90 55, 11, 1650

King, William H., -, -, 1, 260, 20, 60, -, 4500, 160, 650, 120, 110, 60, 12, 1600

Stokely, Stockley L., -, -, 1, 40, -, -, -, 1000, 50, 200, 30, 30, 50, 10, 430

Stokeley, William W., 1, -, -, 30, -, -, -, 1200, 60, 275, 50, 30, 60, 12, 475

Thraul, Mary, 1, -, -, 110, -, -, -, 2500, 60, 300, 50, 60, 150, 30, 600

Donavan, Levan A., -, -, 1, 60, -, 10, -, 1500, 75, 175, 50, 25, 50, 10, 675

Collins, Jonathan C., 1, -, -, 100, -, 15, -, 2500, 100, 1100, 50, 50, 120, 24, 925

Tunnell, Edward, 1, -, -, 220, -, 30, -, 4000, 100, 300, -, -, -, -, 800

Melson, William S. Sr., 1, -, -, 100, -, 10, -, 2000, 75, 275, -, -, -, -, 450

Bailey, Sadie, -, -, 1, 70, -, 10, -, 1600, 75, 250, -, -, -, -, 350

Futcher, John M., 1, -, -, 40, -, 5, -, 1400, 75, 325, -, -, -, -, 400

Fisher, Robert W., 1, -, -, 200, -, 35, -, 4000, 100, 400, -, -, -, -, 1100

Burton, Thomas P., -, -, 1, 125, -, 20, -, 2500, 100, 300, -, -, -, -, 600

Burton, Cathern, 1, -, -, 100, -, 15, -, 2000, 100, 300, -, -, -, -, 600

Palmer, Robert L., -, -, 1, 140, 10, 40, -, 2200, 75, 350, -, -, -, -, 850

Palmer, Samuel P., 1, -, -, 160, 10, 50, -, 2000, 5, 275, -, -, -, -, 900

Cannon, Reece, 1, -, -, 40, -, -, -, 700, 50, 75, -, -, -, -, 250

Downing, James B., -, -, 1, 120, -, 10, -, 2500, 75, 250, 20, 30, 100, 20, 750

Cannon, William, -, -, 1, 60, -, 10, -, 900, 50, 150, -, 90, -, -, 400

Phillips, Edward C., 1, -, -, 200, 33, -, -, 4000, 100, 500, 15, 90, 150, 30, 1400

Paynter, Richard G. Sr., -, -, 1, 100, -, 10, -, 1000, 50, 150, -, 30, 30, 6, 400

White, Sheppard P., 1, -, -, 60, -, -, -, 1200, 75, 250, -, 60, 30, 6, 400

Conwell, Jacob H., 1, -, -, 80, -, 15, -, 2000, 75, 300, -, 60, 150, 30, 600

Morris, James P., 1, -, -, 60, -, -, -, 1200, 50, 225, -, 30, 30, 6, 360

Lynch, Theodore H., -, -, 1, 120, -, -, -, 1600, 75, 350, -, 30, 30, 6, 750

Holland, Mann, -, -, 1, 240, -, 20, -, 4000, 100, 550, 10, 90, 150, 30, 2000

Donovan, Loda, -, -, 1, 160, -, 20, -, -, 100, 400, -, 60, 100, 20, 1900

Burton, James, 1, -, -, 100, -, -, -, 1000, 50, 200, 5, -, 50, 10, 450

Hevelo, John, 1, -, -, 40, -, -, -, 800, 50, 200, -, -, -, 250

Robinson, Orange, 1, -, -, 120, -, 20, -, 1500, 75, 300, -, 30, 100, 20, 550

Lynch, James, 1, -, -, 160, -, 30, -, 2000, 75, 300, -, 30, 100, 20, 600

Hudson, Jane, 1, -, -, 40, -, - -, 800, 50, 75, 15, -, -, -, 200
Marsh, Andrew J., 1, -, -, 110, -, 10, -, 3300, 75, 201, -, 30, 50, 10, 550
Marsh, Sydnham, 1, -, -, 110, -, 10, -, 3000, 75, 201, -, 30, 50, 10, 550
Thompson, William P., 1, -, -, 140, -, 15, -, 5000, 150, 550, -, 60, 150, 30, 110
Holland, John C., 1, -, -, 250, 30, 50, -, 6500, 200, 800, -, 90, 200, 40, 1600
Hart, Thomas, 1, -, -, 105, -, 10, -, 2750, 100, 50, 20, 60, 100, 20, 800
Miller, David, -, -, 1, 80, -, -, -, 1500, 75, 200, -, 30, 100, 20, 375
Prettyman, George, -, -, 1, 100, -, -, -, 2500, 75, 200, -, 30, 120, 24, 425
Stokely, William, A., 1, -, -, 60, -, -, -, 1200, 75, 225, -, 30, 60, 12, 400
Stokeley, Henry B., 1, -, -, 80, -, -, -, 1500, 75, 240, -, 25, 50, 10, 450
Thompson, Sarah M., 1, -, -, 160, -, 20, -, 3000, 100, 425, 10, 60, 150, 30, 850
Burton, Elijah, 1, -, -, 100, -, -, -, 2000, 75, 375, -, 45, 50, 10, 800
Wolfe, Henry D., 1, -, -, 80, -, 10, -, 2000, 100, 375, -, 60, 75, 15, 800
Robinson, Robert B., 1, -, -, 160, -, 30, -, 3000, 100, 450, -, 50, 150, 30, 800
Thompson, Rhodes, 1, -, -, 110, 20, 10, -, 2500, 100,3 50, 10, 30, 100, 20,700
Carpenter, Gideon T., -, -, 1, 105, -, 10, -, 1500, 75, 375, 5, 30, 75, 15, 550
Little, Henry, -, -, 1, 120, -, -, -, 1500, 50, 200, 5, 25, 30, 6, 375
Stokely, Woodman, -, -, 1, 200, 25, 20, -, 3000, 100, 400, -, 40,150, 30, 110
Marsh, Lemuel W., 1, -, -, 105, -, 25, -, 1600, 75, 225, 5, 25, 25, 5, 400
Dodd, William A., 1, -, -, 250, -, 40, -, 5000, 200, 750, 10, 100, 200, 40, 1600
Lynch, Mary E., 1, -, -, 160, -, 10, -, 2500, 100, 350, -, 60, 120, 24, 1150
Hood, Charles H., 1, -, -, 100, -, 10, -, 1200, 50, 200, -, 25, 60, 12, 400
Williard, James, -, 1, -, 240, 40, 30, -, 5000, 150, 800, 20, 120, 225, 45, 1700
Green, John, -, -, 1, 90, -, -, -, 1200, 75, 725, 10, 25, 30, 6, 425
Cord, Blizzard, -, -, 1, 120, -, 20, -, 1500, 75, 250, -, 45, 60, 12, 575
Wright, Return, -, -, 1, 80, -, -, -, 1000, 50, 150, -, 25, 30, 6, 400
Blizzard, Leven A. C., 1, -, -, 80, -, 5, -, 1600, 75, 200, 10, -, -, -, 300
Morris, Richard P., 1, -, -, 75, -, 5, -, 1500, 100, 300, -, 50, 150, 30, 650
Burton, Comfort G., 1, -, -, 45, -, -, -, 900, 50, 200, -, -, -, -, 300
Lodge, Martin V., 1, -, -, 37, -, -, -, 1000, 75, 200, 10, -, 15, 3, 450
Warren, Jacob S., -, -, 1, 140, -, 15, -, 3000, 100, 340, 20, 60, 100, 20, 80
Marvel, Edgar, -, -, 1, 130, -, 10, -, 3500, 100, 300, 10, 60, 100, 120, 1000
Prettyman, Hetty, 1, -, -, 60, -, -, -, 1500, 100, 350, -, 50, 120, 24, 875
Lankford, John N., -, -, 1, 30, -, -, -, 1000, 50, 100, -, -, -, -, 550
Cord, Alfred, -, -, 1, 54, -, 5, -, 2000, 75, 300, -, 30, 100, 20, 600
Johnson, George W., 1, -, -, 50, -, -, -, 1250, 75, 350, 5, 30, 60, 12, 575
Marvell, James H., 1,-, -, 100, -, 40, -, 1000, 50, 200, 10, -, -, -, 375
Walls, John W., -, -, 1, 120, -, -, -, 2500, 100, 350, -, 60, -, -, 1000
Blizzard, Henry C., -, -, 1, 200, -, 100, -, 3000, 100, 300, 20, 60, 100, 20, 950
Lyons, Labon L., 1, -, -, 210, -, 20, -, 3300, 200, 375, 10, 120, 250, -, 1600
Marsh, Erasmus W. W., 1, -, -, 110, -, 25, -, 1500, 75, 300, 5, 30, 60, -, 425
Marsh, Joseph W., 1, -, -, 80, -, 10, -, 2500, 100, 400, -, 60, 100, -, 1000
Lank, Levin, J., 1, -, -, 40, -, -, -, 1000, 50, 200, -, 25, 60, -, 425
Stevens, James, 1, -, -, 40, -, -, -, 1000, 50, 200, -, 25, 60, -, 425
Wright, Robert, -, -, 1, 80, -, 10, -, 1200, 75, 200, 5, 30, 50, -, 425
Drane, Jacob H., -, -, 1, 120, -, 20, -, 1500, 75, 300, -, 30, 75, -, 500

Wilson, John C., -, -, 1, 210, -, 20, -, 4000, 100, 375, -, 60, 100, 20, 950
Coursey, Eliza, 1, -, -, 80, -, -, -, 1250, 50, 200, -, 25, -, -, 400
Wright, Elisha, 1, -, -, 120, -, 10, -, 2500, 100, 300, 10, 50, 300, 20, 600
Wright, Mariah, 1, -, -, 40, -, -, -, 1000, 50, 200, -, 25, 60, 12, 350
Beynum, William H., -, -, 1, 120, -, 15, -, 1500, 75, 200, -, 25, 40, 8, 450
Futcher, Joseph F., 1, -, -, 140, -, 20, -, 3000, 100, 500, 20, 60, 150, 30, 1400
Joseph, Samuel W., 1, -, -, 160, -, 20, -, 3000, 100, 300, -, 60, 100, 20, 900
Drane, Abraham, -, -, 1, 100, -, 100, -, 1500, 75, 150, -, 25, 50, 10, 475
Pride, Wingate, -, -, 1, 200, -, 40, -, 3000, 75, 250, 5, 30, 100, 20, 700
Roach, John C., 1, -, -, 125, -, 10, -, 2500, 100, 350, -, 30, 120, 24, 875
Hopkins, William, 1, -, -, 170, -, 20, -, 3500, 100, 500, 10, 75, 150, 30, 1100
Coverdale, George A., -, -, 1, 240, -, 50, -, 3000, 100, 400, -, 100, 200, 40, 1050
Martin, Robert H., 1, -, -, 160, -, 15, -, 3500, 100, 375, 3, 60, 120, 24, 875
Vesey, John L., 1, -, -, 120, -, 10, -, 2500, 100, 350, -, 60, 100, 20, 828
Warrington, James E., 1, -, -, 125, -, 15, -, 2500, 75, 350, 10, 50, 120, 24, 850
Warrington, David M., 1, -, -, 160, -, 25, -, 3000, 75, 350, 20, 50, 100, 20, 825
Palmer, Thomas D., -, -, 1, 120, -, 10, -, 1200, 50, 175, 5, 25, 60, 10, 675
Steel, George, -, -, 1, 140, -, 25, -, 1500, 50, 175, -, 25, 60, 12, 700
Wilson, James B., -, -, 1, 110, -, 20, -, 1100, 50, 150, 5, 24, 75, 15, 600
Johnson, Erasmus, -, -, 1, 100, -, 10, -, 1000, 50, 175, -, 30, 50, 10, 750
Johnson, John E., -, -, 1, 125, -, 20, -, 2000, 75, 200, 10, 30, 100, 20, 675
Carpenter, William, -, -, 1, 100, -, 25, -, 2500, 75, 200, 15, 30, 100, 20, 600
Lank (Lauk), Hetty, 1, -, -, 140, -, 20, -, 1500, 50, 250, 10, 30, -, -, 675
Reynolds, Elka___, 1, -, -, 200, -, 40, -, 4000, 100, 600, -, 90, 200, 40, 1550
Wilson, Samuel, -, -, 1, 80, -, 5, -, 1200, 50, 200, 10, 25, 50, 10, 500
Carpenter, Lemuel L., -, -, 1, 140, -, 20, -, 2250, 50, 175, 10, 25, 60, 12, 550
Johnson, Nellie, -, -, 1, 120, -, 10, -, 2500, 50, 200, -, 25, 60, 12, 550
Turner, Thomas W. B., 1, -, -, 125, -, 20, -, 3000, 100, 500, -, 90, 200, 40, 1850
Tindale, John, -, -, 1, 200, -, 50, -, 2500, 75, 200, -, 30, 50, 10, 550
Jones, William P., 1, -, -, 160, -, 10, -, 3500, 150, 750, -, 125, 300, 60, 2400
Dell, William A., 1, -, -, 140, -, -, -, 3000, 100, 300, 10, 60, 175, 35, 6900
Wilson, Thomas, B., -, -, 1, 120, -, -, -, 3500, 100, 300, 10, 60, 100, 20, 1250
King, David H., 1, -, -, 125, -, 20, -, 2000, 75, 150, 5, 25, 60, 12, 1050
Carson, Alfred, 1, -, -, 80, -, 10, -, 2000, 75, 200, 15, 60, 40, 8, 900
Lyons, Rodney E., -, -, 1, 130, -, -, -, 3500, 100, 500, 10, 75, 200, 40, 100
Collins, Joseph J., -, -, 1, 120, -, 10, -, 3000, 75, 150, 5, 30, 60, 12, 675
Vesey, Edward T., -, -, 1, 100, -, -, -, 2000, 100, 300, 20, 30, 100, 20, 850
Joseph, John C., -, -, 1, 160, -, 30, -, 3000, 75, 300, 10, 60, 150, 30, 1200
Donavan, James H., -, -, 1, 160, -, 20, -, 3500, 75, 350, 20, 60, 120, 24, 950
Lauk (Lank), John M., -, -, 1, 200, -, 10, -, 1500, 75, 200, 10, 25, 50, 10, 500
Lank, James, -, -, 1, 200, 25, 20, -, 3000, 75, 275, 10, 245, 100, 20, 1050
Wilson, Major E., -, -, 1, 150, 20, 25, -, 2500, 75, 175, 10, 60, 100, 20, 600
Truitt, John H., 1, -, -, 60, -, 10, -, 1000, 50, 150, -, 25, 40, 8, 550
Hill, Theopilus, -, -, 1, 250, -, 40, -, 5000, 100, 400, -, 60, 20, 40, 1375
Paynter, Samuel C., 1, -, -, 100, -, 10, -, 3000, 100, 400, 50, 60, 160, 32, 775
Waples, James W., -, -, 1, 300, -, 30, -, 7500, 100, 900, 60, 120, 240, 48, 1750

Warrington, James A., -, -, 50, 20, 10, -, 4000, 100, 600, 20, 60, 160, 32 1600
King, James A., -, -, 1, 250, 50, 15, -, 5000, 100, 600, 15, 90, 120, 24, 1550
Robinson, David H., -, -, 1, 200, 25, 10, -, 3500, 75, 200, 10, 40, 140, 28, 725
Palmer, Lemuel W., -, -, 1, 115, -, 10, -, 3000, 75, 200, -, 25, 160, 32, 1000
Lodge, Samuel, 1, -, -, 150, -, 20, -, 3000, 100, 300, -, 120, 300, 60, 1500
Donavan, W. Bevers, -, -, 1, 200, -, 20, -, 3000, 225, 600, -, 125, 320, 65, 2050
Morrix, John L., -, -, 1, 400, -, 50, -, 8000, 225, 950, -, 300, 480, 96, 3450
Palmer, William P., -, -, 1, 220, -, 30, -, 4000, 225, 850, -, 250, 390, 75, 1800
Tindale, William, -, -, 1, 100, -, 10, -, 1500, 75, 175, 10, 30,60, 12, 550
Norwood, William, 1, -, -, 120, -, -, -, 2500, 100, 350, 10, 30, 40, 8, 600
Jackson, Alfred, -, -, 1, 60, -, -, -, 900, 50, 125, 5, -, -, -, 250
Maull, Orange, -, -, 1, 80, -, -, -, 1100, 50, 125, 5, -, -, -, 250
Paynter, Thomas, -, -, 1, 50, -, -, -, 8800, 50, 125, 5, -, -, -, 280
White, Orange, -, -, 1, 40, -, -, -, 700, 50, 125, 5, -, -, -, 250
Hevelo, Moses, 1, -, -, 200, 20, 50, -, 2000,75, 250, 10, 30, 100, 20, 700
Cordrey, Eligha, 1, -, -, 75, -, 1, -, 1100, 50, 100, 10, -, -, -, 410
Burch, Jozah, 1, -, -, 68, -, 12, -, 1500, 80, 290, 10, 20, -, 4, 500
Records, Jonathan T., 1, -, -, 75, -, 30, -, 2000, 25, 70, 8, 10, -, 3, 300
Pennel, H. F., 1, -, -, 110, -, 125, -, 2500, 75, 300, 15, 20, 50 4, 500
Boyce, Daniel E., -, -, 1, 725, -, 75, -, 3000, 100, 400, 10, -, 125, 52, 600
Dickerson, L. A., 1, -, -, 150, -, 150, -, 1500, 50, 300, 10, 5, 70, 6, 500
Hearn, Stephan, 1, -, -, 45, -, 20, -, 1500, 50, 100, -, -, 75, 4, 490
Stradling, Moses, 1, -, -, 30, -, 20, -, 500, 12, 125, -, 25, -, -, 600
Hastings, John, -, -, 1, 140, -, 80, -, 2500, 30, 200, 10, 12, 16, 45, 520
Ellis, John H., 1, -, -, 20, -, 7, -, 300, 10, 100, 8, 6, -, -, 360
Calloway, Jobe, 1, -, -, 90, -, 30, -, 1000, 20, 50, 10, 12, 25, 4, 379
Lynch, Richard, 1, -, -, 75, -, 35, -, 1000, 35, 200, 20, 17, 35, 7, 560
Lynch, Samuel, 1, -, -, 75, -, 30, -, 900, 10, 250, 10, 17, 35, 8, 600
James, Wm. L., 1, -, -, 80, -, 50, -, 1510, 200, 300, 12, 15, 40, 12, 550
Parsons, W. W., -, -, 1, 85, -, 143, -, 2000, 28, 100, -, -, -, -, 400
Cormean, Elija W., 1, -, -, 30, -, 120, -, 800, 12, 150, -, -, -, -, 300
Elliott, Mary E., 1, -, -, 50, -, 20, -, 800, 12, 120, 30, -, -, -, 410
King, George, 1, -, -, 132, -, 50, -, 200, 15, 116, -, -, -, -, 450
Elliott, N. J., -, -, 1, 100, -, 50, -, 2000, -, 200, 15, 20, 60, 13, 600
Wiley, John W., -, -, 1, 200, -, -, -, 2500, -, 200, 100, 30, -, -, 700
Baeer, Jacob, 1, -, -, 60, -, 40, -, 1500, 100, 200, 10, 12, 60,17, 600
Parsons, S. H., 1, -, -, 45, -, -, -, 800, 50, 50, -, -, 20, 10, 300
Smith, Martial, 1, -, -, 200, -, 300, -, 10000, 460, 800, 25, 55, -, -, 1500
Hastings, W., 1, -, -, 250, -, 100, -, 7000, 500,800, 30, 200, 500, 104, 1500
Hearn, Edward R., 1, -, -, 100, -, 75, -, 1500, 100, 300, 20, 10, 100, 50, 700
Phillips, John, -, -, 1, 30, -, -, -, 800, 25, 125, 20, 17, 20, 4, 500
Truitt, Sultry, -, -, 1, 33, -, -, -, 1000, 10, 125, -, -, -, -, 495
Britingham, M. H., 1, -, -, 50, -, 15, -, 2000, 25, 175, 5, 35, 100, 52, 600
Elliott, J. S., 1, -, -, 200, -, -, -, 1500, -, 150, -, -, -, -, 560
Hastings, Joseph, 1, -, -, 67, -, 33, -, 10000, 100, 250, 10, 35, 25, 3, 600
Hearn, Kizza, 1, -, -, 40, -, -, 74, 800, -, -, -, -, -, -, 50

Elliott, Josephus, -, -, 1, 50, -, -, 50, 800, 10, 100, -, -, -, -, 30
West, Panter, -, -, 1, 26, -, -, -, 300, 8, 30, -, -, -, -, 50
Elliott, J. W., 1, -, -, 35, -, -, 45, 1100, 10, 100, 8, -, -, -, 100
Ward, L. G., 1, -, -, 60, -, -, 49, 900, 30, 100, 10, -, -, -, 200
Gordy, E. H., 1, -, -, 230, -, -, 40, 300, 100, 75, 10, 75, 100, 52, 500
Ward, Thomas B., 1, -, -, 45, -, -, 58, 1700, 100, 250, 15, -, -, -, 200
Hearn, Wm. C., 1, -, -, 120, -, -, -, 177, 3000, 100, 400, 10, 30, 125, 52, 401
White, S. Wells, 1, -, -, 50, -, -, 30, 80, 100, 125, -, -, -, -, 200
White, J. G., 1, -, -, 11, -, -, -, 1500, -, -, -, -, -, -, 75
James, Wm. F., 1, -, -, 150, -, 50, -, 1500, 75, 240, 14, 14, -, -, 225
Jones, G. W., 1, -, -, 150, -, 50, -, 1500, 75, 277, 14, 14, -, -, 225
Morris, John, 1, -, -, 75, -, 40, -, 1500, 25, 225, 10, 12, -, -, 454
Melson, G. T. B., 1, -, -, 80, -, 100, -, 1800, -, 125, 58, 966, 8, -, 200
Hastings, S. W., -, -, 1, -, -, -, -, -, -, -, -, -, -, -, 200
Ward, William, 1, -, -, 90, -, 54, -, 3000, 20, 230, -, -, 25, -, 30
Ward, S., 1, -, -, 70, -, 80, -, 3000, 95, 350, 10, -, 350, 52, 150
Ward, John, 1, -, -, 187, -, 187, -, 5510, 60, 280, 5, -, -, -, 280
Maddox, N. R., 1, -, -, 47, -, -, -, 400, 30, 150, 10, 6, -, -, 30
Maddox, W. B., 1, -, -, 53, -, -, -, 600, 10, 100, 8, -, -, -, 400
Bacon, Thomas, 1, -, -, 120, -, 35, -, 4500, 100, 870, 25, 150, 100, 20, 1200
Boyce, Watson, -, -, 1, 204, -, 136, -, 6800, -, -, -, -, -, -, -
Bacon, John L., -, -, 1, 57, -, 57, -, 3500, -, -, -, -, -, -, -
Bacon, Samuel, 1, -, -, 125, -, 55, -, 4500, 200, 600, 30, 150, 200, 52, 840
Kinney, E. C., -, -, 1, 140, -, 80, -, 2750, 40, 200, 10, -, -, -, 400
Brown, James, -, -, 1, 55, -, 20, -, 900, -, -, -, -, -, -, 250
Smith, George, -, -, 1, 100, -, 100, -, 3001, -, 40, -, -, -, 6, 150
Riggin, T. H. 1, -, -, 85, -, 4, -, 3115, 50, 175, -, 100, 150, 52, 800
Records, W. B., 1, -, -, 235, -, 135, -, 4700, 200, 500, 25, 150, 225, 50, 1000
Horsey, T. C., 1, -, -, 25, -, -, -, 6000, 50, 150, -, 40, 150, 40, 400
Records, E. W., 1, -, -, 33, -, 17, -, 750, 25, 100, -, 25, -, -, 150
Ellis, Joseph, 1, -, -, 125, -, 87, -, 3180, 100, 200, 20, 20, 200, 48, 200
Collins, J. D. D., 1, -, -, 40, -, 20, -, 1200, 125, 400, 10, 50, 75, 12, 500
Brooks, B. F., 1, -, -, 300, -, 46, -, 8650, 200, 400, -, 201, -, -, 2317
Gordy, E. P., 1, -, -, 212, -, 40, 3000, 125, 250, 25, 200, 150, -, 500
Phillips, G. B., 1, -, -, 106, -, -, -, 1272, 100, 300, 150, 95, 300, 104, 250
Phillips, G. B., 1, -, -, 15, -, -, -, 2500, -, -, -, -, -, -, 130
Phillips, G. B., 1, -, -, 30, -, -, -, 3000, -, -, -, -, -, -, 550
Elliott, F. G., 1, -, -, 60, -, 35, -, 1500, 150, 75, 20, 30, 20, 2, 600
Hearn, Isaac, 1, -, -, 85, 10, 17, 3, 2000, 100, 300, 10, 28, 10, 3, 600
White, W. G., -, -, 1, 35, -, 40, -, 800, 25, 200, 8, 30, 25, 6, 500
Gillis, Gillis B., -, -, 1, 67, 14, 35, 1, 2000, 12, 300, 20, -, -, -, 620
Elliott, Hettie C., 1, -, -, 100, -, 207, -, 2500, 35, 175, 70, 30, 50, 9, 700
Gordy, Harry, 1, -, -, 78, -, 1500, 25, 200, 10, 20, 37, 4, 620
Warrington E. H., 1, -, -, 108, -, 108, -, 3000, 100, 300, -, -, 25, 5, 550
Elliott, Alford, -, -, 1, 100, -, 75, -, 1500, 20, 175, 8, 24, -, -, 535
Swain, Benjamin, -, -, 1, 40, -, 43, -, 800, 30, 125, -, -, -, -, 530

Elliott, W. W., 1, -, -, 75, -, 72, -, 1000, 100, 300, -, -, -, 600
Hearn, Frank, 1, -, -, 100, -, 20, -, 1025, 25, 175, -, -, -, -, 419
Winget, M. W., 1, -, -, 40, 10, 40, 8, 1000, 50, 200, 5, 20, 35, 3, 500
West, James, 1, -, -, 40, -, 30, -, 700, -, 150, 15, 22, 15, 2, 459
Hastings, Hisakiah, 1, -, -, 7, -, 15, -, 2000, 50, 150, 10, 9, -, -, 500
Hastings, William Thomas, 1, -, -, 50, -, 25, -, 600, 25, 125, -, -, -, -, 250
Low, Sebba, 1, -, -, 100, -, 55, -, 2500, 30, 200, 10, 30, 27, 4, 500
McGee, D. W., 1, -, -, 150, -, 50, -, 2500, 20, 330, 18, 23, 60, 15, 800
Oliphan, T. B., 1, -, -, 20, -, -, -, 400, 25, 30, -, 6, -, -, 300
Henry, John, -, -, 1, -, -, -, -, -, -, -, -, -, -, -, 75
Hill, James A., 1, -, -, 150, -, 100, 10, 300, 50, 400, 15, 23, 25, 4, 640
Oliphan, E. M., 1, -, -, 125, -, 87, -, 2500, 110, 300, 11, 60, 40, 7, 600
Lynch, John S., 1, -, -, 100, -, 40, -, 1, 2000, 30, 300, 10, -, -, -, 300
Collier, Z. J., -, -, 1, -, -, -, -, -, -, -, -, -, -, -, -, 150
Carmean, Wm., 1, -, -, 40, -, 12, -, 250, 25, 150, 5, -, -, -, 500
Hastings, Louvenia, 1, -, -, 20, -, 10, -, 400, 10, 125, -, 8, -, -, 400
Elliott, Harry B., 1, -, -, 50, -, 18, -, 800, 25, 150, -, -, -, -, 250
Elliott, Wade J., 1, -, -, 50, -, 3, -, 600, 25, 150, -, -, -, -, 50
Davis, L. J., -, -, 1, 61, -, -, 30, -, 900, 20, 125, 10, 8, -, -, 500
Corden, Eliza, 1, -, -, 125, -, 75, 10, 4000, 30, 400, 10, 50, 30, 8, 500
Lynch, Mary J., 1, -, -, 100, -, 10, -, 1500, 2, 210, 10, -, 35, 4, 500
Simeon, Joba, 1, -, -, 55, -, 10, -, 620, 25, 195, 10, 20, 35, 7, 500
Game, John, 1, -, -, 100, -, 38, -, 2500, 35, 150, 8, 37, 29, 8, 500
Collins, Stephan, -, -, 1, 60, -, 35, -, 800, 25, 210, -, 32, 35, 13, 567
Elliott, David H., -, -, 1, 65, -, 50, -, 1000, 12, 100, 4, -, -, -, 500
Collins, Martin, -, -, 1, 5, -, 24, -, 1000, 100, 200, -, 25, 10, 2, 634
Lecates, John, W., 1, -, -, 50, -, 80, -, 1300, 80, 135, -, 6, 10, 3, 250
Hastings, John H., 1, -, -, 80, -, 50, -, 1950, 100, 250, 10, 10, 75, 52, 500
Gordy, Sallie, 1, -, -, 70, -, 70, -, 2000, 25, 150, 8, -, 50, 37, 504
Hearn, William, -, -, 1, 100, -, 50, -, 800, 50, 175, -, -, -, -, 600
Anderson, J. W., 1, -, -, 300, -, 100, -, 4000, -, 660, 60, 150, 250, 122, 800
Wootten (Wrotten), Elijah, 1, -, -, 100, -, 45, 1000, 50, 160, 10, -, -, -, 500
Horsey, N., 1, -, -, 175, -, 75, -, 10000, 500, 2000, -, 300, 500, 104, 1000
Kinney, Samuel, 1, -, -, 50, -, 25, -, 1300, 35, 100, 10, -, -, -, 139
Collins, Levin A., 1, -, -, 40, -, 40, -, 560, 25, 100, 10, -, -, -, 400
Wright, James H., -, -, 1, 90, -, 20, -, 1500, 75, 175, 10, -, -, -, 350
Cormean, Burton, -, -, 1, 40, -, 30, -, 900, 25, 200, 8, -, -, -, 300
James, Edward H., 1, -, -, 80, -, 20, -, 1500, 30, 75, 5, -, 25, 6, 500
Elliott, Solomon, 1, -, -, 30, -, 10, -, 600, 20, 150, 10, 7, 28, 9, 410
Cormean, Jacob, -, -, 1, 40, -, 27, -, 800, 12, 100, -, -, -, -, 525
Culver, Wm. E., 1, -, -, 60, -, 15, -, 800, 25, 200, 5, -, -, -, 560
Baker, David, -, -, 1, -, -, -, -, -, -, 15, -, -, -, -, 50
Beach, Jonathan G., 1, -, -, 50, -, 15, -, 1200, 200, 350, 10, 21, 71, 10, 200
Dunn, Burton, 1, -, -, 100, -, 39, -, 1500, 100, 125, 20, 10, 50, 5, 400
Reddish, Lewis L., -, -, 1, 75, -, 25, -, 800, 100, 150, -, -, -, -, 300
Callaway, Lerie, -, -, 1, 50, -, 14, -, 700, 20, 60, -, -, -, -, 125

Hastings, Thomas T., -, -, 1, 45, -, 44, -, 1000, -, -, -, -, -, -, 100
Ellis, John A., -, -, 1, -, -, -, -, -, -, -, -, -, -, -, -, 150
Culver, William, -, -, 1, 50, -, 8, -, 1000, 25, 90, 5, -, -, -, 300
Culver, Washington, -, -, 1, 101, -, 20, -, 2000, 30, 150, 5, -, 35, 21, 400
Culver, Selathiel, -, -, 1, 150, -, -, -, 2000, 25, 400, 5, 40, 125, 52, 500
Culver, Wilmore, 1, -, -, 45, -, 33, -, 700, 20, 250, 5, -, -, -, 200
Cordrey, Elizabeth, 1, -, -, 20, -, 20, -, 800, -, -, -, -, -, -, 250
Collins, Charles F. P., -, -, 1, 75, -, 25, -, 1000, 50, 225, 10, -, 10, 3, 200
Henry, George W., -, -, 1, 120, -, 20, -, 800, 75, 125, -, -, 12, 2, 300
Dickerson, C. W., 1, -, -, 65, -, 40, -, 1500, 400, 300, 10, 5, 60, 14, 500
Culver, Minos, -, -, 1, -, -, -, -, -, -, -, -, -, -, -, -, 125
Records, George S., 1, -, -, 80, -, 17, -, 3000, 15, 400, 75, 350, 52, 500
Henry, John H., 1, -, -, 20, -, 5, -, 800, 25, 150, 10, 15, 15, 2, 500
Morris, James H., 1, -, -, 60, -, 40, -, 800, 35, 150, 10, -, 15, 2, 175
Walston, Thomas, 1, -, -, 100, -, 100, -, 2000, 50, 200, 5, -, 112, 30, 300
Collins, Mary A., 1, -, -, 30, -, 30, -, 800, 50, 100, -, -, -, -, 500
Collier, C. W. 1, -, -, 150, -, 70, -, 3700, 100, 300, -, -, 150, 50, 500
Kinking, John T., -, -, 1, 65, -, 100, -, 200, 1100, 200, 10, 8, 25, 3, 550
Ellis, A. D., 1, -, -, 30, -, 107, -, 600, 40, 250, -, -, -, -, 200
Ellis, M. R., -, -, 1, 70, -, -, -, 3400, 44, 300, 11, -, -, -, 300
Callaway, W. B., 1, -, -, 150, -, 140, -, 4000, 100, 400, 25, -, 75, 6, 700
Waller, Johnathan, 1, -, -, 150, -, 235, -, 4000, 200, 300, 20, -, 175, 52, 400
Hastings W. W., 1, -, -, -, -, -, -, -, -, -, -, -, -, -, -, 250
Hearn, Samuel T., 1, -, -, 37, -, 22, -, 1000, 35, 100, 5, 55, 60, 8, 254
Waller, J. _., 1, -, -, 8, -, -, -, 400, 6, 200, 12, -, 50, 6, 110
Waller, H. B., 1, -, -, 60, 28, -, 1500, 40, 250, -, 10, -, -, 250
Collins, John W., -, -, 1, 75, -, 25, -, 1000, 50, 125, -, -, -, -, 300
McKown, John, -, -, 1, 30, -, -, -, 500, 25, 125, -, -, -, -, 500
McGee, Wm. T., 1, -, -, 75, -, 50, -, 900, 45, 170, 10, 97, 25, 3, 300
Elliott, Wm. T., 1, -, -, 12, -, 60, -, 800, 30, 150, -, -, -, -, 250
Wootten, George M., 1, -, -, 60, -, 60, -, 1000, 25, 100, 5, 8, 35, 4, 510
McGee, John W., 1, -, -, 100, -, 115, -, 1500, 75, 150, 25, 50, 190, 52, 600
Vincent, M. W., 1, -, -, 50, -, 75, -, 1000, 100, 200, 10, 60, -, -, 500
Lloyd, Thomas E., -, -, 1, 45, -, 50, -, 800, -, -, -, -, -, -, 116
Cannon, P. Wm., 1, -, -, 100, -, 70, -, 2015, 300, 575, 25, 73, 40, 10, 800
Ellis, James V., -, -, 1, 75, -, 40, -, 2500, 10, 20, -, -, -, -, 100
Bradley, John, -, -, 1, 25, -, 51, -, 600, 20, 110, 5, -, -, -, 200
Adams, Henry, 1, -, -, 230, -, 270, -, 5000, 125, 400, -, -, -, -, 275
Adams, Jacob H., 1, -, -, 200, -, 125, -, 4000, 200, 500, -, 120, -, -, 500
Beach, William, -, -, 1, 60, -, 40, -, 2000, 40, 125, 8, 10, -, -, 400
Kissikin (Kinikin), George, -, -, 1, 40, -, 20, -, 1000, 10, 125, 10, -, 5, -, 350
Bailey, James, -, -, 1, 60, -, 24, -, 1100, 100, 115, -, -, -, -, 410
Ralph, James E., -, -, 1, 95, -, 47, -, 1400, 75, 160, 25, 14, 10, -, 470
Records, Thomas, 1, -, -, 74, -, 30, -, 2000, 100, 75, 13, -, 16, -, 505
Cooper, N. C., -, -, 1, 101, -, 15, -, 2000, -, 150, -, -, -, -, -
Workman, Wm., -, -, 1, 45, -, 115, -, 1110, -, 25, 10, -, -, -, 400

Elliott, T. W., 1, -, -, -, -, -, -, -, -, -, -, -, -, -, 300
Walston, C. -., 1, -, -, 125, -, 100, -, 2000, 75, 200, 10, 75, 150, -, 1000
Twilley, James, E. 1, -, -, 75, -, 50, -, 1100, 40, 125, 8, 24, 25, -, 500
Twilley, Levin, 1, -, -, 150, -, 75, -, 2000, 50, 300, 20, 15, 20, -, 410
Ellis, Joseph, 1, -, -, 70, -, 50, -, 1000, 50, 100, 10, 50, 125, 400
Owens, Josiah E., 1, -, -, 130, -, 220, -, 2000, 100, 250, 20, 18, 260, -, 600
Nicols, J. A., 1, -, -, 85, -, 34, -, 800, 50, 125, 8, -, 25, -, 300
Phillips, Noah, -, -, 1, 40, -, 40, -, 800, 25, 150, 10, -, 20, -, 175
Ellis, Massib__, -, -, 1, 45, -, -, -, -, -, -, -, 10, -, -, -, 200
Body, James K., -, -, 1, 75, -, 75, -, 1000, 20, -, 5, -, -, -, 125
Phillips, Emaline, 1, -, -, 30, -, 200, 10, 5000, 125, -, 20, -, -, -, 400
Spencer, Samuel, 1, -, -, 59, -, -, -, 200, 50, -, 10, -, -, -, 175
Culver, Isaac H., -, -, 1, 150, -, 50, -, 5000, 100, 250, 10, -, -, -, 400
Collins, J. J., 1, -, -, 10, -, -, 300, 10, 60, -, 12, -, -, 200
Kissikin, S. G., -, -, 1, 100, -, 80, 20, 3000, 100, 200, 10, 20, 18, -, 500
Henry, Wm. B., 1, -, -, 30, -, 30, -, 700, 25, 125, 20, 3, 60, -, 300
Henry, John K., 1, -, -, 30, -, -, 5, 500, 25, 100, 5, -, -, -, 300
Collins, Jacob A., 1, -, -, 160, -, 40, -, 3000, 200, 400, 10, 25, 100, -, 150
Bradly, Sarn J., 1, -, -, 60, -, 10, -, 1000, 75, 150, 2, -, -, -, 500
Adam, Henry, 1, -, -, -, -, -, -, -, -, -, -, -, -, -, 50
Moore, Thomas S., 1, -, -, 100, -, -, -, 800, 10, 25, -, -, -, -, 400
Smith, James, 1, -, -, 16, -, 10, -, 400, 10, 200, 50, 20, 20, -, 200
Elzy (Elyz), Charles C., 1, -, -, 15, -, 25, -, 800, 100, 150, 8, -, -, -, 400
Wright, Jesse, 1, -, -, 50, -, -, 38, 1000, 50, 125, 25, -, -, -, 250
Ralph, David, 1, -, -, 75, -, -, 50, 1000, 100, 300, 20, -, -, -, 310
Walston, Lambert, 1, -, -, 40, -, 45, -, 2500, 60, 250, -, 40, 225, -, 800
Ellis, Samuel H., -, -, 1, 65, -, 30, -, 1000, 30, 100, -, -, -, -, -
Collins, John B., -, -, 1, 60, -, 50, -, 1000, 35, 100, 8, -, 25, 3, 400
Culver, Daniel S., 1, -, -, 62, -, 20, 2, 1400, 157, 260, 17, 23, 70, 10, 500
Culver, Handy, 1, -, -, 62, 30, 20, -, 1400, 150, 270, 10, 25, 125, 15, 500
Hastings, Elihu, 1, -, -, 20, 4, 15, 9, 500, 20, 230, -, 400, 20, 200, 50
Hastings, Charles, 1, -, -, 30, 1, 4, -, 350, 10, 50, 10, -, 20, 10, 50
Hastings, Moses, 1, -, -, 30, 1, 4, 9, 350, 10, 50, 8, -, 18, 500, 50
Hearn, Andrew J., 1, -, -, 180, 25, 125, 15, 14000, 100, 225, 200, 35, -, -, 350
Hearn, Joseph J., -, 1, -, -, -, -, -, -, -, 10, -, -, -, -, 40
Evans, Isaac, -, 1, -, -, -, -, -, -, -, -, -, -, -, -, -
Coffin, Leven, W., -, -, 1, 30, -, 50, -, 3000, 100, 175, 15, -, -, -, 100
Ellis, G. W., 1, -, -, 100, 8, 100, -, 6000, 500, 970, 10, 125, -, -, 100
Phillip, William P., -, -, 1, 50, -, 20, -, 1000, 50, 185, 300, -, 400, 2, 200
Kenney, E. W., 1, -, -, 90, 8, 104, 3, 4500, 75, 450, 25, 125, 250, -, 200
Ellis, L. W., 1, -, -, 60, 30, 219, -, 2000, 30, 170, 10, 47, 46, -, 200
Ellis, Wm. _., 1, -, -, 30, 20, 20, -, 900, 10, 150, 8, 30, 5, -, 150
Ellis, M. _., 1, -, -, 150, 25, 50, -, 8000, 600, 1000, 50, 500, 600, -, 2000
Ellis, M. _., 1, -, -, 46, -, -, -, 2000, -, -, 50, -, -, -, -
Ellis, M. _., 1,-, -, 60, -, -, -, -, -, -, -, -, -, -, -
Ellis, M. _., 1, -, -, 60, 2, 43, -, 3500, -, -, -, -, -, -, -

Ellis, M. _., 1, -, -, 50, -, 74, -, 3500, -, -, -, -, -, -, -
Ellis, M. _., 1, -, -, 40, 10, 60, -, 2000, -, -, -, -, -, -, -
Low, E. M. 1, -, -, 120, 18, 80, -, 5000, 700, 1050, 100, 240, 150, -, 1200
Henry, Isaac N., 1, -, -, 100, 20, 15, -, 2000, 150, 390, 10, 53, -, -, 375
Henry, Isaac N., 1, -, -, 12, -, 13, -, 500, -, -, -, -, -, -, 400
Low, James C., 1, -, -, 100, 3, 35, -, 2000, 150, 350, 20, 22, 50, -, 500
Phillips, T. A., 1, -, -, 60, 2, 60, -, 1100, 100, 250, 15, 20, 10, -, 400
Phillips, A. W., 1, -, -, 40, -, 40, -, 1200, 75, 200, 10, 10, 10, -, 320
Phillips, W. R., 1, -, -, 40, -, 40, -, 1400, 75, 300, 10, 10, 5, -, 450
Kenney, George A., 1, -, -, 25, -, -, -, 125, -, -, -, -, -, -, 100
Phillips, Samuel J., 1, -, -, 100, -, 100, -, 2000, 150, 350, 25, -, -, -, 800
Phillips, John E., -, -, 1, 100, -, 50, -, 1000, 50, 100, 10, -, -, -, 400
Phippin, Jos. G., -, 1, -, -, -, -, -, -, -, -, -, 5, -, -, -, 75
Phillips, Harram, C., 1, -, -, 75, -, 20, -, 1000, 50, 200, 25, -, -, -, 500
Cooper, John, 1, -, -, 200, -, 200, -, 4000, 400, 400, 80, 100, 300, -, 500
Cormean, Asbery, C., -, -, 1, 50, -, 90, -, 1000, 30, 150, 5, 32, -, 350
Hill, Clayton J., -, -, 1, 25, -, 50, -, 800, 20, 100, 4, 9, -, -, 400
Collins, John B., -, -, 1, 60, 10, 60, -, 1000, 25, 50, 6, 8, -, -, 520
Kenney, William, 1, -, -, 100, 3, 30, -, 1500, 125, 400, 30, 60, 75, 52, 875
Ellis, James J., 1, -, -, 120, -, 40, -, 2000, 200, 200, 60, 90, 150, 68, 900
Elliott, Joseph D., -, 1, -, 20, 10, 50, 40, 700, -, 100, 505, -, -, -, 100
Ellis, Anzy (Angy), 1, -, -, 30, -, 40, -, 1000, 150, 200, 808, 100, -, -, 200
Mills Jarris (Jarvis), -, -, 1, 125, -, 175, -, -, 125, 325, 10, -, -, -, 500
Bailey, Edward, -, -, 1, 50, -, 53, -, 125, -, 10, -, -, -, 200
Bailey, Edward, 1, -, -, 45, -, 24, -, 1000, -, -, 10, 25, -, -, 51
Cochey, Mary E., 1, -, -, 30, -, 20, -, 500, 25, 100, 5, 3, -, -, 60
Elzey, Charles, 1, -, -, 100, -, 150, -, 2000, 100, 350, 10, 50, 50, -, 500
Phillips, Samuel, 1, -, -, 50, -, 550, -, 10000, 200, 1000, 100, 150, 300, -, 1300
Bradley, Jas. F., 1, -, -, 60, -, 40, -, 1000, -, 100, 251, 25, 10, 51, -, 150
Buell, B., 1, -, -, 100, -, 80, -, 1800, 75, 300, 20, 45, -, -, 510
Bradley, J. A. D., 1, -, -, 80, -, 80, -, 1800, 70, 700, 20, 700, -, -, 403
Bennet, H. G., -, -, 1, 75, -, 65, 1525, 25, 100, 8, -, -, -, 300
Bradly, F. R., 1, -, -, 25, -, 62, -, 609, -, 15, 8, -, -, -, 40
Wright, Levin, -, -, 1, 50, -, 50, -, 700, -, 50, 10, 8, 5, -, 50
Knowles, Cannon, 1, -, -, 50, -, 50, -, 700, 30, 150, 5, 3, -, -, 300
Twilley, Robert, 1, -, -, 150, -, 62, -, 2000, 25, 200, 15, -, -, -, 300
Bradley, _. D., 1, -, -, 40, -, 20, -, 806, -, 40, 220, 18, 5, 25, -, 300
Collins, N. W., 1, -, -, 50, -, 12, -, 910, 20, 100, 10, 15, -, -, 300
Hearn, Wm. G., 1, -, -, 100, -, 75, -, 2500, 100, -, 300, 75, 300, -, 500
Walley, John H., 1, -, -, 30, -, 10, -, 800, 40, 150, 5, 3, -, -, 500
Phillips, F. P., -, -, 1, 90, -, 81, -, 1300, 25, 100, 5, -, 50, -, 200
Phillips, C. W., 1, -, -, -, -, -, -, -, -, -, -, -, -, -, 200
Owens, Jas. C., 1, -, -, 60, -, 82, -, 2000, 50, 310, 10, 10, -, -, 200
Phillips, Joseph W., 1, -, -, 150, -, 50, -, 2000, 75, 400, 10, 43, 150, -, 250
Phillips, John G., -, -, 1, 90, -, 100, -, 2000, 150, -, 300, 25, 50, -, 240
Kinney, Samuel, -, -, 1, 30, -, 25, -, 800, 20, 39, 5, -, -, -, 200

Ellis, John A., -, -, 1, 40, -, 60, -, 850, 30, 150, 8, -, 10, -, 1000
Twilley, R. O., 1, -, -, 75, -, 60, -, 1200, 20, 200, 5, 30, -, -, 200
Kinney, Jacob D., -, -, 1, 50, -, 25, -, 8000, -, 150, 5, -, -, -, 200
Knowles, Wm. J., -, -, 1, 45, -, 40, -, 1000, -, 50, 8, -, -, -, 300
Moore, C. N., 1, -, -, 60, -, 40, -, 1200, 100, 500, 15, 31, 50, -, 400
Gillis, Isaac, 1, -, -, 600, -, 200, -, 14000, 100, 400, 100, 321, 150, -, 1000
Ralph, Wm., 1, -, -, 140, -, 60, -, 4000, 100, 400, 50, 150, 50, -, 1000
Bailey, Josiah, 1, -, -, 25, -, 126, -, 5000, 150, 300, 15, -, -, -, 300
Bradley, R. H. L., 1, -, -, 30, 30, -, 800, 58, 150, 10, -, -, -, 200
Phillips, Melsin, 1, -, -, 40, -, 29, -, 900, 55, 150, 15, -, 30, -, 500
Owens, Hamilton, 1, -, -, 40, -, 60, -, 1000, 35, 350, -, 10, -, -, 400
Owens, Isaac W., 1, -, -, 40, -, 60, -, 1000, 100, 250, 6, 9, -, -, 300
Ellis, James, 1, -, -, 100, -, 50, -, 5000, 201, 400, 10, 50, 100, -, 1000
Horsey, A. J., 1, -, -, 140, -, 40, -, 7200, 750, 1000, 25, 300, 300, 100, 1000
Horsey, G. W., 1, -, -, 170, 30, 30, -, 17625, 400, 900, 25, 125, 300, 104, 1065
Hastings, Levin S., 1, -, -, 50, -, 25, -, 1000, 50, 170, 10, 20, -, -, 500
Serman, Wm. L., 1, -, -, 100, -, 30, -, 2500, 200, 175, 20, 90, 25, 4, 640
Serman, Wm. L., 1, -, -, 25, -, 12, -, 800, -, -, -, -, -, -, 525
Serman, Wm. L., 1, -, -, 70, -, -, -, 1000, -, 170, -, -, -, -, 600
Dunn, Thomas, -, -, 1, 100, -, 140, -, 3000, 100, 300, 25, 70, -, -, 500
Records, Samuel, 1, -, -, 150, -, 30, -, 1800, 50, 350, -, -, -, -, 518
Culver, Charles, 1, -, -, 25, -, 12, -, 600, -, 120, -, -, -, -, 500
Elliott, R. S., 1, -, -, 101, -, 110, -, 1200, 35, 200, 10, 12, -, -, 600
Collins, John B., -, -, 1, 100, -, 75, -, 5000, 50, 350, 20, 30, 50, 6, 600
Callaway, Isaac H., 1, -, -, 125, -, 75, -, 3000, 100, 400, 35, 28, 60, 7, 700
Short, Wm. T., 1, -, -, 36, -, -, -, 500, 25, 175, -, -, -, -, 500
Lockwood, John, -, -, 1, 40, -, -, -, 400, 30, 100, -, -, -, -, 525
Elliott, Elias, 1, -, -, 130, -, 65, -, 1960, 125, 275, 20, 35, 75, 8, 654
Ellis, Elias, 1, -, -, 50, -, -, -, 506, 25, 175, 20, 50, 25, 3, 510
Elliott, Joseph, 1, -, -, 100, -, -, -, 800, -, -, -, -, -, -, 600
Elliott, S. W., 1, -, -, 60, -, 73, -, 1350, 100, 250, 10, 30, 50, 5, 528
West, Stokely, E., -, -, 1, 100, -, 50, -, 1000, 125, 300, 5, 11, 24, 7, 500
King, Wm. C., 1, -, -, 300, -, 200, -, 1810, 150, 275, 15, 30, 75, 70, 700
Boyce, Jas. H., 1, -, -, 33, -, -, -, 2100, 100, 300, 10, 28, 100, 52, 1700
Windsor, Jas. H., 1, -, -, 50, -, -, -, 1000, 25, 75, 10, 30, 100, 52, 1000
Horsey, Charles R., 1, -, -, 36, -, -, -, 400, -, -, -, -, -, -, 300
Andrews, Alford, 1, -, -, 200, -, -, 50, 3000, 150, 300, 25, 61, 100, 50, 1800
Pollitt, L. A., 1, -, -, 125, -, -, 125, 2500, 100, 175, 12, 30, 200, 65, 801
Lecates, James, 1, -, -, 150, -, -, 50, 200, 200, 400, 25, 60, 20, 3, 900
Smith, James T., 1, -, -, 40, -, -, -, 300, 35, 175, -, -, -, -, 500
Hearn, Thomas, -, -, 1, 80, -, -, 50, 1040, 100, 200, 26, 35, 100, 52, 500
Elliott, James R., 1, -, -, 75, -, -, -, 800, 50, 105, 30, -, -, -, 525
Morris, James, 1, -, -, 150, -, -, -, 1000, 100, 310, 20, 30, -, -, 700
Workman, Wm., 1, -, -, 40, -, 30, -, 700, 100, 200, 20, 30, -, 3, 500
Kinney, Jas. B., 1, -, -, 60, -, 27, -, 850, 125, 175, 10, 20, -, 5, 700
Hearn, John, 1, -, -, 45, -, 27, -, 800, 50, 200, -, -, -, -, 500

Lecates, Cary, 1, -, -, 75, -, 25, -, 1000, 100, 300, 20, 30, -, -, 700
Oliphan, T. N., 1, -, -, 30, -, 10, -, 450, -, 125, -, -, -, -, 500
Dunn, Thomas, -, -, 1, 160, -, 30, -, 2000, 50, 300, 20, 30, 25, 6, 600
Dunn, William, -, -, 1, 30, -, 100, -, 400, 20, 125, -, -, -, -, 500
Nea, John, -, -, 1, 50, -, 20, -, 700, 30, 150, 10, 20, 20, 5, 550
Hastings, Eli, -, -, 1, 100, -, 25, -, 1500, 150, 400, -, 60, 75, 18, 650
Hastings, Linn, -, -, 1, 100, -, 25, -, 1500, 150, 301, -, 50, 50, 10, 600
Kenney, Samuel, 1, -, -, 250, -, 250, -, 10000, 250, 750, 50, 300, 200, 104, 1200
Cannon, James S., -, 1, -, 30, -, -, -, 1500, -, -, -, -, -, -, 500
CaClanen, Joseph S., 1, -, -, 79, -, 94, -, 1900, 25, 175, 10, 20, -, -, 700
Ellis, T. H., 1, -, -, 6, -, -, -, 1800, -, 100, -, -, -, -, 300
Bae__, John S., 1, -, -, 43, -, -, -, 1800, -, 10, -, -, -, -, 300
Ellis, James E., 1, -, -, 100, -, 125, -, 800, 125, 200, 20, 60, 100, 52, 800
Culver, Charles E., 1, -, -, 60, -, 40, -, 1000, 100, 225, 12, 20, 48, 3, 500
Parker, B. J., 1, -, -, 40, -, 40, -, 800, 50, 100, 10, -, -, -, 600
High, John, 1, -, -, 100, -, 20, -, 1050, 125, 350, 75, 30, 60, 30, 650
Callaway, J. W. E., 1, -, -, 45, -, -, -, 600, 50, 175, -, -, -, -, 500
Bailey, William, 1, -, -, 20, -, 10, -, 350, 20, 100, 10, 10, 82, 500
Grody, Griffin, 1, -, -, 60, -, 12, -, 750, 50, 125, -, -, -, -, 550
Phillips, W. W., 1, -, -, 80, -, -, -, 800, 30, 200, 10, 30, -, -, 500
Hastings, Edward, -, -, 1, 60, -, -, -, 700, 125, 200, 20, 21, -, -, 600
Callaway, Sharp, 1, -, -, 75, -, 25, -, 800, 50, 100, -, -, -, -, 500
Hearn, George M., -, -, 1, 40, -, 12, -, 475, 35, 200, -, -, -, -, 500
Ellis, Rebecca, 1, -, -, 30, -, 10, -, 400, 10, 100, -, -, -, -, 500
Hastings, Winder, 1, -, -, 100, -, 30, -, 2000, 100, 300, -, -, -, -, 650
Hazzard, Harry, -, -, 1,75, -, -, -, 750, 75, 100, -, -, -, -, 500
Callaway, Handy, -, -, 1, 30, -, -, -, 400, 20, 100, -, -, -, -, 500
Ellis, James, -, -, 1, 80, -, 30, -, 1500, 100, 250, 10, 30, 20, 6, 500
Hearn, W. S., 1, -, -, 22, -, 82, -, 1040, 150, 200, 10, 30, 50, 15, 500
Ellis, James S., 1, -, -, 100, -, 50, -, 1500, 200, 300, 12, 40, 25, 10, 500
Graham, Uriah S., 1, -, -, 140, -, 65, -, 4000, 100, 600, -, 10, -, -, 490
Hitchens, Philip, -, -, 1, 50, -, 67, -, 900, 35, 425, -, 9, -, -, 150
Taylor, Elias, -, -, 1, 75, -, 25, -, 1500, 30, 150, 10, 58, 15, 3, 300
Jones, Ezekiel H., -, -, 1, 150, -, 60, -, 5000, 90, 450, -, 95, 250, 200, 1400
Morgan, S. M. Sr., 1, -, -, 165, -, 185, 50, 5000, 200, 850, 30, 100, 200, 50, 1200
Messick, Nathaniel D., -, -, 1, 100, -, 60, -, 1800, 50, 225, 5, -, -, -, 300
Jones, Jacob M., -, 1, -, 100, -, 100, -, 1500, 75, 400, 10, 33, 80, 30, 500
Simpson, William, 1, -, -, 75, -, 75, -, 1800, 71, 275, 5, 31, -, -, 300
Preden, Stephen, 1, -, -, 4, -, -, -, 100, 10, -, 4, -, 5, 225
Jones, Thomas A., -, -, 1, 175, -, 75, -, 1600, 60, 200, 10, 10, 5, 2, 350
Johnson, Bayard, 1, -, -, 100, -, 60, -, 2500, 110, 400, 20, 45, -, -, 550
Prettyman, Lavinia A., -, -, 1, 160, -, 40, -, 4000, 50, 300, 15, 60, -, -, 600
Adams, Isaac, -, -, 1, 45, -, 150, -, 2000, 40, 260, 10, 15, 1000, 45, 215
Bennett, William, 1, -, -, 3, -, -, -, 300, -, 45, -, -, 6, 1, 70
Conaway, Miles M., -, 1, -, 9, -, 5, -, 500, 30, 75, 5, 7, -, 75
Short, John C., 1, -, -, 150, -, 140, -, 5500, 600, 1500, 100, 110, 275, 45, 1500

Short, Margaret, 1, -, -, 36, -, 7, -, 400, -, 5, -, -, -, 85
Swain, Walter, 1, -, -, 75, -, 25, -, 2000, 50, 250, 10, 37, -, -, 412
Willey, Tilghman L., 1, -, -, 100, -, 50, -, 1500, 100, 425, 5, 58, 25, 3, 625
Swain, Aby, 1, -, -, 45, -, 25, -, 600, -, 45, 5, -, -, -, 100
Short, Priscilla F., 1, -, -, 75, -, 78, -, 1200, 30, 100, 10, 15, -, -, 398
Coverdale, James, -, -, 1, 150, -, 100, -, 2000, 100, 600, 20, 17, -, -, 650
Owens, Edward, 1, -, -, 80, -, 40, -, 2000, 200, 500, 20, 100, 60, 10, 725
Prettyman, Lemuel D., -, -, 1, 70, -, 35, -, 1200, 45, 200, 12, -, -, -, 290
Short, John W., -, -, 1, 50, -, 110, -, 1000, 75, 200, 15, -, -, -, 350
Cornwall, Baptist L., 1, -, -, 60, -, 115, -, 1000, 45, 175, 10, 8, -, -, 285
Isaacs, Noah Sr., 1, -, -, 100, -, 100, -, 5000, 150, 600, 30, 200, 125, 40, 1300
Isaacs, James O., -, 1, -, 75, -, 50, 50, 500, 25, 250, 10, -, 100, 40, 200
Isaacs, Louis S., -, 1, -, 60, -, 40, -, 700, 60, 175, 10, -, -, -, 175
Bryan, Jacob E., -, -, 1, 55, -, 50, -, 1000, 34, 225, 15, 25, -, -, -
Isaacs, Hiram J., 1, -, -, 60, -, 90, -, 1200, 25, 200, 15, 19, -, -, 325
Isaacs, Mary, 1, -, -, 85, -, 15, -, 1000, 125, 550, 12, 25, -, -, 435
Isaacs, Minos, 1, -, -, 75, -, 58, -, 2000, 100, 600, 30, 27, 125, 52, 475
Short, James E., 1, -, -, 50, -, 61, -, 1000, 100, 450, 25, 15, 20, 5, 400
Wilson, Elsey, 1, -, -, 50, -, 50, -, 1000, 75, 250, 30, -, -, -, 300
Wilson, Joseph, 1, -, -, 45, -, 35, -, 1200, 50, 300, 10, 11, 7, 1, 350
Wilson, George, 1, -, -, 60, -, 30, -, 1000, 25, 450, 15, -, 35, 10, 325
Lofland, Cornelius, -, -, 1, 80, -, 75, -, 1800, 35, 325, 10, -, -, -, 300
Simmons, David R., 1, -, -, 65, -, 43, -, 500, 20, 125, 10, 15, 125, 50, 225
Warrington, Alfred A., 1, -, -, 20, -, 30, -, 600, 50, 200, 15, -, 20, 6, 200
Lofland, John P., -, -, 1, 15, -, 300, -, 2500, 25, 175, 20, -, 12, 3, 150
Ryan, James M., 1, -, -, 10, -, 2, -, 300, 58, 300, 10, -, -, -, 75
Macklin, Joseph A., -, -, 1, 75, -, 65, -, 1000, 50, 475, -, 48, 85, 20, 350
Macklin, Emory, 1, -, -, 6, -, 48, -, 1000, 10, 375, 5, 17, 35, 10, 178
Russell, Theodore, -, -, 1, 20, -, 90, -, 800, 10, 125, 8, 5, -, -, 100
Russell, William N., 1, -, -, 75, -, 250, -, 1500, 200, 1000, 5, 20, 20, 4, 400
Prettyman, Josiah, 1, -, -, 100, -, 96, -, 1500, 45, 500, 12, 16, -, -, 400
Matthews, David, -, -, 1, 100, -, 63, -, 1000, 75, 225, 20, -, -, -, 300
Owens, Joshua S., -, -, 1, 90, -, 10, -, 650, 45, 200, 5, 14, -, -, 750
Legates, John, -, -, 1, 60, -, 40, -, 600, 30, 60, 20, -, -, -, 150
Ellegood, Robert C., 1, -, -, 100, -, 25, 25, 2500, 100, 800, 50, 150, 500, 120, 1500
Conaway, Noble, 1, -, -, 100, -, 200, -, 3000, 125, 500, 50, 130, 100, -, 1200
Outten, James, 1, -, -, 19, -, 10, -, 1000, 100, 80, 10, 16, 50, 16, 125
Calhoun, Thomas, 1, -, -, 80, -, 23, -, 2000, 50, 250, 18, 58, 38, 9, 800
Thompson, William H., -, -, 1, 75, -, 5, -, 1000, 40, 200, 5, 30, 10, 3, 600
Jefferson, Elizabeth, -, -, 1, 75, -, 25, -, 800, 20, 150, 5, -, -, -, 200
Matthews, John, 1, -, -, 38, -, 2, -, 700, 30, 155, 5, 29, 8, 2, 328
Matthews, Wingate T., 1, -, -, 4, -, -, -, 300, 5, 24, -, 12, -, -, 100
Hurly, William -, 1, -, 3, -, 1, -, 300, 5, 10, -, -, -, -, 55
Waller, Levin E., -, -, 1, 180, 10, -, -, 1500, 45, 200, 5, 34, -, -, 450
Reynolds, David, -, -, 1, 90, -, 10, -, 650, 43, 200, 5, 15, -, -, 200
Isaacs, Joseph, 1, -, -, 150, -, 100, -, 3500, 100, 554, 15, -, -, -, 400

Bebee, John S., 1, -, -, 125, -, 75, -, 1000, 50, 300, 87, 100, 50, 650
Sharp, Theophilus, -, -, 1, 75, 50, 40, -, 2500, 40, 300, 5, -, 20, 4450
Sharp, Benton, 1, -, -, 150, 24, 50, -, 2500, 35, 300, 12, 6, 50, 50, 375
Satterfield, William M., 1, -, -, 40, -, 112, -, 850, 50, 300, 10, 17, -, -, 300
Stewart, George H., -, -, 1, 20, -, 250, -, 4000, 200, 550, 12, 16, 25, 4, 475
Prettyman, James E., 1, -, -, 108, -, 120, -, 3000, 135, 375, 15, -, 35, 4, 625
Hill, DeWitt C., 1, -, -, 120, -, 120, -, 4000, 500, 350, 15, 35, 150, 55, 675
Staten, Pleasant, -, -, 1, 100, -, 100, 75, 1800, 25, 130, 10, 10, -, -, 355
Callaway, Samuel, -, -, 96, -, 12, -, 2000, 50, 300, 15, 90, 20, 5, 500
Burris, Isaiah C., -, -, 1, 78, -, 12, -, 1500, 40, 175, 4, 30, -, -, 428
Prenable, John, 1, -, -, 85, -, 15, -, 1200, 60, 400, 10, -, 100, 500
Prenable, John, -, -, 1, 30, -, 20, -, 400, -, -, 2, -, -, -, 57
Workman, John W., -, -, 1, 75, -, 25, -, 820, 30, 100, 10, -, -, -, 200
Outten, John, -, -, 1, 15, -, 15, -, 300, 20, 90, 3, -, -, -, 175
Fleetwood, Curtis S., 1, -, -, 91, -, 28, -, 1200, 60, 225, 10, 40, -, -, 500
Jones, Thomas F., -, -, 1, 75, -, 125, -, 500, 70, 300, 25, -, -, -, 375
Spicer, William M. -, -, 1, 60, -, 65, -, 1800, 40, 175 10, -, -, -, 300
Calhoun, John & Silas, 1, -, -, 150, 10, 40, -, 2800, 200, 1000, 10, 110, 225, 65, 1000
Ratcliff, William W., 1, -, -, 125, -, 46, -, 6000, 200, 300, 10, 30,75, 20, 628
Reynolds, James C., -, -, 1, 200, -, 200, -, 3500, 30, 220, 5, -, -, -, 450
Macklin, Margaret A., 1, -, -, 75, -, 350, 5, 3000, 25, 350, 10, 20, -, -, 800
Johnson, William D., -, -, 1, 50, -, 150, -, 800, 15, 120, 5, -, -, -, 225
Hastings, John H., -, -, 1, 70, -, 130, -, 1000, 25, 130, 150, -, -, -, 450
Sharp, James L., -, -, 1, 75, -, 150, -, 1200, 55, 300, 25, -, 200, 60, 400
Dickerson, Burton, -, -, 1, 60, -, 60, -, 1500, 100, 300, 10, 20, -, -, 480
Smith, William H., -, -, 1, 70, -, 75, -, 1000, 150, 700, 12, -, 125, 40, 450
Cooper, George, -, -, 1, 20, -, 60, -, 800, 20, 125, 9, -, -, -, 275
Brown, George F., 1, -, -, 100, -, 80, 3000, 75, 510, -, 10, 100, 35, 900
Lynch, James, -, -, 1, 60, -, 35, -, 600, 20, 125, 5, -, -, -, 178
Short, Samuel T & John M., 1, -, -, 100, -, 100, -, 3000, 75, 200, 10, -, -, -, 400
Walls, John C., -, -, 1, 90, -, 200, -, 3000, 125, 400, 25, 30, 10, 3, 400
Messick, James P., -, 1, -, 25, -, 50, -, 500, 25, 125, 10, -, -, -, 175
Prichard, Theophilus C., 1, -, -, 30, -, 10, -, 800, 225, 600, 25, 30, 125, 50, 375
Fleetwood, Philip H., 1, -, -, 120, -, 8, 10, 1600, 100, 300, 15, 90, 75, 15, 600
Lamden, Sovereign A., 1, -, -, 200, -, 200, -, 3000, 150, 430, 15, 30, 10, 2, 1000
Conaway, John, 1, -, -, 85, -, 80, -, 800, 10, 50, 10, -, -, -, 425
Conaway, Philip, H., -, 1, -, 75, -, 25, -, 800, 20, 125, 10, 10, -, -, 250
Conaway, Joseph P., -, 1, -, 45, -, 25, -, 700, 25, 125, 8, 30, -, -, 200
Hastings, James H., 1, -, -, 145, -, 60, -, 1800, 110, 300, 15, -, 150, 50, 900
Hill, Albert, 1, -, -, 75, -, 25, -, 1000, 100, 250, 10, -, -, -, 450
Fleetwood, William E., -, 1, -, 30, -, 40, -, 400, -, 12, 12, -, 10, 3, 200
Legates, Philip S., -, -, 1, 100, -, 400, -, 1000, 35, 130, -, -, 10, 3, 350
Tindal, Peter, 1, -, -, 120, -, 60, -, 1000, 150, 40, 12, 40, 20, 4, 475
Taylor, William E., -, -, 1, 120, -, 270, -, 2500, 110, 300, 25, 22, 12, 3, 750
Spicer, Charles J., -, -, 1, 90, -, 110, -, 1500, 60, 128, 10, 190, -, -, 475

Smith, George W., 1, -, -, 27, -, 279, -, 1500, 60, 250, 3, 20, 175, 50, 290
Messick, John T., -, -, 1, 50, -, 55, -, 2000, 40, 250, 10, -, 30, 8, 275
Turner, Charles H., -, -, 1, 75, -, 80, -, 900, 20, 130, 10, -, -, -, 300
Elliott, James, 1, -, -, 25, -, 10, 1, 250, 8, 10, 5, -, -, -, 108
Collins, Robert, -, -, 1, 125, -, 150, -, 2000, 40, 220, 5, 30, -, -, 325
Taylor, Harriet A., 1, -, -, 14, -, -, -, 400, 10, 40, 10, -, -, -, 75
Bryan, Waterman, -, -, 1, 53, -, 40, -, 400, 25, 50, 50, -, 8, 2, 150
Tindal, Jordan, 1, -, -, 68, -, 20, -, 600, 30, 100, 10, -, -, -, 375
Teague, Mary S., 1, -, -, 55, -, 5, -, 900, -, 255, -, 48, 10, 200
Messick, George H., -, -, 1, 110, -, -, 500, 65, 55, 25, -, -, -, 225
Sammons, Edward T., 1, -, -, 75, -, 25, -, 800, 75, 600, 10, 45, 12, 3, 375
Elliott, Ananias, -, -, 1, 30, -, -, -, 500, 40, 120, 5, -, -, -, 160
Fleetwood, William, 1, -, -, 145, -, 60, -, 2000, 110, 375, 20, 50, 125, 45, 500
Hill, David H., 1, -, -, 90, -, 10, -, 2000, 110, 350, 10, 30, -, -, 450
Maxwell, Adaline, 1, -, -, 40, -, 5, -, 500, 40, 250, 10, -, -, -, 275
Taylor, Reuben J., -, -, 1, 100, -, 50, -, 1500, 20, 225, 12, -, -, -, 250
Bennett, Nicholas A., -, -, 1, 125, -, 75, -, 1200, 45, 365, 10, -, -, -, 800
Tindal, Ahasuerue, 1, -, -, 100, -, 200, -, 1800, 75, 185, 10, 6, -, -, 450
Mitchell, Denard W., 1, -, -, 75, -, 100, -, 1400, 15, 150, 5, -, 30, 7, 325
Conaway, John W., 1, -, -, 50, -, 25, -, 1200, 100, 400, 10, 36, -, -, 450
Waller, James, -, -, 1, 90, -, 10, -, 1500, 25, 225, -, -, -, -, 375
Fooks, Nathan R., -, -, 1, 70, 40, 40, -, 700, -, 10, -, 30, -, -, 200
Elliott, John H., 1, -, -, 100, -, 50, -, 1200, 50, 350, 12, -, -, -, 600
Dolby, Jackson, 1, -, -, 160, -, 100, -, 3500, 250, 750, 30, 80, 28, 7, 1300
Givens, Julia, 1, -, -, 30, -, 10, -, 500, -, 35, 5, 15, 50, 14, 190
Spicer, Edward W., 1, -, -, 35, -, 5, -, 500, 5, 45, 3, 10, -, -, 100
Tindal, Rev. William J., -, -, 1, 30, -, 5, -, 400, -, 35, 5, 30, 12, 3, 110
Tindal, Benton H., 1, -, -, 100, -, 50, -, 100, 100, 160, 12, -, -, -, 300
Messick, Phillip G., -, -, 1, 60, -, 15, -, 900, 100, 250, 15, -, -, 275
Tindal, Jonatha, 1, -, -, 130, -, 66, -, 1500, 125, 35, 15, 30, -, -, 625
Jones, Charles H., 1, -, -, 60, -, 40, -, 2500, 50, 200, 15, -, 10, 3, 375
Murray, John T., -, -, 1, 75, -, 75, -, 1500, -, 2, 40, -, -, -, 275
Conaway, Minos T., 1, -, -, 85, -, 115, -, 2900, 150, 700, 15, -, -, -, 750
Tindal, George P., -, -, 1, 70, -, 50, -, 1200, -, 110, -, -, 500
Dolby, John C., 1, -, -, 120, -, 40, -, 2500, 20, 550, 15, 75, 50, 12, 600
Dolby, William P., -, 1, -, 65, -, 65, -, 200, 75, 200, 20, 51, 10, 2, 525
Smith, James T., 1, -, -, 100, -, 50, -, 1200, 40, 225, 10, 25, 5, 1, 350
McAlister, Jams, -, -, 1, 100, -, 80, -, 1500, 25, 125, 12, 10, -, -, 200
Morris, William -, -, 1, 50, -, 60, -, 900, 10, 50, 8, 6, -, -, 190
Tindal, John A., 1, -, -, 80, -, 100, -, 3000, 200, 500, 10, -, -, -, 675
Tindal, William J., 1, -, -, 115, -, 115, -, 2000, 10, 175, 8, 30, -, -, 200
Tindal, Sarah A., 1, -, -, 200, -, 100, -, 420, 200, 500, 20, 48, 65, 15, 1000
Fooks, Hetty A., 1, -, -, 200, -, 100, -, 4000, 200, 650, 25, 90, 150, 40, 100
Tindal, James H., -, -, 100, -, 100, -, 2000, 50, 300, 10, -, -, -, 425
Barr, David W., 1, -, -, 125, -, 75, -, 200, 100, 500, 100, -, 150, 200, 675
Carey, Theodore D., -, -, 1, 170, -, 30, -, 2500, 75, 300, 10, -, 5, 1, 875

Elliott, Wingate B., -, -, 1, 95, -, 22, -, 1700, 50, 250, 10, -, -, -, 550
Campbell, Wilson C., -, -, 1, 95, -, 20, -, 1400, 30, 200, 8, -, -, -, 425
Littleton, Isaac T., -, -, 1, 100, -, 25, -, 1500, 35, 200, 10, -, -, -, 600
Hitchens, Gideon, -, -, 1, 125, -, 50, -, 1500, 40, 300, 15, -, 5, 1, 800
Jones, Isaac, -, -, 1,75, -, 5, -, 800, 45, 25, 10, -, -, -, 100
Marvell, Josiah P., 1, -, -, 250, -, 100, -, 7000, 350, 1200, 50, 30, 150, 35, 2500
Sammons, William H., -, -, 1, 77, -, 78, -, 1500, 40, 200, 10, 6, -, -, 275
McIlvain, James, -, -, 1, 180, -, 2, 3,8000, 150, 300, -, 75, 150, 49, 1000
McIlvain, John, -, -, 1, 180, -, -, 2, 8000, 150, 300, -, 100, 175, 52, 1000
Layton, Thomas W., 1, -, -, 140, -, 137, -, 3500, 70, 230, 15, 15, 10, 2, 500
Conaway, Job, 1, -, -, 40, -, 40, -, 1200, 60, 400, 25, 32, 85, 15, 650
Allen, Theodore, -, 1, -, 45, -, 130, 3, 1000, -, 20, 12, -, -, -, 190
Morgan, Henry H., -, -, 1, 175, -, 100, -, 3000, 150, 500, 100, 90, 159, 50, 1700
Todd, Jacob, -, -, 1, 60, -, 100, -, 1200, 12, 10, -, 100, -, -, 450
Butler, Joshua, -, -, 1, 60, -, 200, -, 1500, 25, 125, 5, -, -, -, 575
Lord, David, -, -, 1, 275, -, 75, -, 6000, 550, 650, 25, 65, -, -, 1400
Morery, Jesse H., 1, -, -, 50, -, 12, -, 1000, 25, 250, 10, 30, -, -, 400
Carey, James B., -, -, 1, 100, -, 100, -, 3000, 125, 300, 20, 75, -, -, 1450
Ocheltree, John V., -, -, 1, 150, -, 50, -, 2200, 75, 300, 10, 30, -, -, 825
Elliott, George W., 1, -, -, 100, -, 40, -, 3000, 75, 400, 20, 60, 140, 50, 525
Willey, William H., -, -, 1, 50, -, 10, -, 1500, 40, 210, 20, 60, -, -, 250
Staten, Amos J., 1, -, -, 90, -, 37, -, 2100 100, 400, 10, 35, 60, 15, 650
Willey, Cyrus, 1, -, -, 80, -, 50, -, 1100, 30, 125, 11, 5, 1, 450
Tucker, Bryan R., -, -, 1, 70, -, 90, -, 3000, 20, 560, 10, 125, 125, 55, 1000
Carlisle, William J., 1, -, -, 120, -, 80, -, 7000, 100, 250, 50, 120, 185, 75, 1100
Stewart, Caleb, -, 1, -, 12, -, 45, -, 250, 20, 100, 5, -, -, -, 100
Spanish, Joshua S., 1, -, -, 18, -, 23, -, 600, 30, 125, 10, -, -, -, 100
Vincent, James H., -, -, 1, 100, -, 80, -, 2500, 50, 200, 10, 52, 25, 5, 500
Davis, James H., 1, -, -, 16, -, -, -, 1200, 50, 250, 18, 18, 15, 3, 328
Stewart, Mary A., 1, -, -, 220, -, 100, -, 5000, 125, 550, 50, 112, 100, 20, 1100
Griffith, William, -, -, 1, 24, -, -, -, 800, 50, 300, -, 11, -, -, 270
Sipple, Caleb, -, -, 1, 65, -, 30, -, 1000, 50, 250, -, -, 50, 10, 225
Curry, Albert, 1, -, -, 400, -, 50, -, 25000, 1500, 2145, 200, 125, 1000, 30, 550
Tucker, William -, -, 1, 60, -, 40, -, 1000, 100, 150, 5, 20, -, -, 371
Hays, John T., -, -, 1, 100, -, 10, -, 5000, 100, 500, 15, 75, 125, 40, 1200
Carlisle, William E., -, -, 1, 100, -, -, -, 2000, 150, 300, 10, 50, -, -, 825
Newton, Matthew, S., -, -, 1, 135, -, -, -, 2000, 35, 90, -, -, -, -, 700
Reynols, Sarah, -, 1, -, 40, -, 10, -, 900, 10, 28, 10, 10, -, -, 200
Fowler, John H., 1, -, -, 60, -, 60, -, 800, 50, 200, 10, 17, 20, 5, 300
Owens, John, 1, -, -, 100, -, 60, -, 2000, 100, 450, 20, 55, 100, 52, 475
Hasty, Samuel C., -, -, 1, 70, -, 40, -, 800, 60, 100, 12, -, -, -, 380
Fisher, Alex. Brd., 1, -, -, 100, -, 40, -, 1500, -, 75, 10, 30,75, 40, 800
Paswaters, Thomas, -, -, 1, 75, -, 60, -, 1200, 10, 30, 10, 30, -, -, 300
Messick, Samuel C., -, -, 1, 175, -, 90, -, 2500, 25, 300, 10, 50, -, -, 785
Messick, Miles, 1, -, -, 250, -, 75, -, 12000,300, 1000, 20, 375, 100, 52, 1100
Truit, Miles, -, -, 1, 140, -, 15, -, 1000, 30, 250, 15, 15, -, -, 350

O'Day, James, -, -, 1, 90, -, 140, -, 2000, 50, 100, 10, 25, -, -, 200
Newton, Beaucham P., -, -, 1, 110, -, 40, -, 2000, 100, 250, 10, 30, -, 550
Lyons, John R., -, -, 1, 100, -, 20, -, 2300, 60, 125, -, 42, -, -, 925
Dawson, William E., -, -, 1, 75, -, 175, 2000, 125, 300, 12, -, -, -, 225
Dawson, Albert C., -, -, 1, 75, -, 75, -, 1500, 15, 175, 10, 8, 30, 6, 350
Coates, Thomas, 1, -, -, 50, -, 100, -, 1500, 10, 50, 10, 12, -, -, 378
Duker, Isaac, -, -, 1, 93, -, 7, -, 1000, 40, 200, 5, -, -, -, 400
Sorden, William H., 1, -, -, 20, -, 7, -, 500, 45, 150, -, -, -, -, 160
Carlisle, James, -, -, 1, 75, -, 125, -, 2000, 70, 300, 15, -, 40, 9, 275
Sweeney, Oliver, 1, -, -, 40, -, 40, -, 1200, 25, 50, -, -, 40, 9, 175
Johnson, Charles J., 1, -, -, 28, -, 22, -, 1000, 75, 125, -, 10, -, -, 240
Aswell, John, 1, -, -, 100, -, 33, -, 2500, 100, 165, 20, 25, -, -, 750
Satterfield, John H., 1, -, -, 100, -, 30, -, 3200, 125, 600, 5, -, -, -, 578
Redman, Rev. William W., 1, -, -, 135, -, 25, -, 3000, 20, 100, -, -, -, -, 400
Johnson, John Q., 1, -, -, 100, -, 140, -, 3000, 50, 150, 5, -, -, -, 400
Buell, Roland, 1, -, -, 100, -, 60, -, 4000, 200, 300, 10, 50, 140, 50, 1400
Clifton, James W., 1, -, -, 75, -, 95, -, 1500, 200, 300, -, 35, 100, 20, 450
Clifton, Pemberton, 1, -, -, 48, -, -, -, 4000, 150, 400, -, -, -, -, 400
Guntor, Matthew, 1, -, -, 50, -, 24, 30, 1000, 45, 75, -, -, -, -, 175
Ryan, John W., -, -, 1, 80, -, 40, -, 1000, 30, 120, -, 9, -, -, 325
Ryan, John S., -, -, 1, 18, -, 2, -, 650, 50, 200, 10, 6, 20, 4, 300
Webb, William B., -, -, 1, 175, -, 105, -, 3000, 100, 400, 5, -, -, -, 375
Pettyjohn, James, -, -, 1, 40, -, 60, -, 1200, 15, 125, 5, 10, -, -, 325
Webb, James, 1, -, -, 125, -, 70, -, 3500, 50, 500, 25, 40, 150, 40, 700
Collins, John W., -, -, 1, 85, -, 68, -, 2000, 35, 225, 10, 48, 15, 4, 500
Tatman, Charles P., 1, -, -, 90, -, 53, -, 1800, 150, 550, 15, 50, 175, 38, 800
Owens, Robert B., 1, -, -, 140, -, 80, -, 7500, 150, 600, 300, 150, 50, 12, 2200
Willey, Samuel, 1, -, -, 50, -, 80, -, 1200, 75, 300, 5, 18, -, -, 275
Griffith, Jeremiah C., - -, 1, 100, -, 200, -, 2500, 50, 300, -, 130, -, -, 385
Fowler, Margaret, 1, -, -, 50, -, 80, -, 1000, 75, 350, -, 15, 10, 2, 350
Tucker, Isaac, -, -, 1, 30, -, 100, -, 2000, 15, 60, 10, 10, -, -, 150
Owens, John T., -, -, 1, 70, -, 175, -, 2000, 78, 225, -, 20, -, -, 300
Pettyjohn, William H., -, -, 1, 25, -, 100, -, 1000, 10, 75, -, -, -, -, 100
Scott, John W., -, -, 1, 60, -, 65, -, 1200, 25, 200, 5, -, -, -, 150
Jackson, Nathaniel, -, -, 1, 77, -, 10, -, 1000, 25, 225, -, -, 10, 2, 425
Hayes, Priscilla, 1, -, -, 30, -, 6, -, 200, 10, 85, -, 3, -, -, 60
Carlisle, Eliza J., -, -, 1, 80, -, 100, -, 2500, 40, 300, 15, -, 100, 40, 60
Torbert, William E., -, -, 1, 60, -, 200, -, 2600, 50, 300, -, -, 25, 16, 325
Tatman, John, 1, -, -, 175, -, 106, -, 5000, 200, 600, 20, 150, 50, 100, 1300
Willey, Isaac, -, 1, -, 100, -, 235, -, 1800, 25, 100, -, -, -, -, 300
Willey, Robert J. H., -, -, 1, 110, -, 10, 40, 1600, 40, 200, 15, 30, 8, 2, 550
Willey, Richard, -, -, 1, 95, -, 15, -, 1500, 40, 300, 5, -, -, 500
Johnson, Jordan, -, -, 1, 75, -, 33, -, 1000, 10, 200, -, 23, -, -, 350
Willey, Minos, -, -, 1, 25, -, 8, -, 900, -, 35, -, -, -, -, 275
Owens, Jonathan 1, -, -, 200, -, 60, -, 2000, 25, 350, 20, -, 125, 40, 675

Adams, Bartholonew M., -, -, 1, 200, -, 323, -, 6000, 100, 550, -, 135, 150, 50, 1100

Lyons, John H., -, -, 1, 200, -, 100, -, 5000, 300, 600, 20, -, 400, 10, 1400

Carlisle, Samuel R., 1, -, -, 90, -, 35, -, 3500, 60, 350, 18, 28, -, -, 550

Torbert, Warren, -, -, 1, 100, -, 50, -, 2000, 40, 125, 15, -, -, -, 375

Tucker, Robert, -, -, 1, 150, -, 45, -, 7000, 65, 600, 10, 32, -, -, 825

Lynch, Alexander D., -, -, 1, 150, -, 147, -, 2000, 35, 350, 10, -, -, 650

Coverdale, Tobias, -, -, 1, 85, -, 50, -, 900, 50, 250, 10, -, 12, 2, 325

Owens, Amelia E., 1, -, -, 75, -, 45, -, 1500, 50, 260, 5, 30, 10, 2, 450

Ricards, John T., -, -, 1, 120, -, 180, -, 3000, 125, 400, 15, -, -, -, 300

Dickerson, Sarah M., 1, -, -, 100, -, 60, -, 2200, 60, 275, 10, 19, -, -, 775

Smith, Ezekiel, -, -, 1, 100, -, 50, -, 1800, 15, 115, 15, -, -, -, 325

Paswaters, Clement, -, -, 1, 30, -, 175, -, 1100, 15, 100, 12, -, -, -, 100

Stephens, F. John, -, -, 1, 40, -, 120, -, 1200, 50, 300, 18, 75, 25, 6, 300

Clifton, George, -, -, 1, 36, -, 70, -, 1000, 10, 50, 10, -, -, -, 165

Turner, Louis, -, -, 1, 46, -, 45, -, 800, 30, 110, -, -, 5, 1, 300

Sharp, William W., 1, -, -, 215, -, 245, -, 7000, 200, 800, 25, 100, 150, 35, 2200

Sharp, Joseph L., 1, -, -, 65, -, 43, -, 2000, 50, 375, 10, 28, -, -, 500

Pettyjohn, John, -, -, 1, 35, -, 78, -, 625, 10, 50, 5, -, -, -, 125

Conaway, Elizabeth, 1, -, -, 90, -, 98, -, 3000, 35, 400, 14, 32, 75, 18, 375

Webb, Isaac C., 1, -, -, 125, -, 200, -, 3500, 150, 500, 10, 80, 225, 70, 950

Morgan, Benjamin, -, -, 1, 50, -, 80, -, 900, 20, 125, 8, 17, -, -, 300

Smith, Chalton, 1, -, -, 75, -, 65, -, 2000, 65, 425, 10, -, 150, 33, 1500

Brace, Rev. George V., 1, -, -, 135, -, 65, -, 2500, 150, 455, 10, -, 150, 33, 1500

Banning, Thomas, 1, -, -, 60, -, 110, -, 1000, 50, 175, 30, -, -, -, 475

Banning, John T., 1, -, -, 45, -, 100, -, 1000, 25, 125, 10, 18, -, -, 225

Paswaters, James H., -, -, 1, 75, -, 45, -, 1000, 35, 200, 10, -, -, -, 250

Paswaters, Sebastian, 1, -, -, 68, -, 66, -, 1200, 40, 100, 8, -, 10, 3, 100

Paswaters, Samuel, -, -, 1, 40, -, 160, -, 2000, 45, 75, 10, -, -, -, 1150

Smith, David R., 1, -, -, 120, -, 106, -, 1900, 75, 500, 18, 20, 25, 6, 590

Polk, William A., 1, -, -, 100, -, 300, -, 6000, 150, 500, 5, 25, 175, 58, 625

Smith, Gideon, 1, -, -, 90, -, 30, -, 1600, 100, 225, 20, -, -, -, 300

Lynch, Peter, -, -, 1, 100, -, 100, -, 2000, 12, 75, 5, -, -, -, 528

Wilkins, Isaac, 1, -, -, 55, -, 27, -, 1650, 25, 200, 15, 8, -, -, 200

Dickerson, Charles, -, -, 1, 65, -, 90, -, 1000, 25, 75, 15, -, -, -, 225

Wilkins, John, 1, -, -, 75, -, 25, -, 1500, 100, 450, 15, 6, -, -, 900

Tindal, Isaac N., 1, -, -, 90, -, 65, -, 3500, 200, 800, 5, 40, -, -, 600

Jones, William F., 1, -, -, 175, -, 125, -, 6000, 150, 650, 30, 75, 100, 25, 1178

Joseph, Thomas H., -, -, 1, 105, -, 75, -, 2000, 125, 330, 10, 5, 20, 5, 625

Littleton, Henry, -, -, 1, 75, -, 25, -, 1000, 25, 200, 5, -, -, -, 300

Wilson, David, -, -, 1, 80, -, -, -, 900, 35, 225, 20, -, -, -, 400

Willey, Asbury, -, -, 1, 45, -, 4, -, 900, 20, 175, 5, -, -, -, 400

Littleton, Jesse B., -, -, 1, 55, -, -, -, 1100, 25, 175, 5, -, -, -, 550

Campbell, Simpson, -, -, 1, 65, -, 55, -, 2000, 25, 175, 10, -, -, -, 300

Donahoe, Peter W., -, -, 1, 50, -, 75, -, 1800, 40, 300, 10, -, -, -, 300

Littleton, Thomas, -, -, 1, 20, -, 140, -, 1000, 12, 120, 5, 4, -, -, 100

Hastings, Solomon T., -, -, 1, 125, -, 45, -, 3000, 40, 260, 20, -, -, -, 550
Short, Wingate, 1, -, -, 200, -, 1000, -, 5500, 175, 750, 15, -, 180, 50, 1175
McDowell, Zachariah, 1, -, -, 20, -, 30, -, 500, 20, 75, 5, 6, 6, 1310
Messick, Willard S., 1, -, -, 85, -, 15, -, 900, 10, 95, 6, -, -, -, 175
Carey, John, -, -, 1, 47, -, 60, -, 1500, 25, 125, 15, -, -, -, -, 300
Gumby, John, -, -, 1, 75, -, 25, -, 1100, 15, 75, 10, 20, 20, 5, 325
Smith, William, 1, -, -, 85, -, 35, -, 2000, 100, 400, 12, -, 15, 4, 550
Smith, Sallie A., 1, -, -, 50, -, 70, -, 1500, 20, 250, 10, -, 12, 3, 250
Swain, John, 1, -, -, 50, 65, -, 1200, 40, 400, 10, -, -, -, 495
Macauley, Daniel, H., 1, -, -, 100, -, 100, -, 4000, 150, 75, 15, 50, 30, 8, 1000
Fooks, J. & James, B., 1, -, -, 75, -, 195, -, 3500, 150, 400, 10, 30, 125, 27, 550
Willey, James, -, -, 1, 200, -, 300, 60, 5000, 200, 550, 25, 45, 25, 6, 828
Swain, John B., 1, -, -, 60, -, 40, -, 1000, 30, 300, 10, 15, -, -, 325
Myce(?), David S., 1, -, -, 15, -, 1, 5, -, 7000, 300, 400, -, 200, 400, 130, 2000
Spudler, Sarah, 1, -, -, 130, 1, 70, -, 10000, 200, 600, 100, 120, 200, 33, 1200
Cahal, Elizabeth, 1, -, -, 45, -, -, -, 3000, -, 25, -, 100, -, -, 200
Layton, Elbert W., 1, -, -, 10, ½, -, -, 3000, 25, 200, 25, 75, 150, 24, 600
Willey, Loxley, 1, -, -, 27, -, -, -, 1500, 100, 200, 10, 60, 50, 7, 300
Carn, Peter, 1, -, -, 956, 3, 10, -, 3500, 200, 200, 25, 100, 100, 16, 420
Willen, William 1, -, -, 20, -, -, -, 700, 5, -, 5, 50, 10, 1, 100
Willey, Joseph, 1, -, -, 27, 4, -, -, 3000, 5, -, -, 150, 225, 40, 800
Rawlins, Chas. H., 1, -, -, 8, -, -, -, 500, 50, 75, -, -, 20, 3, 100
Gray, Peter, -, -, ½, 60, 3, -, -, 4000, 100, 460, -, -, 280, 40, 550
Layton, Garret, 1, -, -, 50, 2, 4, -, 7000, 300, 300, 50, 100, 200, 25, 1428
Rust, Wm. C., 1, -, -, 60, -, 63, -, 4600, 100, 450, -, -, 400, 125, 520
Jacobs,Kate, 1, -, -, 152, 5, -, -, 10000, 100, 500, 100, 100, 200, 50, 2160
Lord, Richard, -, 1/3, -, 175, -, 2, 125, -, 6000, 100, 500, 90, 90, 200, 56, 250
Layton, Richard, -, ½, -, 85, 1, 200, 10, 5000, 75, 100, 75, 75, 150, 55, 600
Rickards, Joseph, 1, -, -, 100, 6, 50, -, 15000, 600, 875, 200, 200, 300, 95, 3855
Jacobs, Wm. A., 1, -, -, 170, 4, 100, -, 6000, 150, 700, 84, 84, 275, 70, 1500
Jacobs, Loxley, 1, -, -, 200, 3, 100, -, 15000, 100, 600, 100, 100, 250, 100, 930
Moore, Gilley, -, -, 2/3, 100, 1, 40, -, 3000, 30, 250, -, -, -, -, 300
Wainwright, Wm., -, -, ½, 150, 2, 219, -, 4000, 100, 300, 90, 90, 150, 80, 700
Knox, Samuel, -, -, 1/3, 65, 3, 20, -, 2500, 50, 150, 120, 40, -, -, 225
Meaupou, Adolph, 1, -, -, 60, -, 40, -, 1200, 25, 200, 20, 10, 100, 30, 145
Hamilton, J. M., -, -, 2/5, 200, 1, 50, -, 10000, 100, 100, 10, 70, 60, 32, 968
Gray, William, 1, -, -, 90, -, 30, -, 8000, 200, 500, -, 80, 350, 60, 950
Hamilton,William, -, -, 1/3, 60, -, 64, -, 1000, 60, 60, 10, 15,-, -, 200
Lines, Clement, -, -, 1/3, 100, -, 23, -, 2500, 4, 420, -, 125, 34, 625
Jewell, William -, -, 1/3, 35, -, 8, -, 1000, 20, 25, -, -, 15, 2, 110
Flyerbrand, Frank, 1, -, -, 60, 24, 75, -, 3200, 40, 80, -, 25, -, -, 165
Workman, James, -, -, ½, 60, -, 50, -, 1500, 10, 100, -, 54, -, -, 440
Gorder, Elisha, 1, -, -, 120, -, 50, 12000, 600, 350, 70,70, 250, 48, 750
Richards, Philip, 1, -, -, 200, 6, 175, -, 10000, 500, 500, 50, 300, 600, 110, 2580
Ellegood, Robert, -, -, ½, 40, -, -, -, 500, 10, 150, 10, 20, -, -, 200
Jones, William -, -, 2/3, 60, -, 15, -, 1000, 6, 40, -, -, -, -, 268

Eskridge, George, -, -, 1/3, 60, 2, 40, -, 2000, 100, 250, 10, 30, 40, 7, 360
Baker, Samuel, -, -, 2/5, 60, -, 1, 40, -, 2000, 60, 150, 15, 45, 40, 50, 450
Owens, James, -, -, 1/3, 100, 3, 40, -, 3000, 5, 250, 30, 30, -, -, 550
Lewis, Charles, -, -, 1/3, 80, 2, 45, -, 2500, 100, 250, 20, 60, 100, 60, 500
Moon, Thomas, 1, -, -, 120, 6, 125, -, 6000, 100, 650, 15, 60, 100, 50, 750
Doniphan, Robert, -, -, 1/3, 140, -, 260, -, 7000, 75, 400, 50, -, -, 1140
Cannon, Isaac, -, -, 1/3, -, -, 50, -, 2000, 10, 100, -, -, 30, 10, 280
Swain, Robert, 1, -, -, 90, 1, 30, -, 2000, 170, 325, -, 70, 50, 5, 600
Ray, John, 1, -, -, 75, 3, 125, -, 400, 150, 300, -, 65, 150, 30, 1100
Campbell, John, 1, -, -, 90, 5, 30, -, 5000, 160, 450, 25, 75, 60, 10, 784
Brown, Wilbert, 1, -, -, 80, 3, 5, -, 6000, 50, 140, 25, -, -, -, 840
Kinder, Edgar L., -, -, 2/5, 75, 5, 50, -, 3000, 50, 150, -, 30, 120, 62, 465
Benton, Jacob S., 1, -, -, 200, -, 200,-, 7000, 200, 600, 20, 86, 300, 50, 850
Read, John, 1, -, -, 75, -, 25, -, 2000, 40, 150, 10, 40, 200, 35, 225
Bennet, Henry, 1, -, -, 48, -, 10, 20, 2500, 150, 300, 60, 60, 400, 112, 700
Kinder, David, 1, -, -, 150, 1, 100, -, 3000, 200, 400, 24, 50, 300, 75, 1350
Wright, Rayford, 1, -, -, 100, -, 40, -, 4000, 75, 250, 25, 40, 160, 27, 475
Ross, Saunders S., -, -, 1/3, -, 70, -, 30, -, 2000, 40, 350, 20, -, 30, 6, 330
Benning, John, -, -, 1/3, 60, -, 50, -, 1500, 20,70, 10, -, 10, 2, 200
Sullivan, Henry, -, -, 1/3, 90, -, 65, -, 1500, 25, 145, 10, -, 10, 2, 300
Rankin, Andrew, 1, -, -, 100, -, 32, -, 1880, 75, 300, -, 100, 50, 12, 225
Williams, James, 1, -, -, 15, -, 25, -, 500, 8, 80, -, -, -, -, 140
Williams Jesse, 1, -, -, 15, -, 5, -, 500, 3, 100, -, -, -, -, 70
Corbin, Stephen, 1, -, -, 125, -, 30, -, 2000, 20, 200, 20, 30, 150, 24, 670
Russum, Plymouth, 1, -, -, 40, -, 60, -, 1000, 50, 270, 50, 20, -, -, 242
Corbin, Amos, 1, -, -, 75, -, 40, -, 1500, 80, 200, -, 20, 150, 25, 225
Layton, Jerome, 1, -, -, 75, -, 30, -, 1500, 50, 225, 20, 30, 50, 4, 675
Cannon, Jeremiah, 1, -, -, 80, -, 50, -, 2000, 50, 75, 40, 37, 15, 3, 400
Bradley, Peter, 1, -, -, 35, -, 10, -, 1500, 20, 100, 5, -, -, -, 230
Noble, Jonathan, 1, -, -, 100, 2, 87, -, 6000, 200, 500, 25, 100, -, -, 1060
Corbin, Wm. A., 1, -, -, 100, 3, 25, -, 3000, 100, 350, 20, 60, 125, 25, 560
Hill, _. D., 1, -, -, 100, -, 20, -, 1200, 80, 200, 20, 60, 80, 14, 600
Melson, Benj., 1, -, -, 200, -, 100, -, 6000, 75, 550, 40, 40, 250, 42, 1675
Smith, Lawson, 1, -, -, 130, 1, 20, -, 3000, 60, 300, 15, 15, -, -, 375
Allen, Joseph, -, -, 1/3, 90, 3, 20, -, 1000, 30, 125, 10, -, -, -, 100
Taylor, Isaac, 1, -, -, 30, 1, 2, -, 880, 45, 120, 10, 10, -, -, 200
Williams, John, 1, -, -, 55, 1, 40, -, 2000, 100, 200, 10, 30, -, -, 550
Ellegood, Charles P., -, -, 1/3, 125, -, 75, -, 1800, 30, 150, 40, 5, -, -, 1, -, 150
Coates, Raymond, 1, -, -, 150, 6, 76, -, 8000, 200, 500, 120, -, 100, 33, 1100
Noble, Joseph, 1, -, -, 80, 4, 35, -, 2000, 150, 420, 30, -, 20, 2, 500, -,
Brown, C. C., 1, -, -, 20, 8, 25, -, 2000, 100, 240, 100, 10, 100, 35,675
Brown, Joshua, 1, -, -, 70, 1, 70, -, 1500, 50, 125, -, 10, -, -, 170
Hart, Matilda, 1, -, -, 180, 1, 64, -, 4000, 60, 300, 80, 12, 320, 50, 800
Swain, Cornelius P., 1, -, -, 125, 4, 25, -, 1500, 300, 60, 10, 30, 5, 1275
Culver, George, -, -, 1/3, 120, 6, 300, 30, 5000, 170, 300, 104, 20, 200, 30, 800
Kinder, John, 1, -, -, 154, -, 4, 70, -, 5000, 230, 500, 200, 30, 270, 40, 1400

Hastings, D. C., 1, -, -, 9, -, 1, -, 500, 20, 100, -, 5, -, -, 160
Whitney, J. N., 1, -, -, 100, 4, 100, -, 3000, 75, 150, 25, 30, 50, 8, 328
Banning, A. B., -, -, 1/3, 100, 8, 104, -, 2000, 50, 300, 20, 25, 20, 3, 400
Johnson, H. D., 1, -, -, 50, 1, 14, -, 2500, 100, 250, -, 20, 10, 1, 375
Jacobs, Curtis M., 1, -, -, 200, 4, 115, -, 4500, 125, 300, 10, 15, 100, 5, 550
Melson, Benj., -, -, 1/3, 100, 2, 75, -, 2500, 86, 200, 15, 365, 25, 3,650
Corbin, Amos K., 1, -, -, 80, -, 40, -, 1200, 130, 300, -, 6, -, -, 100
Rust, C. F., 1, -, -, 180, 4, 50, -, 7000, 200, 575, 20, 100, 350, 50, 850
Robinson, James, -, -, 1/3, 70, 2, 30, -, 2500, 30, 175, 10, 30, 100, 5, 600
Wright, Isaac K., 1, -, -, 120, 3, 75, -, 4000, 120, 430, 10, 160, 100, 15, 1100
Wright, Lewis N., 1, -, -, 282, 8, 80, -, 8000, 200, 850, 10, 120, 50, 22, 1600
Burr (Barr), D. F., 1, -, -, 125, 1, 25, 10000, 150, 200, 50, 30, 200, 30, 800
Otwell, Thos., -, 1/3, -, 8, -, -, -, 180, 5, -, -, -, -, -, 60
Williamson, A. M., -, 1/3, -, 40, 1, -, 10, 2000, 5, 50, -, 10, 75, 26, 460
Smith, Lewis, 1, -, -, 75, -, 10, -, 2, -, 1000, 25, 525, -, -, 20, 2, 100
Robinson, Mitchell, 1, -, -, 30, -, 10, -, 886, 25, 65, -, -, -, -, 175
Hill, Robert, 1, -, -, 30, 3, 25, 8, 1200, 170, 10, 35, 10, 5, 425
McMillan, N., 1, -, -, 100, 3, 29, -, 4000, 150, 250, 30, 30, 100, 15, 800
Brown, Daniel, 1, -, -, 45, 1, -, 15, 2000, 20, 150, 10, 30, -, -, 450
Hitch, John, -, -, 2/5, 180, 1, 75, -, 4500, 70, 300, 20, 45, 300, 45, 1040
Langford, Jas. R., 1, -, -, 140, -, 60, -, 4000, 15, 50, -, 75, 180, 50, 650
Cannon, John E., -, -, 2/3, 100, -, 50,-, 2500, 75, 275, 10, 90, 50, 7, 425
Wright, Enos, -, -, ½, 125, 4, 75, -, 4000, 140, 300, 15, 60, 200, 32, 1000
Kinder, Lewis W., 1, -, -, 125, 2, 120, -, 4000, 125, 250, 15, 40, 150, 25, 750
Kinder, Jacob, 1, -, -, 175, 4, 51, -, 4000, 150, 250, 25, 100, 300, 45, 1400
Short, Freeland P., -, -, ½, 66, -, 75, -, 3000, 45, -, 100, 10, 56, 120, 18, 800
Higman, John W., 1, -, -, 80, -, 100, -, 3000, 10, 360, 10, 15, -, -, 440
Coates, William J., 1, -, -, 125, 4,75, -, 5000, 125, 900, 10, 150, 200, 30, 1300
Taylor, Joseph H., -, -, 1/3, 75, -, 100, -, 2000, 125, 75, -, 28, -, -, 210
Ward & Stradley, 1, -, -, 25, 1, 4, -, 1500, 100, 150, -, 30, 100, 15, 500
Swain, Thos. B., 1, -, -, 60, 2, 45, -, 300, 150, 250, 5, 90, 200, 30, 600
Fleetwood, George, 1, -, -, 100, 3, 130, -, 4500, 150, 300, 10, 75, 160, 23, 450
Wright, William H., -, -, 2/5, 250, 5, 500, -, 3000, 250, 500, 20, 35, -, -, 750
Coates, M. A. B., 1, -, -, 150, 2, 80, -, 4000, 150, 500, 25, 100, 300, 33, 900
Pooley, Henry, 1, -, -, 75, ½, 85, -, 1500, 150, 150, 20, 60, 180, 26, 450
Bridge, James, 1, -, -, 130, 2, 80, -, 1000, 75, 45, -, -, 20, 3, 350
Foster, Elbert, 1, -, -, 30, -, 100, -, 1050, 30, -, 100, -, -, -, 120
Cannon, Thos. J., -, -, 1/3, 150, 5, 116, -, 3000, 100, 200, 10, 15, -, -, 230
Russum, E. S., 1, -, -, 10, -, 15, -, 200, 20, 90, 5, -, -, -, 100
Shepherd, Jno, 1, -, -, 52, -, 18, -, 1000, 100, 300, -, 24, -, -, 250
Willoughby, Job, 1, -, -, 20, 5,78, -, 3000, 100, 200, 10, 35, 180, 26, 450
Bullock, John, -, -, 1/3, 140, -, 35, -, 1820, 30, 200,10, 7, -, -, 170
Read, Ezekiel, -, -, 1/3, 50, 7, 50, -, 1500, 50, 162, 10, 36, 10, 2, 400
Evans, Archibald, -, -, 2/3, 30, 2, 114, -, 1000, 5, 40, 15, 16, -, -, 130
Draper, Oliver, 1, -, -, 50, -, 50, -, 800, 50, 175, -, -, -, -, 190
Rathel, Juniper, -, -, 1/3, 60, -, 60, -, 1200, 35, 160, 100, 3, -, -, 180

Odell, John, 1, -, -, 80, 3, -, 4, 3000, 250, 250, -, 90, 60, 9, 980
Frut, George, 1, -, -, 180, 3, -, -, 7000, 350, 500, 20, 100, 200, 32, 1400
Jones, Lavinia, -, -, 2/5, 180, 3, 45, -, 10000, 200, 1100, 50, 150, 150, 21, 2300
McAuley, Chas., -, -, 2/5, 100, 2, 75, -, 6000, 50, 300, 6, 30, 50, 9, 800
Raughley, James, -, -, 2/5, 100, 3, 100, -, 7000, 100, 570, 20, 70, 160, 26, 1000
Willey, Isaac P., 1, -, -, 57, -, 3, -, 2500, 45, 250, 20, 30, 150, 21, 800
Morris, Jeremiah, 1, -, -, 6 8, -, 5, -, 800, 15, 115, 14, 20, 30, 4, 350
Jones, Wm. H., -, -, ½, 250, 3, 150, -, 8000, 225, 600, 30, 100, 100, 17, 1100
McAuley, Ed., C., -, -, 1/3, 120, -, 80, -, 3000, 200, 700, 50, 85, 500, 80, 1500
Richards, John E., 1, -, -, 160, 6, 60, -, 3000, 300, 400, 50, 105, 500, 45, 1800
Lednum, Frank, 1, -, -, 200, 8, 150, -, 7000, 300, 400, 25, 100, 400, 30, 1200
Blades, William, -, 35, -, 160, 4, -, -, 1200, 10, 100, 10, -, -, -, 156
Bissell, Gullaous, 1, -, -, 75, 2, 75, -, 3500, 200, 175, 10, 60, 160, 27, 425
Parker, William, 1, -, -, 100, 4, 20, -, 2000, 75, 354, 10, 16, 140, 20, 100
Parker, William P., 1, -, -, 65, 2, 40, -, 1500, 40, 2000, 5, 15, -, -, 250
Patten, Wellington, 1, -, -, 125, 3, -, 125, 5000, 600, 1000, 30, 100, 500, 75, 950
Wright, C. S., 1, -, -, 100, 3, 69, -, 2000, 40, 300, 10, 84, 140, 20, 800
Adams, E. W., 1, -, -, 170, 2, 30, -, 2000, 200, 600, 25, 50, 260, 40, 900
Higdon, John H., 1, -, -, 33, -, 4, -, 840, 40, 200, 20, 7, -, -, 200
Read, John B., 1, -, -, 105, 3, 5, -, 600, 10, 200, 50, 27, -, -, 200
Hignett, Ellen, 1, -, -, 80, -, 80, -, 1000, 20, 140, 12, 10, 108, 15, 200
Wright, James, -, -, 1/3, 100, -, 20, -, 600, 15, 70, 5, 15, -, -, 85
Webber, William S., -, -, ½, 70, 1, 30, -, 600, 5, 30, 10, 20, -, -, 175
Noble, William, 1, -, -, 40, 2, 10, -, 1200, 30, 200, 19, 30, -, -, 600
Chaffin, Massey, -, -, 1/3, 100, -, 100, -, 2000, 30, 130, 10, 60, 16, 5, 375
Atkinson, Groe, 1, -, -, 90, 1, 70, -, 1000, 25, 130, -, -, 20, 3, 400
Kemp, Thos., 1, -, -, 60, -, 102, -, 800, 50, 250, 20, -, -, -, 60
Adams, Wm. H., 1, -, -, 100, 4, 130, -, 3500, 3000, 1000, 30, 40, 240, 32, 1100
Stayton, Lewis P., -, -, 1/3, 100, -, 200, -, 2000, 25, 125, 15, 56, -, -, 250
Adams, Roger, 1, -, -, 215, 15, 300, -, 12000, 100, 700, 10, 40, 150, 21, 1100
Adams, Daniel, 1, -, -, 100, 2, 80, -, 2000, 125, 420, 40, 100, 200, 30, 950
Vanderting, John, 1, -, -, 120, 2, 10, -, 10000, 50, 300, -, 150, 500, 371, 2600
Weaver, Chas., 1, -, -, 75, 1, 7, -, 3000, 150, 165, -, 65, 150, 21, 1150
Jacobs, John, -, -, 1/3, 225, -, 115, -, 3500, 125, 300, 2, 400, 400, 60, 1400
Willey, John R., -, -, 1/3, 65, -, 100, -, 2000, 45, 150, 25, -, -, -, 237
Dillon, Blades, -, -, ½, 90, 2, 122, -, 3000, 200, 400, 25, 75, 400, 57, 700
Stewart, Wm. N., -, -, 1/3, 114, 1, 184, -, 1500, 30, 225, -, 50, 30, 4, 250
Scott, Jas. B., -, -, 1/3, 100, 2, 50, -, 3000, 75, 3000, 10, 43, 200, 31, 700
Patten, Jacob, 1, -, -, 50, -, 30, -, 1200, 100, 100, -, -, 20, 3, 250
Collison, Jas. A., -, -, 2/5, 150, 3, 70, -, 4000, 150, 250, 10, 30, 50, 7, 500
Jones, James B., -, -, 2/5, 85, 10, 300, -, 4000, 50, 150, -, -, -, -, 275
Foopes, charpes, 1, -, -, 140, 2, 163, -, 3000, 100, 160, 5, 50, 75, 1, 1375
Gardiner, Mahlon M., 1, -, -, 30, -, 32, -, 700, 30,75, -, -, -, -, 140
Hallowell, Thos. E., 1, -, -, 40, 2,88, -, 1500, 40, 150, -, 42, -, -, 370
Allison, Theodore, 1, -, -, 100, 4, 125, -, 2000, 200, 370, 50, 60, -, -, 450
McDowell, Ed., 1, -, -, 6, -, 54, -, 500, 35, 175, -, -, -, -, 125

Knowles, Wm., 1, -, -, 80, 1, 50, -, 2000, 50, 200, 15, 40, 112, 16, 370
Raughley, Samuel, 1, -, -, 85, -, 15, -, 3000, 150, 450, 25, 70, 150, 22, 800
O'Day, Solomon, -, -, 1/3, 125, 1, 162, -, 8000, 250, 400, 20, 75, 260, 40, 1700
Jones, Alexander, 1, -, -, 160, 3, 45, -, 4000, 125, 400, 25, 75, 180, 26, 1000
Lord, James K., -, -, 2/3, 175, 3, 68, -, 4000, 100, 600, 20, 95, 150, 22,750
Collison, Hiram, 1, -, -, 100, 3, 60, -, 1000, 50, 175, 10, 15, 50, 7, 375
Roser, Samuel, 1, -, -, 129, 3, 80, -, 350, 250, 200, -, 45, 150, 21, 400
Bennet, Amos, -, -, 1/3, 90, 1, 58, -, 1800, 50, 380, 10, 30, 100, 14, 275
Pratt, Fredric, -, -, 1/3, 150, 4, 150, -, 2000, 60, 150, 10, 30, -, -, 270
Pratt, Henry, 1, -, -, 60, 28, -, 1000, 40, 300, 10, 28, 20, 3, 380
Dolvins, Wm, H., 1, -, -, 50, -, 50, -, 1200, 75, 230, 20, 40, -, -, 278
Fisher, David W., 1, -, -, 65, 2, 32, -, 2000, 110, 200, 5, 70, 50, 7, 600
Todd, George M., 1, -, -, 50, 1, 25, -, 1000, 20, 160, -, 30, -, -, 125
Davis, Samuel J., -, -, 1/3, 80, -, 5, -, 1600, 20, 250, 20, 42, 40, 6, 675
Bullock, Thos. B., 1, -, -, 100, 1, 153, -, 2000, 50, 120, -, 24, -, -, 140
Kinnamer & Bros., 1, -, -, 65, 1, 442, -, 5000, 200, 560, -, 112, -, -, 1600
Nelson, Columbus, -, -, 2/5, 60, 4, 80, -, 1800, 8, 150, -, 30, 220, 31, 430
Curry, Thomas, 1, -, -, 16, -, 24, -, 700, 10, 70, -, 30, 150, 22, 150
Langendyk, B. H., 1, -, -, 15, 4, -, -, 2500, 40, 100, -, -, 50, 7, 597
Blanchard, George, 1, -, -, 25, 3, 12, -, 2500, 100, 100, 50, -, 100, 14, 600
Blanchard, Mils, 1, -, -, 170, -, 42, -, 800, 70, 1200, 120, 120, 350, 50, 220
Sadler, Wm. T., 1, -, -, 7, -, -, -, 350, 10, 75, -, 10, 25, 3, 110
Hallowell, George, 1, -, -, 85, -, 72, -, 2000, 60, 350, 10, -, 150, 11, 800
Rickards, Jno. R., 1, -, -, 120, -, -, -, 6000, 1000, 1500, 20, 90, 250, 18, 840
Milman, Jouat, -, -, 2/5, 100, 7, 25, -, 10000, 200, 600, 25, 112, 200, 14, 2000
Cannon, John, -, -, 2/5, 240, 5, 60, -, 12000, 300, 1000, 30, 130, 400, 30, 270
Todd, Luther, -, -, 2/5, 120, -, 10, -, 2000, 50, 200, 5, 36, -, -, 800
Ellingsworth, Jno., -, -, ½, 200, 4, 15, -, 13000, 300, 425, 15, 250, 200, 15, 1900
Morris, Ann, E., 1, -, -, 90, 1, 32, -, 2500, 85, 250, 100, 45, 150, 11, 578
Smaller (Smoller, Smeller), Elijah, 1, -, -, 60, -, 15, -, 750, 35, 145, -, 3, -, -, 100
Smeller, Whittington, 1, -, -, 10, 3, 20, -, 2000, 60, 200, -, -, 20, 3, 150
Smith, Geo. M., -, -, 1/3, 135, -, 5, -, 3000, 200, 200, 30, 70, 60, 10, 1800
Stevens, Jno. M., 1, -, -, 90, 4, 12, -, 4000, 200, 600, 20, 70, 50, 8, 500
Willey, Ezekiel, -, -, 2/3, 10, -, -, -, 300, 10, 85, -, -, -, -, 350
Pennewik, Simeon, 1, -, -, 300, 20, 100, -, 8000, 500, 1400, 50, 1000, 500, 834, 200
Lynch, Wm. O., 1, -, -, 90, -, 10, -, 2500, 300, 1000, 25, 125, 300, 50, 1100
Elliott, S. K., -, -, 2/5, 140, 3, 50, -, 5000, 50, 300, 20, 57, 75, 12, 1200
Scott, Jams, -, -, 2/5, 85, -, 25, -, 1000, 50, 500, 10, -, 20, 3, 320
Todd, Ross, -, -, 1/3, 105, -, 15, -, 1000, 20, 100, -, 20, 40, 6, 200
Hamilton, William, -, -, 2/5, 110, -, 30, -, 2500, 25, 250, 10, 36, -, -, 375
Hayes, Isabella, 1, -, -, 90, 2, 20, -, 5000, 100, 700, 20, 75, 400, 70, 1700
Collison, Chas. G., -, -, 2/5, 40, -, 10, -, 1000, 30, 60, -, -, 25, 3, 320
Drake, John, -, -, 2/5, 35, -, -, -, 700, 30, 80, -, 15, -, -, 260
Collison, Jas. M., 1, -, -, 60, 2, 90, -, 2000, 100, 450, 20, 30, 200, 33, 800
Sullivan, James, 1, -, -, 10, -, -, -, 2500, 50, 40, -, 8, 40, 6, 400

Redden, Elijah, -, -, 1/3, 100, 8, 200, -, 3500, 50, 150, 20, 42, 10, 2, 450
Phillips, Eugene, -, 3/5, -, 35, 2, 15, -, 2000, 20, 100, -, 30, -, -, 450
Woodruff, Ann G., 1, -, -, 45, -, -, -, 1000, 100, 75, 20, 40, 250, 70, 450
Stayton, Nehemiah, 1, -, -, 200, 3, 100, -, 6000, 125, 660, 20, 60, 150, 21, 800
Edgell, Wm. H., -, -, 1/3, 75, 3, 79, -, 2400, 50, 250, 15, 31, -, -, 250
Jones, George G., -, -, 2/5, 50, 2, 10, -, 600,75, 20, -, 12, -, -, 350
Gallan, Samuel, 1, -, -, 120, 1, 40, -, 2000, 65, 600, -, 75, -, -, 750
Adams, Nathan & Son, 1, -, -, 80, 3, 76, -, 6000, 40, 350, -, 70, 100, 14, 230
Morris, Wm. W., 1, -, -, 100, 1, 60, -, 5000, 200, 800, 20, 200, 400, 60, 1570
O'Day, Wm. S., -, -, 2/5, 150, -, 250, -, 8000, 60, 300, 5, 30, 100, 15, 450
Frank, John H., -, -, ½, 250, 4, 250, -, 5000, 150, 450, 20, 84, 50, 7, 500
Barwick, Nathan J., 1, -, -, 80, 3, 120, -, 3000, 100, 450, 15, 30, 100, 16, 850
Lord, Andrew, 1, -, -, 51, 1, -, -, 3000, 100, 400, 10, 15, 5, -, 660
Collins, Geo. W., 1, -, -, 150, 6, 50, -, 5000, 400, 700, 50, 150, 300, 45, 1330
Edgwell, William, -, -, 1/3, 150, 6, 110, -, 3000, 100, 600, 10, 45, -, -, 1100
Morris, Hezekiah, 1, -, -, 72, -, 20, -, 2000, 200, 360, -, 32, 80, 12, 450
Collison, William, 1, -, -, 75, 2, 85, -, 2500, 100, 450, 15, 35, -, -, 600
Hamilton, Sarah, 1, -, -, 60, -, 36, -, 1000, 50, 475, 10, 45, -, -, 500
Todd, John H., 1, -, -, 120, 3, 83, -, 6000, 100, 575, 20, 125, 320, 53, 1500
Lynch, John, -, -, 2/5, 100, 18, 40, -, 6000, 100, 450, 20, 70, 120, 17, 1200
Andrews. Wm. G., -, -, 1/3, 175, 2, 85, -, 10000, 110, 350, 20, 60, 150, 22, 1000
Hallowell, George, 1, -, -, 30, 2, 7, -, 1000, 50, 80, -, -, -, -, 300
Wright, Chas. Heirs, 1, -, -, 245, 4, 75, -, 6000, 100, 260, 200, -, 25, 2, 665
Pinder, Isaac, 1, -, -, 120, 12, 20, 34, 3000, 200, 375, 1500, 56, 150, 52, 250
Colbourn, Michael, 1, -, -, 750, 100, 50, -, 1500, 75, 75, -, -, -, -, 150
Rawlins, John M., 1, -, -, 340, 40, 100, -, 1500, 150, 270, -, -, -, -, 390
Rawlins, John, 1, -, -, 130, 50, -, -, 1000, 100, 300, -, -, -, -, 150
Rawlins, William, 1, -, -, 230, 50, 25, -, 1500, 75, 100, -, -, -, -, 250
Wright, Eliza A., 1, -, -, 250, 28, 75, -, 12200, 310, 500, -, -, -, -, 800
Cannon, Cyrus, 1, -, -, 125, 10, -, -, 3000, 150, 250, -, -, -, -, 520
Giles, Thomas B., 1, -, -, 225, 25, 60, -, 10000, 300, 600, 100, 140, 400, 52, 1000
Ross, William H., 1, -, -, 230, 80, -, -, 5000, 200, 575, -, 300, 200, 52, 650
Brown, James, 1, -, -, 350, 40, 105, -, 7000, 300, 1000, 100, 150, 300, 52, 1000
Willey, John, 1, -, -, 20, -, -, -, 600, 100, 200, -, -, -, -, 300
Cannon, Catharine, 1, -, -, 300, 10, 100, -, 1500, 150, 395, -, -, -, -, 350
Morgan, James, 1, -, -, 125, 6, 50, -, 1500, 125, 400, -, -, -, -, 415
Knowles, James, 1, -, -, 130, -, 40, -, 1200, 100, 350, -, -, -, -, 218
Willey, John, 1, -, -, 185, 14, 65, -, 2500, 75, 175, -, -, -, -, 318
Eskridge, James, 1, -, -, 117, 2, 50, -, 2000, 100, 588, -, -, -, -, 250
Cannon, Margaret, 1, -, -, 150, -, 50, -, 1000, 75, 175, -, -, -, -, 300
Prettyman, James, 1, -, -, 200, 10, 100, -, 1000, 100, 580, 25, -, -, -, 336
Nevin, ____, 1, -, -, 300, -, 200, -, 1500, 50, 110, -, -, -, -, 250
Coats, Raymond, 1, -, -, 325, -, 200, -, 1625, 75, 140, -, -, -, -, 100
Starkey, Thomas H., 1, -, -, 73, 5, 10, -, 600, 50, 50, -, -, -, -, 300
Moore, Thomas, 1, -, -, 100, -, 20, -, 1000, 50, 125, -, -, -, -, 150
Hendrick, Sallie, 1, -, -, 251, 25, 80, -, 10000, 200, 576, -, 100, 150, 52, 1000

Ward, Joseph, 1, -, -, 131, -, 40, -, 3500, 100, 375, -, 90, 100, 52, 630
Giles, Isaac, 1, -, -, 700, -, 500, -, 16000, 75, 400, -, -, -, -, 550
Farron, Mc. Joseph, 1, -, -, 400, -, 30, -, 6000, 100, 737, -, -, -, -, 375
Cannon, Elizabeth, 1, -, -, 150, -, 50, -, 2000, 50, 102, -, -, -, -, 250
Cannon, Elizabeth, 1, -, -, 25, -, -, -, 200, 50, 100, -, -, -, -, 125
Cannon, Elizabeth, 1, -, -, 25, -, -, -, 250, 50, 100, -, -, -, -, 125
Short, George, 1, -, -, 59, -, 10, -, 1500, 100, 200, -, -, -, -, 200
Short, George, 1, -, -, 75, -, 15, -, 750, 100, 185, -, -, -, -, 185
Weisman, _____, 1, -, -, 116, -, 30, -, 1500, 75, 210, -, -, -, -, 250
Sharp, Elie R., 1, -, -, 47, -, -, -, 1000, 100, 225, -, -, -, -, 375
Burnside, William, 1, -, -, 98, -, 10, -, 1200, 100, 165, -, -, -, -, 279
Martin, John E., 1, -, -, 179, 50, 40, -, 6000, 150, 150, -, -, -, -, 400
Kinder, Warren, 1, -, -, 30, 30, 50, -, 5000, 100, 300, -, -, -, -, 450
Cannon, Levie, 1, -, -, 14, -, 6, -, 500, -, 35, -, -, -, -, 100
Cannon, Kissiah, 1, -, -, 46, -, 6, -, 1000, 50, 125, -, -, -, -, 150
Ross, William H., 1, -, -, 150, -, 130, -, 7000, 200, 360, 300, 250, -, -, 650
Cannon, William J., 1, -, -, 100, -, 15, -, 4000, 50, 250, -, 75, -, -, 375
Williams, Levin, 1, -, -, 70, -, 20, -, 1500, 105, 130, -, -, -, -, 275
Hooper, Haix, Heirs, 1, -, -, 30, -, 20, -, 500, -, -, -, -, -, -, 75
Houston, Robert, 1, -, -, 56, -, 70, -, 1000, 225, 125, -, 14, -, -, 175
Farron (Farrow), Mc. Joseph, 1, -, -, 40, -, -, -, 500, 75, 400, -, -, -, -, 300
Brown, Robert, 1, -, -, 50, -, 10, -, 500, -, -, -, -, -, -, 75
Wainwright, Elizabeth, 1, -, -, 30, -, -, -, 300, 40, 75, -, -, -, -, 100
Allen, Jesse W., 1, -, -, 100, 20, 20, -, 4000, 100, 250, -, 50, -, -, 390
Brown, Robert, 1, -, -, 107, 30, 40, 20, 3000, 250, 680, -, 75, -, -, 700
Allen, Joseph, 1, -, -, 125, 6, 20, -, 2500, 100, 510, -, -, -, -, 430
Cannon, William H., 1, -, -, 60, 2, 10, -, 1000, 100, 250, -, -, -, -, 200
Fleetwood, James, 1, -, -, 70, 6, 7, -, 1500, 100, 350, -, -, -, -, 275
Allen, John H., 1, -, -, 85, 14, 35, -, 2500, 100, 300, -, 28, -, -, 400
Allen, William H., 1, -, -, 63, 14, -, -, 1500, 100, 375, -, -, -, -, 541
Bradley, John H., 1, -, -, 166, 10, 75, -, 2000, 100, 280, -, 15, 150, 52, 577
Kinder, Warren, 1, -, -, 190, 10, 137, -, 10000, 250, 450, -, 75, 250, 52, 729
Kinder, Warren, 1, -, -, 85, 5, 45, -, 2500, 100, 400, -, -, 200, 52, 403
Kinder, John H., 1, -, -, 158, 42, 100, -, 5000, 175, 415, -, -, 150, 52, 615
Skidmore, Timothy, 1, -, -, 37, -, 30, -, 1500, 75, 175, -, -, -, -, 474
Hurly, Mary, 1, -, -, 49, 1, -, -, 1000, 75, 80, -, -, -, -, 425
Dashields, William, 1, -, -, 200, -, 160, -, 8000, 400, 456, -, 60, 200, 52, 1000
Martin, Halsey, 1, -, -, 155, 20, 50, -, 5000, 200, 300, -, 75, 150, 52, 600
Ford, Samuel, 1, -, -, 150, 10,100, -, 2500, 350, 230, -, -, 100, -, 350
Swoden, Aaron, 1, -, -, 12, -, 1, -, 300, 50, 50, -, -, -, -, 75
Frame, Robert, 1, -, -, 150, 38, 72, -, 3000, 50, 300, -, 15, 100, 52, 500
Frame, Robert, 1, -, -, 75, -, 30, -, 500, -, -, -, -, -, -, 75
Kinder, Warren, 1, -, -, 91, 9, 80, -, 3500, 25, 215, -, 28, -, -, 360
Wright, Isaac, 1, -, -, 75, 10, 65, -, 2500, 50, 300, -, 30, -, -, 500
Neal, Joseph, 1, -, -, 30, 20, 80, -, 1000, 10, 50, -, -, -, -, 100
Spicer, William W., 1, -, -, 56, 19, 25, -, 1500, 100, 465, -, 60, 50, 14, 600

Hallowell, George, 1, -, -, 100, 5, 200, -, 2500, 25, 300, -, 40, -, -, 350
Hallowell, George, 1, -, -, 80, 15, 5, -, 1000, 12, 210, -, 40, 60, 36, 455
Brown, N. H., 1, -, -, 95, 12, 7, -, 1500, 10, 260, -, 14, -, -, 425
Waller, Peter, 1, -, -, 28, 2, 30, -, 1500, 25, 380, -, -, -, -, 200
Brown, James H., 1, -, -, 80, 10, 20, -, 100, 300, 500, -, 50, 150, 52, 500
Cannon, William E., 1, -, -, 165, 25, 100, -, 6000, 200, 600, -, 70, 150, 52, 1000
Brown, Thomas H., 1, -, -, 128, 12, 20, -, 4000, 200, 900, -, 75, 200, 50, 500
Neall, Joseph, 1, -, -, 96, 20, 75, -, 4000, 200, 900, -, 75, 200, 50, 500
Spicer, William W., 1, -, -, 43, 62, 40, -, 6000, 300, 1000, -, 90, 250, 52, 1150
Charles, James, 1, -, -, 190, 50, 240, -, 5000, 100, 700, -, 56, 150, 50, 900
Obilar, Joshua, 1, -, -, 146, 5, 75, -, 2000, 75, 365, 20, 56, 175, 52, 750
Neal, James F., 1, -, -, 60, 5, 50, -, 1500, 100, 325, -, 75, 150, 52, 575
Martin, Edward, 1, -, -, 134, 90, 76, -, 20000, 500, 600, 600, 200, 1000, 52, 3000
Benson, Charles, 1, -, -, 78, 2, 20, - 1000, 10, 125, -, -, 50, -, 150
Obier, Joshua, 1, -, -, 285, 12, 100, 20, 4000, 1000, 800, -, 250, 400, 52, 300
Obier, Joshua, 1, -, -, 30, -, -, -, 500, -, -, -, -, -, -, 150
Fleetwood, William, 1, -, -, 100, -, 40, -, 1000, 100, 150, -, 60, -, -, 300
Obier, Joshua, 1, -, -, 122, 3, 75, -, 2000, 50, 275, 100, 28, 80, 52, 200
Conoway, Curtis, 1, -, -, 206, 24, 30, -, 5000, 150, 850, -, 100, 75, 52, 1000
Waller, Peter, 1, -, -, 50, -, 5, -, 1400, 25, 200, -, -, -, -, 300
Warren, Isaac, 1, -, -, 98, 2, 15, -, 3000, 300, 500, -, 50, 20, 52, 700
Kinder, Jacob A., 1, -, -, -, -, -, -, 500, 50, 300, -, -, -, -, 100
Kinder, Jacob A., 1, -, -, 105, 20, 50, -, 1800, 100, 200, -, -, -, -, 630
Kinder, William F., 1, -, -, 144, 23, 60, -, 4000, 300, 400, -, 113, 300, 52, 600
Gilmer, John, 1, -, -, 89, 9, 40, -, 6000, 300, 375, -, 28, 20, 10, 300
Kinder, Daniel E., 1, -, -, 59, 6, 35, -, 2000, 150, 225, -, 14, 7, 52, 400
Davis Heirs, 1, -, -, 240 -, 100, -, 1800, 10, 100, -, -, -, -, 175
Williams, Newton, 1, -, -, 3, -, -, -, 500, 10, 75, -, -, -, -, 50
Williams, Newton, 1, -, -, 199, 22, 100, -, 3500, 100, 475, -, 60, 100, 52, 500
Obier, Augustus C., 1, -, -, 85, 15, 40, -, 1000, 50, 200, -, 30, 100, 52, 400
Obier, Isaac C., 1, -, -, 34, -, 34, -, 700, 10, 20, -, -, 95, 200
Davis, John M., 1, -, -, 86, - 50, -, 1000, 5, 20, -, -, -, -, -
Morgan, Charles H., 1, -, -, 85, -, 10, -, 100, 50, 225, -, 14, -, -, 457
Lusk, Alfred, 1, -, -, 125, 11, 25, -, 3000, 200, 400, -, 50, 200, 52, 1000
Ruvse(Purse), John J., 1, -, -, 103, 12, 10, -, 1200, 100, 100, 750, -, -, -, 300
Willin, Edward, 1, -, -, 122, 10, 30, -, 1800, 25, 250, -, 90, 60, 24, 500
Darbie, Perry T., 1, -, -, 145, 5, 50, -, 2000, 75, 500, -, 30, -, -, 500
Lecates, Levin, 1, -, -, 93, 15, 40, -, 2000, 100, 200, -, 30, 80, 16, 400
Cottingham, Charles A., 1,-, -, 6, 2, -, -, 1200, -, 50,-, -, 30, 10, 200
Neal, Joseph, 1, -, -, 200, 23, 100, -, 6000, 200, 500, -, 100, 200, 52, 1500
Allen, M. W., 1, -, -, 35, 15, 5, -, 2000, -, -, -, -, -, -, 500
Bell, Henry, 1, -, -, 75, 30, 24, -, 1000, -, -, -, -, -, -, 200
Ward, Samuel, 1, -, -, 80, 5, 55, -, 1500, 50, 300, -, -, -, -, 300
Brown, Hugh Heirs, 1, -, -, 128, 15, 75, -, 2500,75, 400, -, 28, 50, 20, 600
Hearn, Daniel, 1, -, -, 98, 2, 20, -, 2000, 10, 150, -, -, -, -, 175
Clifton, William, 1, -, -, 6, -, -, -, 300, 5, 50, -, -, -, -, 50

Brown, Hugh C., 1, -, -, 57, 6, 32, -, 2500, 115, 250, -, 50, 150, 52, 500
King, Ira, 1, -, -, 85, 16, 7, -, 2000, 15, 150, -, 25, -, -, 100
Jackson, Samuel, 1, -, -, 7, -, 2, -, 250, -, -, -, -, -, -, 25
Wainwright, Joseph, 1, -, -, 35, 5, 10, -, 2500, 100, 150, -, 28, -, -, 500
Dashields, William, 1, -, -, 180, 20, 50, -, 6000, 100, 350, -, 60, -, -, 1000
Williams, David S., 1, -, -, 89, 6, 25, -, 1500, 50, 210, -, 30, 30, 15, 300
Handy, Willard H., 1, -, -, 52, 3, 20, -, 1500, 150, 200, -, 50, -, -, 600
Sears, Henry, 1, -, -, 47, 13, 100, -, 2500, 100, 300, -, 30, 100, 52, 500
Wainwright, Silas C., 1, -, -, 6, -, 4, -, 300, -, -, -, -, -, -, 75
Wainwright, Silas C, 1, -, -, 127, 33, 50, -, 6000, 300, 430, -, 60, 150, 52, 1500
Morris, Robert C., 1, -, -, 19, -, -, -, 500, -, 75, -, -, -, -, 200
Thomas, Charles Heirs, 1, -, -, 120, 40, 140, -, 7000, 300, 600, -, -, 250, 150, 1000
Bounds, Jacob, 1, -, -, 115, 55, 15, -, 3000, 150, 300, -, 60, -, -, 1000
Bryan, David S., 1, -, -, 52, 11, 10, -, 1500, 50, 260, 5, 13, 5, 2, 418
O'Day, James T., 1, -, -, 57, 3, 40, -, 1000, -, -, -, -, -, -, 150
O'Day, James T., 1, -, -, 98, 20, 10, -, 2500, 250, 615, 25, 60, 50, 20, 1000
Houston, William, 1, -, -, 120, 30, 50, -, 6000, 400, 695, 40, 250, 400, 32, 1800
Lankford, Littleton, 1, -, -, 84, -, 15, -, 1500, 20, 30, -, -, -, -, 100
Lankford, Littleton, 1, -, -, 182, 18, 60, -, 2000, 10, 600, -, -, 100, 52, 1000
White, Theodore, 1, -, -, 42, 2, 35, -, 1000, 100, 150, -, 15, -, -, 250
Allen, Robert J., 1, -, -, 37, -, -, -, 500, 24, 230, -, -, 15, 50, 10, 500
Allen, Levin W., 1, -, -, 108, 2, 10, -, 1000, 25, 120, 1, 6, 60, 40, 200
Tull, John A., 1, -, -, 34 -, 4, -, 500, 25, 150, -, -, -, -, 200
Allen, William H., 1, -, -, 71, -, 17, -, 1000, 100, 85, -, 30, -, -, 300
Collins, Josephus, 1, -, -, 77, -, 30, -, 1500, 50, 175, -, 30, 50, -, 400
McKinzie, James S., 1, -, -, 20, -, 80, -, 2000, 50, 175, 20, 5,-, -, 30
Lorens, John H., 1, -, -, 37, -, 5, -, 800, 25, 150, -, 10, -, -, 400
Tull, Peter & William, 1, -, -, 100, -, 25, -, 1000, 50, 50, -, -, -, -, 300
Tull, Robert A., 1, -, -, 38, 2, 5, -, 800, 50, 300, -, 30, -, -, 500
Losen, George, 1, -, -, 100, -, 25, -, 1000, 50, 75, -, -, -, -, 400
Tull, Robert E., 1, -, -, 30, -, 15, -, 500, -, -, -, -, -, -, 100
Tull, Robert A., 1, -, -, 35, -, 24, -, 500, -, -, -, -, -, -, 200
Tull, Samuel E., 1, -, -, 27, -, 25, -, 800, -, 110, -, -, -, -, 300
Rodgers, William E., 1, -, -, 122, 7, 35, -, 9000, 500, 600, -, 80, 100, 52, 2000
Colbourn, Michael, 1, -, -, 110, 20, 10, -, 7000, 200, 325, -, 90, -, -, 1000
Obier, Jesse W., 1, -, -, 91, 7, 15, -, 2000, 150, 300, -, 60, 80, 30, 1600
Hazzard, Lucinda, 1, -, -, 75, -, 25, -, 1500, -, -, -, -, -, -, 350
Reed, Andrew, 1, -, -, 100, -, 40, -, 1500, -, -, -, -, -, -, 300
Allen, M. W., 1, -, -, 80, 10, 5,-, 3000, 150, 335, -, 60, 100, 52, 00
Moore, Tharp, 1, -, -, 144, 18, 50, -, 5000, 400, 300, -, 125, 250, 52, 1300
Foster, George W., 1, -, -, 33, -, 1, -, 500, -, 30, -, -, -, -, 75
Gillis, Edward T., 1, -, -, 100, 95, 10, -, 4000, 200, 500, 200, 60, 200, 52, 1000
Gillis, Edward T., 1, -, -, 75, 14, 12, -, 2000, -, 60, -, -, -, -, 250
Fooks, Benjamin, 1, -, -, 203, 5, 100, 50, 2100, 60, 225, -, -, -, -, 500
Colbourn, Elijah, 1, -, -, 99, 4, 34, -, 1000, 60, 100, -, 12, -, -, 200
Allen, James H., 1, -, -, 70, 3, 35, -, 1000, 10, 50, -, -, -, -, 100

Grayham, John K., 1, -, -, 25, -, 5, -, 300, 10, 25, -, -, -, -, 100
Denbee (Darbie), Perry, 1, -, -, 80, 4, 25, -, 1000, 50, 177, -, -, -, -, 200
Lankford, Jno. P., 1, -, -, 40, -, 20, -, 800, 25, -, -, -, -, -, 200
Cannon, Curtis, 1, -, -, 15, -, 15, -, 600, 20, 150, -, -, -, -, 400
Nicholson, George W. S., 1, -, -, 75, -, 75, -, 1000, -, -, -, -, -, -, 200
Nicholson, George W. S., 1, -, -, 150, -, 300, -, 2000, -, -, -, -, -, -, 500
Nicholson, George W. S., 1, -, -, 1130, -, 150, -, 1000, -, -, -, -, -, -, 1000
Ellis, Elijah J., 1, -, -, 65, 14, 20, -, 2500, 250, 395, -, 60, -, -, 1000
Rickards, William T., 1, -, -, 179, 6, 40, -, 2500, 40, 265, -, -, 100, 25, 500
Dickerson, Manias J., 1, -, -, 217, 8, 75, -, 5000, 250, 500, -, 15, 175, 52, 1000
Huston, William, 1, -, -, 25, -, 125, -, 2000, -, -, -, -, -, -, 100
Huston, Martin Heirs, 1, -, -, 45, -, 45, -, 1000, 100, 100, -, -, -, -, 500
Ellis, George R., 1, -, -, 30, 1, 130, -, 2000, -, -, -, -, -, -, 300
Ellis, William Sr., 1, -, -, 131, 9, 20, -, 7000, 200, 425, -, 119, 250, 5, 4000
Ellis, William Jr., 1, -, -, 6, -, 24, -, 5000, -, -, -, -, -, -, 300
Knight, William W., 1, -, -, 162, 8, 30, -, 5000, 100, 275, -, -, -, -, 1000
Chamberlin, Enoch, Heirs, 1, -, -, 75, 25, 10, -, 3000, 150, 305, -, -, -, -, 700
Cash, William, 1, -, -, 80, -, 20, -, 1200, 25, -, -, 30, -, -, 300
Brown, Thomas H., 1, -, -, 905, 5, 100, -, 1500, 50, 150, -, -, -, -, 200
Fooks, Benjamin, 1, -, -, 35, 25, 260, 30, 3000, 50, 250, -, -, -, -, 500
Reed, Ezekiel Heirs, 1, -, -, 75, 3, 35, -, 3000, 25, 150, -, -, 25, 4, 500
Hall, John W., 1, -, -, 110, -, 1000, -, 1500, 10, 50, -, -, -, -, 100
Wheatly, Jesse R., 1, -, -, 100, -, 30, -, 1000, 10, 100, -, 28, -, -, 200
Lere, George, 1, -, -, 150, -, 50, -, 1000, 50, 215, -, -, -, -, 300
Dickerson, Jasper, 1, -, -, 200, -, 100, -, 3000, -, 25, -, -, -, -, 400
Cannon, Allen, 1, -, -, 150, -, 200, -, 2500, 75, 330, -, -, -, -, 300
McWilliams, Henry, 1, -, -, 100, 10, 50, -, 1500, 75, 350, -, -, -, -, 200
Chamberlin, James L., 1, -, -, 20, -, 10, -, 500, -, 30, -, -, -, -, 100
Dulaney, William W., 1, -, -, 60, -, 20, -, 2000, -, -, -, -, -, -, 800
Dulaney, William W., 1, -, -, 199, 8, 150, -, 20000, 150, 521, -, 60, 200, 52, 1000
Larramore, Thomas, 1, -, -, 7, -, -, -, 1500, 10, 35, -, -, -, -, 100
Blackson, Spencer Heirs, 1, -, -, 30, -, 10, -, 500, -, 25, -, -, -, -, 50
Willis, Charles, 1, -, -, 20, 2, -, -, 1000, -, 100, -, -, -, -, 150
Ross, William H., 1, -, -, 250, 130, 80, -, 20000, -, -, -, -, 1000, -, -, 2500
Ross, James J., 1, -, -, 225, 120, 50, -, 20000, 1000, 2575, -, 500, 5000, 78, 15000
Cannon, Wingate, 1, -, -, 44, -, -, -, 5000, 50, 200, -, -, -, -, 200
Fisher, Sarah, 1, -, -, 14, -, -, -, 500, -, -, -, -, -, -, 150
Halloway, George, 1, -, -, 30, 1, 7, -, 1000, 10, 80, -, -, -, -, 300
Heirs of Chas. Wright, 1, -, -, 45, 45, 75, -, 5000, 100, 260, 20, -, 25, 2, 465
Kinder, Isaac, 1, -, -, 20, 12, 20, -, 3000, 200, 375, 1500, 50, 150, 52, 250
Colbourn, Michael, 1, -, -, 50, 100, 50, -, 1500, 75, 75, -, -, -, -, 150
Wright, Eliza (widow), 1, -, -, 40, 24, 75, -, 10000, 300, 350, -, -, 5, 4, 340
Rawlins, John M., 1, -, -, 40, 60, 150, -, 1500, 150, 270, -, -, -, -, 325
Rawlins, James, 1, -, -, 65, -, -, -, 1000, 100, 200, -, -, -, -, 100
Stevens, William H., 1, -, -, 7, -, -, -, 1500, -, -, -, -, -, -, 200
Darber (Darbie), James, 1, -, -, 50, -, -, -, 5000, 50, 75, -, 60, -, -, 500

Cannon, Ann, 1, -, -, 16, -, -, -, 1000, -, -, -, -, -, -, 200
Stockly, Benjamin, 1, -, -, 28, 9, -, -, 400, 100, 80, -, -, 150, -, 800
Hopkins, Henry L., 1, -, - 40, 15, 2, -, 5000, 100, 200, -, -, -, -, 928
Williams, Jacob, 1, -, -, 112, 3, 20, -, 18000, 500, 600, -, 100, 400, 52, 1000
Martin, Hugh, 1, -, -, 8, -, -, 1000, 50, 100, -, -, -, -, 300

Aaron, 239
Abbert, 240
Abbler, 227
Abbot, 337
Abbott, 211, 222, 324-325, 328, 341, 354-355
Ables, 206
Abott, 224
Abram, 238
Ackerman, 219
Adair, 284
Adam, 376
Adamno, 319
Adams, 233-234, 352, 375, 379, 385, 389, 391
Addicks, 256
Adhern, 342
Adkins, 313, 345-347, 349, 357
Adkinson, 234
Ahern, 265, 294
Aiken, 270
Ake, 360
Albruy, 356
Alcord, 261
Aldrich, 226
Alexander, 218, 225, 240, 265, 272
Alfree, 299, 301
Allaband, 235
Allcorn, 266
Allcott, 220
Allen, 275, 299, 317, 365, 383, 392-394
Allison, 389
Allmond, 254, 258
Allston, 299
Alrich, 274, 279
Alrichs, 289
Alston, 285
Anderson, 198-199, 202-203, 218, 226, 232-233, 237, 240, 243-244, 255, 257-258, 294, 311-312, 374
Andre, 220
Andrew, 222, 315, 318

Andrews, 306, 350, 378, 391
Anthony, 200, 229-230, 2490250
Antney, 246
Appleby, 275-276, 279
Appleman, 218
Appleton, 290, 294-296
Archibald, 388
Argae, 216
Argo, 202, 216, 324
Argoe, 334-336
Armor, 219, 259, 261, 267
Armstrong, 217, 263-263, 268-269, 271, 275, 279, 281, 291, 293, 297, 301
Arnold, 224, 251, 341
Aron, 207, 211
Arrington, 201
Artist, 205
Ash, 282
Ashcroft, 285
Aspril, 288
Aswell, 384
Atkins, 326-328, 330, 353, 356, 362, 365
Atkinson, 211, 389
Attix, 248
Atwell, 286, 293
Atwood, 276
Aurton, 295
Austin, 338-339
Auston, 295
Ayars, 273
Aydelotte, 306, 309, 319
Bacon, 373
Bae_-, 379
Baeer, 372
Bailey, 200, 241, 247, 251, 268, 293, 326, 340, 369, 375, 377-379
Bailio, 317
Baily, 322, 341
Baker, 201, 229, 231, 240, 277, 312, 314-315, 317, 320, 342, 348, 358-360, 364, 369, 374, 387

Bakey, 262
Baldwin, 273
Balis, 351
Ball, 262, 266, 268
Ballen, 279
Bamborough, 243
Bancroft, 235, 263
Bankes, 262
Banks, 257, 272-273, 306-307, 309
Banning, 262, 264, 340, 385, 388
Bannon, 341
Banta, 337
Barber, 282
Barcus, 214, 226, 236
Barens, 214
Barker, 219, 226, 228, 269, 351, 353, 367
Barlow, 223, 253, 265, 270, 293
Barnaby, 275
Barnard, 235
Barnes, 261
Barnett, 239, 287, 297, 306
Barnitt, 236
Barr, 382, 388
Barrett, 205, 237
Barry, 260
Bartholomew, 268, 276
Bartlett, 266, 328
Barton, 364
Bartran, 260
Barwick, 391
Basket, 244
Bastin, 243
Bateman, 240
Batson, 318
Batton, 281, 284
Baudlett, 211
Baum, 210
Baylis, 364
Baynard, 228-230, 233
Baynolds, 295
Baynum, 327
Beach, 374-375
Beachum, 356, 358
Bead, 294
Beardsley, 339

Bearman, 207
Beaston, 291
Beatty, 262
Beauregard, 256
Bebee, 341, 381
Beck, 285, 300
Becklin, 213
Beddle, 244
Bedwell, 210-211, 213, 250, 300
Beers, 208
Beeson, 254-255
Beideman, 334
Belknap, 338
Bell, 203, 206, 218, 251, 269, 393
Bellman, 211
Belville, 286
Bennet, 298, 377, 387, 390
Bennett, 218, 218, 222, 288-289, 300-301, 304-305, 307-308, 311, 313, 332, 334-336, 379, 382
Benning, 387
Bennum, 353, 362
Benson, 223, 227, 229, 245, 249, 265, 280, 321, 361, 393
Benston, 331, 334
Benton, 387
Benwell, 213
Berry, 204, 215-216, 247, 252
Beswick, 222
Bethards, 220, 351
Betts, 217, 234, 307, 314, 322, 324, 329-331, 337, 340, 343-344, 350, 358, 361
Betty, 259
Beynum, 371
Bice, 297
Bickel, 222
Bickle, 220
Bidding, 241
Biddle, 274
Biderman, 259
Biediman, 342
Bigger, 255
Biggs, 273, 280
Bilderback, 251
Biles, 299

Billings, 228
Binge, 289
Binns, 239
Bird, 253-255, 285
Bishop, 198, 206, 209, 216, 305, 310, 315, 317, 349
Bissell, 389
Black, 208, 241, 328-329, 345
Blackiston, 251, 301
Blackson, 395
Blackston, 203
Blackwell, 253, 277
Blade, 228
Bladees, 199
Blades, 229-230, 389
Blair, 332
Blake, 199
Blanchard, 390
Blandy, 277
Blaudan, 232
Bleakley, 198
Blindt, 298
Blizzard, 309, 325-326, 366, 370
Blood, 235-236
Bobbet, 211
Bochwell, 211
Body, 376
Bogart, 339
Boggs, 211, 216, 298
Bolden, 225
Bolton, 276
Bonnell, 278
Bonsall, 253
Bonwill, 211
Boone, 241
Booth, 220, 224, 226, 234, 255
Bosket, 244
Bostick, 243
Bouden, 359
Boughman, 268, 272
Boulden, 283-284, 290
Bounds, 394
Bourke, 248
Bowden, 200, 346, 359
Bowen, 230, 282
Bowers, 204

Bowing, 266
Bowman, 232, 271
Bowton, 314
Boyce, 318-321, 324, 331, 372-372, 378
Boyd, 247, 260, 300
Boyed, 197
Boyer, 208-209, 216, 247
Boyles, 201
Boynton, 240
Boys, 243, 280-281, 284-285
Bozman, 332
Brace, 385
Brackin, 264-265, 267-268
Bradford, 314
Bradley, 222, 241, 243, 375, 377-378, 387, 392
Bradly, 224-225, 376-377
Brady, 286, 291
Braklin, 212
Brannon, 260
Brasure, 307, 314, 349, 357
Brayman, 247
Bredan, 232
Bregator, 284
Bremefle, 264
Brereton, 363
Bridge, 388
Briggs, 301
Brimer, 347
Brinckle, 263
Bringhurst, 258
Brinkley, 333
Brion, 355
Brises, 297
Brittingham, 323-324, 333, 347, 359-361, 372
Brockett, 294
Brockson, 297, 300
Brodaway, 301
Brooks, 279, 281-282, 373
Broom, 202
Brothers, 207, 295
Brown, 226, 228-230, 235-236, 242-243, 265, 271-273, 276, 283-284,

292, 327, 355, 373, 381, 387-388, 391-395
Brumbly, 357
Bruton, 327
Bryan, 247, 285, 319, 327, 331, 345-346, 367, 380, 382, 384
Buchanan, 301
Buckingham, 271
Buckmaster, 250
Buckson, 297
Budd, 294
Buell, 377, 384
Bullock, 229, 321, 388, 390
Bumgust, 269
Bunting, 311-316, 349-350, 356
Burbage, 308
Burch, 372
Burcharel, 243
Burchart, 294
Burchenal, 237
Burgess, 289
Burk, 239
Burnham, 203, 226
Burnitt, 246
Burnside, 392
Burr, 388
Burres, 245
Burris, 212-213, 218, 274-275, 292, 317, 321-322, 339, 342, 344, 381
Burrows, 248, 250
Burry, 220
Burton, 231, 236, 243, 307-308, 322-324, 328-329, 332-333, 336, 343, 345, 348, 351, 354, 362-365, 369-370
Bush, 256
Butabee, 355
Butcher, 199, 355
Butler, 201, 224, 232, 345, 383
Bye, 218
CaClanen, 379
Cady, 349
Cahal, 209-210, 386
Cahall, 231, 233, 243
Cain, 223, 228, 230, 232, 234
Calahan, 213-214

Caldwell, 205, 214
Caleal, 205
Calhoon, 305, 309, 334
Calhoun, 273, 281, 342, 352-354, 380-381
Callahan, 209
Callaway, 228, 230-232, 374, 379, 381
Calloway, 354, 372
Campball, 351
Campbel, 297
Campbell, 206, 253, 272, 279, 315-316, 339, 349, 351, 358, 383, 385, 386
Camper, 230
Campher, 228
Cann, 280, 284, 287
Cannon, 203, 217, 233-234, 259, 278-279, 288, 295, 305, 319-320, 323, 332-333, 244, 346, 350, 356, 361, 367, 369, 375, 387-388, 390-393, 395-396
Carey, 235, 299, 305, 312-315, 327-328, 330-331, 338, 346-347, 358, 362, 368, 382-382, 386
Carlisle, 269, 279, 331, 339, 383-385
Carmean, 353, 365, 374
Carn, 334, 386
Carney, 204, 216, 247-248, 259
Carom, 258
Carpenter, 208, 222, 240, 251, 262-263, 287, 289, 297-298, 324, 326-328, 332, 336, 338-339, 343, 362, 370-371
Carpreseters, 256
Carrey, 203
Carron, 251
Carrow, 200, 204, 206, 288, 301
Carson, 200, 215, 244, 371
Carsons, 210
Carter, 203, 206, 238, 243, 296
Carttrell, 298
Cary, 198-200, 337
Case, 224, 241-242
Casey, 256, 349
Cash, 395

Casho, 278
Caskey, 282
Casperson, 250
Cassady, 254
Casson, 211
Catlin, 227, 236, 356
Catts, 202, 283
Caulk, 202, 235, 237, 293-294
Causey, 339
Cavender, 284-285, 287, 292, 301
Caverly, 219
Cewden, 275
Chadwick, 299
Chaffin, 389
Chalk, 274
Chamber, 270
Chamberlain, 265, 309-310
Chamberlin, 395
Chambers, 270, 277, 281
Chandler, 257, 259-260, 262-263, 268-269, 310, 330, 349
Chapman, 220
Charles, 235, 242, 393
Charnean, 353
Charnece, 353
Chase, 318, 353, 368
Chaxelle, 262
Chaytor, 255
Chears, 293
Cheasman, 222
Cheffin, 238
Chemberg, 245
Chericks, 305
Chillas, 273
Chipman, 219, 320-321
Christopher, 310, 351
Churchman, 279
Cirwithen, 332, 334
Clair, 254-255, 266
Clairnan, 267
Clark, 211-213, 217, 224, 227-228, 236, 241-242, 245, 247-249, 253, 268, 273, 275, 279, 284-286, 305, 343, 365-366
Clarke, 261
Clarkson, 282

Claten, 293
Clay, 279
Clayton, 205, 249-250, 255, 291-292, 294, 296, 302
Claytor, 249
Cleacie, 285
Cleave, 286
Cleaver, 236, 269, 288, 290
Cleaves, 231, 281, 285, 287-290
Clement, 261
Clements, 212, 216
Clendaniel, 214, 219, 324, 328, 331-332, 336-337, 339-343, 364
Clendenin, 282
Clifton, 222, 241, 318, 324, 328, 334, 3370339, 342, 384-385, 393
Clinton, 259
Clogg, 310, 350
Closs, 255
Clothier, 292
Cloud, 255-257, 260-261, 269
Clouds, 216
Clough, 244
Clow, 206
Cluff, 279
Clump, 243, 245
Coarsey, 242
Coates, 384, 387-388
Coats, 391
Cobb, 305
Cochey, 377
Cochran, 231, 287, 290-292, 295-296
Coe, 250
Coffen, 334
Coffin, 210, 307, 311, 314, 327, 335, 346, 352, 366, 376
Cohee, 237-238, 245
Cohol, 209
Cohoon, 207
Colborn, 218
Colbourn, 220, 391, 394-395
Colburn, 285
Coldwell, 243-244
Cole, 199-200, 215, 220, 241, 247, 254, 282

Colescott, 285
Colison, 232
Collier, 308, 374-375
Collins, 201-202, 206, 220, 225, 232-233, 250, 269, 272, 295, 297-298, 300, 306, 310, 313, 318-319, 324, 3290332, 336, 339, 349, 353, 357-359, 364-365, 369, 371, 373, 375-377, 382, 384, 391, 394
Collison, 232-234, 366, 389-391
Colman, 226
Colt, 210
Compton, 286
Conard, 201
Conaway, 216, 315, 319-320, 322-323, 328, 343-344, 354-356, 379-383, 385
Concannon, 260
Concealor, 248
Condright, 198
Condry, 360-361
Congo, 283
Coning, 262
Conkey, 277
Conley, 260
Conly, 205, 281
Connard, 199
Connell, 269
Conner, 202, 239, 242-243, 265, 335, 338
Conners, 260
Connor, 217, 219, 235
Conoway, 393
Conquest, 351
Conrow, 255
Conwell, 236, 239, 326, 329-330, 369
Conyer, 242
Cooch, 281
Cook, 204, 236-237, 244, 248
Cooney, 260
Cooper, 200, 204, 216, 226-228, 236, 238, 246, 298, 339, 355, 366, 375, 377, 381
Copes, 365
Coppage, 211

Corbet, 286
Corbin, 387-388
Corbit, 289, 294
Cord, 370
Corden, 357, 374
Cordray, 226, 233
Cordrey, 225, 372, 375
Cordry, 360
Cormean, 323, 360, 374, 372, 377
Cormer, 318
Cornegys, 209
Cornish, 229
Cornly, 281
Cornwall, 380
Corsa, 220
Corsdon, 333
Corse, 208, 339
Cosden, 212
Cottingham, 393
Coughter, 355
Coulborn, 317-318
Coulter, 325-326, 334
Counselor, 270
Course, 209
Coursey, 371
Cousin, 198
Covell, 211
Coverdale, 210, 215, 219, 222, 238, 340, 371, 380, 395
Cowgill, 201-202, 214, 236
Cox, 205, 212, 224, 239
Coyle, 278
Coznegys, 205
Cozzens, 256
Cradick, 241
Craig, 205, 209, 248, 327, 330
Crammer, 205
Crampfield, 358
Cranston, 264, 266, 272
Crapper, 334
Crawford, 250, 296
Crawly, 209
Credin, 268
Cresson, 256
Cripps, 268
Crockett, 257, 272, 291

Croes, 278
Croft, 289-290
Croker, 248
Cropper, 329
Crosland, 291-292
Cross, 259, 274
Crossan, 275, 277
Crosser, 267
Crossland, 284
Crow, 199, 283
Cubbage, 220, 238
Cudney, 216
Cullen, 202, 241, 356
Culler, 202, 210, 224
Culver, 317-318, 374-376, 378-379, 387
Cummins, 197-201, 216
Cummons, 301
Cunningham, 212-213, 282, 284-285
Currinder, 272
Curry, 383, 390
Curtis, 224, 226
Cusick, 260
CuyKendall, 331
Dager, 234
Daisey, 348, 358-358
Dalby, 345
Daldvan, 350
Dale, 291, 305
Daley, 298
Dalson, 293
Daniel, 332, 343
Daniels, 292-293
Danner, 203
Danzenbaker, 256
Darber, 395
Darbie, 393, 395
Darby, 243, 336
Darling, 214, 237
Darnell, 204
Dary, 366
Dasey, 305-307, 309, 312, 314
Dashields, 392, 394
Dashper, 272
David, 210, 216, 249-250, 298

Davidson, 274-275, 283, 310, 329, 344, 349, 363, 366-367
Davis, 201, 203, 205, 214, 217-218, 220, 223, 237, 239, 241-242, 244, 247, 251, 260, 272-273, 276, 280, 283, 285, 292, 296-297, 300, 306, 311-313, 327, 332, 334-336, 339, 349-351, 355, 366, 374, 383, 390, 393
Davos, 244
Dawes, 202
Dawnham, 235
Dawson, 250, 281, 338, 384
Day, 223, 253, 256, 352, 355-356
Dayett, 281, 284
Deakins, 296
Deakyne, 296, 298-301
Dean, 225, 233, 273, 280
Dear, 247
Deats, 249
Deffadoffer, 225
Degan, 290
DeHitt, 217
Dehority, 241
Del Beet Sugar Co., 256
Delany, 205
Dell, 371
Demar, 214
Dempsey, 280, 289
Denbee, 395
Denison, 265, 270
Denner, 213
Denney, 198-199, 202-203, 214-216, 289
Dennis, 208, 323, 358
DePont DeNemours, 263
Deputy, 277, 286, 332, 338, 340
Derborough, 235
Derby, 236
Derickson, 226, 228, 230, 253, 266-267, 270-271, 292, 304-313, 341, 350
Derley, 256
Derma, 295
Derrah, 206
Derrick, 257

Derrickson, 265-266, 297-298, 352
Desch, 197
Dewese, 269
Dickenson, 223
Dickerson, 219, 226, 242, 287, 293, 308, 314, 320, 323, 325, 338-339, 341-343, 352-353, 355, 372, 375, 381, 385, 395
Dickinson, 286-287, 326
Dickson, 201, 240, 251, 324
Diehl, 275-276, 289
Dill, 212-213, 229, 236, 238, 243-246, 299, 345
Dillon, 389
Dilworth, 260, 263, 288
Dingle, 349
Dixon, 218, 255, 258, 269
Dixson, 260
Dobly, 323
Dod, 244
Dodd, 208, 306, 324-325, 353, 368-370
Dodson, 295
Dohaney, 293
Doherty, 258
Dolby, 308, 319, 345, 382
Dole, 305
Dolvins, 390
Donahoe, 385
Donaphon, 243
Donavan, 369, 371-371
Donaway, 359-360
Doniphan, 387
Donley, 300
Donnell, 279
Donoghoue, 261
Donoho, 297
Donohoue, 279
Donophon, 227
Donovan, 225, 250, 300-301, 324-325, 328, 331, 333, 337, 339-340, 352, 354-355, 369
Doody, 238
Dorman, 223, 229, 328, 366
Dorsey, 219-220
Dougherty, 253, 258, 260, 263, 275

Douglass, 327
Douthard, 337
Dowed, 226
Downes, 208
Downey, 276
Downham, 201, 204, 241, 244
Downing, 351, 369
Downs, 198, 204, 211, 248, 251, 299, 357-361
Doy, 223
Drain, 362
Drake, 390
Drane, 368-371
Draper, 227, 229-230, 235-236, 246, 278, 333-336, 388
Dreden, 317
Dreggis, 211
Driggars, 207
Driggers, 207
Driggis, 207, 240
Driggs, 201
Droolinger, 284
Dubois, 343
Duckey, 283
Duffey, 257
Duggan, 279
Duhadaway, 211
Duker, 225, 232, 384
Dukes, 225, 312-313, 319, 348, 350, 360
Dulaney, 395
Dulin, 251
Duncan, 266
Dunham, 227
Dunn, 374, 378-379
Dunning, 289
DuPont DeNemours & Co., 258
Dupont, 259, 263
Durham, 206, 247-248, 297
Durross, 267
Dutton, 286, 325, 328-330, 354
Dyer, 203
Dyson, 259
Dyton, 227
Earvin, 246
Eashum, 306, 315

Eastburn, 270-271, 274, 278
Eater, 271
Eaton, 290
Ebright, 253
Eccles, 202
Eckles, 273
Ector, 265
Edgell, 391
Edgelle, 294
Edgemoor Iron Co., 258
Edgwell, 391
Edmunds, 239
Edmundson, 276
Edwards, 227, 241, 246, 255, 273
Egee, 272
Eggleton, 277
Eisenbrey, 231
Eliason, 280, 291
Elison, 291
Ellegood, 380, 386-387
Ellensworth, 338
Eller, 319
Ellingsworth, 300, 324-325, 348, 390
Elliott, 218, 258, 279, 319, 321, 323,
352, 372-374, 376-278, 382-383, 390
Ellis, 240, 250, 294, 305, 351, 373-
373, 375-379, 395
Ellison, 280, 284, 288
Ely, 259, 262
Elyz, 376
Elzey, 377
Elzy, 376
Emerson, 204, 241
Emmerson, 216
Emory, 204, 219, 242, 276
English, 318, 320, 323
Enich, 205
Ennis, 214-215, 248, 256, 298, 312,
326, 333, 343-344, 361, 363
Eskridge, 320, 387, 391
Esra, 209
Estlin, 274
Evan, 222, 231
Evans, 198, 221, 234-236, 241, 248,
263, 277, 290, 304-309, 311-315,

317, 319, 334, 345, 354, 361, 376,
388
Everett, 198, 235
Everson, 264, 273
Evert, 206
Everts, 204
Evits, 238
Ewing, 216, 257
Ewinge, 356
Faison, 201
Faley, 238
Farrell, 198, 297
Farris, 247
Farron, 392
Farrow, 202, 206, 211, 392
Faucet, 356
Faucett, 236
Faulkner, 206, 208, 239, 282
Faust, 268
Fay, 341
Fell, 264, 269-270
Felton, 339
Fenimore, 214, 287, 294, 297
Fenn, 261
Fennemore, 299-300
Fenner, 338
Fennimore, 299
Fergison, 268
Ferguson, 302
Ferrens, 231
Ferrie, 268
Ferris, 262, 281
Fiddaman, 331
Fie, 297
Fields, 329
Finn, 250
Finter, 297
Finton, 278
Fisher, 212, 217, 226, 233, 235, 268-
269, 278-279, 286, 314, 316, 325,
327, 330, 332, 343, 369, 383, 390,
395
Fitzgerald, 222, 333, 338
Fitzsimmons, 273
Flannery, 273

Fleetwood, 224, 340, 356, 381-382, 388, 392-393
Fleming, 225, 232, 244, 290, 339
Flicks, 259
Flinn, 264, 267, 274
Flitt, 263
Fluharty, 251
Flyerbrand, 386
Foard, 289, 291
Fols, 273
Fooks, 382, 386, 394-395
Foopes, 389
Foote, 266, 270-272
Foracre, 272
Ford, 198-200, 205, 209, 213, 247, 256, 261, 266, 280-281, 300, 392
Fordham, 284, 286
Foreacre, 218-219
Foreman, 265
Forsythe, 275
Forum, 248
Forwood, 253-254, 256
Foskey, 360-361
Fosque, 347, 350
Foster, 227, 388, 394
Foulk, 253-254
Fountain, 205, 242, 335
Fow, 275
Fowler, 203, 383-384
Fox, 215, 274, 330-331
Frail, 275
Fraim, 257
Frame, 348, 350, 365-366, 32
Frances, 300, 306
Francis, 296
Franer, 244
Frank, 391
Franklin, 228
Frasher, 237
Frasure, 247
Frazer, 281, 284
Frazier, 198-199, 217, 238, 240, 278, 368
Frear, 236-237
Fredell, 243-244
Frederick, 260

Freeman, 314
Freeny, 319
Freeston, 215
French, 221-222
Frismuth, 277
Frut, 389
Fry, 285
Fulton, 204
Furbish, 244
Furey, 256
Furguson, 206
Furman, 304, 309, 311-312
Furnisk, 244
Futcher, 369
Futcher, 371
GA__, 250
Gallan, 391
Galleger, 287
Gamble, 261
Game, 374
Ganrice, 293
Gardiner, 389
Gardner, 219, 298, 301
Garman, 285
Garnett, 267
Garrison, 198, 201, 215, 247
Gartin, 201
Garton, 239
Gates, 343
Gatt, 293
Gattis, 293
Gay, 280
Gebhart, 265
Gehbart, 267
George, 199, 215, 276, 281
Gerker, 249
Gesford, 294
Gettz, 259
Gibbs, 205-206, 212, 296, 366
Gibson, 202
Gier, 210
Gilchrist, 331
Giles, 322, 391-392
Gillace, 336
Gillice, 260
Gillis, 373, 378, 394

Gilman, 341
Gilmer, 393
Ginn, 293, 296
Ginng, 297
Givens, 382
Givin, 201
Gleeson, 342
Godfrey, 305, 316
Godwin, 242, 246, 304, 310
Gold, 287
Golding, 262
Goldsborough, 199, 201, 250
Goldstone, 294
Gomery, 210
Good, 248
Gooden, 212-213, 236-237, 244
Gooding, 264
Goodley, 256
Goodpastur, 256
Goodwin, 198
Gorder, 386
Gordon, 266, 290
Gordy, 322-323, 356, 359-361, 373-374
Goslee, 364-365
Goslin, 233
Gott, 289
Gould, 284, 291
Gow, 277
Grace, 267
Grady, 249, 263
Graff, 256
Graham, 206, 231, 237, 244, 248, 251, 268, 289, 321, 379
Grant, 212
Grasly, 214
Graves, 257, 259-260, 262, 267-268
Gravner, 358
Gray, 236, 238, 274, 285, 291, 300, 307-310, 313, 315, 318, 331-331, 340, 350-351, 359, 386
Grayham, 395
Greeman, 274
Green, 213-215, 235, 237-238, 242, 261, 263, 283, 287, 289, 291, 294, 313, 353-354, 364, 367, 370

Greenage, 205
Greenfield, 259
Greenley, 300
Greenlley, 245
Greenly, 211, 337, 354
Greenwalt, 272, 279
Greenwell, 209
Greer, 207, 246, 293, 298
Greewell, 244
Gregg, 262-263, 267, 277
Grells, 244
Grennell, 243
Grier, 224, 239
Grieves, 298
Griffin, 198, 201, 268
Griffith, 199, 209-210, 219-220, 225, 235, 247, 259, 319, 338, 340, 342, 367, 383-384
Grills, 244
Grimes, 281, 285
Grinby, 357
Griswold, 223
Grody, 379
Groom, 369
Grose, 276
Grove, 279, 334
Groves, 269, 281, 334
Grubb, 255-256, 276
Gruman, 202
Gruwell, 237-239
Guessford, 248
Guest, 255
Guinn, 294
Gullett, 207, 246
Gum, 351
Gumby, 386
Gunby, 359
Gundy, 322
Gunly, 322
Guntor, 384
Gurton, 295
Guthrie, 273, 277
Guy, 280
Habbart, 254
Hacker, 306
Hackett, 209

Haines, 280
Halcomb, 274
Haley, 262-263
Hall, 205, 208-209, 221-222, 234, 242, 261, 282, 303-304, 306-308, 313, 323, 331, 339, 349, 395
Hallowell, 389-391, 393
Haman, 271
Hamilton, 228, 233, 256, 279, 290, 299, 386, 390-391
Hamm, 215-216
Hammond, 205-206, 217-220, 225, 228, 240, 333, 340, 342
Hanby, 254-256
Hance, 285-286
Hancock, 204, 316, 353
Hand, 253, 257
Hands, 223
Handy, 214, 315, 394
Hanger, 207
Hanna, 266, 270-271
Hannafee, 300
Hanniford, 332
Hanson, 235, 277, 296
Hansor, 365
Hardcastle, 247
Harden, 343
Hardesty, 231-232
Haregrove, 234
Hargadine, 207, 237, 241
Hargardine, 237
Harigan, 260, 267
Harkins, 257
Harkness, 270, 272-273
Harkstein, 283
Harman, 296, 322-323, 364
Harmer, 210, 242
Harmin, 322
Harmon, 327, 352, 365-366, 369
Harnet, 208
Harnon, 231
Harper, 215-216
Harras, 270
Harrass, 244

Harrington, 202-203, 212, 214, 216-218, 221, 223-224, 226, 228-230, 232-234, 240-243, 245, 337
Harris, 234-235, 250, 259, 293, 356, 363
Harrison, 315-316, 368
Harriss, 224, 311
Hart, 363-364, 368, 370, 387
Hartnet, 208
Hartup, 299
Harvey, 254, 257
Hastings, 310, 316-323, 349, 357, 359, 361, 372-3779, 381, 386, 388
Hasty, 383
Hatfield, 211, 224, 356
Hatt, 290
Haughey, 253
Hausee, 368
Hauthorn, 278
Hawkins, 205, 248
Hawthorn, 279
Hay, 205
Hayden, 297
Hayes, 203, 218, 220, 276, 384, 390
Haylett, 266
Hays, 383
Hazard, 301
Hazel, 214, 248
Hazell, 249, 336
Hazzard, 330-331, 336, 349-350, 360, 362-363, 379, 394
Heald, 267
Heall, 361
Healy, 257
Heard, 207, 245-246
Hearn, 308, 317-318, 322, 352, 360-361, 373-379, 393
Heather, 329
Heaveloe, 320, 329
Heavelow, 329
Hefnal, 206
Heinholt, 250
Heistand, 241
Helem, 351
Hellems, 340
Helm, 304, 306

Hemmons, 340
Hemping, 230
Henderson, 220
Hendrick, 391
Hendricks, 225
Hendrickson, 261, 269
Hendricson, 216, 259
Henell, 368
Henry, 219, 294, 374-377
Hepburn, 249
Herrick, 287, 292
Hetsh__, 216
Hevaloe, 339
Hevelo, 369, 372
Hevern, 203, 300
Hews, 210
Heyd, 243
Heye, 207
Hibithia, 346
Hickey, 242
Hickman, 222, 310-313, 315-316,
336, 357, 358-359, 367-368
Hicks, 228
Higdon, 389
Higgins, 218, 273, 285, 289
High, 379
Highfield, 266, 268
Higman, 332, 335, 388
Hignett, 389
Hill, 217-219, 222, 224, 226-227,
248, 250, 286, 294, 299, 302, 320,
326, 333-334, 339, 343, 371, 374,
377, 381-382, 387-388
Hillingsworth, 264
Hillyard, 250
Hinkson, 257
Hinsley, 212
Hinsly, 221
Hipple, 261
Hirons, 204, 211
Hitch, 217, 242, 317, 321-323, 388
Hitchens, 219, 310, 318-319, 322-
323, 347-348, 350, 357, 359, 361,
368, 379, 383
Hitchins, 321
Hobbs, 200, 364, 367

Hobs, 232
Hobson, 261-262, 269, 295
Hockster, 229, 240
Hodson, 241
Hoey, 225
Hoffecker, 198, 250, 265, 291
Hogg, 284
Holcomb, 276
Holden, 214, 243, 245
Holder, 243
Holfield, 225
Holister, 319
Holland, 212, 221, 272, 326-328,
330, 369-370
Hollett, 281
Hollinger, 222
Hollingsworth, 268
Hollis, 219
Hollitt, 299
Holloway, 305, 311-312, 315, 317,
395
Holmes, 217, 260
Holstein, 343
Holston, 215
Holt, 203, 210, 239, 305-307, 321,
338
Holtson, 350
Holtz, 199
Homewood, 278
Honey, 213
Hood, 324, 326, 350, 364, 368, 370
Hooker, 210
Hooper, 269, 392
Hoopes, 260, 266, 269
Hoops, 268
Hoover, 240
Hope, 210
Hopkins, 222, 227-232, 233-234,
246, 270, 285, 287-288, 294, 317,
319, 325, 331, 346, 350, 353, 371,
396
Hornby, 257
Horsey, 212, 240, 374, 378
Horton, 340
Hosea, 357
Hosey, 347

Hossinger, 280
Houlton, 292
Houston, 287, 291, 329, 339, 343, 347, 350-351, 368, 392, 394
Howard, 310, 369
Hubbard, 212, 229
Huber, 283
Hudson, 200, 221, 223, 241, 250, 288, 304, 308-309, 311-317, 323, 327-328, 331-331, 336-337, 340, 342-343, 347-349, 352-353, 357, 359-360, 365, 370
Huelser, 369
Huelson, 369
Huggins, 273, 301
Hughes, 227, 246, 260
Hughs, 206
Hukill, 289
Hulett, 268
Humphreys, 261, 332
Hunter, 257, 362
Hurd, 208, 211-212, 228
Hurdle, 362, 367
Hurlock, 251
Hurly, 214, 380, 392
Hurst, 274
Husband, 262
Husbands, 236, 249, 257-258
Hushebick, 281
Huston, 320, 395
Hutchens, 212
Hutchins, 209, 212
Hutchinson, 198, 202, 211, 247, 249, 286
Hutson, 208, 213
Hutton, 248
Hybedo, 208
Hyde, 266
Hydom, 222
Hynson, 216
Ingraham, 340
Ingrahm, 326
Ingram, 222, 295, 321, 333, 336, 343, 348-350
Insley, 219
Ireland, 207

Irons, 264, 283
Isaacs, 293, 380
Isleib, 219
Jackson, 202, 204-206, 221-222, 235, 237-240, 242, 251, 260, 269, 271, 274, 276, 330-331, 335, 372, 384, 394
Jacobs, 202, 205-206, 217-218, 232, 248, 270, 312, 386, 388-389
Jacquott, 270
James, 203, 206, 210, 283, 292, 318-319, 323, 373-374, 381
Jamison, 286, 288
Jance, 285
Janvier, 273, 281, 286, 290
Jarman, 234
Jarrel, 243
Jarrell, 224, 244, 287
Jarves, 245
Jefferis, 267
Jefferson, 198, 215, 309, 324, 331, 335, 337, 344, 355, 380
Jenkins, 220, 234
Jerman, 198, 200, 209, 322, 360
Jermen, 359
Jerrell, 299
Jessup & Moore Paper Co., 259
Jestace, 305
Jester, 220-222, 227-228, 230, 232, 237, 267, 279, 285, 324-325, 341-342
Jestur, 359
Jewel, 206
Jewell, 229-230, 386
Johnis, 208
Johnson, 197, 199, 205, 207, 209, 212-213, 217, 219, 222, 224-226, 232-233, 235-236, 239, 241-242, 244, 251, 255, 272, 278, 283, 301-302, 305, 309-310, 315-316, 319, 324-326, 329, 331, 334, 337-338, 340, 342-344, 346, 350-351, 357, 362-367, 370-371, 379, 381, 384, 388
Johnston, 202, 277, 343
Joines, 350

Jolly, 208
Jones, 198-201, 204-205, 208-211, 226, 228, 231, 233-234, 237, 241, 248-250, 266, 278-282, 286-287, 289, 291-295, 298-299, 301, 306-309, 311, 319, 322, 324, 330, 333, 335-337, 354, 356-358, 361, 371, 373, 279, 381-383, 386, 389-391
Jordan, 263, 268, 274-275
Joseph, 327, 344, 351, 362-367, 371, 385
Joster, 287
Journey, 259
Jump, 219, 246
Justice, 266
Kalas, 228
Kallock, 347
Kane, 246, 260, 262
Kanely, 296
Kapple, 294
Kates, 228
Kearns, 270
Kee, 245
Keegan, 277
Keeley, 282
Keen, 298, 301
Keer, 278
Keiffer, 300
Keiser, 199
Keith, 214-215
Keller, 295
Kelley, 231, 237, 241, 274, 286
Kellum, 256
Kelly, 261, 270
Kemp, 202, 244-245, 389
Kendell, 283
Keniether, 283
Kennedy, 301
Kenney, 219, 376-377, 379
Kent County Almshouse, 236
Kent, 235, 263
Kenton, 227
Kerney, 265
Kersey, 208-209, 211, 240
Kerson, 332
Kesler, 218

Keyes, 288
Kign, 369
Kilgore, 332
Killen, 226-228, 241, 244
Killgore, 263
Kimey, 203
Kimmey, 198, 229
Kinder, 387-388, 392-393, 395
Kinekin, 345
King, 206, 238, 254, 281, 287, 293, 296, 327-328, 355, 361, 371-372, 378, 394
Kinikin, 375
Kinking, 375
Kinnamer, 390
Kinney, 235, 373-374, 377-378
Kinny, 318
Kirbin, 239
Kirby, 199, 221
Kirk, 211, 253, 274
Kirkley, 298
Kissikin, 375-376
Klair, 260, 266, 268, 270
Knapp, 212, 241
Knight, 203, 239, 395
Kniverse, 210
Knopp, 212
Knotts, 203, 249, 275, 290
Knowles, 268, 320, 377-378, 390-391
Knox, 232, 246, 306, 386
Kollock, 347
Kooper, 245
Kuley, 284
Kyle, 278
Lachus, 289
Ladd, 291
Ladoux, 202
Laffer, 203
Lafferty, 203, 215, 240, 248
Laffety, 267
Lamb, 216, 293
Lambden, 322
Lamborn, 267, 269
Lambson, 277
Lamden, 381

Lan, 275
Landerman, 255
Landers, 275
Landis, 225, 293
Lane, 234, 238
Langendyk, 390
Langford, 388
Lank, 318-319, 327, 330, 364, 370-371
Lankford, 370, 394-395
Lany, 360
Lapham, 249
Laramore, 226-227, 229
Lark, 282
Larned, 204
Larramore, 395
Lasser, 203
Lathbery, 309
Latman, 293
Lator, 245
Latta, 297
Lattamus, 299
Lattemous, 296
Laubach, 211
Lauk, 319, 330, 371
Laurance, 289
Lauterwasser, 300
Law, 262, 311
Lawrence, 235
Laws, 211-212, 216, 224, 284, 333
Lawson, 363, 366-367
Layman, 239
Layton, 231-233, 239, 304-305, 309, 339, 346, 348, 356, 383, 386-387
Lea, 278
Leach, 246, 254, 257, 260, 268, 271
Leatherm(an), 297
LeCarpentier, 263
Lecates, 355, 374, 378-379, 393
Lecompt, 223, 287, 339
Lecount, 251
Lednum, 389
Lee, 282, 293, 300
LeFevre, 277, 280
Legal, 204
Legar, 207

Legark, 289
Legates, 224-225, 232, 357, 380-381
Legats, 223
LeKites, 314, 361
Lemans, 244
Lentz, 274, 277
Lere, 395
Lester, 296
Leverage, 248, 334
Levick, 201
Lewes, 234
Lewis, 200, 202, 204, 218, 220, 224, 231, 236, 242, 248, 265, 278, 293, 298, 308-309, 324, 359, 387
Lightcop, 299
Lighter, 232
Lim, 283
Limberger, 200
Linch, 291
Lindale, 207, 235, 39
Lindall, 332, 336
Lindell, 284
Lindle, 222, 325, 327, 329
Lindsey, 280
Lines, 386
Lingo, 235, 318, 331, 356, 362, 364-366
Lion, 342
Lippinscott, 255
Lister, 245
Little, 268, 270-271, 370
Littleton, 304
Littleton, 307-308, 346-347, 352, 356, 360, 383, 385
Litz, 223
Livingston, 209
Lloyd, 236, 375
Loatman, 199
Lockard, 272
Lockerman, 298, 301
Lockwood, 292, 295, 313, 316
Lodge, 235, 239, 254-255, 339, 368, 370, 372
Loffland, 329-330, 342

Lofland, 217, 249, 276, 285, 299, 324, 330, 339, 341-342, 344, 355, 380
Lofton, 315
Logan, 200, 267
Lollis, 242
Long, 226, 304, 311, 315-316, 324, 348-351, 356
Longfellow, 226, 245
Longland, 289
Longlellow, 245
Longrel, 228
Loose, 236, 249
Loper, 203, 222, 236
Lord, 207, 289-290, 338, 383, 386, 390-391
Lore, 287
Lorens, 394
Lorres, 244
Losen, 394
Loughead, 254, 282
Louk, 275
Low, 347, 374, 377
Lowber, 215, 234, 240, 242-243
Lowe, 201, 318, 357
Lower, 255
Lowther, 260-261
Loyd, 320
Loyed, 233
Ludlow, 288
Lum, 278, 283
Lumford, 266
Lusk, 393
Lutz, 265
Lyman, 331
Lynam, 257, 263-264, 266-267, 271, 279
Lynch, 199, 217, 222, 230-231, 240, 244, 261, 267, 273, 304-305, 307-313, 314, 317, 321, 337, 341, 343, 350, 352, 354, 364, 369-370, 372, 374, 381, 385, 390-391
Lynde, 356
Lyons, 224, 353, 368, 370-371, 384-385
Maberry, 250-251

Macauley, 386
MacDonald, 273
Mace, 261
Mackey, 260, 277
Macklin, 218, 221-222, 326, 336, 338, 340-341, 343, 355, 380-381
Macollister, 360
Macy, 250, 294
Maddis, 295
Maddox, 373
Magee, 325-326
Mahaffy, 254
Mahan, 251, 284, 293
Mahon, 282
Mahony, 255
Mailley, 290
Male, 336
Malin, 259
Maloney, 223, 231, 235, 293, 297
Manlove, 242
Manship, 330-331
Marcey, 286
Mare, 250
Maree, 279
Margargargle, 268
Margaugle, 267
Marine, 229
Mariner, 344
Mark, 240
Marker, 244, 354
Marnes, 210
Marrim, 300
Marsh, 362-363, 370
Marshall, 204, 212, 269, 330, 332-333, 368
Marten, 293
Martin, 203, 220, 226, 243, 249, 325-326, 328, 330, 352-353, 371, 392-393, 396
Martine, 255
Marvel, 204, 210, 212-213, 218, 225, 305-306, 326, 344, 348, 352, 354, 362, 365-366, 370, 383
Marvell, 204, 370
Marvil, 320
Marvill, 343, 351

Masmon, 199
Mason, 217, 220, 243, 269, 331, 336, 343
Massey, 211, 240, 251, 317, 320, 346, 354, 364
Masten, 217, 221, 226-228, 343
Mastin, 207, 224
Mathews, 285, 323, 349
Mathis, 346
Matthews, 297, 321-323, 358, 360, 380
Mattiford, 300
Matts, 296
Maull, 328, 367-368, 372
Maxwell, 241, 382
May, 204, 261
Mayberry, 211
McAlaster, 255
McAlister, 382
McAuley, 389
McBride, 237, 264, 279
McCabe, 307, 311-316, 348-349
McCalister, 288
McCall, 254, 286
McCallion, 260
McCannon, 270
McCarter, 293
McCarty, 267
McCauley, 225
McClain, 203, 358
McClar, 266
McClean, 288
McCleavy, 224
McCloskey, 282
McColley, 217-219, 221-222, 333-334, 363
McCollom, 260
McCollough, 263
McColly, 332
McComerick, 310
McConaughey, 282
McCormick, 269, 271
McCoy, 274, 276, 283, 285, 295-296
McCracken, 284
McCray, 350
McCullough, 262

McDermott, 259
Mcdill, 282
McDonald, 248, 258, 261
McDowell, 312, 36, 389
McElwee, 279
McFaden, 361
McFann, 216
McFarlan, 272
McFarland, 275-276
McGee, 303-305, 312, 314, 317, 322, 349, 365, 367, 374-375
McGill, 240
McGilligan, 240, 258
McGinnis, 236, 238
McGonigal, 216
McGovern, 266
McGowan, 277
McGrellis, 260
McGrier, 274
McIlvain, 242, 368, 383
McIlvaine, 241
McIlvane, 362
McIntire, 280
McIntosh, 240
McIntyre, 258, 284
McKay, 293
McKee, 207, 271, 275-276
McKenney, 313
McKeowan, 278
McKeowon, 277
McKimm, 257
McKin, 356
McKinzie, 394
McKnatt, 229
McKown, 375
Mclain, 206
McLane, 297
Mclany, 198
McLaughlin, 278
Mclen, 269
Mclevee, 268
Mclewrie, 267
McMahan, 274
McMillan, 388
McMullen, 281, 288
McMullin, 273, 288

McNamee, 299
McNeal, 315
McNutt, 216
McQueen, 238
McSorley, 255
McSough, 269
McVaugh, 269
McVay, 294
McWhorter, 285, 288, 292
McWilliams, 395
McWorter, 288
Mearks, 208
Mears, 344, 346
Meaupou, 386
Medhurst, 340
Medill, 282
Megee, 354
Meggenson, 254
Megget, 282
Melson, 206, 306, 309, 319, 345, 360, 369, 373, 387-388
Melven, 231
Melvin, 207, 217-218, 227, 229, 238, 242, 245
Menden, 266
Mensch, 237
Meredith, 222, 232, 237-238
Mereith, 244-245
Merrikan, 232
Merrkan, 231
Mesius, 322
Messick, 223, 233, 304, 318-319, 32, 323, 325, 335-336, 345-346, 352-354, 363, 367, 379, 381-383, 386
Metcalf, 368
Meyers, 273
Middleton, 250, 299, 301, 348, 355
Mier, 272
Mifflin, 234-235
Milbey, 324, 330
Milbourne, 208
Milbrone, 296
Milburn, 241
Milby, 342
Miles, 274
Mill, 245

Miller, 200, 203, 210, 212, 215, 245, 253-254, 258, 260, 265, 268, 276, 278, 281-282, 309-310, 315, 338, 343, 363-366, 370
Millington, 199
Millman, 353
Mills, 219-220, 222, 317-318, 321, 333, 377
Milman, 335-336, 341-343, 390
Milten, 220-221
Milver, 366
Minehart, 266
Minner, 226-227, 231, 243, 245-246
Minnie, 273
Minones, 202
Minos, 209
Mirch, 326
Mirely, 274
Mitchell, 197, 266, 268-269, 272, 275, 278, 292, 304, 306, 309, 316, 323, 347-348, 357-359, 361, 382
Miten, 241
Mitten, 241
Moffat, 300
Mohler, 244
Money, 283, 299-300
Monro, 299
Monroe, 345
Montage, 214
Montague, 213, 235
Montgomery, 280
Moody, 281-282, 288
Moon, 206, 387
Moor, 245, 315
Moore, 198-201, 204, 206-207, 209, 213-215, 218, 220, 236-237, 242-243, 246-247, 249, 262-264, 266, 269-271, 273, 275-276, 278, 281, 289, 293-294, 299-300, 305, 310, 315, 317-318, 320-322, 324, 328, 344, 347-350, 357-359, 376, 378, 386, 391, 394
More, 239
Morery, 383
Morflight, 212

Morgan, 216, 223, 228, 232, 247, 276, 285, 318, 320-322, 332-333, 336, 342, 379, 383, 391, 393
Morgin, 293
Morice, 227
Morrel, 238
Morris, 200-201, 205, 212, 216-217, 221, 223, 226-228, 230, 234, 277, 306, 311, 314, 319, 322, 324, 326, 329-330, 337, 341, 343, 350-353, 359-360, 366-370, 373, 375, 378, 382, 389-391, 394
Morrison, 270-272, 275-276, 278-280, 282, 338
Morrix, 372
Morrow, 260-261, 278
Morton, 233, 258, 278
Moseley, 207, 221, 325
Mosely, 206-207, 210, 221
Mosley, 207
Moss, 268
Mote, 271, 277
Mouseley, 254, 256, 259
Mousely, 276
Mousley, 253, 262
Muller, 209, 261
Mumford, 316, 345, 350
Munce, 204
Murphey, 283, 291
Murphy, 234, 240, 248, 259, 261-262, 278, 320, 341
Murray, 281-283, 304-306, 313-316, 364, 382
Murrey, 240, 243, 261, 272
Murry, 238, 318, 347
Musgrove, 349
Mustard, 201, 362-363
Myce, 386
Myers, 250, 256, 281
Nailor, 298-300
Nancy, 290
Nandain, 296
Naudain, 279
Naylor, 257-258, 295, 329
Nea, 379
Neal, 392-393

Neall, 361, 393
Needles, 220-221, 243-244
Neff, 300
Negendank, 263, 271-272
Neil, 224, 229
Nelson, 223, 288, 390
Neuson, 318
Nevin, 391
Newcomb, 199
Newkirk, 274
Newlin, 264, 266
Newlon, 273
Newman, 283
Newnum, 206
Newsum, 206
Newton, 200, 208, 383-384
Niblet, 323
Niblett, 360-361
Nichols, 207, 262, 283
Nicholson, 253, 261, 318, 361, 395
Nickerson, 205, 214, 257
Nickols, 225
Nicols, 376
Nivin, 274
Noble, 387, 389
Noland, 299
Norcross, 337
Norris, 216, 284
Norvel, 223
Norwood, 362, 366, 372
Nowell, 223, 247
Nowland, 287
Null, 254
Numbers, 249, 251
Numering, 296
O'Day, 384, 390-391, 394
O'Rourke, 282-283
Oakes, 258, 262
Obier, 393-394
Obilar, 393
Ocheltree, 383
Oday, 233
Odell, 389
Ogram, 277
Ograne, 277
Oldham, 278

Oliphan, 374, 379
Oliver, 205, 277, 314, 330
Olophant, 347, 359
Oneal, 296, 317-319, 321-322
Onel, 293
Orr, 255
Ortist, 205
Oskins, 255, 267
Ottwell, 346, 388
Outten, 231, 381
Outter, 319
Outwell, 323
Ovil, 266
Owens, 218, 238, 320, 376-378, 380, 383-385, 387
Paige, 341
Paisly, 220
Palmatory, 200
Palmer, 242, 257, 260, 279, 281, 327, 331, 361, 371-372
Paon, 204, 212
Paor, 204
Parcks, 333
Pardee, 202
Parker, 205, 209, 273, 298, 320, 322, 330, 345-346, 350, 357, 359, 367, 379, 389
Parkes, 346
Parkhurst, 351
Parkin, 261
Parkinson, 243
Parkr, 348
Parmer, 319
Parris, 229, 233
Parry, 257
Parson, 284
Parsons, 306, 346-348, 350, 357-358, 374
Parus, 229
Parvis, 203, 213, 233, 297
Paschall, 254
Paskins, 203
Passmore, 261
Paswaters, 234, 342, 383, 385
Patten, 202, 213-214, 389
Patterson, 237

Pattison, 257
Paulsbury, 208
Paw__ly, 212
Paxon, 285
Paxson, 271
Paynter, 368-369, 371-372
Peach, 275
Peak, 203
Pearace, 296
Pearce, 275, 293
Pearse, 199, 211
Pearson, 206-208, 210-211, 226
Peck, 243, 274, 354
Peckaham, 224
Penise, 213
Pennel, 358, 361, 372
Pennewik, 390
Pennington, 287, 290
Pennock, 270-271
Peoples, 265, 267
Pepper, 231, 313, 328-329, 331, 351-352, 354, 356
Perkins, 240, 254-257, 318
Perry, 209, 216, 242, 257, 297, 299, 301, 327, 362
Peters, 233, 235, 271, 275, 279
Petildemange, 256, 258, 265
Pettyjohn, 325, 330, 336, 338, 353-355, 367, 384-385
Pfleegon, 236
Pharo, 227, 287, 292
Philips, 202, 204, 213, 267-268, 298, 305
Phillips, 219, 236, 247, 254-255, 318-320, 345-346, 348, 350, 356, 360-361, 362, 372-373, 376-379, 391
Pickery, 237
Pierce, 231, 253-257, 331, 335
Pierson, 245, 265-267, 301
Pinder, 246, 391
Piper, 335
Pitcher, 228
Plankinton, 261
Pleasanton, 198, 215, 289
Plum, 223
Plummer, 288, 317, 365

Pogue, 278
Polite, 310, 316
Polk, 290-292, 330, 385
Pollitt, 378
Polyte, 309
Pond, 230
Ponder, 328
Pool, 289
Poole, 269, 274, 307, 309, 350
Pooley, 388
Poor, 294
Pordham, 284
Porter, 226-227, 229, 233, 275, 280-282
Postles, 201-203, 217, 234-235, 239
Potter, 258, 333
Powel, 205, 214
Powell, 204-205, 211, 224, 272, 275, 282, 297, 300, 323
Pratt, 209, 212, 240, 250-251, 390
Prattis, 242
Preden, 379
Prenable, 381
Prettyman, 225-226, 239, 317, 325-326, 330-333, 336-337, 341, 343-345, 362, 365, 370, 379, 380, 381, 391
Price, 214, 229-230, 247, 263, 280, 284, 288, 292, 200
Prichard, 381
Pride, 325, 343, 355, 371
Priestly, 202
Primrose, 256
Prince, 256
Prior, 248, 300
Pritchard, 279
Pritchell, 240
Pronerel, 204
Proud, 236
Pryor, 301
Pugh, 262, 265, 274
Pullen, 219
Purnell, 232, 337
Purse, 393
Pusey, 304, 306, 308, 321, 324, 357, 359-360

Pyle, 262, 271, 277
Quigley, 264
Quil, 270
Quillan, 243
Quillen, 217, 317
Quillin, 223, 304, 306, 308-309, 312
Quillon, 214, 217
Racine, 280-281
Ralph, 375-376, 378
Ralston, 225, 268
Rambo, 254, 277, 282
Ramsey, 257
Rankin, 270, 272, 277, 387
Ranson, 201
Rash, 202, 206-209, 213, 219, 232, 248
Ratcliff, 381
Rathel, 388
Ratledge, 292
Raughley, 197, 199-201, 229-231, 233-234, 389-390
Raughly, 198, 214, 224
Raush, 235
Rawley, 238
Rawlins, 386, 391, 395
Ray, 295, 387
Raymond, 202, 215, 238
Read, 387-389
Readie, 343
Reardon, 265
Records, 286, 299, 319, 345, 372-373, 375, 378
Redden, 220, 233, 391
Reddish, 374
Redifer, 219
Redman, 384
Reece, 223, 268
Reed, 204, 211, 223, 227-228, 236, 239-240, 245, 250, 281, 284, 294, 297, 300, 324-325, 329-330, 341, 394-395
Rees, 250
Reese, 223, 250, 282
Reeves, 285
Register, 199, 207
Reinholt, 250

Remley, 249
Rench, 199
Ressum, 387
Reston, 337
Reville, 207
Revis, 274
Revuse, 393
Reybold, 285-286
Reynolds, 199, 201, 205, 234, 240, 244, 246, 270, 277, 281, 283, 288, 295, 298-299, 327-329, 334, 339, 355, 364, 37, 380-381
Reynols, 383
Rheins, 250-251
Rhine, 247
Rhodes, 339
Rias, 203
Ricard, 319
Ricards, 385
Rice, 198, 241
Richards, 220-222, 235, 239, 272, 274, 295, 317, 328, 356, 386, 389
Richardson, 205, 215, 218, 264, 271, 297, 368
Rickard, 233
Rickards, 214, 225, 234-235, 288, 303-304, 306-311, 314-315, 363, 386, 390, 395
Rickets, 349-350
Ridder, 245
Riddle, 263, 282
Rider, 337
Ridgley, 210
Ridgway, 204, 207
Riggin, 320, 373
Riggs, 198, 219, 301
Right, 244
Righter, 257, 274
Riley, 288, 290, 293
Rimenter, 279
Rion, 340, 342
Risler, 337
Ritchie, 261
Rittenhouse, 276
Roach, 220, 239, 328-329, 334-336, 344, 353, 371

Robbins, 220, 239, 329-330, 355
Robert, 315
Roberts, 205, 208-209, 275, 294, 297, 300, 335, 357, 356
Robinson, 198-199, 210, 236, 248, 250, 259, 270, 272, 278, 305, 324, 328, 363-364, 369-370, 388
Robson, 285
Roccass, 199
Roch, 244
Rodgers, 394
Rodney, 322-323
Roe, 200, 204, 241
Rogers, 242, 273, 277, 304, 313-314, 317, 333, 344-346, 351, 367
Rohuehencane, 266
Roland, 258
Rolston, 259
Romer, 294
Roo, 200
Rose, 274
Roser, 390
Ross, 226, 231, 250, 275, 387, 391-392, 395
Rothwel, 296
Rothwell, 263, 279, 291, 301
Rotthouse, 257
Roughley, 240
Roush, 247
Rouss, 245
Rov, 200
Roy, 283
Rubencane, 266
Rudy, 219
Runner, 279
Rush, 248
Russ, 243
Russel, 233, 239, 242, 340, 342, 344, 367
Russell, 270, 275, 280, 329-330, 355-356, 380
Russum, 388
Rust, 234, 281, 306, 324, 327-328, 353, 365-367, 386, 388
Ruth, 278
Ryan, 380, 384

Rycott, 262
Ryon, 232, 314, 316
Sackett, 343
Sadler, 390
Sallomus, 295
Salmond, 368
Salmons, 228, 231, 324, 352, 355
Sammon, 383
Sammons, 251, 299, 382
Samons, 342
Sampson, 224, 358, 360
Samuel, 268
Sanders, 216, 218, 220, 247-248, 296, 323
Sapp, 203, 210, 217-218, 226-228, 245, 281-282
Sarde, 235
Satterfield, 219, 235, 240-241, 381, 384
Satterthwaite, 272
Saulesbury, 353
Saulsbery, 233
Saulsbury, 208, 240
Savage, 314, 351, 358-359
Savin, 251
Saxton, 204, 214, 237-238, 301, 342
Sayle, 272
Sayres, 278
Sayton, 342
Scagg, 335
Scanling, 267
Scanlon, 238
Sceine, 337
Schafee, 199
Schalinger, 246
Schwenderman, 268
Scott, 209, 213, 218, 220-223, 225-226, 228, 233, 244, 246, 259, 294, 322, 332, 345-346, 351, 355, 363, 384, 389-390
Scott__, 208
Scotten, 212
Scotter, 208, 212
Scout, 200
Scrofield, 224
Seal, 262, 269

Sears, 394
Seemons, 300
Seeney, 280
Segark, 289
Seims, 294
Selby, 360
Sellers, 291
Selletos, 276
Senn, 214
Sentman, 269
Seny, 294
Serman, 378
Serven, 219
Serwithia, 329
Sester, 285
Sesterhenn, 266
Severson, 201
Sevil, 298
Seward, 212, 214-215
Sewerson, 199
Shabinger, 246
Shahan, 204, 206, 210, 235, 249
Shaher, 210
Shakespeare, 265, 267
Shannon, 279
Shanof, 287
Sharp, 226, 228, 233, 326-328, 332, 353-354, 381, 385, 392
Sharpe, 254
Sharpless, 263, 269
Sharpley, 257
Shaurlot, 246
Shaw, 228
Sheats, 201
Sheldon, 276, 284
Shellece, 229
Shelten, 231
Shepherd, 336, 341, 388
Shepperd, 278, 283
Sherard, 208
Sherdon, 293
Sherman, 326-327
Sherwood, 224
Shetzline, 280
Shew, 341-342
Shiles, 32, 320

Shillcuts, 243
Shipley, 258
Shivler, 298
Shockley, 225, 228, 238, 280, 314, 331-335, 337-339, 342-343, 346
Shoot, 355
Short, 199, 214, 220-221, 223-224, 306, 310, 317, 322, 331, 333, 341, 343-348, 350, 352, 354, 357-360, 379-381, 386, 388, 392
Shorts, 198, 205, 207-208, 211, 249
Showell, 309, 351
Shubert, 338
Shutts, 244
Siere, 206
Silbey, 198
Silcox, 282, 296
Silver, 274
Simeon, 374
Simkins, 290
Simmons, 206, 265, 337, 380
Simon, 258, 277
Simpler, 309, 317, 330-331, 363, 366-367
Simples, 328, 330
Simpson, 200, 231-232, 241, 276, 342, 344, 379
Simsler, 307
Sinex, 251
Singer, 278
Sipkle, 245
Sipple, 230, 243, 245, 383
Sirse, 211
Skaggs, 293, 295
Skase, 211
Skidmore, 392
Slack, 276, 282
Slaughter, 199-200, 202-203, 207-209, 211, 216, 228, 239
Slay, 213, 235
Slaymaker, 201, 235
Slayton, 224
Sloan, 256
Small, 337, 339
Smaller, 390
Smally, 279

Smeed, 298
Smeller, 390
Smith, 202, 204, 206, 209, 214-217, 225-227, 229-234, 236, 238-239, 245, 247, 261, 265-266, 274, 277, 286, 325, 331, 337-338, 340, 344, 347, 352, 355-356, 358, 372-373, 376, 378, 381-382, 385-388, 390
Smithers, 215
Smoller, 390
Smoot, 220
Smyth, 296
Snech, 216
Snell, 219
Snow, 197
Snyder, 200, 259, 264
Sockrider, 340
Sockwriter, 309-310
Soffle, 220
Sollis, 242
Sommers, 369
Sorden, 384
Southgate, 264
Sowder, 268
Sowdon, 275
Spanish, 383
Sparklin, 243
Sparks, 210, 288, 300
Speakman, 198
Spear, 198
Spearman, 298
Spence, 207, 232, 234, 240, 261
Spenceneer, 209
Spencer, 217, 241, 244, 283, 332, 376
Spicer, 317-320, 325, 354, 381-382, 392-393
Spocer, 354
Sport, 354
Sprague, 337
Springer, 259, 261, 265-266
Spruance, 199-200, 250
Spudler, 386
Squibb, 253
Staats, 296-299
Staber, 210

Stacling, 198
Stafford, 230, 272, 275-276
Stapleford, 225
Stapleton, 276
Starkey, 391
Starr, 259, 279
Staten, 381, 383
Stayton, 298, 342, 389, 391
Ste__, 211
Stean, 345, 351
Stedman, 277
Steel, 213, 227, 278, 303-305,m307,
354, 362, 364, 371
Steele, 238, 273
Steelman, 363
Stepens, 316
Stephens, 209, 311, 315-316, 385
Stephenson, 198, 300-301, 308, 312,
316, 327
Stergus, 355
Sterling, 286
Steskey, 294
Stevens, 208, 210-211, 234, 249-
250, 280, 289-290, 339-370, 390,
395
Stevenson, 220, 239-240, 242, 335,
338, 365
Steward, 277
Stewart, 247, 267, 281, 283-284,
337, 340, 380-381, 383, 389
Stidham, 259, 264
Stinson, 272
Stirling, 259
Stockeley, 368
Stockley, 200-201, 356, 362
Stockly, 396
Stockwell, 216
Stoekel, 344
Stokeley, 369-370
Stokely, 368-370
Stokes, 223
Stoops, 273, 276
Stout, 215, 239, 242, 338
Stradley, 388
Stradling, 372
Stradly, 280, 300

Street, 365-366
Stroud, 247, 279, 282
Stroup, 265, 275
Struts, 250
Stuart, 263, 353
Stubbs, 236
Stubs, 230
Stuckert, 285
Sturdivant, 211
Sturgis, 361
Stutz, 237
Sudars, 234
Sudler, 213
Sulivan, 202
Sullivan, 238, 243, 282-283, 387,
390
Summers, 204
Surgeon, 247
Surman, 313
Surrat, 282
Sutton, 265, 273-274, 277
Swain, 229, 232, 288, 317, 321, 339,
354-355, 373, 380, 386-388
Swaine, 333
Swainey, 267
Swallow, 213
Swan, 285
Swayne, 262
Sweatman, 294
Sweeney, 355, 384
Swoden, 392
Sylvester, 246, 249
Syskle, 245
Taggart, 255
Taglibet, 210
Talley, 252-255, 257-261
Tam, 351
Tatman, 225, 290, 293, 296, 339-
340, 384
Tatnall, 265, 275
Tatum, 265
Taylor, 202-203, 206, 209, 212, 214-
215, 217, 221, 240, 248-249, 266,
268, 270-275, 277, 294, 305, 312-
313, 316, 321, 346, 351, 379, 381-
382, 387-388

Teaf, 277
Teague, 382
Teal, 252
Tease, 333
Teet, 207
Temple, 246, 271
Templeman, 287
Tharp, 223-224, 232
Thistlewood, 217, 223
Thomas, 204, 208, 210, 213-224, 221-222, 231, 237, 241, 294, 300, 349, 368, 394
Thompson, 201, 205-207, 209, 212-213, 221, 224, 262, 267, 269, 276, 278, 281, 301, 319, 343, 346, 348-350, 265, 370, 380
Thornton, 284
Thoroughgood, 365-366
Thorp, 226, 232, 234
Thouroughgood, 351
Thraul, 369
Thrawley, 230
Thurston, 337
Tignor, 251
Timmons, 312, 315, 347, 349-350, 358
Tindal, 381-382, 385
Tindale, 371-372
Tingle, 310, 312-313, 316, 348-349, 359
Tinley, 227, 237, 295
Titter, 281
Titus, 334
Todd, 205, 214, 336, 383, 390-391
Tolbert, 219-221
Toll, 331
Tolstone, 213
Tomas, 314
Tomlin, 258
Tomlinson, 200, 217-218, 345-346, 348, 351
Tootle, 228
Torbert, 219, 305, 319, 354, 384-385
Toulson, 290

Townsend, 205, 219, 237, 243, 288, 294-297, 304, 307, 312-313, 317, 333, 358, 365
Toy, 263
Tracy, 309
Trader, 218
Trane, 261
Transul, 339
Travis, 229-230
Traynor, 259
Tremble, 225
Tribbitt, 222
Tribit, 246
Trice, 230-231
Trimble, 256
Trinder, 271
Truax, 198-201, 299, 301
Truit, 383
Truitt, 203, 222, 309, 313, 318-319, 322-324, 327, 332, 335-336, 338-340, 342-343, 346, 350, 356-357, 359-361, 371
Trumpeller, 201
Trusty, 283
Tubbs, 307, 315
Tuck, 293
Tucker, 224, 231, 383-385
Tull, 394
Tunnel, 309
Tunnell, 303, 306-307, 310, 356, 369
Turner, 198-199, 211, 217, 242, 294, 309-310, 312, 357, 371, 382, 385
Twaddell, 259
Tweed, 270, 277
Twilley, 376-378
Tybont, 275
Tybout, 275
Tyre, 311
Tyree, 356
Underwood, 249
Urian, 251
Urias, 251
Vail, 287-288, 290, 292
Valentine, 277
Van Burkalow, 234
Vance, 267, 285

VandenLehe, 273
Vandergrift, 264, 281, 288-290, 294
Vanderting, 389
Vandever, 269
Vandyke, 293, 295-296
Vane, 199, 275
Vanhekil, 288
Vanhekle, 289
VanKirk, 331
Vanpelt, 295
Vansant, 228, 284, 287
VanVorst, 219
Vaughan, 214
Vaughn, 215-216, 318, 321, 329, 362
Vauls, 339-340
Veale, 254
Veasey, 326, 341, 362
Veazey, 283-284
Vent, 326
Vernon, 256, 260
Vernor, 256
Vesey, 371
Vessels, 363
Vible, 205
Viccurs, 310, 313
Vickers, 349, 366
Vickry, 246
Victor, 333
Vincent, 213, 233, 236, 239, 322-323, 334, 375, 383
Viney, 257
Vinyard, 217-218
Violet, 210
Virden, 221, 326, 365
Virdin, 209, 212-213
Voshal, 212-213
Voshall, 204, 206, 208, 210, 212-214
Voshel, 246
Voshell, 198, 203, 222, 239, 248-249, 251
Voss, 231
Vought, 272
Vreland, 331
Wadkins, 221, 223
Wadsley, 297
Waggamon, 348, 350

Wainwright, 386, 392, 394
Wales, 218
Walhearter, 237
Walker, 204, 223, 244, 270-273, 291-292, 294, 297-299, 304, 305, 323, 325-327, 331, 368
Walkins, 289
Wall, 215
Wallace, 201, 206-207
Waller, 318, 320-321, 375, 380, 382, 393
Walley, 377
Walls, 213, 222, 225, 230, 325-326, 333, 335, 337, 340-341, 343, 346, 349-350, 352, 362, 365-366, 370, 381
Walson, 295
Walston, 253, 375-376
Walter, 304, 312, 314
Walters, 263, 283, 289
Walther, 275
Walton, 203, 273, 282-283, 338
Waples, 231, 336, 356, 363, 366, 368, 371
Ward, 215, 232, 248, 260, 281, 284, 351, 355, 361, 373, 388, 392-393
Ware, 273
Warington, 357
Warl, 210
Warran, 293
Warren, 217, 226, 245-246, 248, 253, 270, 291, 298, 324-325, 333, 338, 341-341, 352, 354, 370, 393
Warrington, 308, 311, 320, 322, 326-328, 330, 332, 346, 353, 361-362, 364-365, 369, 371-373, 380
Waston, 340
Waters, 274
Watson, 218, 220-222, 297, 307, 320, 327, 332-334, 336, 338, 340, 342, 347, 353, 360
Weatherly, 334
Weaver, 389
Webb, 201, 220-223, 236, 247, 290, 292, 295, 318, 331, 338, 340, 343, 346, 363, 384-385

Webber, 254, 389
Webster, 208, 253, 258, 300
Weeks, 224
Weer, 253
Weisman, 392
Welch, 203, 230, 246, 266, 295, 327, 331, 369
Weld, 255
Weldin, 254-258, 274, 276, 295, 312
Weller, 251
Wells, 232, 236, 301, 306, 316, 333, 348, 357, 359-360, 362, 364
Welsh, 271, 340, 342
Welston, 208
Weslon, 208
West, 205
West, 205, 251, 259, 305, 307-308, 310, 319, 323, 326, 335, 344, 346-347, 349, 355, 357-358, 360-361, 368, 373-374, 378
Wetherby, 207
Whaley, 322-323
Whaly, 321
Wharton, 202, 210, 219, 237, 308, 312, 340
Whatly, 319
Wheatley, 273
Wheatly, 210, 219, 321, 395
Wheatman, 198, 201, 206
Whelby, 209
Whitaker, 214, 242, 289
White, 210, 261, 265, 274-277, 295, 318, 327, 329, 360-361, 369, 372-373, 394
Whiteman, 268, 270-273
Whitfield, 275
Whitlock, 287, 291, 294
Whitney, 388
Whitnock, 207
Whitten, 279
Whittington, 251
Wiatt, 246
Wickett, 208
Wiggin, 295
Wilbourn, 306, 312
Wilcutts, 218, 239

Wilds, 246-247, 249
Wiley, 207, 280, 321, 372
Wilgus, 307-308, 311
Wilkins, 333, 336, 352-353, 385
Wilkinson, 213, 253, 255, 337-338, 357-348, 351
Willard, 301
Willcutts, 234
Willen, 386
Willey, 206, 232, 355, 380, 383-384, 386, 389-391
Williams, 198-200, 204, 206, 210, 213-216, 219, 231, 233, 238, 242, 247, 257, 261, 284, 290-291, 295, 304, 306-308, 313, 315, 320, 329, 335, 340-341, 344-345, 347-349, 354, 387, 392-394, 394, 396
Williamson, 233, 388
Williard, 368, 370
Willin, 353, 356, 393
Willis, 210, 223, 225, 242, 274, 395
Willits, 292
Willoughby, 388
Wills, 282
Wilson, 202-203, 208, 215-216, 218-219, 227, 233-234, 240, 254, 256-260, 262-263, 267-268, 277, 280, 283, 290, 292, 296, 317, 322, 325-326, 328, 335, 343, 352-354, 364-365, 367, 371, 380, 385
Wiltbank, 253, 329, 331
Wiltbanks, 369
Wimbrow, 348
Winant, 223
Winbrow, 348
Windal, 200
Window, 318
Windson, 321, 351
Windsor, 378
Wingate, 355
Winget, 374
Wingett, 349, 350
Wistar, 256
Wiswell, 331
Wix, 228
Wolcott, 228

Wolfe, 324, 368, 370
Wollaston, 271-272
Woman, 226
Wood, 255, 261-262, 292, 352
Woodall, 204, 331
Woodhull, 201
Woodkeeper, 297
Woodrow, 218, 280
Woodruff, 219, 391
Woods, 287
Woodward, 261-262, 264, 267, 269, 282, 286
Woolford, 348
Woollens, 262
Wooten, 358
Wootens, 230
Wootten, 356-360, 374-375
Workman, 232, 321-322, 325, 329, 343, 353-354, 358, 361, 375, 378, 381, 386
Worral, 280
Worrall, 272

Worrell, 272
Worth, 268
Wright, 201, 211-212, 233, 239, 247, 251, 263, 277, 279-280, 283, 300, 317, 356, 362, 366, 370-371, 374, 376-377, 387-389, 391-392, 395
Wroten, 234, 339, 343
Wrotten, 374
Wryatt, 323
Wyatt, 223, 2260229, 240, 360
WynKoop, 334
Yarnall, 273
Yates, 251
Yearsley, 266-267
Yerkes, 271
Yolt, 210
York, 203, 216
Young, 256, 300, 334-336
Zaylon, 212
Zebley, 264
Zerratt, 214

Made in the USA
Charleston, SC
03 December 2009